T0323621

LONDON MATHEMATICAL SOCIETY LECTURE NOTE SERIES

Managing Editor: Professor NJ. Hitchin, Mathematical Institute,
University if Oxford, 24–29 St Giles, Oxford OX1 3LB, United Kingdom

All the titles listed below can be obtained from good booksellers or from Cambridge
University Press. For a complete series listing visit
http://publishing.cambridge.org/stm/mathematics/lmsn/

Photo by M. Gualtieri.

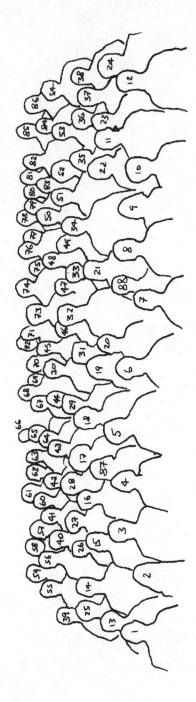

Key to symposium picture

1. N. Hitchin
2. S.T. Tsou
3. G. Segal
4. D. McDuff
5. D. Sullivan
6. E. Lupercio
7. I. Galvez
8. S. Kallel
9. N. Wahl
10. U. Tillmann
11. N. Bass

12. G. Powell
13. S. Majid
14. P. Boalch
15. R. Cohen
16. A. Vina
17. ??
18. A. Ranicki
19. S. Dean
20. K. Marathe
21. J. McKay
22. E. Mann

23. Mrs. Teichner
24. D. Husemoller
25. V. Toledano Laredo
26. I. Smith
27. M. Hyland
28. E. Beggs
29. M. Dunajski
30. S. Rancaniere
31. S. Schaefer–Nameki
32. S. Galatius
33. M. Weiss

34. J. Anderson
35. K. Erdmann
36. F. Kirwan
37. J. Sawon
38. D. Freed
39. R. Picken
40. C. Teleman
41. E. Rees
42. D. Quillen
43. S. Garoufalidis
44. E. Getzler

45. J. Roberts
46. T. Perutz
47. T. Hodge
48. H. Fegan
49. S. Willerton
50. I. Leary
51. P. Turner
52. ??
53. D. Indelicato
54. H. Khudaverdian
55. M. Mackaay

56. P. Teichner
57. R. Marsh
58. H. Skarke
59. J. Morava
60. Y. Kim
61. ??
62. J. Drummond
63. M. Jacob
64. K. Hannabus
65. I. Zois
66. G. Luke

67. C.-F. Bödigheimer
68. M. Kontsevich
69. ??
70. J. Klein
71. M. Anton
72. S. Kerstan
73. J. Baez
74. K. Feldman
75. ??
76. F. Clarke
77. M. Y. Mo

78. ??
79. T. Riley
80. H. Segerman
81. A. Ramirez
82. A. Stacey
83. T. Haire
84. V. Manuilov
85. C. Stretch
86. T. Voronov
87. C. Rietsch
88. C. Douglas

London Mathematical Society Lecture Note Series: 308

Topology, Geometry and Quantum Field Theory

Proceedings of the 2002 Oxford Symposium in the honour of the 60th birthday of Graeme Segal

Edited by

Ulrike Tillmann
University of Oxford

CAMBRIDGE
UNIVERSITY PRESS

CAMBRIDGE UNIVERSITY PRESS
Cambridge, New York, Melbourne, Madrid, Cape Town, Singapore,
São Paulo, Delhi, Dubai, Tokyo, Mexico City

Cambridge University Press
The Edinburgh Building, Cambridge CB2 8RU, UK

Published in the United States of America by Cambridge University Press, New York

www.cambridge.org
Information on this title: www.cambridge.org/9780521540490

© Cambridge University Press 2004

First published 2004

A catalogue record for this publication is available from the British Library

ISBN 978-0-521-54049-0 Paperback

Contents

vii

Preface

The Symposium on Topology, Geometry and Quantum Field Theory took place in Oxford during the week 24–29 June 2002. Graeme Segal's sixtieth birthday was celebrated at a special dinner; his mathematics throughout the meeting. These proceedings reflect the scientific excitement at the Symposium, which more than 140 physicists, geometers and topologists attended.

The Symposium was financially supported by the EPSRC, the London Mathematical Society and Oxford University. I would like to thank all the speakers, participants and everyone who helped with the organization for making the Symposium such a success. Special thanks are due to the contributors to these proceedings and the referees. Brenda Willoughby not only offered technical help in putting this volume together but also typed Part II more than ten years ago. Finally and foremost, I would like to thank Graeme Segal for his consent (if reluctant) to holding this Symposium in honour of his birthday, and publishing his influential manuscript as part of these proceedings.

Ulrike Tillmann
Oxford, June 2003

Participants

Rui Albuquerque, Warwick
Jorgen E. Andersen, Aarhus
Marian F. Anton, Sheffield
Nils A. Baas, Trondheim
John Baez, Riverside
John Barrett, Nottingham
Eliezer Batista, QMW-London
Edwin Beggs, Swansea
Philip Boalch, Strasbourg
C.-F. Boedigheimer, Bonn
Simon Brain, Oxford
Mark Brightwell, Aarhus
Philip Candelas, Oxford
John Cardy, Oxford
Catarina Carvalho, Oxford
Rogerio Chaves, Oxford
Francis Clarke, Swansea
Ralph Cohen, Stanford
Michael Crabb, Aberdeen
Xenia De La Ossa, Oxford
Sarah Dean, Harvard
Gustav Delius, York
Robbert Dijkgraaf, Amsterdam
Christopher Douglas, MIT
James Drummond, KC-London
Maciej Dunajski, Oxford
Vivien Easson, Oxford
Karin Erdmann, Oxford
Alon Faraggi, Oxford

Thomas Fawcett, Oxford
Howard Fegan, Lehigh
Konstantin Feldman, Edinburgh
Joel Fine, Imperial-London
Neil Firth, Oxford
Bogdan Florea, Oxford
Dan Freed, Austin
Edward Frenkel, Berkeley
Soren Galatius, Aarhus
Imma Galvez, Lille
Manuel Garcia Islas, Nottingham
Stavros Garoufalidis, Warwick
Elizabeth Gasparin, New Mexico
Ezra Getzler, Northwestern
Veronique Godin, Stanford
Michele Grassi, Pisa
John Greenlees, Sheffield
Ian Grojnowski, Cambridge
Marco Gualtieri, Oxford
Timothy James Haire, Cambridge
Mark Hale, QMW-London
Keith Hannabuss, Oxford
Nigel Hitchin, Oxford
Thomas Hodge, Imperial-London
Luke Hodgkin, KC-London
Mike Hopkins, MIT
Dale Husemoller, MPI-Bonn
Martin Hyland, Cambridge
Davide Indelicato, Zürich

Martin Jakob, Neuchatel
Shabnam Kadir, Oxford
Sadok Kallel, Lille
Ayse Kara, Trieste
Sven Kerstan, KC-London
Hovhannes Khudaverdian, UMIST
Yunhyong Kim, Trieste
Rob Kirby, Berkeley
Frances Kirwan, Oxford
John Klein, Wayne State
Maxim Kontsevich, IHES
Andrzej Kozlowski, Toyama
Ian Leary, Southampton
Glenys Luke, Oxford
Ernesto Lupercio, Madison
Marco Mackaay, Nottingham
Shahn Majid, QMW-London
Elizabeth Mann, Oxford
Vladimir Manuilov, Moscow
Kishore Marathe, CUNY
Alexander Markos, McGill
Robert Marsh, Leicester
Stephen Marshall, Oxford
Lionel Mason, Oxford
Dusa McDuff, Stony Brook
John McKay, Concordia
Man Yue Mo, Oxford
Greg Moore, Rutgers
Jack Morava, Johns Hopkins
Jeff Murugan, Oxford
Nikita Nekrasov, IHES
Kyungho Oh, Missouri
Jonathan Palmer, Oxford
Tim Perutz, Imperial-London
Roger Picken, Lisbon
Adam Piggott, Oxford
Daniel Pons, Imperial-London
Geoffrey Powell, Paris 13
Dan Quillen, Oxford
Sebastian Racaniev, Strasbourg

Antonio Ramirez, Stanford
Jose J. Ramon-Mari, Durham
Andrew Ranicki, Edinburgh
Elmer Rees, Edinburgh
Dean Rickles, Leeds
Konstanze Rietsch, Oxford
Tim Riley, Oxford
Justin Roberts, San Diego
Julius Ross, Imperial-London
John de Sa, Oxford
Paolo Salvatore, Rome
Justin Sawon, Oxford
S. Schafer-Nameki, Cambridge
Simon Scott, KC-London
Graeme Segal, Oxford
Matteo Semplice, Oxford
Harald Skarke, Oxford
Ivan Smith, Polytechnique-Paris
Marc Ulrich Sommers, Warwick
Andrew Stacey, Stanford
Stephan Stolz, Notre Dame
Christopher Stretch, Ulster
Dennis Sullivan, CUNY
Wilson Sutherland, Oxford
Tim Swift, Bristol
David Swinarski, Oxford
Peter Teichner, San Diego
Constantin Teleman, Cambridge
Richard Thomas, Oxford
Ulrike Tillmann, Oxford
Valerio Toledano Laredo, Paris 6/7
Andrea Tomatis, Imperial-London
Tsou Sheung Tsun, Oxford
Paul Turner, Heriot-Watt
Andres Vina, Oviedo
Theodore Voronov, UMIST
Nathalie Wahl, Northwestern
Michael Weiss, Aberdeen
Simon Willerton, Edinburgh
George Wilson, Imperial-London

Frederik Witt, Oxford

Edward Witten, IAS-Princeton

Peter Woit, Columbia

Nicholas Woodhouse, Oxford

Ioannis Zois, IHES

Introduction

SIR MICHAEL ATIYAH

It is a great pleasure for me to write this introduction to the volume celebrating Graeme Segal's 60th birthday. Graeme was one of my first Ph.D. students but he rapidly moved on to become a collaborator and colleague. Over the years we have written a number of joint papers, but the publications are merely the tidemarks of innumerable discussions. My own work has been subtly influenced by Graeme's point of view: teacher and student can and do interchange roles, each educating the other.

Graeme has a very distinctive style. For him brevity is indeed the soul of wit, arguments should be elegant and transparent, lengthy calculations are a sign of failure and algebra should be kept firmly in its place. He only publishes when he is ready, when he is satisfied with the final product. At times this perfectionist approach means that his ideas, which he generously publicizes, get absorbed and regurgitated by others in incomplete form. But his influence is widely recognized, even when the actual publication is long-delayed.

Topology has always been at the heart of Graeme's interests, but he has interpreted this broadly and found fruitful pastures as far away as theoretical physics. There was a time when such deviation from the strict path of pure topology was deemed a misdemeanour, particularly when the field into which Graeme deviated was seen as less than totally rigorous. But time moves on and subsequent developments have fully justified Graeme's 'deviance'. He is one of a small number of mathematicians who have had an impact on theoretical physicists.

Of all his works I would single out his beautiful book on Loop Groups, written jointly with his former student Andrew Pressley. In a difficult area, straddling algebra, geometry, analysis and physics the book manages to maintain a coherent outlook throughout, and it does so with style. It is a real

1

treasure, a worthy successor in its way to Hermann Weyl's *The Classical Groups*.

Perhaps we should look forward to another book in the same mould – in time for the 70th birthday?

Edinburgh, June 2003

Part I

Contributions

1

A variant of K-theory: $K_\pm$

MICHAEL ATIYAH and MICHAEL HOPKINS

University of Edinburgh and MIT.

1 Introduction

Topological K-theory [2] has many variants which have been developed and exploited for geometric purposes. There are real or quaternionic versions, 'Real' K-theory in the sense of [1], equivariant K-theory [14] and combinations of all these.

In recent years K-theory has found unexpected application in the physics of string theories [6] [12] [13] [16] and all variants of K-theory that had previously been developed appear to be needed. There are even variants, needed for the physics, which had previously escaped attention, and it is one such variant that is the subject of this paper.

This variant, denoted by $K_\pm(X)$, was introduced by Witten [16] in relation to 'orientifolds'. The geometric situation concerns a manifold X with an involution τ having a fixed sub-manifold Y. On X one wants to study a pair of complex vector bundles (E^+, E^-) with the property that τ interchanges them. If we think of the virtual vector bundle $E^+ - E^-$, then τ takes this into its negative, and $K_\pm(X)$ is meant to be the appropriate K-theory of this situation.

In physics, X is a 10-dimensional Lorentzian manifold and maps $\Sigma \to X$ of a surface Σ describe the world-sheet of strings. The symmetry requirements for the appropriate Feynman integral impose conditions that the putative K-theory $K_\pm(X)$ has to satisfy.

The second author proposed a precise topological definition of $K_\pm(X)$ which appears to meet the physics requirements, but it was not entirely clear how to link the physics with the geometry.

In this paper we elaborate on this definition and also a second (but equivalent) definition of $K_\pm(X)$. Hopefully this will bring the geometry and physics closer together, and in particular link it up with the analysis of Dirac operators.

5

Although $K_\pm(X)$ is defined in the context of spaces with involution it is rather different from Real K-theory or equivariant K-theory (for $G = Z_2$), although it has superficial resemblances to both. The differences will become clear as we proceed, but at this stage it may be helpful to consider the analogy with cohomology. Equivariant cohomology can be defined (for any compact Lie group G), and this has relations with equivariant K-theory. But there is also 'cohomology with local coefficients', where the fundamental group $\pi_1(X)$ acts on the abelian coefficient group. In particular for integer coefficients Z the only such action is via a homomorphism $\pi_1(X) \to Z_2$, i.e. by an element of $H^1(X; Z_2)$ or equivalently a double-covering $\tilde{X}$ of X.

This is familiar for an unoriented manifold with $\tilde{X}$ its oriented double-cover. In this situation, if say X is a compact n-dimensional manifold, then we do not have a fundamental class in $H^n(X; Z)$ but in $H^n(X; \tilde{Z})$ where $\tilde{Z}$ is the local coefficient system defined by $\tilde{X}$. This is also called 'twisted cohomology'.

Here $\tilde{X}$ has a fixed-point-free involution τ and, in such a situation, our group $K_\pm(\tilde{X})$ is the precise K-theory analogue of twisted cohomology. This will become clear later.

In fact K-theory has more sophisticated twisted versions. In [8] Donovan and Karoubi use Wall's graded Brauer group [15] to construct twistings from elements of $H^1(X; Z_2) \times H^3(X; Z)_{\text{torsion}}$. More general twistings of K-theory arise from automorphisms of its classifying space, as do twistings of equivariant K-theory. Among these are twistings involving a general element of $H^3(X; Z)$ (i.e., one which is not necessarily of finite order). These are also of interest in physics, and have recently been the subject of much attention [3] [5] [9]. Our $K_\pm$ is a twisted version of equivariant K-theory,[1] and this paper can be seen as a preliminary step towards these other more elaborate versions.

2 The first definition

Given a space X with involution we have two natural K-theories, namely $K(X)$ and $K_{Z_2}(X)$ – the ordinary and equivariant theories respectively. Moreover we have the obvious homomorphism

$$\phi : K_{Z_2}(X) \to K(X) \tag{2.1}$$

---

[1] It is the twisting of equivariant K-theory by the non-trivial element of $H^1_{Z_2}(\text{pt}) = Z_2$. From the point of view of the equivariant graded Brauer group, $K_\pm(X)$ is the K-theory of the graded cross product algebra $C(X) \otimes M \rtimes Z_2$, where $C(X)$ is the algebra of continuous functions on X, and M is the graded algebra of 2×2-matrices over the complex numbers, graded in such a way that (i, j) entry has degree $i + j$. The action of Z_2 is the combination of the geometric action given on X and conjugation by the permutation matrix on M.

which 'forgets' about the Z_2-action. We can reformulate this by introducing
the space $(X \times Z_2)$ with the involution $(x, 0) \to (\tau(x), 1)$. Since this action is
free we have

$$K_{Z_2}(X \times Z_2) \cong K(X)$$

and (2.1) can then be viewed as the natural homomorphism for K_{Z_2} induced
by the projection

$$\pi : X \times Z_2 \to X. \tag{2.2}$$

Now, whenever we have such a homomorphism, it is part of a long exact
sequence (of period 2) which we can write as an exact triangle

$$\begin{array}{ccc} K_{Z_2}^*(X) & \xrightarrow{\phi} & K^*(X) \\ & \nwarrow & \swarrow \delta \\ & K_{Z_2}^*(\pi) & \end{array} \tag{2.3}$$

where $K^* = K^0 \oplus K^1$, δ has degree 1 mod 2 and the relative group $K_{Z_2}^*(\pi)$
is just the relative group for a pair, when we replace π by a Z_2-homotopically
equivalent inclusion. In this case a natural way to do this is to replace the X
factor on the right of (2.2) by $X \times I$ where $I = [0, 1]$ is the unit interval with
τ being reflection about the mid-point $\frac{1}{2}$. Thus, explicitly

$$K_{Z_2}^*(\pi) = K_{Z_2}^*(X \times I, X \times \partial I) \tag{2.4}$$

where ∂I is the (2-point) boundary of I.

We now take the group in (2.4) (with the degree shifted by one) as our definition of $K_{\pm}^*(X)$. It is then convenient to follow the notation of [1] where
$R^{p,q} = R^p \oplus R^q$ with the involution changing the sign of the first factor,
and we use K-theory with compact supports (so as to avoid always writing the
boundary). Then our definition of $K_{\pm}$ becomes

$$K_{\pm}^0(X) = K_{Z_2}^1(X \times R^{1,0}) \cong K_{Z_2}^0(X \times R^{1,1}) \tag{2.5}$$

(and similarly for K^1).

Let us now explain why this definition fits the geometric situation we began
with (and which comes from the physics). Given a vector bundle E we can
form the pair (E, τ^*E) or the virtual bundle

$$E - \tau^*E.$$

Under the involution, E and τ^*E switch and the virtual bundle goes into its
negative. Clearly, if E came from an equivariant bundle, then $E \cong \tau^*E$ and
the virtual bundle is zero. Hence the virtual bundle depends only the element

defined by E in the cokernel of ϕ, and hence by the image of E in the next term of the exact sequence (2.3), i.e. by

$$\delta(E) \in K_{\pm}^0(X).$$

This explains the link with our starting point and it also shows that one cannot always define $K_{\pm}(X)$ in terms of such virtual bundles on X. In general the exact sequence (2.3) does not break up into short exact sequences and δ is not surjective.

At this point a physicist might wonder whether the definition of $K_{\pm}(X)$ that we have given is the right one. Perhaps there is another group which is represented by virtual bundles. We will give two pieces of evidence in favour of our definition, the first pragmatic and the second more philosophical.

First let us consider the case when the involution τ on X is trivial. Then $K_{Z_2}^*(X) = R(Z_2) \otimes K^*(X)$ and $R(Z_2) = Z \oplus Z$ is the representation ring of Z_2 and is generated by the two representations:

$$1 \quad \text{(trivial representation)}$$
$$\rho \quad \text{(sign representation)}.$$

The homomorphism ϕ is surjective with kernel $(1 - \rho)K^*(X)$ so $\delta = 0$ and

$$K_{\pm}^0(X) \cong K^1(X). \tag{2.6}$$

This fits with the requirements of the physics, which involves a switch from type IIA to type IIB string theory. Note also that it gives an extreme example when ∂ is not surjective.

Our second argument is concerned with the general passage from physical (quantum) theories to topology. If we have a theory with some symmetry then we can consider the quotient theory, on factoring out the symmetry. Invariant states of the original theory become states of the quotient theory but there may also be new states that have to be added. For example if we have a group G of geometric symmetries, then closed strings in the quotient theory include strings that begin at a point x and end at $g(x)$ for $g \in G$. All this is similar to what happens in topology with (generalized) cohomology theories, such as K-theory. If we have a morphism of theories, such as ϕ in (2.1) then the third theory we get fits into a long-exact sequence. The part coming from $K(X)$ is only part of the answer – other elements have to be added. In ordinary cohomology where we start with cochain complexes the process of forming a quotient theory involves an ordinary quotient (or short exact sequence) at the level of cochains. But this becomes a long exact sequence at the cohomology level. For K-theory the analogue is to start with bundles over small open sets and at this level we can form the naïve quotients, but the K-groups arise when we impose the matching conditions to get bundles, and then we end up with long exact sequences.

It is also instructive to consider the special case when the involution is free so that we have a double covering $\tilde{X} \to X$ and the exact triangle (2.3), with $\tilde{X}$ for X, becomes the exact triangle

$$
\begin{array}{ccc}
K^*(X) & \overset{\phi}{\to} & K^*(\tilde{X}) \\
\nwarrow & & \swarrow \delta \\
& K^*_{Z_2}(L) &
\end{array}
\tag{2.7}
$$

Here L is the real line bundle over X associated to the double covering $\tilde{X}$ (or to the corresponding element of $H^1(X, Z_2)$), and we again use compact supports. Thus (for $q = 0, 1 \mod 2$)

$$
K^q_{\pm}(\tilde{X}) = K^{q+1}(L).
\tag{2.8}
$$

If we had repeated this argument using equivariant cohomology instead of equivariant K-theory we would have ended up with the twisted cohomology mentioned earlier, via a twisted suspension isomorphism

$$
H^q(X, \tilde{Z}) = H^{q+1}(L).
\tag{2.9}
$$

This shows that, for free involutions, $K_{\pm}$ **is precisely the analogue of twisted cohomology,** so that, for example, the Chern character of the former takes values in the rational extension of the latter.

3 Relation to Fredholm operators

In this section we shall give another definition of $K_{\pm}$ which ties in naturally with the analysis of Fredholm operators, and we shall show that this definition is equivalent to the one given in Section 2.

We begin by recalling a few basic facts about Fredholm operators. Let H be complex Hilbert space, $\mathcal{B}$ the space of bounded operators with the norm topology and $\mathcal{F} \subset \mathcal{B}$ the open subspace of Fredholm operators, i.e. operators A so that ker A and coker A are both finite-dimensional. The index defined by

$$
\text{index } A = \dim \ker A - \dim \operatorname{coker} A
$$

is then constant on connected components of $\mathcal{F}$. If we introduce the adjoint A^* of A then

$$
\operatorname{coker} A = \ker A^*
$$

so that

$$
\text{index } A = \dim \ker A - \dim \ker A^*.
$$

More generally if we have a continuous map

$$f : X \to \mathcal{F}$$

(i.e. a family of Fredholm operators, parametrized by X), then one can define

$$\text{index } f \in K(X)$$

and one can show [2] that we have an isomorphism

$$\text{index} : [X, \mathcal{F}] \cong K(X) \qquad (3.1)$$

where [,] denotes homotopy classes of maps. Thus $K(X)$ has a natural definition as the 'home' of indices of Fredholm operators (parametrized by X): it
gives the complete homotopy invariant.

Different variants of K-theory can be defined by different variants of (3.1).
For example real K-theory uses real Hilbert space and equivariant K-theory
for G-spaces uses a suitable H-space module of G, namely $L_2(G) \otimes H$. It is
natural to look for a similar story for our new groups $K_\pm(X)$. A first candidate
might be to consider Z_2-equivariant maps

$$f : X \to \mathcal{F}$$

where we endow $\mathcal{F}$ with the involution $A \to A^*$ given by taking the adjoint
operator. Since this switches the role of kernel and cokernel it acts as -1 on
the index, and so is in keeping with our starting point.

As a check we can consider X with a trivial involution, then f becomes a
map

$$f : X \to \widehat{\mathcal{F}}$$

where $\widehat{\mathcal{F}}$ is the space of self-adjoint Fredholm operators. Now in [4] it is shown
that $\widehat{\mathcal{F}}$ has three components

$$\widehat{\mathcal{F}}_+, \widehat{\mathcal{F}}_-, \widehat{\mathcal{F}}_*$$

where the first consists of A which are essentially positive (only finitely many
negative eigenvalues), the second is given by essentially negative operators.
These two components are trivial, in the sense that they are contractible, but
the third one is interesting and in fact [4]

$$\widehat{\mathcal{F}}_* \sim \Omega \mathcal{F} \qquad (3.2)$$

where Ω denotes the loop space. Since

$$[X, \Omega \mathcal{F}] \cong K^1(X)$$

this is in agreement with (2.6) – though to get this we have to discard the two trivial components of $\widehat{\mathcal{F}}$, a technicality to which we now turn.

Lying behind the isomorphism (3.1) is Kuiper's Theorem [11] on the contractibility of the unitary group of Hilbert spaces. Hence to establish that our putative definition of $K_\pm$ coincides with the definition given in Section 2 we should expect to need a generalization of Kuiper's Theorem incorporating the involution $A \to A^*$ on operators. The obvious extension turns out to be false, precisely because $\widehat{\mathcal{F}}$, the fixed-point set of $*$ on $\mathcal{F}$, has the additional contractible components. There are various ways we can get round this but the simplest and most natural is to use 'stabilization'. Since $H \cong H \oplus H$ we can always stabilize by adding an additional factor of H. In fact Kuiper's Theorem has two parts in its proof:

(1) The inclusion $U(H) \to U(H \oplus H)$ defined by $u \to u \oplus 1$ is homotopic to the constant map.
(2) This inclusion is homotopic to the identity map given by the isomorphism $H \cong H \oplus H$.

The proof of (1) is an older argument (sometimes called the 'Eilenberg swindle'), based on a correct use of the fallacious formula

$$1 = 1 + (-1 + 1) + (-1 + 1) \dots$$
$$= (1 + -1) + (1 + -1) + \dots$$
$$= 0.$$

The trickier part, and Kuiper's contribution, is the proof of (2).

For many purposes, as in K-theory, the stronger version is a luxury and one can get by with the weaker version (1), which applies rather more generally. In particular (1) is consistent with taking adjoints (i.e. inverses in $U(H)$), which is the case we need.

With this background explanation we now introduce formally our second definition, and to distinguish it temporarily from $K_\pm$ as defined in Section 2, we put

$$\mathcal{K}_\pm(X) = [X, \mathcal{F}]_*^s \tag{3.3}$$

where $*$ means we use Z_2-maps compatible with $*$ and s means that we use **stable** homotopy equivalence. More precisely the Z_2-maps

$$f : X \to \mathcal{F}(H) \quad g : X \to \mathcal{F}(H)$$

are called stably homotopic if the 'stabilized' maps

$$f^s : X \to \mathcal{F}(H \oplus H) \quad g^s : X \to \mathcal{F}(H \oplus H)$$

given by $f^s = f \oplus J$, $g^s = g \oplus J$ are homotopic, where J is a fixed (essentially unique) automorphism of H with

$$J = J^*, \ J^2 = 1, \quad +1 \text{ and } -1 \text{ both of infinite multiplicity.} \qquad (3.4)$$

Note that under such stabilization the two contractible components $\widehat{\mathcal{F}}_+$ and $\widehat{\mathcal{F}}_-$ of $\widehat{\mathcal{F}}(H)$ both end up in the interesting component $\widehat{\mathcal{F}}_*$ of $\widehat{\mathcal{F}}(H \oplus H)$.

The first thing we need to observe about $\mathcal{K}_\pm(X)$ is that it is an abelian group. The addition can be defined in the usual way by using direct sums of Hilbert spaces. Moreover we can define the negative degree groups $\mathcal{K}_\pm^{-n}(X)$ (for $n \geq 1$) by suspension (with trivial involution on the extra coordinates), so that

$$\mathcal{K}_\pm^{-n}(X) = \mathcal{K}_\pm(X \times S^n, X \times \infty).$$

However, at this stage we do not have the periodicity theorem for $\mathcal{K}_\pm(X)$. This will follow in due course after we establish the equivalence with $K_\pm(X)$. As we shall see our construction of (4.2) is itself closely tied to the periodicity theorem.

Our aim in the subsequent sections will be to show that there is a natural isomorphism

$$\mathcal{K}_\pm(X) \cong K_\pm(X). \qquad (3.5)$$

This isomorphism will connect us up naturally with Dirac operators and so should tie in nicely with the physics.

4 Construction of the map

Our first task is to define a natural map

$$K_\pm(X) \to \mathcal{K}_\pm(X). \qquad (4.1)$$

We recall from (2.5) that

$$K_\pm(X) = K_{Z_2}(X \times R^{1,1})$$
$$= K_{Z_2}(X \times S^2, \ X \times \infty)$$

where S^2 is the 2-sphere obtained by compactifying $R^{1,1}$, and ∞ is the added point. Note that Z_2 now acts on S^2 by a **reflection,** so that it reverses its orientation.

Thus to define a map (4.1) it is sufficient to define a map

$$K_{Z_2}(X \times S^2) \to \mathcal{K}_\pm(X). \qquad (4.2)$$

This is where the Dirac operator enters. Recall first that, if we ignore involutions, there is a basic map

$$K(X \times S^2) \to [X, \mathcal{F}] \cong K(X) \qquad (4.3)$$

which is the key to the Bott periodicity theorem. It is given as follows. Let D be the Dirac operator on S^2 from positive to negative spinors and let V be a complex vector bundle on $X \times S^2$, then we can extend, or couple, D to V to get a family D_V of elliptic operators along the S^2-fibres. Converting this, in the usual way, to a family of (bounded) Fredholm operators we get the map (4.3).

We now apply the same construction but keeping track of the involutions. The new essential feature is that Z_2 reverses the orientation of S^2 and hence takes the Dirac operator D into its adjoint D^*. This is precisely what we need to end up in $\mathcal{K}_{\pm}(X)$ so defining (4.2).

Remark 4.1. Strictly speaking the family D_V of Fredholm operators does not act in a fixed Hilbert space, but in a bundle of Hilbert spaces. The problem can be dealt with by adding a trivial operator acting on a complementary bundle W (so that $W + V$ is trivial).

5 Equivalence of definition

Let us sum up what we have so far. We have defined a natural homomorphism

$$K_{\pm}(X) \to \mathcal{K}_{\pm}(X)$$

and we know that this is an isomorphism for spaces X with trivial involution – both groups coinciding with $K^1(X)$. Moreover, if for general X, we ignore the involutions, or equivalently replace X by $X \times \{0, 1\}$, we also get an isomorphism, both groups now coinciding with $K^0(X)$.

General theory then implies that we have an isomorphism for all X. We shall review this general argument.

Let A, B be representable theories, defined on the category of Z_2-spaces, so that

$$A(X) = [X, \mathcal{A}]$$
$$B(X) = [X, \mathcal{B}]$$

where [,] denotes homotopy classes of Z_2-maps into the classifying spaces of $\mathcal{A}$, $\mathcal{B}$ of the theories. A natural map $A(X) \to B(X)$ then corresponds to a Z_2-map $\mathcal{A} \to \mathcal{B}$. Showing that A and B are isomorphic theories is equivalent to showing that $\mathcal{A}$ and $\mathcal{B}$ are Z_2-homotopy equivalent.

If we forget about the involutions then isomorphism of theories is the same as ordinary homotopy equivalence. Restricting to spaces X with trivial involution corresponds to restricting to the fixed-point sets of the involution on A and B.

Now there is a general theorem in homotopy theory [10] which asserts (for reasonable spaces including Banach manifolds such as $\mathcal{F}$) that, if a Z_2-map $A \to B$ is both a homotopy equivalence ignoring the involution and for the fixed-point sets, then it is a Z_2-homotopy equivalence. Translated back into the theories A, B it says that the map $A(X) \to B(X)$ is an isomorphism provided it holds for spaces X with the trivial Z_2-action, and for Z_2-spaces X of the form $Y \times \{0, 1\}$.

This is essentially the situation we have here with

$$A = K_\pm \qquad B = \mathcal{K}_\pm.$$

Both are representable. The representability of the first

$$K_\pm(X) \cong K_{Z_2}(X \times R^{1,1})$$

arises from the general representability of K_{Z_2}, the classifying space being essentially the double loop space of $\mathcal{F}(H \otimes C^2)$ with an appropriate Z_2-action. The second is representable because

$$\mathcal{K}_\pm(X) = [X, \mathcal{F}]^s_* = [X, \mathcal{F}_s]_* \qquad (5.1)$$

where $\mathcal{F}^s$ is obtained by stabilizing $\mathcal{F}$. More precisely

$$\mathcal{F}^s = \lim_{n \to \infty} \mathcal{F}_n$$

where $\mathcal{F}_n = \mathcal{F}(H \otimes C^n)$ and the limit is taken with respect to the natural inclusions, using J of (3.4) as a base point. The assertion in (5.1) is easily checked and it simply gives two ways of looking at the stabilization process.

We have thus established the equivalence of our two definitions $K_\pm$ and $\mathcal{K}_\pm$.

6 Free involutions

We shall now look in more detail at the case of free involutions and, following the notation of Section 1, we shall denote the free Z_2-space by $\tilde{X}$ and its quotient by X.

The reason for introducing the stabilization process in Section 3 concerned fixed points. We shall now show that, for free involutions, we can dispense with stabilization. Let

$$\mathcal{F} \to \mathcal{F}^s$$

be the natural inclusion of $\mathcal{F}$ in the direct limit space. This inclusion is a Z_2-map and a homotopy equivalence, though **not** a Z_2-homotopy equivalence (because of the fixed points). Now given the double covering $\tilde{X} \to X$ we can form the associated fibre bundles $\mathcal{F}_X$ and $\mathcal{F}_X^s$ over X with fibres $\mathcal{F}$ and $\mathcal{F}^s$. Thus

$$\mathcal{F}_X = \tilde{X} \times_{Z_2} \mathcal{F} \qquad \mathcal{F}_X^s = \tilde{X} \times_{Z_2} \mathcal{F}^s$$

and we have an inclusion

$$\mathcal{F}_X \to \mathcal{F}_X^s$$

which is fibre preserving. This map is a homotopy equivalence on the fibres and hence, by a general theorem [7] (valid in particular for Banach manifolds) a fibre homotopy equivalence. It follows that the homotopy classes of sections of these two fibrations are isomorphic. But these are the same as

$$\left[\tilde{X}, \mathcal{F}\right]_* \text{ and } \left[\tilde{X}, \mathcal{F}\right]_*^s = \mathcal{K}_\pm(\tilde{X}).$$

This show that, for a free involution, we can use $\mathcal{F}$ instead of $\mathcal{F}^s$. Moreover it gives the following simple description of $K_\pm(\tilde{X})$

$$K_\pm(\tilde{X}) = \text{Homotopy classes of sections of } \mathcal{F}_X. \qquad (6.1)$$

This is the K-theory analogue of twisted cohomology described in Section 1. A corresponding approach to the higher twist of K-theory given by an element of $H^3(X; Z)$ will be developed in [3].

7 The real case

Everything we have done so far extends, with appropriate modifications, to real K-theory. The important difference is that the periodicity is now 8 rather than 2 and that, correspondingly, we have to distinguish carefully between **self-adjoint** and **skew-adjoint** Fredholm operators. Over the complex numbers multiplication by i converts one into the other, but over the real numbers there are substantial differences.

We denote by $\mathcal{F}^1(R)$ the interesting component of the space of real self-adjoint Fredholm operators $\widehat{\mathcal{F}}(R)$ on a real Hilbert space (discarding two contractible components as before). We also denote by $\mathcal{F}^{-1}(R)$ the space of skew-adjoint Fredholm operators. Then in [4] it is proved that

$$[X, \mathcal{F}^1(R)] \cong KO^1(X) \qquad (7.1)$$

$$[X, \mathcal{F}^{-1}(R)] \cong KO^{-1}(X) \cong KO^7(X) \qquad (7.2)$$

showing that these are essentially different groups.

Using (7.1), stabilizing, and arguing precisely as before, we define

$$KO_{\pm}(X) = KO^1_{Z_2}(X \times R^{1,0}) \cong KO_{Z_2}(X \times R^{1,7})$$
$$\mathcal{KO}_{\pm}(X) = [X, \mathcal{F}(R)]^s_*$$

where (in (2.5)) the mod 2 periodicity of K has been replaced by the mod 8 periodicity of KO. But we cannot now just use the Dirac operator on S^2 because this is not real. Instead we have to use the Dirac operator on S^8, which then gives us our map

$$KO_{\pm}(X) \to [X, \mathcal{F}(R)]^s_*. \tag{7.3}$$

The same proof as before establishes the isomorphism of (7.3), so that

$$KO_{\pm}(X) \cong \mathcal{KO}_{\pm}(X)$$

and more generally for q modulo 8

$$KO^q_{\pm}(X) \cong \mathcal{KO}^q_{\pm}(X). \tag{7.4}$$

In [4] there is a more systematic analysis of Fredholm operators in relation to Clifford algebras and using this it is possible to give more explicit descriptions of $KO^q_{\pm}(X)$, for all q, in terms of Z_2-mappings into appropriate spaces of Fredholm operators. This would fit in with the different behaviour of the Dirac operator in different dimensions (modulo 8).

References

[1] Atiyah, M. F., *K*-theory and reality, *Quart. J. Math. Oxford Ser.* (2) 17 (1966), 367–386. MR 34 #6756.

[2] Atiyah, M. F., *K-theory*, Lecture notes by D. W. Anderson, W. A. Benjamin, Inc., New York-Amsterdam, 1967. MR 36 #7130.

[3] Atiyah, M. F. and G. B. Segal, Twisted *K*-theory, to appear.

[4] Atiyah, M. F. and I. M. Singer, Index theory for skew-adjoint Fredholm operators, *Inst. Hautes Études Sci. Publ. Math.* (1969), no. 37, 5–26. MR 44 #2257.

[5] Bouwknegt, Peter, Alan L. Carey, Varghese Mathai, Michael K. Murray, and Danny Stevenson, Twisted *K*-theory and *K*-theory of bundle gerbes, *Comm. Math. Phys.* 228 (2002), no. 1, 17–45. MR 1 911 247.

[6] Diaconescu, Duiliu-Emanuel, Gregory Moore, and Edward Witten, E_8 Gauge theory, and a derivation of *K*-theory from *M*-theory, hep-th/0005090.

[7] Dold, Albrecht, Partitions of unity in the theory of fibrations, *Ann. of Math.* (2) 78 (1963), 223–255. MR 27 #5264.

[8] Donovan, P. and M. Karoubi, Graded Brauer groups and *K*-theory with local coefficients, *Inst. Hautes Études Sci. Publ. Math.* (1970), no. 38, 5–25. MR 43 #8075.

[9] Freed, Daniel S., Mike J. Hopkins, and Constantin Teleman, Twisted equivariant *K*-theory with complex coefficients, arXiv:math.AT/0206257.

[10] James, I. M. and G. B. Segal, On equivariant homotopy type, *Topology* 17 (1978), no. 3, 267–272. MR 80k:55045.

[11] Kuiper, Nicolaas H., The homotopy type of the unitary group of Hilbert space, *Topology* 3 (1965), 19–30. MR 31 #4034.

[12] Minasian, Ruben and Gregory Moore, *K*-theory and Ramond–Ramond charge, *J. High Energy Phys.* (1997), no. 11, Paper 2, 7 pp. (electronic). MR 2000a:81190.

[13] Moore, Gregory and Edward Witten, Self-duality, Ramond–Ramond fields and *K*-theory, *J. High Energy Phys.* (2000), no. 5, Paper 32, 32. MR 2001m:81254.

[14] Segal, Graeme, Equivariant *K*-theory, *Inst. Hautes Études Sci. Publ. Math.* (1968), no. 34, 129–151. MR 38 #2769.

[15] Wall, C. T. C., Graded Brauer groups, *J. Reine Angew. Math.* 213 (1963/1964), 187–199. MR 29 #4771.

[16] Witten, Edward, D-branes and *K*-theory, *J. High Energy Phys.* (1998), no. 12, Paper 19, 41 pp. (electronic). MR 2000e:81151.

Email: atiyah@maths.ed.ac.uk and mjh@math.mit.edu.

2

Two-vector bundles and forms
of elliptic cohomology

NILS A. BAAS, BJØRN IAN DUNDAS and
JOHN ROGNES

NTNU, Trondheim, NTNU, Trondheim and University of Oslo.

Dedicated to Graeme Segal on the occasion of his 60th birthday

1 Introduction

The work to be presented in this paper has been inspired by several of Professor Graeme Segal's papers. Our search for a geometrically defined elliptic cohomology theory with associated elliptic objects obviously stems from his Bourbaki seminar [28]. Our readiness to form group completions of symmetric monoidal categories by passage to algebraic K-theory spectra derives from his *Topology* paper [27]. Our inclination to invoke 2-functors to the 2-category of 2-vector spaces generalizes his model for topological K-theory in terms of functors from a path category to the category of vector spaces. We offer him our admiration.

Among all generalized (co-)homology theories, a few hold a special position because they are, in some sense, geometrically defined. For example, de Rham cohomology of manifolds is defined in terms of cohomology classes of closed differential forms, topological K-theory of finite CW complexes is defined in terms of equivalence classes of complex vector bundles, and complex bordism is defined in terms of bordism classes of maps from stably complex manifolds. The geometric origin of these theories makes them particularly well suited to the analysis of many key problems. For example, Chern–Weil theory associates differential forms related to the curvature tensor to manifolds with a connection, whose de Rham cohomology classes are the Chern classes of the tangent bundle of the manifold. The Atiyah–Segal index theory [2] associates formal differences of vector bundles to parametrized families of Fredholm operators, arising e.g. from complexes of elliptic pseudo-differential operators, and their isomorphism classes live in topological K-theory. Moduli spaces of isomorphism classes of solutions to e.g. Yang–Mills gauge-theoretic problems can generically yield maps from suitably structured manifolds, with well-defined bordism classes in the corresponding form of bordism homology. On the other hand, Quillen's theorem that the coefficient ring $\pi_*(MU)$ for complex bordism theory is the Lazard ring that corepresents (commutative one-dimensional) formal group laws has no direct manifold-geometric interpretation, and may seem to be a fortuitous coincidence in this context.

From the chromatic point of view of stable homotopy theory, related to the various periodicity operators v_n for $n \geq 0$ that act in many cohomology theories, these three geometrically defined cohomology theories detect an increasing amount of information. De Rham cohomology or real singular cohomology sees only rational phenomena, because for each prime p multiplication by $p = v_0$ acts invertibly on $H^*(X; \mathbb{R})$. Topological K-theory only picks up Bott periodic phenomena, because multiplication by the Bott class $u \in \pi_2(KU)$ acts invertibly on $KU^*(X)$, and $u^{p-1} = v_1$ for each prime p. Complex bordism $MU_*(X)$ instead detects all levels of periodic phenomena. We can say that real cohomology, topological K-theory and complex bordism have *chromatic filtration* 0, 1 and ∞, respectively. A precise interpretation of this is that the spectra $H\mathbb{R}$ and KU are Bousfield local with respect to the Johnson–Wilson spectra $E(n)$ for $n = 0$ and 1, respectively, while MU is not $E(n)$-local for any finite n. Traditionally, an elliptic cohomology theory is a complex oriented Landweber exact cohomology theory associated to the formal group law of an elliptic curve. It will have chromatic filtration 2 when the elliptic curve admits a supersingular specialization, and so any cohomology theory of chromatic filtration 2 might loosely be called a form of elliptic cohomology. However, the formal group law origin of traditional elliptic cohomology is not of a directly geometric nature, and so there has been some lasting interest in finding a truly geometrically defined form of elliptic cohomology.

It is the aim of the present paper to introduce a geometrically defined cohomology theory that is essentially of chromatic filtration 2, or, more precisely, a connective form of such a theory. It therefore extends the above list of distinguished cohomology theories one step beyond topological K-theory, to a theory that will detect v_2-periodic phenomena, but will ignore the complexity of all higher v_n-periodicities for $n \geq 3$.

The theory that we will present is represented by the algebraic K-theory spectrum $K(\mathcal{V})$ of the Kapranov–Voevodsky 2-category of 2-vector spaces [15]. A 2-vector space is much like a complex vector space, but with all occurrences of complex numbers, sums, products and equalities replaced by finite-dimensional complex vector spaces, direct sums, tensor products and coherent isomorphisms, respectively. It is geometrically defined in the sense that the 0th cohomology group $K(\mathcal{V})^0(X)$ of a space X can be defined in terms of equivalence classes of *2-vector bundles* over X (or, more precisely, over the total space Y of a Serre fibration $Y \to X$ with acyclic homotopy fibers, i.e., an acyclic fibration) (cf. Theorem 4.10). A 2-vector bundle over X is a suitable bundle of categories, defined much like a complex vector bundle over X, but subject to the same replacements as above. The previously studied notion of a gerbe over X with band $\mathbb{C}^*$ is a special case of a 2-vector bundle, corresponding in the same way to a complex line bundle.

We conjecture in 5.1 that the spectrum $K(\mathcal{V})$ is equivalent to the algebraic K-theory spectrum $K(ku)$ of the connective topological K-theory spectrum ku, considered as a 'brave new ring', i.e., as an $\mathbb{S}$-algebra. This is a special case of a more general conjecture, where for a symmetric bimonoidal category $\mathcal{B}$ (which is a generalization of a commutative semi-ring) we compare the category of finitely generated free modules over $\mathcal{B}$ to the category of finitely generated free modules over the commutative $\mathbb{S}$-algebra $\mathbb{A} = \mathrm{Spt}(\mathcal{B})$ (which is a generalization of a commutative ring) associated to $\mathcal{B}$. The conjecture amounts to a form of 'positive thinking', asserting that for the purpose of forming algebraic K-theory spectra it should not matter whether we start with a semiring-like object (such as the symmetric bimonoidal category $\mathcal{B}$) or the ring-like object given by its additive Grothendieck group completion (such as the commutative $\mathbb{S}$-algebra $\mathbb{A}$). This idea originated with Marcel Bökstedt, and we are indebted to him for suggesting this approach. We have verified the conjecture in the case of actual commutative semi-rings, interpreted as symmetric bimonoidal categories that only have identity morphisms, and view this as strong support in favor of the conjecture.

Continuing, we know that $K(ku)$, or rather a spectrum very closely related to it, is essentially of chromatic filtration 2. For connective spectra, such as all those arising from algebraic K-theory, there is a more appropriate and flexible variation of the chromatic filtration that we call the *telescopic complexity* of the spectrum; cf. Definition 6.1. For example, integral and real cohomology have telescopic complexity 0, connective and periodic topological K-theory have telescopic complexity 1, and traditional elliptic cohomology has telescopic complexity 2.

It is known, by direct non-trivial calculations [3], that $K(\ell_p^\wedge)$ has telescopic complexity 2, where $\ell_p^\wedge$ is the connective p-complete Adams summand of topological K-theory and $p \geq 5$. The use of the Adams summand in place of the full connective p-complete topological K-theory spectrum $ku_p^\wedge$, as well as the hypothesis $p \geq 5$, are mostly technical assumptions that make the calculations manageable, and it seems very likely that also $K(ku_p^\wedge)$ will have telescopic complexity 2 for any prime p. It then follows from [9], if we assume the highly respectable Lichtenbaum–Quillen conjecture for $K(\mathbb{Z})$ at p, that also $K(ku)$ has telescopic complexity 2. In this sense we shall allow ourselves to think of $K(ku)$, and conjecturally $K(\mathcal{V})$, as a connective form of elliptic cohomology.

The definition of a 2-vector bundle is sufficiently explicit that it may carry independent interest. In particular, it may admit notions of *connective structure* and *curving*, generalizing the notions for gerbes [8, §5.3], such that to each 2-vector bundle $\mathcal{E}$ over X with connective structure there is an associated virtual vector bundle H over the free loop space $\mathcal{L}X = \mathrm{Map}(S^1, X)$, generalizing the *anomaly line bundle* for gerbes [8, §6.2]. If $\mathcal{E}$ is equipped with a curving, there probably arises an *action functional* for oriented compact surfaces over X (ibid.), providing a construction of an *elliptic object* over X in the sense of Segal [28]. Thus 2-vector bundles over X (with extra structure) may have naturally associated elliptic objects over X. However, we have not yet developed this theory properly, and shall therefore postpone its discussion to a later joint paper, which will also contain proofs of the results announced in the present paper. Some of the basic ideas presented here were sketched by the first author in [4].

The paper is organized as follows. In Section 2 we define a charted 2-vector bundle of rank n over a space X with respect to an open cover $\mathcal{U}$ that is indexed by a suitably partially ordered set $\mathcal{J}$. This corresponds to a Steenrod-style definition of a fiber bundle, with standard fiber the category $\mathcal{V}^n$ of n-tuples of finite-dimensional complex vector spaces, chosen trivializations over the chart domains in $\mathcal{U}$, gluing data that compare the trivializations over the intersection of two chart domains and coherence isomorphisms that systematically relate the two possible comparisons that result over the intersection of three chart domains. We also discuss when two such charted 2-vector bundles are to be viewed as equivalent, i.e., when they define the same abstract object.

In Section 3 we think of a symmetric bimonoidal category $\mathcal{B}$ as a generalized semi-ring, and make sense of the algebraic K-theory $K(\mathcal{B})$ of its 2-category of finitely generated free 'modules' $\mathcal{B}^n$. We define the weak equivalences $\mathcal{B}^n \to \mathcal{B}^n$ to be given by a monoidal category $\mathcal{M} = GL_n(\mathcal{B})$ of *weakly invertible* matrices over $\mathcal{B}$, cf. Definition 3.6, in line with analogous constructions for simplicial rings and $\mathbb{S}$-algebras [33]. It is a key point that we allow $GL_n(\mathcal{B})$ to contain more matrices than the strictly invertible ones, of which there are too few to yield an interesting theory. We also present an explicit bar construction $B\mathcal{M}$ that is appropriate for such monoidal categories. Our principal example is the symmetric bimonoidal category $\mathcal{V}$ of finite-dimensional complex vector spaces under direct sum and tensor product, for which the modules $\mathcal{V}^n$ are the 2-vector spaces of Kapranov and Voevodsky.

In Section 4 we bring these two developments together, by showing that the equivalence classes of charted 2-vector bundles of rank n over a (reasonable) space X is in natural bijection (Theorem 4.5) with the homotopy classes of maps from X to the geometric realization $|BGL_n(\mathcal{V})|$ of the bar construction on the monoidal category of weakly invertible $n \times n$ matrices over $\mathcal{V}$. The group of homotopy classes of maps from X to the algebraic K-theory space $K(\mathcal{V})$ is naturally isomorphic (Theorem 4.10) to the Grothendieck group completion of the abelian monoid of virtual 2-vector bundles over X, i.e., the 2-vector bundles $\mathcal{E} \downarrow Y$ over spaces Y that come equipped with an acyclic fibration $a : Y \to X$. Hence the contravariant homotopy functor represented by $K(\mathcal{V})$ is geometrically defined, in the sense that virtual 2-vector bundles over X are the (effective) cycles for this functor at X.

In Section 5 we compare the algebraic K-theory of the generalized semi-ring $\mathcal{B}$ to the algebraic K-theory of its additive group completion. To make sense of the latter as a ring object, as is necessary to form its algebraic K-theory, we pass to structured ring spectra, i.e., to the commutative $\mathbb{S}$-algebra $\mathbb{A} = \mathrm{Spt}(\mathcal{B})$. We propose that the resulting algebraic K-theory spectra $K(\mathcal{B})$ and $K(\mathbb{A})$ are weakly equivalent (Conjecture 5.1), and support this assertion by confirming that it holds true in the special case of a discrete symmetric bimonoidal category $\mathcal{B}$, i.e., a commutative semi-ring in the usual sense. In the special case of 2-vector spaces the conjecture asserts that $K(\mathcal{V})$ is the algebraic K-theory $K(ku)$ of connective topological K-theory ku viewed as a commutative $\mathbb{S}$-algebra.

In Section 6 we relate the spectrum $K(ku)$ to the algebraic K-theory spectrum $K(\ell_p^\wedge)$ of the connective p-complete Adams summand $\ell_p^\wedge$ of $ku_p^\wedge$. The latter theory $K(\ell_p^\wedge)$ is known (Theorem 6.4, [3]) to have telescopic complexity 2, and this section makes it plausible that also the former theory $K(ku)$ has telescopic complexity 2, and hence is a connective form of elliptic cohomology. Together with Conjecture 5.1 this says that (a) the generalized cohomology theory represented by $K(ku)$ is geometrically defined, because its 0th cohomology group, which is then represented by $K(\mathcal{V})$, is defined in terms of formal differences of virtual 2-vector bundles, and (b) that it has telescopic complexity 2, meaning that it captures one more layer of chromatic complexity than topological K-theory does.

2 Charted two-vector bundles

Definition 2.1. Let X be a topological space. An *ordered open cover* $(\mathcal{U}, \mathcal{J})$ of X is a collection $\mathcal{U} = \{U_\alpha \mid \alpha \in \mathcal{J}\}$ of open subsets $U_\alpha \subset X$, indexed by a partially ordered set $\mathcal{J}$, such that

(i) the U_α cover X in the sense that $\bigcup_\alpha U_\alpha = X$, and
(ii) the partial ordering on $\mathfrak{I}$ restricts to a total ordering on each finite subset $\{\alpha_0, \ldots, \alpha_p\}$ of $\mathfrak{I}$ for which the intersection $U_{\alpha_0, \ldots, \alpha_p} = U_{\alpha_0} \cap \cdots \cap U_{\alpha_p}$ is non-empty.

The partial ordering on $\mathfrak{I}$ makes the nerve of the open cover $\mathcal{U}$ an ordered simplicial complex, rather than just a simplicial complex. We say that $\mathcal{U}$ is a *good cover* if each finite intersection $U_{\alpha_0, \ldots, \alpha_p}$ is either empty or contractible.

Definition 2.2. Let X be a topological space, with an ordered open cover $(\mathcal{U}, \mathfrak{I})$, and let $n \in \mathbb{N} = \{0, 1, 2, \ldots\}$ be a non-negative integer. A *charted 2-vector bundle* $\mathcal{E}$ of rank n over X consists of

(i) an $n \times n$ matrix

$$E^{\alpha\beta} = (E_{ij}^{\alpha\beta})_{i,j=1}^n$$

of complex vector bundles over $U_{\alpha\beta}$, for each pair $\alpha < \beta$ in $\mathfrak{I}$, such that over each point $x \in U_{\alpha\beta}$ the integer matrix of fiber dimensions

$$\dim(E_x^{\alpha\beta}) = (\dim E_{ij,x}^{\alpha\beta})_{i,j=1}^n$$

is invertible, i.e., has determinant ± 1, and
(ii) an $n \times n$ matrix

$$\phi^{\alpha\beta\gamma} = (\phi_{ik}^{\alpha\beta\gamma})_{i,k=1}^n : E^{\alpha\beta} \cdot E^{\beta\gamma} \xrightarrow{\ \cong\ } E^{\alpha\gamma}$$

of vector bundle isomorphisms

$$\phi_{ik}^{\alpha\beta\gamma} : \bigoplus_{j=1}^n E_{ij}^{\alpha\beta} \otimes E_{jk}^{\beta\gamma} \xrightarrow{\ \cong\ } E_{ik}^{\alpha\gamma}$$

over $U_{\alpha\beta\gamma}$, for each triple $\alpha < \beta < \gamma$ in $\mathfrak{I}$, such that
(iii) the diagram

$$
\begin{array}{ccc}
E^{\alpha\beta} \cdot (E^{\beta\gamma} \cdot E^{\gamma\delta}) & \xrightarrow{\ \underline{\alpha}\ } & (E^{\alpha\beta} \cdot E^{\beta\gamma}) \cdot E^{\gamma\delta} \\
{\scriptstyle id \cdot \phi^{\beta\gamma\delta}} \big\downarrow & & \big\downarrow {\scriptstyle \phi^{\alpha\beta\gamma} \cdot id} \\
E^{\alpha\beta} \cdot E^{\beta\delta} \xrightarrow{\ \phi^{\alpha\beta\delta}\ } E^{\alpha\delta} & \xleftarrow{\ \phi^{\alpha\gamma\delta}\ } & E^{\alpha\gamma} \cdot E^{\gamma\delta}
\end{array}
$$

of vector bundle isomorphisms over $U_{\alpha\beta\gamma\delta}$ commutes, for each chain $\alpha < \beta < \gamma < \delta$ in $\mathfrak{I}$.

Here $\underline{\alpha}$ denotes the (coherent) natural associativity isomorphism for the matrix product $\cdot$ derived from the tensor product $\otimes$ of vector bundles. We call the $n \times n$ matrices $E^{\alpha\beta}$ and $\phi^{\alpha\beta\gamma}$ the *gluing bundles* and the *coherence isomorphisms* of the charted 2-vector bundle $\mathcal{E} \downarrow X$, respectively.

Definition 2.3. Let $\mathcal{E}$ and $\mathcal{F}$ be two charted 2-vector bundles of rank n over X, with respect to the same ordered open cover $(\mathcal{U}, \mathcal{I})$, with gluing bundles $E^{\alpha\beta}$ and $F^{\alpha\beta}$ and coherence isomorphisms $\phi^{\alpha\beta\gamma}$ and $\psi^{\alpha\beta\gamma}$, respectively. An *elementary change of trivializations* $(T^\alpha, \tau^{\alpha\beta})$ from $\mathcal{E}$ to $\mathcal{F}$ is given by

(i) an $n \times n$ matrix $T^\alpha = (T^\alpha_{ij})^n_{i,j=1}$ of complex vector bundles over U_α, for each α in $\mathcal{I}$, such that over each point $x \in U_\alpha$ the integer matrix of fiber dimensions $\dim(T^\alpha_x)$ has determinant ± 1, and

(ii) an $n \times n$ matrix of vector bundle isomorphisms

$$\tau^{\alpha\beta} = (\tau^{\alpha\beta}_{ij})^n_{i,j=1} : F^{\alpha\beta} \cdot T^\beta \xrightarrow{\cong} T^\alpha \cdot E^{\alpha\beta}$$

over $U_{\alpha\beta}$, for each pair $\alpha < \beta$ in $\mathcal{I}$, such that

(iii) the diagram

$$
\begin{array}{ccc}
F^{\alpha\beta} \cdot F^{\beta\gamma} \cdot T^\gamma & \xrightarrow{id \cdot \tau^{\beta\gamma}} F^{\alpha\beta} \cdot T^\beta \cdot E^{\beta\gamma} \xrightarrow{\tau^{\alpha\beta} \cdot id} & T^\alpha \cdot E^{\alpha\beta} \cdot E^{\beta\gamma} \\
{\scriptstyle \psi^{\alpha\beta\gamma} \cdot id} \downarrow & & \downarrow {\scriptstyle id \cdot \phi^{\alpha\beta\gamma}} \\
F^{\alpha\gamma} \cdot T^\gamma & \xrightarrow{\hspace{3cm} \tau^{\alpha\gamma} \hspace{3cm}} & T^\alpha \cdot E^{\alpha\gamma}
\end{array}
$$

(natural associativity isomorphisms suppressed) of vector bundle isomorphisms over $U_{\alpha\beta\gamma}$ commutes, for each triple $\alpha < \beta < \gamma$ in $\mathcal{I}$.

Definition 2.4. Let $(\mathcal{U}, \mathcal{I})$ and $(\mathcal{U}', \mathcal{I}')$ be two ordered open covers of X. Suppose that there is an order-preserving *carrier function* $c \colon \mathcal{I}' \to \mathcal{I}$ such that for each $\alpha \in \mathcal{I}'$ there is an inclusion $U'_\alpha \subset U_{c(\alpha)}$. Then $(\mathcal{U}', \mathcal{I}')$ is a *refinement* of $(\mathcal{U}, \mathcal{I})$.

Let $\mathcal{E}$ be a charted 2-vector bundle of rank n over X with respect to $(\mathcal{U}, \mathcal{I})$, with gluing bundles $E^{\alpha\beta}$ and coherence isomorphisms $\phi^{\alpha\beta\gamma}$. Let

$$c^* E^{\alpha\beta} = E^{c(\alpha)c(\beta)} | U'_{\alpha\beta}$$

for $\alpha < \beta$ in $\mathcal{I}'$ and

$$c^* \phi^{\alpha\beta\gamma} = \phi^{c(\alpha)c(\beta)c(\gamma)} | U'_{\alpha\beta\gamma}$$

for $\alpha < \beta < \gamma$ in $\mathcal{I}'$, be $n \times n$ matrices of vector bundles and vector bundle isomorphisms over $U'_{\alpha\beta}$ and $U'_{\alpha\beta\gamma}$, respectively. Then there is a charted 2-vector bundle $c^* \mathcal{E}$ of rank n over X with respect to $(\mathcal{U}', \mathcal{I}')$, with gluing bundles $c^* E^{\alpha\beta}$ and coherence isomorphisms $c^* \phi^{\alpha\beta\gamma}$. We say that $c^* \mathcal{E}$ is an *elementary refinement* of $\mathcal{E}$.

More generally, two charted 2-vector bundles of rank n over X are said to be *equivalent 2-vector bundles* if they can be linked by a finite chain of elementary changes of trivializations and elementary refinements. (This is the notion of equivalence that appears to be appropriate for our representability Theorem 4.5.)

Remark 2.5. A charted 2-vector bundle of rank 1 consists of precisely the data defining a *gerbe* over X with band $\mathbb{C}^*$, as considered e.g. by Giraud [10], Brylinski [8] and Hitchin [13, §1]. There is a unitary form of the definition above, with Hermitian gluing bundles and unitary coherence isomorphisms, and a unitary 2-vector bundle of rank 1 is nothing but a gerbe with band $U(1)$. In either case, the set of equivalence classes of $\mathbb{C}^*$-gerbes or $U(1)$-gerbes over X is in natural bijection with the third integral cohomology group $H^3(X; \mathbb{Z})$ [8, 5.2.10].

Definition 2.6. Let $\mathcal{E} \downarrow X$ be a charted 2-vector bundle of rank n, with notation as above, and let $a: Y \to X$ be a map of topological spaces. Then there is a charted 2-vector bundle $a^*\mathcal{E} \downarrow Y$ of rank n obtained from $\mathcal{E}$ by *pullback* along a. It is charted with respect to the ordered open cover $(\mathcal{U}', \mathcal{J})$ with $\mathcal{U}' = \{U'_\alpha = a^{-1}(U_\alpha) \mid \alpha \in \mathcal{J}\}$. It has gluing bundles $a^*E^{\alpha\beta}$ obtained by pullback of the matrix of vector bundles $E^{\alpha\beta}$ along $a: U'_{\alpha\beta} \to U_{\alpha\beta}$, and coherence isomorphisms $a^*\phi^{\alpha\beta\gamma}$ obtained by pullback of the matrix of vector bundle isomorphisms $\phi^{\alpha\beta\gamma}$ along $a: U'_{\alpha\beta\gamma} \to U_{\alpha\beta\gamma}$. By definition there is then a *map* of charted 2-vector bundles $\hat{a}: a^*\mathcal{E} \to \mathcal{E}$ covering $a: Y \to X$.

Definition 2.7. Let $\mathcal{E} \downarrow X$ and $\mathcal{F} \downarrow X$ be charted 2-vector bundles with respect to the same ordered open cover $(\mathcal{U}, \mathcal{J})$ of X, of ranks n and m, with gluing bundles $E^{\alpha\beta}$ and $F^{\alpha\beta}$ and coherence isomorphisms $\phi^{\alpha\beta\gamma}$ and $\psi^{\alpha\beta\gamma}$, respectively. Their *Whitney sum* $\mathcal{E} \oplus \mathcal{F} \downarrow X$ is then the charted 2-vector bundle of rank $(n + m)$ with gluing bundles given by the $(n + m) \times (n + m)$ matrix of vector bundles

$$\begin{pmatrix} E^{\alpha\beta} & 0 \\ 0 & F^{\alpha\beta} \end{pmatrix}$$

and coherence isomorphisms given by the $(n + m) \times (n + m)$ matrix of vector bundle isomorphisms

$$\begin{pmatrix} \phi^{\alpha\beta\gamma} & 0 \\ 0 & \psi^{\alpha\beta\gamma} \end{pmatrix}: \begin{pmatrix} E^{\alpha\beta} & 0 \\ 0 & F^{\alpha\beta} \end{pmatrix} \cdot \begin{pmatrix} E^{\beta\gamma} & 0 \\ 0 & F^{\beta\gamma} \end{pmatrix} \xrightarrow{\cong} \begin{pmatrix} E^{\alpha\gamma} & 0 \\ 0 & F^{\alpha\gamma} \end{pmatrix}.$$

There is an elementary change of trivializations from $\mathcal{E} \oplus \mathcal{F}$ to $\mathcal{F} \oplus \mathcal{E}$ given by the $(n + m) \times (n + m)$ matrix

$$T^{\alpha} = \begin{pmatrix} 0 & I_m \\ I_n & 0 \end{pmatrix}$$

for each α in $\mathfrak{J}$, and identity isomorphisms $\tau^{\alpha\beta}$. Here I_n denotes the identity $n \times n$ matrix, with the trivial rank 1-vector bundle in each diagonal entry and zero bundles elsewhere.

3 Algebraic K-theory of two-vector spaces

Let $(\mathcal{B}, \oplus, \otimes, \underline{0}, \underline{1})$ be a *symmetric bimonoidal category*, with sum and tensor functors

$$\oplus, \otimes \colon \mathcal{B} \times \mathcal{B} \longrightarrow \mathcal{B}$$

and zero and unit objects $\underline{0}, \underline{1}$ in $\mathcal{B}$. These satisfy associative, commutative and distributive laws, etc., up to a list of natural isomorphisms, and these isomorphisms are *coherent* in the sense that they fulfil a (long) list of compatibility conditions, as presented by Laplaza in [17, §1]. We say that $\mathcal{B}$ is a *bipermutative category* if the natural isomorphisms are almost all identity morphisms, except for the commutative laws for $\oplus$ and $\otimes$ and the left distributive law, and these in turn fulfil the (shorter) list of compatibility conditions listed by May in [20, §VI.3].

Suppose that $\mathcal{B}$ is *small*, i.e., that the class of objects of $\mathcal{B}$ is in fact a set. Let $\pi_0(\mathcal{B})$ be the set of path components of the geometric realization of $\mathcal{B}$. (Two objects of $\mathcal{B}$ are in the same path component if and only if they can be linked by a finite chain of morphisms in $\mathcal{B}$.) Then the sum and tensor functors induce sum and product pairings that make $\pi_0(\mathcal{B})$ into a commutative semi-ring with zero and unit. We can therefore think of the symmetric bimonoidal category $\mathcal{B}$ as a kind of generalized commutative semi-ring. Conversely, any commutative semi-ring may be viewed as a discrete category, with only identity morphisms, which is then a symmetric bimonoidal category.

The additive Grothendieck group completion $\mathrm{Gr}(\pi_0(\mathcal{B}))$ of the commutative semi-ring $\pi_0(\mathcal{B})$ is a commutative ring. Likewise, the geometric realization $|\mathcal{B}|$ can be group completed with respect to the symmetric monoidal pairing induced by the sum functor $\oplus$, and this group completion can take place at the categorical level, say by Quillen's construction $\mathcal{B}^{-1}\mathcal{B}$ [11] or its generalization $\mathcal{B}^{+} = E\mathcal{B} \times_{\mathcal{B}} \mathcal{B}^2$ due to Thomason [31, 4.3.1]. However, the tensor functor $\otimes$ does not readily extend to $\mathcal{B}^{-1}\mathcal{B}$, as was pointed out by Thomason [32]. So $\mathcal{B}^{-1}\mathcal{B}$ is a symmetric monoidal category, but usually not a symmetric bimonoidal category.

Example 3.1. Let $\mathcal{V}$ be the topological bipermutative category of finite-dimensional complex vector spaces, with set of objects $\mathbb{N} = \{0, 1, 2, \dots\}$ with $d \in \mathbb{N}$ interpreted as the complex vector space $\mathbb{C}^d$, and morphism spaces

$$
\mathcal{V}(d, e) = \begin{cases} U(d) & \text{if } d = e, \\ \emptyset & \text{otherwise} \end{cases}
$$

from d to e. The sum functor $\oplus$ takes (d, e) to $d + e$ and embeds $U(d) \times U(e)$ into $U(d + e)$ by the block sum of matrices. The tensor functor $\otimes$ takes (d, e) to de and maps $U(d) \times U(e)$ to $U(de)$ by means of the left lexicographic ordering, which identifies $\{1, \dots, d\} \times \{1, \dots, e\}$ with $\{1, \dots, de\}$. Both of these functors are continuous. The zero and unit objects are 0 and 1, respectively.

In this case, the semi-ring $\pi_0(\mathcal{V}) = \mathbb{N}$ is that of the non-negative integers, with additive group completion $\mathrm{Gr}(\mathbb{N}) = \mathbb{Z}$. The geometric realization $|\mathcal{V}| = \coprod_{d \geq 0} BU(d)$ is the classifying space for complex vector bundles, while its group completion $|\mathcal{V}^{-1}\mathcal{V}| \simeq \mathbb{Z} \times BU$ classifies virtual vector bundles. The latter space is the infinite loop space underlying the spectrum $ku = \mathrm{Spt}(\mathcal{V})$ that represents connective complex topological K-theory, which is associated to either of the symmetric monoidal categories $\mathcal{V}$ or $\mathcal{V}^{-1}\mathcal{V}$ by the procedure of Segal [27], as generalized by Shimada and Shimakawa [29] and Thomason [31, 4.2.1].

Definition 3.2. Let $(\mathcal{B}, \oplus, \otimes, \underline{0}, \underline{1})$ be a symmetric bimonoidal category. The category $M_n(\mathcal{B})$ of $n \times n$ *matrices over* $\mathcal{B}$ has objects the matrices $V = (V_{ij})_{i,j=1}^n$ with entries that are objects of $\mathcal{B}$, and morphisms the matrices $\phi = (\phi_{ij})_{i,j=1}^n$ with entries that are morphisms in $\mathcal{B}$. The source (domain) of ϕ is the matrix of sources of the entries ϕ_{ij}, and similarly for targets (codomains).

There is a *matrix multiplication functor*

$$
M_n(\mathcal{B}) \times M_n(\mathcal{B}) \xrightarrow{\quad\cdot\quad} M_n(\mathcal{B})
$$

that takes two matrices $U = (U_{ij})_{i,j=1}^n$ and $V = (V_{jk})_{j,k=1}^n$ to the matrix $W = U \cdot V = (W_{ik})_{i,k=1}^n$ with

$$
W_{ik} = \bigoplus_{j=1}^n U_{ij} \otimes V_{jk}
$$

for $i, k = 1, \dots, n$. In general, we need to make a definite choice of how the n-fold sum is to be evaluated, say by bracketing from the left. When the direct sum functor is strictly associative, as in the bipermutative case, the choice does not matter.

The *unit object* I_n of $M_n(\mathcal{B})$ is the $n \times n$ matrix with unit entries $\underline{1}$ on the diagonal and zero entries $\underline{0}$ everywhere else.

Proposition 3.3. $(M_n(\mathcal{B}), \cdot, I_n)$ *is a monoidal category.*

In other words, the functor $\cdot$ is associative up to a natural associativity isomorphism

$$\underline{\alpha}: U \cdot (V \cdot W) \overset{\cong}{\longrightarrow} (U \cdot V) \cdot W$$

and unital with respect to I_n up to natural left and right unitality isomorphisms. These are coherent, in the sense that they fulfil a list of compatibility conditions, including the Mac Lane–Stasheff pentagon axiom. The proof of the proposition is a direct application of Laplaza's first coherence theorem from [17, §7].

Definition 3.4. Let B be a commutative semi-ring with additive Grothendieck group completion the commutative ring $A = \mathrm{Gr}(B)$. Let $M_n(A)$ and $M_n(B)$ be the multiplicative monoids of $n \times n$ matrices with entries in A and B, respectively, and let $GL_n(A) \subset M_n(A)$ be the subgroup of invertible $n \times n$ matrices with entries in A, i.e., those whose determinant is a unit in A. Let the submonoid $GL_n(B) \subset M_n(B)$ be the pullback in the diagram

$$
\begin{array}{ccc}
GL_n(B) & \longrightarrow & GL_n(A) \\
\downarrow & & \downarrow \\
M_n(B) & \longrightarrow & M_n(A) \, .
\end{array}
$$

Example 3.5. When $B = \mathbb{N}$ and $A = \mathbb{Z}$, $GL_n(\mathbb{N}) = M_n(\mathbb{N}) \cap GL_n(\mathbb{Z})$ is the monoid of $n \times n$ matrices with non-negative integer entries that are invertible as integer matrices, i.e., have determinant ± 1. It contains the elementary matrices that have entries 1 on the diagonal and in one other place, and 0 entries elsewhere. This is a larger monoid than the subgroup of units in $M_n(\mathbb{N})$, which only consists of the permutation matrices.

Definition 3.6. Let $\mathcal{B}$ be a symmetric bimonoidal category. Let $GL_n(\mathcal{B}) \subset M_n(\mathcal{B})$ be the full subcategory with objects the matrices $V = (V_{ij})_{i,j=1}^n$ whose matrix of path components $[V] = ([V_{ij}])_{i,j=1}^n$ lies in the submonoid $GL_n(\pi_0(\mathcal{B})) \subset M_n(\pi_0(\mathcal{B}))$. We call $GL_n(\mathcal{B})$ the category of *weakly invertible $n \times n$ matrices over $\mathcal{B}$*.

Corollary 3.7. $(GL_n(\mathcal{B}), \cdot, I_n)$ *is a monoidal category.*

Definition 3.8. Let $(\mathcal{M}, \cdot, e)$ be a monoidal category, and write $[p] = \{0 < 1 < \cdots < p\}$. The *bar construction* $B\mathcal{M}$ is a simplicial category $[p] \mapsto B_p\mathcal{M}$. In simplicial degree p the category $B_p\mathcal{M}$ has objects consisting of

  (i) triangular arrays of objects $M^{\alpha\beta}$ of $\mathcal{M}$, for all $\alpha < \beta$ in $[p]$, and
 (ii) isomorphisms

$$\mu^{\alpha\beta\gamma} : M^{\alpha\beta} \cdot M^{\beta\gamma} \xrightarrow{\;\cong\;} M^{\alpha\gamma}$$

   in $\mathcal{M}$, for all $\alpha < \beta < \gamma$ in $[p]$, such that
(iii) the diagram of isomorphisms

$$
\begin{array}{ccc}
M^{\alpha\beta} \cdot (M^{\beta\gamma} \cdot M^{\gamma\delta}) & \xrightarrow{\;\underline{\alpha}\;} & (M^{\alpha\beta} \cdot M^{\beta\gamma}) \cdot M^{\gamma\delta} \\
{\scriptstyle id \cdot \mu^{\beta\gamma\delta}} \downarrow & & \downarrow {\scriptstyle \mu^{\alpha\beta\gamma} \cdot id} \\
M^{\alpha\beta} \cdot M^{\beta\delta} \xrightarrow{\mu^{\alpha\beta\delta}} & M^{\alpha\delta} & \xleftarrow{\mu^{\alpha\gamma\delta}} M^{\alpha\gamma} \cdot M^{\gamma\delta}
\end{array}
$$

commutes, for all $\alpha < \beta < \gamma < \delta$ in $[p]$.

Here $\underline{\alpha}$ is the associativity isomorphism for the monoidal operation $\cdot$ in $\mathcal{M}$.

The morphisms in $B_p\mathcal{M}$ from one object $(M_0^{\alpha\beta}, \mu_0^{\alpha\beta\gamma})$ to another $(M_1^{\alpha\beta}, \mu_1^{\alpha\beta\gamma})$ consist of a triangular array of morphisms $\phi^{\alpha\beta} : M_0^{\alpha\beta} \to M_1^{\alpha\beta}$ in $\mathcal{M}$ for all $\alpha < \beta$ in $[p]$, such that the diagram

$$
\begin{array}{ccc}
M_0^{\alpha\beta} \cdot M_0^{\beta\gamma} & \xrightarrow{\mu_0^{\alpha\beta\gamma}} & M_0^{\alpha\gamma} \\
{\scriptstyle \phi^{\alpha\beta} \cdot \phi^{\beta\gamma}} \downarrow & & \downarrow {\scriptstyle \phi^{\alpha\gamma}} \\
M_1^{\alpha\beta} \cdot M_1^{\beta\gamma} & \xrightarrow{\mu_1^{\alpha\beta\gamma}} & M_1^{\alpha\gamma}
\end{array}
$$

commutes, for all $\alpha < \beta < \gamma$ in $[p]$.

To allow for degeneracy operators f in the following paragraph, let $M^{\alpha\alpha} = e$ be the unit object of $\mathcal{M}$, let $\mu^{\alpha\alpha\beta}$ and $\mu^{\alpha\beta\beta}$ be the left and right unitality isomorphisms for $\cdot$, respectively, and let $\phi^{\alpha\alpha}$ be the identity morphism on e.

The simplicial structure on $B\mathcal{M}$ is given as follows. For each order-preserving function $f : [q] \to [p]$ let the functor $f^* : B_p\mathcal{M} \to B_q\mathcal{M}$ take the object $(M^{\alpha\beta}, \mu^{\alpha\beta\gamma})$ of $B_p\mathcal{M}$ to the object of $B_q\mathcal{M}$ that consists of the triangular array of objects $M^{f(\alpha)f(\beta)}$ for $\alpha < \beta$ in $[q]$ and the isomorphisms $\mu^{f(\alpha)f(\beta)f(\gamma)}$ for $\alpha < \beta < \gamma$ in $[q]$.

Each monoidal category $\mathcal{M}$ can be rigidified to an equivalent *strict monoidal category* $\mathcal{M}_s$, i.e., one for which the associativity isomorphism and the left and right unitality isomorphisms are all identity morphisms [18, XI.3.1]. The usual strict bar construction for $\mathcal{M}_s$ is a simplicial category $[p] \mapsto \mathcal{M}_s^p$, and corresponds in simplicial degree p to the full subcategory of $B_p\mathcal{M}_s$ where all the isomorphisms $\mu^{\alpha\beta\gamma}$ are identity morphisms.

Proposition 3.9. *The bar construction $B\mathcal{M}$ is equivalent to the strict bar construction $[p] \mapsto \mathcal{M}_s^p$ for any strictly monoidal rigidification $\mathcal{M}_s$ of $\mathcal{M}$.*

This justifies calling $B\mathcal{M}$ the bar construction. The proof is an application of Quillen's theorem A [22] and the coherence theory for monoidal categories.

Definition 3.10. Let $\operatorname{Ar}\mathcal{M} = \operatorname{Fun}([1], \mathcal{M})$ be the *arrow category* of $\mathcal{M}$, with the morphisms of $\mathcal{M}$ as objects and commutative square diagrams in $\mathcal{M}$ as morphisms. There are obvious *source* and *target functors* $s, t \colon \operatorname{Ar}\mathcal{M} \to \mathcal{M}$. Let $\operatorname{Iso}\mathcal{M} \subset \operatorname{Ar}\mathcal{M}$ be the full subcategory with objects the isomorphisms of $\mathcal{M}$.

Lemma 3.11. *Let $(\mathcal{M}, \cdot, e)$ be a monoidal category. The category $B_2\mathcal{M}$ is the limit of the diagram*

$$\mathcal{M} \times \mathcal{M} \xrightarrow{\;\cdot\;} \mathcal{M} \xleftarrow{\;s\;} \operatorname{Iso}\mathcal{M} \xrightarrow{\;t\;} \mathcal{M}.$$

For $p \geq 2$ each object or morphism of $B_p\mathcal{M}$ is uniquely determined by the collection of its 2-faces in $B_2\mathcal{M}$, which is indexed by the set of monomorphisms $f \colon [2] \to [p]$.

Consider the symmetric bimonoidal category $\mathcal{B}$ as a kind of generalized semi-ring. The sum and tensor operations in $\mathcal{B}$ make the product category $\mathcal{B}^n$ a generalized (right) module over $\mathcal{B}$, for each non-negative integer n. The collection of $\mathcal{B}$-module homomorphisms $\mathcal{B}^n \to \mathcal{B}^n$ is encoded in terms of (left) matrix multiplication by the monoidal category $M_n(\mathcal{B})$, and we shall interpret the monoidal subcategory $GL_n(\mathcal{B})$ as a category of *weak equivalences* $\mathcal{B}^n \xrightarrow{\sim} \mathcal{B}^n$. This motivates the following definition.

Definition 3.12. Let $\mathcal{B}$ be a symmetric bimonoidal category. The *algebraic K-theory* of the 2-category of (finitely generated free) modules over $\mathcal{B}$ is the loop space

$$K(\mathcal{B}) = \Omega B\Big(\coprod_{n \geq 0} |BGL_n(\mathcal{B})|\Big).$$

Here $|BGL_n(\mathcal{B})|$ is the geometric realization of the bar construction on the monoidal category $GL_n(\mathcal{B})$ of weakly invertible $n \times n$ matrices over $\mathcal{B}$. The block sum of matrices $GL_n(\mathcal{B}) \times GL_m(\mathcal{B}) \to GL_{n+m}(\mathcal{B})$ makes the coproduct $\coprod_{n \geq 0} |BGL_n(\mathcal{B})|$ a topological monoid. The looped bar construction ΩB provides a group completion of this topological monoid.

When $\mathcal{B} = \mathcal{V}$ is the category of finite-dimensional complex vector spaces, the (finitely generated free) modules over $\mathcal{V}$ are called *2-vector spaces*, and $K(\mathcal{V})$ is the algebraic K-theory of the 2-category of 2-vector spaces.

Let $GL_\infty(\mathcal{B}) = \operatorname{colim}_n GL_n(\mathcal{B})$ be the infinite stabilization with respect to block sum with the unit object in $GL_1(\mathcal{B})$, and write $B = \pi_0(\mathcal{B})$ and $A = \operatorname{Gr}(B)$. Then $K(\mathcal{B}) \simeq \mathbb{Z} \times |BGL_\infty(\mathcal{B})|^+$ by the McDuff–Segal group completion theorem [21]. Here the superscript '+' refers to Quillen's plus-construction with respect to the (maximal perfect) commutator subgroup of $GL_\infty(A) \cong \pi_1|BGL_\infty(\mathcal{B})|$; cf. Proposition 5.3 below.

4 Represented two-vector bundles

Let X be a topological space, with an ordered open cover $(\mathcal{U}, \mathcal{I})$. Recall that all morphisms in $\mathcal{V}$ are isomorphisms, so $\operatorname{Ar} GL_n(\mathcal{V}) = \operatorname{Iso} GL_n(\mathcal{V})$.

Definition 4.1. A *represented 2-vector bundle* $\mathcal{E}$ of rank n over X consists of:

(i) a *gluing map*
$$g^{\alpha\beta} : U_{\alpha\beta} \longrightarrow |GL_n(\mathcal{V})|$$
for each pair $\alpha < \beta$ in $\mathcal{I}$, and

(ii) a *coherence map*
$$h^{\alpha\beta\gamma} : U_{\alpha\beta\gamma} \longrightarrow |\operatorname{Ar} GL_n(\mathcal{V})|$$
satisfying $s \circ h^{\alpha\beta\gamma} = g^{\alpha\beta} \cdot g^{\beta\gamma}$ and $t \circ h^{\alpha\beta\gamma} = g^{\alpha\gamma}$ over $U_{\alpha\beta\gamma}$, for each triple $\alpha < \beta < \gamma$ in $\mathcal{I}$, such that

(iii) the *2-cocycle condition*
$$h^{\alpha\gamma\delta} \circ (h^{\alpha\beta\gamma} \cdot id) \circ \underline{\alpha} = h^{\alpha\beta\delta} \circ (id \cdot h^{\beta\gamma\delta})$$
holds over $U_{\alpha\beta\gamma\delta}$ for all $\alpha < \beta < \gamma < \delta$ in $\mathcal{I}$.

There is a suitably defined notion of *equivalence* of represented 2-vector bundles, which we omit to formulate here, but cf. Definitions 2.3 and 2.4.

Definition 4.2. Let $E(d) = EU(d) \times_{U(d)} \mathbb{C}^d \downarrow BU(d)$ be the universal $\mathbb{C}^d$-bundle over $BU(d)$. There is a universal $n \times n$ matrix
$$E = (E_{ij})_{i,j=1}^n$$

of Hermitian vector bundles over $|GL_n(\mathcal{V})|$. Over the path component

$$|GL_n(\mathcal{V})_D| = \prod_{i,j=1}^{n} BU(d_{ij})$$

for $D = (d_{ij})_{i,j=1}^n$ in $GL_n(\mathbb{N})$, the (i, j)th entry in E is the pullback of the universal bundle $E(d_{ij})$ along the projection $|GL_n(\mathcal{V})_D| \to BU(d_{ij})$.

Let $|\mathrm{Ar}\, U(d)|$ be the geometric realization of the arrow category $\mathrm{Ar}\, U(d)$, where $U(d)$ is viewed as a topological groupoid with one object. Each pair $(A, B) \in U(d)^2$ defines a morphism from C to $(A, B) \cdot C = BCA^{-1}$ so

$$|\mathrm{Ar}\, U(d)| \cong EU(d)^2 \times_{U(d)^2} U(d)$$

equals the Borel construction for this (left) action of $U(d)^2$ on $U(d)$. There are source and target maps $s, t \colon |\mathrm{Ar}\, U(d)| \to BU(d)$, which take the 1-simplex represented by a morphism (A, B) to the 1-simplices represented by the morphisms A and B, respectively. By considering each element in $U(d)$ as a unitary isomorphism $\mathbb{C}^d \to \mathbb{C}^d$ one obtains a universal unitary vector bundle isomorphism $\phi(d) \colon s^*E(d) \xrightarrow{\cong} t^*E(d)$.

There is a universal $n \times n$ matrix of unitary vector bundle isomorphisms

$$\phi \colon s^*E \cong t^*E$$

over $|\mathrm{Ar}\, GL_n(\mathcal{V})|$. Over the path component

$$|\mathrm{Ar}\, GL_n(\mathcal{V})_D| = \prod_{i,j=1}^{n} |\mathrm{Ar}\, U(d_{ij})|$$

for D as above, the (i, j)th entry in ϕ is the pullback of the universal isomorphism $\phi(d_{ij})$ along the projection $|\mathrm{Ar}\, GL_n(\mathcal{V})_D| \to |\mathrm{Ar}\, U(d_{ij})|$.

Lemma 4.3. *Let $\mathcal{E}$ be a represented 2-vector bundle with gluing maps $g^{\alpha\beta}$ and coherence maps $h^{\alpha\beta\gamma}$. There is an associated charted 2-vector bundle with gluing bundles*

$$E^{\alpha\beta} = (g^{\alpha\beta})^*(E)$$

over $U_{\alpha\beta}$ and coherence isomorphisms

$$\phi^{\alpha\beta\gamma} = (h^{\alpha\beta\gamma})^*(\phi) \colon E^{\alpha\beta} \cdot E^{\beta\gamma} = (g^{\alpha\beta} \cdot g^{\beta\gamma})^*(E) \xrightarrow{\cong} (g^{\alpha\gamma})^*(E) = E^{\alpha\gamma}$$

over $U_{\alpha\beta\gamma}$. This association induces a bijection between the equivalence classes of represented 2-vector bundles and the equivalence classes of charted 2-vector bundles of rank n over X.

Definition 4.4. Let 2-Vect$_n(X)$ be the set of equivalence classes of 2-vector bundles of rank n over X. For path-connected X let

$$2\text{-Vect}(X) = \coprod_{n \geq 0} 2\text{-Vect}_n(X).$$

Whitney sum (Definition 2.7) defines a pairing that makes 2-Vect(X) an abelian monoid.

Theorem 4.5. *Let X be a finite CW complex. There are natural bijections*

$$2\text{-Vect}_n(X) \cong [X, |BGL_n(\mathcal{V})|]$$

and

$$2\text{-Vect}(X) \cong [X, \coprod_{n \geq 0} |BGL_n(\mathcal{V})|].$$

To explain the first correspondence, from which the second follows, we use the following construction.

Definition 4.6. Let $(\mathcal{U}, \mathcal{I})$ be an ordered open cover of X. The *Mayer–Vietoris blow-up* $MV(\mathcal{U})$ of X with respect to $\mathcal{U}$ is the simplicial space with p-simplices

$$MV_p(\mathcal{U}) = \coprod_{\alpha_0 \leq \cdots \leq \alpha_p} U_{\alpha_0, \ldots, \alpha_p}$$

with $\alpha_0 \leq \cdots \leq \alpha_p$ in $\mathcal{I}$. The ith simplicial face map is a coproduct of inclusions $U_{\alpha_0, \ldots, \alpha_p} \subset U_{\alpha_0, \ldots, \hat{\alpha}_i, \ldots, \alpha_p}$, and similarly for the degeneracy maps. The inclusions $U_{\alpha_0, \ldots, \alpha_p} \subset X$ combine to a natural map $e \colon |MV(\mathcal{U})| \to X$, which is a (weak) homotopy equivalence.

Sketch of proof of Theorem 4.5. By Lemma 3.11, a simplicial map $g \colon MV(\mathcal{U}) \to |BGL_n(\mathcal{V})|$ is uniquely determined by its components in simplicial degrees 1 and 2. The first of these is a map

$$g_1 \colon MV_1(\mathcal{U}) = \coprod_{\alpha \leq \beta} U_{\alpha\beta} \longrightarrow |B_1 GL_n(\mathcal{V})| = |GL_n(\mathcal{V})|$$

which is a coproduct of gluing maps $g^{\alpha\beta} \colon U_{\alpha\beta} \to |GL_n(\mathcal{V})|$. The second is a map

$$g_2 \colon MV_2(\mathcal{U}) = \coprod_{\alpha \leq \beta \leq \gamma} U_{\alpha\beta\gamma} \to |B_2 GL_n(\mathcal{V})|.$$

The simplicial identities and Lemma 3.11 imply that g_2 is determined by g_1 and a coproduct of coherence maps $h^{\alpha\beta\gamma} \colon U_{\alpha\beta\gamma} \to |\text{Ar}\, GL_n(\mathcal{V})|$. Hence such a simplicial map g corresponds bijectively to a represented 2-vector bundle of rank n over X.

Any map $f: X \to |BGL_n(\mathcal{V})|$ can be composed with the weak equivalence $e: |MV(\mathcal{U})| \to X$ to give a map of spaces $fe: |MV(\mathcal{U})| \to |BGL_n(\mathcal{V})|$, which is homotopic to a simplicial map g if $\mathcal{U}$ is a good cover, and for reasonable X any open ordered cover can be refined to a good one. The homotopy class of f corresponds to the equivalence class of the represented 2-vector bundle determined by the simplicial map g.   □

Remark 4.7. We wish to interpret the 2-vector bundles over X as (effective) 0-cycles for some cohomology theory at X. Such theories are group-valued, so a first approximation to the 0th cohomology group at X could be the Grothendieck group $Gr(2\text{-Vect}(X))$ of formal differences of 2-vector bundles over X. The analogous construction for ordinary vector bundles works well to define topological K-theory, but for 2-vector bundles this algebraically group completed functor is not even representable, like in the case of the algebraic K-theory of a discrete ring. We thank Haynes Miller for reminding us of this issue.

Instead we follow Quillen and perform the group completion at the space level, which leads to the algebraic K-theory space

$$K(\mathcal{V}) = \Omega B\left(\coprod_{n \geq 0} |BGL_n(\mathcal{V})|\right)$$

$$\simeq \mathbb{Z} \times |BGL_\infty(\mathcal{V})|^+$$

from Definition 3.12. But what theory does this loop space represent? One interpretation is provided by the theory of virtual flat fibrations, presented by Karoubi in [16, Ch. III], leading to what we shall call virtual 2-vector bundles. Another interpretation could be given using the homology bordism theory of Hausmann and Vogel [12].

Definition 4.8. Let X be a space. An *acyclic fibration* over X is a Serre fibration $a: Y \to X$ such that the homotopy fiber at each point $x \in X$ has the integral homology of a point, i.e., $\tilde{H}_*(\text{hofib}_x(a); \mathbb{Z}) = 0$. A map of acyclic fibrations from $a': Y' \to X$ to $a: Y \to X$ is a map $f: Y' \to Y$ with $af = a'$.

A *virtual 2-vector bundle* over X is described by an acyclic fibration $a: Y \to X$ and a 2-vector bundle $\mathcal{E} \downarrow Y$. We write $\mathcal{E} \downarrow Y \xrightarrow{a} X$. Given a map $f: Y' \to Y$ of acyclic fibrations over X there is an induced 2-vector bundle $f^*\mathcal{E} \downarrow Y'$. The virtual 2-vector bundles described by $\mathcal{E} \downarrow Y \xrightarrow{a} X$

and $f^*\mathcal{E} \downarrow Y' \xrightarrow{a'} X$ are declared to be *equivalent* as virtual 2-vector bundles over X.

Lemma 4.9. *The abelian monoid of equivalence classes of virtual 2-vector bundles over X is the colimit*

$$\operatorname*{colim}_{a:\, Y \to X} 2\text{-Vect}(Y)$$

where $a: Y \to X$ ranges over the category of acyclic fibrations over X. Its Grothendieck group completion is isomorphic to the colimit

$$\operatorname*{colim}_{a:\, Y \to X} \text{Gr}(2\text{-Vect}(Y)).$$

The functor $Y \mapsto 2\text{-Vect}(Y)$ factors through the homotopy category of acyclic fibrations over X, which is directed.

The following result says that formal differences of virtual 2-vector bundles over X are the geometric objects that constitute cycles for the contravariant homotopy functor represented by the algebraic K-theory space $K(\mathcal{V})$. Compare [16, III.3.11]. So $K(\mathcal{V})$ represents sheaf cohomology for the topology of acyclic fibrations, with coefficients in the abelian presheaf $Y \mapsto \text{Gr}(2\text{-Vect}(Y))$ given by the Grothendieck group completion of the abelian monoid of equivalence classes of 2-vector bundles.

Theorem 4.10. *Let X be a finite CW complex. There is a natural group isomorphism*

$$\operatorname*{colim}_{a:\, Y \to X} \text{Gr}(2\text{-Vect}(Y)) \cong [X, K(\mathcal{V})]$$

where $a: Y \to X$ ranges over the category of acyclic fibrations over X. Restricted to $\text{Gr}(2\text{-Vect}(X))$ (with $a = id$) the isomorphism extends the canonical monoid homomorphism $2\text{-Vect}(X) \cong [X, \coprod_{n \geq 0} |BGL_n(\mathcal{V})|] \to [X, K(\mathcal{V})]$.

Remark 4.11. The passage to sheaf cohomology would be unnecessary if we replaced $\mathcal{V}$ by a different symmetric bimonoidal category $\mathcal{B}$ such that each $\pi_0(GL_n(\mathcal{B}))$ is abelian. This might entail an extension of the category of vector spaces to allow generalized vector spaces of arbitrary real, or even complex, dimension, parallel to the inclusion of the integers into the real or complex numbers. Such an extension is reminiscent of a category of representations of a suitable C^*-algebra, but we know of no clear interpretation of this approach.

5 Algebraic K-theory of topological K-theory

Is the contravariant homotopy functor $X \mapsto [X, K(\mathcal{V})] = K(\mathcal{V})^0(X)$ part of a cohomology theory, and, if so, what is the spectrum representing that theory?

The topological symmetric bimonoidal category $\mathcal{V}$ plays the role of a generalized commutative semi-ring in our definition of $K(\mathcal{V})$. Its additive group completion $\mathcal{V}^{-1}\mathcal{V}$ correspondingly plays the role of a generalized commutative ring. This may be tricky to realize at the level of symmetric bimonoidal categories, but the connective topological K-theory spectrum $ku = \mathrm{Spt}(\mathcal{V})$ associated to the additive topological symmetric monoidal structure of $\mathcal{V}$ is an E_∞ ring spectrum, and hence a commutative algebra over the sphere spectrum $\mathbb{S}$.

The algebraic K-theory of an $\mathbb{S}$-algebra $\mathbb{A}$ can on the one hand be defined as the Waldhausen algebraic K-theory [34] of a category with cofibrations and weak equivalences, with objects the finite cell $\mathbb{A}$-modules, morphisms the $\mathbb{A}$-module maps and weak equivalences the stable equivalences. Alternatively, it can be defined as a group completion

$$K(\mathbb{A}) = \Omega B\Big(\coprod_{n \geq 0} B\widehat{GL}_n(\mathbb{A})\Big)$$

where $\widehat{GL}_n(\mathbb{A})$ is essentially the topological monoid of $\mathbb{A}$-module maps $\mathbb{A}^n \to \mathbb{A}^n$ that are stable equivalences. The former definition produces a spectrum, so the space $K(\mathbb{A})$ is in fact an infinite loop space, and its deloopings represent a cohomology theory.

The passage from modules over the semi-ring object $\mathcal{V}$ to modules over the ring object ku corresponds to maps $|GL_n(\mathcal{V})| \to \widehat{GL}_n(ku)$ and a map $K(\mathcal{V}) \to K(ku)$.

Conjecture 5.1. *There is a weak equivalence $K(\mathcal{V}) \simeq K(ku)$. More generally, $K(\mathcal{B}) \simeq K(\mathbb{A})$ for each symmetric bimonoidal category $\mathcal{B}$ with associated commutative $\mathbb{S}$-algebra $\mathbb{A} = \mathrm{Spt}(\mathcal{B})$.*

Remark 5.2. The conjecture asserts that the contravariant homotopy functor $X \mapsto [X, K(\mathcal{V})]$ with 0-cycles given by the virtual 2-vector bundles over X is the 0th cohomology group for the cohomology theory represented by the spectrum $K(ku)$ given by the algebraic K-theory of connective topological K-theory. We consider the virtual 2-vector bundles over X to be sufficiently geometric objects (like complex vector bundles), that this cohomology theory then admits as geometric an interpretation as the classical examples of de Rham cohomology, topological K-theory and complex bordism.

As a first (weak) justification of this conjecture, recall that to the eyes of algebraic K-theory the block sum operation $(g, h) \mapsto \left[\begin{smallmatrix} g & 0 \\ 0 & h \end{smallmatrix}\right]$ is identified with the stabilized matrix multiplication $(g, h) \mapsto \left[\begin{smallmatrix} gh & 0 \\ 0 & I \end{smallmatrix}\right]$, where I is an identity matrix. The group completion in the definition of algebraic K-theory adjoins inverses to the block sum operation, and thus also to the stabilized matrix multiplication. In particular, for each elementary $n \times n$ matrix $e_{ij}(V)$ over $\mathcal{B}$ with (i, j)th off-diagonal entry equal to an object V of $\mathcal{B}$, the inverse matrix $e_{ij}(-V)$ is formally adjoined as far as algebraic K-theory is concerned. Hence the formal negatives $-V$ in $\mathcal{B}^{-1}\mathcal{B}$ are already present, in this weak sense.

A stronger indication that the conjecture should hold true is provided by the following special case. Recall that a commutative semi-ring is the same as a (small) symmetric bimonoidal category that is *discrete*, i.e., has only identity morphisms.

Proposition 5.3. *Let B be a commutative semi-ring, with additive Grothendieck group completion $A = \mathrm{Gr}(B)$. The semi-ring homomorphism $B \to A$ induces a weak equivalence*

$$BGL_\infty(B) \xrightarrow{\;\simeq\;} BGL_\infty(A)$$

and thus a weak equivalence $K(B) \simeq K(A)$. In particular, there is a weak equivalence $K(\mathbb{N}) \simeq K(\mathbb{Z})$.

A proof uses the following application of Quillen's Theorem B [22].

Lemma 5.4. *Let $f \colon M \to G$ be a monoid homomorphism from a monoid M to a group G. Write $mg = f(m) \cdot g$. Let $Q = B(*, M, G)$ be the category with objects $g \in G$ and morphisms $(m, g) \in M \times G$ from mg to g*

$$mg \xrightarrow{(m,g)} g \,.$$

Then there is a fiber sequence up to homotopy

$$|Q| \longrightarrow BM \xrightarrow{Bf} BG \,.$$

Sketch of proof of Proposition 5.3. Applying lemma 5.4 to the monoids $M_n = GL_n(B)$ and groups $G_n = GL_n(A)$ we obtain categories Q_n for each natural number n. There are stabilization maps $i \colon Q_n \to Q_{n+1}$, $M_n \to M_{n+1}$ and $G_n \to G_{n+1}$, with (homotopy) colimits Q_∞, M_∞ and G_∞, and a quasi-fibration

$$|Q_\infty| \longrightarrow BGL_\infty(B) \longrightarrow BGL_\infty(A) \,.$$

It suffices to show that each stabilization map $i: |Q_n| \to |Q_{n+1}|$ is weakly null-homotopic, because then $|Q_\infty|$ is weakly contractible.

For each full subcategory $K \subset Q_n$ with finitely many objects, the restricted stabilization functor $i|K$ takes g to $i(g) = \begin{bmatrix} g & 0 \\ 0 & 1 \end{bmatrix}$. It receives a natural transformation from a functor $j: K \to Q_{n+1}$ that maps g to $j(g) = \begin{bmatrix} g & v \\ 0 & 1 \end{bmatrix}$ for some column vector $v = v(g)$ with positive entries in B. The trick is to construct $v(g)$ inductively for the finite set of objects g of K, so that $v(mg)$ is sufficiently positive with respect to $m \cdot v(g)$ for all morphisms $mg \to g$ in K.

Furthermore, the finiteness of K ensures that there is a row vector w with entries in B and an object $h = \begin{bmatrix} I_n & 0 \\ -w & 1 \end{bmatrix}$ of Q_{n+1} such that there is a natural transformation from j to the constant functor to h. These two natural transformations provide a homotopy from $i|K$ to a constant map. As K was arbitrary with finitely many objects, this means that i is weakly null-homotopic. $\square$

Remark 5.5. If there exists a symmetric bimonoidal category $\mathcal{W}$ and a functor $\mathcal{V} \to \mathcal{W}$ of symmetric bimonoidal categories that induces an additive equivalence from $\mathcal{V}^{-1}\mathcal{V}$ to $\mathcal{W}$, then most likely the line of argument sketched above in the case of commutative semi-rings can be adapted to the symmetric bimonoidal situation. This could provide one line of proof toward Conjecture 5.1. Similar remarks apply for a general symmetric bimonoidal category $\mathcal{B}$ in place of $\mathcal{V}$.

6 Forms of elliptic cohomology

In this section we shall view the algebraic K-theory $K(\mathbb{A})$ of an $\mathbb{S}$-algebra $\mathbb{A}$ as a spectrum, rather than as a space.

We shall argue that the algebraic K-theory $K(ku)$ of the connective topological K-theory spectrum ku is a connective form of elliptic cohomology, in the sense that it detects homotopy theoretic phenomena related to v_2-periodicity, much like how topological K-theory detects phenomena related to v_1-periodicity (which is really the same as Bott periodicity) and how rational cohomology detects phenomena related to v_0-periodicity. Furthermore, from this point of view the homotopy type of $K(ku)$ is robust with respect to changes in the interpretation of the phrase 'algebraic K-theory of topological K-theory'.

We first introduce a filtration of the class of spectra that is related to the *chromatic filtration* given by the property of being Bousfield local with respect to some Johnson–Wilson theory $E(n)$ (cf. Ravenel [24, §7]), but is more appropriate for the connective spectra that arise from algebraic K-theory. Our notion is also more closely linked to aspects of v_n-periodicity than to being $E(n)$-local.

Let p be a prime, $K(n)$ the nth Morava K-theory at p and F a p-local finite CW spectrum. The least number $0 \le n < \infty$ such that $K(n)_*(F)$ is non-trivial is called the *chromatic type* of F. (Only contractible spectra have infinite chromatic type.) By the Hopkins–Smith periodicity theorem [14, Thm. 9], F admits a v_n-*self map* $v \colon \Sigma^d F \to F$ such that $K(m)_*(v)$ is an isomorphism for $m = n$ and zero for $m \ne n$. The v_n-self map is sufficiently unique for the mapping telescope

$$ v^{-1} F = \mathrm{Tel}\big(F \xrightarrow{\ v\ } \Sigma^{-d} F \xrightarrow{\ v\ } \ldots \big) $$

to be well-defined up to homotopy. The class of all p-local finite CW spectra of chromatic type $\ge n$ is closed under weak equivalences and the formation of homotopy cofibers, desuspensions and retracts, so we say that the full subcategory that it generates is a *thick* subcategory. By the Hopkins–Smith thick subcategory theorem [14, Thm. 7], any thick subcategory of the category of p-local finite CW spectra has this precise form, for a unique number $0 \le n \le \infty$.

Definition 6.1. Let X be a spectrum, and let $\mathcal{T}_X$ be the full subcategory of p-local finite CW spectra F for which the localization map

$$ F \wedge X \longrightarrow v^{-1} F \wedge X $$

induces an isomorphism on homotopy groups in all sufficiently high degrees. Then $\mathcal{T}_X$ is a thick subcategory, hence consists of the spectra F of chromatic type $\ge n$ for some unique number $0 \le n \le \infty$. We call this number $n = \mathrm{telecom}(X)$ the *telescopic complexity* of X. (This abbreviation is due to Matthew Ando.)

Lemma 6.2. *If Y is the k-connected cover of X, for some integer k, then X and Y have the same telescopic complexity.*

Let $X \to Y \to Z$ be a cofiber sequence and

$$ m = \max\{\mathrm{telecom}(X), \mathrm{telecom}(Y)\}. $$

If $\mathrm{telecom}(X) \ne \mathrm{telecom}(Y)$ *then* $\mathrm{telecom}(Z) = m$, *otherwise* $\mathrm{telecom}(Z) \le m$.

If Y is a (de-)suspension of X then X and Y have the same telescopic complexity.

If Y is a retract of X then $\mathrm{telecom}(Y) \le \mathrm{telecom}(X)$.

If X is an $E(n)$-local spectrum then X has telescopic complexity $\le n$.

Examples 6.3. (1) Integral, rational, real and complex cohomology ($H\mathbb{Z}$, $H\mathbb{Q}$, $H\mathbb{R}$ or $H\mathbb{C}$) all have telescopic complexity 0.

(2) Connective or periodic, real or complex topological K-theory (ko, ku, KO or KU) all have telescopic complexity 1. The étale K-theory $K^{et}(R)$ of a ring $R = \mathcal{O}_{F,S}$ of S-integers in a local or global number field has telescopic complexity 1, and so does the algebraic K-theory $K(R)$ if the Lichtenbaum–Quillen conjecture holds for the ring R.

(3) An Ando–Hopkins–Strickland [1] *elliptic spectrum* (E, C, t) has telescopic complexity ≤ 2, and the telescopic complexity equals 2 if and only if the elliptic curve C over $R = \pi_0(E)$ has a supersingular specialization over some point of $\mathrm{Spec}(R)$.

(4) The Hopkins–Mahowald–Miller *topological modular forms* spectra tmf and TMF have telescopic complexity 2.

(5) The Johnson–Wilson spectrum $E(n)$ and its connective form, the Brown–Peterson spectrum $BP\langle n\rangle$, both have telescopic complexity n.

(6) The sphere spectrum $\mathbb{S}$ and the complex bordism spectrum MU have infinite telescopic complexity.

Let $V(1)$ be the four-cell Smith–Toda spectrum with $BP_*(V(1)) = BP_*/(p, v_1)$. For $p \geq 5$ it exists as a commutative ring spectrum. It has chromatic type 2, and there is a v_2-self map $v: \Sigma^{2p^2-2}V(1) \to V(1)$ inducing multiplication by the class $v_2 \in \pi_{2p^2-2}V(1)$. We write $V(1)_*(X) = \pi_*(V(1) \wedge X)$ for the $V(1)$-homotopy groups of X, which are naturally a graded module over $P(v_2) = \mathbb{F}_p[v_2]$.

Let $X_{(p)}$ and $X_p^\wedge$ denote the p-localization and p-completion of a spectrum X, respectively. The first Brown–Peterson spectrum $\ell = BP\langle 1\rangle$ is the connective p-local Adams summand of $ku_{(p)}$, and its p-completion $\ell_p^\wedge$ is the connective p-complete Adams summand of $ku_p^\wedge$. These are all known to be commutative $\mathbb{S}$-algebras.

The spectrum $TC(\ell_p^\wedge)$ occurring in the following statement is the *topological cyclic homology* of $\ell_p^\wedge$, as defined by Bökstedt, Hsiang and Madsen [5]. The theorem is proved in [3, 0.3] by an elaborate but explicit computation of its $V(1)$-homotopy groups, starting from the corresponding $V(1)$-homotopy groups of the topological Hochschild homology $THH(\ell_p^\wedge)$.

Theorem 6.4 (Ausoni–Rognes). *Let $p \geq 5$. The algebraic K-theory spectrum $K(\ell_p^\wedge)$ of the connective p-complete Adams summand $\ell_p^\wedge$ has telescopic complexity 2. More precisely, there is an exact sequence of $P(v_2)$-modules*

$$0 \to \Sigma^{2p-3}\mathbb{F}_p \to V(1)_*K(\ell_p^\wedge) \xrightarrow{\ \ \mathrm{trc}\ \ } V(1)_*TC(\ell_p^\wedge) \to \Sigma^{-1}\mathbb{F}_p \to 0$$

and an isomorphism of $P(v_2)$-modules

$$V(1)_* TC(\ell_p^\wedge) \cong P(v_2) \otimes$$

$$\left(E(\partial, \lambda_1, \lambda_2) \oplus E(\lambda_2)\{\lambda_1 t^d \mid 0 < d < p\} \oplus E(\lambda_1)\{\lambda_2 t^{dp} \mid 0 < d < p\} \right).$$

Here ∂, t, λ_1 and λ_2 have degrees -1, -2, $2p-1$ and $2p^2-1$, respectively. Hence $V(1)_* TC(\ell_p^\wedge)$ is free of rank $(4p+4)$ over $P(v_2)$, and agrees with its v_2-localization in sufficiently high degrees.

Since $K(\ell_p^\wedge)$ has telescopic complexity 2, it has a chance to detect v_2-periodic families in $\pi_* V(1)$. This is indeed the case. Let $\alpha_1 \in \pi_{2p-3} V(1)$ and $\beta_1' \in \pi_{2p^2-2p-1} V(1)$ be the classes represented in the Adams spectral sequence by the cobar 1-cycles $h_{10} = [\bar{\xi}_1]$ and $h_{11} = [\bar{\xi}_1^p]$, respectively. There are maps $V(1) \to v_2^{-1} V(1) \to L_2 V(1)$, and Ravenel [23, 6.3.22] computed

$$\pi_* L_2 V(1) \cong P(v_2, v_2^{-1}) \otimes E(\zeta)\{1, h_{10}, h_{11}, g_0, g_1, h_{11}g_0 = h_{10}g_1\}$$

for $p \geq 5$. Hence $\pi_* L_2 V(1)$ contains twelve v_2-periodic families. The telescope conjecture asserted that $v_2^{-1} V(1) \to L_2 V(1)$ might be an equivalence, but this is now considered to be unlikely [19]. The following detection result can be read off from [3, 4.8], and shows that $K(\ell_p^\wedge)$ detects the same kind of homotopy theoretic phenomena as $E(2)$ or an elliptic spectrum.

Proposition 6.5. *The unit map $\mathbb{S} \to K(\ell_p^\wedge)$ induces a $P(v_2)$-module homomorphism $\pi_* V(1) \to V(1)_* K(\ell_p^\wedge)$ which takes 1, α_1 and β_1' to 1, $t\lambda_1$ and $t^p \lambda_2$, respectively. Hence $V(1)_* K(\ell_p^\wedge)$ detects the v_2-periodic families in $\pi_* V(1)$ generated by these three classes.*

Turning to the whole connective p-complete topological K-theory spectrum $ku_p^\wedge$, there is a map $\ell_p^\wedge \to ku_p^\wedge$ of commutative $\mathbb{S}$-algebras. It induces a natural map $K(\ell_p^\wedge) \to K(ku_p^\wedge)$, and there is a transfer map $K(ku_p^\wedge) \to K(\ell_p^\wedge)$ such that the composite self-map of $K(\ell_p^\wedge)$ is multiplication by $(p-1)$. Hence the composite map is a p-local equivalence.

Lemma 6.6. *The algebraic K-theory spectrum $K(ku_p^\wedge)$ of connective p-complete topological K-theory $ku_p^\wedge$ contains $K(\ell_p^\wedge)$ as a p-local retract, hence has telescopic complexity ≥ 2.*

Most likely $K(ku_p^\wedge)$ also has telescopic complexity exactly 2. It may be possible to prove this directly by computing $V(1)_* TC(ku_p^\wedge)$, by similar methods as in [3], but the algebra involved for $ku_p^\wedge$ is much more intricate than it was for the Adams summand. Some progress in this direction has recently been made by Ausoni.

The following consequence of a theorem of the second author [9, p. 224] allows us to compare the algebraic K-theory of $ku_p^\wedge$ to that of the integral spectra ku and $K(\mathbb{C})$.

Theorem 6.7 (Dundas). *Let $\mathbb{A}$ be a connective $\mathbb{S}$-algebra. The commutative square*

$$
\begin{array}{ccc}
K(\mathbb{A}) & \longrightarrow & K(\mathbb{A}_p^\wedge) \\
\downarrow & & \downarrow \\
K(\pi_0(\mathbb{A})) & \longrightarrow & K(\pi_0(\mathbb{A}_p^\wedge))
\end{array}
$$

becomes homotopy Cartesian after p-completion.

We apply this with $\mathbb{A} = ku$ or $\mathbb{A} = K(\mathbb{C})$. Also in the second case $\mathbb{A}_p^\wedge \simeq ku_p^\wedge$, by Suslin's theorem on the algebraic K-theory of algebraically closed fields [30]. Then $\pi_0(\mathbb{A}) = \mathbb{Z}$ and $\pi_0(\mathbb{A}_p^\wedge) = \mathbb{Z}_p$. It is known that $K(\mathbb{Z}_p)$ has telescopic complexity 1, by Bökstedt–Madsen [6], [7] for p odd and by the third author [26] for $p = 2$. It is also known that $K(\mathbb{Z})$ has telescopic complexity 1 for $p = 2$, by Voevodsky's proof of the Milnor conjecture and Rognes–Weibel [25]. For p odd it would follow from the Lichtenbaum–Quillen conjecture for $K(\mathbb{Z})$ at p that $K(\mathbb{Z})$ has telescopic complexity 1, and this now seems to be close to a theorem by the work of Voevodsky, Rost and Positselski.

Proposition 6.8. *Suppose that $K(\mathbb{Z})$ has telescopic complexity 1 at a prime $p \geq 5$. Then $K(ku)$ and $K(K(\mathbb{C}))$ have the same telescopic complexity as $K(ku_p^\wedge)$, which is ≥ 2.*

More generally it is natural to expect that $K(K(R))$ has telescopic complexity 2 for each ring of S-integers $R = \mathcal{O}_{F,S}$ in a local or global number field F, including the initial case $K(K(\mathbb{Z}))$. A discussion of such a conjecture has been given in lectures by the third author, but should take place in the context of étale covers or Galois extensions of commutative $\mathbb{S}$-algebras, which would take us too far afield here.

The difference between the connective and periodic topological K-theory spectra ku and KU may also not affect their algebraic K-theories greatly.

There is a (localization) fiber sequence

$$K(\mathcal{C}^{KU}(ku)) \to K(ku) \to K(KU)$$

where $\mathcal{C}^{KU}(ku)$ is the category of finite cell ku-module spectra that become contractible when induced up to KU-modules [34, 1.6.4]. Such spectra have finite Postnikov towers with layers that are induced from finite cell $H\mathbb{Z}$-module spectra via the map $ku \to H\mathbb{Z}$, and so it is reasonable to expect that a generalized form of the dévissage theorem in algebraic K-theory applies to identify $K(\mathcal{C}^{KU}(ku))$ with $K(\mathbb{Z})$.

Proposition 6.9. *If there is a fiber sequence* $K(\mathbb{Z}) \to K(ku) \to K(KU)$ *and* $K(\mathbb{Z})$ *has telescopic complexity 1, at a prime* $p \geq 5$*, then the algebraic K-theory spectrum* $K(KU)$ *of the periodic topological K-theory spectrum KU has the same telescopic complexity as* $K(ku)$*, which is* ≥ 2*.*

Remark 6.10. Unlike traditional elliptic cohomology, the spectrum $K(ku)$ is not complex orientable. For the unit map of $K(\mathbb{Z})$ detects $\eta \in \pi_1(\mathbb{S})$ and factors as $\mathbb{S} \to K(ku) \to K(\mathbb{Z})$, where the first map is the unit of $K(ku)$ and the second map is induced by the map $ku \to H\mathbb{Z}$ of $\mathbb{S}$-algebras. Hence the unit map for $K(ku)$ detects η and cannot factor through the complex bordism spectrum MU, since $\pi_1(MU) = 0$. This should not be perceived as a problem, however, as e.g. also the topological modular forms spectrum tmf is not complex orientable. It seems more likely that $K(\overline{KU})$ can be complex oriented, where $\overline{KU}$ is an 'algebraic closure' of KU in the category of commutative $\mathbb{S}$-algebras.

References

[1] Ando, M., M. J. Hopkins and N. P. Strickland. Elliptic spectra, the Witten genus and the theorem of the cube. *Invent. Math.*, 146(3): 595–687, 2001.

[2] Atiyah, M. F. and G. B. Segal. The index of elliptic operators. II. *Ann. of Math. (2)*, 87: 531–545, 1968.

[3] Ausoni, Christian and John Rognes. Algebraic K-theory of topological K-theory. *Acta Math.*, 188(1): 1–39, 2002.

[4] Baas, Nils A. A general topological functor construction. 1998. NTNU preprint, Trondheim.

[5] Bökstedt, M., W. C. Hsiang and I. Madsen. The cyclotomic trace and algebraic K-theory of spaces. *Invent. Math.*, 111(3): 465–539, 1993.

[6] Bökstedt, M. and I. Madsen. Topological cyclic homology of the integers. *Astérisque*, (226): 7–8, 57–143, 1994. K-theory (Strasbourg, 1992).

[7] Bökstedt, M. and I. Madsen. Algebraic K-theory of local number fields: the unramified case. In *Prospects in Topology (Princeton, NJ, 1994)*, volume 138 of *Ann. of Math. Stud.*, pages 28–57. Princeton Univ. Press, Princeton, NJ, 1995.

[8] Brylinski, Jean-Luc. *Loop Spaces, Characteristic Classes and Geometric Quantization*, volume 107 of *Progress in Mathematics*. Birkhäuser Boston Inc., Boston, MA, 1993.

[9] Dundas, Bjørn Ian. Relative K-theory and topological cyclic homology. *Acta Math.*, 179(2): 223–242, 1997.

[10] Giraud, Jean. *Cohomologie non abélienne*. Springer-Verlag, Berlin, 1971. Die Grundlehren der mathematischen Wissenschaften, Band 179.

[11] Grayson, Daniel. Higher algebraic K-theory. II (after Daniel Quillen). In *Algebraic K-theory (Proc. Conf., Northwestern Univ., Evanston Ill., 1976)*, pages 217–240. *Lecture Notes in Math.*, Vol. 551. Springer, Berlin, 1976.

[12] Hausmann, Jean-Claude and Pierre Vogel. The plus construction and lifting maps from manifolds. In *Algebraic and Geometric Topology (Proc. Sympos. Pure Math., Stanford Univ., Stanford, Calif, 1976), Part 1, Proc. Sympos. Pure Math.*, XXXII, pages 67–76. Amer. Math. Soc., Providence, R.I., 1978.

[13] Hitchin, Nigel. Lectures on special Lagangian submanifolds. In *Winter School on Mirror Symmetry, Vector Bundles and Lagrangian Submanifolds (Cambridge, MA, 1999)*, volume 23 of *AMS/IP Stud. Adv. Math.*, pages 151–182. Amer. Math. Soc., Providence, RI, 2001.

[14] Hopkins, Michael J. and Jeffrey H. Smith. Nilpotence and stable homotopy theory. II. *Ann. of Math. (2)*, 148(1): 1–49, 1998.

[15] Kapranov, M. M. and V. A. Voevodsky. 2-categories and Zamolodchikov tetrahedra equations. In *Algebraic Groups and Their Generalizations: Quantum and Infinite-Dimensional Methods (University Park, PA, 1991)*, volume 56 of *Proc. Sympos. Pure Math.*, pages 177–259. Amer. Math. Soc., Providence, RI, 1994.

[16] Karoubi, Max. Homologie cyclique et K-théorie. *Astérisque* (149): 147, 1987.

[17] Laplaza, Miguel L. Coherence for distributivity. In *Coherence in Categories*, pages 29–65. *Lecture Notes in Math.*, Vol. 281. Springer, Berlin, 1972.

[18] Macl, Lane, Saunders. *Categories for the Working Mathematician. Graduate Texts in Mathematics*, Vol. 5. Springer-Verlag, New York, 1971.

[19] Mahowald, Mark, Douglas Ravenel and Paul Shick. The triple loop space approach to the telescope conjecture. In *Homotopy Methods in Algebraic Topology (Boulder, CO, 1999)* volume 271 of *Contemp. Math.*, pages 217–284. Amer. Math. Soc., Providence, RI, 2001.

[20] May, J. Peter. E_∞ *Ring Spaces and* E_∞ *Ring Spectra*. With contributions by Frank Quinn, Nigel Ray, and Jørgen Tornehave, *Lecture Notes in Mathematics*, Vol. 577. Springer-Verlag, Berlin, 1977.

[21] McDuff, D. and G. Segal. Homology fibrations and the 'group-completion' theorem. *Invent. Math.*, 31(3): 279–284, 1975/76.

[22] Quillen, Daniel. Higher algebraic K-theory. I. In *Algebraic K-theory, I: Higher K-theories (Proc. Conf., Battelle Memorial Inst., Seattle, Wash., 1972)*, pages 85–147. *Lecture Notes in Math.*, Vol. 341. Springer, Berlin, 1973.

[23] Ravenel, Douglas C. *Complex Cobordism and Stable Homotopy Groups of Spheres*, volume 121 of *Pure and Applied Mathematics*. Academic Press Inc., Orlando, FL, 1986.

[24] Ravenel, Douglas C. *Nilpotence and Periodicity in Stable Homotopy Theory*, volume 128 of *Annals of Mathematics Studies*. Princeton University Press, Princeton, NJ, 1992. Appendix C by Jeff Smith.

[25] Rognes, J. and C. Weibel. Two-primary algebraic K-theory of rings of integers in number fields, *J. Amer. Math. Soc.*, 13(1): 1–54, 2000. Appendix A by Manfred Kolster.

[26] Rognes, John. Algebraic K-theory of the two-adic integers. *J. Pure Appl. Algebra*, 134(3): 287–326, 1999.

[27] Segal, Graeme. Categories and cohomology theories. *Topology*, 13: 293–312, 1974.

[28] Segal, Graeme. Elliptic cohomology (after Landweber-Stong, Ochanine, Witten, and others). *Astérisque* (161–162): Exp. No. 695, 4, 187–201 (1989), 1988. Séminaire Bourbaki, Vol. 1987/88.

[29] Shimada, Nobuo and Kazuhisa Shimakawa. Delooping symmetric monoidal categories. *Hiroshima Math. J.*, 9(3): 627–645, 1979.

[30] Suslin, A. On the K-theory of algebraically closed fields. *Invent. Math.*, 73(2): 241–245, 1983.

[31] Thomason, R. W. Homotopy colimits in the category of small categories. *Math. Proc. Cambridge Philos. Soc.*, 85(1): 91–109, 1979.

[32] Thomason, R. W. Beware the phony multiplication on Quillen's $a^{-1}a$. *Proc. Amer. Math. Soc.*, 80(4): 569–573, 1980.

[33] Waldhausen, Friedhelm. Algebraic K-theory of topological spaces. I. In *Algebraic and Geometric Topology (Proc. Sympos. Pure Math., Stanford Univ., Stanford, Calif., 1976), Part 1, Proc. Sympos. Pure Math.*, XXXII, pages 35–60. Amer. Math. Soc., Providence, R.I., 1978.

[34] Waldhausen, Friedhelm. Algebraic K-theory of spaces. In *Algebraic and Geometric Topology (New Brunswick, N.J., 1983)*, volume 1126 of *Lecture Notes in Math.*, pages 318–419. Springer, Berlin, 1985.

E-mail: baas@math.ntnu.no and dundas@math.ntnu.no and rognes@math.uio.no

3

Geometric realization of the Segal–Sugawara construction

DAVID BEN-ZVI and EDWARD FRENKEL
University of Chicago and University of California, Berkeley.

To Graeme Segal on his 60th birthday

Abstract

We apply the technique of localization for vertex algebras to the Segal–Sugawara construction of an 'internal' action of the Virasoro algebra on affine Kac–Moody algebras. The result is a lifting of twisted differential operators from the moduli of curves to the moduli of curves with bundles, with arbitrary decorations and complex twistings. This construction gives a uniform approach to a collection of phenomena describing the geometry of the moduli spaces of bundles over varying curves: the KZB equations and heat kernels on non-abelian theta functions, their critical level limit giving the quadratic parts of the Beilinson–Drinfeld quantization of the Hitchin system, and their infinite level limit giving a Hamiltonian description of the isomonodromy equations.

1 Introduction

1.1 Uniformization

Let G be a complex connected simply connected simple algebraic group with Lie algebra $\mathfrak{g}$, and X a smooth projective curve over $\mathbb{C}$. The geometry of G-bundles on X is intimately linked to representation theory of the affine Kac–Moody algebra $\widehat{\mathfrak{g}}$, the universal central extension of the loop algebra $L\mathfrak{g} = \mathfrak{g} \otimes \mathcal{K}$, where $\mathcal{K} = \mathbb{C}((t))$. More precisely, let $G(\mathcal{O})$, where $\mathcal{O} = \mathbb{C}[[t]]$, denote the positive half of the loop group $G(\mathcal{K})$. There is a principal $G(\mathcal{O})$-bundle over the moduli space (stack) $\mathrm{Bun}_G(X)$ of G-bundles on X, which carries an infinitesimally simply transitive action of $L\mathfrak{g}$. This provides an infinitesimal 'uniformization' of the moduli of G-bundles. Moreover, this uniformization lifts to an infinitesimal action of $\widehat{\mathfrak{g}}$ on the 'determinant' line bundle

We would like to thank Tony Pantev, Spencer Bloch, Sasha Beilinson, Dennis Gaitsgory and Matthew Emerton for useful discussions, and Roman Fedorov for comments on an early draft.

$\mathcal{C}$ on $\mathrm{Bun}_G(X)$, whose sections give the non-abelian versions of the spaces of theta functions.

The geometry of the moduli space $\mathfrak{M}_{g,1}$ of smooth pointed curves of genus g is similarly linked to the Virasoro algebra Vir, the universal central extension of the Lie algebra $\mathrm{Der}\,\mathcal{K}$ of derivations of $\mathcal{K}$ (or vector fields on the punctured disc). Let $\mathrm{Aut}\,\mathcal{O}$ denote the group of automorphisms of the disc. Then there is a principal $\mathrm{Aut}\,\mathcal{O}$-bundle over $\mathfrak{M}_{g,1}$ which carries an infinitesimally transitive action of $\mathrm{Der}\,\mathcal{K}$. This uniformization lifts to an infinitesimal action of the Virasoro algebra on the Hodge line bundle $\mathcal{H}$ on $\mathfrak{M}_{g,1}$.

These uniformizations can be used, following [BB, BS], to construct *localization functors* from representations of the Virasoro and Kac–Moody algebras to sheaves on the corresponding moduli spaces. Using these localization functors, one can describe sheaves of modules over (twisted) differential operators, as well as sheaves on the corresponding (twisted) cotangent bundles over the moduli spaces. The sheaves of twisted differential operators and twisted symbols (functions on twisted cotangent bundles) themselves are particularly easy to describe in this fashion: they are the localizations of the vacuum modules of the respective Lie algebras.

The Virasoro algebra acts on the Kac–Moody algebra via its natural action by derivations on the space $\mathcal{K}$ of Laurent series. Therefore we form the semidirect product $\widetilde{\mathfrak{g}} = Vir \ltimes \widehat{\mathfrak{g}}$, which uniformizes the moduli stack $\mathrm{Bun}_{G,g,1}$ classifying G-bundles over varying pointed curves of genus g. There is now a two-parameter family of line bundles $\mathcal{L}_{k,c}$ on which the (two-parameter) central extension $\widetilde{\mathfrak{g}}$ acts, and the corresponding two-parameter families of sheaves of twisted differential operators and twisted symbols. Thus it is natural to look to the structure of $\widetilde{\mathfrak{g}}$ for information about the variation of G-bundles over the moduli of curves.

1.2 The Segal–Sugawara construction

The interaction between the variation of curves and that of bundles on curves is captured by a remarkable feature of affine Kac–Moody algebras, best seen from the perspective of the theory of vertex algebras. The Segal–Sugawara construction presents the action of the Virasoro algebra on $\widehat{\mathfrak{g}}$ as an *internal* action, by identifying infinite quadratic expressions in the Kac–Moody generators which satisfy the Virasoro relations. More precisely, the Virasoro algebra embeds in a certain completion of $U\widehat{\mathfrak{g}}$, and hence acts on all *smooth* representations of $\widehat{\mathfrak{g}}$ of non-critical levels. From the point of view of vertex algebras, the construction simply involves the identification of a conformal structure, i.e., a certain distinguished vector, in the vacuum module over $\widehat{\mathfrak{g}}$.

Our objective in this paper is to draw out the different geometric consequences of this construction in a simple uniform fashion. Our approach involves four steps:

- We consider the Segal–Sugawara construction as defining a $G(\mathcal{O})$–
  invariant embedding of the vacuum module of the Virasoro algebra into
  the vacuum module of the Virasoro and Kac–Moody algebra $\widetilde{\mathfrak{g}}$.
- Twists of the vacuum modules form sheaves of algebras on the relevant
  moduli spaces, and the above construction gives rise to a homomorphism of these sheaves of algebras.
- The sheaves of twisted differential operators on the moduli spaces are
  quotients of the above algebras, and the vertex algebraic description of
  the Segal–Sugawara operators guarantees that the homomorphism descends to a homomorphism between sheaves of differential operators.
- All of the constructions vary flatly with respect to the Kac–Moody level
  and the Virasoro central charge and possess various 'classical' limits
  which may be described in terms of the twisted cotangent bundles to
  the respective moduli spaces.

Our main result is the following theorem. Let $\Pi : \mathrm{Bun}_{G,g} \to \mathfrak{M}_g$ denote the
projection from moduli of curves and bundles to moduli of curves, $k, c \in \mathbb{C}$
(the level and charge), $\mu_{\mathfrak{g}} = h^\vee \dim \mathfrak{g}$ and $c_k = c - \dfrac{k \dim \mathfrak{g}}{k + h^\vee}$. We denote
by $\mathcal{D}_{k,c}$, $\mathcal{D}_{k/\mathfrak{M}_g}$, and $\mathcal{D}_{c_k}$ respectively the corresponding sheaves of twisted
differential operators on $\mathrm{Bun}_{G,g}$, relative to $\mathfrak{M}_g$, and on $\mathfrak{M}_g$. The sheaves of
twisted symbols (functions on twisted cotangent bundles) with twists λ, μ on
$\mathfrak{M}_g$ and $\mathrm{Bun}_{G,g}$ are denoted by $\mathcal{O}(T^*_{\lambda,\mu}\mathrm{Bun}_{G,g})$ and $\mathcal{O}(T^*_{\lambda\mu}\mathfrak{M}_g)$, respectively.

Theorem 1.1.
(1) *Let $k, c \in \mathbb{C}$ with k not equal to minus the dual Coxeter number $h^\vee$. There
    is a canonical homomorphism of sheaves of algebras on $\mathfrak{M}_g$*

$$\mathcal{D}_{c_k} \longrightarrow \Pi_* \mathcal{D}_{k,c}$$

and an isomorphism of twisted $\mathcal{D}$-modules on $\mathrm{Bun}_{G,g}$

$$\mathcal{D}_{k/\mathfrak{M}_g} \underset{\mathcal{O}}{\otimes} \Pi^* \mathcal{D}_{c_k} \cong \mathcal{D}_{k,c}.$$

(2) *Let $c \in \mathbb{C}$, $k = -h^\vee$. There is an algebra homomorphism*

$$\mathcal{O}(T^*_{\mu_{\mathfrak{g}}}\mathfrak{M}) \longrightarrow \Pi_* \mathcal{D}_{-h^\vee,c}.$$

(3) *For every λ, $\mu \in \mathbb{C}$ there is a (λ-Poisson) homomorphism*

$$\mathcal{O}(T^*_{\lambda\mu}\mathfrak{M}_g) \longrightarrow \Pi_* \mathcal{O}(T^*_{\lambda,\mu}\mathrm{Bun}_{G,g}).$$

1.2.1

Moreover, the homomorphisms (2) and (3) are suitably rescaled limits of (1). In fact we prove a stronger result, valid for moduli of curves and bundles with arbitrary 'decorations'. Namely, we consider moduli of curves with an arbitrary number of marked points, jets of local coordinates and jets of bundle trivializations at these points. (For the sake of simplicity of notation, we work with the case of a single marked point or none – the multipoint extension is straightforward.)

1.3 Applications

The homomorphism (1) allows us to lift vector fields on $\mathfrak{M}_g$ to differential operators on $\mathrm{Bun}_{G,g}$, which are first order along the moduli of curves and second order along moduli of bundles. In the special case of $k \in \mathbb{Z}_+$ and no decorations, this gives a direct algebraic construction of the heat operators acting on non-abelian theta functions (the global sections of $\mathcal{L}_{k,c}$ along $\mathrm{Bun}_G(X)$) constructed by different means in [ADW, Hi2, BK, Fa1]. Our approach via vertex algebras makes clear the compatibility of this heat equation with the projectively flat connection on the bundle of conformal blocks coming from conformal field theory ([TUY, FrS]) – the Knizhnik–Zamolodchikov–Bernard (KZB) equations, or higher genus generalization of the Knizhnik–Zamolodchikov equations, [KZ]. This compatibility has also been established in [La]. The important feature of our proof which follows immediately from part (1) of the above theorem is that the heat operators depend on two *complex* parameters k and c, allowing us to consider various limits. This answers some of the questions raised by Felder in his study of the general KZB equations over the moduli space of pointed curves for arbitrary complex levels (see [Fel], §6).

The geometric significance of the two classical limit statements (2), (3) comes from the identification of the twisted cotangent bundles of $\mathfrak{M}_g$ and $\mathrm{Bun}_G(X)$ as the moduli spaces $T_1^*\mathfrak{M}_g = \mathcal{P}\mathrm{roj}_g$ of algebraic curves with projective structures and $T_1^*\mathrm{Bun}_G(X) = \mathcal{C}\mathrm{onn}_G(X)$ of G–bundles with holomorphic connections, respectively. As the parameter k approaches the *critical level* $-h^\vee$, the component of the heat kernels along the moduli of curves drops out, and we obtain commuting global second-order operators on the moduli $\mathrm{Bun}_G(X)$ (for a fixed curve X) from linear functions on the space of projective structures on X. Thus we recover the quadratic part of the Beilinson–Drinfeld quantization of the Hitchin Hamiltonians (see [BD1]) in part (2). When we take k, c to infinity in different fashions, we obtain the classical limits in part (3). For $\lambda, \mu = 0$ (i.e., as the symbols of our operators) we recover the quadratic part of the Hitchin integrable system [Hi1], namely, the Hamiltonians corresponding to the trace of the square of a Higgs field. For $\lambda \neq 0$ we obtain a canonical (non-affine) decomposition of the twisted cotangent bundle $T^*_{\lambda,\mu} \operatorname{Bun}_G = \mathcal{E} \times \operatorname{Conn}^{\lambda,\mu}_{G,g}$, the space of *extended connections* ([BS, BZB2]), into a product over $\mathfrak{M}_g$ of the spaces of projective structures and flat connections. This recovers a construction of Bloch–Esnault and Beilinson [BE]. (This decomposition is also described from the point of view of kernel functions in [BZB1], where it is related to the Klein construction of projective connections from theta functions and used to give a conjectural coordinate system on the moduli of bundles.)

For $\lambda \neq 0$, the homomorphism (3) gives a Hamiltonian action of the vector fields on $\mathfrak{M}_g$ on the moduli space of flat G-bundles. This gives a time-dependent Hamiltonian system (see [M]), or flat symplectic connection, on the moduli $\operatorname{Conn}_{G,g}$ of flat connections, with times given by the moduli of decorated curves. By working with arbitrary level structures, we obtain a similar statement for moduli spaces of connections with *arbitrary* (regular or irregular) singularities. We show that this is in fact a Hamiltonian description of the equations of isomonodromic deformation of meromorphic connections with respect to variation of the decorated curve. In particular, since all of the maps of the theorem come from a single multi-parametric construction, Theorem 1.2 immediately implies the following picture suggested by the work of [LO] (see also [I]), with arrows representing degenerations:

Quadratic Hitchin System $\longleftarrow$ Isomonodromy Equations

$\uparrow$ $\uparrow$

Quadratic Beilinson–Drinfeld System $\longleftarrow$ Heat Operators/KZ Equations

Namely, the isospectral flows of the Hitchin system arise as a degeneration of the isomonodromy equations as $\lambda \to 0$ (see also [Kr]), and the isomonodromy Hamiltonians appear as a degeneration of the KZB equations, previously proved for the Schlesinger equations (isomonodromy for regular singularities in genus zero) in [Res] (see also [Har]).

1.3.1 Outline of the paper.

We begin in Section 2 with a review of the formalism of localization and the description of twisted differential operators and symbols from vacuum modules following [BB]. We then apply this formalism in the special case when the Lie algebras are the affine Kac–Moody and the Virasoro algebras. First, we review the necessary features of these Lie algebras in Section 3. Theorem 3.3.1 provides the properties of the Segal–Sugawara construction at the level of representations. Next, we describe in Section 4 the spaces on which representations of these Lie algebras localize. These are the moduli stacks of curves and bundles on curves. In Section 5 we describe the implications of the Segal–Sugawara constructions for the sheaves of differential operators on these moduli spaces. Finally, we consider the classical limits of the localization and their applications to the isomonodromy questions in Section 6.

2 Localization

In this section we review the localization construction for representations of Harish–Chandra pairs as modules over algebras of twisted differential operators or sheaves on twisted cotangent bundles. In particular we emphasize the localization of vacuum representations, which give the sheaves of twisted differential operators or twisted symbols (functions on twisted cotangent bundles) themselves.

2.1 Deformations and limits

We first review the standard pattern of deformations and limits of algebras (see [BB]).

First recall that by the Rees construction, a filtered vector space $B = \bigcup_i B_i$ has a canonical deformation to its associated graded gr B over the affine line $\mathbb{A}^1 = \text{Spec } \mathbb{C}[\lambda]$, with the fiber at the point λ isomorphic to B for $\lambda \neq 0$, and to gr B for $\lambda = 0$. The corresponding $\mathbb{C}[\lambda]$-module is just the direct sum $\bigoplus_i B_i$, on which λ acts by mapping each B_i to B_{i+1}.

Let $\mathfrak{g}$ be any Lie algebra. The universal enveloping algebra $U\mathfrak{g}$ has the canonical PBW filtration. The Rees construction then gives us a one-parameter family of algebras $U_\lambda \mathfrak{g}$, with the associated graded algebra $U_0 \mathfrak{g} = \text{Sym } \mathfrak{g}$ being the symmetric algebra. The λ-deformation rescales the bracket by λ, so that taking the λ-linear terms defines the standard Poisson bracket on the Sym $\mathfrak{g} = \mathbb{C}[\mathfrak{g}^*]$.

Let

$$0 \to \mathbb{C}\mathbf{1} \to \widehat{\mathfrak{g}} \to \mathfrak{g} \to 0$$

be a one-dimensional central extension of $\mathfrak{g}$. We define a two-parameter family of algebras

$$U_{k,\lambda}\widehat{\mathfrak{g}} = U_\lambda\widehat{\mathfrak{g}}/(\mathbf{1} - k \cdot 1)$$

which specializes to $U_\lambda\mathfrak{g}$ when $k = 0$. When $\lambda = 1$ we obtain the family $U_k\widehat{\mathfrak{g}}$ of level k enveloping algebras of $\widehat{\mathfrak{g}}$, whose representations are the same as representations of $\widehat{\mathfrak{g}}$ on which $\mathbf{1}$ acts by $k \cdot \mathrm{Id}$. This family may be extended to $(k, \lambda) \in \mathbb{P}^1 \times \mathbb{A}^1$, by defining a $\mathbb{C}[k, k^{-1}]$-algebra

$$\widehat{U} = \oplus \left(\frac{\lambda}{k}\right)^i (U_{k,\lambda\widehat{\mathfrak{g}}})_{\leq i}$$

and setting

$$U_{\infty,\lambda\widehat{\mathfrak{g}}} = \widehat{U}/k^{-1}\widehat{U}.$$

Equivalently, we consider the $\mathbb{C}[k^{-1}]$-lattice Λ inside $U_{k,\lambda}$ (for $k \in \mathbb{A}^1 \setminus 0$) generated by 1 and $\overline{x} = \frac{\lambda}{k}x$ for $x \in \mathfrak{g}$. The algebra $U_{\infty,\lambda\widehat{\mathfrak{g}}}$ is then identified with $\Lambda/k^{-1}\Lambda$. (Note that the algebras $U_{k,\lambda\widehat{\mathfrak{g}}}$ may also be obtained from the standard λ-deformation of the filtered algebras $U_k\widehat{\mathfrak{g}} = U_{k,1\widehat{\mathfrak{g}}}$.)

Let us choose a vector space splitting $\widehat{\mathfrak{g}} \cong \mathfrak{g} \oplus \mathbb{C}\mathbf{1}$, so that the bracket in $\widehat{\mathfrak{g}}$ is given by a two-cocycle $\langle \cdot, \cdot \rangle$ on $\mathfrak{g}$

$$[x, y]_{\widehat{\mathfrak{g}}} = [x, y]_{\mathfrak{g}} + \langle x, y \rangle \cdot \mathbf{1}$$

for $x, y \in \mathfrak{g} \subset \widehat{\mathfrak{g}}$. If we let $\overline{x}$ denote $\frac{\lambda}{k}x$ for $x \in \mathfrak{g} \subset \widehat{\mathfrak{g}}$, the algebra $U_{k,\lambda\widehat{\mathfrak{g}}}$ may be described as generated by elements $\overline{x}$, $x \in \mathfrak{g}$ and $\mathbf{1}$, with the relations

$$[\overline{x}, \overline{y}] = \frac{\lambda}{k}\left(\overline{[x, y]_{\mathfrak{g}}} + \lambda\langle x, y \rangle \cdot \mathbf{1}\right), \qquad [\mathbf{1}, \overline{x}] = 0.$$

In particular, in the limit $k \to \infty$ we obtain a commutative algebra

$$U_{\infty,\lambda\widehat{\mathfrak{g}}} \cong \mathrm{Sym}_\lambda \widehat{\mathfrak{g}} = \mathrm{Sym}\,\widehat{\mathfrak{g}}/(1 - \lambda)$$

with a Poisson structure.

The algebra $\mathrm{Sym}_\lambda \widehat{\mathfrak{g}}$ is nothing but the algebra of functions $\mathbb{C}[\widehat{\mathfrak{g}}_\lambda^*]$ on the hyperplane in $\widehat{\mathfrak{g}}^*$ consisting of functionals, whose value on $\mathbf{1}$ is λ. The functions on $\widehat{\mathfrak{g}}^*$ form a Poisson algebra, with the Kirillov–Kostant–Poisson bracket determined via the Leibniz rule by the Lie bracket on the linear functionals, which are elements of $\widehat{\mathfrak{g}}$ itself. Since $\mathbf{1}$ is a central element, $\mathrm{Sym}_\lambda \widehat{\mathfrak{g}}$ inherits a Poisson bracket as well.

Thus, we have defined a two-parameter family of 'twisted enveloping algebras' $U_{k,\lambda\widehat{\mathfrak{g}}}$ for $(k, \lambda) \in \mathbb{P}^1 \times \mathbb{A}^1$ specializing to $U_k\widehat{\mathfrak{g}}$ for $\lambda = 1$ and to the Poisson algebras $\mathrm{Sym}_\lambda \widehat{\mathfrak{g}}$ for $k = \infty$. The algebras $U_{k,0\widehat{\mathfrak{g}}}$ with $\lambda = 0$ are all isomorphic (as Poisson algebras) to the symmetric algebra $\mathrm{Sym}\,\mathfrak{g}$.

2.1.1 Harish-Chandra pairs.

We now suppose that the Lie algebra $\mathfrak{g}$ is part of a Harish-Chandra pair $(\mathfrak{g}, K)$. Thus K is an algebraic group, and we are given an embedding $\mathfrak{k} = \mathrm{Lie}\,K \subset \mathfrak{g}$ and an action of K on $\mathfrak{g}$ compatible with the adjoint action of K on $\mathfrak{k}$ and the action of $\mathfrak{k}$ on $\mathfrak{g}$. Suppose further that we have a central extension $\widehat{\mathfrak{g}}$, split over $\mathfrak{k}$. Then the action of K lifts to $\widehat{\mathfrak{g}}$ and $(\widehat{\mathfrak{g}}, K)$ is also a Harish-Chandra pair.

It follows that the subalgebra of $U_{k,\lambda}\widehat{\mathfrak{g}}$ generated by the elements $\overline{x}$ for $x \in \mathfrak{k}$ is isomorphic to $U_{\lambda/k}\mathfrak{k}$, degenerating to $\mathrm{Sym}\,\mathfrak{k}$ for $\lambda = 0$ or $\lambda = \infty$. Moreover, the action of $\mathfrak{k}$ on $U_{k,\lambda}\widehat{\mathfrak{g}}$ given by bracket with $x = \frac{k}{\lambda}\overline{x}$ preserves the lattice Λ and thus is well-defined in the limit $k = \infty$. (In this limit the action is generated by Poisson bracket with $\mathfrak{k} \subset \mathrm{Sym}\,\mathfrak{k} \subset \mathrm{Sym}_\lambda\widehat{\mathfrak{g}}$.) In particular $\mathfrak{k}$ acts on $U_{k,\lambda}\widehat{\mathfrak{g}}$ for all $(k,\lambda) \in \mathbb{P}^1 \times \mathbb{A}^1$, and so it makes sense to speak of $(U_{k,\lambda}\widehat{\mathfrak{g}}, K)$-modules.

2.2 Twisted differential operators and twisted cotangent bundles

We review briefly the notions of sheaves of twisted differential operators and of twisted symbols (functions on twisted cotangent bundles) following [BB].

Let M be a smooth variety equipped with a line bundle $\mathcal{L}$. The Atiyah sequence of $\mathcal{L}$ is an extension of Lie algebras

$$0 \to \mathcal{O}_M \to \mathcal{T}_{\mathcal{L}} \xrightarrow{a} \Theta_M \to 0$$

where $\mathcal{T}_{\mathcal{L}}$ is the Lie algebroid of infinitesimal automorphisms of $\mathcal{L}$, Θ_M is the tangent sheaf, and the anchor map a describes the action of $\mathcal{T}_{\mathcal{L}}$ on its subsheaf $\mathcal{O}_M$.

We can generalize the construction of the family of algebras $U_{k,\lambda}\widehat{\mathfrak{g}}$ from Section 2.1 to the Lie algebroid $\mathcal{T}_{\mathcal{L}}$. We first introduce the sheaf of unital associative algebras $\mathcal{D}_k$ generated by the Lie algebra $\mathcal{T}_{\mathcal{L}}$, with $1 \in \mathcal{O}_M \subset \mathcal{T}_{\mathcal{L}}$ identified with k times the unit. Due to this identification, $\mathcal{D}_k$ is naturally a filtered algebra. Thus we have the Rees λ-deformation $\mathcal{D}_{k,\lambda}$ from $\mathcal{D}_k$ to its associated graded algebra at $\lambda = 0$. Specializing to $\lambda = 0$ we obtain the symmetric algebra $\mathrm{Sym}\,\Theta_M$ with its usual Poisson structure, independently of k. We may also define a limit of the algebras $\mathcal{D}_{k,\lambda}$ as k goes to infinity. In order to do so we introduce a $\mathbb{C}[k,k^{-1}]$–algebra $\mathcal{D}_{\lambda,k}$ as

$$\mathcal{D}_{\lambda,k} = \bigoplus_{i=0}^{\infty} \left(\frac{\lambda}{k}\right)^i \cdot \mathcal{D}_{\leq i}(\mathcal{L}^{\otimes k})$$

and define the limit algebra

$$\mathcal{D}_{\infty,\lambda} = \mathcal{D}_{\lambda,k} / k^{-1} \cdot \mathcal{D}_{\lambda,k}.$$

This algebra is commutative, and hence inherits a Poisson structure from the deformation process.

2.2.1 Twisted differential operators.

Consider the sheaf $\mathcal{D}(\mathcal{L})$ of differential operators acting on sections of $\mathcal{L}$. This is a sheaf of filtered associative algebras, and the associated graded sheaf is identified with the sheaf of symbols (functions on the cotangent bundle) as Poisson algebras. Similarly, the sheaf of differential operators $\mathcal{D}(\mathcal{L}^{\otimes k})$ for any integer k is also a sheaf of filtered associative unital $\mathcal{O}$-algebras, whose associated graded is commutative and isomorphic (as a Poisson algebra) to the sheaf of functions on T^*M. The sheaf $\mathcal{D}_k$ defined above for an arbitrary $k \in \mathbb{C}$ shares these properties, which describe it as a *sheaf of twisted differential operators*. For $k \in \mathbb{Z}$ there is a canonical isomorphism of sheaves of twisted differential operators $\mathcal{D}_k \cong \mathcal{D}(\mathcal{L}^{\otimes k})$. In particular, the subsheaf $\mathcal{D}(\mathcal{L}^{\otimes k})_{\leq 1}$ of differential operators of order at most one is identified with the subsheaf $(\mathcal{D}_k)_{\leq 1}$, which is nothing but $\mathcal{T}_{\mathcal{L}}$ with the extension rescaled by k. The sheaf $\mathcal{D}_k$ for general k may also be described as a subquotient of the sheaf of differential operators on the total space of $\mathcal{L}^\times$, the complement of the zero section in $\mathcal{L}$ (namely, as a quantum reduction by the natural action of $\mathbb{C}^\times$).

2.2.2 Twisted cotangent bundles.

We may also associate to $\mathcal{L}$ a *twisted cotangent bundle* of M. This is the affine bundle $\mathcal{C}onn\,\mathcal{L}$ of connections on $\mathcal{L}$, which is a torsor for the cotangent bundle T^*M. For $\lambda \in \mathbb{C}$ we have a family of twisted cotangent bundles $T^*_\lambda M$, which for $\lambda \in \mathbb{Z}$ are identified with the T^*M-torsors $\mathcal{C}onn\,\mathcal{L}^{\otimes \lambda}$ of connections on $\mathcal{L}^{\otimes \lambda}$. Let $\mathcal{L}^\times$ denote the principal $\mathbb{C}^\times$-bundle associated to the line bundle $\mathcal{L}$. Then the space $T^*_\lambda M$ is identified with the Hamiltonian reduction (at $\lambda \in \mathbb{C} = (\mathrm{Lie}\,\mathbb{C}^\times)^*$) of the cotangent bundle $T^*\mathcal{L}^\times$ by the action of $\mathbb{C}^\times$. In particular, $T^*_\lambda M$ carries a canonical symplectic structure for any λ.

Thus the *twisted symbols*, i.e., functions on $T^*_\lambda M$ (pushed forward to M) form a sheaf of $\mathcal{O}_M$-Poisson algebras $\mathcal{O}(T^*_\lambda M)$. Its subsheaf of *affine* functions $\mathcal{O}(T^*_\lambda M)_{\leq 1}$ on the affine bundle $T^*_\lambda M$ forms a Lie algebra under the Poisson bracket. It is easy to check that there is an isomorphism of sheaves of Poisson algebras $\mathcal{D}_{\infty,\lambda} \cong \mathcal{O}(T^*_\lambda M)$, identifying $\mathcal{O}(T^*_\lambda M)_{\leq 1}$ with $\mathcal{T}_{\mathcal{L}}$ (with the bracket rescaled by λ).

Thus to a line bundle we have associated a two–parameter family of algebras $\mathcal{D}_{k,\lambda}$ for $(k, \lambda) \in \mathbb{P}^1 \times \mathbb{A}^1$. This specializes for $k \in \mathbb{Z}$, $\lambda = 1$ to the differential operators $\mathcal{D}(\mathcal{L}^{\otimes k})$, for $\lambda = 0$ and k arbitrary to symbols $\mathcal{O}(T^*M)$, and for $k = \infty$, $\lambda \in \mathbb{Z}$ to the Poisson algebra of twisted symbols, $\mathcal{O}(\mathcal{C}onn\,\mathcal{L}^{\otimes \lambda})$.

2.3 Localization

In this section we combine the algebraic picture of Section 2.1 with the geometric picture of Section 2.2 following [BB, BD1] (see also [FB], Ch. 16).

Let $\mathfrak{g}, \widehat{\mathfrak{g}}, K$ be as in Section 2.1. We will consider the following geometric situation: $\mathfrak{M}$ is a smooth scheme equipped with an action of $(\mathfrak{g}, K)$, in other words, an action of the Lie algebra $\mathfrak{g}$ on $\widehat{\mathfrak{M}}$ which integrates to an algebraic action of the algebraic group K. Let $\mathfrak{M} = \widehat{\mathfrak{M}}/K$ be the quotient, which is a smooth algebraic stack, so that $\pi : \widehat{\mathfrak{M}} \to \mathfrak{M}$ is a K-torsor. Thus if the $\mathfrak{g}$ action on $\widehat{\mathfrak{M}}$ is infinitesimally transitive, then given $\widehat{x} \in \widehat{\mathfrak{M}}$ and $x \in \mathfrak{M}$ its K-orbit, the formal completion of $\mathfrak{M}$ at x is identified with the double quotient of the formal group $\exp(\mathfrak{g})$ of $\mathfrak{g}$: by the formal group $\exp(\mathfrak{k})$ of $\mathfrak{k}$ on one side, and the formal stabilizer of $\widehat{x}$ on the other.

Let $\mathcal{L}$ denote a K-equivariant line bundle on $\widehat{\mathfrak{M}}$ and the corresponding line bundle on $\mathfrak{M}$. We assume that the action of $\mathfrak{g}$ on $\widehat{\mathfrak{M}}$ lifts to an action of $\widehat{\mathfrak{g}}$ on $\mathcal{L}$, so that $\mathbf{1}$ acts as the identity. For $k, \lambda \in \mathbb{P}^1 \times \mathbb{A}^1$, we have an algebra $U_{k,\lambda}\widehat{\mathfrak{g}}$ from Section 2.1 and a sheaf of algebras $\mathcal{D}_{k,\lambda}$ on $\widehat{\mathfrak{M}}$ from Section 2.2. The action of $\widehat{\mathfrak{g}}$ on $\mathcal{L}$ matches up these definitions: we have an algebra homomorphism

$$\mathcal{O}_{\widehat{\mathfrak{M}}} \otimes U_{k,\lambda}\widehat{\mathfrak{g}} \longrightarrow \mathcal{D}_{k,\lambda}, \qquad (k, \lambda) \in \mathbb{P}^1 \times \mathbb{A}^1.$$

This may also be explained by reduction. Namely, $\widehat{\mathfrak{g}}$ acts on the total space $\mathcal{L}^\times$ of the $\mathbb{C}^\times$-bundle associated to $\mathcal{L}$. Therefore $U\widehat{\mathfrak{g}}$ maps to differential operators on $\mathcal{L}^\times$. It also follows that $\widehat{\mathfrak{g}}$ acts in the Hamiltonian fashion on $T^*\mathcal{L}^\times$, so that $\mathrm{Sym}\,\widehat{\mathfrak{g}}$ maps in the Poisson fashion to symbols on $\mathcal{L}^\times$. The actions of $U_k\widehat{\mathfrak{g}} \to \mathcal{D}_k$ and $\mathrm{Sym}_\lambda\,\widehat{\mathfrak{g}} \to \mathcal{O}(T_\lambda^*\widehat{\mathfrak{M}})$ are then obtained from the quantum (respectively, classical) Hamiltonian reduction with respect to the $\mathbb{C}^\times$ action on $\mathcal{L}^\times$ and the moment values k (respectively, λ).

2.3.1 Localization functor.

The action of the Harish–Chandra pair $(\widehat{\mathfrak{g}}, K)$ on $\widehat{\mathfrak{M}}$ and $\mathcal{L}$ may be used to construct localization functors from $(\widehat{\mathfrak{g}}, K)$-modules to sheaves on $\mathfrak{M}$. Let M be a $U_{k,\lambda}\widehat{\mathfrak{g}}$-module with a compatible action of K. We then consider the sheaf of $\mathcal{D}_{k,\lambda}$-modules

$$\widehat{\Delta}(M) = \mathcal{D}_{k,\lambda} \underset{U_{k,\lambda}\widehat{\mathfrak{g}}}{\otimes} M$$

on $\widehat{\mathfrak{M}}$, whose fibers are the coinvariants of M under the stabilizers of the $\widehat{\mathfrak{g}}$-action on $\widehat{\mathcal{L}}^\times$. (By our hypotheses, these lift to $\widehat{\mathfrak{g}}$ the stabilizers of the $\mathfrak{g}$-action on $\widehat{\mathfrak{M}}$.) The sheaf $\widehat{\Delta}(M)$ is a K-equivariant $\mathcal{D}_{k,\lambda}$-module, and so it descends to a $\mathcal{D}_{k,\lambda}$-module $\Delta_{k,\lambda}(M)$ on $\mathfrak{M}$, the localization of M

$$\Delta_{k,\lambda}(M) = (\pi_*(\mathcal{D}_{k,\lambda} \underset{U_{k,\lambda}\widehat{\mathfrak{g}}}{\otimes} M))^K.$$

Equivalently, we may work directly on $\mathfrak{M}$ by twisting. Let

$$\mathcal{M} = \widehat{\mathfrak{M}} \underset{K}{\times} M = \pi_*(\mathcal{O}_{\widehat{\mathfrak{M}}} \otimes M)^K$$

denote the twist of M by the K-torsor $\widehat{\mathfrak{M}}$ over $\mathfrak{M}$ (i.e., the vector bundle on $\mathfrak{M}$ associated to the principal bundle $\widehat{\mathfrak{M}}$ and representation M). Then there is a surjection

$$\mathcal{M} \twoheadrightarrow \Delta_{k,\lambda}(M) \tag{2.1}$$

and $\Delta_{k,\lambda}(M)$ is the quotient of $\mathfrak{M}$ by the twist $(\mathfrak{g}_{\text{stab}})_{\widehat{\mathfrak{M}}} \subset (\mathfrak{g})_{\widehat{\mathfrak{M}}}$ of the sheaf of stabilizers.

Remark 2.1. The localization functor $\Delta_{k,\lambda}$ 'interpolates' between the localization

$$\Delta_k : (U_k\widehat{\mathfrak{g}}, K) - \text{mod} \quad \longrightarrow \quad \mathcal{D}_k - \text{mod}$$

as modules for twisted differential operators, and the classical localization

$$\overline{\Delta}_\lambda : (\text{Sym}_\lambda \widehat{\mathfrak{g}}, K) - \text{mod} \quad \longrightarrow \quad \mathcal{O}(T_\lambda^*\mathfrak{M}) - \text{mod}$$

as quasicoherent sheaves on $T_\lambda^*\mathfrak{M}$. The latter assigns to a module over the commutative ring $\text{Sym}_\lambda \widehat{\mathfrak{g}}$ a quasicoherent sheaf on $T_\lambda^*\widehat{\mathfrak{M}}$ via the embedding of $\text{Sym}_\lambda \widehat{\mathfrak{g}}$ into global twisted symbols. This K-equivariant sheaf descends to $\mathfrak{M}$, where it becomes a module over the sheaf of twisted symbols $\mathcal{O}(T_\lambda^*\mathfrak{M})$ and therefore we obtain a sheaf on the twisted cotangent bundle.

2.4 Localizing the vacuum module

The fundamental example of a $(\mathfrak{g}, K)$-module is the *vacuum module*

$$V_{\mathfrak{g},K} = \text{Ind}_{U\mathfrak{k}}^{U\mathfrak{g}} \mathbb{C} = U\mathfrak{g}/(U\mathfrak{g} \cdot \mathfrak{k})$$

(we will denote it simply by V when the relevant Harish-Chandra pair $(\mathfrak{g}, K)$ is clear). The vector in $V_{\mathfrak{g},K}$ corresponding to $1 \in \mathbb{C}$ is denoted by $|0\rangle$ and referred to as the vacuum vector. It is a cyclic vector for the action of $\mathfrak{g}$ on $V_{\mathfrak{g},K}$. The vacuum representation has the following universality property: given a $\mathfrak{k}$-invariant vector $m \in M$ in a representation of $\mathfrak{g}$, there exists a unique $\mathfrak{g}$-homomorphism $m_* : V_{\mathfrak{g},K} \to M$ with $m_*(|0\rangle) = m$.

An important feature of the localization construction for Harish-Chandra modules is that the sheaf of twisted differential operators $\mathcal{D}_k$ itself arises as the localization $\Delta(V_{\mathfrak{g},K})$ of the vacuum module. This is a generalization of the description of differential operators on a homogeneous space G/K as sections of the twist

$$\mathcal{V}_{\mathfrak{g},K} = \mathcal{P}_K \underset{K}{\times} V_{\mathfrak{g},K}$$

by the K-torsor $\mathcal{P}_K = G$ over G/K. Informally, the general statement follows by applying reduction from G/K to $\mathfrak{M}$, which is (formally) a quotient of the formal homogeneous space $\exp(\mathfrak{g})/\exp(\mathfrak{k})$.

Recall that $\pi : \widehat{\mathfrak{M}} \to \mathfrak{M}$ denotes a K-torsor with compatible $\mathfrak{g}$-action. We assume K is connected and affine. In particular, the functor π_* is exact.

Proposition 2.2.

(1) *The twist* $\mathcal{V} = \widehat{\mathfrak{M}} \underset{K}{\times} V$, *its subsheaf of invariants* $\mathcal{V}^K = \mathcal{O}_{\mathfrak{M}} \otimes V^K$, *and the localization* $\Delta(V)$ *are sheaves of algebras on* $\mathfrak{M}$.

(2) *The localization* $\Delta(V)$ *is naturally identified with the sheaf* $\mathcal{D}$ *of differential operators on* $\mathfrak{M}$.

(3) *The natural maps*

$$\mathcal{O}_{\mathfrak{M}} \otimes V^K \longrightarrow \mathcal{V} \longrightarrow \Delta(V) = \mathcal{D}_{\mathfrak{M}}$$

are algebra homomorphisms.

(4) *For any* $(\mathfrak{g}, K)$-*module* M, *the sheaves*

$$\mathcal{O}_{\mathfrak{M}} \otimes M^K \longrightarrow \mathcal{M} \longrightarrow \Delta(M)$$

are modules over the corresponding algebras in (3).

Proof. The vector space $V^K = [U\mathfrak{g}/U\mathfrak{g} \cdot \mathfrak{k}]^{\mathfrak{k}}$ is naturally an algebra, isomorphic to the quotient $N(I)/I$ of the normalizer $N(I)$ of the ideal $I = U\mathfrak{g} \cdot \mathfrak{k}$ by I. It is also isomorphic to the algebra $(\mathrm{End}_{\mathfrak{g}}(V))^{\mathrm{opp}}$ (with the opposite multiplication), thanks to the universality property of V, and thus to the algebra of endomorphisms of the functor of K-invariants (with the opposite multiplication). Hence $\mathcal{O}_{\mathfrak{M}} \otimes V^K$ is an $\mathcal{O}_{\mathfrak{M}}$-algebra.

The sheaf $\mathcal{O}_{\widehat{\mathfrak{M}}} \otimes U\mathfrak{g}$ is a sheaf of algebras, where $U\mathfrak{g}$ acts on $\mathcal{O}_{\widehat{\mathfrak{M}}}$ as differential operators, and so is $\mathcal{A} = \pi_*(\mathcal{O}_{\widehat{\mathfrak{M}}} \otimes U\mathfrak{g})$. Let $\mathcal{K}$ be the sheaf of $\mathcal{O}_{\mathfrak{M}}$-Lie algebras $\pi_*(\mathcal{O}_{\widehat{\mathfrak{M}}} \otimes \mathfrak{k})$ (note that the action of $\mathfrak{k}$ on $\mathcal{O}_{\widehat{\mathfrak{M}}}$ is $\pi^{-1}(\mathcal{O}_{\mathfrak{M}})$-linear). Then we have

$$\mathcal{V} = [\pi_*(\mathcal{O}_{\widehat{\mathfrak{M}}} \otimes V)]^K$$
$$= [\pi_*(\mathcal{O}_{\widehat{\mathfrak{M}}} \otimes V)]^{\mathcal{K}}$$
$$= [\mathcal{A}/\mathcal{A} \cdot \mathcal{K}]^{\mathcal{K}}$$

which is thus a sheaf of algebras (again as the normalizer of an ideal modulo that ideal).

On the other hand, the localization

$$\Delta(V) = [\pi_*(\mathcal{D}_{\widehat{\mathfrak{M}}} \otimes_{U\mathfrak{g}} V)]^K$$
$$= [\pi_*(\mathcal{D}_{\widehat{\mathfrak{M}}}/\mathcal{D}_{\widehat{\mathfrak{M}}} \cdot \mathcal{K})]^{\mathcal{K}}$$
$$= \mathcal{D}_{\mathfrak{M}}$$

since $\mathcal{K}$ is the sheaf of *vertical* vector fields. This is in fact the common description of differential operators on a quotient as the quantum hamiltonian (BRST) reduction of differential operators upstairs. The obvious maps are compatible with this definition of the algebra structures. This proves parts (1)–(3) of the proposition.

Finally, for part (4), observe that the $U\mathfrak{g}$-action on M descends to a $\mathcal{V}$-action on $\mathcal{M}$, since the ideal $\mathcal{K}$ acts trivially on the $\mathcal{K}$-invariants $\mathcal{M} = \pi_*(\mathcal{O}_{\widehat{\mathfrak{M}}} \otimes M)^{\mathcal{K}}$. Equivalently, the Lie algebroid $\widehat{\mathfrak{M}} \underset{K}{\times} \mathfrak{g}$ on $\mathfrak{M}$ acts on the twist $\mathcal{M}$, and the action descends to the quotient algebroid $\widehat{\mathfrak{M}} \underset{K}{\times} \mathfrak{g}/\mathfrak{k}$, which generates $\mathcal{V}$. The compatibility with the actions of invariants on $\mathcal{M}^K$ and of the localization $\mathcal{D}$ on the $\mathcal{D}$-module $\Delta(W)$ follow $\square$

Corollary 2.3. *The space of invariants V^K is naturally a subalgebra of the algebra $\Gamma(\mathfrak{M}, \mathcal{D})$ of global differential operators on $\mathfrak{M}$.*

2.4.1 Changing the vacuum.

Let $\overline{K} \subset K$ be a normal subgroup. Then a minor variation of the proof establishes that the twist of $\overline{K}$-invariants $V^{\overline{K}} \hookrightarrow \mathcal{V}$ is a subalgebra. Note moreover that $\mathcal{V}^{\overline{K}}$ depends only on the induced $K/\overline{K}$-torsor $\widehat{\mathfrak{N}} = (K/\overline{K})_{\widehat{\mathfrak{M}}}$

$$\mathcal{V}^{\overline{K}} = \widehat{\mathfrak{N}} \times_{K/\overline{K}} V^{\overline{K}}.$$

Now suppose $(\mathfrak{a}, K)$ is a sub-Harish-Chandra pair of $(\mathfrak{g}, K)$ (i.e. $\mathfrak{k} \subset \mathfrak{a} \subset \mathfrak{g}$). Then we have the two vacuum modules $V = V_{\mathfrak{g},K} \supset V_{\mathfrak{a},K}$ and the module $V_{\mathfrak{g},\mathfrak{a}} = \operatorname{Ind}_{\mathfrak{a}}^{\mathfrak{g}} \mathbb{C}$, which is not literally the vacuum module of a Harish-Chandra pair unless $\mathfrak{a}$ integrates to an algebraic group.

By Proposition 2.2, the twist $\mathcal{V}_{\mathfrak{a},K}$ and the localization $\Delta(V_{\mathfrak{a},K})$ – with respect to the Harish-Chandra pair $(\mathfrak{a}, K)$ – are sheaves of algebras, and act on $\mathcal{V}$ and $\Delta(V)$ respectively. We then have:

Lemma 2.4. *The quotients $\mathcal{V}/(\mathcal{V}_{\mathfrak{a},K} \cdot \mathcal{V})$ and $\Delta(V)/(\Delta(V_{\mathfrak{a},K}) \cdot \Delta(V))$ are isomorphic, respectively, to the twist $\mathcal{V}_{\mathfrak{g},\mathfrak{a}}$ and the localization $\Delta(V_{\mathfrak{g},\mathfrak{a}})$ with respect to the Harish-Chandra pair $(\mathfrak{g}, K)$.*

2.4.2 Twisted and deformed vacua.

In general, we define the twisted, classical and deformed vacuum representations $V_k(\widehat{\mathfrak{g}})$, $\overline{V}_\lambda(\widehat{\mathfrak{g}})$ and $V_{k,\lambda}$ of $U_k\widehat{\mathfrak{g}}$, $\operatorname{Sym}_\lambda \widehat{\mathfrak{g}}$ and $U_{k,\lambda}\widehat{\mathfrak{g}}$, respectively, as the induced representations

$$V_k = U_k\widehat{\mathfrak{g}} \underset{U\mathfrak{k}}{\otimes} \mathbb{C}$$

$$\overline{V}_\lambda = \operatorname{Sym}_\lambda \widehat{\mathfrak{g}} \underset{\operatorname{Sym}\mathfrak{k}}{\otimes} \mathbb{C}$$

$$V_{k,\lambda} = U_{k,\lambda}\widehat{\mathfrak{g}} \underset{U_{k,\lambda}\widehat{\mathfrak{k}}}{\otimes} \mathbb{C}$$

where we use the observation (Section 2.1.1) that $U_k\widehat{\mathfrak{k}}$ is isomorphic to $U\mathfrak{k}$ for

$k \neq \infty$ and to Sym $\mathfrak{k}$ for $k = \infty$. All of these representations carry compatible K-actions.

We then have a general localization principle, describing twisted differential operators and twisted symbols on $\mathfrak{M}$ in terms of the corresponding vacuum representations. The proof of Proposition 2.2 generalizes immediately to the family $V_{k,\lambda}$ over $\mathbb{P}^1 \times \mathbb{A}^1$. This is deduced formally from the analogous statement for homogeneous spaces G/K, twisted by an equivariant line bundle. In that case the algebra of functions on the cotangent space $T^*G/K|_{1K} = (\mathfrak{g}/\mathfrak{k})^*$ is identified with

$$\overline{V}_0 = \text{Sym}(\mathfrak{g}/\mathfrak{k}) = \text{Sym}\,\mathfrak{g} \underset{\text{Sym}\,\mathfrak{k}}{\otimes} \mathbb{C}$$

and sections of the twist of $\overline{V}_0$ give the sheaf of functions on T^*G/K. Note that $\overline{V}_\lambda$ is a (commutative) algebra; however, it only becomes Poisson after twisting or localization. To summarize, we obtain the following:

Proposition 2.5. *The localization $\Delta_{k,\lambda}(V_{k,\lambda})$ is canonically isomorphic to the sheaf of algebras $\mathcal{D}_{k,\lambda}$, for $(k, \lambda) \in \mathbb{P}^1 \times \mathbb{A}^1$. In particular $\Delta_k(V_k) \cong \mathcal{D}_k$, and $\overline{\Delta}_\lambda(\overline{V}_\lambda) \cong \mathcal{O}(T^*_\lambda \mathfrak{M})$ as Poisson algebras.*

3 Virasoro and Kac–Moody algebras

In this section we introduce the Lie algebras to which we wish to apply the formalism of localization outlined in the previous sections. These are the affine Kac–Moody algebras, the Virasoro algebra and their semi-direct product. We describe the Segal–Sugawara construction which expresses the action of the Virasoro algebra on an affine algebra as an 'internal' action. We interpret this construction in terms of a homomorphism between vacuum representations of the Virasoro and Kac–Moody algebras, and identify the critical and classical limits of these homomorphisms. In the subsequent sections we will use the localization of this construction to describe sheaves of differential operators on the moduli spaces of curves and bundles on curves.

3.1 Virasoro and Kac–Moody algebras

Let $\mathfrak{g}$ be a simple Lie algebra. It carries an invariant bilinear form $(\cdot, \cdot)$ normalized in the standard way so that the square length of the maximal root is equal to 2. We choose bases $\{J^a\}$, $\{J_a\}$ dual with respect to this bilinear form. The affine Kac–Moody Lie algebra $\widehat{\mathfrak{g}}$ is a central extension

$$0 \to \mathbb{C}\mathbf{K} \to \widehat{\mathfrak{g}} \to L\mathfrak{g} \to 0$$

of the loop algebra $L\mathfrak{g} = \mathfrak{g} \otimes \mathcal{K}$ of $\mathfrak{g}$, with (topological) generators $\{J_n^a = J^a \otimes t^n\}_{n \in \mathbb{Z}}$ and relations

$$[J_n^a, J_m^b] = [J^a, J^b]_{n+m} + (J^a, J^b)\delta_{n,-m}\mathbf{K}. \tag{3.2}$$

The central extension splits over the Lie subalgebra $\mathfrak{g}(\mathcal{O}) \subset \mathfrak{g}(\mathcal{K})$ (topologically spanned by $J_n^a, n \geq 0$), so that the affine proalgebraic group $G(\mathcal{O})$ acts on $\widehat{\mathfrak{g}}$. Thus we have a Harish-Chandra pair $(\widehat{\mathfrak{g}}, G(\mathcal{O}))$. We denote by

$$V_k = V_k(\widehat{\mathfrak{g}}) = U_k\widehat{\mathfrak{g}} \underset{U\mathfrak{g}(\mathcal{O})}{\otimes} \mathbb{C}$$

the corresponding vacuum module (see Section 2.4). More generally, let $\mathfrak{m} \subset \mathcal{O}$ denote the maximal ideal $t\mathbb{C}[[t]]$. Let $\mathfrak{g}_n(\mathcal{O}) = \mathfrak{g} \otimes \mathfrak{m}^n \subset \mathfrak{g}(\mathcal{O})$ denote the congruence subalgebra of level n, and $G_n(\mathcal{O}) \subset G(\mathcal{O})$ the corresponding algebraic group, consisting of loops which equal the identity to order n. We denote by

$$V_k^n = V_k^n(\widehat{\mathfrak{g}}) = U_k\widehat{\mathfrak{g}} \underset{U\mathfrak{g}_n(\mathcal{O})}{\otimes} \mathbb{C}$$

the vacuum module for $(\widehat{\mathfrak{g}}, G_n(\mathcal{O}))$.

Let Vir denote the Virasoro Lie algebra. This is a central extension

$$0 \to \mathbb{C}\mathbf{C} \to \text{Vir} \to \text{Der}\,\mathcal{K} \to 0$$

of the Lie algebra of derivations of the field $\mathcal{K}$ of Laurent series. It has (topological) generators $L_n = -t^{n+1}\partial_t, n \in \mathbb{Z}$, and the central element $\mathbf{C}$, with relations

$$[L_n, L_m] = (n - m)L_{n+m} + \frac{1}{12}(n^3 - n)\delta_{n,-m}\mathbf{C}. \tag{3.3}$$

The central extension splits over the Lie subalgebra $\text{Der}\,\mathcal{O} \subset \text{Der}\,\mathcal{K}$, topologically spanned by $L_n, n \geq -1$. Consider the module

$$\text{Vir}_c = U_c\text{Vir} \underset{U\,\text{Der}\,\mathcal{O}}{\otimes} \mathbb{C}.$$

Since $\text{Der}\,\mathcal{O}$ is not the Lie algebra of an affine group scheme, Vir_c is not strictly speaking a vacuum module of a Harish-Chandra pair.

The affine group scheme $\text{Aut}\,\mathcal{O}$ of all changes of coordinates on the disc fixes the closed point $t = 0$, so that $L_{-1} = -\partial_t$ is not in its Lie algebra, which we denote by $\text{Der}\,_1\mathcal{O}$. More generally, we set

$$\text{Der}\,_n(\mathcal{O}) = \mathfrak{m}^n\text{Der}\,\mathcal{O} \cong t^n\mathbb{C}[[t]]\partial_t$$

and let $\text{Aut}_n(\mathcal{O}) \subset \text{Aut}_0(\mathcal{O}) = \text{Aut}\,\mathcal{O}$ be the corresponding algebraic subgroups (for $n \geq 0$). Then we have Harish-Chandra pairs $(\text{Vir}, \text{Aut}_n\,\mathcal{O})$ and the corresponding vacuum modules Vir_c^n (thus we may denote $\text{Vir}_c = \text{Vir}_c^0$).

The Virasoro algebra acts on $\widehat{\mathfrak{g}}$, via its action as derivations of $\mathcal{K}$. In terms of our chosen bases this action is written as follows

$$[L_n, J_m^a] = -m J_{n+m}^a.$$

Let $\widetilde{\mathfrak{g}}$ be the resulting semidirect product Lie algebra $\mathrm{Vir} \ltimes \widehat{\mathfrak{g}}$. Thus $\widetilde{\mathfrak{g}}$ has topological generators

$$J_m^a \quad (m \in \mathbb{Z}, J^a \in \mathfrak{g}), \qquad L_n \quad (n \in \mathbb{Z}), \qquad \mathbf{K}, \mathbf{C}$$

with relations as above.

The central extension splits over the semidirect product $\widetilde{\mathfrak{g}}_+ = \mathrm{Der}\, \mathcal{O} \ltimes \mathfrak{g}(\mathcal{O})$. We let

$$V_{k,c}(\widetilde{\mathfrak{g}}) = \mathrm{Ind}_{\widetilde{\mathfrak{g}}_+ \oplus \mathbb{C}\mathbf{1}}^{\widetilde{\mathfrak{g}}} \mathbb{C}_{k,c} = U\widetilde{\mathfrak{g}} \underset{U(\widetilde{\mathfrak{g}}_+ \oplus \mathbb{C}\mathbf{1})}{\otimes} \mathbb{C}_{k,c}$$

where $\mathbb{C}_{k,c}$ is the one-dimensional representation of $\widetilde{\mathfrak{g}}_+ \oplus \mathbb{C}\mathbf{K} \oplus \mathbb{C}\mathbf{C} \subset \widetilde{\mathfrak{g}}$ on which $\widetilde{\mathfrak{g}}_+$ acts by zero and $\mathbf{K}, \mathbf{C}$ act by k, c. Let

$$\widetilde{\mathfrak{g}}_n(\mathcal{O}) = \mathrm{Der}\, {}_{2n}(\mathcal{O}) \ltimes \mathfrak{g}_n(\mathcal{O}) = \mathfrak{m}^{2n} \mathrm{Der}\, \mathcal{O} \ltimes \mathfrak{m}^n \mathfrak{g}(\mathcal{O}).$$

Remark 3.1. This normalization above, by which we pair $\mathfrak{g}_n(\mathcal{O})$ with $\mathrm{Der}\, {}_{2n}(\mathcal{O})$, is motivated by the Segal–Sugawara construction, cf. Proposition 3.3: the Virasoro generators will be constructed from *quadratic* expressions in $\widehat{\mathfrak{g}}$. Thus in geometric applications of the Segal–Sugawara operators the orders of trivializations or poles along the Virasoro (moduli of curves) directions will turn out to be double those along the Kac–Moody (moduli of bundles) directions (for example the quadratic Hitchin hamiltonians double the order of the pole of a Higgs field).

3.1.1

The corresponding representation

$$V_{k,c}^n = U_{k,c}\widetilde{\mathfrak{g}} \underset{U\widetilde{\mathfrak{g}}_n(\mathcal{O})}{\otimes} \mathbb{C}$$

agrees with $V_{k,c}$ for $n = 0$, while for $n > 0$ it is identified as the vacuum module for the Harish-Chandra pair $(\widetilde{\mathfrak{g}}, \widetilde{G}_n(\mathcal{O}))$ where $\widetilde{G}_n(\mathcal{O}) = \mathrm{Aut}_{2n}\, \mathcal{O} \ltimes G_n(\mathcal{O})$. We will also denote by $\widetilde{G}(\mathcal{O})$ the semidirect product $\mathrm{Aut}\, \mathcal{O} \ltimes G(\mathcal{O})$.

The pattern of deformations and limits from Section 2.1 applies to the Kac–Moody and Virasoro central extensions, giving rise to two families of algebras which we denote $U_{k,\lambda}\widehat{\mathfrak{g}}$ and $U_{c,\mu}\mathrm{Vir}$. Thus for $\lambda = \mu = 1$ and $k = c = 0$, $U_{k,\lambda}\widehat{\mathfrak{g}} = U\mathfrak{g}(\mathcal{K})$ and $U_{c,\mu}\mathrm{Vir} = U\,\mathrm{Der}\,\mathcal{K}$, etc. For $k = c = \infty$ we obtain the Poisson algebras $\mathrm{Sym}_\lambda \widehat{\mathfrak{g}}$ and $\mathrm{Sym}_\mu \mathrm{Vir}$.

We also have classical vacuum modules $\overline{V}_\lambda(\widehat{\mathfrak{g}})$ for $(\operatorname{Sym}\widehat{\mathfrak{g}}, G(\mathbb{O}))$ and $\overline{V}_\mu(\text{Vir})$ for $(\operatorname{Sym}\text{Vir}, \operatorname{Aut}\mathbb{O})$ and the interpolating families $V_{k,\lambda}(\widehat{\mathfrak{g}})$ and $V_{c,\mu}(\text{Vir})$.

3.2 Vertex algebras

The vacuum representations V_k, Vir_c and $V_{k,c}$ have natural structures of vertex algebras, and V_k^n, Vir_c^n and $V_{k,c}^n$ are modules over the respective vertex algebras (see [FB] for the definition of vertex algebras and in particular the Virasoro and Kac–Moody vertex algebras).

The vacuum vector $|0\rangle$ plays the role of the unit for these vertex algebras. Moreover, all three vacuum modules carry natural Harish-Chandra actions of $(\operatorname{Der}\mathbb{O}, \operatorname{Aut}\mathbb{O})$: these are defined by the natural action of $\operatorname{Der}\mathbb{O}$ on $\widehat{\mathfrak{g}}$, Vir and $\widetilde{\mathfrak{g}}$, preserving the positive halves and hence giving rise to actions on the corresponding representations. These actions give rise to a grading operator L_0 and a translation operator $T = L_{-1} = -\partial_z \in \operatorname{Der}\mathbb{O}$. The vertex algebra V_k is generated by the fields

$$J^a(z) = Y(J_{-1}^a|0\rangle, z) = \sum_{n\in\mathbb{Z}} J_n^a z^{-n-1}$$

associated to the vectors $J_{-1}^a|0\rangle \in V_k$, which satisfy the operator product expansions

$$J^a(z)J^b(w) = \frac{k(J^a, J^b)}{(z-w)^2} + \frac{[J^a, J^b](w)}{z-w} + \cdots,$$

where the ellipses denote regular terms. This may be seen as a shorthand form for the defining relations (3.2). This structure extends to the family $V_{k,\lambda}$. Explicitly, introduce the $\mathbb{C}[\lambda, k^{-1}]$-lattice Λ in $V_{k,\lambda}$ generated by monomials in $\overline{J}_n^a = \frac{\lambda}{k}J_n^a$. These satisfy relations

$$[\overline{J}_n^a, \overline{J}_m^b] = \frac{\lambda}{k}([\overline{J^a, J^b}]_{n+m} + \lambda(J^a, J^b))\delta_{n,-m},$$

or in shorthand

$$[\overline{J}, \overline{J}] = \frac{\lambda}{k}(\overline{J} + \lambda(\cdots))$$

The vertex algebra structure is defined by replacing the J operators by the rescaled versions $\overline{J}$. In particular we see that we recover the commutative vertex algebra $\overline{V}_\lambda(\mathfrak{g})$ when $k = \infty$. Recall from [FB], Ch. 15, that a *commutative vertex algebra* is essentially the same as a unital commutative algebra with derivation and grading. Thus the classical vacuum representations $\overline{V}_\lambda(\mathfrak{g})$, $\overline{\text{Vir}}_\mu$ and $\overline{V}_{\lambda,\mu}(\mathfrak{g})$ are naturally commutative vertex algebras.

Moreover, the description as a degeneration endows $\overline{V}_\lambda(\mathfrak{g})$ with a vertex Poisson algebra structure. The Poisson structure comes with the following relations

$$\{\overline{J}_n^a, \overline{J}_m^b\} = \overline{[J^a, J^b]}_{n+m} + \lambda(J^a, J^b)\delta_{n,-m}.$$

Likewise, the vertex algebra Vir_c is generated by the field

$$T(z) = Y(L_{-2}|0\rangle, z) = \sum_{n\in\mathbb{Z}} L_n z^{-n-2} \tag{3.4}$$

satisfying the operator product expansion

$$T(z)T(w) = \frac{c/2}{(z-w)^4} + \frac{2T(w)}{(z-w)^2} + \frac{\partial_w T(w)}{z-w} + \cdots \tag{3.5}$$

encapsulating the relations (3.3). The μ-deformation $\mathrm{Vir}_{c,\mu}$ is straightforward. The commutation relations for the generators $\overline{L}_n = \frac{\mu}{c}L_n$ read

$$[\overline{L}_n, \overline{L}_m] = \frac{\mu}{c}(n-m)\overline{L}_{n+m} + \frac{\mu^2}{c}\frac{1}{12}(n^3-n)\delta_{n,-m}$$

or simply

$$[\overline{L}, \overline{L}] = \frac{\mu}{c}(\overline{L} + \mu(\cdots)).$$

In the limit $c = \infty$ we obtain a vertex Poisson algebra structure on $\overline{V}_\mu(\mathfrak{g})$. (As a commutative algebra with derivation it is freely generated by the single vector $\overline{L}_{-2}|0\rangle$.) The Poisson operators $\overline{L}$ have relations

$$\{\overline{L}, \overline{L}\} = \overline{L} + \mu(\cdots).$$

The space $V_{k,c}$ also carries a natural vertex algebra structure, so that $\mathrm{Vir}_c \subset V_{k,c}(\mathfrak{g})$ is a vertex subalgebra, complementary to $V_k(\widehat{\mathfrak{g}}) \subset V_{k,c}(\mathfrak{g})$, which is a vertex algebra ideal. In particular, the vertex algebra is generated by the fields $J^a(z)$ and $T(z)$, with the additional relation

$$T(z)J^a(w) = \frac{J^a(w)}{(z-w)^2} + \frac{\partial_w J^a(w)}{z-w} + \cdots. \tag{3.6}$$

We may combine the above deformations into a two-parameter deformation of $V_{k,c}$. We introduce the $\mathbb{C}[k^{-1}, c^{-1}]$-lattice Λ in $V_{k,c}$ generated by monomials in $\overline{J}_n^a = \frac{\lambda}{k}J_n^a$ and $\overline{L}_n = \frac{\mu}{c}L_n$. These satisfy relations

$$[\overline{J}, \overline{J}] = \frac{\lambda}{k}(\overline{J} + \lambda), \qquad [\overline{L}, \overline{L}] = \frac{\mu}{c}(\overline{L} + \mu), \qquad [\overline{L}, \overline{J}] = \frac{\mu}{c}\overline{J}.$$

If we impose $\lambda = \mu = 0$, we obtain the deformation of the enveloping vertex algebra $V_{k,c}$ to the symmetric vertex Poisson algebra associated to $\widetilde{\mathfrak{g}}$.

We will denote this limit vertex Poisson algebra by $\overline{V}_{0,0}(\mathfrak{g})$. We may also specialize μ to 0, making the $\overline{L}$ generators central. Thus in this limit we have a noncommutative vertex structure on the Kac–Moody part, while the Virasoro part degenerates to a vertex Poisson algebra. Note however that if we keep μ nonzero but specialize $\lambda = 0$, the Kac–Moody generators become commutative but *not* central. Hence the Kac–Moody part does not acquire a vertex Poisson structure, and we do not obtain an action of the Kac–Moody vertex Poisson algebra on $V_{k,c}$ in this limit.

In order to obtain a vertex Poisson structure, we need a vertex subalgebra which becomes central, together with a 'small parameter' or direction of deformation, in which to take the linear term. The 'generic limit' we consider is obtained by letting $k, c \to \infty$, but with the constraint that our small parameter is

$$\frac{\lambda}{k} = \frac{\mu}{c}.$$

In other words, we take terms linear in either one of these ratios. In this limit, the entire vertex algebra becomes commutative, as can be seen from the relations above. Thus in the limit we obtain a vertex Poisson algebra, generated by $\overline{L}, \overline{J}$ with relations

$$\{\overline{J}, \overline{J}\} = \overline{J} + \lambda \quad \{\overline{L}, \overline{L}\} = \overline{L} + \mu \quad \{\overline{L}, \overline{J}\} = \overline{J}.$$

The resulting vertex Poisson algebra will be denoted by $\overline{V}_{\lambda,\mu}(\mathfrak{g})$, for $\lambda, \mu \in \mathbb{C}$.

3.3 The Segal–Sugawara vector

To any vertex algebra V we associate a Lie algebra $U(V)$ topologically spanned by the Fourier coefficients of vertex operators from V (see [FB]). This Lie algebra acts on any V-module. In the case of the Kac–Moody vertex algebra V_k, the Lie algebra $U(V_k)$ belongs to a completion of the enveloping algebra $U_k\widehat{\mathfrak{g}}$. An important fact is that it contains inside it a copy of the Virasoro algebra (if $k \neq -h^\vee$). In vertex algebra terminology, this means that the Kac–Moody vacuum modules are *conformal* vertex algebras.

The conformal structure for $V_{k,c}$ is automatic. Let $\omega_V = L_{-2}|0\rangle \in V_{k,c}(\mathfrak{g})$. This is a conformal vector for the vertex algebra $V_{k,c}(\mathfrak{g})$: the field $T(z)$ it generates satisfies the operator product expansion (3.5), and hence its Fourier coefficients L_n (see (3.4)) give rise to a Virasoro action. The operator L_0 is the grading operator and L_{-1} is the translation operator T on $V_{k,c}$. The action of Der $\mathcal{O}$ induced by the L_n ($n \geq -1$) preserves the Kac–Moody part $V_k \subset V_{k,c}$, but the negative L_ns take us out of this subspace.

For level k not equal to minus the dual Coxeter number $h^\vee$ of $\mathfrak{g}$, the vertex algebra V_k itself carries a Virasoro action, and in fact a conformal structure, given by the Segal–Sugawara vector. This means that we have a conformal vector $\omega_S \in V_k(\widehat{\mathfrak{g}}) \subset V_{k,c}(\mathfrak{g})$ such that the corresponding field satisfies the Virasoro operator product expansion (3.5). This conformal vector is given by

$$\omega_S = \frac{S_{-2}}{k + h^\vee}, \qquad S_{-2} = \frac{1}{2} \sum_a J_1^a J_{a,-1}. \tag{3.7}$$

The corresponding Virasoro algebra has central charge $\dfrac{k \dim \mathfrak{g}}{k + h^\vee}$. In other words

$$Y(\omega_S, z) = \sum_{n \in \mathbb{Z}} L_n^S z^{-n-2} \qquad (L_n^S \in \operatorname{End} V_{k,c}(\mathfrak{g}))$$

and we have

$$[L_n^S, J_m^a] = -m J_{n+m}^a \tag{3.8}$$

$$[L_n^S, L_m^S] = (n - m) L_{n+m}^S + \frac{1}{12} \frac{k \dim \mathfrak{g}}{k + h^\vee} (n^3 - n) \delta_{n,m}. \tag{3.9}$$

We will continue to take the vector ω_V as the conformal vector for $V_{k,c}$, and use the notation L_n for the coefficients of the corresponding field $T(z)$. Due to the above commutation relations, ω_S is *not* a conformal vector for $V_{k,c}$, because L_0^S and L_{-1}^S do not act as the grading and translation operators on the Virasoro generators.

Note that the commutator $[L_n, J_m^a]$ coincides with the right-hand side of formula (3.8), and therefore $[L_n, L_m^S]$ is given by the right-hand side of formula (3.9).

The above construction of the Virasoro algebra action on V_k is nontrivial from the Lie algebra point of view. Indeed, the operators L_n^S are given by infinite sums of quadratic expressions in the J_n^a, which are nonetheless well-defined as operators on V_k

$$Y(\omega_S, z) = \frac{1}{2(k + h^\vee)} \sum_a \; : J^a(z) J_a(z) :$$

so that

$$L_n^S = \frac{1}{2(k + h^\vee)} \sum_a \sum_m \; : J_m^a J_{a,n-m} :$$

$$= \frac{1}{2(k + h^\vee)} \sum_a \left(\sum_{m<0} J_m^a J_{a,n-m} + \sum_{m \geq 0} J_{a,n-m} J_m^a \right).$$

In fact, since this action is given by a vertex operator in V_k, it follows immediately that any module over the vertex algebra $V_k, k \neq -h^\vee$, carries a compatible action of the Virasoro algebra. This may also be expressed using completions of the enveloping algebras of $\widehat{\mathfrak{g}}$ and $\widetilde{\mathfrak{g}}$. These are the completions which act on all smooth representations, i.e., those in which every vector is stabilized by a deep enough congruence subgroup $\widetilde{G}_n(\mathcal{O}), n \geq 0$. Namely, we define a completion of $U_{k,c}\widetilde{\mathfrak{g}}$ as the inverse limit

$$\widehat{U}_{k,c}\widetilde{\mathfrak{g}} = \varprojlim\ U_{k,c}\widetilde{\mathfrak{g}}/(U_{k,c}\widetilde{\mathfrak{g}} \cdot \widetilde{\mathfrak{g}}_n(\mathcal{O})).$$

This is a complete topological algebra, since the multiplication on $U_{k,c}(\widetilde{\mathfrak{g}})$ is continuous in the topology defined by declaring the left ideals generated by $\widetilde{\mathfrak{g}}_n(\mathcal{O})$ to be base of open neighborhoods of 0. We define a completion of $U_k\widehat{\mathfrak{g}}$ in the same way. These completions contain the Lie algebras $U(V_{k,c})$ and $U(V_k)$, and in particular for $k \neq -h^\vee$ they contain the Virasoro algebra generated by the Segal–Sugawara operators $L_n^S, n \in \mathbb{Z}$. Hence any smooth representation of $\widetilde{\mathfrak{g}}$ or $\widehat{\mathfrak{g}}$ of level $k \neq -h^\vee$ inherits a Virasoro action. In particular, the algebra $\widehat{U}_{k,c}(\widetilde{\mathfrak{g}})$ acts on the vacuum modules $V_{k,c}^n$

Proposition 3.2. *For any $k \neq -h^\vee$, the Segal–Sugawara operators $L_m^S, m \in \mathbb{Z}$, define (Vir, $\mathrm{Aut}_{2n}\ \mathcal{O}$)-action of central charge*

$$c = \frac{k \dim \mathfrak{g}}{k + h^\vee}$$

on the vacuum modules V_k^n and $V_{k,c}^n$. Together with the action of $(\widehat{\mathfrak{g}}, G_n(\mathcal{O}))$, this (Vir, $\mathrm{Aut}_{2n}\ \mathcal{O}$)-action combines into a $(\widetilde{\mathfrak{g}}, \widetilde{G}_n(\mathcal{O}))$-action on V_k^n.

Proof. The action of the L_m^S given by the Segal–Sugawara operators is well-defined on V_k^n because V_n^k is a smooth $\widehat{\mathfrak{g}}$-module. Next, we claim $L_m^S \cdot |0\rangle_n = 0$ for $m \geq 2n - 1$. For (precisely) such m, for each term: $J_{m-i}^a J_{a,i}$: at least one of the two factors lies in $\widetilde{\mathfrak{g}}_n(\mathcal{O})$, either immediately annihilating $|0\rangle_n$ or first passing through the other factor to annihilate $|0\rangle_n$, leaving a commutator of degree m which also annihilates $|0\rangle_n$.

Observe that $\mathfrak{g}_n(\mathcal{O})$ acts locally nilpotently on V_k^n, $\mathrm{Der}_{2n}\mathcal{O}$ acts locally nilpotently on $\widehat{\mathfrak{g}}/\mathfrak{g}_n(\mathcal{O})$ and $\mathrm{Der}_{2n}\mathcal{O}$ annihilates $|0\rangle_n$. This shows that $\mathrm{Der}_{2n}\mathcal{O}$ acts locally nilpotently on V_k^n. It follows that this action may be exponentiated to the pro-unipotent group $\mathrm{Aut}_{2n}(\mathcal{O})$. The arguments for $V_{k,c}^n$ are identical.

The fact that the Virasoro action on V_k^n defined by the Segal–Sugawara operators is compatible with the action of $\widehat{\mathfrak{g}}$ follows from commutation relations (3.8). $\qquad\square$

3.3.1 The Segal–Sugawara singular vector.

We now have two Virasoro actions on the representations $V_{k,c}^n$: one given by the operators L_m and one given by the operators L_m^S. Moreover, both actions have the same commutation relations with the Kac–Moody generators J_n^a. Their difference now defines a third Virasoro action, which has the crucial feature that it commutes with $\widehat{\mathfrak{g}}$. Indeed, setting $\mathbb{S}_m = L_m - L_m^S$, we have

$$[\mathbb{S}_l, \mathbb{S}_m] = [L_l, L_m] + ([L_l^S, L_m^S] - [L_l, L_m^S] - [L_l^S, L_m])$$

$$= [L_l, L_m] - [L_l^S, L_m^S]$$

$$= (l - m)\mathbb{S}_{l-m} + \frac{c_k}{12}(l^3 - l)\delta_{l,-m}$$

and

$$[\mathbb{S}_m, J_l^a] = 0$$

where we have introduced the notation

$$c_k = c - \frac{k \dim \mathfrak{g}}{k + h^\vee} \tag{3.10}$$

for the central charge of the $\mathbb{S}_m$.

These operators are also defined from the action of a vertex operator. Define the Segal–Sugawara singular vector $\mathbb{S} \in V_{k,c}$ as the difference $\mathbb{S} = \omega_V - \omega_S$, for $k \neq -h^\vee$. The corresponding field

$$\mathbb{S}(z) = \sum_{m \in \mathbb{Z}} \mathbb{S}_m z^{-m-2}$$

generates the action of the $\mathbb{S}_m$s on $V_{k,c}^n$. The crucial property of $\mathbb{S}$ is that it is a singular vector for the Kac–Moody action, i.e., it is $\mathfrak{g}(\mathcal{O})$-invariant

$$J_n^a \cdot \mathbb{S} = 0, \quad n \geq 0.$$

In what follows we consider $\mathrm{Vir}_{c_k}^{2n}$ as a $\widetilde{\mathfrak{g}}$-module, where $\widehat{\mathfrak{g}} \subset \widetilde{\mathfrak{g}}$ acts by zero. By Proposition 3.2, V_k^n is also a $\widetilde{\mathfrak{g}}$-module for $k \neq -h^\vee$. Therefore their tensor product is a $\widetilde{\mathfrak{g}}$-module.

Proposition 3.3. *Let $k, c \in \mathbb{C}$ with $k \neq -h^\vee$.*

(1) *The action of the $\mathbb{S}_m$ on $V_{k,c}^n$ defines an embedding $\mathbb{S}_{k,c}^n : \mathrm{Vir}_{c_k}^{2n} \longrightarrow V_{k,c}^n$ of Vir-modules.*

(2) *$\mathbb{S}_{k,c}^n$ is a homomorphism of $\widetilde{G}_n(\mathcal{O})$-modules with respect to the standard action of $\widetilde{G}_n(\mathcal{O})$ on $V_{k,c}^n$ and the trivial action of $G_n(\mathcal{O})$ on $\mathrm{Vir}_{c_k}^{2n}$.*

(3) *There is an isomorphism of* $(\widetilde{\mathfrak{g}}, \widetilde{G}_n(\mathcal{O}))$*-modules*

$$\mathfrak{s}_{k,c} : V_k^n \underset{\mathbb{C}}{\otimes} \mathrm{Vir}_{c_k}^{2n} \longrightarrow V_{k,c}^n$$

such that $\mathfrak{s}_{k,c}(|0\rangle \otimes v) = \mathbb{S}_{k,c}^n(v)$.

Proof. The morphism $\mathbb{S}_{k,c}^n$ is uniquely determined by the requirement that it intertwine the Virasoro action on $\mathrm{Vir}_{c_k}^{2n}$ with the action of the Virasoro algebra generated by the $\mathbb{S}_m$s on $V_{k,c}^n$, which annihilate the vacuum vector for $m \geq 2n - 1$. (In particular, for $n = 0$, ω_V is sent to the Segal–Sugawara singular vector $\mathbb{S} = \mathbb{S}_{-2} \cdot |0\rangle_n$.) To show that this is an embedding, note that for $m < 2n - 1$ the operators L_m act freely on $V_{k,c}^n$, hence so do the $\mathbb{S}_m$s, which have the same leading term with respect to the filtration on $V_{k,c}^n$ by the order in the L_m operators.

The Segal–Sugawara operators commute with the action of $\widehat{\mathfrak{g}}$. It follows that the subspace $\mathbb{S}_{k,c}(\mathrm{Vir}_{c_k}^{2n})$ of $V_{k,c}^n$ generated by the action of the $\mathbb{S}_m$s on $|0\rangle_n$ is $\mathfrak{g}_n(\mathcal{O})$-invariant, and hence $G(\mathcal{O})$-invariant. Therefore any vector $v \in \mathbb{S}_{k,c}^n(\mathrm{Vir}_{c_k}^{2n}) \subset V_{k,c}^n$ determines a unique $\widehat{\mathfrak{g}}$-homomorphism $v_* : V_k^n \to V_{k,c}^n$ with $v_*(|0\rangle_n) = v$. Hence we have a natural embedding of $\mathbb{S}_{k,c}^n(\mathrm{Vir}_{c_k}^{2n})$ in $\mathrm{Hom}_{\widehat{\mathfrak{g}}}(V_k^n, V_{k,c}^n)$, and therefore a $\widehat{\mathfrak{g}}$-homomorphism

$$\mathfrak{s}_{k,c}^n : V_k^n \otimes \mathrm{Vir}_{c_k}^{2n} \longrightarrow V_{k,c}^n. \tag{3.11}$$

The fact that this is a homomorphism of Vir-modules, and therefore of $\widetilde{\mathfrak{g}}$-modules, of this map is immediate from the formula $\mathbb{S}_n = L_n - L_n^S$ for the Virasoro action, where L_n^S and L_n denote the Virasoro actions on V_k^n and $V_{k,c}^n$, respectively. In particular, we see that the map $\mathbb{S}_{k,c}^n$, identified with the inclusion $\mathrm{Vir}_{c_k}^{2n} \to |0\rangle \otimes \mathrm{Vir}_{c_k}^{2n}$, followed by $\mathfrak{s}_{k,c}^n$, is a homomorphism of $\widetilde{G}_n(\mathcal{O})$-modules, since the actions of $\widetilde{G}_n(\mathcal{O})$ fix the vector $|0\rangle \subset V_k^n$.

We claim that the map $\mathfrak{s}_{k,c}^n$ in (3.11) is an isomorphism. By the Poincaré–Birkhoff–Witt Theorem, $V_{k,c}^n$ has a basis of monomials of the form

$$J_{m_1}^{a_1} \cdots J_{m_k}^{a_l} L_{n_1} \cdots L_{n_l} |0\rangle_n \qquad (m_i < n, \ n_j < 2n - 1)$$

where we choose an ordering on the Kac–Moody and Virasoro generators. It follows that the same holds with the L_{n_j} replaced by $\mathbb{S}_{m_j}$. The map $\mathfrak{s}_{k,c}^n$ acts as follows

$$(J_{m_1}^{a_1} \cdots J_{m_k}^{a_l} |0\rangle) \otimes (L_{n_1} \cdots L_{n_l} |0\rangle) \mapsto J_{m_1}^{a_1} \cdots J_{m_k}^{a_k} \mathbb{S}_{n_1} \cdots \mathbb{S}_{n_l} |0\rangle_n.$$

Hence it is indeed an isomorphism as claimed. $\square$

3.4 Limits of Segal–Sugawara

We would like to describe the behavior of the homomorphisms $\mathbb{S}^n_{k,c}$, or equivalently, of the vector $\mathbb{S} \in V^n_{k,c}$ which generates them, as we vary the parameters k, c. In order to describe the different limits, it is convenient to introduce the parameters λ, μ and consider the full four-parameter family $V_{k,c,\lambda,\mu}$ of vertex algebras. The vector $\mathbb{S} = \omega_V - \omega_S$ is a well-defined element of $V_{k,c,\lambda,\mu}$ for $k \in \mathbb{C} \setminus -h^\vee$, $c \in \mathbb{C}$ and $\lambda, \mu \in \mathbb{C} \setminus 0$. It has a first-order pole when $k = -h^\vee$, since the Segal–Sugawara central charge $c_k = c - \dfrac{k \dim \mathfrak{g}}{k + h^\vee}$ does. It also has a second-order pole along $\lambda = 0$, since it contains a term quadratic in the J generators; a first-order pole along $\mu = 0$ and $c = \infty$, since it is first order in the L generators; and a first-order pole along $k = \infty$, since it is quadratic in the Js but divided by $k + h^\vee$. Thus in these limits it is necessary to normalize $\mathbb{S}$ by its leading term to obtain well-defined, non-zero limits of the map $\mathbb{S}_{k,c}$.

3.4.1 Critical level.

We would like to specialize the vector $\mathbb{S}$ and morphism $\mathbb{S}_{k,c}$ to the critical level $k = -h^\vee$. Introduce the rescaled operators

$$\overline{\mathbb{S}}_m = (k + h^\vee)\mathbb{S}_m = (k + h^\vee)(L_m - L^S_m)$$

which are generated by the vertex operator $\overline{\mathbb{S}}(z)$ associated to

$$\overline{\mathbb{S}} = (k + h^\vee)\mathbb{S} = (k + h^\vee)L_{-2} - S_{-2}.$$

Thus when $k = -h^\vee$, the vector $\overline{\mathbb{S}}$ is well-defined and equal to $-S_{-2}$. The operators $\overline{\mathbb{S}}_m$ satisfy

$$[\overline{\mathbb{S}}_l, \overline{\mathbb{S}}_m] = (k + h^\vee)\left((l - m)\overline{\mathbb{S}}_{l+m} + \frac{(k + h^\vee)(k \dim \mathfrak{g})}{12}(l^3 - l)\delta_{l,-m}\right). \quad (3.12)$$

Let us introduce the notation

$$\mu_\mathfrak{g} = h^\vee \dim \mathfrak{g}.$$

We see that as k approaches the critical level $-h^\vee$, as the central charge of the Virasoro action of the $\mathbb{S}_m$ becomes infinite, the renormalized operators $\overline{\mathbb{S}}_m$ become commuting elements, and moreover satisfy the Poisson relations of the classical Virasoro algebra

$$U_{\infty,\mu_\mathfrak{g}}\mathrm{Vir} = \mathrm{Sym}_{\mu_\mathfrak{g}}\mathrm{Vir}.$$

Proposition 3.4. *The action of the operators $\overline{\mathbb{S}}_m$ defines a $G(\mathbb{O})$–invariant homomorphism $\overline{\mathbb{S}}^n_{-h^\vee,c} : \overline{\mathrm{Vir}}^{2n}_{\mu_\mathfrak{g}} \longrightarrow V^n_{-h^\vee,c}$ of $\mathrm{Sym}_{\mu_\mathfrak{g}} \mathrm{Vir}$-modules. For*

$n = 0$, $\overline{\mathbb{S}}_{-h^\vee,c}$ *defines a homomorphism of vertex Poisson algebras* $\overline{\mathrm{Vir}}_{\mu_{\mathfrak{g}}} \to \mathcal{Z}(V_{-h^\vee,c}(\mathfrak{g}))$ *to the center of the Virasoro and Kac–Moody vertex algebra.*

Proof. The morphism $\overline{\mathbb{S}}^n_{-h^\vee,c}$ is defined by the universal mapping property of the vacuum module $\overline{\mathrm{Vir}}^n_{\mu_{\mathfrak{g}}}$ of $\mathrm{Sym}_{\mu_{\mathfrak{g}}} \mathrm{Vir}$. The centrality of $\overline{\mathbb{S}}$ (and hence the map) for $k = -h^\vee$ follows from the fact that the commutators of $\overline{\mathbb{S}}_n = (k+h^\vee)\mathbb{S}_n$ (for $k \neq -h^\vee$) with the L_n and J_n^a are divisible by $k+h^\vee$. That this is a morphism of vertex Poisson algebras is immediate from the commutation relation (3.12). □

3.4.2 Infinite limit.

Now we would like to study the 'generic' classical limit of the Segal–Sugawara construction in $\overline{V}_{\lambda,\mu}$. In order to do so we approach the plane $c = k = \infty$ along the direction $\dfrac{\lambda}{k} = \dfrac{\mu}{c}$, which is the direction we used to define the vertex Poisson structure on $\overline{V}_{\lambda,\mu}(\mathfrak{g})$. The Segal–Sugawara operators are rescaled as follows: $\overline{\mathbb{S}}_m = \frac{\lambda^2}{k}\mathbb{S}_m$. These are the Fourier modes of the vertex operator $\overline{\mathbb{S}}(z)$ associated to the vector

$$\overline{\mathbb{S}} = \frac{\lambda^2}{k}\mathbb{S} \in V_{k,c,\lambda,\mu}$$

which is regular for $k \neq 0, -h^\vee$ and c, λ, μ arbitrary. In terms of the regular elements $\overline{L}$ and $\overline{J}$, we have

$$\overline{\mathbb{S}} = \lambda\overline{L}_{-2} - \frac{1}{2(k+h^\vee)}\sum_a \overline{J}^a_{-1}\overline{J}_{a,-1} + \cdots.$$

Thus when $\lambda = 0$ the linear term drops out and we recover the symbol, $-\dfrac{1}{2(k+h^\vee)}\overline{S}_{-2}$. The commutation relations for the $\overline{\mathbb{S}}_m$ are as follows

$$[\overline{\mathbb{S}}_l, \overline{\mathbb{S}}_m] = \frac{\lambda^4}{k^2}[\mathbb{S}_l, \mathbb{S}_m] \tag{3.13}$$

$$= \lambda\frac{\lambda}{k}((l-m)\overline{\mathbb{S}}_{l-m} + \frac{\lambda\mu}{12}(l^3-l)\delta_{l,-m} + \cdots). \tag{3.14}$$

We have used the relation $\mu = \dfrac{\lambda c}{k}$, and that

$$\frac{\lambda^2}{k}\frac{k\dim\mathfrak{g}}{k+h^\vee} = \frac{\lambda^2\dim\mathfrak{g}}{k+h^\vee}$$

vanishes in the limit $k \to \infty$. Therefore in this limit the $\overline{\mathbb{S}}_m$s satisfy the relations of the Virasoro Poisson algebra $\mathrm{Sym}_{\lambda\mu} \mathrm{Vir}$, with the bracket rescaled by λ. We will refer to a morphism which is a homomorphism after rescaling by λ as a λ-*homomorphism*. We therefore obtain the following analogue of Proposition 3.4:

Proposition 3.5. *The action of the Segal–Sugawara operators* $\overline{\mathbb{S}}_m$ *defines a* λ-*homomorphism* $\overline{\mathbb{S}}^n_{\lambda,\mu} : \overline{\text{Vir}}^{2n}_{\lambda\mu} \longrightarrow \overline{V}^n_{\lambda,\mu}$ *of* $\text{Sym}_{\lambda\mu}$ Vir-*modules. The image is* $G(\mathcal{O})$-*invariant, and* $\overline{\mathbb{S}}_{\lambda,\mu}$ *is a* λ-*homomorphism of vertex Poisson algebras.*

4 Moduli spaces

In this section we describe the spaces on which representations of the Virasoro and Kac–Moody algebras and their semidirect product localize. These are the moduli spaces of curves, of bundles on a fixed curve and bundles on varying curves, respectively. We will consider these localization functors following the general formalism outlined in Section 2. Note that the above moduli spaces are not algebraic varieties, but algebraic *stacks*. However, as explained in [BB, BD1], the localization formalism is applicable to them because they are 'good' stacks, i.e., the dimensions of their cotangent stacks are equal to the twice their respective dimensions.

4.1 Moduli of bundles

Let X be a smooth projective curve over $\mathbb{C}$. Denote by $\text{Bun}_G(X)$ the moduli stack of principal G-bundles on X, and $\mathfrak{P}$ the tautological G-bundle on the product $X \times \text{Bun}_G(X)$. (Its restriction to $X \times \{\mathcal{P}\}$, for $\mathcal{P} \in \text{Bun}_G(X)$, is identified with $\mathcal{P}$.)

Given $x \in X$, we denote by $\text{Bun}_G(X, x, n)$ the moduli stack of G-bundles with an nth-order jet of trivialization at x, and by $\widehat{\text{Bun}}_G(X, x)$ the moduli stack of G-bundles with trivializations on the formal neighborhood of x (the latter moduli stack is in fact a *scheme* of infinite type). For now we fix a formal coordinate t at x, so that the complete local ring $\mathcal{O}_x$ is identified with $\mathcal{O} = \mathbb{C}[[t]]$. Later we will vary this coordinate by the $\text{Aut}\,\mathcal{O}$-action. The group scheme $G(\mathcal{O})$ acts on $\widehat{\text{Bun}}_G(X, x)$ by changing the formal trivialization, making $\widehat{\text{Bun}}_G(X, x) \to \text{Bun}_G(X)$ into a $G(\mathcal{O})$-torsor. More generally $\widehat{\text{Bun}}_G(X, x) \to \text{Bun}_G(X, x, n)$ is a $G_n(\mathcal{O})$-torsor.

Theorem 4.1. (Kac–Moody Uniformization.)

(1) *The* $G_n(\mathcal{O})$ *action on the moduli space* $\widehat{\text{Bun}}_G(X, x)$ *extends to a formally transitive action of the Harish-Chandra pair* $(\mathfrak{g}(\mathcal{K}), G_n(\mathcal{O}))$.

(2) *([BL1, BL2, DSi]) For G semisimple, the action of the ind-group $G(\mathcal{K})$ on $\widehat{\mathrm{Bun}}_G(X, x)$ is transitive, and there are isomorphisms*

$$\widehat{\mathrm{Bun}}_G(X, x) \simeq G(\mathcal{K})_{\mathrm{out}} \backslash G(\mathcal{K}), \quad \mathrm{Bun}_G(X) \simeq G(\mathcal{K})_{\mathrm{out}} \backslash G(\mathcal{K}) / G(\mathcal{O}).$$

4.1.1 Line bundles on $\mathrm{Bun}_G(X)$.

We refer to [Sor] for a detailed discussion of the line bundles on $\mathrm{Bun}_G(X)$. They are classified by integral invariant forms on $\mathfrak{g}$, which also label the Kac–Moody central extensions of LG. The action of $\widehat{\mathfrak{g}}$ on $\widehat{\mathrm{Bun}}_G(X, x)$ lifts with level one to the line bundle $\mathcal{C}$ given by the corresponding invariant form. This line bundle may be defined by using Theorem 4.1 from the action of the Kac–Moody *group* $\widehat{G(\mathcal{K})}$ (the central extension splits over $G(X \setminus x)$ and hence gives rise to a line bundle on $G(X \setminus x) \backslash \widehat{G(\mathcal{K})} = \widehat{\mathrm{Bun}}_G(X, x)$, which descends to $\mathrm{Bun}_G(X)$).

For example, if $G = SL_n$, the line bundle $\mathcal{C}$ may be identified with the determinant of the cohomology of the universal vector bundle $\mathfrak{E} = \mathfrak{P} \underset{SL_n}{\times} \mathbb{C}^n$ over $X \times \mathrm{Bun}_{SL_n}(X)$, $\det R\pi_{2*}\mathfrak{E}$ (where $\pi_2 : X \times \mathrm{Bun}_G(X) \to \mathrm{Bun}_G(X)$). This identification however is not canonical (it is not valid for bundles over varying curves, see Section 4.3). More generally, for any simple algebraic group G powers of $\mathcal{C}$ can be defined as determinant line bundles associated to representations of G.

4.1.2 Localization.

For every n the triple

$$(\widehat{\mathfrak{M}}, \mathfrak{M}, \mathcal{L}) = (\widehat{\mathrm{Bun}}_G(X, x), \mathrm{Bun}_G(X, x, n), \mathcal{C})$$

defined above carries a transitive Harish-Chandra action of $(\widehat{\mathfrak{g}}, G_n(\mathcal{O}))$ as in Section 2.3. Therefore we have localization functors from $(\widehat{\mathfrak{g}}, G_n(\mathcal{O}))$-modules to twisted $\mathcal{D}$-modules on $\mathrm{Bun}_G(X, x, n)$ (we denote these functors by Δ as before). In particular, according to Proposition 2.2 (2), for the vacuum module V_k^n, the sheaf $\Delta(V_k^n)$ on $\mathrm{Bun}_G(X, x, n)$ is just the corresponding sheaf of twisted differential operators, which we denote uniformly by $\mathcal{D}_k$. Furthermore, the twist $\mathcal{V}_k^n = \mathcal{V}_{\widehat{\mathfrak{g}}, G_n(\mathcal{O})}$ is a sheaf of algebras on $\mathrm{Bun}_G(X, x, n)$, and we have a surjective homomorphism $\mathcal{V}_k^n \to \mathcal{D}_k$.

The corresponding classical vacuum representations $\overline{V}_k^n$ localize to give the Poisson sheaves of functions on the twisted cotangent bundles $T_\lambda^* \mathrm{Bun}_G(X, x, n) = \mathrm{Conn}\, \mathcal{C}^\lambda$ corresponding to $\mathcal{C}$. Recall that the cotangent space $T_\mathcal{P}^* \mathrm{Bun}_G(X)$ at a bundle $\mathcal{P}$ is the space of $\mathfrak{g}_\mathcal{P}$-valued differentials $H^0(X, \mathfrak{g}_\mathcal{P} \otimes \Omega)$. Therefore we obtain

Proposition 4.2 ([Fa1, BS]). *The twisted cotangent bundle $T_1^* \operatorname{Bun}_G(X, x, n)$ is canonically identified, as a torsor over $T^* \operatorname{Bun}_G(X, x, n)$, with the moduli stack $\operatorname{Conn}_G(X, x, n)$ of bundles with connections with a pole of order at most n at x.*

4.2 Moduli of curves

Let $\mathfrak{M}_g$ denote the moduli stack of smooth projective curves of genus g, and $\pi : \mathfrak{X}_g \to \mathfrak{M}_g$ the universal curve. The stack $\mathfrak{X}_g$ is identified with the moduli stack $\mathfrak{M}_{g,1}$ of *pointed* genus g curves. More generally, we denote by $\mathfrak{M}_{g,1,n}$ the moduli stack of pointed curves with an nth order jet of coordinate at the marked point, and $\widehat{\mathfrak{M}}_{g,1} = \mathfrak{M}_{g,1,\infty}$ the moduli scheme of curves with a marked point and formal coordinate (i.e. classifying triples (X, x, z), where $(X, x) \in \mathfrak{M}_{g,1}$ and z is a formal coordinate on X at x). The group scheme $\operatorname{Aut} \mathcal{O}$ acts on $\widehat{\mathfrak{M}}_{g,1}$ by changing the coordinate z, making $\widehat{\mathfrak{M}}_{g,1}$ into an $\operatorname{Aut} \mathcal{O}$–torsor over $\mathfrak{M}_{g,1}$ and an $\operatorname{Aut}_n \mathcal{O}$–torsor over $\mathfrak{M}_{g,1,n}$.

For any family of curves $\pi : X \to S$, we have the Hodge line bundle $\mathcal{H}$ on S, defined by

$$\mathcal{H} = \det R\pi_* \omega_{X/S},$$

the determinant of the cohomology of the canonical line bundle of X over S. By abuse of notation we will denote by $\mathcal{H}$ the Hodge line bundle of an arbitrary family of curves. Over the moduli stack $\mathfrak{M}$, Mumford has shown that (for $g > 1$) $\mathcal{H}$ generates the Picard group $\operatorname{Pic} \mathfrak{M} \cong \mathbb{Z} \cdot \mathcal{H}$.

Theorem 4.3. [BS, TUY, ADKP, K] (Virasoro Uniformization.) *The $\operatorname{Aut} \mathcal{O}$ action on the moduli space $\widehat{\mathfrak{M}}_{g,1}$ of pointed, coordinatized curves extends to a formally transitive action of the Harish-Chandra pair $(\operatorname{Vir}, \operatorname{Aut} \mathcal{O})$ (of level 0). The action of Vir lifts to an action with central charge -2 on the line bundle $\mathcal{H}$.*

4.2.1 Localization.

The pattern of localization from Section 2.3 applies directly to the $\operatorname{Aut}_n \mathcal{O}$–bundle $\widehat{\mathfrak{M}}_{g,1} \to \mathfrak{M}_{g,1,n}$ and the Harish-Chandra pairs $(\operatorname{Vir}, \operatorname{Aut}_n \mathcal{O})$. Therefore we have localization functors from $(\operatorname{Vir}, \operatorname{Aut}_n \mathcal{O})$–modules to twisted $\mathcal{D}$–modules on $\mathfrak{M}_{g,1,n}$. In particular, the localization of the vacuum module Vir_c^n gives us the sheaf of twisted differential operators on $\mathfrak{M}_{g,1,n}$, denoted by $\mathcal{D}_c$, and their classical versions give us Poisson algebras of functions on twisted cotangent bundles. Furthermore, the twist $\operatorname{Vir}_c^n = \mathcal{V}_{\operatorname{Vir}, \operatorname{Aut}_n(\mathcal{O})}$ is a sheaf of algebras on $\mathfrak{M}_{g,1,n}$, and we have a surjective homomorphism $\operatorname{Vir}_c^n \to \mathcal{D}_c$.

A similar picture holds for the fibration $\widehat{\mathfrak{M}}_{g,1} \to \mathfrak{M}_g$ and the pair (Vir, Der $\mathcal{O}$), except that Der $\mathcal{O}$ does not integrate to a group (only to an ind-group, see [FB], Ch. 5), and $\widehat{\mathfrak{M}}_{g,1} \to \mathfrak{M}_g$ is not a principal bundle, so that the definition of localization does not carry over directly. Nevertheless, we obtain the desired description of differential operators on $\mathfrak{M}_g$ by first localizing Vir_c on $\mathfrak{M}_{g,1}$ (or $\mathfrak{M}_{g,1,n}$), using the fact that the corresponding $\mathcal{D}$-modules descend along the projection $\pi : \mathfrak{M}_{g,1} \to \mathfrak{M}_g$:

Proposition 4.4. *The localization of the Virasoro module* Vir_c *on* $\mathfrak{M}_{g,1}$ *is isomorphic to the pullback* $\pi^* \mathcal{D}_c$ *of the sheaf of twisted differential operators on* $\mathfrak{M}_g$.

Proof. By Lemma 2.4, the localization $\Delta(\mathrm{Vir}_c)$ is the quotient of $\Delta(\mathrm{Vir}_c^0) = \mathcal{D}_c$ by the action of the partial vacuum representation $\Delta(V_{\mathrm{Der}\,\mathcal{O},\mathrm{Der}_0\mathcal{O}})$. However, the latter is readily identified as the sheaf of *relative* differential operators, $\Delta(V_{\mathrm{Der}\,\mathcal{O},\mathrm{Der}_0\mathcal{O}}) = \mathcal{D}_{c/\mathfrak{M}_g}$. Indeed the action of Der $\mathcal{O}$ on $\widehat{\mathfrak{M}}_g$ is free and generates the relative vector fields on the universal curve – the $\widehat{\mathfrak{M}}_g$ twist $(\mathrm{Der}\,\mathcal{O}/\mathrm{Der}_0\mathcal{O})_{\widehat{\mathfrak{M}}_g}$ is precisely the relative tangent sheaf of $\mathfrak{M}_{g,1}$ over $\mathfrak{M}_g$. But the quotient of the sheaf of differential operators by the ideal generated by vertical vector fields is the pullback of the sheaf of differential operators downstairs, whence the proposition. $\qquad\square$

4.2.2

Recall (see, e.g., [FB]) that the space of projective structures on X is a torsor over the quadratic differentials $\mathrm{H}^0(X, \Omega^{\otimes 2})$, which is the cotangent fiber of $\mathfrak{M}_g$ at x. The Virasoro uniformization of $\mathfrak{M}_g$, together with the canonical identification of projective structures on the punctured disc with a hyperplane in Vir*, give us the following:

Corollary 4.5. ([BS]) *There is a canonical identification of the twisted cotangent bundles* $T_\lambda^* \mathfrak{M}_g = \mathcal{P}\mathrm{roj}_g$ *(for $\lambda = 12$).*

4.2.3

Similarly the twisted cotangents to the moduli $\mathfrak{M}_{g,1,n}$ of curves with marked points and level structures are identified with the moduli $\mathcal{P}\mathrm{roj}_{g,1,n}^\lambda$ of λ-projective structures with poles at the corresponding points.

4.3 Moduli of curves and bundles

The discussions of the moduli spaces $\mathfrak{M}_g$ and $\mathrm{Bun}_G(X)$ above may be generalized to the situation where we vary both the curve and the bundle on it. Let

$\text{Bun}_{G,g}$ denote the moduli stack of pairs $(X, \mathcal{P})$, where X is a smooth projective curve of genus g, and $\mathcal{P}$ is a principal G-bundle on X. Thus we have a projection

$$\Pi : \text{Bun}_{G,g} \longrightarrow \mathfrak{M}_g$$

with fiber over X being the moduli stack $\text{Bun}_G(X)$. Let $\text{Bun}_{G,g,1}$ denote the moduli stack classifying G-bundles on pointed curves, i.e., the pullback of the universal curve to $\text{Bun}_{G,g}$

$$\text{Bun}_{G,g,1} = \mathfrak{M}_{g,1} \underset{\mathfrak{M}_g}{\times} \text{Bun}_G \xrightarrow{\pi_G} \text{Bun}_{G,g} .$$

We denote by $\Pi^0 : \text{Bun}_{G,g,1} \to \mathfrak{M}_{g,1}$ the map forgetting the bundles. We let $\text{Bun}_{G,g,1,n}$ denote the moduli stack of quintuples $(X, x, \mathcal{P}, z, \tau)$ consisting of a G-bundle $\mathcal{P}$ with an nth-order jet of trivialization τ on a pointed curve (X, x) with $2n$th-order jet of coordinate z. We let $\Pi^n : \text{Bun}_{G,g,1,n} \to \mathfrak{M}_{g,1,2n}$ denote the map forgetting $(\mathcal{P}, \tau)$. For $n = \infty$ we obtain the moduli scheme $\widehat{\text{Bun}}_{G,g}$ of bundles with formal trivialization on pointed curves with formal coordinates.

There is a natural two-parameter family of line bundles on $\text{Bun}_{G,g}$. Namely, there is a Hodge bundle $\mathcal{H}$ associated to the family of curves $\pi_G : \text{Bun}_{G,g,1} \to \text{Bun}_{G,g}$ (which is the Π pullback of the Hodge bundle on $\mathfrak{M}_g$). The universal principal G-bundle $\mathfrak{P}$ lives on this universal curve $\text{Bun}_{G,g,1}$, so we may also consider the line bundle $\mathcal{C}$ associated to the principal bundle $\mathfrak{P}$. The bundle $\mathcal{C}$ is trivial on the section triv : $\mathfrak{M}_g \to \text{Bun}_{G,g}$ sending a curve to the trivial G-bundle. Let $\mathcal{L}_{k,c} = \mathcal{C}^{\otimes k} \otimes \mathcal{H}^{\otimes c}$. For simply connected G this assignment gives an identification $\text{Pic}(\text{Bun}_{G,g}) \cong \mathbb{Z} \oplus \mathbb{Z}$ (see [La]).

For $G = SL_n$ we may consider the determinant bundle $\det R\pi_{2*}\mathfrak{E}$ as before, where π_2 is the projection from the universal curve and $\mathfrak{E} = \mathfrak{P} \times_{SL_n} \mathbb{C}^n$ is the universal vector bundle. The determinant of the cohomology of the trivial rank n bundle gives the nth power of the Hodge line bundle, while for fixed curve the determinant bundle may be identified with $\mathcal{C}$. Thus we have the Riemann–Roch identification

$$\det R\pi_{2*}\mathfrak{E} \cong \mathcal{C} \otimes \mathcal{H}^{\otimes n} = \mathcal{L}_{1,n}.$$

(In this case the determinant and Hodge bundles also span the Picard group.) In general we have determinant line bundles $\mathcal{L}_\rho$ for any representation $\rho : G \to SL_n$.

4.3.1 Extended connections.

For each $\lambda, \mu \in \mathbb{C}$, we have the corresponding twisted cotangent bundle $T^*_{\lambda,\mu} \text{Bun}_{G,g}$, which is $\mathcal{C}\text{onn}(\mathcal{L}_{\lambda,\mu})$ when λ, μ are integral. Following

[BZB2, BZB1], we will refer to points of $\mathcal{E}x\mathcal{C}onn_{G,g}^{\lambda,\mu} = T_{\lambda,\mu}^* \operatorname{Bun}_{G,g}$ as (λ, μ)-*extended connections*. Using the short exact sequence of cotangent bundles

$$0 \to \Pi^* T^* \mathfrak{M}_g \to T^* \operatorname{Bun}_{G,g} \to T_{/\mathfrak{M}}^* \operatorname{Bun}_{G,g} \to 0$$

and the description (see Proposition 4.1.2) of the relative twisted cotangent bundle of $\operatorname{Bun}_{G,g}$, we obtain an affine projection

$$\mathcal{E}x\mathcal{C}onn_{G,g}^{\lambda,\mu} \to \mathcal{C}onn_{G,g}^\lambda \qquad (4.15)$$

with fibers affine spaces over quadratic differentials. Here $\mathcal{C}onn_{G,g}$ denotes the moduli space of curves equipped with G-bundles and λ-connections, which is identified as the *relative* twisted cotangent bundle

$$\mathcal{C}onn_{G,g}^\lambda \cong \mathcal{C}onn_{/\mathfrak{M}}\mathcal{L}_{\lambda,\mu}$$

for any $\mu \in \mathbb{C}$. (Note that $\mathcal{L}_{0,\mu} = \Pi^*\mathcal{H}_\mu$ has a canonical connection relative to $\mathfrak{M}$, so that $\mathcal{C}onn_{/\mathfrak{M}}\mathcal{L}_{\lambda,\mu} \cong \mathcal{C}onn_{/\mathfrak{M}}\mathcal{L}_{\lambda,0}$ for any μ.)

Remark: kernel functions 4.6. See also [BZB2] where a concrete description is given of the identification of projective structures, connections and extended connections with the twisted cotangent bundles of the moduli of curves and bundles using kernel functions along the diagonal (in particular the Szegö kernel), in the spirit of [BS].

4.3.2

The group schemes $\widetilde{G}_n(\mathcal{O})$ act on $\widehat{\operatorname{Bun}}_{G,g}$, changing the coordinate z and trivialization τ. This action makes $\widehat{\operatorname{Bun}}_{G,g}$ into a $\widetilde{G}_n(\mathcal{O})$-torsor over $\operatorname{Bun}_{G,g,1,n}$.

Theorem 4.7. (Virasoro–Kac–Moody Uniformization.) *The $\widetilde{G}_n(\mathcal{O})$-action on the moduli stack $\widehat{\operatorname{Bun}}_{G,g}$ extends to a formally transitive action of the Harish-Chandra pair $(\widetilde{\mathfrak{g}}, \widetilde{G}_n(\mathcal{O}))$ (of level and central charge 0). The action of $\widetilde{\mathfrak{g}}$ lifts to an action with level k and central charge c on the line bundle $\mathcal{L}_{k,c}$.*

4.3.3 Localization

The pattern of localization from Section 2.3 again applies directly to $\operatorname{Bun}_{G,g,1,n}$ and the Harish-Chandra pairs $(\widetilde{\mathfrak{g}}, \widetilde{G}_n(\mathcal{O}))$. Therefore we have localization functors from $(\widetilde{\mathfrak{g}}, \widetilde{G}_n(\mathcal{O}))$-modules to twisted $\mathcal{D}$-modules on $\operatorname{Bun}_{G,g,1,n}$. In particular, the localization of the vacuum module $V_{k,c}^n$ gives us the sheaf of twisted differential operators on $\operatorname{Bun}_{G,g,1,n}$, denoted by $\mathcal{D}_{k,c}$. As in Section 4.2.1, we would like to show that the corresponding sheaves descend along the projection $\pi_G : \operatorname{Bun}_{G,g,1} \to \operatorname{Bun}_{G,g}$ to the moduli of curves and bundles itself. The argument of Proposition 4.2.1 carries over directly to this setting, and we obtain:

Proposition 4.8. *The localization of the Virasoro–Kac–Moody module $V_{k,c}$ on* $\mathrm{Bun}_{G,g,1}$ *is isomorphic to the pullback $\pi_G^* \mathcal{D}_{k,c}$ of the sheaf of twisted differential operators on* $\mathrm{Bun}_{G,g}$.

Note that we now have a *two*-parameter family of line bundles $\mathcal{L}_{k,c}$, and hence the pattern of deformations of Section 2.1 can be 'doubled', to match up with the picture of the Virasoro–Kac–Moody vertex algebra $V_{k,c}$. Thus, we introduce deformation parameters λ, μ coupled to the level and charge k, c. This defines a four-parameter family of algebras $\mathcal{D}_{\lambda,\mu}(\mathcal{L}_{k,c})$, to which the analogous quasi-classical localization statements apply.

5 The Segal–Sugawara homomorphism

In this section we apply the techniques of Lie algebra localization and vertex algebra conformal blocks to the Segal–Sugawara construction from Section 3. We interpret the result as a homomorphism between sheaves of twisted differential operators on the moduli stacks introduced in the previous section, and as heat operators on spaces of non-abelian theta functions. Various classical limits of this construction will be considered in the next section.

5.1 Homomorphisms between sheaves of differential operators

Let $k, c \in \mathbb{C}$ with $k \neq -h^\vee$. Let $\mathcal{V}_{k,c}^n = \mathcal{V}_{\widetilde{\mathfrak{g}}, \widetilde{G}_n(\mathcal{O})}$ be the sheaf on $\mathrm{Bun}_{G,g,1,n}$ obtained as the twist of the vacuum module $V_{k,c}^n$ over $(\widetilde{\mathfrak{g}}, \widetilde{G}_n(\mathcal{O}))$ following the construction of Section 2.4. By Proposition 2.4, this is a sheaf of algebras, equipped with a surjective homomorphism to the sheaf of (k, c)-twisted differential operators $\Delta(\mathcal{V}_{k,c}^n) = \mathcal{D}_{k,c}$.

According to Corollary 2.3, the subspace of invariants $(V_{k,c}^n)^{\widetilde{G}_n(\mathcal{O})}$ give rise to *global* differential operators on $\mathrm{Bun}_{G,g,1,n}$. Unfortunately, for $k \neq -h^\vee$ the space $(V_{k,c}^n)^{\widetilde{G}_n(\mathcal{O})}$ is one-dimensional, spanned by the vacuum vector. However, the Segal–Sugawara construction provides us with a large space of invariants for the smaller group $G_n(\mathcal{O}) \subset \widetilde{G}_n(\mathcal{O})$. Namely, by Proposition 3.3.1, the $G_n(\mathcal{O})$-invariants contain a copy of the Virasoro vacuum module $S^n : \mathrm{Vir}_{c_k}^{2n} \to (V_{k,c}^n)^{G_n(\mathcal{O})}$. We will use this fact to obtain a homomorphism $\mathcal{D}_{c_k} \to \Pi_*^n \mathcal{D}_{k,c}$ of sheaves on $\mathfrak{M}_{g,1,2n}$. The first step is the following assertion.

Consider $\mathrm{Vir}_{c_k}^{2n}$ as a $\widetilde{\mathfrak{g}}$-module by letting $\widehat{\mathfrak{g}}$ act by zero. Then the twist $\mathrm{Vir}_{c_k}^{2n} = \mathcal{V}_{\widetilde{\mathfrak{g}}, \widetilde{G}_n(\mathcal{O})}$ becomes a sheaf of algebras on $\mathfrak{M}_{g,1,2n}$.

Proposition 5.1. *There is a homomorphism of sheaves of algebras on* $\mathfrak{M}_{g,1,2n}$

$$\mathbb{S}^n_{k,c} : \mathrm{Vir}^{2n}_{c_k} \longrightarrow \Pi^n_* \mathcal{V}^n_{k,c}.$$

Proof. Note that the map $\mathbb{S}^n : \mathrm{Vir}^{2n}_{c_k} \to (V^n_{k,c})^{G_n(\mathcal{O})}$ is a homomorphism of $\widetilde{G}_n(\mathcal{O})$-modules (where the subgroup $G_n(\mathcal{O})$ of $\widetilde{G}_n(\mathcal{O})$ acts by zero). Hence its gives rise to a homomorphism of the corresponding twists by the $\widetilde{G}_n(\mathcal{O})$-torsor $\widehat{\mathrm{Bun}}_{G,g} \to \mathrm{Bun}_{G,g,1,n}$

$$\widehat{\mathrm{Bun}}_{G,g} \underset{\mathrm{Bun}_{G,g,1,n}}{\times} \mathrm{Vir}^{2n}_{c_k} \longrightarrow \widehat{\mathrm{Bun}}_{G,g} \underset{\mathrm{Bun}_{G,g,1,n}}{\times} (V^n_{k,c})^{G_n(\mathcal{O})} \longrightarrow$$

$$\longrightarrow \mathcal{V}^n_{k,c} = \widehat{\mathrm{Bun}}_{G,g} \underset{\mathrm{Bun}_{G,g,1,n}}{\times} V^n_{k,c}. \qquad (5.16)$$

The first two sheaves are pullbacks from $\mathfrak{M}_{g,1,2n}$. Indeed, since the actions of $\widetilde{G}_n(\mathcal{O})$ on $\mathrm{Vir}^{2n}_{c_k}$ and $(V^n_{k,c})^{G_n(\mathcal{O})}$ factor through $\mathrm{Aut}_{2n} \mathcal{O}$, their twists depend only on the associated $\mathrm{Aut}_{2n} \mathcal{O}$-torsor, which is nothing but the Π^n-pullback of the $\mathrm{Aut}_{2n} \mathcal{O}$-torsor $\pi_{2n} : \widetilde{\mathfrak{M}}_{g,1} \to \mathfrak{M}_{g,1,2n}$. Therefore the maps (5.16) may be written as

$$(\Pi^n)^* \mathrm{Vir}^{2n}_{c_k} \longrightarrow (\Pi^n)^* (V^n_{k,c})^{G_n(\mathcal{O})} \longrightarrow \mathcal{V}^n_{k,c}.$$

By adjunction, we obtain the following maps on $\mathfrak{M}_{g,1,2n}$

$$\mathrm{Vir}^{2n}_{c_k} \longrightarrow (V^n_{k,c})^{G_n(\mathcal{O})} \longrightarrow \Pi^n_* \mathcal{V}^n_{k,c}.$$

We claim that these sheaves are algebras and these maps are algebra homomorphisms. The third term is the pushforward of an algebra, hence an algebra. The first term is the twist of the vacuum module for $(\mathrm{Vir}, \mathrm{Aut}_{2n} \mathcal{O})$, so its algebra structure comes from Proposition 2.2. Next, note that the Harish-Chandra pair $(\widetilde{\mathfrak{g}}, \widetilde{G}_n(\mathcal{O}))$ acts on $\widehat{\mathfrak{M}}_{g,1}$ through the quotient $(\mathrm{Vir}, \mathrm{Aut}_{2n} \mathcal{O})$, so that the $\mathrm{Aut}_{2n} \mathcal{O}$-twist of the $G_n(\mathcal{O})$-invariants in $V^n_{k,c}$ may be rewritten in terms of the semidirect product $\widetilde{G}_n(\mathcal{O})$

$$(\mathcal{V}^n_{k,c})^{G_n(\mathcal{O})} = \left[\pi_{n*}(\mathcal{O}_{\widehat{\mathfrak{M}}_{g,1}} \otimes V^n_{k,c}) \right]^{\widetilde{G}_n(\mathcal{O})}.$$

Thus as in Proposition 2.4 the sheaves $(\mathcal{V}^n_{k,c})^{G_n(\mathcal{O})}$ on $\mathfrak{M}_{g,1,2n}$ or $\mathrm{Bun}_{G,g,1,n}$ are naturally sheaves of algebras. This structure is clearly compatible with that on $\Pi^n_* \mathcal{V}^n_{k,c}$.

Finally, the maps $\mathrm{Vir}^{2n}_{c_k} \to V^n_{k,c}$ are induced from the homomorphism $U(\mathrm{Vir}_{c_k}) \to \widehat{U}_{k,c}\widetilde{\mathfrak{g}}$ into the completion of the enveloping algebra of $\widetilde{\mathfrak{g}}$. This homomorphism maps the subalgebra $U\mathrm{Der}_{2n} \mathcal{O}$ to the left ideal generated by the Lie subalgebra $\widetilde{\mathfrak{g}}_n(\mathcal{O})$. Hence we obtain a homomorphism of the corresponding vacuum modules, because $V^n_{k,c}$ can be defined as the quotient of the completed algebra $\widehat{U}_{k,c}\widetilde{\mathfrak{g}}$ by the left ideal generated by the Lie subalgebra $\widetilde{\mathfrak{g}}_n(\mathcal{O})$. It follows that the map $\mathrm{Vir}^{2n}_{c_k} \longrightarrow (\mathcal{V}^n_{k,c})^{G_n(\mathcal{O})}$ above respects algebra structures. More precisely, this map comes from the above homomorphism of enveloping algebras by passing to the normalizers of ideals on both sides, hence it remains a homomorphism. $\qquad\qquad\square$

5.1.1

By Proposition 5.1, we have a diagram of algebra homomorphisms on $\mathfrak{M}_{g,1,2n}$

$$
\begin{array}{ccc}
\mathrm{Vir}^n_{c_k} & \xrightarrow{\ \mathbb{S}^n_{k,c}\ } & \Pi^n_*\mathcal{V}^n_{k,c} \\
\downarrow & & \downarrow \\
\mathcal{D}_{c_k} & & \Pi^n_*\mathcal{D}_{k,c}
\end{array}
$$

We wish to show that the homomorphism $\mathbb{S}^n_{k,c}$ descends to the sheaves of twisted differential operators. It will then automatically be a homomorphism of *algebras* of differential operators.

Theorem 5.2. *The homomorphism* $\mathbb{S}^n_{k,c}$ *of descends to a homomorphism of algebras*

$$
\mathbb{S}^n_{k,c} : \mathcal{D}_{c_k} \longrightarrow \Pi^n_*\mathcal{D}_{k,c}.
$$

5.1.2 Spaces of coinvariants

Equivalently, we need to show that up on $\mathrm{Bun}_{G,g,1,n}$, the morphism $(\Pi^n)^*\mathrm{Vir}^n_{c_k} \to \mathcal{V}^n_{k,c}$ descends to $(\Pi^n)^*\mathcal{D}_{c_k} \to \mathcal{D}_{k,c}$, since the morphism on $\mathfrak{M}_{g,1,2n}$ is obtained from the former by adjunction.

If we apply the Virasoro and Kac–Moody localization functor Δ on $\mathrm{Bun}_{G,g,1,n}$ to the $(\widetilde{\mathfrak{g}}, \widetilde{G}_n(\mathcal{O}))$-modules $\mathrm{Vir}^{2n}_{c_k}$ and $V^n_{k,c}$, we obtain the desired sheaves $(\Pi^n)^*\mathcal{D}_{c_k}$ and $\mathcal{D}_{k,c}$. However, the embedding $\mathrm{Vir}^{2n}_{c_k} \to V_{k,c}$ of Proposition 3.3 is *not* a homomorphism of $\widetilde{\mathfrak{g}}$-modules: it intertwines the Virasoro action on $\mathrm{Vir}^n_{c_k}$ with the action on $V_{k,c}$ of the Virasoro algebra generated by the $\mathbb{S}_n$'s, not the L_ns. Because of that, it is not immediately clear that the map $\mathrm{Vir}^{2n}_{c_k} \to V_{k,c}$ gives rise to a morphism of sheaves $(\Pi^n)^*\mathcal{D}_{c_k} \to \mathcal{D}_{k,c}$. In order to prove that, we must use Proposition 5.3 below to pass from Lie algebra coinvariants to vertex algebra coinvariants.

Let us recall some results from [FB] on the spaces of (twisted) coinvariants of vertex algebras.

Let V be a conformal vertex algebra with a compatible $\widehat{\mathfrak{g}}$-structure (see [FB], Section 6.1.3). This means that V carries an action of the Harish-Chandra pair $(\widetilde{\mathfrak{g}}, \widetilde{G}(\mathcal{O}))$, such that the action of the Lie algebra $\widetilde{\mathfrak{g}}$ is generated by Fourier coefficients of vertex operators. In particular, our vacuum modules $V_{k,c}$ and Vir_c are examples of such vertex algebras.

Given a vertex algebra of this type, we define its space of twisted coinvariants as in [FB], Section 8.5.3. Namely, let R be a local $\mathbb{C}$-algebra and $(X, x, \mathcal{P})$ an R-point of $\mathrm{Bun}_{G,g}$, i.e., a pointed curve (X, x) and a G-bundle on X, all defined over $\mathrm{Spec}\, R$. Then X carries a natural $\widetilde{G}(\mathcal{O})$-bundle $\widehat{\mathcal{P}}$, whose fiber over $y \in X$ consists of pairs (z, t), where z is a formal coordinate at y and t is a trivialization of $\mathcal{P}$ over the formal disc around y.

Set

$$\mathcal{V}^{\mathcal{P}} = \widehat{\mathcal{P}} \underset{\widetilde{G}(\mathcal{O})}{\times} V.$$

This vector bundle carries a flat connection, and we define the sheaf $h(\mathcal{V}^{\mathcal{P}} \otimes \Omega)$ as the sheaf of zeroth de Rham cohomology of $\mathcal{V}^{\mathcal{P}} \otimes \Omega$. The vertex algebra structure on V makes this sheaf into a sheaf of Lie algebras. In particular, the Lie algebra $U^{\mathcal{P}}(\mathcal{V}_x)$ of sections of $h(\mathcal{V}^{\mathcal{P}} \otimes \Omega)$ over the punctured disc $D_x^{\times}$ is isomorphic to the Lie algebra $U(V)$ topologically spanned by the Fourier coefficients of all vertex operators from V. It acts on $\mathcal{V}_x^{\mathcal{P}}$, the fiber of $\mathcal{V}^{\mathcal{P}}$ at x.

Let $U_{X\backslash x}^{\mathcal{P}}(\mathcal{V}_x)$ be the image of the Lie algebra $\Gamma(X\backslash x, h(\mathcal{V}^{\mathcal{P}} \otimes \Omega))$ in $U^{\mathcal{P}}(\mathcal{V}_x) = \Gamma(D_x^{\times}, h(\mathcal{V}^{\mathcal{P}} \otimes \Omega))$. The space $H^{\mathcal{P}}(X, x, V)$ of twisted coinvariants of V is by definition the quotient of $\mathcal{V}_x^{\mathcal{P}}$ by the action of $U_{X\backslash x}^{\mathcal{P}}(\mathcal{V}_x)$.

Let $\mathcal{A}^{\mathcal{P}}$ be the Atiyah algebroid of infinitesimal symmetries of $\mathcal{P}$. We have the exact sequence

$$0 \to \mathfrak{g}^{\mathcal{P}} \to \mathcal{A}^{\mathcal{P}} \to \Theta_X \to 0$$

where $\mathfrak{g}^{\mathcal{P}}$ is the sheaf of sections of the vector bundle $\mathcal{P} \underset{G}{\times} \mathfrak{g}$.

When $V = V_{k,c}$, the Lie algebra $U^{\mathcal{P}}(\mathcal{V}_{k,c,x})$ contains as a Lie subalgebra a canonical central extension of the Lie algebra $\Gamma(D_x^{\times}, \mathcal{A}^{\mathcal{P}})$ (it becomes isomorphic to $\widetilde{\mathfrak{g}}$ if we choose a formal coordinate at x and a trivialization of $\mathcal{P}|_{D_x}$). Also, the Lie algebra $U_{X\backslash x}^{\mathcal{P}}(\mathcal{V}_{k,c,x})$ contains

$$\widetilde{\mathfrak{g}}_{\mathrm{out}}^{\mathcal{P}} = \Gamma(X\backslash x, \mathcal{A}^{\mathcal{P}})$$

as a Lie subalgebra (it is isomorphic to $\mathrm{Vect}(X\backslash x) \ltimes \mathfrak{g} \otimes \mathbb{C}[X\backslash x]$).

Likewise, in the case when $V = \mathrm{Vir}_c$, $U^{\mathcal{P}}(\mathrm{Vir}_{c,x})$ contains the Virasoro algebra, and $U_{X\backslash x}^{\mathcal{P}}(\mathrm{Vir}_{c,x})$ contains the Lie algebra $\mathrm{Vect}(X\backslash x)$ of vector fields on $X\backslash x$. The homomorphism $\mathrm{Vir}_{c_k} \to V_{k,c}$ of vertex algebra induces injective homomorphisms of Lie algebras

$$U^{\mathcal{P}}(\mathrm{Vir}_{c_k,x}) \hookrightarrow U^{\mathcal{P}}(\mathcal{V}_{k,c,x})$$

$$U_{X\backslash x}^{\mathcal{P}}(\mathrm{Vir}_{c_k,x}) \hookrightarrow U_{X\backslash x}^{\mathcal{P}}(\mathcal{V}_{k,c,x})$$

(since the action of $\widehat{\mathfrak{g}}$ on Vir_c is trivial, the Lie algebras on the left do not depend on $\mathcal{P}$). Note that though the Segal–Sugawara vertex operator is quadratic in the generating fields of $V_{k,c}$, the elements of $U_{X\backslash x}^{\mathcal{P}}(\mathrm{Vir}_{c_k,x})$ (and other elements of $U_{X\backslash x}^{\mathcal{P}}(V_x)$) cannot be expressed in terms of the Lie subalgebra $\widetilde{\mathfrak{g}}_{\mathrm{out}}^{\mathcal{P}}$ of $U_{X\backslash x}^{\mathcal{P}}(V_{k,c,x})$. Nevertheless, we have the following result which is proved in [FB], Theorem 8.3.3 (see also Remark 8.3.10).

5.3 Proposition. *For any smooth $\widetilde{\mathfrak{g}}^{\mathcal{P}}$-module M (which is then automatically a $U^{\mathcal{P}}(V_{k,c,x})$-module), the space of coinvariants of M by the action of $U_{X\backslash x}^{\mathcal{P}}(V_{k,c,x})$ is equal to the space of coinvariants of M by the action of $\widetilde{\mathfrak{g}}_{\mathrm{out}}^{\mathcal{P}}$.*

Proof of Theorem 5.1.1. Fix an R-point $(X, x, z, \mathcal{P}, t)$ of $\mathrm{Bun}_{G,g,1,n}$, i.e., a pointed curve (X, x), a $2n$-jet of coordinate z at x, a G-bundle $\mathcal{P}$ and an n-jet of trivialization t of $\mathcal{P}$ at x, all defined over the spectrum of some local $\mathbb{C}$-algebra R. Let v be a vector in the fiber of the sheaf $\mathrm{Vir}_{c_k}^{2n}$ over the R-point (X, x, z) of $\mathfrak{M}_{g,1,2n}$, which lies in the kernel of the surjection $\mathrm{Vir}_{c_k}^{2n}|_R \to \mathcal{D}_{c_k}|_R$. In order to prove the theorem, it is sufficient to show that the image of v in the fiber of the sheaf $V_{k,c}^n$ over the R-point $(X, x, z, \mathcal{P}, t)$ of $\mathrm{Bun}_{G,g,1,n}$ belongs to the kernel of the surjection $V_{k,c}^n|_R \to \mathcal{D}_{k,c}|_R$.

But according to [FB], Lemmas 16.2.9 and 16.3.6, the kernel of the map $\mathrm{Vir}_{c_k}^{2n}|_R \to \mathcal{D}_{c_k}|_R$ is spanned by all vectors of the form $s \cdot A$, where $A \in \mathrm{Vir}_{c_k}^{2n}|_R$ and $s \in \mathrm{Vect}(X\backslash x)$ (so that $\mathcal{D}_{c_k}|_R$ is the space of coinvariants of $\mathrm{Vir}_{c_k}^{2n}|_R$ with respect to $\mathrm{Vect}(X\backslash x)$). Likewise, the kernel of the map $V_{k,c}^n|_R \to \mathcal{D}_{k,c}|_R$ is spanned by all vectors of the form $f \cdot B$, where $B \in V_{k,c}^n|_R$ and $f \in \widetilde{\mathfrak{g}}_{\mathrm{out}}^{\mathcal{P}}$ (so that $\mathcal{D}_{k,c}|_R$ is the space of coinvariants of $V_{k,c}^n|_R$ with respect to $\widetilde{\mathfrak{g}}_{\mathrm{out}}^{\mathcal{P}}$).

So we need to show that the image of a vector of the above form $s \cdot A$ in $V_{k,c}^n|_R$ is in the image of the Lie algebra $\widetilde{\mathfrak{g}}_{\mathrm{out}}^{\mathcal{P}}$. But by Proposition 5.3, the space of coinvariants of $V_{k,c}^n|_R$ with respect to $\widetilde{\mathfrak{g}}_{\mathrm{out}}$ is equal to the space of coinvariants of $V_{k,c}^n|_R$ with respect to $U_{X\backslash x}^{\mathcal{P}}(V_{k,c,x})$. This implies the statement of the theorem, because according to the discussion of Section 5.1.2, $U_{X\backslash x}^{\mathcal{P}}(V_{k,c,x})$ contains the image of $\mathrm{Vect}(X\backslash x) \subset U_{X\backslash x}(\mathrm{Vir}_{c_k,x})$ as a Lie subalgebra. $\qquad\square$

5.1.3 Descent to $\mathfrak{M}_g$.

The homomorphism of Theorem 5.2 may be used to describe differential operators on the moduli of unmarked curves and bundles, $\Pi : \mathrm{Bun}_{G,g} \to \mathfrak{M}_g$. To do so we first localize the corresponding 'vacuum' representations Vir_{c_k} and $V_{k,c}$ on $\mathrm{Bun}_{G,g,1}$ and $\mathfrak{M}_{g,1}$ and then descend.

Corollary 5.4. *There is a canonical homomorphism* $\mathcal{D}_{c_k} \longrightarrow \Pi_*\mathcal{D}_{k,c}$ *of sheaves of algebras on* $\mathfrak{M}_g$.

Proof. We will construct a morphism of sheaves $\Pi^*\mathcal{D}_{c_k} \to \mathcal{D}_{k,c}$ on $\mathrm{Bun}_{G,g}$, which gives rise to the desired morphism on $\mathfrak{M}_g$ by adjunction. This map furthermore is constructed by descent from $\mathrm{Bun}_{G,g,1}$. By Proposition 4.4 and Proposition 4.8, the sheaves $\Delta_{c_k}(\mathrm{Vir}_{c_k})$ and $\Delta_{k,c}(V_{k,c})$ on $\mathfrak{M}_{g,1}$ and $\mathrm{Bun}_{G,g,1}$ are identified with $\pi^*\mathcal{D}_{c_k}$ and $\pi_G^*\mathcal{D}_{k,c}$. Moreover these identifications are *horizontal* with respect to the relative connections inherited by the pullback sheaves. We now need to construct a map between the pullback of $\pi^*\mathcal{D}_{c_k}$ to $\mathrm{Bun}_{G,g,1}$ and $\pi_G^*\mathcal{D}_{k,c}$ which is flat relative to π_G and hence descends to $\mathrm{Bun}_{G,g}$.

This morphism $\mathbb{S}_{k,c}$ may be constructed directly following Theorem 5.2, or by applying the change of vacuum isomorphism Lemma 2.4 to the homomorphism $\mathbb{S}_{k,c}^1$. The connection along the curve $\pi_G : \mathrm{Bun}_{G,g,1} \to \mathrm{Bun}_{G,g}$ is deduced (by passing to subquotients) from the action of $L_{-1} \in \mathrm{Der}\,\mathcal{O}$ on the vacuum modules Vir_{c_k} and $V_{k,c}$, which are compatible under the Segal–Sugawara homomorphism (see Proposition 3.3). Hence the morphism is horizontal along this curve and descends to $\mathrm{Bun}_{G,g,1}$.

The morphism $\mathbb{S}_{k,c}$ on $\mathfrak{M}_g$ is a homomorphism, since it is obtained by reduction (by change of vacuum, see Lemma 2.4) from the corresponding homomorphism $\mathbb{S}_{k,c}^1$ on $\mathfrak{M}_{g,1}$. $\square$

5.2 Tensor product decomposition

By applying localization to the isomorphism of $(\tilde{\mathfrak{g}}, \tilde{G}(\mathcal{O}))$-modules

$$\mathfrak{s}_{k,c} : V_k \otimes \mathrm{Vir}_{c_k}^{2n} \xrightarrow{\sim} V_{k,c}$$

of Proposition 3.3, we obtain an isomorphism

$$\Delta\mathfrak{s}_{k,c} : \Delta(V_k \otimes \mathrm{Vir}_{c_k}^{2n}) \xrightarrow{\sim} \Delta(V_{k,c}).$$

However, the functors of coinvariants and localization are not tensor functors. Since V_k is a $\hat{\mathfrak{g}}$-submodule of $V_{k,c}$, we obtain a natural map $\Delta(V_k) \to \Delta(V_{k,c})$. The localization $\Delta(V_k)$ is naturally identified with the sheaf $\mathcal{D}_{k/\mathfrak{M}} \subset \mathcal{D}_{k,c}$ of *relative* differential operators. We now use the homomorphism $\mathbb{S}_{k,c}$ to lift the Virasoro operators to $\mathcal{D}_{k,c}$:

Theorem 5.5. *There is an isomorphism of sheaves on* $\mathfrak{M}_{g,1,2n}$

$$\Pi_*^n\mathcal{D}_{k/\mathfrak{M}} \underset{\mathcal{O}}{\otimes} \mathcal{D}_{c_k} \cong \Pi_*^n\mathcal{D}_{k,c}$$

compatible with the inclusions of the two factors as subalgebras of $\Pi_*^n\mathcal{D}_{k,c}$.

Proof. Consider the isomorphism of $(\widetilde{\mathfrak{g}}, \widetilde{G}_n(\mathcal{O}))$-modules

$$\mathfrak{s}^n_{k,c} : V^n_k \underset{\mathbb{C}}{\otimes} \mathrm{Vir}^{2n}_{c_k} \longrightarrow V^n_{k,c}$$

of Proposition 3.3. Twisting by the $\widetilde{G}_n(\mathcal{O})$-torsor $\widehat{\mathrm{Bun}}_{G,g} \to \mathrm{Bun}_{G,g,1,n}$ we obtain an isomorphism

$$\mathcal{V}^n_k \underset{\mathcal{O}}{\otimes} \mathrm{Vir}^{2n}_{c_k} \longrightarrow \mathcal{V}^n_{k,c}$$

of sheaves. This morphism restricts to $\mathbb{S}^n_{k,c}$ on $|0\rangle \otimes \mathrm{Vir}^{2n}_{c_k}$, and hence descends to an isomorphism

$$\Pi^n_* \mathcal{V}^n_k \underset{\mathcal{O}}{\otimes} \mathrm{Vir}^{2n}_{c_k} \longrightarrow \Pi^n_* \mathcal{V}^n_{k,c}.$$

Thus we see that $\Pi^n_* \mathcal{V}^n_{k,c}$ is generated by its two subalgebras $\Pi^n_* \mathcal{V}^n_k$ and $\mathrm{Vir}^n_{c_k}$.

The inclusion $V_k \to V_{k,c}$ of $\widehat{\mathfrak{g}}$-modules gives rise to a natural algebra inclusion

$$\Delta(V^n_k) = \mathcal{D}_{k/\mathfrak{M}} \longrightarrow \Delta(V^n_{k,c}) = \mathcal{D}_{k,c}$$

from the sheaf of twisted differential operators *relative* to the moduli of curves $\mathfrak{M}_{g,1,2n}$ to the full sheaf of twisted differential operators. We define the map

$$\Pi^n_* \mathcal{D}_{k/\mathfrak{M}} \underset{\mathcal{O}}{\otimes} \mathcal{D}_{c_k} \to \Pi^n_* \mathcal{D}_{k,c}$$

as one generated by the homomorphisms from $\Pi^n_* \mathcal{D}_{k/\mathfrak{M}}$ and $\mathcal{D}_{c_k}$ to $\Pi^n_* \mathcal{D}_{k,c}$.

This map is surjective because it comes from the composition

$$\mathcal{V}^n_k \underset{\mathcal{O}}{\otimes} \mathrm{Vir}^{2n}_{c_k} \to \mathcal{V}^n_{k,c} \to \Delta(V^n_{k,c})$$

which is surjective because the first map is an isomorphism by Proposition 3.3 and the second map is surjective by definition. It remains to check that this map is injective. We have a commutative diagram

$$
\begin{array}{ccc}
\mathcal{V}^n_k \underset{\mathcal{O}}{\otimes} \mathrm{Vir}^{2n}_{c_k} & \overset{\sim}{\longrightarrow} & \mathcal{V}^n_{k,c} \\
\downarrow & & \downarrow \\
\Delta(V^n_k) \underset{\mathcal{O}}{\otimes} \Delta(\mathrm{Vir}^{2n}_{c_k}) & \longrightarrow & \Delta(V^n_{k,c})
\end{array}
$$

For an R-point of $\mathrm{Bun}_{G,g,1,n}$ as in the proof of Theorem 5.2, the kernel of the right vertical map over $\mathrm{Spec}\, R$ is the image of the action of the Lie algebra $\widetilde{\mathfrak{g}}^{\mathcal{P}}_{\mathrm{out}}$ in $\mathcal{V}^n_{k,c}|_R \simeq \mathcal{V}_k|_R \otimes \mathrm{Vir}^{2n}_{c_k}|_R$. But

$$\widetilde{\mathfrak{g}}^{\mathcal{P}}_{\mathrm{out}} \cdot (\mathcal{V}_k|_R \otimes \mathrm{Vir}^{2n}_{c_k}|_R) \subset (\mathfrak{g}^{\mathcal{P}}_{\mathrm{out}} \cdot \mathcal{V}_k|_R) \otimes \mathrm{Vir}^{2n}_{c_k}|_R + \mathcal{V}_k|_R \otimes (\widetilde{\mathfrak{g}}^{\mathcal{P}}_{\mathrm{out}} \cdot \mathrm{Vir}^{2n}_{c_k}|_R)$$

and so the right-hand side is in the kernel of the projection $\mathcal{V}_k|_R \otimes \mathrm{Vir}^{2n}_{c_k}|_R$

$\rightarrow (\Delta(V_k^n) \otimes \Delta(\text{Vir}_{c_k}^{2n}))|_R$. This proves that our map is injective and hence an isomorphism. $\square$

Remark 5.6. The theorem may be applied to describe differential operators on the moduli of curves and bundles without level structure, following Corollary 5.4. However, it is known ([BD1]) that for $k \neq -h^\vee$, there are no non-constant global twisted differential operators on $\text{Bun}_G(X)$, in other words, $\Pi_* \mathcal{D}_{k/\mathfrak{M}} = \mathcal{O}_{\mathfrak{M}_g}$. Thus in this case the pushforward of twisted differential operators $\mathcal{D}_{k,c}$ to $\mathfrak{M}_g$ is simply identified with the differential operators $\mathcal{D}_{c_k}$ on $\mathfrak{M}_g$.

5.3 Heat operators and projectively flat connections

Let $\pi : M \rightarrow S$ be a smooth projective morphism with connected fibers and $\mathcal{L} \rightarrow M$ a line bundle. Let $\mathcal{D}_\mathcal{L}$ be the sheaf of twisted differential operators on $\mathcal{L}$.

Definition 5.7. A heat operator on $\mathcal{L}$ relative to π is a lifting of the identity map $\text{id} : \Theta_S \rightarrow \Theta_S$ to a sheaf homomorphism $\mathcal{H} : \Theta_S \rightarrow \pi_* \mathcal{D}_{\mathcal{L}, \leq 1_S}$ to differential operators, which are of order one along S, such that the corresponding map $\Theta_S \rightarrow \pi_* \mathcal{D}_{\mathcal{L}, \leq 1_S}/\mathcal{O}_S$ is a Lie algebra homomorphism.

Suppose that the sheaf $\pi_* \mathcal{L}$ is locally free, i.e., is a sheaf of sections of a vector bundle on S. Then a heat operator gives rise to a projectively flat connection on this vector bundle (see [W, BK, Fa1]).

Consider the morphism $\Pi : \text{Bun}_{G,g} \rightarrow \mathfrak{M}_g$. Recall that $\mathcal{C} = \mathcal{L}_{k,0}$ is the line bundle on $\text{Bun}_{G,g}$ whose restriction to each $\text{Bun}_G(X)$ is the ample generator of the Picard group of $\text{Bun}_G(X)$. For any $k \in \mathbb{Z}_+$ the sheaf $\Pi_* \mathcal{C}^k$ is locally free, and its fiber at a curve $X \in \mathfrak{M}_g$ is the vector space $\text{H}^0(\text{Bun}_G(X), \mathcal{C}^k)$ of *non-abelian theta functions* of weight k. It is well-known that the corresponding vector bundle (which we will also denote by $\Pi_* \mathcal{C}^k$), possesses a projectively flat connection [Hi2, Fa1, BK]. This connection may be constructed using a heat operator on $\mathcal{C}^k$.

Theorem 5.7. ([BK]) *For any $k \geq 0$ the sheaf $\Pi_* \mathcal{C}^k$ on $\mathfrak{M}_g$ possesses a unique flat projective connection given by a heat operator on $\mathcal{C}^k$.*

5.3.1

The existence and uniqueness of this projective connection are deduced for $\mathcal{L}$ satisfying the vanishing of the composition

$$\Theta_S \rightarrow R^1 \pi_* \mathcal{D}_\mathcal{L} \rightarrow R^1 \pi_* \mathcal{D}_{\mathcal{L}/S}$$

and the identification $\pi_* \mathcal{D}_{\mathcal{L}/S} = \mathcal{O}_S$. The connection was also constructed by Hitchin [Hi2] and Faltings [Fa1] by different means.

The connection on non-abelian theta functions has been explicitly identified in [La] with the connection on the (dual of) the space of conformal blocks for the basic integrable representation $L_k(\widehat{\mathfrak{g}})$ of $\widehat{\mathfrak{g}}$ at level $k \in \mathbb{Z}_+$. The latter connection, known as the KZB or WZW connection (and specializing to the Knizhnik–Zamolodchikov connection in genus zero), is defined following the general procedure of [FB], Ch. 16. Namely, for any conformal vertex algebra V we defined a twisted $\mathcal{D}$-module of coinvariants on $\mathfrak{M}_g$. Its fibers are the spaces of coinvariants $H(X, x, V)$ (see Section 5.1.2), and the action of the sheaf $\mathcal{D}_c$ of differential operators (where c is the central charge of V) comes from the action of the Virasoro algebra on V. In our case, we take as V the module $L_k(\widehat{\mathfrak{g}})$. For any positive integer k this module is a conformal vertex algebra with central charge $c(k) = k \dim \mathfrak{g}/(k + h^\vee)$, with the conformal structure defined by the Segal–Sugawara vector. In this case the sheaf of coinvariants is locally free as an $\mathcal{O}$-module, and so it is the sheaf of sections of a vector bundle with a projectively flat connection. The sheaf of sections of its dual vector bundle (whose fibers are the spaces of conformal blocks of $L_k(\widehat{\mathfrak{g}})$) is therefore also a twisted $\mathcal{D}$-module, more precisely, a $\mathcal{D}_{-c(k)}$-module. The corresponding projectively flat connection on the bundle of conformal blocks is the KZB connection.

On the other hand, in Theorem 5.2 we produced homomorphisms $\mathbb{S}_{k,0}^n$: $\mathcal{D}_{-c(k)} \longrightarrow \Pi_*^n \mathcal{D}_{k,0}$. Their restrictions $\mathbb{S}_{k,0}|_{\leq 1}$ to $(\mathcal{D}_{-c(k)})_{\leq 1}$ give us heat operators on $\mathcal{C}^k = \mathcal{L}_{k,0}$. But the sheaf $\Pi_* \mathcal{C}^k$ is isomorphic to the sheaf of conformal blocks on $\mathfrak{M}_{g,1}$ corresponding to $L_k(\widehat{\mathfrak{g}})$ [BL1, KNR, Fa2, Te]. Under this identification, the projectively flat connection on $\Pi_* \mathcal{C}^k$ obtained from the heat operators $\mathbb{S}_{k,0}|_{\leq 1}$ tautologically coincides with the KZB connection, because both connections are constructed by applying the Segal–Sugawara construction. Thus, we obtain.

Proposition 5.8.

(1) *For every $k \in \mathbb{Z}_+$, the heat operator defining the projectively flat connection on the sheaf $\Pi_* \mathcal{C}^k$ of non-abelian theta functions over $\mathfrak{M}_g$ is given by the restriction of the Segal–Sugawara map $\mathbb{S}_{k,0}|_{\leq 1}$: $(\mathcal{D}_{-c(k)})_{\leq 1} \to \Pi_* \mathcal{D}_{k,0}$.*

(2) *For any n, $\mathbb{S}_{k,0}^n|_{\leq 1}$ gives heat operators defining a projectively flat connection on $\Pi_*^n \mathcal{L}_{k,c(k)}$ over $\mathfrak{M}_{g,1,2n}$.*

(3) *Under the identification between $\Pi_* \mathcal{C}^k$ and the sheaf of conformal blocks on $\mathfrak{M}_{g,1}$ corresponding to the integrable representation of $\widehat{\mathfrak{g}}$ of level k and highest weight 0, the KZB connection on the sheaf of conformal blocks is given by the heat operators $\mathbb{S}_{k,0}|_{\leq 1}$ on $\mathcal{C}^k$.*

In the same way we obtain heat operators for any level structure (where the morphisms Π^n are no longer projective). Note also that we can replace $\mathcal{C}^k$ by $\mathcal{L}_{k,c}$ for any integer c, which have isomorphic restrictions to $\mathrm{Bun}_G(X)$ for any X, and obtain analogous projectively flat connections.

6 Classical limits

In this section we describe the limits of the (suitably rescaled) Segal–Sugawara operators at the critical level $k = -h^\vee$, and in the classical limit $k, c \to \infty$. In the former case the algebra of twisted differential operators $\mathcal{D}_{c_k}$ degenerates into the Poisson algebra of functions on the space of curves with projective structure, and the construction gives twisted differential operators on $\mathrm{Bun}_{G,g}$ which are *vertical* (i.e. preserve $\mathrm{Bun}_G(X)$ for fixed X) and *commute*. We identify these operators with the quadratic part of the Beilinson–Drinfeld quantization of the Hitchin system. When the level and charge become infinite, both sides of the construction become commutative (Poisson) algebras, and the Segal–Sugawara construction is interpreted as a map from the moduli of extended connections to the moduli of projective structures or quadratic differentials. We interpret this map as defining a symplectic connection over the moduli of pointed curves on the moduli spaces of connections with arbitrary poles. We also sketch the interpretation of this connection as a new Hamiltonian form of the equations of isomonodromic deformation.

6.1 The critical level

Recall that we assume throughout that the group G is simply connected.

By the general formalism of localization (Proposition 2.2), we have an algebra homomorphism $(V_k^n)^{G(\mathcal{O})} \to \Gamma(\mathrm{Bun}_G(X, x, n), \mathcal{D}_k)$ from the $G(\mathcal{O})$-invariants of the vacuum (which are the endomorphisms of the vacuum representation) to global differential operators on the moduli of bundles. For general k this gives only scalars, but at the critical level $k = -h^\vee$ the space $V_k^{G(\mathcal{O})}$ becomes very large. In [FF] (see also [Fr1, Fr2]), Feigin and Frenkel identify the algebra $V_{-h^\vee}^{G(\mathcal{O})}$ canonically with the ring of functions $\mathbb{C}[\mathcal{O}p_{G^\vee}(D)]$ on the space of *opers* on the disc, for the Langlands dual group $G^\vee$ of G. Opers, introduced in [BD1] by Beilinson and Drinfeld, are $G^\vee$-bundles equipped with a Borel reduction and a connection, which satisfies a strict form of Griffiths transversality. For $G^\vee = PSL_2$ (so $G = SL_2$), opers are identified with projective structures $\mathcal{P}\mathrm{roj}\,(X)$. In fact for arbitrary G there is a natural projection $\mathcal{O}p_{G^\vee}(X) \to \mathcal{P}\mathrm{roj}\,(X)$, which identifies the space of opers with the affine space for the vector space $\mathrm{Hitch}_G(X)$ induced from the affine space $\mathcal{P}\mathrm{roj}\,(X)$

for quadratic differentials $H^0(X, \Omega_X^{\otimes 2}) \subset \mathrm{Hitch}_G(X)$. (Similar considerations apply to any level vacuum representations V_k^n, replacing regular opers by opers with singularities, which form an affine space for the meromorphic version of the Hitchin space.)

Thus there is a homomorphism

$$\mathbb{C}[\mathcal{O}p_{G^\vee}(D)] \longrightarrow \Gamma(\mathrm{Bun}_G(X), \mathcal{D}_{-h^\vee}).$$

Beilinson and Drinfeld show that this homomorphism factors through functions on *global* opers $\mathbb{C}[\mathcal{O}p_{G^\vee}(X)]$, is independent of the choice of $x \in X$ used in localization, and gives rise to an *isomorphism*

$$\mathbb{C}[\mathcal{O}p_{G^\vee}(X)] \cong \Gamma(\mathrm{Bun}_G(X), \mathcal{D}_{-h^\vee}).$$

$$\mathcal{O}(\mathcal{O}p_{G^\vee}(\mathfrak{X})) \to \Pi_* \mathcal{D}_{k/\mathfrak{M}}.$$

We wish to compare the restriction of this homomorphism to the subalgebra $\mathbb{C}[\mathcal{P}\mathrm{roj}\,(X)]$ of $\mathbb{C}[\mathcal{O}p_{G^\vee}(X)]$ with the critical level limit of the Segal–Sugawara construction. Let $\mathcal{O}p_{G^\vee} \to \mathfrak{M}_g$ denote the moduli stack of curves with $G^\vee$-opers.

Recall the notation $\mu_\mathfrak{g} = h^\vee \dim \mathfrak{g}$.

Theorem 6.1.

(1) *For $c \in \mathbb{C}$, the homomorphism $(k + h^\vee)\mathbb{S}_{k,c}^n$ is regular at $k = -h^\vee$, defining an algebra homomorphism*

$$\overline{\mathbb{S}}_{-h^\vee,c}^n : \mathcal{O}(T_{\mu_\mathfrak{g}}^* \mathfrak{M}_{g,1,2n}) \longrightarrow \Pi_*^n \mathcal{D}_{-h^\vee,c}.$$

(2) *The homomorphism $\overline{\mathbb{S}}_{-h^\vee,c}$ is the restriction of the Beilinson–Drinfeld homomorphism to projective structures: we have a commutative diagram*

$$
\begin{array}{ccc}
\mathcal{O}(\mathcal{P}\mathrm{roj}_g^{\mu_\mathfrak{g}}) & \xrightarrow{\overline{\mathbb{S}}_{-h^\vee,c}} & \Pi_* \mathcal{D}_{-h^\vee,c} \\
\downarrow & & \uparrow \\
\mathcal{O}(\mathcal{O}p_{G^\vee}) & \xrightarrow{\mathrm{BD}} & \Pi_* \mathcal{D}_{-h^\vee/\mathfrak{M}_g}
\end{array}
$$

Proof. As described in Section 3.4.1 (see Proposition 3.4), the rescaled Segal–Sugawara operators $\overline{\mathbb{S}}_m = (k + h^\vee)\mathbb{S}_m$ are regular at $k = -h^\vee$, and define a homomorphism of vertex algebras. We may now repeat the constructions of Section 5 leading to Theorem 5.2 for the rescaled Sugawara operators. Note that the classical vacuum representations $\overline{\mathrm{Vir}}_{\mu_\mathfrak{g}}$ are (commutative) vertex algebras – the vertex Poisson structure is not used in the definition of coinvariants – and for commutative vertex algebras, the comparison of coinvariants for a generating set and the algebra it generates is obvious. Identifying the localization of $\overline{\mathrm{Vir}}_{\mu_{\mathfrak{g}}}^{-2n}$ with $\mathcal{O}(T^*_{\mu_{\mathfrak{g}}}\mathfrak{M}_{g,1,2n}) = \mathcal{O}(\mathcal{P}\mathrm{roj}_{g,1,2n}^{\mu_{\mathfrak{g}}})$, we obtain the first assertion.

Moreover, by Section 3.4.1, the critical Segal–Sugawara homomorphism factors as follows

$$\overline{\mathbb{S}}^n_{-h^\vee,c} : \overline{\mathrm{Vir}}_{\mu_{\mathfrak{g}}}^{-2n} \hookrightarrow (V^n_{-h^\vee})^{G_n(\mathcal{O})} \hookrightarrow (V^n_{-h^\vee,c})^{G_n(\mathcal{O})}$$

(since $\overline{\mathbb{S}} = -S_{-2}$ lands inside $V_{-h^\vee} \subset V_{-h^\vee,c}$). It follows that the localized map on twisted differential operators also factors through the localization of $V_{-h^\vee}$, which is the sheaf of relative differential operators. Thus the critical Segal–Sugawara construction is part of the localization of the $G(\mathcal{O})$-invariants in the vacuum representation, giving the Beilinson–Drinfeld operators.   □

6.2 The infinite level

Recall from Section 4 (see Proposition 4.2) that the $\mathcal{C}$-twisted cotangent bundle of the moduli stack of bundles $\mathrm{Bun}_G(X)$ is isomorphic to the moduli stack $\mathcal{C}\mathrm{onn}\,_G(X)$ of bundles with regular connections on X, while the twisted cotangent bundle of the moduli of bundles with n-jet of trivialization $\mathrm{Bun}_G(X, x, n)$ is the moduli $\mathcal{C}\mathrm{onn}\,_G(X, x, n)$ of connections having at most nth-order poles at x. Similarly, we may consider moduli $\mathrm{Bun}_G(X, x_1, \ldots, x_m, n_1, \ldots, n_m)$ of bundles with several marked points and jets of coordinates, whose twisted cotangent bundles are identified with moduli of connections with poles of the corresponding orders. As elsewhere, we restrict to the one-point case for notational simplicity though all constructions carry over to the multipoint case in a straightforward fashion.

By virtue of their identification as twisted cotangent bundles, the moduli of meromorphic connections on a fixed curve carry canonical (holomorphic) symplectic structures. As we vary X and x, the moduli stack $\mathcal{C}\mathrm{onn}\,_{G,g,1,n}$ forms a relative twisted cotangent bundle to $\mathrm{Bun}_{G,g,1,n}$ over $\mathfrak{M}_{g,1,2n}$. To obtain a symplectic variety, we consider the *absolute* twisted cotangent bundles of $\mathrm{Bun}_{G,g,1,n}$. The twisted cotangent bundle corresponding to the line bundle $\mathcal{L}_{\lambda,\mu}$ is the moduli of *extended connections* $\mathcal{C}\mathrm{onn}\,_{\mathrm{Bun}_{G,g,1,n}}(\mathcal{L}_{\lambda,\mu}) = \mathcal{E}\mathrm{x}\mathcal{C}\mathrm{onn}_{G,g,1,n}^{\lambda,\mu}$ defined in Section 4.3. In the limit of infinite level the sheaves of twisted differential operators on $\mathrm{Bun}_{G,g,1,n}$ and $\mathfrak{M}_{g,1,2n}$ degenerate to the commutative (and hence Poisson) algebra of functions on the moduli of extended connections and projective structures, respectively. Therefore it is convenient to reinterpret the infinite limit of the Segal–Sugawara homomorphism as a morphism between these moduli spaces (rather than a homomorphism between the corresponding algebras of functions), and examine its Poisson properties.

As in Section 3.4.2, we let $k, c \to \infty$ by introducing auxiliary parameters λ, μ with $\dfrac{\lambda}{k} = \dfrac{\mu}{c}$. Then the homomorphism $\dfrac{\lambda^2}{k} \mathbb{S}^n_{k,c}$ is regular as $k, c \to \infty$, as is (for $\lambda \neq 0$) the isomorphism $\mathfrak{s}^n_{k,c} : V^n_k \otimes \mathrm{Vir}^{2n}_{c_k} \to V^n_{k,c}$. We thus obtain classical limits of Theorem 5.2 and Theorem 5.5.

Let $\mathrm{Quad}_{g,1,n} = T^* \mathfrak{M}_{g,1,n}$ and $\mathrm{Higgs}_{G,g,1,n} = T^*_{/\mathfrak{M}_{g,1,2n}} \mathrm{Bun}_{G,g,1,n}$ denote the moduli stacks of curves equipped with a quadratic differential and bundles with Higgs field, having at most nth-order pole at the marked point, respectively. Let $\mathrm{Hitch}_{G,g,1,n}$ denote the target of the Hitchin map for $\mathrm{Higgs}_{G,g,1,n}$, i.e.

$$\mathrm{Hitch}_{G,g,1,n} = \bigoplus_{i=1}^{\ell} \Gamma(X, \Omega(nx)^{d_i+1})$$

where $\ell = \mathrm{rank}\,\mathfrak{g}$ and d_i is the ith exponent of $\mathfrak{g}$.

Recall from Section 3.4.2 that a commutative algebra homomorphism of Poisson algebras (and the corresponding morphism of spaces) is called λ-*Poisson* if it rescales the Poisson bracket by λ.

Theorem 6.2.

(1) *For every* $\lambda, \mu \in \mathbb{C}$ *there is a* λ-*Poisson homomorphism*

$$\overline{\mathbb{S}}^n_{\lambda,\mu} : \mathcal{O}(T^*_{\lambda\mu}\mathfrak{M}_{g,1,2n}) \longrightarrow \Pi^n_* \mathcal{O}(T^*_{\lambda,\mu} \mathrm{Bun}_{G,g,1,n})$$

Equivalently, for every $\lambda, \mu \in \mathbb{C}$ *there is a* λ-*Poisson map*

$$\Pi^n_{\lambda,\mu} : \mathcal{E}x\mathcal{C}onn^{\lambda,\mu}_{G,g,1,n} \to \mathcal{P}roj^{\lambda\mu}_{g,1,2n}$$

lifting $\Pi^n : \mathrm{Bun}_{G,g,1,n} \to \mathfrak{M}_{g,1,2n}$.

(2) *For* $\lambda \neq 0$, *there is a canonical (non-affine) product decomposition over the moduli of curves*

$$\mathcal{E}x\mathcal{C}onn^{\lambda,\mu}_{G,g,1,n} \cong \mathcal{C}onn^{\lambda}_{G,g,1,n} \underset{\mathfrak{M}_{g,1,2n}}{\times} \mathcal{P}roj^{\lambda\mu}_{g,1,2n}.$$

(3) *For* $\lambda = \mu = 0$, $\Pi^n_{\lambda,\mu}$ *factors through the quadratic Hitchin map*

$$
\begin{array}{ccc}
T^*\mathrm{Bun}_{G,g,1,n} & \overset{\Pi^n_{0,0}}{\longrightarrow} & \mathrm{Quad}_{g,1,2n} \\
\downarrow & & \uparrow \\
\mathrm{Higgs}_{G,g,1,n} & \overset{\mathrm{Hitch}}{\longrightarrow} & \mathrm{Hitch}_{G,g,1,n}
\end{array}
$$

Proof. As in Theorem 6.1, the construction of the classical Segal–Sugawara homomorphism $\overline{\mathbb{S}}_{\lambda,\mu}$ is identical to the proof of Theorem 5.2, appealing to Proposition 3.5 for the description of the rescaled vertex algebra homomorphism. The localization of $\overline{\mathrm{Vir}}^{2n}_{\mu}$ is $\mathcal{O}(T^*_{\mu}\mathfrak{M}_{g,1,2n})$ and that of $\overline{V}^n_{\lambda,\mu}$ is

$\mathcal{O}(T^*_{\lambda,\mu}\mathrm{Bun}_{G,g,1,2n})$, while the Poisson bracket is rescaled by λ, by Proposition 3.4.2. The homomorphism $\overline{\mathbb{S}}^n_{\lambda,\mu}$ of $\mathcal{O}_{\mathrm{Bun}_{G,g}}$-algebras defines, upon taking Spec over the moduli of curves, a morphism $T^*_{\lambda,\mu}\mathrm{Bun}_{G,g,1,n} \to T^*_{\lambda\mu}\mathfrak{M}_{g,1,2n}$ relative to $\mathfrak{M}_{g,1,2n}$, hence the geometric reformulation.

The morphism in (2) is given in components by the natural projection (4.15) from extended connections to connections and the map $\Pi^n_{\lambda,\mu}$. The spaces $\mathcal{E}\mathrm{x}\mathcal{C}\mathrm{onn}^{\lambda,\mu}_{G,g,1,n}$ and $\mathcal{P}\mathrm{roj}^{\lambda\mu}_{g,1,2n} \underset{\mathfrak{M}_{g,1,2n}}{\times} \mathcal{C}\mathrm{onn}^{\lambda}_{G,g,1,n}$ are both torsors for quadratic differentials over $\mathcal{C}\mathrm{onn}_{G,g,1,n}$. It follows from the explicit form of the classical Segal–Sugawara vector that the resulting map of torsors over the space of connections is an isomorphism (up to rescaling by λ). Equivalently, the assertion (2) is the classical limit of Theorem 5.5 and may be proved identically, replacing $\mathfrak{s}^n_{k,c}$ by its rescaled version.

For $\lambda = \mu = 0$, the image of $\overline{\mathbb{S}}^n_{0,0} : \overline{\mathrm{Vir}}^{2n}_0 \to \overline{V}^n_{0,0}$ lies strictly in $\overline{V}^n_0 \subset \overline{V}^n_{0,0}$. It follows that the image of $\mathcal{O}(T^*\mathfrak{M}_{g,1,2n}) = \mathcal{O}(\mathrm{Quad}_{g,1,2n})$ in $\mathcal{O}(T^*\mathrm{Bun}_{G,g,1,n})$ consists of functions which are pulled back from $T^*_{/\mathfrak{M}_{g,1,2n}}\mathrm{Bun}_{G,g,1,n}$. Therefore the map $\Pi^n_{0,0}$ descends to $T^*_{/\mathfrak{M}_{g,1,2n}}\mathrm{Bun}_{G,g,1,n}$. Restricting to fibers over a fixed curve X we obtain a map $T^*\mathrm{Bun}_G(X,x,n) \to T^*\mathfrak{M}_{g,1,2n}|_X = H^0(X,\Omega^2(2nx))$.

The quadratic Hitchin map is the map $T^*\mathrm{Bun}_G(X,x,n) \to H^0(X,\Omega^2(2nx))$ sending a Higgs bundle $(\mathcal{P},\eta)$, $\eta \in H^0(X,\mathfrak{g}_\mathcal{P} \otimes \Omega(nx))$ to $\frac{1}{2}\mathrm{tr}\,\eta^2$. Using the pairing on $L\mathfrak{g}$, this is identified with the map defined by the Segal–Sugawara vector

$$S_{-2} = \frac{1}{2}\sum_a J^a_{-1}J_{a,-1}$$

as desired. □

Remark 6.3. In [BZB1], the product decomposition (2) is described concretely in terms of kernel functions along the diagonal (the necessary translation from vertex algebra language to kernel language is described in Chapter 7 of [FB]). Moreover, it is related, for $\mu \neq 0$, to the classical constructions of projective structures on Riemann surfaces using theta functions due to Klein and Wirtinger. Composing the projection $\Pi_{\lambda,\mu}$ with the canonical meromorphic sections of the twisted cotangent bundles over $\mathrm{Bun}_{G,g}$ given by nonabelian theta functions (see also [BZB2]), this gives a construction of interesting rational maps from the moduli of bundles to the spaces of projective structures, and, more generally, opers.

Now we discuss various applications of the above results.

6.2.1 The Bloch–Esnault and Beilinson connection.

Let $\mathfrak{X} \to S$ be a smooth family of curves, and E a vector bundle on $\mathfrak{X}$. The vector bundle E defines a line bundle $\mathcal{C}(E)$ on S, the c_2-line bundle defined in [De] (whose Chern class is the pushforward of $c_2(E)$) which differs from the determinant of cohomology of E by the nth power of the Hodge line bundle on S. In fact, this line bundle is identified canonically with the pullback of $\mathcal{C}$ to S from the map $S \to \mathrm{Bun}_{G,g}$ classifying E and X (see [BK]). Bloch–Esnault and Beilinson [BE] construct a connection on $\mathcal{C}(E)$ from a regular connection $\nabla_{\mathfrak{X}/S}$ on E relative to S. It is easy to see that this connection can be recovered from Theorem 6.2 (2) (more precisely, its straightforward version for $G = GL_n$ and unpointed curves). Let $\lambda = 1$, $\mu = 0$, so that $\mathcal{L}_{\lambda,\mu} = \mathcal{C}$ and $\mathcal{C}\mathrm{onn}\,(\mathcal{C}) = \mathcal{E}\mathrm{x}\mathcal{C}\mathrm{onn}_{G,g}^{1,0}$ over $\mathrm{Bun}_{G,g}$. We have a decomposition $\mathcal{C}\mathrm{onn}\,(\mathcal{C}) = \mathcal{C}\mathrm{onn}_{G,g} \underset{\mathfrak{M}_g}{\times} T^*\mathfrak{M}_g$, and hence a canonical lifting from $\mathcal{C}\mathrm{onn}_{G,g}$ to $\mathcal{C}\mathrm{onn}\,(\mathcal{C})$, lying over the zero section of $T^*\mathfrak{M}_g$. When pulled back to S by the classifying map $S \to \mathcal{C}\mathrm{onn}_{G,g}$ of $(E, \nabla_{\mathfrak{X}/S})$, this gives a connection on $\mathcal{C}(E)$ over S. The compatibility with the construction of Bloch–Esnault and Beilinson follows for example from the uniqueness of the splitting $\mathcal{C}\mathrm{onn}_{G,g} \to \mathcal{E}\mathrm{x}\mathcal{C}\mathrm{onn}_{G,g}^{1,0}$, due to the absence of one–forms on $\mathfrak{M}_g$.

6.2.2 The Segal–Sugawara symplectic connection.

Suppose $f : N \to M$ is a smooth Poisson map of symplectic varieties. It then follows that N carries a flat symplectic connection over M, i.e., a Lie algebra lifting of vector fields on M to vector fields on N (defining a foliation on N transversal to f), which preserve the symplectic form on fibers (see [GLS] for a discussion of symplectic connections). Namely, we represent local vector fields on M by Hamiltonian functions, pull back to functions on N and take the symplectic gradient. The Poisson property of f guarantees the flatness and symplectic properties of the connection. (In the algebraic category, this defines the structure of $\mathcal{D}$-scheme or crystal of schemes on N over M, compatible with the symplectic structure.)

The morphism $\Pi_{1,\mu}^n$ of Theorem 6.2 is such a Poisson morphism of symplectic varieties, and hence defines a symplectic connection (or crystal structure) on $\mathcal{E}\mathrm{x}\mathcal{C}\mathrm{onn}_{G,g,1,n}^{1,\mu}$ over $\mathcal{P}\mathrm{roj}_{g,1,2n}^{\mu}$. When $\mu = 0$ we may reduce this connection as follows. First we restrict to the zero section $\mathfrak{M}_{g,1,2n} \hookrightarrow \mathcal{P}\mathrm{roj}_{g,1,2n}^{0}$ of the cotangent bundle to obtain a connection on $\mathcal{E}\mathrm{x}\mathcal{C}\mathrm{onn}_{G,g,1,n}^{1,0}$ over $\mathfrak{M}_{g,1,2n}$. Next, the connection respects the product decomposition Theorem 6.2,(2) relative to $\mathfrak{M}_{g,1,2n}$, so that we obtain a flat symplectic connection on the moduli space $\mathfrak{Conn}_{G,g,1,n}$ of G–bundles with connections over the moduli $\mathfrak{M}_{g,1,2n}$ of decorated curves (with respect to the relative symplectic structure):

Corollary 6.4. *The projection* $\Pi_{1,0}^{n}$ *defines a flat symplectic connection on the moduli stack* $\mathfrak{Conn}_{G,g,1,n}$ *over* $\mathfrak{M}_{g,1,2n}$.

6.2.3 Time-dependent Hamiltonians.

The structure of flat symplectic connection on a relatively symplectic $P \to M$ variety can also be encoded in a closed two-form Ω on P, restricting to the symplectic form on the fibers. The connection is then defined by the null-foliation of Ω. If P is locally a product, then this structure can be encoded in the data of Hamiltonian functions on P that are allowed to depend on the 'times' M. Thus, following [M], we may consider the data of such a form as the general structure of a time-dependent (or non–autonomous) Hamiltonian system. In our case $P = \mathfrak{Conn}_{G,g,1,n}$, we thus have three equivalent formulations of a non-autonomous Hamiltonian system, with times given by the moduli $\mathfrak{M}_{g,1,2n}$ of decorated curves: the Hamiltonian functions $\overline{\mathbb{S}}_{1,0}^{n}$ (or more precisely the functions on $\mathfrak{Conn}_{G,g,1,n}$ obtained from local vector fields on $\mathfrak{M}_{g,1,2n}$, considered as linear functions on $T^{*}\mathfrak{M}_{g,1,2n}$); the symplectic connection induced by $\Pi_{1,0}^{n}$; and the two-form Ω on $\mathfrak{Conn}_{G,g,1,n}$ obtained by restricting the symplectic form on $\mathcal{E}\mathrm{x}\mathfrak{Conn}_{G,g,1,n}^{1,0}$ under the embedding (as in Section 6.2.1) along the zero section of $T^{*}\mathfrak{M}_{g,1,2n}$.

6.3 Isomonodromy

In this final section, we describe the algebraic definition of isomonodromic deformation of arbitrary meromorphic connections over the moduli of decorated curves and sketch its identification with the Segal–Sugawara symplectic connection (as well as the compatibility with the analytic iso-Stokes connections of [JMU, Ma, Bo]). We thus obtain an algebraic time-dependent Hamiltonian description of the isomonodromy equations. It also follows that the isomonodromy Hamiltonians are classical limits of the heat operators defining the KZB equations, and are non-autonomous deformations of the quadratic Hitchin Hamiltonians.

6.3.1 Isomonodromic deformation.

The moduli spaces of G-bundles with regular connection $\mathfrak{Conn}_{G,g}$ carry a flat connection (crystal structure) over the moduli of curves, namely the connection of isomonodromic deformation or non-abelian Gauss–Manin connection (see [Si]). This connection is a manifestation of the topological description of connections with regular singularities on a Riemann surface as representations of the fundamental group, which does not change under holomorphic deformations. Namely, given a family $\mathfrak{X} \to S$ of Riemann surfaces and a bundle with (holomorphic, hence flat) connection on one fiber X, there is a unique extension of the connection to nearby fibers so that the monodromy representation does not change. The families of connections relative to S one obtains this way are uniquely characterized as those families which admit an *absolute* flat connection over $\mathfrak{X}$, the total space of the family. This is captured algebraically in the crystalline interpretation of flat connections: a flat connection on a variety admits a unique flat extension to an arbitrary nilpotent thickening of the variety. Thus algebraically one defines families of flat connections relative to a base to be isomonodromic if they may be extended to an absolute flat connection. It follows that the moduli spaces of flat connections on varieties carry a crystal structure over the deformation space of the underlying variety – the non-abelian Gauss–Manin connection of Simpson [Si].

More generally, there is an algebraic connection (crystal or $\mathcal{D}$-scheme structure) on the moduli stack of $\widehat{\mathcal{C}\mathrm{onn}}_{G,g,1,n}$ of meromorphic connections over curves with formal coordinates, which is the pullback of $\mathcal{C}\mathrm{onn}_{G,g,1,n}$ to the moduli of pointed curves with coordinates $\widehat{\mathfrak{M}}_{g,1}$. This connection is defined by 'fixing' connections around their poles and deforming them isomonodromically on the complement, combining the crystalline description of isomonodromy and the Virasoro uniformization of the moduli of curves. Namely, given a family $(\widetilde{X}, \widetilde{x}, \widetilde{t}) \in \widehat{\mathfrak{M}}_{g,1}(S)$ of pointed curves with formal coordinate over S, the spectrum of an Artinian local ring, and a connection $(\mathcal{P}, \nabla)$ on the special fiber X with pole at x of order at most n, we must produce a canonical extension of $(\mathcal{P}, \nabla)$ to $\widetilde{X}$. As explained in [FB], Ch. 15, p. 282, any such deformation $(\widetilde{X}, \widetilde{x})$ of pointed curves is given by 'regluing' $X \backslash x$ and the formal disc D_x around x using the action of $\mathrm{Aut}\,\mathcal{K}(S)$.

More precisely, we fix an identification of $\widetilde{X} \setminus \widetilde{x}$ with $(X \setminus x) \times S$ and 'glue' it to $D_x \times S$ using an automorphism of the punctured disc over S. Now, given $(\mathcal{P}, \nabla)$, we define an (absolute) flat connection on $\widetilde{X}$ by extending our connection $(\mathcal{P}, \nabla)$ trivially (as a product, with the trivial connection along the second factor) on to $(X \setminus x) \times S$ and on to $D_x \times S$. Note that since the connection on the nilpotent thickening $\widetilde{X} \setminus \widetilde{x} \simeq (X \setminus x) \times S$ of $X \setminus x$ is flat, it is *uniquely* determined by $(\mathcal{P}, \nabla)$, independently of the trivialization of the deformation – is the isomonodromic deformation of $(\mathcal{P}, \nabla)|_{X \backslash x}$ over S (in particular the stabilizer of $(\widetilde{X}, \widetilde{x})$ in $\mathrm{Aut}\,\mathcal{K}(S)$ does not change the deformed connection.) This defines the isomonodromy connection on $\widehat{\mathfrak{M}}_{g,1}$.

Proposition 6.5. *The Segal–Sugawara symplectic connection on* $\mathcal{C}\mathrm{onn}_{G,g,1,n}$ *over* $\mathfrak{M}_{g,1,2n}$ *pulls back to the isomonodromy connection on* $\widehat{\mathcal{C}\mathrm{onn}}_{G,g,1,n}$.

Proof. The compatibility between the two connections follows from the local description of the isomonodromy connection through the action of Aut $\mathcal{K}$ on meromorphic connections, and the description of the classical Segal–Sugawara operators as the corresponding Hamiltonians. Namely, recall from [FB] that the ind-scheme $\mathrm{Conn}_G(D^\times)$ of connections on the trivial G-bundle on the punctured disc is identified with $\widehat{\mathfrak{g}}_1^*$, the level one hyperplane in the dual to the affine Kac–Moody algebra. Moreover, this identification is equivariant with respect to the natural actions of $G(\mathcal{K})$ and Aut $\mathcal{K}$, describing the transformation of connections under gauge transformations and changes of coordinates. These actions are Hamiltonian, in an appropriately completed sense. Namely, the Fourier coefficients of fields from the vertex Poisson algebra $\overline{V}_1(\mathfrak{g})$ form the Lie subalgebra of *local functionals* on $\widehat{\mathfrak{g}}_1^*$ in the Poisson algebra of all functionals on $\widehat{\mathfrak{g}}_1^*$. By the Segal–Sugawara construction, this Lie algebra contains as Lie subalgebra Der $\mathcal{K}$ (as the Fourier coefficients $\overline{L}_n$ of the classical limit of the conformal vector), thereby providing Hamiltonians for the action of Aut $\mathcal{K}$. The construction of the isomonodromy connection on $\widehat{\mathfrak{M}}_{g,1}$ above is expressed in local coordinates (as a flow on meromorphic connections on the disc at x) by this action of Aut $\mathcal{K}$ on connections. Likewise, the Segal–Sugawara connection is defined by Hamiltonian functions, which are reductions of the classical Segal–Sugawara operators to $\mathrm{Conn}_{G,g,1,n}$. Thus the compatibility of the two connections is a consequence of the local statement. $\qquad\square$

6.3.2 Remark: analytic isomonodromy and Stokes data. In the complex analytic setting, one can extend to irregular connections the description of regular-singular connections by topological monodromy data, by introducing Stokes data describing the transitions between asymptotic fundamental solutions to the connection in different sectors. One can thereby define an 'iso-Stokes' generalization of the isomonodromy equations, i.e., an analytic symplectic connection on moduli of irregular holomorphic connections (see [JMU, Ma] and recently [Bo, Kr]). Thus, one has an isomonodromic deformation with more 'times' than just the moduli of pointed curves (one may also vary the most singular term of the connection).

It is easy to see our algebraic isomonodromic connection (with respect to motions of pointed curves) agrees with the analytic one. Since we can choose the gluing transformations in Aut $\mathcal{K}$ used to describe a given deformation to be *convergent* rather than formal, it follows that we are not altering the isomorphism type of the connection on a small analytic disc around x, and hence all Stokes data are automatically preserved. Thus it follows that the iso-Stokes connection on $\mathrm{Conn}_{G,g,1,n}$ is in fact algebraic.

6.3.2 Conclusion: KZB, isomonodromy and Hitchin.

It follows from Proposition 6.5 and Theorem 6.2 that the moduli spaces of meromorphic connections carry algebraic isomonodromy equations, which form a time-dependent Hamiltonian system. Moreover, the isomonodromy Hamiltonians are non-autonomous deformations of the quadratic Hitchin Hamiltonians (see also [Kr]). Most significantly, it follows that the heat operators $\mathbb{S}^n_{k,c}$ defining the KZB connection on the bundles of conformal block quantize the isomonodromy Hamiltonians $\overline{\mathbb{S}}^n_{1,0}$. This makes precise the picture developed in [I] and [LO] of the non-stationary Schrödinger equations defining the KZB connection on conformal blocks as quantizations of the isomonodromy equations or non-autonomous Hitchin systems. In genus zero we thus generalize the result of [Res, Har] that the KZ connection on spaces of conformal blocks on the n-punctured sphere, viewed as a system of multi-time-dependent Schrödinger equations, quantizes the Schlesinger equation, describing isomonodromic deformation of connections with regular singularities on the sphere, which itself is a time-dependent deformation of the Gaudin system (the Hitchin system corresponding to the n-punctured sphere). In genus one we obtain a similar relation between KZB equations, the elliptic form of the Painlevé equations and the elliptic Calogero–Moser system as in [LO].

References

[ADKP]   Arbarello, E., C. deConcini, V. Kac and C. Procesi, Moduli spaces of curves and representation theory, *Comm. Math. Phys.* 117 (1988) 1–36.

[ADW]   Axelrod, S., S. Della Pietra and E. Witten, Geometric quantization of Chern–Simons gauge theory, *J. Diff. Geom.* 33 (1991) 787–902.

[BL1]   Beauville, A. and Y. Laszlo, Conformal blocks and generalized theta functions, *Comm. Math. Phys.* 164 (1993) 385–419.

[BL2]   Beauville, A. and Y. Laszlo, Un lemme de descente, *C. R. Acad. Sci. Paris Sr. I Math.* 320 (1995) 335–340.

[BB]   Beilinson, A. and J. Bernstein. A proof of Jantzen conjectures, in *Advances in Soviet Mathematics*, Vol. 16, Part 1, pp. 1–50, AMS, 1993.

[BD1]   Beilinson, A. and V. Drinfeld, Quantization of Hitchin's integrable system and hecke eigensheaves, Preprint, available at www.math.uchicago.edu/~benzvi.

[BD2]   Beilinson, A. and V. Drinfeld, Chiral algebras, Preprint, available at www.math.uchicago.edu/~benzvi.

[BFM]   Beilinson, A., B. Feigin and B. Mazur, Introduction to algebraic field theory on curves, unpublished manuscript.

[BG]   Beilinson, A. and V. Ginzburg, Infinitesimal structure of moduli spaces of G-bundles, Int. Math. Res. Notices, *Duke Math. J.* 4 (1992) 63–74.

[BK]   Beilinson, A. and D. Kazhdan, Flat projective connection, unpublished manuscript, 1991.

[BS]   Beilinson, A. and V. Schechtman, Determinant bundles and Virasoro algebras, *Comm. Math. Phys.* 118 (1988) 651–701.

[BPZ]   Belavin, A., A. Polyakov and A. Zamolodchikov, Infinite conformal
          symmetries in two-dimensional quantum field theory, *Nucl. Phys.* B241
          (1984) 333–380.
[BZB1]   Ben-Zvi, D. and I. Biswas, Opers and theta functions, Preprint
          math.AG/0204301, to appear in *Adv. Math.*
[BZB2]   Ben-Zvi, D. and I. Biswas, Theta functions and Szegö kernels, Preprint
          math.AG/0211441, to appear in IMRN.
[BE]   Bloch, S. and H. Esnault, Relative algebraic differential characters, Preprint
          math.AG/9912015.
[Bo]   Boalch, P. Symplectic manifolds and isomonodromic deformations, *Adv.
          Math.* 163 (2001) 137–205.
[De]   Deligne, P., Le déterminant de la cohomologie. Current trends in
          arithmetical algebraic geometry (Arcata, Calif., 1985), 93–177, *Contemp.
          Math.* 67, Providence, RI, 1987.
[DSi]   Drinfeld, V. and C. Simpson, B-structures on G-bundles and local
          triviality, *Math. Res. Lett.* 2 (1995) 823–829.
[EFK]   Etingof, P., I. Frenkel and A. Kirillov Jr., *Lectures on Representation
          Theory and Knizhnik–Zamolodchikov Equations.* AMS Publications 1998.
[Fa1]   Faltings, G. Stable G-bundles and projective connections, *J. Alg. Geom.* 2
          (1993) 507–568.
[Fa2]   Faltings, G. A proof for the Verlinde formula. *J. Alg. Geom.* 3 (1994)
          347–374.
[FF]   Feigin, B. and E. Frenkel, Affine Kac–Moody algebras at the critical level
          and Gelfand–Dikii algebras. Infinite analysis, Part A, B (Kyoto, 1991),
          pp. 197–215, *Adv. Ser. Math. Phys.*, 16, World Sci. Publishing, River Edge,
          NJ, 1992.
[Fel]   Felder, G. The KZB equations on Riemann surfaces. *Quantum Symmetries,
          Les Houches Session LXIV.* A. Connes, K. Gawedzki and J. Zinn-Justin
          (eds.). Elsevier (1998) 687–726.
[Fr1]   Frenkel, E. *Affine algebras, Langlands duality and Bethe ansatz. XIth
          International Congress of Mathematical Physics* (Paris, 1994), 606–642,
          Internat. Press, Cambridge, MA, 1995.
[Fr2]   Frenkel, E. Lectures on Wakimoto modules, opers and the critical level,
          Preprint math.QA/0210029.
[FB]   Frenkel, E. and D. Ben-Zvi, *Vertex Algebras and Algebraic Curves.
          Mathematical Surveys and Monographs* 88, American Mathematical
          Society Publications (2001).
[FrS]   Friedan, D. and S. Shenker, The analytic geometry of two-dimensional
          conformal field theory, *Nucl. Phys.* B281 (1987) 509–545.
[GLS]   Guillemin, B., E. Lerman and S. Sternberg, *Symplectic Fibrations and
          Multiplicity Diagrams.* Cambridge University Press, Cambridge, 1996.
[Har]   Harnad, J. Quantum isomonodromic deformations and the
          Knizhnik–Zamolodchikov equations. Symmetries and integrability of
          difference equations (Estérel, PQ, 1994), 155–161, *CRM Proc. Lecture
          Notes*, 9, Amer. Math. Soc., Providence, RI, 1996.
[Hi1]   Hitchin, N. Stable bundles and integrable systems, *Duke Math. J.* 54 (1990)
          91–114.
[Hi2]   Hitchin, N. Projective connections and geometric quantizations, *Comm.
          Math. Phys.* 131 (1990) 347–380.
[I]   Ivanov, D. Knizhnik–Zamolodchikov–Bernard equations as a quantization
          of nonstationary Hitchin system, Preprint hep-th/9610207.

[JMU]   Jimbo, M., T. Miwa and K. Ueno, Monodromy preserving deformation of linear ordinary differential equations with rational coefficients I, Physica 2D (1981) 306–352.

[KZ]   Knizhnik, V. and A. Zamolodchikov, Current algebra and Wess–Zumino model in two dimensions, *Nucl. Phys.* B 247 (1984) 83–103.

[K]   Kontsevich, M. The Virasoro algebra and Teichmüller spaces, *Functional Anal. Appl.* 21 (1987) 156–157, 78–79.

[Kr]   Krichever, I. Isomonodromy equations on algebraic curves, canonical transformations and Whitham equations, Preprint hep-th/0112096.

[KNR]   Kumar, S., M.S. Narasimhan and A. Ramanathan, Infinite Grassmannians and moduli spaces of *G*-bundles, *Math. Ann.* 300 (1994) 41–75.

[La]   Laszlo, Y. Hitchin's and WZW connections are the same, *J. Diff. Geom* 49 (1998) 547–576.

[LMB]   Laumon, G. and L. Moret–Bailly, Champs Algébriques, Preprint Université Paris 11 (Orsay) 1992. Champs algébriques. (French) Ergebnisse der Mathematik und ihrer Grenzgebiete. 3. Folge, 39. Springer–Verlag, Berlin, 2000.

[LO]   Levin, A. and M. Olshanetsky, Hierarchies of isomonodromic deformations and Hitchin systems, in *Moscow Seminar in Mathematical Physics*, 223–262, Amer. Math. Soc. Transl. Ser. 2, 191, Amer. Math. Soc., Providence, RI, 1999.

[Ma]   Malgrange, B. Sur les déformations isomonodromiques, in *Mathématique et Physique* (Séminaire E.N.S. 1979–1982), Birkhüser Boston (1983) 401–426.

[M]   Manin, Yu.I. Sixth Painlevé equation, universal elliptic curve, and mirror of $\mathbb{P}^2$, in *Geometry of Differential Equations*, Amer. Math. Soc. Trans. Ser. 2, 186 (1998) 131–151.

[Res]   Reshetikhin, N. The Knizhnik–Zamolochikov system as a deformation of the isomonodromy problem, *Lett. Math. Phys.* 26 (1992) 167–177.

[Si]   Simpson, C. The Hodge filtration on nonabelian cohomology, *Proc. Symp. Pure Math.* 62 (1997) 217–281.

[Sor]   Sorger, C. Lectures on moduli of principal *G*-bundles over algebraic curves. Notes, ICTP Trieste School on Algebraic Geometry, 1999. (Available at www.ictp.trieste.it/ pub_off/lectures/)

[Te]   Teleman, C. Borel-Weil-Bott theory on the moduli stack of *G*-bundles over a curve, *Invent. Math.* 134 (1998) 1–57.

[TUY]   Tsuchiya, A., K. Ueno, and Y. Yamada, Conformal field theory on universal family of stable curves with gauge symmetries, *Adv. Stud. Pure Math.* 19, 1989.

[W]   Welters, G. Polarized abelian varieties and the heat equations, *Compositio. Math.* 49 (1983) 173–194.

Email: benzvi@math.utexas.edu and frenkel@math.berkeley.edu.

4

Differential isomorphism and equivalence of algebraic varieties

YURI BEREST and GEORGE WILSON

Cornell University and Imperial College London.

To Graeme Segal for his sixtieth birthday

1 Introduction

Let X be an irreducible complex affine algebraic variety, and let $\mathcal{D}(X)$ be the ring of (global, linear, algebraic) differential operators on X (we shall review the definition in Section 2). This ring has a natural filtration (by order of operators) in which the elements of order zero are just the ring $\mathcal{O}(X)$ of regular functions on X. Thus, if we are given $\mathcal{D}(X)$ together with its filtration, we can at once recover the variety X. But now suppose we are given $\mathcal{D}(X)$ just as an abstract noncommutative $\mathbb{C}$-algebra, without filtration; then it is not clear whether or not we can recover X. We shall call two varieties X and Y *differentially isomorphic* if $\mathcal{D}(X)$ and $\mathcal{D}(Y)$ are isomorphic.

The first examples of nonisomorphic varieties with isomorphic rings of differential operators were found by Levasseur, Smith and Stafford (see [LSS] and Section 9 below). These varieties arise in the representation theory of simple Lie algebras; they are still the only examples we know in dimension > 1 (if we exclude products of examples in lower dimensions). For curves, on the other hand, there is now a complete classification up to differential isomorphism; the main purpose of this article is to review that case. The result is very strange. It turns out that for curves, $\mathcal{D}(X)$ determines X (up to isomorphism) except in the very special case when X is homeomorphic to the affine line $\mathbb{A}^1$ (we call such a curve a *framed curve*). There are uncountably many nonisomorphic framed curves (we can insert arbitrarily bad cusps at any finite number of points of $\mathbb{A}^1$). However, the differential isomorphism classes of framed curves are classified by a single non-negative integer n. This invariant n seems to us the most interesting character in our story: it appears in many guises, some of which we describe in Section 8.

The authors were partially supported by the National Science Foundation (NSF) grant DMS 00-71792 and an A. P. Sloan Research Fellowship. The second author is grateful to the Mathematics Department of Cornell University for its hospitality during the preparation of this article.

We can also ask to what extent X is determined by the Morita equivalence class of $\mathcal{D}(X)$: we call two varieties X and Y *differentially equivalent* if $\mathcal{D}(X)$ and $\mathcal{D}(Y)$ are Morita equivalent (as $\mathbb{C}$-algebras). A complete classification of curves up to differential equivalence is not available; however, it is known that the differential equivalence class of a *smooth* affine curve X consists of all the curves homeomorphic to X. In particular, all framed curves are differentially equivalent to each other: that is one reason why the invariant n which distinguishes them has to be somewhat unusual. In dimension > 1, there are already some interesting results about differential equivalence; we include a (very brief) survey in Section 9, where we also mention some generalizations of our questions to non-affine varieties.

At the risk of alienating some readers, we point out that most of the interest in this paper is in *singular* varieties. For smooth varieties it is a possible conjecture that differential equivalence implies isomorphism: indeed, that is true for curves. However, in dimension > 1 the conjecture would be based on no more than lack of counterexamples.

Our aim in this article has been to provide a readable survey, suitable as an introduction to the subject for beginners; most of the material is already available in the literature. For the convenience of readers who are experts in this area, we point out a few exceptions to that rule: Theorem 8.7 is new, and perhaps Theorem 3.3; also, the formulae (7.1) and (8.3) have not previously appeared explicitly.

2 Generalities on differential operators

We first recall the definition of (linear) differential operators, in a form appropriate for applications in algebraic geometry (see [G]). If A is a (unital associative) commutative algebra over (say) $\mathbb{C}$, the filtered ring

$$\mathcal{D}(A) = \bigcup_{r \geq 0} \mathcal{D}^r(A) \subset \mathrm{End}_{\mathbb{C}}(A)$$

of differential operators on A may be defined inductively as follows. First, we set $\mathcal{D}^0(A) := A$ (here the elements of A are identified with the corresponding multiplication operators); then, by definition, a linear map $\theta : A \to A$ belongs to $\mathcal{D}^r(A)$ if

$$\theta a - a\theta \in \mathcal{D}^{r-1}(A) \text{ for all } a \in A.$$

The elements of $\mathcal{D}^r(A)$ are called *differential operators of order* $\leq r$ on A. The commutator of two operators of orders r and s is an operator of order at most $r + s - 1$; it follows that the *associated graded algebra*

$$\operatorname{gr} \mathcal{D}(A) := \bigoplus_{r \geq 0} \mathcal{D}^r(A)/\mathcal{D}^{r-1}(A)$$

is commutative (we set $\mathcal{D}^{-1}(A) := 0$).

Slightly more generally, we can define the ring $\mathcal{D}_A(M)$ of differential operators on any A-module M: the operators of order zero are the A-linear maps $M \to M$, and operators of higher order are defined inductively just as in the special case above (where $M = A$).

Example 2.1. If $A = \mathbb{C}[z_1, \ldots, z_m]$, then $\mathcal{D}(A) = \mathbb{C}[z_i, \partial/\partial z_i]$ is the mth *Weyl algebra* (linear differential operators with polynomial coefficients).

Example 2.2. Similarly, if $A = \mathbb{C}(z_1, \ldots, z_m)$, then $\mathcal{D}(A) = \mathbb{C}(z_i)[\partial/\partial z_i]$ is the algebra of linear differential operators (in m variables) with rational coefficients.

The definition of $\mathcal{D}(A)$ makes sense for an arbitrary $\mathbb{C}$-algebra A; however, in this paper we shall use it only in the cases when A is either the coordinate ring $\mathcal{O}(X)$ of an irreducible affine variety X, or the field $\mathbb{K} \equiv \mathbb{C}(X)$ of rational functions on such a variety. Let us consider first the latter case. If we choose a transcendence basis $\{z_1, \ldots, z_m\}$ for $\mathbb{K}$ over $\mathbb{C}$ (where $m = \dim X$), then there are (unique) $\mathbb{C}$-derivations $\partial_1, \ldots, \partial_m$ of $\mathbb{K}$ such that $\partial_i(z_j) = \delta_{ij}$, and each element of $\mathcal{D}^r(\mathbb{K})$ has a unique expression in the form

$$\theta = \sum_{|\alpha| \leq r} f_\alpha \partial^\alpha$$

(with $f_\alpha \in \mathbb{K}$), as in Example 2.2 above, in which X is the affine space $\mathbb{A}^m$. In particular, $\mathcal{D}(\mathbb{K})$ is generated by $\mathcal{D}^1(\mathbb{K})$, as one would expect, and an element of $\mathcal{D}^1(\mathbb{K})$ is just the sum of a derivation and a multiplication operator. Indeed, it is easy to show that this last fact is true for an arbitrary algebra A.

The case where the ring A is $\mathcal{O}(X)$ is more subtle; in this case $\mathcal{D}(A)$ is denoted by $\mathcal{D}(X)$ and is called the *ring of differential operators on* X. Thus the mth Weyl algebra (see Example 2.1) is the ring of differential operators on $\mathbb{A}^m$. In general one does not have global coordinates on X, as in this example; nevertheless, if X is *smooth*, the structure of $\mathcal{D}(X)$ is still well understood.

Proposition 2.3. *Let X be a smooth (irreducible) affine variety. Then*

(i) $\mathcal{D}(X)$ *is a simple (left and right) Noetherian ring without zero divisors;*

(ii) $\mathcal{D}(X)$ *is generated as a $\mathbb{C}$-algebra by finitely many elements of $\mathcal{D}^1(X)$;*

(iii) *the associated graded algebra* $\operatorname{gr} \mathcal{D}(X)$ *is canonically isomorphic to* $\mathcal{O}(T^*X)$*;*

(iv) $\mathcal{D}(X)$ *has global (that is, homological) dimension equal to* $\dim X$*.*

If X is singular, the situation is less clear. We can still consider the ring $\Delta(X)$ of ($\mathbb{C}$-linear) operators on $\mathcal{O}(X)$ generated by the multiplication operators and the derivations of $\mathcal{O}(X)$; however, in general, $\Delta(X)$ is smaller than $\mathcal{D}(X)$. Our main reason to prefer $\mathcal{D}(X)$ to $\Delta(X)$ is the following. Each differential operator on $\mathcal{O}(X)$ has a unique extension to a differential operator (of the same order) on $\mathbb{K}$, so we may view $\mathcal{D}(X)$ as a subalgebra of $\mathcal{D}(\mathbb{K})$. Furthermore, a differential operator on $\mathbb{K}$ which preserves $\mathcal{O}(X)$ is a differential operator on $\mathcal{O}(X)$ (this last statement would in general not be true for $\Delta(X)$). Thus we have:

Proposition 2.4. *Let* X *be an affine variety with function field* $\mathbb{K}$*. Then*

$$\mathcal{D}(X) = \{D \in \mathcal{D}(\mathbb{K}) : D.\mathcal{O}(X) \subseteq \mathcal{O}(X)\}.$$

For the purposes of the present paper we could well take this as the definition of $\mathcal{D}(X)$. It follows from Proposition 2.4 that $\mathcal{D}(X)$ is without zero divisors also for (irreducible) singular varieties X.

Example 2.5. Let X be the rational curve with coordinate ring $\mathcal{O}(X) := \mathbb{C}[z^2, z^3]$ (thus X has just one simple cusp at the origin). Then $\Delta(X)$ is generated by $\mathcal{O}(X)$ and the derivations $\{z^r \partial : r \geq 1\}$ (we set $\partial := \partial/\partial z$). But $\mathcal{D}^2(X)$ contains the operators $\partial^2 - 2z^{-1}\partial$ and $z\partial^2 - \partial$, neither of which belongs to $\Delta(X)$.

To obtain a concrete realization of $\mathcal{D}_A(M)$ similar to that in Proposition 2.4, we need to suppose that M is embedded as an A-submodule of some $\mathbb{K}$-vector space; to fix ideas, we formulate the result in the case that will concern us, where M has rank 1.

Proposition 2.6. *Suppose* $M \subset \mathbb{K}$ *is a (nonzero)* A*-submodule of* $\mathbb{K}$*. Then*

$$\mathcal{D}_A(M) = \{D \in \mathcal{D}(\mathbb{K}) : D.M \subseteq M\}.$$

Notes.

1. To part (iii) of Proposition 2.3 we should add that the commutator on $\mathcal{D}(X)$ induces on $\operatorname{gr} \mathcal{D}(X)$ the canonical Poisson bracket coming from the symplectic structure of T^*X; that is, $\mathcal{D}(X)$ is a *deformation quantization* of $\mathcal{O}(T^*X)$.

2. For singular varieties, the rings $\Delta(X)$ and $\mathcal{D}(X)$ have quite different properties: for example, $\Delta(X)$ is simple if *and only if* X is smooth (cf. Theorem 3.2 below). It follows that if X is smooth, then $\Delta(X)$ is never isomorphic, or even Morita equivalent, to $\Delta(Y)$ for any singular variety Y. Thus the present paper would probably be very short and dull if we were to work with $\Delta(X)$ rather than with $\mathcal{D}(X)$.

3. Nakai (cf. [Na]) has conjectured that $\mathcal{D}(X) = \Delta(X)$ if *and only if* X is smooth. The conjecture has been proved for curves (see [MV]) and, more generally, for varieties with smooth normalization (see [T]). In [Be] and [R] it is shown that Nakai's conjecture would imply the well known *Zariski–Lipman conjecture*: if the module of derivations of $\mathcal{O}(X)$ is projective, then X is smooth.

4. If X is singular, then in general $\mathcal{D}(X)$ may have quite bad properties. In [BGG] it is shown that if X is the cone in $\mathbb{A}^3$ with equation $x^3 + y^3 + z^3 = 0$, then $\mathcal{D}(X)$ is not a finitely generated algebra, nor left or right Noetherian. In this example X is a normal variety, and has only one singular point (at the origin). In [SS], Section 7, it is shown that if X is a variety of dimension ≥ 2 with smooth normalization and isolated singularities, then $\mathcal{D}(X)$ is right Noetherian but not left Noetherian.

5. In the situation of Proposition 2.6, it may happen that the ring $B := \mathcal{D}_A^0(M)$ is larger than A. In that case the ring $\mathcal{D}_A(M) \subset \mathcal{D}(\mathbb{K})$ would not change if we replaced A by B; thus there is no loss of generality if we restrict attention to modules M for which $B = A$. We call such A-modules *maximal*.

6. Of course, all the statements in this section (and, indeed, in most of the other sections) would remain true if we replaced $\mathbb{C}$ by any algebraically closed field of characteristic zero. If we work over a field of positive characteristic, the above definition of differential operators is still generally accepted to be the correct one, but some of the properties of the rings $\mathcal{D}(X)$ are very different: for example, $\mathcal{D}(X)$ is not Noetherian, or finitely generated, or without zero divisors (see, for example, [Sm]). In particular, in positive characteristic $\mathcal{D}(\mathbb{A}^1)$ is not at all like the Weyl algebra.

7. A convenient reference for this section is the last chapter of the book [MR], where one can find proofs of all the facts we have stated (except for Proposition 2.6, whose proof is similar to that of Proposition 2.4).

3 Differential equivalence of curves

From now on until Section 9, X will be an affine *curve*, probably singular. In this case the problems mentioned in Section 2, Note 4 do not occur.

Proposition 3.1. *Let X be an (irreducible) affine curve. Then $\mathcal{D}(X)$ is a (left and right) Noetherian ring, and is finitely generated as a $\mathbb{C}$-algebra.*

However, the associated graded ring $\operatorname{gr} \mathcal{D}(X)$ is in general not a Noetherian ring (and hence not a finitely generated algebra either). The following theorem of Smith and Stafford shows that for our present purposes there is a very stark division of curve singularities into 'good' and 'bad'.

Theorem 3.2. *Let X be an affine curve, and let $\tilde{X}$ be its normalization. Then the following are equivalent:*

(i) *The normalization map $\pi : \tilde{X} \to X$ is bijective.*
(ii) *The algebras $\mathcal{D}(\tilde{X})$ and $\mathcal{D}(X)$ are Morita equivalent.*
(iii) *The ring $\mathcal{D}(X)$ has global dimension 1 (that is, the same as $\mathcal{D}(\tilde{X})$).*
(iv) *The ring $\mathcal{D}(X)$ is simple.*
(v) *The algebra $\operatorname{gr} \mathcal{D}(X)$ is finitely generated.*
(vi) *The ring $\operatorname{gr} \mathcal{D}(X)$ is Noetherian.*

Perhaps the most striking thing about Theorem 3.2 is that the 'good' singularities (from our present point of view) are the *cusps* (as opposed to double points, or higher-order multiple points). If X has even one double point, the ring $\mathcal{D}(X)$ is somewhat wild; whereas if X has only cusp singularities, no matter how 'bad', then $\mathcal{D}(X)$ is barely distinguishable from the ring of differential operators on the smooth curve $\tilde{X}$.

Theorem 3.2 does not address the question of when two *smooth* affine curves are differentially equivalent. However, the answer to that is very simple.

Theorem 3.3. *Let X and Y be smooth affine curves. Then $\mathcal{D}(X)$ and $\mathcal{D}(Y)$ are Morita equivalent (if and) only if X and Y are isomorphic.*

Theorems 3.2 and 3.3 together determine completely the differential equivalence class of a smooth curve X: it consists of all curves obtained from X by pinching a finite number of points to (arbitrarily bad) cusps.

Notes.

1. Apparently, not much is known about the differential equivalence class of a curve with multiple points. From Theorem 3.2 one might guess that if $\pi : Y \to X$ is regular surjective of degree one, then X and Y are differentially equivalent if and only if π is bijective. However, in [SS] (5.8) there is a counterexample to the 'if' part of this statement. The paper [CH2] contains some curious results about the Morita equivalence class of $\mathcal{D}(A)$ when A is the *local* ring at a multiple point of a curve.

2. Another natural question that is not addressed by Theorem 3.2 is: what is the global dimension of $\mathcal{D}(X)$ if X has multiple points? In [SS] it is proved that if the singularites are all *ordinary* multiple points, then the answer is 2; but for more complicated singularities it seems nothing is known.

3. We have not found Theorem 3.3 stated explicitly in the literature, but it is an easy consequence of the results of [CH1] and [M-L]: we will sketch a proof in Section 6, Note 5.

4. Proposition 3.1 is proved in [SS] and (also in the case of a reducible (but reduced) curve) in [M].

5. We refer to [SS] for the proofs of the various assertions in Theorem 3.2. Here we mention only that a key role is played by the space

$$P \equiv \mathcal{D}(\tilde{X}, X) := \{D \in \mathcal{D}(\mathbb{K}) : D.\mathcal{O}(\tilde{X}) \subseteq \mathcal{O}(X)\}. \qquad (3.1)$$

Clearly, P is a right ideal in $\mathcal{D}(\tilde{X})$ and a left ideal in $\mathcal{D}(X)$; the Morita equivalence in Theorem 3.2 is defined by tensoring with the bimodule P. Another notable property of P is the following: each of the statements in Theorem 3.2 is equivalent to the condition

$$P.\mathcal{O}(\tilde{X}) = \mathcal{O}(X). \qquad (3.2)$$

The formulae (3.1) and (3.2) provide the starting point for the theory of Cannings and Holland which we explain in Section 6; there P is replaced by an arbitrary right ideal in $\mathcal{D}(\tilde{X})$.

4 Differential isomorphism of curves

We now turn to our main question, concerning differential isomorphism. We begin by sketching the history of this subject.

To our knowledge, the papers [St], [Sm] are the first that explicitly pose the question: does $\mathcal{D}(X) \simeq \mathcal{D}(Y)$ imply $X \simeq Y$? In [St], Stafford proved that this is true if X is the affine line $\mathbb{A}^1$ (in which case $\mathcal{D}(X)$ is the Weyl algebra), and also if X is the plane curve with equation $y^2 = x^3$, that is, the rational curve obtained from $\mathbb{A}^1$ by introducing a *simple cusp* at the origin. The first general result in the subject is due to L. Makar-Limanov (see [M-L]). His idea was as follows. Recall that if we take the commutator $(\mathrm{ad} f)L := fL - Lf$ of a function $f \in \mathcal{O}(X)$ with an operator $L \in \mathcal{D}(X)$ of order n, then we get an operator of order at most $n - 1$ (indeed, this is essentially the definition of $\mathcal{D}(X)$, see Section 2 above). It follows that $(\mathrm{ad} f)^{n+1}L = 0$, so that f is a (locally) *ad-nilpotent* element of $\mathcal{D}(X)$. If it happens (as seems likely) that the set $\mathcal{N}(X)$ of all ad-nilpotent elements of $\mathcal{D}(X)$ coincides with $\mathcal{O}(X)$, then we have a purely ring-theoretical description of $\mathcal{O}(X) \subset \mathcal{D}(X)$, namely, it is the unique maximal abelian ad-nilpotent subalgebra (for short: *mad subalgebra*) of $\mathcal{D}(X)$. So in this way $\mathcal{D}(X)$ determines X. Makar-Limanov's main remark was the following.

Lemma 4.1. *Let* $\mathbb{K}$ *be the function field of a curve, and let* $D \in \mathcal{D}(\mathbb{K})$ *have positive order. Let* $N \subset \mathbb{K}$ *be the set of elements of* $\mathbb{K}$ *on which* D *acts ad-nilpotently. Then there is an element* q *in some finite extension field of* $\mathbb{K}$ *such that* $N \subseteq \mathbb{C}[q]$.

If now X is a curve such that $\mathcal{N}(X) \neq \mathcal{O}(X)$, that is, such that $\mathcal{N}(X)$ contains an operator of positive order, then it follows from Lemma 4.1 that $\mathcal{O}(X) \subseteq \mathbb{C}[q]$ for suitable q. Equivalently:

Theorem 4.2. *If* $\mathcal{N}(X) \neq \mathcal{O}(X)$, *then the normalization* $\tilde{X}$ *of* X *is isomorphic to* $\mathbb{A}^1$.

In his thesis (see [P1]), P. Perkins refined this result.

Theorem 4.3. *Let* X *be an affine curve. Then* $\mathcal{N}(X) \neq \mathcal{O}(X)$ *if and only if:*

(i) *$\tilde{X}$ is isomorphic to* $\mathbb{A}^1$*; and*
(ii) *the normalization map* $\pi : \tilde{X} \to X$ *is bijective.*

In other words, the differential isomorphism class of a curve X consists just of (the class of) X itself, except, possibly, when X has the properties (i) and (ii) above.

For short, we shall call a curve with these two properties a *framed curve*. More precisely, by a framed curve we shall mean a curve X together with a regular bijective map $\pi : \mathbb{A}^1 \to X$: the choice of 'framing' (that is, of the isomorphism $\tilde{X} \simeq \mathbb{A}^1$) is fairly harmless, because any two choices differ only by an automorphism $z \mapsto az + b$ of $\mathbb{A}^1$. The two curves considered by Stafford are certainly framed curves: Stafford's results do not contradict those of Perkins, because, although the rings $\mathcal{D}(X)$ in these examples have many ad-nilpotent elements not in $\mathcal{O}(X)$, their mad subalgebras are all isomorphic, so we can still extract $\mathcal{O}(X)$ (up to isomorphism) from $\mathcal{D}(X)$. For a while it might have seemed likely that the situation is similar for any framed curve; but counterexamples were found by Letzter [L] and by Perkins [P2]. The following example of Letzter is perhaps the simplest and most striking. Let X and Y be the curves with coordinate rings

$$\mathcal{O}(X) = \mathbb{C} + z^4\mathbb{C}[z]; \quad \mathcal{O}(Y) = \mathbb{C}[z^2, z^5].$$

Each of X and Y is obtained from $\mathbb{A}^1$ by introducing a single cusp at the origin; X and Y are clearly not isomorphic. Indeed, we have $\mathcal{O}(X) \subset \mathcal{O}(Y)$, so the singularity of X is strictly 'worse' than that of Y. Nevertheless, Letzter proved that X and Y are differentially isomorphic. This example, and others in [P2], [L], shows that the problem of classifying *framed* curves up to differential isomorphism is nontrivial.

This problem was solved completely in the thesis [K] of K. Kouakou. The simplest way to state his result is as follows. For each $n \geq 0$, let X_n denote the curve with coordinate ring

$$\mathcal{O}(X_n) := \mathbb{C} + z^{n+1}\mathbb{C}[z]. \tag{4.1}$$

(Thus the curves considered by Stafford are $X_0 \equiv \mathbb{A}^1$ and X_1, while the curve X in Letzter's example above is X_3.)

Theorem 4.4. *[Kouakou] Every framed curve X is differentially isomorphic to one of the above curves X_n.*

On the other hand, Letzter and Makar-Limanov (see [LM]) have proved the following:

Theorem 4.5. *No two of the curves X_n are differentially isomorphic to each other.*

It follows that each framed curve X is differentially isomorphic to exactly one of the special curves X_n: we shall call this number n the *differential genus* of X, and denote it by $d(X)$.

Notes.

1. Of course, this is very unsatisfactory as a *definition* of the differential genus, because it does not make sense until after we have proved the two nontrivial Theorems 4.4 and 4.5. In Section 8 we discuss several more illuminating ways to define $d(X)$. We use the term 'genus' because $d(X)$ is in some ways reminiscent of the arithmetic genus of a curve: it turns out that it is a sum of local contributions from each singular point, so it simply counts the cusps of our framed curve with appropriate weights. In Section 8 we shall explain how to calculate these weights: here we just mention that the weight of a *simple* (that is, of type $y^2 = x^3$) cusp is equal to 1, so if all the cusps of X are simple, then $d(X)$ is just the number of cusps.

2. Recall from Theorem 3.2 that the algebras $\mathcal{D}(X)$ (for X a framed curve) are all Morita equivalent to each other: thus the invariant $d(X)$ that distinguishes them must be fairly subtle.

3. Makar-Limanov's Lemma 4.1 (in a slightly disguised form) plays a basic role also in the theory of bispectral differential equations (compare the proof in [M-L] with similar arguments in [DG] or [W1]).

4. There is no convenient reference where the reader can find a complete proof of Kouakou's theorem: Kouakou's thesis has never been published, and the (different) proof in [BW1] is mostly omitted. The proof that we shall explain in the next three sections amplifies the sketch given in [W3]: it is not the most elementary possible, but it seems to us the most natural available at present.

5 The adelic Grassmannian

It is actually easier to prove a more general theorem than Theorem 4.4, as follows. Let X be a framed curve, and let $\mathcal{L}$ be any rank 1 torsion-free coherent sheaf over X; it corresponds to a rank 1 torsion-free $\mathcal{O}(X)$-module M. Then we have the ring $\mathcal{D}_{\mathcal{L}}(X) \equiv \mathcal{D}_{\mathcal{O}(X)}(M)$ of differential operators on (global) sections of $\mathcal{L}$. If $\mathcal{L} = \mathcal{O}_X$ is the sheaf of regular functions on X, then $\mathcal{D}_{\mathcal{L}}(X)$ is just the ring $\mathcal{D}(X)$ discussed previously. Generalizing Theorem 4.4, we have the following:

Theorem 5.1. *Every algebra* $\mathcal{D}_{\mathcal{L}}(X)$ *is isomorphic to one of the algebras* $\mathcal{D}(X_n)$.

Of course, Theorem 4.5 shows that the integer n in this assertion is unique: we call it the *differential genus of the pair* $(X, \mathcal{L})$ and denote it by $d_{\mathcal{L}}(X)$.

The reason Theorem 5.1 is easier to prove than Theorem 4.4 is that the space of pairs $(X, \mathcal{L})$ has a large group of symmetries that preserves the isomorphism class of the algebra $\mathcal{D}_{\mathcal{L}}(X)$ (but does not preserve the subset of pairs of the form $(X, \mathcal{O}_X)$). In fact the isomorphism classes of these pairs form the *adelic Grassmannian* $\mathrm{Gr}^{\mathrm{ad}}$, a well-studied space that occurs in at least two other contexts, namely, in the theory of the Kadomtsev–Petviashvili hierarchy (cf. [Kr]) and in the problem of classifying bispectral differential operators (see [DG], [W1]). The adelic Grassmannian is a subspace[1] of the much larger Grassmannian Gr studied in [SW]. We recall the definition of $\mathrm{Gr}^{\mathrm{ad}}$. For each $\lambda \in \mathbb{C}$, we choose a λ-*primary* subspace of $\mathbb{C}[z]$, that is, a linear subspace V_λ such that

$$(z - \lambda)^N \mathbb{C}[z] \subseteq V_\lambda \text{ for some } N.$$

We suppose that $V_\lambda = \mathbb{C}[z]$ for all but finitely many λ. Let $V = \bigcap_\lambda V_\lambda$

---

[1] It is perhaps the most interesting Grassmannian not mentioned explicitly in [SW].

(such a space V is called *primary decomposable*) and, finally, let

$$W = \prod_\lambda (z - \lambda)^{-k_\lambda} V \subset \mathbb{C}(z)$$

where k_λ is the codimension of V_λ in $\mathbb{C}[z]$. By definition, $\mathrm{Gr}^{\mathrm{ad}}$ consists of all $W \subset \mathbb{C}(z)$ obtained in this way. The correspondence between points of $\mathrm{Gr}^{\mathrm{ad}}$ and pairs $(X, \mathcal{L})$ is a special case of the construction explained in [SW]. Given W, we obtain $(X, \mathcal{L})$ by setting

$$\mathcal{O}(X) := \{f \in \mathbb{C}[z] : fW \subseteq W\}$$

and W is then the rank 1 $\mathcal{O}(X)$-module corresponding to $\mathcal{L}$. Conversely, given $(X, \mathcal{L})$, we let W be the space of global sections of $\mathcal{L}$, regarded as a subspace of $\mathbb{C}(z)$ by means of a certain distinguished rational trivialization of $\mathcal{L}$ (implicitly described above).

Proposition 5.2. *This construction defines a bijection between* $\mathrm{Gr}^{\mathrm{ad}}$ *and the set of isomorphism classes of pairs* $(X, \mathcal{L})$, *where* X *is a framed curve and* $\mathcal{L}$ *is a maximal rank 1 torsion-free sheaf over* X.

'Maximal' here means that the $\mathcal{O}(X)$-module corresponding to $\mathcal{L}$ is maximal in the sense of Note 5, Section 2.

Example 5.3. If X_n is the curve defined by (4.1), then $\mathcal{O}(X_n)$ is 0-primary, and the corresponding point of $\mathrm{Gr}^{\mathrm{ad}}$ is $W_n = z^{-n} \mathcal{O}(X_n)$. More generally, let $\Lambda \subset \mathbb{N}$ be any (additive) semigroup obtained from $\mathbb{N}$ by deleting a finite number of positive integers, and let $\mathcal{O}(X)$ be the subring of $\mathbb{C}[z]$ spanned by $\{z^i : i \in \Lambda\}$. Such a curve X is called a *monomial curve*; the corresponding point of $\mathrm{Gr}^{\mathrm{ad}}$ is $z^{-m} \mathcal{O}(X)$, where m is the number of elements of $\mathbb{N} \setminus \Lambda$.

Example 5.4. If X has simple cusps at the (distinct) points $\lambda_1, \ldots, \lambda_r \in \mathbb{C}$, then $\mathcal{O}(X)$ consists of all polynomials whose first derivatives vanish at these points, and the corresponding point of $\mathrm{Gr}^{\mathrm{ad}}$ is

$$W = \prod_{i=1}^r (z - \lambda_i)^{-1} \mathcal{O}(X).$$

More generally, if in addition we choose $\alpha_1, \ldots, \alpha_r \in \mathbb{C}$, then

$$V = \{f \in \mathbb{C}[z] : f'(\lambda_i) = \alpha_i f(\lambda_i) \text{ for } 1 \leq i \leq r\}$$

is primary decomposable, and the corresponding point of $\mathrm{Gr}^{\mathrm{ad}}$ is

$$W = \prod_{i=1}^r (z - \lambda_i)^{-1} V.$$

In the pairs $(X, \mathcal{L})$ here, the curve X is the same as before, and as we vary the parameters α_i we get the various line bundles $\mathcal{L}$ over X.

The rings $\mathcal{D}_{\mathcal{L}}(X)$ that interest us are easy to describe in terms of $\mathrm{Gr}^{\mathrm{ad}}$. If $W \in \mathrm{Gr}^{\mathrm{ad}}$, we define the *ring of differential operators on* W by

$$\mathcal{D}(W) := \{D \in \mathbb{C}(z)[\partial] : D.W \subseteq W\}$$

(as in Section 2, the dot denotes the natural action of differential operators on functions). Proposition 2.6 shows:

Proposition 5.5. *Let* $W \in \mathrm{Gr}^{\mathrm{ad}}$ *correspond to the pair* $(X, \mathcal{L})$ *as in Proposition 5.2. Then there is a natural identification*

$$\mathcal{D}(W) \simeq \mathcal{D}_{\mathcal{L}}(X).$$

It remains to discuss the symmetries of $\mathrm{Gr}^{\mathrm{ad}}$. Some of them are fairly obvious. First, we have the commutative group Γ of the *KP flows*: it corresponds to the action $(X, \mathcal{L}) \mapsto (X, L \otimes \mathcal{L})$ of the Jacobian (that is, the group of line bundles L over X) on the space of pairs $(X, \mathcal{L})$. If $W^{\mathrm{an}} \supset W$ is the space of analytic sections of $\mathcal{L}$, then Γ is the group of maps of the form $W^{\mathrm{an}} \mapsto e^{p(z)} W^{\mathrm{an}}$, where p is a polynomial. Another fairly evident symmetry is the *adjoint involution* c defined by

$$c(W) = \{f \in \mathbb{C}(z) : \mathrm{res}_{\infty} f(z)g(z)dz = 0 \text{ for all } g \in W\}.$$

Like the KP flows, c is just the restriction to $\mathrm{Gr}^{\mathrm{ad}}$ of a symmetry of the Grassmannian Gr of [SW]. A more elusive symmetry of $\mathrm{Gr}^{\mathrm{ad}}$ is the *bispectral involution* b introduced in [W1]; it does not make sense on Gr, and does not have a simple description in terms of the pairs $(X, \mathcal{L})$. It can be characterized by the formula

$$\psi_{bW}(x, z) = \psi_W(z, x)$$

where ψ is the *stationary Baker function* of W (see, for example, [SW]). Let $\varphi = bc$, and let G be the group of symmetries of $\mathrm{Gr}^{\mathrm{ad}}$ generated by Γ and φ. In view of Proposition 5.5, Theorems 4.4, 4.5 and 5.1 are all consequences of:

Theorem 5.6

(i) *Let* $V, W \in \mathrm{Gr}^{\mathrm{ad}}$. *Then* $\mathcal{D}(V)$ *and* $\mathcal{D}(W)$ *are isomorphic if and only if* V *and* W *belong to the same* G-*orbit in* $\mathrm{Gr}^{\mathrm{ad}}$.

(ii) *Each orbit contains exactly one of the points* W_n *from Example 5.3.*

Although it is possible to formulate a proof of Theorem 5.6 within our present context, the proof will appear more natural if we use two alternative descriptions of Gr^{ad}: we explain these in the next sections. First, in Section 6 we shall see that Gr^{ad} can be identified with the space of ideals in the Weyl algebra $\mathcal{D}(\mathbb{A}^1)$: the ring $\mathcal{D}(W)$ then becomes the endomorphism ring of the corresponding ideal, and G becomes the automorphism group of the Weyl algebra. Part (i) of Theorem 5.6 then turns into a theorem of Stafford (see [St]). In Section 7 we explain how Gr^{ad} decomposes into the union of certain finite-dimensional varieties $\mathcal{C}_n$ that have a simple explicit description in terms of matrices; part (ii) of Theorem 5.6 then follows from the more precise assertion that these spaces $\mathcal{C}_n$ are exactly the G-orbits. Since the action of G also has a simple description in terms of matrices, part (ii) of the Theorem becomes a problem in linear algebra.

Notes.

1. The fact that the action of $\Gamma \subset G$ preserves the isomorphism class of $\mathcal{D}(W)$ is almost trivial. Indeed, if $g \in \Gamma$ is given (as above) by multiplication by $e^{p(z)}$, then $\mathcal{D}(gW) = e^{p(z)}\mathcal{D}(W)e^{-p(z)}$. It follows that $\mathcal{D}(gW)$ is even isomorphic to $\mathcal{D}(W)$ as a *filtered* algebra. Thus the (filtered) isomorphism class of $\mathcal{D}_{\mathcal{L}}(X)$ depends only on the orbit of the Jacobian of X in the space of rank 1 torsion-free sheaves; for example, if $\mathcal{L}$ is locally free, then $\mathcal{D}_{\mathcal{L}}(X)$ is isomorphic to $\mathcal{D}(X)$. A direct proof that φ preserves the isomorphism class of $\mathcal{D}(W)$ is also not too difficult: it follows from the facts that $\mathcal{D}(bW)$ and $\mathcal{D}(cW)$ are *anti*-isomorphic to $\mathcal{D}(W)$ (cf. [BW2], Sections 7 and 8). We regard the main assertions in Theorem 5.6 to be part (ii) and the 'only if' statement in part (i).

2. The spaces W_n are fixed by b, so b induces an involutory anti-automorphism on each of the rings $\mathcal{D}(X_n)$. Thus Theorem 5.1 shows that the distinction between isomorphism and anti-isomorphism in the preceding note was immaterial.

3. If $\mathcal{L}$ is not locally free, then in general $\mathcal{D}_{\mathcal{L}}(X)$ is not isomorphic to $\mathcal{D}(X)$ (see Example 8.4 below).

4. Details of the proof of Proposition 5.2 can be found in [W1]; see also [CH4], 1.4 and [E], p. 945.

6 The Cannings–Holland correspondence

In this section we explain a different realization of Gr^{ad} (due to Cannings and Holland) as the space of ideals in the Weyl algebra. Let $A := \mathbb{C}[z, \partial]$ from now on denote the (first) Weyl algebra, and let $\mathfrak{I}$ be the set of nonzero right ideals of A. Let $\mathfrak{S}$ be the set of all linear subspaces of $\mathbb{C}[z]$. If $V, W \in \mathfrak{S}$ (or, later, also if V and W are subspaces of $\mathbb{C}(z)$) we set

$$\mathcal{D}(V, W) := \{D \in \mathbb{C}(z)[\partial] : D.V \subseteq W\}. \tag{6.1}$$

We define maps $\alpha : \mathfrak{S} \to \mathfrak{I}$ and $\gamma : \mathfrak{I} \to \mathfrak{S}$ as follows. If $V \in \mathfrak{S}$, we set

$$\alpha(V) := \mathcal{D}(\mathbb{C}[z], V) \tag{6.2}$$

and if $I \in \mathfrak{I}$, we set

$$\gamma(I) := \{D.\mathbb{C}[z] : D \in I\}. \tag{6.3}$$

Theorem 6.1.
  (i) *We have $\alpha\gamma(I) = I$ if and only if $I \cap \mathbb{C}[z] \neq \{0\}$.*
 (ii) *We have $\gamma\alpha(V) = V$ if and only if V is primary decomposable.*
(iii) *The maps α and γ define inverse bijections between the set of primary decomposable subspaces of $\mathbb{C}[z]$ and the set of right ideals of A that intersect $\mathbb{C}[z]$ nontrivially.*
(iv) *If V and W are primary decomposable and $I := \alpha(V)$, $J := \alpha(W)$ are the corresponding (fractional) ideals, then*

$$\mathcal{D}(V, W) = \{D \in \mathbb{C}(z)[\partial] : DI \subseteq J\} \simeq \mathrm{Hom}_A(I, J).$$

Example 6.2. Let I_n be the right ideal

$$I_n := z^{n+1}A + \prod_{r=1}^{n}(z\partial - r)\, A.$$

The second generator kills $z, z^2, \ldots, z^n$, so we find that $\gamma(I_n) = \mathcal{O}(X_n)$.

The assertions (iii) and (iv) in Theorem 6.1 follow at once from (i) and (ii). Now, not every right ideal of A intersects $\mathbb{C}[z]$ nontrivially; but every ideal is isomorphic (as right A-module) to one with this property (see [St], Lemma 4.2). Furthermore, two such ideals I, J are isomorphic if and only if $pI = qJ$ for some polynomials $p(z), q(z)$. On the other hand, two primary decomposable subspaces V, W determine the same point of $\mathrm{Gr}^{\mathrm{ad}}$ if and only if $pV = qW$ for some polynomials $p(z), q(z)$; and the bijections α and γ are clearly compatible with multiplication by polynomials. Let $\mathcal{R}$ denote the set of isomorphism classes of nonzero right ideals of A (equivalently, of finitely generated torsion-free rank 1 right A-modules). Combining the remarks above with Theorem 6.1, we get the following:

Theorem 6.3.

(i) *The maps defined by the formulae (6.2) and (6.3) define inverse bijections*

$$\alpha : \mathrm{Gr}^{\mathrm{ad}} \to \mathcal{R} \quad \text{and} \quad \gamma : \mathcal{R} \to \mathrm{Gr}^{\mathrm{ad}}.$$

(ii) *For $V, W \in \mathrm{Gr}^{\mathrm{ad}}$, there is a natural identification*

$$\mathcal{D}(V, W) \simeq \mathrm{Hom}_A(\alpha(V), \alpha(W)).$$

As a special case of (ii), we see that if $W \in \mathrm{Gr}^{\mathrm{ad}}$ and $I := \alpha(W)$ is the corresponding ideal in A, then the algebra $\mathcal{D}(W) \equiv \mathcal{D}(W, W)$ is identified with $\mathrm{End}_A(I)$. On the other hand, if W corresponds to the pair $(X, \mathcal{L})$, then according to Proposition 5.5, $\mathcal{D}(W)$ is just the algebra $\mathcal{D}_{\mathcal{L}}(X)$ that interests us. In this way Theorem 6.3 translates any question about the algebras $\mathcal{D}_{\mathcal{L}}(X)$ into a question about ideals in the Weyl algebra. It remains to give the translation into these terms of the group G of symmetries of $\mathrm{Gr}^{\mathrm{ad}}$. Note that if σ is an automorphism of A and I is finitely generated torsion-free rank 1 A-module, then $\sigma_*(I)$ is a module of the same type: thus the automorphism group $\mathrm{Aut}(A)$ acts naturally on $\mathcal{R}$.

Theorem 6.4. *Under the bijection α, the action of the group Γ of KP flows corresponds to the action on $\mathcal{R}$ induced by the automorphisms $D \mapsto e^{p(z)} D e^{-p(z)}$ of A; while the map φ corresponds to the map on $\mathcal{R}$ induced by the formal Fourier transform $(z \mapsto \partial, \partial \mapsto -z)$ of A.*

Now, if σ is an automorphism of (any algebra) A, and M is any A-module, then it is trivial that $\mathrm{End}_A(M) \simeq \mathrm{End}_A(\sigma_* M)$. Thus Theorem 6.4 makes the 'if' part of Theorem 5.6(i) transparent.

Notes.

1. According to Dixmier (see [D]), the automorphisms mentioned in Theorem 6.4 generate the full automorphism group of A; thus we may identify our symmetry group G with $\mathrm{Aut}(A)$.

2. There are two routes available to prove the 'only if' part of Theorem 5.6(i). If we use Dixmier's theorem, we can simply note that it translates into a known theorem of Stafford (see [St]): if I and J are two ideal classes of A, then their endomorphism rings are isomorphic (if and) only if I and J belong to the same orbit of $\mathrm{Aut}(A)$ in $\mathcal{R}$. Alternatively, after we have classified the orbits, this fact will follow from Theorem 4.5 (whose proof in [LM] does not use Stafford's theorem, nor Dixmier's).

3. To get an idea of the depth of Stafford's theorem, let us give a proof (following [CH3]) of a crucial special case: if I is an ideal of A whose endomorphism ring is isomorphic to $\text{End}_A(A) = A$, then $I \simeq A$. Let $(X, \mathcal{L})$ be the pair corresponding to I; then $\mathcal{D}_\mathcal{L}(X)$ is isomorphic to A, hence $\mathcal{O}(X)$ is isomorphic to a mad subalgebra of A. Another (nontrivial) theorem of Dixmier (see [D]) says that all the mad subalgebras of A are isomorphic to $\mathbb{C}[z]$; hence $X \simeq \mathbb{A}^1$ and $\mathcal{L}$ is the trivial line bundle (because this is the only rank 1 torsion-free sheaf over $\mathbb{A}^1$). According to Theorem 6.3, it follows that $I \simeq A$. The general case of Stafford's theorem is a relatively formal consequence of this special case (see [St], Corollary 3.2).

4. If we introduce the category $\mathfrak{P}$ with objects the primary decomposable subspaces of $\mathbb{C}[z]$ and morphisms $\mathcal{D}(V, W)$, then we could summarize Theorem 6.1 by saying that we have an *equivalence of categories* between $\mathfrak{P}$ and the category of ideals in A (regarded as a full subcategory of the category of right A-modules).

5. Theorems 6.1 and 6.3 remain true (*mutatis mutandis*) if we replace the Weyl algebra by the ring of differential operators on any smooth affine curve (see [CH1]). Using this fact, we can sketch a proof of Theorem 3.3. Suppose that X and Y are smooth affine curves such that $\mathcal{D}(X)$ is Morita equivalent to $\mathcal{D}(Y)$. Since these are Noetherian domains, that means that $\mathcal{D}(Y)$ is isomorphic to the endomorphism ring of some ideal in $\mathcal{D}(X)$, and hence to $\mathcal{D}(V)$ for some primary decomposable subspace V of $\mathcal{O}(X)$. This in turn is isomorphic to some ring $\mathcal{D}_\mathcal{L}(X')$, where X' is a curve with bijective normalization $X \to X'$. If Y is not isomorphic to $\mathbb{A}^1$, then Theorem 4.2 shows that $\mathcal{D}(Y)$ has only one mad subalgebra. The same is therefore true of $\mathcal{D}_\mathcal{L}(X')$; extracting these mad subalgebras gives $\mathcal{O}(Y) \simeq \mathcal{O}(X')$, hence $Y \simeq X'$. Since Y is smooth, this implies $X = X'$, hence $Y \simeq X$. Finally, if Y is isomorphic to $\mathbb{A}^1$, then $\mathcal{D}(Y)$, and hence also $\mathcal{D}_\mathcal{L}(X')$, has more than one mad subalgebra, so Lemma 4.1 implies that $X \simeq \mathbb{A}^1$.

6. Theorem 6.1 is proved in [CH1]; Theorem 6.4 is proved in [BW2].

7. A different view of the construction of Cannings and Holland, and some further generalizations, can be found in [BGK2].

7 The Calogero–Moser spaces

Our third realization of Gr^{ad} involves the *Calogero–Moser spaces* $\mathcal{C}_n$. For each $n \geq 0$, let $\tilde{\mathcal{C}}_n$ be the space of pairs (X, Y) of complex $n \times n$ matrices such that

$$[X, Y] + I \text{ has rank } 1$$

and let $\mathcal{C}_n := \tilde{\mathcal{C}}_n/\mathrm{GL}(n, \mathbb{C})$, where the action of $g \in \mathrm{GL}(n, \mathbb{C})$ is by simultaneous conjugation: $(X, Y) \mapsto (gXg^{-1}, gYg^{-1})$. One can show that $\mathcal{C}_n$ is a smooth irreducible affine variety of dimension $2n$ ($\mathcal{C}_0$ is supposed to be a point).

Theorem 7.1. *There is a natural bijection*

$$\beta : \mathcal{C} := \bigsqcup_{n \geq 0} \mathcal{C}_n \to \mathrm{Gr}^{\mathrm{ad}}$$

such that:

(i) *the action on $\mathrm{Gr}^{\mathrm{ad}}$ of the multiplication operators $e^{\{p(z)\}} \in \Gamma$ corresponds to the maps $(X, Y) \mapsto (X + p'(Y), Y)$ on $\mathcal{C}_n$;*

(ii) *the action of φ on $\mathrm{Gr}^{\mathrm{ad}}$ corresponds to the map $(X, Y) \mapsto (-Y, X)$ on $\mathcal{C}_n$;*

(iii) *the action of the group G on each $\mathcal{C}_n$ is transitive.*

It follows from part (iii) of this theorem that the spaces $\beta(\mathcal{C}_n)$ are the orbits of G in $\mathrm{Gr}^{\mathrm{ad}}$. To complete the proof of Theorem 5.6 we have only to check that $\beta^{-1}(W_n)$ belongs to $\mathcal{C}_n$: that is done in Example 8.2 below.

The decomposition of $\mathrm{Gr}^{\mathrm{ad}}$ in Theorem 7.1 was originally obtained using ideas from the theory of integrable systems (see [W2]). Here we sketch a different method. In view of Theorem 6.3, it is enough to see why the space $\mathcal{R}$ of ideals in the Weyl algebra should decompose into the finite-dimensional spaces $\mathcal{C}_n$. That can be understood by analogy with the corresponding commutative problem, namely, to describe the space $\mathcal{R}_0$ of isomorphism classes of ideals in $A_0 := \mathbb{C}[x, y]$. This problem is easy, because each ideal class in A_0 has a unique representative of finite codimension; hence $\mathcal{R}_0$ decomposes into the disjoint union of the *point Hilbert schemes* $\mathrm{Hilb}_n(\mathbb{A}^2)$ (that is, the spaces of ideals of codimension n) for $n \geq 0$. It is elementary that $\mathrm{Hilb}_n(\mathbb{A}^2)$ can be identified with the space of pairs (X, Y) of commuting $n \times n$ matrices possessing a cyclic vector (see [N], 1.2); thus $\mathrm{Hilb}_n(\mathbb{A}^2)$ is the commutative analogue of the Calogero–Moser space $\mathcal{C}_n$. Because the Weyl algebra has no nontrivial ideals of finite codimension, it is not immediately clear how to adapt this discussion to the noncommutative case; however, there is a less elementary point of view which generalizes more easily. We may regard an ideal of A_0 as a rank 1 torsion-free sheaf over $\mathbb{A}^2$; it has a unique extension to a torsion-free sheaf over the projective plane $\mathbb{P}^2$ trivial over the line at infinity. The classification of ideals by pairs of matrices can then be regarded as a (trivial) special case of Barth's classification of framed bundles (of any rank) over $\mathbb{P}^2$ (see [N], Ch. 2). In a similar way, an ideal of the Weyl algebra determines a rank 1 torsion-free sheaf over a suitably defined quantum projective plane $\mathbb{P}_q^2$; these can then be classified much as in the commutative case.

Notes.

1. Let us try to give something of the flavour of the noncommutative projective geometry needed to carry out the plan sketched above (see, for example [A], [AZ] for more details). Let $X \subseteq \mathbb{P}^N$ be a projective variety, and let $A = \oplus_{k \geq 0} A_k$ be its (graded) homogeneous coordinate ring. To any quasicoherent sheaf $\mathcal{M}$ over X we can assign the graded A-module

$$M := \bigoplus_{k \in \mathbb{Z}} H^0(X, \mathcal{M}(k)).$$

A theorem of Serre (see [S]) states that this defines an equivalence between the category of quasicoherent sheaves over X and a certain quotient of the category of graded A-modules (we have to divide out by the so-called *torsion modules*, in which each element is killed by some A_k). Thus many results about projective varieties can be formulated in a purely algebraic way, in terms of graded A-modules; in this form the theory makes sense also for a noncommutative graded ring A. The coordinate ring of the space $\mathbb{P}_q^2$ referred to above is the ring of noncommutative polynomials in three variables x, y, z of degree 1, where z commutes with everything, but $[x, y] = z^2$. It turns out that the homological properties of this ring are similar to those of the commutative graded ring $\mathbb{C}[x, y, z]$; in particular, the classification of bundles (of any rank) over $\mathbb{P}_q^2$ is similar to that of bundles over $\mathbb{P}^2$ (see [KKO]).

2. The idea of using $\mathbb{P}_q^2$ to classify the ideals in the Weyl algebra is due to L. Le Bruyn (see [LeB]). However, Le Bruyn's chosen extension of an ideal in A to a sheaf over $\mathbb{P}_q^2$ was in general not trivial over the line at infinity, so he did not obtain the decomposition of $\mathcal{R}$ into the Calogero–Moser spaces. That was done in [BW3] and (in a different way) in [BGK1].

3. The connection between the spaces $\mathrm{Hilb}_n(\mathbb{A}^2)$ and $\mathcal{C}_n$ is actually much closer than we have indicated: $\mathrm{Hilb}_n(\mathbb{A}^2)$ is a hyperkähler variety, and $\mathcal{C}_n$ is obtained by deforming the complex structure of $\mathrm{Hilb}_n(\mathbb{A}^2)$ within the hyperkähler family. See [N], Ch. 3, especially 3.45.

4. The assertions (i) and (ii) in Theorem 7.1 are proved in [BW2] (using the original construction of β), and in [BW3] (using the construction sketched above). The fact that the two constructions agree is also proved in [BW3].

5. Parts (i) and (ii) of Theorem 7.1 reduce the proof of part (iii) (transitivity of the G-action) to an exercise in linear algebra. Unfortunately, the exercise seems to be quite difficult, and the published solution in [BW2] strays outside

elementary linear algebra at one point (see Lemma 10.3 in [BW2]). P. Etingof has kindly pointed out to us that transitivity also follows easily from the fact that the functions $(X, Y) \mapsto \operatorname{tr}(X^k)$ and $(X, Y) \mapsto \operatorname{tr}(Y^k)$ generate $\mathcal{O}(\mathcal{C}_n)$ as a Poisson algebra (see [EG], 11.33).

6. In [BW3], Section 5 we have given an elementary construction of the map $\mathcal{R} \to \mathcal{C}$, in a similar spirit to the elementary treatment of the commutative case. It turns out that the inverse map $\mathcal{C} \to \mathcal{R}$ can also be written down explicitly, as follows. Let $(X, Y) \in \mathcal{C}_n$, and choose column and row vectors v, w such that $[X, Y] + I = vw$. Define[2]

$$\kappa := 1 - w(Y - zI)^{-1}(X - \partial I)^{-1}v$$

(thus κ belongs to the quotient field of the Weyl algebra A). Then the (fractional) right ideal

$$\det(Y - zI)\, A + \kappa \det(X - \partial I)\, A \subset \mathbb{C}(z)[\partial] \tag{7.1}$$

represents the class in $\mathcal{R}$ corresponding to (X, Y). Using these formulae, it is possible to give a completely elementary proof that $\mathcal{R}$ decomposes into the spaces $\mathcal{C}_n$. More details will appear elsewhere.

8 The invariant n

Theorem 7.1 assigns to each $W \in \mathrm{Gr}^{\mathrm{ad}}$ a non-negative integer n, namely, the index of the 'stratum' $\mathcal{C}_n$ containing $\beta^{-1}(W)$. Using Proposition 5.2 and Theorem 6.3, we may equally well regard n as an invariant of a pair $(X, \mathcal{L})$, or of an ideal (class) in the Weyl algebra A. In this section we discuss various descriptions of this invariant. The first two begin with an ideal class in A.

n **as a Chern class.** We return to the quantum projective plane $\mathbb{P}_q^2$ explained at the end of Section 7. Let M be an ideal class of A, and let $\mathcal{M}$ denote its unique extension to a sheaf over $\mathbb{P}_q^2$ trivial over the line at infinity. Then we claim that

$$n = \dim_{\mathbb{C}} H^1(\mathbb{P}_q^2, \mathcal{M}(-1)). \tag{8.1}$$

To see that, we need to give more details of the construction of the map $\mathcal{R}$ to $\mathcal{C}$. Recall that the homogeneous coordinate ring of $\mathbb{P}_q^2$ has three generators x, y, z. It turns out that multiplication by z induces an isomorphism

$$H^1(\mathbb{P}_q^2, \mathcal{M}(-2)) \to H^1(\mathbb{P}_q^2, \mathcal{M}(-1) := V.$$

---

[2] We get this formula by combining Remark 5.4 in [BW3] with formula (3.5) in [W2].

If we use this isomorphism to identify these spaces, then multiplication by x and y gives us a pair (X, Y) of endomorphisms of V: this is the point of $\mathcal{C}$ associated with M. Obviously, the size of the matrices (X, Y) is given by (8.1).

Note. By analogy with the commutative case (see [N], Ch. 2), we would like to interpret n as the second Chern class $c_2(\mathcal{M})$. However, at the time of writing, Chern classes have not yet been discussed in noncommutative projective geometry.

n as a codimension. Again, let M be an ideal of A. By [St], Lemma 4.2, we may suppose that M intersects $\mathbb{C}[z] \subset A$ nontrivially; let I be the ideal in $\mathbb{C}[z]$ generated by the leading coefficients of the operators in M, and let $p(z)$ be a generator of I. Then $p^{-1}M \subset \mathbb{C}(z)[\partial]$ is a fractional ideal representing the class of M. Define a map $D \mapsto D_+$ from $\mathbb{C}(z)[\partial]$ to A by

$$\left(\sum_i f_i \partial^i \right)_+ = \sum_i (f_i)_+ \partial^i.$$

Here f_+ denotes the polynomial part of a rational function f (that is, the polynomial such that $f - f_+$ vanishes at infinity). Then we claim that

$$n \text{ is the codimension of } (p^{-1}M)_+ \text{ in } A.$$

A proof can be found in [BW3], Section 6, where it is shown that the quotient space $A/(p^{-1}M)_+$ can be identified with the (Čech) cohomology group on the right of (8.1).

Note. The special representative for an ideal class that we used in this subsection is the same one as is given by the formula (7.1). It is the unique representative of the form $\mathcal{D}(\mathbb{C}[z], W)$ with $W \in \mathrm{Gr}^{\mathrm{ad}}$ (cf. Theorem 6.3).

The differential genus of a framed curve. The following characterization of n was one of the main results of [W2].

Theorem 8.1. *Let $W \in \mathrm{Gr}^{\mathrm{ad}}$. Then the integer n that we have associated to W is equal to the dimension of the open cell in $\mathrm{Gr}^{\mathrm{ad}}$ containing W.*

This theorem leads easily to a simple formula for calculating n in concrete examples (cf. [PS], 7.4). Recall from Section 5 that W is constructed from a family of λ-primary subspaces $V_\lambda \subseteq \mathbb{C}[z]$ (one for each $\lambda \in \mathbb{C}$, and almost all of them equal to $\mathbb{C}[z]$). In terms of these V_λ, we can calculate n as follows. First, we have $n = \sum_\lambda n_\lambda$, where n_λ depends only on V_λ (and is zero if $V_\lambda = \mathbb{C}[z]$). To find n_λ, let

$$r_0 < r_1 < r_2 < \ldots \tag{8.2}$$

be the numbers r such that V_λ contains a polynomial that vanishes exactly to order r at λ. For large i we have $r_i = g + i$, where g is the number of 'gaps' (non-negative integers that do not occur) in the sequence (8.2). Then we have

$$n_\lambda = \sum_{i \geq 0} (g + i - r_i). \tag{8.3}$$

Example 8.2. For the 0-primary space $V := \mathcal{O}(X_n)$ defined by (4.1), the sequence (8.2) is

$$0 < n + 1 < n + 2 < \ldots$$

whence $g = n$, and the right-hand side of (8.3) is equal to n.

This calculation completes the proof of Theorem 5.6(ii), and shows that we can identify the number n associated with a pair $(X, \mathcal{L})$ with the differential genus $d_\mathcal{L}(X)$ introduced in Section 5.

Example 8.3. If Y_r is the curve with coordinate ring $\mathcal{O}(Y_r) := \mathbb{C}[z^2, z^{2r+1}]$, then again $\mathcal{O}(Y_r)$ is 0-primary, and the sequence (8.2) is

$$0 < 2 < 4 < \ldots < 2r < 2r + 1 < \ldots .$$

Hence $g = r$, and $d(Y_r) = r + (r - 1) + \ldots + 2 + 1 = r(r + 1)/2$.

In particular, $d(Y_2) = 3$ so Y_2 is differentially isomorphic to X_3, in agreement with G. Letzter (see [L]).

Example 8.4. Here is the simplest example to show that in general $d_\mathcal{L}(X)$ depends on $\mathcal{L}$, not just on X. Let V be the 0-primary space spanned by $\{z^i : i \neq 2, 3\}$. Then the the sequence (8.2) is

$$0 < 1 < 4 < 5 < \ldots$$

whence $n = 4$. Clearly, V is a maximal module over the ring $\mathcal{O}(X_3)$, and thus corresponds to a maximal torsion-free (but not locally free) sheaf $\mathcal{L}$ over X_3. For this sheaf $\mathcal{L}$ we therefore have $d_\mathcal{L}(X_3) = 4$, and the ring $\mathcal{D}_\mathcal{L}(X_3)$ is isomorphic to $\mathcal{D}(X_4)$.

The Letzter–Makar–Limanov invariant. Next, we describe the invariant originally used in [LM] to distinguish the rings $\mathcal{D}(X_n)$. We return temporarily to the case of any affine curve X, with normalization $\tilde{X}$ and function field $\mathbb{K}$; as usual (see Proposition 2.4), we view $\mathcal{D}(X)$ and $\mathcal{D}(\tilde{X})$ as subalgebras of $\mathcal{D}(\mathbb{K})$. In general, $\mathcal{D}(X)$ is not contained in $\mathcal{D}(\tilde{X})$; however, the associated graded algebra $\operatorname{gr}\mathcal{D}(X)$ is always contained in $\operatorname{gr}\mathcal{D}(\tilde{X})$ (see [SS], 3.11). In the case that most concerns us when $\tilde{X} = \mathbb{A}^1$, this simply means that the *leading* coefficient of each operator in $\mathcal{D}(X)$ is a polynomial (although the other coefficients may be rational functions, as we saw in Example 2.5). Continuing Theorem 3.2, we have:

Theorem 8.5. *Each of the conditions in Theorem 3.2 is equivalent to:*

$$\operatorname{gr}\mathcal{D}(X) \ \text{has finite codimension in} \ \operatorname{gr}\mathcal{D}(\tilde{X}).$$

In our case, when $\tilde{X} = \mathbb{A}^1$ and X is a framed curve, $\operatorname{gr}\mathcal{D}(X)$ is a subalgebra of finite codimension in $\mathbb{C}[z, \zeta]$; we call its codimension the *Letzter–Makar–Limanov invariant* of X, and denote it by $LM(X)$. The definition of $LM(X)$ uses the standard filtration on $\mathcal{D}(X)$; nevertheless, in [LM] it is proved that it depends only on the isomorphism class of the algebra $\mathcal{D}(X)$; that is, if X and Y are differentially isomorphic framed curves, then $LM(X) = LM(Y)$. On the other hand, it is not hard to calculate that $LM(X_n) = 2n$ (see [LM], Section 5). Combined with Theorem 5.6, that gives:

Theorem 8.6. *Let X be any framed curve. Then $2\,d(X) = LM(X)$.*

Notes.

1. Theorem 8.5 is proved (though not explicitly stated) in [SS], 3.12.

2. In [LM] the rings $\mathcal{D}_{\mathcal{L}}(X)$ (for $\mathcal{L} \neq \mathcal{O}_X$) are not considered; however, it is not hard to extend the discussion to include that case. Thus we can define the invariant $LM(\mathcal{D}(W))$ for any $W \in \operatorname{Gr}^{\mathrm{ad}}$, and Theorem 5.6 shows that it is equal to $2n$.

3. It is possible to prove directly (that is, without using Theorem 5.6) that $LM(\mathcal{D})$ is twice the number n defined by (8.1). The interested reader may see [B].

All our descriptions of n so far have been specific to our particular situation. It is natural to ask whether n is a special case of some general invariant of rings that is able to distinguish between different Morita equivalent domains. Our last two subsections are attempts in that direction.

Pic and Aut. Let $\mathcal{D}$ momentarily be any domain (associative algebra without zero divisors) over $\mathbb{C}$. The following idea for obtaining subtle invariants of the isomorphism class of $\mathcal{D}$ is due to Stafford (see [St]). Consider the group[3] $\text{Pic}(\mathcal{D})$ of all Morita equivalences of $\mathcal{D}$ with itself, that is, of all self-equivalences of the category Mod-$\mathcal{D}$ of (say right) $\mathcal{D}$-modules. Each such equivalence is given by tensoring with a suitable $\mathcal{D}$-bimodule, so we may also think of $\text{Pic}(\mathcal{D})$ as the group of all invertible $\mathcal{D}$-bimodules. Each automorphism of $\mathcal{D}$ induces a self-equivalence of Mod-$\mathcal{D}$, so there is a natural map

$$\omega : \text{Aut}(\mathcal{D}) \to \text{Pic}(\mathcal{D}). \tag{8.4}$$

Although the group $\text{Pic}(\mathcal{D})$ is a Morita invariant of $\mathcal{D}$, the automorphism group and the map ω are not.

We return to our case, where $\mathcal{D}$ is one of the algebras $\text{End}_A(I)$ (or $\mathcal{D}_{\mathcal{L}}(X)$). In general, the kernel of ω consists of the *inner* automorphisms of $\mathcal{D}$; in our case these are trivial, so ω is injective. For the Weyl algebra A, Stafford showed that ω is an isomorphism. We thus have a natural inclusion

$$\text{Aut}(\mathcal{D}) \hookrightarrow \text{Pic}(\mathcal{D}) \simeq \text{Pic}(A) = \text{Aut}(A)$$

(the isomorphism from $\text{Pic}(\mathcal{D})$ to $\text{Pic}(A)$ is defined by tensoring with the $\mathcal{D}$-A-bimodule I). Recalling that the group $\text{Aut}(A)$ acts transitively on $\mathcal{C}_n$, one can calculate that the isotropy group of the point in $\mathcal{C}_n$ corresponding to I is exactly this subgroup $\text{Aut}(\mathcal{D})$. It follows that we have a natural bijection

$$\mathcal{C}_n \simeq \text{Pic}(\mathcal{D})/\text{Aut}(\mathcal{D})$$

so it is tempting to claim that our invariant n is given by

$$2n = \dim_{\mathbb{C}} \text{Pic}(\mathcal{D})/\text{Aut}(\mathcal{D}). \tag{8.5}$$

The flaw in this is that the structure of algebraic variety on the quotient 'space' in (8.5) has been imposed *a posteriori*, and has not been extracted intrinsically from the algebra $\mathcal{D}$.

Note. In view of the above, we may hope that there should be (at least for some algebras $\mathcal{D}$) a natural structure of (infinite-dimensional) algebraic group on $\text{Pic}(\mathcal{D})$ for which $\text{Aut}(\mathcal{D})$ would be a closed subgroup. In our case, we can identify $\text{Pic}(\mathcal{D})$ with $\text{Aut}(A)$, which does indeed have a natural structure of algebraic group; however, for this structure $\text{Aut}(\mathcal{D})$ is not a closed subgroup (see [BW2], Section 11 for more details).

---

[3] More properly, we should write $\text{Pic}_{\mathbb{C}}(\mathcal{D})$ to indicate that we consider only equivalences that commute with multiplication by scalars. For a similar reason, we should write $\text{Aut}_{\mathbb{C}}$ too.

Mad subalgebras. The idea behind our final description of n is very simple, namely: n should measure the 'number' of mad subalgebras of $\mathcal{D}(X)$. Let us formulate a precise statement. For each $W \in \mathrm{Gr}^{\mathrm{ad}}$ with invariant n, we may choose an isomorphism

$$\phi : \mathcal{D}(W) \to \mathcal{D}(X_n).$$

Since $\mathcal{D}^0(W)$ is a mad subalgebra of $\mathcal{D}(W)$, $B := \phi(\mathcal{D}^0(W))$ is a mad subalgebra of $\mathcal{D}(X_n)$. Furthermore, ϕ extends to an isomorphism of quotient fields, in particular, it maps $z \in \mathbb{C}(z)[\partial]$ to some element $u := \phi(z)$ in the quotient field of B. Clearly, $\mathbb{C}[u]$ is the integral closure of B. According to [LM], the integral closure $\overline{B}$ of *any* mad subalgebra B is isomorphic to $\mathbb{C}[u]$: we shall call a choice of generator for $\overline{B}$ a *framing* of B. Thus the above isomorphism ϕ gives us a framed mad subalgebra (B, u) of $\mathcal{D}(X_n)$. Any two choices of ϕ differ only by an automorphism of $\mathcal{D}(X_n)$, so the *class* (modulo the action of $\mathrm{Aut}\,\mathcal{D}(X_n)$) of the framed mad subalgebra we have obtained depends only on W. Moreover (cf. Section 5, Note 1), if we replace W by gW, where g belongs to the group Γ of KP flows, then conjugation by g defines an isomorphism of $\mathcal{D}(gW)$ with $\mathcal{D}(W)$ which is the identity on $\mathcal{D}^0$, so the isomorphism

$$\mathcal{D}(gW) \to \mathcal{D}(W) \xrightarrow{\phi} \mathcal{D}(X_n)$$

defines the same framed mad subalgebra as ϕ. It follows that we have constructed a well-defined map

$$\mathcal{C}_n / \Gamma \to \{\text{classes of framed mad subalgebras in } \mathcal{D}(X_n)\}. \qquad (8.6)$$

Theorem 8.7. *The map (8.6) is a set-theoretical bijection.*

We will explain the proof elsewhere. Since the (categorical) quotient $\mathcal{C}_n /\!/ \Gamma$ is n-dimensional, we should like to interpret n as the dimension of the 'space' of (classes of framed) mad subalgebras of $\mathcal{D}(X_n)$. However, the word 'space' here is open to even more serious objections than in the preceding subsection.

Notes.

1. In the definition of a framed mad subalgebra (B, u) we did not assume *a priori* that the curve $\mathrm{Spec}\,B$ was free of multiple points (indeed, that was not proved in [LM]). This momentary inconsistency of terminology is resolved by Theorem 8.7, which asserts (*inter alia*) that every mad subalgebra B arises from the construction described above; in particular, that $\mathrm{Spec}\,B$ is a framed curve as defined earlier.

2. In the case $n = 0$, the left-hand side of (8.6) is a point, so Theorem 8.7 becomes a well-known result of Dixmier: in the Weyl algebra there is only one class of mad subalgebras (see [D]).

9 Higher dimensions

The examples of Levasseur, Smith and Stafford. Let $\mathfrak{g}$ be a simple complex Lie algebra, and let O be the closure of the minimal nilpotent orbit in $\mathfrak{g}$. Let $\mathfrak{g} = \mathfrak{n}_- \oplus \mathfrak{h} \oplus \mathfrak{n}_+$ be a triangular decomposition of $\mathfrak{g}$; then $O \cap \mathfrak{n}_+$ breaks up into several irreducible components X_i. In [LSS] it is shown that in some cases the ring $\mathcal{D}(X_i)$ can be identified with $U(\mathfrak{g})/J$, where J is a certain distinguished completely prime primitive ideal of $U(\mathfrak{g})$ (the *Joseph ideal*). The examples of differential isomorphism arise in the case $\mathfrak{g} = \mathfrak{so}(2n, \mathbb{C})$ (with $n \geq 5$), because in that case there are two nonisomorphic components X_1 and X_2 of this kind. They can be described quite explicitly: X_1 is the quadric cone $\sum z_i^2 = 0$ in $\mathbb{C}^{2n-2}$, and X_2 is the space of skew-symmetric $n \times n$ matrices of rank ≤ 2. In contrast to what we saw for curves, these spaces X_1 and X_2 are quite different topologically.

Morita equivalence. There are several papers that study differential *equivalence* in dimension > 1. In view of Theorem 3.2, attention has focused on the question of when a variety X is differentially equivalent to its normalization $\tilde{X}$. Of course, in dimension > 1 the normalization is not necessarily smooth: in [J1] there are examples of differential equivalence in which $\tilde{X}$ is not smooth (they can be thought of as generalizations of the monomial curves of Example 5.3). Another point that does not arise for curves is that the condition that X be *Cohen–Macaulay* plays an important role (we recall that every curve is Cohen–Macaulay). For example, a theorem of Van den Bergh states that if $\mathcal{D}(X)$ is simple, then X must be Cohen–Macaulay (see [VdB], Theorem 6.2.5). For varieties with *smooth* normalization, there are good generalizations of at least some parts of Theorem 3.2. For example, piecing together various results scattered through the literature, we can get the following.

Theorem 9.1. *Let X be an (irreducible) affine variety with smooth normalization $\tilde{X}$. Then the following are equivalent.*

(i) *The normalization map $\pi : \tilde{X} \to X$ is bijective and X is Cohen–Macaulay.*

(ii) *The algebras $\mathcal{D}(\tilde{X})$ and $\mathcal{D}(X)$ are Morita equivalent.*

(iii) *The ring $\mathcal{D}(X)$ is simple.*

Beautiful examples are provided by the varieties of *quasi-invariants* of finite reflection groups (see [BEG], [BC]): here $\tilde{X}$ is the affine space $\mathbb{A}^m$, so these examples are perhaps the natural higher-dimensional generalizations of our framed curves.

References for the proof of Theorem 9.1. For the implication '(1) $\Rightarrow$ (2)' in Theorem 9.1 we are relying on the recent preprint [BN] (at least in dimension > 2: for surfaces it was proved earlier in [HS]). For the rest, the implication '(2) $\Rightarrow$ (3)' is trivial, and the fact that $\mathcal{D}(X)$ simple implies X Cohen–Macaulay is the theorem of Van den Bergh mentioned above. The only remaining assertion in Theorem 9.1 is that if $\mathcal{D}(X)$ is simple then π is bijective. Suppose $\mathcal{D}(X)$ is simple. Then by [SS], 3.3, $\mathcal{D}(X)$ is isomorphic to the endomorphism ring of the right $\mathcal{D}(\tilde{X})$-module P defined by (3.1); the dual basis lemma then implies that P is a projective $\mathcal{D}(\tilde{X})$-module. It now follows from [CS], Theorem 3.1 that $\mathcal{D}(X)$ is a *maximal order*, then from [CS], Corollary 3.4 that π is bijective.

Non-affine varieties. In this paper we have considered only affine varieties. However, the problem of differential *equivalence* has an obvious generalization to arbitrary (for example, projective) varieties X. Namely: on X we have the *sheaf* $\mathcal{D}_X$ of differential operators (whose sections over an affine open set Spec A are the ring $\mathcal{D}(A)$), and given two varieties X and Y, we can ask whether the categories of $\mathcal{O}$-quasicoherent sheaves of modules over $\mathcal{D}_X$ and $\mathcal{D}_Y$ are equivalent. For X affine, the global section functor gives an equivalence between the categories of $\mathcal{D}_X$-modules and of $\mathcal{D}(X)$-modules, so we recover our original problem. The available evidence (namely [SS] and [BN]) suggests that results about the affine case carry over to this more general situation.

The question of differential isomorphism does not make sense for sheaves; however, we can always consider the ring $\mathcal{D}(X)$ of global sections of $\mathcal{D}_X$, and ask when $\mathcal{D}(X)$ and $\mathcal{D}(Y)$ are isomorphic. In general, $\mathcal{D}(X)$ may be disappointingly small: for example, if X is a smooth projective curve of genus > 1, then we have no global vector fields, so $\mathcal{D}(X) = \mathbb{C}$. Probably the question is a sensible one only if X is close to being a $\mathcal{D}$-*affine* variety (for which the global section functor still gives an equivalence between $\mathcal{D}_X$-modules and $\mathcal{D}(X)$-modules). As far as we know, there are not yet any papers on this subject: however, the question of Morita eqivalence of rings $\mathcal{D}(X)$ has been studied in [HoS] (where $X = \mathbb{P}^1$, this being the only $\mathcal{D}$-affine smooth projective curve); and in [J2] (where X is a weighted projective space). We should like to state one of the results of [HoS], since it is very close to our framed curves.

Let X be a 'framed projective curve', that is, we have a bijective normalization map $\mathbb{P}^1 \to X$. Then Holland and Stafford show that the rings $\mathcal{D}(X)$ (for X singular) are all Morita equivalent to each other, but *not* to $\mathcal{D}(\mathbb{P}^1)$. A key point is that, although $\mathbb{P}^1$ is $\mathcal{D}$-affine, the singular curves X are not.

References

[A]   Artin, M., Geometry of quantum planes, in Azumaya Algebras, Actions and Groups, *Contemporary Math.* **124**, Amer. Math. Soc., Providence, 1992, pp. 1–15.

[AZ]   Artin, M. and J. Zhang, Noncommutative projective schemes, *Adv. in Math.* **109** (1994), 228–287.

[BGK1]   Baranovsky, V., V. Ginzburg and A. Kuznetsov, Quiver varieties and a noncommutative $\mathbb{P}^2$, *Compositio Math.* **134** (2002), 283–318.

[BGK2]   Baranovsky, V., V. Ginzburg and A. Kuznetsov, Wilson's Grassmannian and a noncommutative quadric, *Internat. Math. Res. Notices* **21** (2003), 1155–1197.

[Be]   Becker, J., Higher derivations and integral closure, *Amer. J. Math.* **100**(3) (1978), 495–521.

[BN]   Ben-Zvi, D. and T. Nevins, Cusps and $\mathcal{D}$-modules, arXiv:math.AG/0212094.

[B]   Berest, Yu., A remark on Letzter–Makar–Limanov invariants, in Proceedings of ICRA X (Toronto, August 2002), to appear in Fields Institute Communication Series 40 (2004).

[BC]   Berest, Yu. and O. Chalykh, Quasi-invariants of complex reflection groups, in preparation.

[BEG]   Berest, Yu., P. Etingof and V. Ginzburg, Cherednik algebras and differential operators on quasi-invariants, *Duke Math. J.* **118**(2) (2003), 279–337.

[BW1]   Berest, Yu. and G. Wilson, Classification of rings of differential operators on affine curves, *Internat. Math. Res. Notices* **2** (1999), 105–109.

[BW2]   Berest, Yu. and G. Wilson, Automorphisms and ideals of the Weyl algebra, *Math. Ann.* **318**(1) (2000), 127–147.

[BW3]   Berest, Yu. and G. Wilson, Ideal classes of the Weyl algebra and noncommutative projective geometry (with an Appendix by M. Van den Bergh), *Internat. Math. Res. Notices* **26** (2002), 1347–1396.

[BGG]   Bernstein, J.N., I. M. Gel'fand and S.I. Gel'fand, Differential operators on the cubic cone, *Russian Math. Surveys* **27** (1972), 169–174.

[CH1]   Cannings, R.C. and M.P. Holland, Right ideals of rings of differential operators, *J. Algebra* **167** (1994), 116–141.

[CH2]   Cannings, R.C. and M.P. Holland, Differential operators, n-branch curve singularities and the n-subspace problem, *Trans. Amer. Math. Soc.* **347** (1995), 1439–1451.

[CH3]   Cannings, R.C. and M.P. Holland, Etale covers, bimodules and differential operators, *Math. Z.* **216**(2) (1994), 179–194.

[CH4]   Cannings, R.C. and M.P. Holland, Limits of compactified Jacobians and $\mathcal{D}$-modules on smooth projective curves, *Adv. Math.* **135**(2) (1998), 287–302.

[CS]   Chamarie, M. and J.T. Stafford, When rings of differential operators are maximal orders, *Math. Proc. Cambridge Philos. Soc.* **102**(3) (1987), 399–410.

[D]   Dixmier, J., Sur les algèbres de Weyl, *Bull. Soc. Math. France* **96** (1968), 209–242.

[DG]   Duistermaat, J.J. and F.A. Grünbaum, Differential equations in the spectral parameter, *Commun. Math. Phys.* **103** (1986), 177–240.

[E]   El boufi, B., Idéaux à droite réflexifs dans l'algèbre des opérateurs différentiels, *Comm. Algebra* **24**(3) (1996), 939–947.

[EG]   Etingof, P. and V. Ginzburg, Symplectic reflection algebras, Calogero–Moser space, and deformed Harish–Chandra homomorphism, *Invent. Math.* **147** (2002), 243–348.

[G]   Grothendieck, A., Eléments de Géométrie Algébrique IV, *Publ. Math.* **32**, IHES, Paris, 1967.

[HS]   Hart, R. and S.P. Smith, Differential operators on some singular surfaces, *Bull. London Math. Soc.* **19**(2) (1987), 145–148.

[HoS]   Holland, M. and J.T. Stafford, Differential operators on rational projective curves, *J. Algebra* **147**(1) (1992), 176–244.

[J1]   Jones, A.G., Some Morita equivalences of rings of differential operators, *J. Algebra* **173**(1) (1995), 180–199.

[J2]   Jones, A.G., Non-normal 𝒟-affine varieties with injective normalization, *J. Algebra* **173**(1) (1995), 200–218.

[KKO]   Kapustin, A., A. Kuznetsov and D. Orlov, Noncommutative instantons and twistor transform, *Commun. Math. Phys.* **220** (2001), 385–432.

[K]   Kouakou, K.M., Isomorphismes entre algèbres d'opérateurs différentielles sur les courbes algébriques affines, Thèse, Univ. Claude Bernard Lyon-I, 1994.

[Kr]   Krichever, I.M., On rational solutions of the Kadomtsev–Petviashvili equation and integrable systems of Calogero type, *Funct. Anal. Appl.* **12** (1978), 59–61.

[LeB]   Le Bruyn, L., Moduli spaces of right ideals of the Weyl algebra, *J. Algebra* **172** (1995), 32–48.

[L]   Letzter, G., Non-isomorphic curves with isomorphic rings of differential operators, *J. London Math. Soc.* **45**(2) (1992), 17–31.

[LM]   Letzter, G. and L. Makar-Limanov, Rings of differential operators over rational affine curves, *Bull. Soc. Math. France* **118** (1990), 193–209.

[LSS]   Levasseur, T., S.P. Smith and J.T. Stafford, The minimal nilpotent orbit, the Joseph ideal, and differential operators, *J. Algebra* **116** (1988), 480–501.

[M-L]   Makar-Limanov, L., Rings of differential operators on algebraic curves, *Bull. London Math. Soc.* **21** (1989), 538–540.

[MR]   McConnell, J.C. and J.C. Robson, Noncommutative Noetherian Rings, *Graduate Studies in Mathematics* **30**, American Mathematical Society, Providence, RI, 2001.

[MV]   Mount, K. and O.E. Villamayor, On a conjecture of Y. Nakai, *Osaka J. Math.* **10** (1973), 325–327.

[M]   Muhasky, J., The differential operator ring of an affine curve, *Trans. Amer. Math. Soc.* **307**(2) (1988), 705–723.

[Na]   Nakai, Y., High order derivations I, *Osaka J. Math.* **7** (1970), 1–27.

[N]   Nakajima, H. Lectures on Hilbert schemes of points on surfaces, *University Lecture Series*, vol. 18, American Mathematical Society, Rhode Island, 1999.

[P1]   Perkins, P., Commutative subalgebras of the ring of differential operators on a curve, *Pacific J. Math.* **139**(2) (1989), 279–302.

[P2]   Perkins, P., Isomorphisms of rings of differential operators on curves, *Bull. London Math. Soc.* **23** (1991), 133–140.

[PS]   Pressley, A. and G. Segal, *Loop Groups*, Clarendon Press, Oxford, 1986.

[R]   Rego, C.J., Remarks on differential operators on algebraic varieties, *Osaka J. Math.* **14** (1977), 481–486.

[SW]   Segal, G. and G. Wilson, Loop groups and equations of KdV type, *Publ. Math. IHES* **61** (1985), 5–65.

[S]   Serre, J.-P., Faisceaux algébriques cohérents, *Ann. Math.* **61** (1955), 197–278.

[Sm]   Smith, S.P., Differential operators on commutative algebras, *Lecture Notes in Math.* **1197**, Springer, Berlin, 1986, 165–177.

[SS]   Smith, S.P. and J.T., Stafford, Differential operators on an affine curve, *Proc. London Math. Soc.* (3) **56** (1988), 229–259.

[St]   Stafford, J.T., Endomorphisms of right ideals of the Weyl algebra, *Trans. Amer. Math. Soc.* **299** (1987), 623–639.

[T]   Traves, W.N., Nakai's conjecture for varieties smoothed by normalization, *Proc. Amer. Math. Soc.* **127** (1999), 2245–2248.

[VdB]   Van den Bergh, M. Differential operators on semi-invariants for tori and weighted projective spaces, *Lecture Notes in Math.* **1478**, Springer, Berlin, 1991, 255–272.

[W1]   Wilson, G., Bispectral commutative ordinary differential operators, *J. Reine Angew. Math.* **442** (1993), 177–204.

[W2]   Wilson, G., Collisions of Calogero-Moser particles and an adelic Grassmannian (with an Appendix by I.G. Macdonald), *Invent. Math.* **133** (1998), 1–41.

[W3]   Wilson, G., Bispectral symmetry, the Weyl algebra and differential operators on curves, *Proceedings of the Steklov Institute of Mathematics* **225** (1999), 141–147.

Email: berest@math.cornell.edu and g.wilson@imperial.ac.uk.

5

A polarized view of string topology

RALPH L. COHEN and VÉRONIQUE GODIN

Stanford University

Dedicated to Graeme Segal on the occasion of his 60th birthday

Abstract

Let M be a closed, connected manifold, and LM its loop space. In this paper we describe closed string topology operations in $h_*(LM)$, where h_* is a generalized homology theory that supports an orientation of M. We will show that these operations give $h_*(LM)$ the structure of a unital, commutative Frobenius algebra without a counit. Equivalently they describe a positive boundary, two-dimensional topological quantum field theory associated to $h_*(LM)$. This implies that there are operations corresponding to any surface with p incoming and q outgoing boundary components, so long as $q \geq 1$. The absence of a counit follows from the nonexistence of an operation associated to the disk, D^2, viewed as a cobordism from the circle to the empty set. We will study homological obstructions to constructing such an operation, and show that in order for such an operation to exist, one must take $h_*(LM)$ to be an appropriate homological pro-object associated to the loop space. Motivated by this, we introduce a prospectrum associated to LM when M has an almost complex structure. Given such a manifold its loop space has a canonical polarization of its tangent bundle, which is the fundamental feature needed to define this prospectrum. We refer to this as the 'polarized Atiyah-dual' of LM. An appropriate homology theory applied to this prospectrum would be a candidate for a theory that supports string topology operations associated to any surface, including closed surfaces.

1 Introduction

Let M^n be a closed, oriented manifold of dimension n, and let LM be its free loop space. The 'string topology' theory of Chas–Sullivan [3] describes a rich

The first author was partially supported by a grant from the NSF.
The authors are grateful to J. Morava, G. Segal, and D. Sullivan for many inspiring conversations about the topics of this paper. They are also grateful to the referee for many useful suggestions.

127

structure in the homology and equivariant homology of LM. The most basic operation is an intersection-type product

$$\circ : H_q(LM) \times H_r(LM) \to H_{q+r-n}(LM)$$

that is compatible with both the intersection product in the homology of the manifold, and the Pontrjagin product in the homology of the based loop space, $H_*(\Omega M)$. Moreover this product structure extends to a Batalin–Vilkovisky algebra structure on $H_*(LM)$, and an induced Lie algebra structure on the equivariant homology, $H_*^{S^1}(LM)$. More recently Chas and Sullivan [4] described a Lie bialgebra structure on the rational reduced equivariant homology, $H_*^{S^1}(LM, M; \mathbb{Q})$, where M is embedded as the constant loops in LM.

These string topology operations and their generalizations are parameterized by combinatorial data related to fat graphs used in studying Riemann surfaces [3][4][19][20][5]. The associated field theory aspects of string topology is a subject that is still very much under investigation. In this paper we contribute to this investigation in the following two ways.

Recall that a two-dimensional topological quantum field theory associates to an oriented compact one manifold S a vector space, A_S, and to any oriented cobordism Σ between S_1 and S_2 a linear map, $\mu_\Sigma : A_{S_1} \to A_{S_2}$. Such an assignment is required to satisfy various well-known axioms, including a gluing axiom. Recall also that if $A = A_{S^1}$, such a TQFT structure is equivalent to a Frobenius algebra structure on A [17][10][1]. This is a unital, commutative algebra structure, $\mu : A \otimes A \to A$, together with a counit (or trace map) $\theta : A \to k$, so that the composition $\theta \circ \mu : A \otimes A \to A \to k$ is a nondegenerate form. From the TQFT point of view, the unit in the algebra, $u : k \to A$ is the operation corresponding to a disk D^2 viewed as a cobordism from the empty set to the circle, and the counit $\theta : A \to k$ is the operation corresponding to the disk viewed as a cobordism from the circle to the empty set. Without the counit θ a Frobenius algebra is equivalent to a unital, commutative algebra A, together with a cocommutative coalgebra structure, $\Delta : A \to A \otimes A$ without counit, where Δ is a map of A-modules. From the TQFT point of view, a non-counital Frobenius algebra corresponds to a 'positive boundary' TQFT, in the sense that operations μ_Σ are defined only when each component of the surface Σ has a *positive* number of outgoing boundary components.

Let h^* be a multiplicative generalized cohomology theory whose coefficient ring, $h^*(point)$ is a graded field (that is, every nonzero homogeneous element is invertible). Besides usual cohomology with field coefficients, natural examples of such theories are periodic K-theory with field coefficients, or any of the Morava K-theories. Any multiplicative generalized cohomology theory gives rise to such a theory by appropriately localizing the coefficient ring.

Let h_* be the associated generalized homology theory. Let M^n be a closed, n-dimensional manifold which is oriented with respect to this theory. Our first result, which builds on work of Sullivan [19], is that string topology operations can be defined to give a two-dimensional positive boundary TQFT, with $A_{S^1} = h_*(LM)$.

Theorem 1.1. *The homology of the free loop space $h_*(LM)$ has the structure of a Frobenius algebra without counit. The ground field of this algebra structure is the coefficient field, $h_* = h_*(point)$.*

The construction of the TQFT operations corresponding to a surface Σ will involve studying spaces of maps from a fat graph Γ_Σ associated to the surface to M, and viewing that space as a finite codimension submanifold of a $(LM)^p$, where p is the number of incoming boundary components of Σ. We will show that this allows the construction of a Thom collapse map for this embedding, which will in turn define a push-forward map $\iota_! : h_*(LM)^{\otimes p} \to h_*(Map(\Gamma_\Sigma; M))$. The operation μ_Σ will then be defined as the composition $\rho_{out} \circ \iota_! : h_*(LM)^{\otimes p} \to h_*(Map(\Gamma_\Sigma; M) \to h_*(LM)^{\otimes q}$ where ρ_{out} is induced by restricting a map from Γ_Σ to its outgoing boundary components.

The second goal of this paper is to investigate the obstructions to constructing a homological theory applied to the loop space which supports the string topology operations, *and* permits the definition of a counit in the Frobenius algebra structure, or, equivalently, would eliminate the 'positive boundary' requirement in the TQFT structure. Let $h_*^{mid}(LM)$ be such a conjectural theory. In some sense this would represent a 'middle dimensional', or 'semi-infinite' homology theory associated to the loop space, because of the existence of a nonsingular form $h_*^{mid}(LM) \otimes h_*^{mid}(LM) \to k$ analogous to the interesection form on the middle-dimensional homology of an even-dimensional oriented manifold.

We will see that defining a counit would involve the construction of a push-forward map for the embedding of constant loops $M \hookrightarrow LM$. Unlike the embeddings described above, this has *infinite* codimension. We will argue that this infinite dimensionality will force the use of an inverse limit of homology theories, or a 'pro-homology theory' associated to the loop space. Using previous work of the first author and Stacey [8], we will show that there are obstructions to the construction of such a pro-object unless M has an almost complex structure. In this case the tangent bundle of the loop space has a canonical complex polarization, and we will use it to define the 'polarized loop space', $LM_\pm$. This space fibers over LM, where an element in the fiber over $\gamma \in LM$ is a representative of the polarization of the tangent space $T_\gamma LM$. We will examine various properties of $LM_\pm$, including its equivariant properties. We will show that the pullback of the tangent bundle TLM over $LM_\pm$ has a filtration that will allow us to define a prospectrum $LM_\pm^{-TLM}$, which we call the (polarized) 'Atiyah dual' of LM. We will end by describing how the application of an appropriate equivariant homology theory to this prospectrum should be a good candidate for studying further field theory properties of string topology. This will be the topic of future work, which will be joint with J. Morava and G. Segal.

The paper is organized as follows. In Section 1 we will describe the type of fat graphs needed to define the string topology operations. These are chord diagrams of the sort introduced by Sullivan [19]. We will define the topology of these chord diagrams using categorical ideas of Igusa [12][13]. Our main technical result, which we will need to prove the invariance of the operations, is that the space of chord diagrams representing surfaces of a particular diffeomorphism type is connected. In Section 2 we define the string topology operations and prove Theorem 1. The operations will be defined using a homotopy theoretic construction (the 'Thom collapse map') generalizing what was done by the first author and Jones in [5]. In Section 3 we describe the obstructions to the existence of a counit or trace in the Frobenius algebra structure. Motivated by these observations, in Section 4 we describe the 'polarized Atiyah dual' of the loop space of an almost complex manifold, and study its properties.

2 Fat graphs and Sullivan chord diagrams

Recall from [15][18] that a fat graph is a graph whose vertices are at least trivalent, and where the edges coming into each vertex come equipped with a cyclic ordering. Spaces of fat graphs have been used by many different authors as an extremely effective tool in studying the topology and geometry of moduli spaces of Riemann surfaces. The essential feature of a fat graph is that when it is thickened, it produces a surface with boundary, which is well defined up to homeomorphism.

For our purposes, the most convenient approach to the space of fat graphs is the categorical one described by Igusa [12][13]. In [12] (Chapter 8) he defined a category $\mathcal{F}at_n(g)$ as follows. The objects of $\mathcal{F}at_n(g)$ are fat graphs (with no lengths assigned to the edges), and the morphisms are maps of fat graphs $f : \Gamma_1 \to \Gamma_2$ (i.e maps of the underlying simplicial complexes that preserve the cyclic orderings) satisfying the following properties:

(a) The inverse image of any vertex is a tree.

(b) The inverse image of an open edge is an open edge.

Igusa proved that the geometric realization $|\mathcal{F}at_n(g)|$ is homotopy equivalent to the classifying space, $B\mathcal{M}_{g,n}$, where $\mathcal{M}_{g,n}$ is the mapping class group of genus g surfaces with n marked, ordered points (Theorem 8.6.3 of [12]). In this theorem $n \geq 1$ for $g \geq 1$ and $n \geq 3$ for $g = 0$. He also proved (Theorem 8.1.17) that $|\mathcal{F}at_n(g)|$ is homotopy equivalent to the space of *metric fat graphs,* which we denote $\mathcal{F}_n(g)$, which is a simplicial space made up of fat graphs with appropriate metrics. These spaces are closely related the simplicial sets studied by Culler and Vogtmann [9] and Kontsevich [14]. See [12] Chapter 8 for details.

Following [9] there are 'boundary cycles' associated to a fat graph Γ defined as follows. Pick an edge and choose an orientation on it. Traversing that edge in the direction of its orientation leads to a vertex. Proceed with the next edge emanating from that vertex in the cyclic ordering, and give it the orientation pointing away from that vertex. Continuing in this way, one traverses several edges, eventually returning to the original edge, with the original orientation. This yields a 'cycle' in the set of oriented edges and represents a boundary component. One partitions the set of all oriented edges into cycles, which enumerate the boundary components of the surface represented by Γ. The cycle structure of the oriented edges completely determines the combinatorial data of the fat graph.

In a metric fat graph each boundary cycle has an orientation and a metric on it. Hence introducing a marked point for each boundary cycle would yield a parameterization of the boundary components. Notice that it is possible that two marked points lie on the same edge, and indeed a single point on an edge might have a 'double marking' since a single edge with its two orientations might lie in two different boundary cycles.

We call the space of metric fat graphs representing surfaces of genus g with n boundary components, that come equipped with marked points on the the boundary cycles, $\mathcal{F}_n^\mu(g)$. This is the space of *marked* metric fat graphs. Using Igusa's simplicial set construction, one sees that $\mathcal{F}_n^\mu(g)$ can be topologized so that the projection map that forgets the markings

$$p : \mathcal{F}_n^\mu(g) \to \mathcal{F}_n(g) \tag{2.1}$$

is a quasifibration whose fiber is the space of markings on a fixed fat graph, which is homeomorphic to the torus $(S^1)^n$. The topology of the space of marked metric fat graphs is studied in detail in [11] with applications to specific combinatorial calculations.

For the purposes of constructing the string topology operations, we will use a particular type of fat graph due to Sullivan (Figure 1).

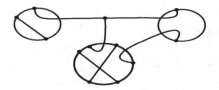

Figure 1. Sullivan chord diagram of type (1;3,3)

Definition 2.1. A 'Sullivan chord diagram' of type $(g; p, q)$ is a fat graph representing a surface of genus g with $p + q$ boundary components, that consists of a disjoint union of p disjoint closed circles together with the disjoint union of connected trees whose endpoints lie on the circles. The cyclic orderings of the edges at the vertices must be such that each of the p disjoint circles is a boundary cycle. These p circles are referred to as the incoming boundary cycles, and the other q boundary cycles are referred to as the outgoing boundary cycles.

The ordering at the vertices in the diagrams that follow are indicated by the clockwise cyclic ordering of the plane. Also in a Sullivan chord diagram, the vertices and edges that lie on one of the p disjoint circles will be referred to as circular vertices and circular edges respectively. The others will be referred to as ghost vertices and edges.

To define the topology on the space of metric chord diagrams, we first need to define the space of metric fat graphs, $\mathcal{F}_{p,q}(g)$, of genus g, with $p+q$-ordered boundary cycles, with the first p distinguished as incoming, and the remaining q distinguished as outgoing. Igusa's simplicial construction of $\mathcal{F}_n(g)$ defines a model of $\mathcal{F}_{p,q}(g)$ as the geometric realization of a simplicial set. Moreover this space is homotopy equivalent to the realization of the nerve of the category $\mathcal{F}at_{p,q}(g)$ defined as above, with the additional feature that the objects come equipped with an ordering of the boundary cycles, with the first p distinguished as incoming cycles. The morphisms must preserve this structure.

Consider the space of 'metric chord diagrams', $\mathcal{CF}_{p,q}(g)$ defined to be the subspace of the metric fat graphs $\mathcal{F}_{p,q}(g)$ whose underlying graphs are Sullivan chord diagrams of type $(g; p, q)$. So if $\mathcal{CF}at_{p,q}(g)$ is the full subcategory of $\mathcal{F}at_{p,q}(g)$ whose objects are chord diagrams of type $(g; p, q)$, then Igusa's argument shows that the space of metric chord diagrams $\mathcal{CF}_{p,q}(g)$ is homotopy equivalent to the realization of the nerve of the category $\mathcal{CF}at_{p,q}(g)$.

Given a metric chord diagram $c \in \mathcal{CF}_{p,q}(g)$, there is an associated metric fat graph, $S(c)$, obtained from c by collapsing all ghost edges. There is an induced cyclic ordering on the vertices of $S(c)$ so that the collapse map

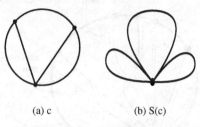

(a) c   (b) S(c)

Figure 2. Collapsing of ghost edges

$\pi : c \to S(c)$ is a map of fat graphs in $\mathcal{F}at_{p,q}(g)$. Figure 2 describes this collapse map.

We will define a marking of a Sullivan chord diagram c to be a marking (i.e a choice of point) on each of the boundary cycles of the associated fat graph $S(c)$. We let $\mathcal{CF}^{\mu}_{p,q}(g)$ denote the space of all marked metric chord diagrams. Like with the full space of marked fat graphs, this space can be topologized in a natural way so that the projection map that forgets the markings

$$p : \mathcal{CF}^{\mu}_{p,q}(g) \leftarrow \mathcal{CF}_{p,q}(g) \tag{2.2}$$

is a quasifibration, with fiber over a metric chord diagram c equivalent to a torus $(S^1)^{p+q}$. Again, the topology of these spaces of marked chord diagrams will be studied in detail in [11].

The space of marked, metric chord diagrams $\mathcal{CF}^{\mu}_{p,q}(g)$ will be used in the next section to parameterize the string topology operations. Its topology, however, is far from understood. It is a proper subspace of a space homotopy equivalent to the classifying space of the mapping class group, and thus moduli space. However very little is known about the topology of this subspace. We make the following conjecture, which would say that the parameterizing spaces of string topology operations are homotopy equivalent to moduli spaces of curves, thus potentially leading to a conformal field theory type structure.

Conjecture. The inclusion $\mathcal{CF}^{\mu}_{p,q}(g) \hookrightarrow \mathcal{CF}_{p,q}(g)$ is a homotopy equivalence, and in particular $\mathcal{CF}^{\mu}_{p,q}(g)$ is homotopy equivalent to the classifying space $B\mathcal{M}^{p+q}_g$, where $\mathcal{M}^{p+q}_g$ is the mapping class group of a surface of genus g with $p + q$-ordered boundary components, and the diffeomorphisms preserve the boundary components pointwise.

For the purposes of this paper we will need the following property of these spaces.

Theorem 2.2. *The space $\mathcal{CF}^{\mu}_{p,q}(g)$ is path connected.*

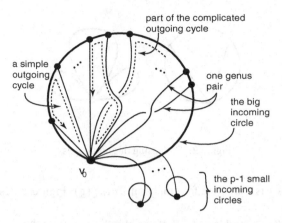

Figure 3. Γ_0

Proof. Because of quasifibration (2.2), it suffices to prove that the space of *unmarked* metric chord diagrams $\mathcal{CF}_{p,q}(g)$ is connected. However as remarked earlier, this space is homotopy equivalent to the nerve of the category $\mathcal{CF}at_{p,q}(g)$.

Now a morphism between objects a and b in a category determines a path from a to b in the geometric realization of its nerve. Since a morphism in $\mathcal{CF}at_{p,q}(g)$ collapses trees to vertices, we refer to such morphisms as 'collapses'. Now by reversing orientation, a morphism from b to a also determines a path from a to b in the geometric realization. We refer to such a morphism as an 'expansion' from a to b. Therefore to prove this theorem it suffices to build, for any chord diagram in our space, a sequence of collapses and and expansions from it to a fixed chord diagram Γ_0. In the following diagrams, dashed lines will represent boundary cycles.

We will choose our basepoint Γ_0 as in Figure 3. In Γ_0, $p-1$ of the incoming circles contain only one vertex. There is also a distinguished vertex v_o in the pth incoming circle (which we refer to as the 'big circle'). Moreover $q-1$ outgoing components share the same structure : they can be traced by going from v_o along a chord edge whose other vertex also lies on the big circle, going along the next circle edge in the cyclic ordering, and then going along the next chord edge back to v_o. In the last outgoing boundary component positive genus is produced by pairs of chord edges twisted, as shown. (These pairs add two generators in the fundamental group of the surface but do not affect the number of boundary components, therefore they 'create genus'.) Notice also that except for v_o all of the vertices of Γ_0 are trivalent. So, in Γ_0 the complexity is concentrated in the big incoming circle, the last outgoing boundary circle, and one vertex v_o.

The ordering of the boundary cycles in Γ_0 is given by making the first incoming cycle the one containing v_o. The ordering of the other incoming boundary components follows the cyclic ordering at v_o (so that in Figure 3 the circle on the right will come second and the one on the left last.) Similarly the cyclic ordering at v_o will give us an ordering of the outgoing boundary components (in which the complicated boundary cycle is last).

To prove the theorem we start from any chord diagram (object) in $\mathcal{CF}at_{p,q}(g)$ and get to Γ_0 by a sequence of collapses and expansions. In our figures, the arrows follow the direction of the corresponding morphism in our category $\mathcal{CF}at_{p,q}(g)$. Note that since, in a Sullivan chord diagram, the incoming boundaries are represented by disjoint circles, a chord edge between two circular vertices cannot be collapsed. Remember also that the ghost edges need to form a disjoint union of trees. Hence if both vertices of a circular edge are part of the same tree of ghost edges (same connected component of the ghost structure), this circular edge cannot be collapsed. We will call an edge 'essential' if it cannot be collapsed. That is, it is either a circular edge and its collapse would create a non-trivial cycle among the ghost edges, or it is a chord edge between two circular vertices.

Throughout this proof, letters from the beginning of the alphabet will be used to label edges that are on the verge of being collapsed, and letters from the end of the alphabet will be used to label edges that have just been created, via an expansion. We will start by assuming that all nonessential edges have been collapsed.

The first step will be to find a path to a chord diagram with a distinguished vertex v_o, the only one with more than three edges emanating from it. Choose v_o to be any vertex on the first incoming boundary cycle. For any vertex v, other than v_o, having more than three edges, we will 'push' the edges of v toward v_o by a sequence of expansions and collapses. This is done as follows.

Since all edges are essential, the vertices of any circular edge are part of the same connected component of the ghost structure. We can therefore choose a path γ from v to v_o contained completely in the ghost structure. Following Figure 4, we can push the edges of v a step closer to v_o. Repeating this process completes this step.

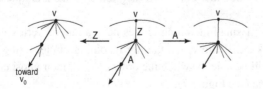

Figure 4. Pushing edges

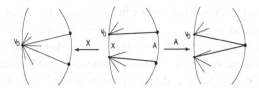

Figure 5. Getting rid of one edge

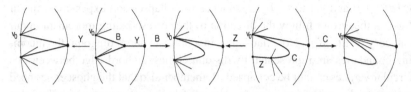

Figure 6. Pushing edges towards v_o

We now have a distinguished vertex v_o, which is the only vertex with more than one ghost edge. Note that there is a unique ordering of the edges emanating from v_o that is compatible with the cyclic order such that one of the circular edges of v_o is first in the order, and the other circular edge is last in the order. We will think of this ordering of edges as passing from left to right.

We will next simplify the incoming circles. In this step all of the incoming cycles but the first will be brought down to one edge. Notice first that all ghost edges have v_o as one of their vertices. A ghost edge between two other vertices would be in itself a connected component of the ghost structure. But a circular edge joining two different components can be collapsed without creating a 'ghost cycle'. Since all the edges of our graphs are essential, we know that all ghost edges have v_o as a vertex.

Now take any 'small' incoming boundary circle containing more than one edge. As seen in Figure 5, the addition of an edge X close to v_o, renders A nonessential and it can be collapsed.

After this has been done, there will be three non-trivalent vertices. The procedure shown in Figure 6 brings this number back down to one. Observe that in this procedure there is no risk of collapsing an essential edge. Repeating this process reduces the number of vertices on these incoming circles down to one per circle. Notice that our chord diagram still has a unique non-trivalent vertex v_o.

Now the $p - 1$ small incoming circles have a unique vertex with a unique ghost edge linking it with v_o on the big incoming circle. Using their simple structure we will be able to switch the order at v_o of their ghost edges and any other ghost edge (see Figure 7).

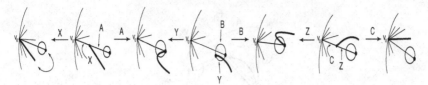

Figure 7. Switching an incoming circle and an edge

The next step is to simplify the structure of $q - 1$ outgoing boundary components (all but the last one). This will be achieved by lowering the number of edges involved in the tracing of these boundary cycles to a minimum (three edges for most). We will use the term 'clean' to refer to this simple form.

Now each of the outgoing boundaries has an edge on the big circle. (The cleaned boundaries will only have one such edge.) Assume by induction that the first $k - 1$ outgoing boundaries have already been cleaned. Assume also that these cleaned boundaries have been pushed to the left of the big incoming circle, meaning that unique incoming edge is situated to the left of all the incoming edges associated to the uncleaned boundaries (no uniqueness here) and that their ghost edges at v_o are attached left of all other ghost edges. Assume that the cleaned boundaries have been ordered. Assume also that at least two outgoing boundaries still need to be cleaned. Firstly, we will clean the next outgoing boundary w_k and, secondly, we will push it left to its proper position.

Since all the clean boundaries are to the left, we will have two successive circular edges, A_k and A_s, on our big circle such that A_k is part of w_k the next outgoing boundary component to be cleaned, and A_s belongs to w_s with $s > k$. We will argue the case where A_k is to the left of A_s. The other situation is argued similarly.

Let v be the common circular vertex of A_k and A_m and let B_k be the ghost edge coming into v. Notice that the cycles representing both w_k and w_m include B_k (with different orientations). We now push all the extra edges involved in the tracing of w_k to the right of B_k and hence into w_m. Since all the ghost edges involved in the cycle representing w_k start at v_o and end on the big circle, the cycle traced by w_k is $(B_{k-1}, \dots, C_{m-1}, E_m, D_m, C_m, B_k, A_k)$ where the Es and the Cs are ghost edges from v_o to the big circle, and the Ds are circular edges on the big circle. See Figure 8. In Figure 9 the start of B_k is glided along the edges C_m D_m and E_m. (B_k is thickened on each of the pictures to help visualize this process.) After these glidings w_k will not include E_m, D_m and C_m. We can repeat this process until only B_k, A_k and B_{k-1} are left in the tracing of w_k.

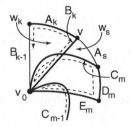

Figure 8.

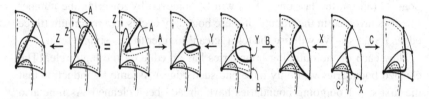

Figure 9. Skipping three edges

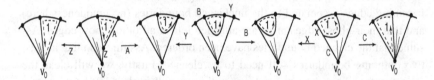

Figure 10. Moving boundary 1 to the left

To make sure we do not interfere with this boundary cycle when making
subsequent rearrangements, we will move it to the left of all the remaining
uncleaned boundaries. Follow Figure 10 to see how to switch this boundary
(labelled 1) with the one directly on the left of it. By induction we can clean
all of the outgoing boundaries but the last one.

We are now very close to our goal. We have the incoming and the outgoing
boundaries in the right order and in the right form. The only issue rests with
the last outgoing boundary component. The one that includes all the 'genus
creating' edges. To finally reach Γ_0, we need to untangle these into twisted
pairs. This is done by induction on the number of pairs of such edges left to
untangle.

Choose M to be first genus creating edge (in the ordering at v_o) that has not
yet been paired in our inductive process. Our first step is to find an edge that
'crosses it', meaning that it starts on the right of M at v_o and ends up on the left
of M on the big circle. Take P to be the next edge at v_o. If P ends up on the left
of M, we have our edge and we are ready to apply the second step. If this is not

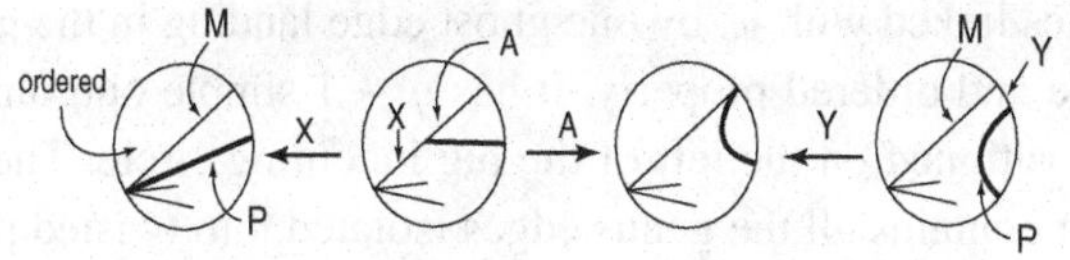

Figure 11. Moving P away

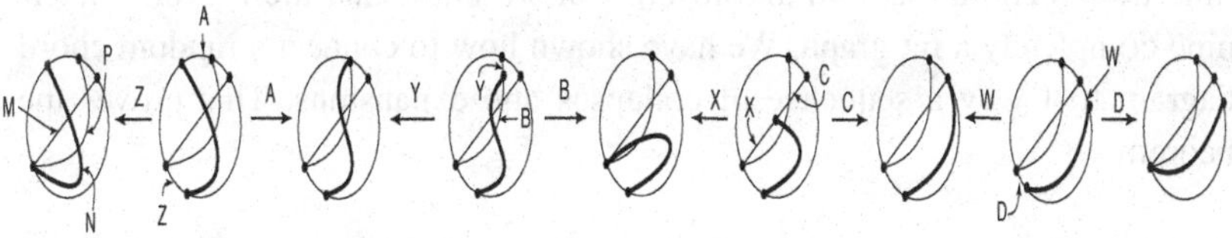

Figure 12. Separating the pair M-P from N

the case, we'll move P along M as shown in Figure 11 and consider the next edge. Since both orientation of M are part of the last boundary component, the tracing of this component moves from the edges on the right of M to the edges on the left of M. This implies that there is at least one edge that starts on the right of M at v_o and ends on the left of M on the big circle. So by moving through the edges P we will find one of these crossing edges and this step will be completed.

Now we have our pair of edges M and P. But we would like to have M and P completely separate from the other edges as in Γ_0. Any edge N that is intertwined with M and P needs to be moved. Since everything on the left of M and P at v_o is already in proper order, this N will always start on the right of M and P at v_o (and end up on the big circle). But there are still two ways that N might be intertwined with our pair: it could either end up between M and P or on the left of these edges. Figure 12 shows how to glide the edge N along first M and then P. A similar operation would get rid of the edges landing between M and P.

Before restarting these steps for isolating the next pair, we need to bring a lot of edges back to v_o. For example in Figure 11 P ends up completely disconnected from v_o and from the rest of the ghost structure. To achieve this we will first collapse all nonessential edges and then we will reduce the number of edges on the non-v_o vertices down to three as done in Figure 4. Note that M and P can be kept isolated while all of this is done.

After this process, all the genus edges are paired and twisted properly. We can finally put the ghost edges connecting the small circles to v_o in their correct position. Our chord diagram has now one big incoming circle and a special vertex v_o, the only non-trivalent vertex. It has $p - 1$ small 'one-vertexed' incoming circles linked with v_o by one ghost edge landing in the last outgoing boundary cycle and ordered properly. It has $q - 1$ simple outgoing boundary components positioned on the left of the big incoming circle. The last outgoing component contains all the genus edges isolated into twisted pairs and the ghost edges linking the big incoming circles with the small ones. This means that the cycles associated to the different boundary components are exactly the same in this chord diagram and in Γ_0. But we know that these cycles determine completely a fat graph. We have shown how to connect a random chord diagram to Γ_0 by a sequence of collapses and expansions. This proves the theorem.

$\square$

3 The Thom collapse map and string topology operations

In this section we use fat graphs to define the string operations, and will prove Theorem 1.1 stated in the introduction.

Let LM denote the space of piecewise smooth maps, $\gamma : S^1 \to M$. Let Σ be an oriented surface of genus g with $p + q$ boundary components.

For $c \in \mathcal{CF}^\mu_{p,q}(g)$, let $Map_*(c, M)$ be the space of continuous maps $f : c \to M$, smooth on each edge, which is constant on each ghost edge. Equivalently, this is the full space of maps $S(c) \to M$ where $S(c)$ is the marked metric fat graph described in the previous section (obtained from c by collapsing each ghost component to a point). Since each ghost component is a tree and therefore contractible, this mapping space is homotopy equivalent to the space of all continuous maps, $c \to M$, which in turn is homotopy equivalent to the smooth mapping space, $Map(\Sigma, M)$. Furthermore, the markings on $S(c)$ induce parameterizations of the incoming and outgoing boundary cycles of c, so restriction to these boundary cycles induces a diagram

$$(LM)^q \xleftarrow{\ \rho_{out}\ } Map_*(c, M) \xrightarrow{\ \rho_{in}\ } (LM)^p. \qquad (3.1)$$

Since $Map_*(c, M)$ is the same as the space of all continuous maps $Map(S(c), M)$, it is clear that the restriction to the incoming boundary components

$$\rho_{in} : Map_*(c, M) \to (LM)^p$$

is an *embedding* of infinite-dimensional manifolds, but it has finite codimension. We now consider its normal bundle.

Let $v(c)$ be the collection of circular vertices of a chord diagram c. Let $\sigma(c)$ be the collection of vertices of the associated graph $S(c)$. The projection map $\pi : c \to S(c)$ determines a surjective set map, $\pi_* : v(c) \to \sigma(c)$. For a vertex $v \in \sigma(c)$, we define the *multiplicity*, $\mu(v)$, to be the cardinality of the preimage, $\#\pi^{-1}(v)$. Let $M^{\sigma(c)}$ and $M^{v(c)}$ be the induced mapping spaces from these vertex collections. Then π induces a diagonal map

$$\Delta_c : M^{\sigma(c)} \to M^{v(c)}.$$

The normal bundle of this diagonal embedding is the product bundle

$$\nu(\Delta_c) = \prod_{v \in \sigma(c)} (\mu(v) - 1)TM \to \prod_{v \in \sigma(c)} M = M^{\sigma(c)}.$$

Here $k \cdot TM$ is the k-fold Whitney sum of the tangent bundle. Since $\sum_{v \in \sigma(c)} \mu(v) = v(c)$, the fiber dimension of this bundle is $(v(c) - \sigma(c))n$. An easy exercise verifies that $(v(c) - \sigma(c)) = -\chi(\Sigma_c)$, minus the Euler characteristic of a surface represented by c.

Now remember that the markings of the incoming boundary components of $S(c)$ define parameterizations of the incoming boundary components of c, since these cycles only consist of circular edges. Now using these parameterizations we can identify $(LM)^p$ with $Map(c_1 \sqcup \cdots \sqcup c_p, M)$, where $c_1, \cdots, c_p$ are the p incoming boundary cycles of c. Consider the evaluation map $e_c : (LM)^p \to M^{v(c)}$ defined on an element of $\gamma \in (LM)^p$ by evaluating γ on the circular vertices. Similarly, define

$$e_c : Map_*(c, M) \to M^{\sigma(c)}$$

by evaluating a map $f : S(c) \to M$ on the vertices. These evaluation maps are fibrations, and notice that the following is a pull-back square

$$
\begin{array}{ccc}
Map_*(c, M) & \xrightarrow{\ \rho_{in}\ } & (LM)^p \\
\ \ \downarrow{\scriptstyle e_c} & & \ \ \downarrow{\scriptstyle e_c} \\
M^{\sigma(c)} & \xrightarrow[\ \Delta_c\ ]{} & M^{v(c)}
\end{array}
\qquad (3.2)
$$

By taking the inverse image of a tubular neighborhood of the embedding Δ_c, one has the following consequence.

Lemma 3.1. $\rho_{in} : Map_*(c, M) \hookrightarrow (LM)^p$ *is a codimension* $-\chi(\Sigma_c)n$ *embedding, and has an open neighborhood $\nu(c)$ diffeomorphic to the total space of the pullback bundle, $e_c^*(\nu(\Delta_c)) = e_c^*(\prod_{v \in \sigma(c)}(\mu(v) - 1)TM)$. The fiber of this bundle over a map $f : c \to M$ is therefore given by*

$$\nu(c)_f = \bigoplus_{v \in \sigma(c)} \bigoplus_{(\mu(v)-1)} T_{f(v)}M.$$

where $\bigoplus_{(\mu(v)-1)} T_{f(v)} M$ *refers to taking the direct sum of* $\mu(v) - 1$ *copies of* $T_{f(v)} M$.

Let $Map_*(c, M)^{\nu(c)}$ be the Thom space of this normal bundle. This result allows us to define a Thom collapse map $\tau : (LM)^p \to Map_*(c, M)^{\nu(c)}$ defined, as usual, to be the identity inside the tubular neighborhood, and the basepoint outside the tubular neighborhood.

Now let h^* be a generalized cohomology theory as before. By the above description of the bundle $\nu(c)$ we see that since M is h^*-oriented, the bundle $\nu(c)$ is h^*-oriented. This defines a Thom isomorphism

$$t : h_*(Map_*(c, M)^{\nu(c)}) \cong h_{*+\chi(\Sigma_c)n}(Map_*(c, M)).$$

Now, since we are assuming that the coefficient ring $h_* = h_*(point)$ is a graded field, the Kunneth spectral sequence collapses, and hence

$$h_*(X \times Y) \cong h_*(X) \otimes_{h_*} h_*(Y).$$

From now on we take all tensor products to be over the ground field h_*. We can therefore make the following definitions:

Definition 3.2. Fix $c \in \mathcal{CF}^\mu_{p,q}(g)$.

(a) Define the push-forward map $(\rho_{in})_! : h_*(LM)^{\otimes p} \to h_{*+\chi(\Sigma_c)n}(Map_*(c, M))$ to be the composition

$$(\rho_{in})_! : h_*(LM)^{\otimes p} \cong h_*((LM)^p) \xrightarrow{\tau_*} h_*(Map_*(c, M)^{\nu(\Delta_c)})$$

$$\xrightarrow[\cong]{t} h_{*+\chi(\Sigma_c)n}(Map_*(c, M)).$$

(b) Define the operation $\mu_c : h_*(LM)^{\otimes p} \to h_*(LM)^{\otimes q}$ to be the composition

$$\mu_c : h_*(LM)^{\otimes p} \xrightarrow{(\rho_{in})_!} h_{*+\chi(\Sigma_c)n}(Map_*(c, M))$$

$$\xrightarrow{(\rho_{out})_*} h_{*+\chi(\Sigma_c)n}((LM)^q).$$

In order to use these operations to prove Theorem 1.1 we will first need to verify the following.

Theorem 3.3. *The operations* $\mu_c : h_*(LM)^{\otimes p} \to h_*(LM)^{\otimes q}$ *do not depend on the choice of marked metric chord diagram* $c \in \mathcal{CF}^\mu_{p,q}(g)$. *In other words, they only depend on the topological type* $(g; p,q)$ *of the chord diagram.*

Proof. We show that if $\gamma : [0, 1] \to \mathcal{CF}^{\mu}_{p,q}(g)$ is a path of chord diagrams, then $\mu_{\gamma(0)} = \mu_{\gamma(1)}$. By the connectivity of $\mathcal{CF}^{\mu}_{p,q}(g)$ (Theorem 2.2), this will prove the theorem. To do this we parameterize the construction of the operation. Namely, let

$$Map_*(\gamma, M) = \{(t, f) : t \in [0, 1], \; f \in Map_*(\gamma(t), M)\}.$$

Then there are restriction maps to the incoming and outgoing boundaries, $\rho_{in} : Map_*(\gamma, M) \to (LM)^p$, and $\rho_{out} : Map_*(\gamma, M) \to (LM)^q$. Let $p : Map_*(\gamma, M) \to [0, 1]$ be the projection map.

Then Lemma 3.1 implies the following.

Lemma 3.4. *The product $\rho_{in} \times p : Map_*(\gamma, M) \hookrightarrow (LM)^p \times [0, 1]$ is a codimension $-\chi(\Sigma_c)n$ embedding, and has an open neighborhood $\nu(\gamma)$ diffeomorphic to the total space of the vector bundle whose fiber over $(t, f) \in Map_*(\gamma, M)$ is given by*

$$\nu(\gamma)_{(t,f)} = \bigoplus_{v \in \sigma(\gamma(t))} \bigoplus_{(\mu(v)-1)} T_{f(v)}M.$$

This allows us to define a Thom collapse map, $\tau : (LM)^p \times [0, 1] \to (Map_*(\gamma, M))^{\nu(\gamma)}$ which defines a homotopy between the collapse maps $\tau_0 : (LM)^p \to Map_*(\gamma(0), M)^{\nu(\gamma(0))} \hookrightarrow Map_*(\gamma, M)^{\nu(\gamma)}$ and $\tau_1 : (LM)^p \to Map_*(\gamma(1), M)^{\nu(\gamma(1))} \hookrightarrow Map_*(\gamma, M)^{\nu(\gamma)}$.

One can then define the push-forward map

$$(\rho_{in})_! : h_*((LM)^p \times [0, 1]) \xrightarrow{\;\tau_*\;} h_*((Map_*(\gamma, M))^{\nu(\gamma)})$$

$$\xrightarrow[\cong]{\;t\;} h_{*+\chi \cdot n}((Map_*(\gamma, M)))$$

and then an operation

$$\mu_\gamma = (\rho_{out})_* \circ (\rho_{in})_! : h_*((LM)^p \times [0, 1]) \to h_{*+\chi \cdot n}((Map_*(\gamma, M)))$$

$$\to h_{*+\chi \cdot n}((LM)^q).$$

The restriction of this operation to $h_*((LM)^p \times \{0\}) \hookrightarrow h_*((LM)^p \times [0, 1])$ is, by definition, $\mu_{\gamma(0)}$, and the restriction to $h_*((LM)^p \times \{1\}) \hookrightarrow h_*((LM)^p \times [0, 1])$ is $\mu_{\gamma(1)}$. This proves that these two operations are equal. $\square$

Now that we have Theorem 3.3 we can introduce the notation $\mu_{p,q}(g)$ to stand for $\mu_c : h_*(LM)^{\otimes p} \to h_*(LM)^{\otimes q}$ for any Sullivan chord diagram $c \in \mathcal{CF}^{\mu}_{p,q}(g)$. $\mu_{p,q}(g)$ is an operation that lowers the total degree by $(2g - 2 + p + q)n$.

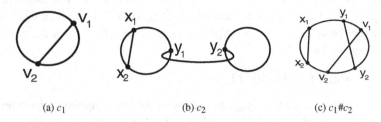

(a) c_1 (b) c_2 (c) $c_1 \# c_2$

Figure 13. Gluing c_1 and c_2

Remark. The above argument is easily modified to show that any element $\alpha \in h_*(\mathcal{CF}_{p,q}^\mu(g))$ defines a string topology operation $\mu_{p,q}(g)(\alpha)$. The operations we are dealing with correspond to the class $1 \in h_0(\mathcal{CF}_{p,q}^\mu(g))$.

In order to complete the proof of Theorem 1.1, by the correspondence between two-dimensional TQFT's and Frobenius algebras [10][1], it suffices to show that these operations respect the gluing of surfaces.

Theorem 3.5. $\mu_{q,r}(g_2) \circ \mu_{p,q}(g_1) = \mu_{p,r}(g_1 + g_2 + q - 1) : h_*(LM)^{\otimes p} \to h_*(LM)^{\otimes q} \to h_*(LM)^{\otimes r}$.

Proof. Let $c_1 \in \mathcal{CF}_{p,q}^\mu(g_1)$ and $c_2 \in \mathcal{CF}_{q,r}^\mu(g_2)$. Notice that we can glue c_1 to c_2 to obtain a Sullivan chord diagram in $c_1 \# c_2 \in \mathcal{CF}_{p,r}^\mu(g_1 + g_2 + q - 1)$ in the following way.

Identify the outgoing boundary circles of c_1 with the incoming boundary circles of c_2 using the parameterizations, and input the vertices and ghost edges of c_2 into the diagram c_1 using these identifications. Figure 13 gives an example of this gluing procedure with $c_1 \in \mathcal{CF}_{1,2}^\mu(0)$, $c_2 \in \mathcal{CF}_{2,2}^\mu(0)$, and $c_1 \# c_2 \in \mathcal{CF}_{1,2}^\mu(1)$. For clarity the vertices have been labeled in these diagrams, both before and after gluing.

Note. We are not claiming that this gluing procedure is continuous, or even well-defined. The ambiguity in definition occurs if, when one identifies the outgoing boundary circle of c_1 with an incoming boundary circle of c_2, a circular vertex x of c_2 coincides with a circular vertex v of c_1 that lies on a ghost edge in the boundary cycle. Then there is an ambiguity over whether to place x at v or at the other vertex of the ghost edge. However for our purposes, we can make any such choice, since the operations that two such glued surfaces define are equal, by Theorem 3.3.

Notice that the parameterizations give us maps of the collapsed fat graphs

$$\phi_1 : S(c_1) \to S(c_1 \# c_2) \quad \text{and} \quad \phi_2 : S(c_2) \to S(c_1 \# c_2).$$

These induce a diagram of mapping spaces

$$Map_*(c_2, M) \xleftarrow{\ \bar{\phi}_2\ } Map_*(c_1 \# c_2, M) \xrightarrow{\ \bar{\phi}_1\ } Map_*(c_1, M)$$

The next two lemmas follow from a verification of the definitions of the mapping spaces and the maps ϕ_i.

Lemma 3.6. $\bar{\phi}_1 : Map_*(c_1 \# c_2, M) \to Map_*(c_1, M)$ *is an embedding, whose image has a neighborhood diffeomorphic to the total space of the bundle $\bar{\phi}_2^*(\nu(c_2))$, where $\nu(c_2) \to Map_*(c_2, M)$ is the normal bundle of $\rho_{in} : Map_*(c_2, M) \hookrightarrow (LM)^p$ described in Lemma 3.1.*

This allows the definition of a Thom collapse map $\tau_{\phi_1} : Map_*(c_1, M) \to Map_*(c_1 \# c_2, M)^{\bar{\phi}_2^*(\nu(c_2))}$ and therefore a push-forward map in homology, $(\bar{\phi}_1)_! : h_*(Map_*(c_1, M)) \to h_{*+\chi(c_2)n}(Map_*(c_1 \# c_2, M))$.

Lemma 3.7. *The following diagram commutes*

$$
\begin{array}{ccccc}
(LM)^r & \xrightarrow{\quad = \quad} & (LM)^r & & \\
{\scriptstyle \rho_{out}(2)}\big\uparrow & & \big\uparrow{\scriptstyle \rho_{out}(1\#2)} & & \\
Map_*(c_2, M) & \xleftarrow{\ \bar{\phi}_2\ } & Map_*(c_1 \# c_2, M) & \xrightarrow{\ \rho_{in}(1\#2)\ } & (LM)^p \\
{\scriptstyle \rho_{in}(2)}\big\downarrow & & \big\downarrow{\scriptstyle \bar{\phi}_1} & & \big\downarrow{\scriptstyle =} \\
(LM)^q & \xleftarrow[\ \rho_{out}(1)\ ]{} & Map_*(c_1, M) & \xrightarrow[\ \rho_{in}(1)\ ]{} & (LM)^p.
\end{array}
$$

The indexing of the restriction maps corresponds to the indexing of the chord diagrams in the obvious way.

By the naturality of the Thom collapse map, and therefore the homological pushout construction, we therefore have the following corollary.

Corollary 3.8. 1. $(\rho_{in}(1\#2))_! = (\bar{\phi}_1)_! \circ (\rho_{in}(1))_! : h_*((LM)^p) \to h_{*+\chi(c_1 \# c_2) \cdot n}(Map_*(c_1 \# c_2, M))$
2. $(\bar{\phi}_2)_* \circ (\bar{\phi}_1)_! = (\rho_{in}(2))_! \circ (\rho_{out}(1))_* : h_*(Map(c_1, M)) \to h_{*+\chi(c_2) \cdot n}(Map_*(c_2, M))$
3. $(\rho_{out}(1\#2))_* = (\rho_{out}(2))_* \circ (\bar{\phi}_2)_* : h_*(Map_*(c_1 \# c_2, M)) \to h_*((LM)^r)$.

We may now complete the proof of Theorem 3.5. We have

$$
\begin{aligned}
\mu_{p,r}(g_1 + g_2 + q - 1) = \mu_{c_1 \# c_2} &= (\rho_{out}(1\#2))_* \circ (\rho_{in}(1\#2))_! \\
&= (\rho_{out}(1\#2))_* \circ (\bar{\phi}_1)_! \circ (\rho_{in}(1))_! \\
&= (\rho_{out}(2))_* \circ (\bar{\phi}_2)_* \circ (\bar{\phi}_1)_! \circ (\rho_{in}(1))_! \\
&= (\rho_{out}(2))_* \circ (\rho_{in}(2))_! \circ (\rho_{out}(1))_* \circ (\rho_{in}(1))_! \\
&= \mu_{c_2} \circ \mu_{c_1} \\
&= \mu_{q,r}(g_2) \circ \mu_{p,q}(g_1). \qquad \square
\end{aligned}
$$

As observed in [1], a Frobenius algebra without counit is the same thing as a positive boundary topological quantum field theory. We have now verified that the string topology operations define such a theory for any generalized cohomology theory h^* satisfying the conditions described above. Recall that it was observed in [3] [5], that the unit in the algebra structure of $h_*(M)$ is the fundamental class, $[M] \in h_n(M) \hookrightarrow h_n(LM)$, where the second map is induced by the inclusion of the manifold in the loop space as the constant loops, $\iota : M \hookrightarrow LM$. Thus $h_*(LM)$ is a unital Frobenius algebra without a counit. This proves Theorem 1.1.

4 Capping off boundary components: issues surrounding the unit and counit

The unit in the Frobenius algebra stucture can be constructed in the same way as the other string topology operations as follows.

Consider the disk D^2 as a surface with zero incoming boundary component and one outgoing boundary component. A graph c_D that represents D^2 can be taken to be a point (i.e a single vertex). Formally, the restriction to the zero incoming boundary components is the map

$$
\rho_{in} : M = Map_*(c_D, M) \to Map(\emptyset, M) = point.
$$

The push-forward map in this setting

$$
(\rho_{in})_! : h_*(point) \to h_{*+n}(M)
$$

is the h_*-module map defined by sending the generator to $[M] \in h_n(M)$. The restriction to the outgoing boundary component is the map

$$
\rho_{out} : M = Map_*(c_D, M) \to LM
$$

which is given by $\iota : M \hookrightarrow LM$. Thus the unit is given by the h_* module homomorphism

$$\mu_{D^2} = (\rho_{out})_* \circ (\rho_{in})_! = \iota_* \circ (\rho_{in})_! : h_* \to h_n(M) \to h_n(LM)$$

which sends the generator to the fundamental class.

The issue of the existence (or nonexistence) of a *counit* in the Frobenius algebra structure given by Theorem 1.1 is formally the same (or dual) to the existence of a unit, but is geometrically much more difficult and subtle. Namely, for this operation one must consider D^2 as a surface with one incoming boundary, and zero outgoing boundary components. In this setting the roles of the restriction maps ρ_{in} and ρ_{out} are reversed, and one obtains the diagram

$$Map(\emptyset, M) \xleftarrow{\rho_{out}} Map_*(c_D, M) \xrightarrow{\rho_{in}} LM$$

or, equivalently

$$point \xleftarrow{\epsilon} M \xrightarrow{\iota} LM.$$

where $\epsilon : M \to point$ is the constant map.

Now notice that in this case, unlike when any of the other fat graphs were considered, the embedding of $Map_*(c_D, M) \hookrightarrow LM$ (i.e $\iota : M \hookrightarrow LM$) is of *infinite* codimension. Therefore to find a theory h_* that supports a counit in the Frobenius algebra structure of $h_*(LM)$, one needs to be able to define a push-forward map for this infinite codimensional embedding. Now in their work on genera of loop spaces, [2], Ando and Morava argued that if one has a theory where this push-forward map exists, one would need that the Euler class of the normal bundle $e(\nu(\iota)) \in h^*(M)$ is invertible. So let us now consider this normal bundle.

The embedding of M as the constant loops in LM is S^1-equivariant where S^1 acts triviallly on M. When M is a simply connected almost complex manifold, the normal bundle has the following description (see [2], for example).

Lemma 4.1. *The normal bundle $\nu(\iota) \to M$ of the embedding $\iota : M \hookrightarrow LM$ is equivariantly isomorphic to the direct sum*

$$\nu(\iota) \cong \bigoplus_{k \neq 0} TM \otimes_{\mathbb{C}} \mathbb{C}(k)$$

where $\mathbb{C}(k)$ is the one-dimensional representation of S^1 of weight k.

This says that the Euler class of the normal bundle will have the formal description

$$e(\nu(\iota)) = \prod_{k \neq 0} e(TM \otimes \mathbb{C}(k)). \tag{4.1}$$

Thus a theory $h_*(LM)$ that supports a counit in a Frobenius algebra structure should have the following properties:

(1) h_* should be an S^1-equivariant theory in order to take advantage of the different equivariant structures of the summands $TM \otimes \mathbb{C}(k)$.

(2) h_* should be a 'pro-object' – an inverse system of homology groups, so that it can accommodate this infinite product.

5 The polarized loop space and its Atiyah dual

Motivated by these homological requirements, in this section we show that the loop space of an almost complex manifold has a natural equivariant pro-object (a 'prospectrum') associated to it. The ideas for the constructions in this section stem from conversations with Graeme Segal. Throughout this section we assume that M is a simply connected, oriented, closed n-manifold.

Let $-TM$ be the virtual bundle (K-theory class) given by the opposite of the tangent bundle. Let M^{-TM} be its Thom spectrum. We refer to M^{-TM} as the 'Atiyah dual' of M_+ because of Atiyah's well-known theorem stating that M^{-TM} is equivalent to the Spanier Whitehead dual of M_+. (Here M_+ is M together with a disjoint basepoint.) This gives M^{-TM} the structure of a ring spectrum, whose multiplication $m : M^{-TM} \wedge M^{-TM} \to M^{-TM}$ is dual to the diagonal $\Delta : M \to M \times M$. When one applies homology and the Thom isomorphism, this multiplication realizes the intersection product ($\cap$), meaning that the following diagram commutes

$$
\begin{array}{ccc}
H_*(M^{-TM}) \otimes H_*(M^{-TM}) & \xrightarrow{\ m_*\ } & H_*(M^{-TM}) \\
t \downarrow \cong & & \cong \downarrow t \\
H_{*+n}(M) \otimes H_{*+n}(M) & \xrightarrow[\cap]{} & H_{*+n}(M).
\end{array}
$$

Here $H_*(M^{-TM})$ refers to the spectrum homology of M^{-TM}.

It is therefore natural to expect that an appropriate pro-object that carries the string topology operations, including a counit (i.e a 2-dimensional TQFT, or Frobenius algebra structure), would be a prospectrum model for the Atiyah dual of the loop space, LM^{-TLM}.

In studying homotopy theoretic aspects of symplectic Floer homology, Jones and Segal used pro-spectra associated to certain infinite-dimensional bundles [6]. The construction was the following. If $E \to X$ is an infinite-dimensional bundle with a filtration by finite-dimensional subbundles

$$
\cdots \hookrightarrow E_i \hookrightarrow E_{i+1} \hookrightarrow \cdots E
$$

such that $\bigcup_i E_i$ is a dense subbundle of E, then one can define the prospectrum X^{-E} to be the inverse system

$$\cdots \leftarrow X^{-E_{i-1}} \xleftarrow{\;u_i\;} X^{-E_i} \xleftarrow{\;u_{i+1}\;} X^{-E_{i+1}} \leftarrow \cdots$$

where $u_j : X^{-E_j} \to X^{-E_{j-1}}$ is the map defined as follows. Let $e_j : E_{j-1} \hookrightarrow E_j$ be the inclusion. Assume for simplicity that E_j is embedded in a large-dimensional trivial bundle, and let $E_j^{\perp}$ and $E_{j-1}^{\perp}$ be the corresponding orthogonal complements. One then has an induced inclusion of complements, $e_j^{\perp} : E_j^{\perp} \to E_{j-1}^{\perp}$. The induced map of Thom spaces then defines a map of Thom spectra, $u_j : X^{-E_j} \to X^{-E_{j-1}}$. A standard homotopy theoretic technique allows one to define this map of Thom spectra even if E_j is not embeddable in a trivial bundle, by restricting E_j to finite subcomplexes of X where it is.

Under the assumption that M is an almost complex manifold of dimension $n = 2m$, then we are dealing with an infinite-dimensional vector bundle (TLM) whose structure group is the loop group $LU(m)$. In [8], Cohen and Stacey studied obstructions to finding an appropriate filtration of an infinite-dimensional $LU(m)$ bundle. In particular for $TLM \to LM$ it was proved that if such a filtration (called a 'Fourier decomposition' in [8]) exists, then the holonomy of any unitary connection on TM, $h : \Omega M \to U(m)$ is null homotopic. This 'homotopy flat' condition is far too restrictive for our purposes, but we can get around this problem by taking into account the canonical *polarization* of the tangent bundle TLM of an almost complex manifold.

Recall that a *polarization* of a Hilbert space E is an equivalence class of decomposition, $E = E_+ \oplus E_-$, where two such decompositions $E_+ \oplus E_- = E_+' \oplus E_-'$ are equivalent if the composition $E_+ \hookrightarrow E \to E_+'$ is Fredholm, and $E_+ \hookrightarrow E \to E_-'$ is compact (see [16] for details). The restricted general linear group of a polarized space $GL_{res}(E)$ consists of all elements of $GL(E)$ that preserve the polarization.

A polarized vector bundle $\zeta \to X$ is one where every fiber is polarized, and the structure group reduces to the restricted general linear group. If M^{2m} is an almost complex manifold, and $\gamma \in LM$, then the tangent space $T_\gamma LM$ is the space of L^2 vector fields of M along γ, and the operator

$$j\frac{d}{d\theta} : T_\gamma LM \to T_\gamma LM$$

is a self-adjoint Fredholm operator. Here $\frac{d}{d\theta}$ is the covariant derivative, and j is the almost complex structure. The spectral decomposition of $j\frac{d}{d\theta}$ polarizes the bundle TLM according to its positive and negative eigenspaces. The structure group in this case is $GL_{res}(L^2(S^1, \mathbb{C}^m))$, where the loop space $L^2(S^1, \mathbb{C}^m)$ is polarized according to the Fourier decompostion. That is, we write

$$L^2(S^1, \mathbb{C}^n) = H_+ \oplus H_-$$

where $H_+ = Hol(D^2, \mathbb{C}^n)$ is the space of holomorphic maps of the disk, and H_- is the orthogonal complement.

For a polarized space E, recall from [16] that the *restricted Grassmannian* $Gr_{res}(E)$ consists of closed subspaces $W \subset E$ such that the projections $W \hookrightarrow E \rightarrow E_+$ is Fredholm, and $W \hookrightarrow E \rightarrow E_-$ is Hilbert–Schmidt. In the case under consideration, the tangent space $T_\gamma LM$, is a $L\mathbb{C}$- module, and therefore a module over the Laurent polynomial ring, $\mathbb{C}[z, z^{-1}]$. Define $Gr_{res}^0(T_\gamma LM) \subset Gr_{res}(T_\gamma LM)$ to be the subspace

$$Gr_{res}^0(T_\gamma LM) = \{W \in Gr_{res}(T_\gamma LM) : zW \subset W\}.$$

For M^{2m} a simply connected almost complex manifold, we can then define the *polarized loop space* $LM_\pm$ to be the space

$$LM_\pm = \{(\gamma, W) : \gamma \in LM, \ W \in Gr_{res}^0(T_\gamma LM)\}. \tag{5.1}$$

We now consider the S^1-equivariance properties of $LM_\pm$. The following theorem will be an easy consequence of the results of [16], chapter 8.

Theorem 5.1. *The natural projection* $p : LM_\pm \rightarrow L(M^{2m})$ *is an S^1-equivariant fiber bundle with fiber diffeomorphic to the based loop space, $\Omega U(m)$. The S^1-fixed points of $LM_\pm$ form a bundle over M with fiber the space of group homomorphisms, $Hom(S^1, U(m))$.*

Proof. Let $\gamma \in LM$. The tangent space, $T_\gamma LM = \Gamma_{S^1}(\gamma^*(TM))$, is the space of L^2 sections of the pullback of the tangent bundle over the circle. The S^1-action on LM differentiates to make the tangent bundle TLM an S^1-equivariant bundle. If $\sigma \in T_\gamma LM$, and $t \in S^1$, then $t\sigma \in T_{t\gamma}LM$ is defined by $t\sigma(s) = \sigma(t + s)$. Since this action preserves the polarization, it induces an action

$$S^1 \times LM_\pm \rightarrow LM_\pm \tag{5.2}$$

$$t \times (\gamma, W) \rightarrow (t\gamma, tW) \tag{5.3}$$

where $tW = \{t\sigma \in T_{t\gamma}LM : \sigma \in W\}$. The fact that the projection map $p : LM_\pm \rightarrow LM$ is an S^1-equivariant bundle is clear. The fiber of this bundle can be identified with $Gr_{res}^0(L^2(S^1, \mathbb{C}^m))$, which was proved in [16] to be diffeomorphic to $\Omega U(m)$. The induced action on $\Omega U(m)$ was seen in [16] (chapter 8) to be given as follows. For $t \in S^1 = \mathbb{R}/\mathbb{Z}$, and $\omega \in \Omega U(m)$, $t \cdot \omega(s) = \omega(s+t)\omega(t)^{-1}$. The fixed points of this action are the group homomorphisms, $Hom(S^1, U(m))$. The theorem now follows. $\square$

Remark. Since the group homomorphisms, $Hom(S^1, U(m))$ are well understood, one can view the above theorem as saying that the equivariant homotopy type of $LM_{\pm}$ is directly computable in terms of the equivariant homotopy type of LM.

By this theorem, the pullback of the tangent bundle, $p^*TLM \to LM_{\pm}$ is an S^1-equivariant bundle. Our final result implies that even though one cannot generally find a prospectrum modeling the Atiyah dual LM^{-TLM}, one can find a pro-spectrum model of the 'polarized Atiyah dual', $LM_{\pm}^{-TLM}$.

The following theorem says that one can build up the bundle $p^*(TLM) \to LM_{\pm}$ by finite-dimensional subbundles.

Theorem 5.2. *There is a doubly graded collection of finite dimensional, S^1-equivariant subbundles of $p^*(TLM) \to LM_{\pm}$*

$$E_{i,j} \to LM_{\pm}, \quad i < j$$

satisfying the following properties:

(1) *There are inclusions of subbundles*

$$E_{i,j} \hookrightarrow E_{i-1,j} \quad \text{and} \quad E_{i,j} \hookrightarrow E_{i,j+1}$$

   such that $\bigcup_{i,j} E_{i,j}$ is a dense subbundle of $p^*(TLM)$.
(2) *The subquotients*

$$E_{i-1,j}/E_{i,j} \quad \text{and} \quad E_{i,j+1}/E_{i,j}$$

   are m-dimensional S^1-equivariant complex vector bundles that are nonequivariantly isomorphic to the pullback of the tangent bundle $\tilde{p}^*TM$, where $\tilde{p} : LM_{\pm} \to M$ is the composition of $p : LM_{\pm} \to LM$ with the map $e_1 : LM \to M$ that evaluates a loop at the basepoint $1 \in S^1$.

Remark. Such a filtration is a 'Fourier decomposition' of the loop bundle $p^*(TLM) \to LM_{\pm}$ as defined in [8]

Proof. We first define certain infinite-dimensional subbundles $E_i \subset p^*(TLM) \to LM_{\pm}$. Define the fiber over $(\gamma, W) \in LM_{\pm}$ to be

$$(E_i)_{(\gamma, W)} = z^{-i} W \subset T_{\gamma}(LM).$$

We note that E_i is an equivariant subbundle, with the property that $zE_i \subset E_i$. Furthermore there is a filtration of subbundles

$$\cdots \hookrightarrow E_i \hookrightarrow E_{i+1} \hookrightarrow \cdots p^*(TLM)$$

with $\bigcup_i E_i$ a dense subbundle of $p^*(TLM)$. Notice that for $j > i$, the subquotient E_j/E_i has fiber at (γ, W) given by $z^{-j}W \cap (z^{-i}W)^\perp$ where $(z^{-i}W)^\perp \subset T_\gamma LM$ is the orthogonal complement of $z^{-i}W$. For $j - i = 1$, an easy argument (done in [8]) gives that the composition

$$z^{-j}W \cap (z^{-(j-1)}W)^\perp \hookrightarrow T_\gamma LM \xrightarrow{e_1} T_{\gamma(1)}(M) \qquad (5.4)$$

is an isomorphism. For $i < j$ we define the bundle $E_{i,j} \to LM_\pm$ to be the quotient E_j/E_i. It is the vector bundle whose fiber over (γ, W) is $z^{-j}W \cap (z^{-i}W)^\perp$. By (5.4), the subquotient of the bundle $E_{j-1,j}$ is (nonequivariantly) isomorphic to the pullback of the tangent bundle $TM \to M$ under the composition $LM_\pm \xrightarrow{p} LM \xrightarrow{e_1} M$. In general the bundle $E_{i,j}$ is nonequivariantly isomorphic to the Whitney sum of $j - i$ copies of $\tilde{p}^*(TM)$.

Now since $z^{-j}W \cap (z^{-i}W)^\perp$ is a subspace of both $z^{-(j+1)}W \cap (z^{-i}W)^\perp$ and of $z^{-j}W \cap (z^{-(i-1)}W)^\perp$, we have inclusions $E_{i,j} \hookrightarrow E_{i,j+1}$ and $E_{i,j} \hookrightarrow E_{i-1,j}$. Clearly $\bigcup_{i,j} E_{i,j}$ is a dense subbundle of $p^*(TLM)$. The theorem follows. $\qquad\qquad\qquad\qquad\qquad\qquad\qquad\qquad\qquad\qquad\qquad\qquad\qquad\qquad\square$

Since the bundles $E_{i,j} \to LM_\pm$ are finite-dimensional S^1-equivariant bundles, we can construct the Thom spectrum of the S^1-equivariant virtual bundle, $-E_{i,j}$, which we denote by $(LM_\pm)^{-E_{i,j}}$. Notice the inclusions of bundles $E_{i,j} \hookrightarrow E_{i,j+1}$ and $E_{i,j} \hookrightarrow E_{i-1,j}$ induce maps of virtual bundles, $\tau_{i,j} : -E_{i,j+1} \to -E_{i,j}$ and $\sigma_{i,j} : -E_{i-1,j} \to -E_{i,j}$, which yields an inverse system of S^1-equivariant spectra

$$
\begin{array}{ccc}
\vdots & & \vdots \\
\Big\downarrow{\scriptstyle\sigma_{i-1,j}} & & \Big\downarrow{\scriptstyle\sigma_{i-1,j+1}} \\
(LM_\pm)^{-E_{i-1,j}} \xleftarrow{\ \tau_{i-1,j}\ } & (LM_\pm)^{-E_{i-1,j+1}} \xleftarrow{\ \tau_{i-1,j+1}\ } \cdots \\
\Big\downarrow{\scriptstyle\sigma_{i,j}} & & \Big\downarrow{\scriptstyle\sigma_{i,j+1}} \\
(LM_\pm)^{-E_{i,j}} \xleftarrow{\ \tau_{i,j}\ } & (LM_\pm)^{-E_{i,j+1}} \xleftarrow{\ \tau i,j+1\ } \cdots \\
\Big\downarrow{\scriptstyle\sigma_{i+1,j}} & & \Big\downarrow{\scriptstyle\sigma_{i+1,j+1}} \\
\vdots & & \vdots
\end{array}
$$

This system defines a pro-object in the category of S^1-equivariant spectra that we call the polarized Atiyah dual, $LM_{\pm}^{-TLM}$. If one applies an equivariant homology theory to this prospectrum, one gets a pro-object in the category of graded abelian groups. Notice that in cohomology, the structure maps $\tau_{i,j}$ and $\sigma_{i,j}$ will induce multiplication by the equivariant Euler classes of the orthogonal complement bundles of these inclusions. As seen above, these orthogonal complement bundles are nonequivariantly isomorphic to the pull back of the tangent bundle, TM. However, they have different equivariant structures. In future work we will study those equivariant cohomology theories for which these Euler classes are units, with the goal being to prove that such theories, when applied to this prospectrum, support the string topology operations including one that corresponds to a disk viewed as a cobordism from a circle to the empty set. By gluing, this will allow the construction of string topology operations for closed surfaces as well as surfaces with a positive number of outgoing boundary components.

References

[1] Abrams, L., Two dimensional topological quantum field theories and Frobenius algebras, *J. Knot theory ramifications* **5** (1996), 569–587.

[2] Ando, M. and J. Morava, A renormalized Riemann-Roch formula and the Thom isomorphism for the free loop space, *Topology, Geometry, and Algebra: Interactions and New Directions* (Stanford, CA, 1999), 11–36, *Contemp. Math.*, **279**, Amer. Math. Soc., Providence, RI, 2001.

[3] Chas, M. and D. Sullivan, String topology, to appear in *Annals of Math.*, preprint: math.GT/9911159 (1999).

[4] Chas, M. and D. Sullivan, Closed string operators in topology leading to Lie bialgebras and higher string algebra, preprint: Math.GT/0212358 (2002).

[5] Cohen, R.L. and J.D.S. Jones, A homotopy theoretic realization of string topology, *Math. Annalen*, published online: DOI 10.1007/s00208-002-0362-0 (2002). Preprint: math.GT0107187.

[6] Cohen, R.L., J.D.S. Jones, and G. Segal, Floer's infinite dimensional Morse theory and homotopy theory, *Floer Memorial Volume*, Birkhauser Verlag Prog. in Math. vol. **133** (1995), 287–325.

[7] Cohen, R.L. J.D.S. Jones, and J. Yan, The loop homology algebra of spheres and projective spaces, to appear in Proc. of Conference on Algebraic Topology, Skye, Scotland, 2001. Preprint: math.AT/0210353.

[8] Cohen, R.L. and A. Stacey, Fourier decompositions of loop bundles, preprint: math/0210351, 2002.

[9] Culler, M. and K. Vogtmann, Moduli of graphs and automorphisms of free groups, *Invent. Math.* 84 (1986), no.1, 91–119.

[10] Dijkgraaf, R., A geometric approach to two dimensional conformal field theory, Ph.D. thesis, Univ. of Utrecht, (1989).

[11] Godin, V., Stanford University Ph.D. thesis, in preparation.

[12] Igusa, K., *Higher Franz-Reidemeister Torsion, AMS/IP Studies in Adv. Math.*, vol 31, International Press, 2002.

[13] Igusa, K., Combinatorial Miller–Morita–Mumford classes and Witten cycles, preprint: math.GT/0207042 (2002).

[14] Kontsevich, M., Intersection theory on the moduli space of curves and the matrix Airy function, *Comm. Math. Phys.* **(147)** (1992), no.1, 1–23.

[15] Penner, R., The decorated Teichmuller space of punctured surfaces, *Comm. Math. Phys.* 113 (1987), 299–339.

[16] Pressley, A. and G. Segal, *Loop Groups, Oxford Math. Monographs*, Clarendon Press (1986).

[17] Segal, G., Two dimensional conformal field theories and modular functors, in IXth Int'l. Congress on Math. Physics (1988), 22–37.

[18] Strebel, K., *Quadratic Differentials*, Springer Verlag, Berlin, 1984.

[19] Sullivan, D., Closed string operators and Feynman graphs, in preparation.

[20] Voronov, A., Notes on universal algebra, to appear in proceedings of Stony Brook conf. on Graphs and Patterns in Mathematics and Physics, June 2001. Preprint: math.QA/0111009.

Email: ralph@math.stanford.edu and vero@math.stanford.edu.

6

Random matrices and Calabi–Yau geometry

ROBBERT DIJKGRAAF

University of Amsterdam

Abstract

I review certain more mathematical aspects of recent work done in collaboration with C. Vafa [1, 2, 3] that has led to a direct connection between the theory of random matrices and string theory invariants associated to (non-compact) Calabi-Yau manifolds.

1 String invariants of Calabi–Yau manifolds

1.1 Moduli and periods

Let X be a Calabi–Yau three-fold, i.e. a Kähler manifold of complex dimension three with vanishing first Chern class $c_1(X) = 0$. We will assume for the moment that X is compact, although we will relax this condition later. Such a CY manifold has a unique (up to scalars) holomorphic $(3, 0)$ form denoted as Ω.

Let $\mathcal{M}$ be the moduli space of inequivalent complex structures on the topological manifold X. It carries a natural line bundle $\mathcal{L}$, whose fiber is given by the complex line $H^{3,0}(X)$. The total space of $\mathcal{L}$ parameterizes pairs (X, Ω) up to equivalence.

The moduli space $\mathcal{M}$ has an elegant description in terms of the period map of Hodge theory. Recall that the vector space $H^3(X, \mathbf{C})$ carries a natural symplectic structure ω given by the intersection form

$$\omega(\alpha, \beta) = \int_X \alpha \wedge \beta.$$

The period map

$$\pi : \mathcal{M} \to \mathbf{P}(H^3(X, \mathbf{C}))$$

assigns to a given complex structure the complex line generated by the holomorphic $(3,0)$ form Ω

$$\pi(X) = H^{3,0}(X) \cong \mathbf{C} \cdot \Omega.$$

155

A standard result is the following:

Theorem[4]. *The period map π is injective and its image in $H^3(X)$ is a Lagrangian cone.*

This implies that in canonical coordinates (q_i, p_i) on $H^3(X)$ with

$$\omega = dp_i \wedge dq_i$$

the image is given by the graph of the differential $d\mathcal{F}_0$ for some function $\mathcal{F}_0(\mathbf{q})$, with $\mathbf{q} = (q_1, \ldots, q_n)$. To be a bit more concrete, let A_i, B_j be a canonical symplectic basis of $H_3(X)$, i.e. a basis in which the only non-vanishing intersection is

$$A_i \cap B_j = \delta_{ij}.$$

This basis of $H_3(X)$ induces local symplectic coordinates (q_i, p_i) on $H^3(X)$. In terms of these coordinates the period map for a given complex structure $X \in \mathcal{M}$ takes the form

$$\int_{A_i} \Omega = q_i(X), \qquad \int_{B_i} \Omega = p_i(X).$$

These periods now satisfy the following properties. First of all the variables q_i can be seen as homogeneous local coordinates on $\mathcal{M}$. Second, the dual periods p_i can be expressed as

$$p_i = \frac{1}{2\pi i} \frac{\partial \mathcal{F}_0}{\partial q_i}.$$

This is usually referred to as special geometry in the physics literature (see e.g. [5]). The *prepotential* $\mathcal{F}_0(\mathbf{q})$ is homogeneous of degree two and should therefore more geometrically be thought of as a section of the line bundle $\mathcal{L}^{\otimes 2}$ over the moduli space $\mathcal{M}$.

The proof of all this follows essentially from Griffith's transversality of the variation of Hodge structures. The variation of the holomorphic $(3, 0)$ form contains at most a $(2, 1)$ form, that is

$$\frac{\partial \Omega}{\partial q_i} \in H^{3,0} \oplus H^{2,1}.$$

Therefore, using the 'Riemann bilinear identities', we immediately have the integrability condition

$$\frac{\partial p_j}{\partial q_i} - \frac{\partial p_i}{\partial q_j} = \int_X \frac{\partial \Omega}{\partial q_i} \wedge \frac{\partial \Omega}{\partial q_j} = 0.$$

Similarly, the homogeneity of $\mathcal{F}_0(\mathbf{q})$ is obtained by the relation

$$\int_X \Omega \wedge \frac{\partial \Omega}{\partial q_i} = 0.$$

1.2 Remarks

(1) A Lagrangian in a symplectic vector space, locally given by the graph of $d\mathcal{F}_0$, can be regarded as the classical limit of a quantum state. Quantization will associate to this Lagrangian a full quantum wave-function, that in the semi-classical WKB approximation is given by

$$\Psi(\mathbf{q}) \sim e^{\mathcal{F}_0(\mathbf{q})/\hbar}.$$

More precisely, the wave-function has an expansion of the form

$$\Psi(\mathbf{q}) = \exp \sum_{g \geq 0} \hbar^{g-1} \mathcal{F}_g(\mathbf{q}).$$

In the present context one often writes

$$\hbar = g_s^2$$

with g_s the string coupling constant. The function $\mathcal{F}_0(\mathbf{q})$ should therefore be regarded as the first of an infinite series of invariants $\mathcal{F}_g(\mathbf{q})$ associated to the quantization of the family of CYs. According to the formalism of (topological) string theory, the quantum invariants $\mathcal{F}_g$ are related to maps $\Sigma_g \to X$, with Σ_g a Riemann surface of genus g. In this so-called topological B-model these maps are in general 'almost constant'. That is, only completely degenerate nodal rational curves contribute to $\mathcal{F}_g$ [6] [7].

The general definition of $\mathcal{F}_{g \geq 2}$ is complicated [7], but the leading correction $\mathcal{F}_1$ has an elegant definition in terms of the Ray–Singer analytic torsion of X, a particular combination of determinants of Laplacians

$$\mathcal{F}_1 = \sum_{0 \leq p,q \leq 3} (-1)^{p+q} pq \log \det{}' \Delta_{p,q}.$$

Here $\Delta_{p,q} = \partial\bar{\partial} + \bar{d}d$ is the Laplacian on (p, q) forms.

(2) Mirror symmetry will relate the invariants $\mathcal{F}_g$ of the B-model to the Gromov–Witten invariants of a mirror CY threefold $\widehat{X}$ (the so-called A-model). The coordinates $\mathbf{q}$ on the moduli space $\mathcal{M}_X$ are related to the natural variable $\mathbf{t} \in H^{1,1}(\widehat{X}; \mathbf{C})$ that parameterizes the complexified Kähler forms on

$\widehat{X}$. The general relation between **q** and **t** is very nontrivial. In terms of this new coordinate t we have an expansion

$$\mathcal{F}_g(\mathbf{t}) = \sum_d GW_{g,d} \exp(-d\mathbf{t})$$

with $GW_{g,d}$ the Gromov–Witten invariant of $\widehat{X}$ for genus g and degree $d \in H_2(\widehat{X})$ (roughly, the 'number' of holomorphic curves on $\widehat{X}$, or, more precisely, the pairing with the virtual fundamental class of the Kontsevich moduli space of stable maps).

1.3 Non-compact Calabi–Yau

It has proven very interesting to generalize this formalism to non-compact Calabi–Yau manifolds. Instead of giving a general analysis we will confine ourselves here to an instructive class of examples given by an affine hypersurface X in $\mathbf{C}^4$, given by an equation of the general form

$$X : f(u, v, x, y) = 0.$$

Such an affine hypersurface always has $c_1 = 0$, and a canonical choice for an holomorphic (3,0)-form is given by

$$\Omega = \frac{du \wedge dv \wedge dx \wedge dy}{df}.$$

If we assume that the function f is homogeneous of degree one (where the coordinates are given (possibly fractional) degrees d_i), then it has been proven [8] that X can carry a Calabi–Yau metric that looks conical at infinity if

$$\sum d_i > 1.$$

We will be mainly interested in families of the form

$$F(x, y) - uv = 0$$

where this relation is always satisfied. To such a family there is naturally an associated affine planar curve given by

$$C : F(x, y) = 0.$$

If we consider X as a fibration over the (x, y)-plane, with as fiber the rational curve $uv = F$, then the curve C is the locus on which this fiber degenerates.

It is not difficult to show that in this case the periods of the holomorphic three-form on the threefold X, which here takes the form

$$\Omega = \frac{du \wedge dx \wedge dy}{u}$$

reduce to the periods of the associated one-form

$$\eta = ydx$$

on the curve C, roughly by integrating first along the fiber as in

$$\int \Omega = \int \frac{du}{u} dxdy = \int_{F \geq 0} dxdy = \int_{F=0} ydx.$$

1.4 Conifold

A relevant example of such a non-compact CY is the case of the 'conifold', given by

$$x^2 + y^2 - uv = q. \tag{1.1}$$

Here q is a parameter. For $q = 0$ this describes a conical singularity. For $q \neq 0$ the smooth manifold is actually diffeomorphic to T^*S^3. In this case there are two relevant homology 3-cycles: the A-cycle is represented by the zero-section S^3; the dual B-cycle is one of the fibers.

The associated Riemann surface is in this case the rational curve

$$x^2 + y^2 = q.$$

The period of the meromorphic one-form ydx around the A-cycle is easily evaluated to be

$$\frac{1}{2\pi i} \oint_A ydx = q.$$

The period around the non-compact period B diverges logarithmically (a common problem for all these non-compact CY spaces) and has to be regularized by a cut-off point Λ (with $|\Lambda| \gg |q|$). This then gives

$$\int_B ydx = \int_0^\Lambda y(x)dx \sim \frac{1}{2}q^2 \log(q/\Lambda).$$

2 Random matrix theory

The theory of random matrices (see for example [9]) was initiated by Wigner in his study of the spectral properties of Hermitian matrices of large rank. Instead of actually diagonalizing a particular matrix he considered a statistical ensemble of matrices and only made probabilistic statements about the distribution of eigenvalues. In the classical case this distribution is given by a Gaussian weight – the so-called Gaussian ensemble.

2.1 Wigner's Gaussian ensemble

More precisely, let $\mathcal{H}_N$ be the space of $N \times N$ Hermitian matrices. One can think of $\mathcal{H}_N$ as the Lie algebra of $U(N)$ equipped with the adjoint action of $U(N)$. The probability density on the linear space of these matrices Φ is then taken to be the $U(N)$-invariant measure

$$d\mu(\Phi) = \frac{1}{V_N} e^{-\frac{1}{g_s} \mathrm{Tr} \Phi^2} d\Phi.$$

This measure is conveniently normalized by the volume of the unitary group

$$V_N = \mathrm{vol}(U(N))$$

as computed in the induced measure from $\mathcal{H}_N$.

A Hermitian matrix Φ can be diagonalized

$$\Phi = U \cdot \begin{pmatrix} \lambda_1 & \cdots & 0 \\ \vdots & \ddots & \vdots \\ 0 & \cdots & \lambda_N \end{pmatrix} \cdot U^{-1}$$

and has an associated spectral density

$$\rho(x) = \frac{1}{N} \sum \delta(x - \lambda_I).$$

We want to take the limit $N \to \infty$ such that the average density in the ensemble becomes a smooth function. In order to do this one has to simultaneously scale the parameter $g_s \to 0$ such that the combination

$$q = g_s N$$

(known in the physics literature as the 't Hooft coupling) remains finite. The classical result of Wigner is that for the Gaussian ensemble in this large N limit the eigenvalues spread out from their classical locus $\lambda_I = 0$ and form a 'cut' $[-\sqrt{q}, \sqrt{q}]$ along the x-axis. More precisely, the limiting eigenvalue distribution is given by the so-called semi-circle law

$$\rho(x) = \frac{2}{\pi q} \sqrt{q - x^2}.$$

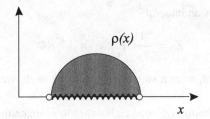

We will see in a moment how this (half of a) circle is related to the *complex* algebraic curve

$$x^2 + y^2 = q$$

and more generally the conifold geometry (1.1).

2.2 Ribbon graphs and matrix models

There is a beautiful connection between matrix integrals and combinatorics of ribbon diagrams that originates in the work of 't Hooft [10]. This connection naturally emerges if one considers more general ensembles, given by measures of the form

$$\frac{1}{V_N} e^{-\frac{1}{g_s} \mathrm{Tr}\, W(\Phi)} d\Phi$$

with $W(\Phi)$ a general polynomial, say of degree $n + 1$

$$W(\Phi) = \sum_{k=1}^{n+1} \frac{1}{k} t_k \Phi^k.$$

One way to analyze this case is to consider it as a perturbation of the quadratic Gaussian. To this end one introduces the expectation values in the Gaussian ensemble

$$\langle f(\Phi) \rangle = \frac{\int d\mu(\Phi) \cdot f(\Phi)}{\int d\mu(\Phi) \cdot 1}.$$

We can now write the general matrix integral

$$Z = \frac{1}{\mathrm{vol}\, U(N)} \int d\Phi \, \exp\left[-\frac{1}{g_s} \mathrm{Tr}\, W(\Phi) \right] \tag{2.1}$$

in terms of the expectation value

$$Z = Z_0 \cdot \left\langle \exp\left(-\frac{1}{g_s} \sum_{k \geq 3} \frac{1}{k} t_k \mathrm{Tr}\, \Phi^k \right) \right\rangle.$$

Here Z_0 is the Gaussian integral

$$Z_0 = \frac{1}{V_N} \int_{\mathcal{H}_N} e^{-\frac{1}{gs}\mathrm{Tr}\,\Phi^2} d\Phi$$

and we assumed that, after a suitable redefinition of Φ, we have $t_1 = 0$ and $t_2 = 1$.

Now we can use combinatorical techniques to evaluate the expectation values as a perturbative series in the couplings t_k using Wick's theorem. Recall that a ribbon graph is a graph with a cyclic ordering of edges at each vertex. Such an orientation can be induced by drawing the graph on the plane and thinking of the edges as two-dimensional strips (ribbons) which have an (infinitesimal) thickness. For example, there are two possibilities to realize a 'two-loop' graph as an (orientable) ribbon graph

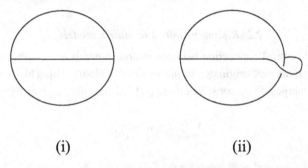

(i)   (ii)

By 'fattening' such a graph we obtain a topological surface with boundaries. We will assume the ribbon graph and therefore this surface to be oriented. Its genus will be denoted as g, and h will be the number of holes (boundary components). For example, the previous two graphs correspond to the surfaces

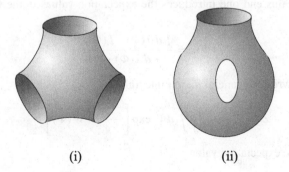

(i)   (ii)

of $(g, h) = (0, 3)$ and $(1, 1)$ respectively. The Euler number, of both the graph and the surface of course, is $\chi = 2 - 2g - h$. Let m_k furthermore be the number of vertices of valency k (i.e. the vertices at which k edges end). Then there is the following combinatorial standard result [11]

$$\left\langle \exp \sum_{k\geq 3} \frac{t_k}{\lambda} \operatorname{Tr} \Phi^k \right\rangle \sim \sum_{\text{ribbon graphs } \Gamma} \frac{1}{|\operatorname{Aut}(\Gamma)|} g_s^{-\chi(\Gamma)} N^{h(\Gamma)} \prod_{k\geq 3} t_k^{m_k(\Gamma)}.$$

Note that the only dependence on the rank N of the matrix is through the factor N^h, with h the number of holes. Combining this with the dependence on the string coupling g_s, which keeps track of the Euler number through the factor g_s^{2g-2+h}, we see that this expression behaves well in the 't Hooft limit

$$N \to \infty, \quad g_s \to 0, \qquad g_s N = q \text{ fixed}.$$

We simply have to rewrite the factor $N^h g_s^{2g-2+h}$ as $q^h g_s^{2g-2}$, and then we can freely sum over the number of holes h. As always in combinatorics, the logarithm of this partition function is expressed as a sum over *connected* surfaces (or graphs)

$$\log Z = \sum_{g\geq 0} \mathcal{F}_g(q; t_k) \lambda^{2g-2}.$$

That is, $\mathcal{F}_g$ is given by the sum over all connected surfaces of genus g.

3 Periods and matrix integrals

We will now proceed to explain how this interpretation of random matrix integrals in terms of Riemann surfaces can be related to string invariant of CY geometries.

3.1 Critical points

Up to now we considered the stationary phase approximation of the matrix integral

$$Z = \frac{1}{\operatorname{vol} U(N)} \int d\Phi \, \exp\left[-\frac{1}{g_s} \operatorname{Tr} W(\Phi) \right]$$

around the critical point $\Phi = 0$, viewing the polynomial $W(\Phi)$ as a small perturbation of the quadratic form Φ^2. However, there are of course more critical points where such an approximation becomes more involved. If we assume that the polynomial

$$W'(x) = \prod_{i=1}^{n} (x - a_i)$$

is non-degenerate, so that the polynomial $W(x)$ has n distinct critical points at $x = a_1, \ldots, a_n$, then the critical point of the matrix function $\mathrm{Tr}\, W(\Phi)$ are naturally labeled by partitions of N into n parts. Indeed, one can diagonalize Φ as

$$
\Phi = U \cdot \begin{pmatrix} \lambda_1 & \cdots & 0 \\ \vdots & \ddots & \vdots \\ 0 & \cdots & \lambda_N \end{pmatrix} \cdot U^{-1}
$$

and consider the function $\mathrm{Tr}\, W(\Phi)$ as a function of these eigenvalues λ_I. Then at a critical point each of the eigenvalues λ_I will have to satisfy the equation

$$
W'(\lambda_I) = 0.
$$

Therefore the stationary points of the matrix integral are given by the various distributions of the N eigenvalues λ_I among the n critical points a_i of the polynomial W'. Let

$$
N = N_1 + \cdots + N_n
$$

be such a distribution with

$$
N_i = \#\{\text{eigenvalues } \lambda_I = a_i\}.
$$

We will consider the limit where these 'filling factions' N_i tend to infinity, while keeping the quantities

$$
q_i = g_s N_i
$$

finite. We will still write

$$
q = \sum q_i = g_s N.
$$

In the large N_i limit the critical points of the matrix integral will now form a continuum moduli space labeled by the real numbers

$$
\mathbf{q} = (q_1, \ldots, q_n).
$$

We will see that the quantities that we compute will turn out to be *analytic* functions in the variables q_i, and therefore we can consider the q_i in the end to be complex valued. So, locally, the moduli space associated to the matrix integral is given by an open domain in $\mathbf{C}^n$.

We will now define the quantities $\mathcal{F}_g(\mathbf{q})$ through the stationary phase approximation

$$
\log Z \sim \sum_{g \geq 0} g_s^{2g-2} \mathcal{F}_g(\mathbf{q})
$$

of the matrix integral around the critical point $\mathbf{q}$.

To really make sense of these stationary phase approximations it is necessary to consider a generalized matrix integral, in which one does not integrate over *Hermitian* matrices, but over a particular contour in the space of *complex* matrices. If this contour is in a generic position, the corresponding matrices will still be diagonalizable, but now with complex eigenvalues. So, alternatively, one can think in terms of a contour integral in the space of complexified eigenvalues. The contour can then be picked in such a way that the saddle-point expansions make sense.

The generating functions again have a combinatorial interpretation, in this case as a sum over *colored* ribbon graphs. The holes of the graph carry here a label $i \in \{1, \dots, n\}$. The edges therefore carry two labels. The Feynman rules are very simple to derive from the potential W and are given in [12].

3.2 Matrices and Calabi–Yau manifolds

We now have the following main result that relates random matrix integrals and string theory invariants of Calabi–Yau geometries.

Theorem. (1) *The functions $\mathcal{F}_g(q)$ that appear in the saddle-point approximation of the matrix integral (2.1) are the quantum prepotentials associated to the family of CY geometries*

$$X : \quad y^2 + W'(x)^2 + f(x) - uv = 0. \tag{3.1}$$

This result requires some explanation. The function $W'(x)$ must be considered here as a fixed polynomial. The above family of geometries (3.1) is therefore encoded in the deformation $f(x)$, which is by definition a polynomial of degree $n - 1$

$$f(x) = \sum_{k=0}^{n-1} b_k x^k.$$

As we will see in a moment, the coefficients b_k are in one-to-one correspondence with the moduli q_i of the matrix integral. This particular set of deformations of the *singular* three-fold (a natural generalization of the conifold)

$$y^2 + W'(x)^2 - uv = 0$$

are picked out because they preserve the behavior of the holomorphic three-form Ω at infinity. More precisely, for these kinds of deformations we have a bound

$$\int \delta\Omega \leq c \cdot \int \Omega.$$

As we explained before, the periods of Ω on a hypersurface X of the type (3.1) can be written in terms of periods on the associated plane curve, which here takes the form of a hyperelliptic curve

$$C : y^2 + W'(x)^2 + f(x) = 0 \qquad (3.2)$$

doubly branched cover of the x-plane. Let A_i, B_i now be a basis of homology cycles for this curve. On the Riemann surface they can be chosen as

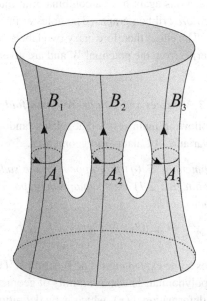

In the x-plane they encircle and cross respectively the branch cuts

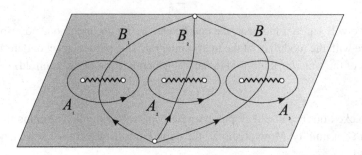

We now claim that the leading genus zero contribution to the matrix model

integral $\mathcal{F}_0(q)$ is encoded through the period maps of the CY geometry

$$q_i = \frac{1}{2\pi i} \oint_{A_i} y dx$$

and

$$\frac{\partial \mathcal{F}_0}{\partial q_i} = \int_{B_i} y dx.$$

Furthermore, the eigenvalue density is directly obtained from the one-form $y dx$ through the relation

$$\rho(x) = \text{disc Im} \frac{y(x)}{\pi q}.$$

This last result is therefore a direct generalization of Wigner's observation that results in the Gaussian case

$$W(\Phi) = \frac{1}{2}\Phi^2.$$

The eigenvalue profile is now given not by a half of a circle, but by a half of a hyperelliptic curve

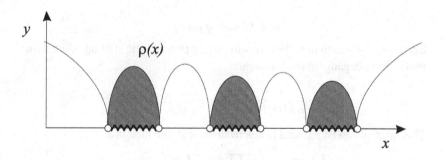

3.3 *Higher genus invariants*

One can also compute explicitly the higher genus corrections $\mathcal{F}_g$ for $g > 0$ in the matrix model, and confirm that they indeed compute the higher quantum invariants of the CY. For example, the genus one contribution can be expressed in terms of the geometry of the hyperelliptic curve as [13] [14]

$$\mathcal{F}_1(\mathbf{q}) = -\frac{1}{2}\log\det A - \frac{1}{12}D.$$

Here D is the discriminant of the hyperelliptic curve C given by (3.2). That is, if we write the curve as

$$y^2 = \prod_{i=1}^{2n}(x - x_i)$$

with x_i the branch points, then

$$D = \prod_{i<j}^{2n}(x_i - x_j)^2.$$

The matrix A is given by the period integrals

$$A_{ij} = \oint_{A_i} \frac{x^{j-1}}{y}dx.$$

More importantly, we can express $\mathcal{F}_1$ as

$$\mathcal{F}_1 = -\frac{1}{2}\log\det\Delta$$

with Δ the scalar determinant on C. This realizes the expected relation with the analytic torsion on the CY, here suitably reduced to the underlying curve C.

3.4 Sketch of proof

Let us briefly sketch how these results can be proven. The starting point is the (holomorphic, gauged) matrix integral

$$Z = \frac{1}{\mathrm{vol}\,U(N)}\int d\Phi\,\exp\left[\frac{1}{g_s}\mathrm{Tr}\,W(\Phi)\right].$$

This integral can be reduced to eigenvalues, with the measure

$$\prod d\Phi_{ij} = dU\prod_I d\lambda_I\prod_{I<J}(\lambda_I - \lambda_J)^2.$$

Here dU is the Haar measure on $U(N)$ and the last factor is the famous Vandermonde determinant. This gives the following integral over the eigenvalues

$$Z = \int\prod d\lambda_I\prod_{I<J}(\lambda_I - \lambda_J)\exp\sum_I\frac{1}{g_s}W(\lambda_I) \tag{3.3}$$

$$= \int\prod d\lambda_I\exp\frac{1}{g_s}\mathcal{S}(\lambda_1,\ldots,\lambda_N) \tag{3.4}$$

with 'effective action'

$$\mathcal{S}(\lambda_1, \ldots, \lambda_N) = \sum_I W(\lambda_I) + 2g_s \sum_{I<J} \log(\lambda_I - \lambda_J).$$

The saddle-point equations now read

$$\frac{\partial \mathcal{S}}{\partial \lambda_I} = W'(\lambda_I) + 2g_s \sum_{J \neq I} \frac{1}{\lambda_I - \lambda_J} = 0.$$

It proves convenient to think of adding an extra eigenvalue $\lambda_{N+1} = x$ to the matrix and to study the dependence on x. To this end introduce the notation

$$y(x) := \frac{\partial \mathcal{S}}{\partial x} = W'(x) + 2g_s \sum_I \frac{1}{x - \lambda_I}.$$

The differential $d\mathcal{S} = y(x)dx$ is a meromorphic one-form on the complex x-plane. It has a single pole at the position of each eigenvalue. It is closely related to the matrix resolvent

$$\omega(x) = \frac{1}{N} \sum_I \frac{1}{x - \lambda_I} = \frac{1}{N} \mathrm{Tr} \frac{1}{x - \Phi}.$$

In fact, in terms of the resolvent and the 't Hooft coupling $q = g_s N$ we have

$$y(x) = W'(x) + 2q\omega(x).$$

It is now a standard result in the theory of random matrices that the resolvent $\omega(x)$ and therefore also the variable $y(x)$ satisfy a differential equation (the loop equation) that reduces to an algebraic equation in the large N (genus zero) limit.

This is proven by simply squaring the resolvent which gives

$$\omega(x)^2 = \frac{1}{N^2} \left(\sum_I \frac{1}{x - \lambda_I} \right)^2 \tag{3.5}$$

$$= \frac{1}{N^2} \sum_{I \neq J} \frac{1}{(x - \lambda_I)(x - \lambda_J)} + \frac{1}{N^2} \sum_I \frac{1}{(x - \lambda_I)^2} \tag{3.6}$$

$$= \frac{1}{N^2} \sum_{I \neq J} \frac{2}{(x - \lambda_I)(\lambda_I - \lambda_J)} - \frac{1}{N} \omega'(x). \tag{3.7}$$

In the large N limit we can now drop the last term, since $\omega(x)$ and therefore also $\omega'(x)$ are normalized in such a way that they remain finite in this limit.

Now we can further manipulate the resulting expression as

$$\omega(x)^2 \sim \frac{1}{N^2 g_s} \sum_I \frac{-W'(\lambda_I)}{x - \lambda_I} \tag{3.8}$$

$$= \frac{1}{q} \left[\sum_I \frac{1}{N} \frac{W'(x) - W'(\lambda_I)}{x - \lambda_I} - W'(x) \sum_I \frac{1}{N} \frac{1}{x - \lambda_I} \right] \tag{3.9}$$

$$= \frac{1}{4q^2} f(x) - \frac{1}{q} \omega(x) W'(x) \tag{3.10}$$

where the quantum correction is defined by

$$f(x) := 4q \sum_I \frac{1}{N} \frac{W'(x) - W'(\lambda_I)}{x - \lambda_I}.$$

This is by inspection a polynomial of degree $n - 1$. Rewriting this all in terms of $y(x)$ we recover the curve

$$y^2 = W'(x)^2 + f(x).$$

The relations with the period map are now easily obtained. First of all, we recall that y has single poles at the location of the eigenvalues. Therefore the period around a branch cut measures the number of eigenvalues

$$\frac{1}{2\pi i} \oint_{A_i} y(x) dx = g_s N_i = q_i.$$

Integrating $y dx = dS$ over the noncompact B_i cycles, corresponds to adding or removing an eigenvalue. Therefore we have

$$\int_{B_i} y dx = \int_{a_i}^{\Lambda} dS = \frac{\partial \mathcal{F}_0}{\partial q_i}.$$

4 Generalizations

The relation between random matrices and Calabi–Yau geometries goes much further than in the simple example considered above. For example, multi-matrix models of quiver type [2] naturally give rise to geometries of the form

$$F(x, y) - uv = 0$$

with $F(x, y)$ a general polynomial.

More involved two-matrix models lead to the geometries discussed in [15], which are hypersurfaces that are essentially complex surfaces, not curves, of the form

$$F(x, y, u) - v^2 = 0.$$

In fact, one can make a bold conjecture that all solvable matrix models (that is, matrix models that can be reduced in some way to eigenvalues) will necessarily reduce to CY three-folds. Perhaps, even non-solvable matrix models can still be interpreted as a kind of non-commutative CY spaces. It would be very interesting to explore this deep connection further.

Acknowledgements

Apart from warm wishes on the occasion of his birthday, I would like to appreciate my intellectual debt to Graeme Segal, who has formed to a large extend my interactions between physics and mathematics over the past years. I would also like to thank the Mathematical Institute at Oxford University and in particular Ulrike Tillmann for organizing such an excellent meeting and for offering me kind hospitality, allowing me to complete the work [1] during the meeting.

The research described here is partly supported by FOM and the CMPA grant of the University of Amsterdam.

References

[1] Dijkgraaf, R. and C. Vafa, Matrix models, topological strings, and supersymmetric gauge theories [hep-th/0206255].
[2] Dijkgraaf, R. and C. Vafa, On geometry and matrix models [hep-th/0207106].
[3] Dijkgraaf, R. and C. Vafa, A perturbative window into non-perturbative physics [hep-th/0208048].
[4] Tian, G., Smoothness of the universal deformation space of compact Calabi–Yau manifolds and its Peterson–Weil metric, in S.-T. Yau (ed.), *Mathematical Aspects of String Theory* (World Scientific, 1987); Smoothing 3-folds with trivial canonical bundles and ordinary double points, in S.-T. Yau (ed.), *Essays on Mirror Manifolds* (International Press, Hong Kong, 1992). A. Todorov, 'Geometry of Calabi-Yau', preprint 1986.
[5] Freed, D.S., Special Kaehler manifolds, *Commun. Math. Phys.* **203**, 31 (1999) [arXiv:hep-th/9712042].
[6] Witten, E., Mirror manifolds and topological field theory, in S.-T. Yau (ed.), *Essays on Mirror Manifolds* (International Press, Hong Kong, 1992).
[7] Bershadsky, M., S. Cecotti, H. Ooguri and C. Vafa, Kodaira–Spencer theory of gravity and exact results for quantum string amplitudes, *Commun. Math. Phys.* **165**, 311 (1994) [arXiv:hep-th/9309140].
[8] Tian, G. and S.-T. Yau, Complete Kähler manifolds with zero Ricci curvature II, *Inv. Math* **106** (1991) 27.
[9] Di Francesco, P., P. Ginsparg and J. Zinn-Justin, 2-D Gravity and random matrices, *Phys. Rept.* **254**, 1 (1995) [arXiv:hep-th/9306153].
[10] Hooft, G. 't A Planar diagram theory for strong interactions, *Nucl. Phys.* B **72**, 461 (1974).
[11] Brezin, E., C. Itzykson, G. Parisi and J.B. Zuber, Planar Diagrams, *Commun. Math. Phys.* **59** (1978) 35.

[12] Dijkgraaf, R., S. Gukov, V.A. Kazakov, C. Vafa, Perturbative Analysis of Gauged Matrix Models, [arXiv:hep-th/0210238].

[13] Dijkgraaf, R., A. Sinkovics and M. Temurhan, Matrix models and gravitational corrections, [arXiv:hep-th/0211241].

[14] Klemm, A., M. Marino and S. Theisen, Gravitational corrections in supersymmetric gauge theory and matrix models, *JHEP* **0303** (2003) 051 [arXiv:hep-th/0211216].

[15] Cachazo, F., S. Katz and C. Vafa, Geometric transitions and $N = 1$ quiver theories, [arXiv:hep-th/0108120].

Email: rhd@science.uva.nl

7

A survey of the topological properties of symplectomorphism groups

DUSA McDUFF*

State University of New York at Stony Brook

for Graeme Segal

Abstract

The special structures that arise in symplectic topology (particularly Gromov–Witten invariants and quantum homology) place as yet rather poorly understood restrictions on the topological properties of symplectomorphism groups. This article surveys some recent work by Abreu, Lalonde, McDuff, Polterovich and Seidel, concentrating particularly on the homotopy properties of the action of the group of Hamiltonian symplectomorphisms on the underlying manifold M. It sketches the proof that the evaluation map $\pi_1(\mathrm{Ham}(M)) \to \pi_1(M)$ given by $\{\phi_t\} \mapsto \{\phi_t(x_0)\}$ is trivial, as well as explaining similar vanishing results for the action of the homology of $\mathrm{Ham}(M)$ on the homology of M. Applications to Hamiltonian stability are discussed.

1 Overview

The special structures that arise in symplectic topology (particularly Gromov–Witten invariants and quantum homology) place as yet rather poorly understood restrictions on the topological properties of symplectomorphism groups. This article surveys some recent work on this subject. Throughout (M, ω) will be a closed (ie compact and without boundary), smooth symplectic manifold of dimension $2n$, unless it is explicitly mentioned otherwise. Background information and more references can be found in [24] [23] [27].

The **symplectomorphism group** $\mathrm{Symp}(M, \omega)$ consists of all diffeomorphisms $\phi : M \to M$ such that $\phi^*(\omega) = \omega$, and is equipped with the C^∞-topology, the topology of uniform convergence of all derivatives. We will sometimes contrast this with the C^0 (i.e. compact-open) topology. The (path)

* Partially supported by NSF grant DMS 0072512.

connected component containing the identity is denoted $\text{Symp}_0(M, \omega)$. (Note that Symp is locally path connected.) This group Symp_0 contains an important normal subgroup called the **Hamiltonian group** $\text{Ham}(M, \omega)$ whose elements are the time-1 maps of Hamiltonian flows. These are the flows $\phi_t^H, t \in [0, 1]$, that at each time t are tangent to the symplectic gradient X_t^H of the function $H_t : M \to \mathbb{R}$, i.e.

$$\dot{\phi}_t^H = X_t^H, \qquad \omega(X_t^H, \cdot) = -dH_t.$$

When $H^1(M, \mathbb{R}) = 0$ the groups Ham and Symp_0 coincide. In general, there is a sequence of groups and inclusions

$$\text{Ham}(M, \omega) \hookrightarrow \text{Symp}_0(M, \omega) \hookrightarrow \text{Symp}(M, \omega) \hookrightarrow \text{Diff}^+(M)$$

where Diff^+ denotes the orientation preserving diffeomorphisms. Our aim is to understand and contrast the properties of these groups.

We first give an overview of basic results on the group Symp_0. Then we describe results on the Hamiltonian group, showing how a vanishing theorem for its action on $H_*(M)$ implies various stability results. Finally, we sketch the proof of this vanishing theorem. It relies on properties of the Gromov–Witten invariants for sections of Hamiltonian fiber bundles over S^2, that can be summarized in the statement (essentially due to Floer and Seidel) that there is a representation of $\pi_1(\text{Ham}(M, \omega))$ into the automorphism group of the quantum homology ring of M. The proof of the vanishing of the evaluation map $\pi_1(\text{Ham}(M)) \to \pi_1(M)$ is easier: it relies on a "stretching the neck" argument, see Lemma 3.2 below. A different but also relatively easy proof of this fact may be found in [23].

1.1 Basic facts

We begin by listing some fundamentals.

Dependence on the cohomology class of ω

The groups $\text{Symp}(M, \omega)$ and $\text{Ham}(M, \omega)$ depend only on the diffeomorphism class of the form ω. In particular, since Moser's argument implies that any path $\omega_t, t \in [0, 1]$, of *cohomologous* forms is induced by an isotopy $\psi_t : M \to M$ of the underlying manifold (i.e. $\psi_t^*(\omega_t) = \omega_0, \psi_0 = id$), the groups do not change their topological or algebraic properties when ω_t varies along such a path. However, as first noticed by Gromov (see Proposition 1.3 below), changes in the cohomology class $[\omega]$ can cause significant changes in the homotopy type of these groups.

Stability properties of Symp(M) *and* Symp$_0$(M)

By this we mean that if G denotes either of these groups, there is a C^1-neighbourhood $\mathcal{N}(G)$ of G in Diff(M) that deformation retracts on to G. This follows from the Moser isotopy argument mentioned above. In the case $G = $ Symp(M), take

$$\mathcal{N}(\text{Symp}) = \{\phi \in \text{Diff}(M) : (1 - t)\phi^*(\omega) + t\omega \text{ is nondegenerate for } t \in [0, 1]\}.$$

By Moser, one can define for each such ϕ a unique isotopy ψ_t (that depends smoothly on $\phi^*(\omega)$) such that $\psi_t^*(t\phi^*(\omega) + (1 - t)\omega) = \omega$ for all t. Hence $\phi \circ \psi_1 \in$ Symp(M). Similarly, when $G = $ Symp$_0$(M) one can take $\mathcal{N}(G)$ to be the identity component of $\mathcal{N}(\text{Symp})$. Note also that these neighborhoods are uniform with respect to ω. For example, given any compact subset K of Symp$_0$(M, ω) there is a C^∞-neighbourhood $\mathcal{N}(\omega)$ of ω in the space of all sympletic forms such that K may be isotoped into Symp$_0$(M, ω') for all $\omega' \in \mathcal{N}(\omega)$. These statements, that we sum up in the rubric **symplectic stability**, exhibit the fiabbiness, or lack of local invariants, of symplectic geometry.

The above two properties are 'soft', i.e. they depend only on the Moser argument. By way of contrast, the next result is 'hard' and can be proved only by using some deep ideas, either from variational calculus (Ekeland–Hofer), generating functions/wave fronts (Eliashberg, Viterbo) or J-holomorphic curves (Gromov).

The group Symp(M, ω) *is* C^0-*closed in* Diff(M)

This celebrated result of Eliashberg and Ekeland–Hofer is known as **symplectic rigidity** and is the basis of symplectic topology. The proof shows that even though one uses the first derivatives of ϕ in saying that a diffeomorphism ϕ preserves ω, there is an invariant $c(U)$ (called a *symplectic capacity*) of an open subset of a symplectic manifold that is continuous with respect to the Hausdorff metric on sets and that is preserved by a diffeomorphism ϕ if and only if $\phi^*(\omega) = \omega$. (When n is even, one must slightly modify the previous statement to rule out the case $\phi^*(\omega) = -\omega$). There are several ways to define a suitable invariant c. Perhaps the easiest is to take Gromov's width

$$c(U) = \sup\{\pi r^2 : B^{2n}(r) \text{ embeds symplectically in } U\}.$$

Here $B^{2n}(r)$ is the standard ball of radius r in Euclidean space $\mathbb{R}^{2n}$ with the usual symplectic form $\omega_0 = \sum_i dx_{2i-1} \wedge dx_{2i}$.

It is unknown whether the identity component Symp$_0$(M) is C^0-closed in Diff(M). In fact this may well not hold. For example, it is quite possible that the group $\mathrm{Symp}^c(\mathbb{R}^{2n})$ (of compactly supported symplectomorphisms of Euclidean space is disconnected when $n > 2$. (When $n = 2$ this group is contractible by Gromov [8].) Hence for some closed manifold M there might be an element in $\mathrm{Symp}(M)\backslash\mathrm{Symp}_0(M)$ that is supported in a Darboux neighbourhood U (i.e. an open set symplectomorphic to an open ball in Euclidean space). Such an element would be in the C^0-closure of $\mathrm{Symp}_0(M)$ since by conformal rescaling in U one could isotop it to have support in an arbitrarily small neighbourhood of a point in U.

We discuss related questions for the group $\mathrm{Ham}(M)$ in Section 2 below. Though less is known about the above questions, some very interesting new features appear. Before doing that we shall give a brief summary of what is known about the homotopy groups of $\mathrm{Symp}(M)$.

1.2 The homotopy type of $\mathrm{Symp}(M)$

In dimension 2 it follows from Moser's argument that $\mathrm{Symp}(M, \omega)$ is homotopy equivalent to Diff^+. Thus $\mathrm{Symp}(S^2)$ is homotopy equivalent to the rotation group $\mathrm{SO}(3)$; $\mathrm{Symp}_0(T^2)$ is homotopy equivalent to an extension of $\mathrm{SL}(2, \mathbb{Z})$ by T^2; and for higher genus the symplectomorphism group is homotopy equivalent to the mapping class group. In dimensions 4 and above, almost nothing is known about the homotopy type of Diff^+. On the other hand, there are some very special 4-manifolds for which the (rational) homotopy type of Symp is fully understood. The following results are due to Gromov [8]. Here σ_Y denotes (the pullback to the product of) an area form on the Riemann surface Y with total area 1.

Proposition 1.1. (Gromov)

 (i) $\mathrm{Symp}^c(\mathbb{R}^4, \omega_0)$ *is contractible:*
 (ii) $\mathrm{Symp}(S^2 \times S^2, \sigma_{S^2} + \sigma_{S^2})$ *is homotopy equivalent to the extension of* $\mathrm{SO}(3) \times \mathrm{SO}(3)$ *by* $\mathbb{Z}/2\mathbb{Z}$ *where this acts by interchanging the factors;*
(iii) $\mathrm{Symp}(\mathbb{C}P^2, \omega_{\mathrm{FS}})$ *is homotopy equivalent to* $\mathrm{PU}(3)$, *where* ω_{FS} *is the Fubini–Study Kähler form.*

It is no coincidence that these results occur in dimension 4. The proofs use J-holomorphic spheres, and these give much more information in dimension 4 because of positivity of intersections.

In Abreu [1] and Abreu–McDuff [5] these arguments are extended to other symplectic forms and (some) other ruled surfaces. Here are the main results, stated for convenience for the product manifold $\Sigma \times S^2$ (though there are similar results for the nontrivial S^2 bundle over Σ.) Consider the following family[1] of symplectic forms on $M_g = \Sigma_g \times S^2$ (where g is genus(Σ))

$$\omega_\mu = \mu \sigma_\Sigma + \sigma_{S^2}, \qquad \mu > 0.$$

Denote by G_μ^g the subgroup

$$G_\mu^g := \mathrm{Symp}(M_g, \omega_\mu) \cap \mathrm{Diff}_0(M_g)$$

of the group of symplectomorphisms of (M_g, ω_μ). When $g > 0$, μ ranges over all positive numbers. However, when $g = 0$ there is an extra symmetry – interchanging the two spheres gives an isomorphism $G_\mu^0 \cong G_{1/\mu}^0$ – and so we take $\mu \geq 1$. Although it is not completely obvious, there is a natural homotopy class of maps from G_μ^g to $G_{\mu+\varepsilon}^g$ for all $\varepsilon > 0$. To see this, let

$$G_{[a,b]}^g = \bigcup_{\mu \in [a,b]} \{\mu\} \times G_\mu^g \subset \mathbb{R} \times \mathrm{Diff}(M_g).$$

It is shown in [5] that the inclusion $G_b^g \to G_{[a,b]}^g$ is a homotopy equivalence. Therefore we can take the map $G_\mu^g \to G_{\mu+\varepsilon}^g$ to be the composite of the inclusion $G_\mu^g \to G_{[\mu,\mu+\varepsilon]}^g$ with a homotopy inverse $G_{[\mu,\mu+\varepsilon]}^g \to G_{\mu+\varepsilon}^g$. Another, more geometric definition of this map is given in [22].

Proposition 1.2. *As $\mu \to \infty$, the groups G_μ^g tend to a limit G_∞^g that has the homotopy type of the identity component $\mathcal{D}_0^g$ of the group of fiberwise diffeomorphisms of $M_g = \Sigma_g \times S^2 \to \Sigma_g$.*

Proposition 1.3. *When $\ell < \mu \leq \ell + 1$ for some integer $\ell \geq 1$*

$$H^*\big(G_\mu^0, \mathbb{Q}\big) = \Lambda(t, x, y) \otimes \mathbb{Q}[w_\ell],$$

where $\Lambda(t, x, y)$ is an exterior algebra over $\mathbb{Q}$ with generators t of degree 1, and x, y of degree 3 and $\mathbb{Q}[w_\ell]$ is the polynomial algebra on a generator w_ℓ of degree 4ℓ.

In the above statement, the generators x, y come from $H^*(G_1^0) = H^*(\mathrm{SO}(3) \times \mathrm{SO}(3))$ and t corresponds to an element in $\pi_1(G_\mu^0)$, $\mu > 1$ found by Gromov in [8]. Thus the subalgebra $\Lambda(t, x, y)$ is the pullback of $H^*(\mathcal{D}_0^0, \mathbb{Q})$ under the map $G_\mu^0 \to \mathcal{D}_0^0$. The other generator w_ℓ is fragile, in the sense that the corresponding element in homology disappears (i.e. becomes null homologous) when μ increases. It is dual to an element in $\pi_{4\ell}$ that is a higher order Samelson product and hence gives rise to a relation (rather than a new generator) in the cohomology of the classifying space. Indeed, when

---

[1] Using results of Taubes and Li–Liu, Lalonde–McDuff show in [14] that these are the *only* symplectic forms on $\Sigma \times S^2$ up to diffeomorphism.

$\ell < \mu \le \ell + 1$

$$H^*\left(BG_\mu^0\right) \cong \frac{\mathbb{Q}[T, X, Y]}{\{T(X - Y)\ldots(\ell^2 X - Y) = 0\}}$$

where the classes T, X, Y have dimensions 2, 4, 4 respectively and are the deloopings of t, x, y.

Anjos [2] calculated the full homotopy type of G_μ^0 for $1 < \mu \le 2$. Her results has been sharpened in Anjos–Granja [3] where it is shown that this group has the homotopy type of the pushout of the following diagram in the category of topological groups

$$\text{SO}(3) \quad \overset{\text{diag}}{\to} \quad \text{SO}(3) \times \text{SO}(3)$$
$$\downarrow$$
$$S^1 \times \text{SO}(3).$$

Thus G_μ^0 is a amalgamated free product of two compact subgroups, $\text{SO}(3) \times \text{SO}(3)$, which is the automorphism group of the product almost complex structure, and $S^1 \times \text{SO}(3)$. The latter appears as the automorphism group of the other integrable almost complex structure with Kähler form ω_μ, namely the Hirzebruch structure on $\mathcal{P}(L_2 \oplus \mathbb{C})$ where the line bundle $L_2 \to \mathbb{C}P^1$ has Chern number 2. As mentioned in [3], this description has interesting parallels with the structure of some Kac–Moody groups.

McDuff [22] proves that the homotopy type of G_μ^0 is constant on all intervals $(\ell - 1, \ell], \ell > 1$. However, their full homotopy type for $\mu > 2$ is not yet understood, and there are only partial results when $g > 0$. Apart from this there is rather little known about the homotopy type of Symp(M). There are some results due to Pinsonnault [26] and Lalonde–Pinsonnault [19] on the one point blow up of $S^2 \times S^2$ showing that the homotopy type of this group also depends on the symplectic area of the exceptional divisor. Also Seidel [31, 30] has done some very interesting work on the symplectic mapping class group $\pi_0(\text{Symp}(M))$ for certain 4-manifolds, and on the case $M = \mathbb{C}P^m \times \mathbb{C}P^n$.

2 The Hamiltonian group

Now consider the Hamiltonian subgroup Ham(M). It has many special properties: it is the commutator subgroup of $\text{Symp}_0(M)$ and is itself a simple group (Banyaga). It also supports a biinvariant metric, the Hofer metric, which gives rise to an interesting geometry. Its elements also have remarkable dynamical properties. For example, according to Arnold's conjecture (finally proven by Fukaya–Ono and Liu–Tian based on work by Floer and Hofer–Salamon) the number of fixed points of $\phi \in$ Ham may be estimated as

$$\#\text{Fix}\,\phi \geq \sum_k \text{rank}\, H^k(M, \mathbb{Q})$$

provided that the fixed points are all nondegenerate, i.e. that the graph of ϕ is transverse to the diagonal.

Many features of this group are still not understood, and it may not even be C^1-closed in Symp_0. Nevertheless, we will see that there are some analogs of the stability properties discussed earlier for Symp. Also the action of $\text{Ham}(M)$ on M has special properties.

2.1 Hofer geometry

Because the elements of the Hamiltonian group are generated by functions H_t, the group itself supports a variety of interesting functions. First of all there is the Hofer norm [10] that is usually defined as follows

$$\|\phi\| := \inf_{\phi_1^H = \phi} \int_0^1 \left(\max_{x \in M} H_t(x) - \min_{x \in M} H_t(x) \right) dt.$$

Since this is constant on conjugacy classes and symmetric (i.e. $\|\phi\| = \|\phi^{-1}\|$), it gives rise to a biinvariant metric $d(\phi, \psi) := \|\psi\phi^{-1}\|$ on $\text{Ham}(M, \omega)$. There are still many open questions about this norm – for example, it is not yet known whether it is always unbounded: for a good introduction see Polterovich's lovely book [27].

Recently, tools (based on Floer homology) have been developed that allow one to define functions on Ham or its universal cover $\widetilde{\text{Ham}}$ by picking out special elements of the action spectrum $\text{Spec}(\tilde{\phi})$ of $\tilde{\phi} \in \widetilde{\text{Ham}}$. This spectrum is defined as follows. Choose a normalized time periodic Hamiltonian H_t that generates $\tilde{\phi}$, i.e. so that the following conditions are satisfied

$$\int H_t \omega^n = 0, t \in \mathbb{R}, \quad H_{t+1} = H_t, t \in \mathbb{R}, \quad \tilde{\phi} = \tilde{\phi}^H := (\phi_1^H, \{\phi_t^H\}_{t \in [0,1]}).$$

Denote by $\tilde{\mathcal{L}}(M)$ the cover of the space $\mathcal{L}(M)$ of contractible loops x in M whose elements are pairs (x, u), where $u : D^2 \to M$ restricts to x on $\partial D^2 = S^1$. Then define the action functional $\mathcal{A}_H : \tilde{\mathcal{L}}(M) \to \mathbb{R}$ by setting

$$\mathcal{A}_H(x, u) = \int_0^1 H_t(x_t)dt - \int_{D^2} u^*(\omega).$$

The critical points of $\mathcal{A}_H$ are precisely the pairs (x, u) where x is a contractible 1-periodic orbit of the flow ϕ_t^H. Somewhat surprisingly, it turns out that the set of critical values of $\mathcal{A}_H$ depends only on the element $\tilde{\phi}^H \in \widetilde{\text{Ham}}$ defined by the flow $\{\phi_t^H\}_{t\in[0,1]}$; in other words, these values depend only on the homotopy class of the path ϕ_t^H rel endpoints. Thus we set

$$\text{Spec}(\tilde{\phi}^H) := \{\text{all critical values of } \mathcal{A}_H\}.$$

There are variants of the Hofer norm that pick out certain special homologically visible elements from this spectrum: see for example Schwarz [28] and Oh [25].

Even more interesting is a recent construction by Entov–Polterovich [7] that uses these spectral invariants to define a nontrivial continuous and homogeneous **quasimorphism** μ on $\widetilde{\text{Ham}}(M, \omega)$, when M is a monotone manifold such as $\mathbb{C}P^n$ that has semisimple quantum homology ring. A quasimorphism on a group G is a map $\mu : G \to \mathbb{R}$ that is a bounded distance away from being a homomorphism, i.e. there is a constant $c = c(\mu) > 0$ such that

$$|\mu(gh) - \mu(g) - \mu(h)| < c, \quad g, h \in G.$$

It is called homogeneous if $\mu(g^m) = m\mu(g)$ for all $m \in \mathbb{Z}$, in which case it restricts to a homomorphism on all abelian subgroups. Besides giving information about the bounded cohomology of G, quasimorphisms can be used to investigate the commutator lengths and dynamical properties of its elements. The example constructed by Entov–Polterovich extends the Calabi homomorphism defined on the subgroups $\widetilde{\text{Ham}}_U$ of elements with support in sufficiently small open sets U. Moreover, in the case of $\mathbb{C}P^n$, it vanishes on $\pi_1(\text{Ham})$ and so descends to the Hamiltonian group Ham (which incidentally equals Symp_0 since $H^1(\mathbb{C}P^n) = 0$.) It is not yet known whether $\widetilde{\text{Ham}}(M)$ or Ham (M) supports a nontrivial quasimorphism for every M. Note that these groups have no nontrivial homomorphisms to $\mathbb{R}$ because they are perfect.

2.2 Relation between Ham and Symp_0

The relation between Ham and Symp_0 is best understood via the **Flux homomorphism**. Let $\widetilde{\text{Symp}_0}(M)$ denote the universal cover of $\text{Symp}_0(M)$. Its elements $\tilde{\phi}$ are equivalence classes of paths $\{\phi_t\}_{t\in[0,1]}$ starting at the identity, where $\{\phi_t\} \sim \{\phi_t'\}$ iff $\phi_1 = \phi_1'$ and the paths are homotopic rel endpoints. We define

$$\text{Flux}(\tilde{\phi}) = \int_0^1 [\omega(\dot{\phi}_t, \cdot)] \in H^1(M, \mathbb{R}).$$

That this depends only on the homotopy class of the path ϕ_t (rel endpoints) is a consequence of the following alternative description: the value of the cohomology class Flux($\tilde{\phi}$) on a 1-cycle $\gamma : S^1 \to M$ is given by the integral

$$\text{Flux}(\tilde{\phi})(\gamma) = \int_{\tilde{\phi}.(\gamma)} \omega \tag{1}$$

where $\tilde{\phi}_*(\gamma)$ is the 2-chain $I \times S^1 \to M : (t, s) \mapsto \phi_t(\gamma(s))$. Thus Flux is well defined. It is not hard to check that it is a homomorphism.

One of the first results in the theory is that the rows and columns in the following commutative diagram are short exact sequences of groups. (For a proof see [24, Chapter 10].)

$$
\begin{array}{ccccc}
\pi_1(\text{Ham}(M)) & \longrightarrow & \pi_1(\text{Symp}_0(M)) & \xrightarrow{\text{Flux}} & \Gamma_\omega \\
\downarrow & & \downarrow & & \downarrow \\
\widetilde{\text{Ham}}(M) & \longrightarrow & \widetilde{\text{Symp}_0}(M) & \xrightarrow{\text{Flux}} & H^1(M, \mathbb{R}) \\
\downarrow & & \downarrow & & \downarrow \\
\text{Ham}(M) & \longrightarrow & \text{Symp}_0(M) & \xrightarrow{\text{Flux}} & H^1(M, \mathbb{R})/\Gamma_\omega.
\end{array} \tag{2}
$$

Here Γ_ω is the so-called **flux group**. It is the image of $\pi_1(\text{Symp}_0(M))$ under the flux homomorphism.

It is easy to see that $\text{Ham}(M)$ is C^1-closed in $\text{Symp}_0(M)$ if and only if Γ_ω is a discrete subgroup of $H^1(M, \mathbb{R})$.

Question 2.1. *Is the subgroup Γ_ω of $H^1(M, \mathbb{R})$ always discrete?*

The hypothesis that Γ_ω is always discrete is known as the **Flux conjecture**. One might think it would always hold by analogy with symplectic rigidity. In fact it does hold in many cases, for example if (M, ω) is Kähler or from (1) above if $[\omega]$ is integral, but we do not yet have a complete understanding of this question. One consequence of Corollary 2.3 is that the rank of Γ_ω is always bounded above by the first Betti number (see Lalonde–McDuff–Polterovich [17] [18]; some sharper bounds are found in Kedra [11]), but the argument does not rule out the possibility that Γ_ω is indiscrete for certain values of $[\omega]$. Thus, for the present one should think of $\text{Ham}(M)$ as a leaf in a foliation of $\text{Symp}_0(M)$ that has codimension equal to the first Betti number of M.

2.3 Hamiltonian stability

When Γ_ω is discrete, the stability principle extends: there is a C^1-neighbourhood of $\text{Ham}(M, \omega)$ in $\text{Diff}(M)$ that deformation retracts into $\text{Ham}(M, \omega)$. Moreover, if this discreteness were uniform with respect to ω (which would hold if (M, ω) were Kähler), then the groups $\text{Ham}(M, \omega)$ would have the same stability with respect to variations in ω as do Symp_0 and Symp.

To be more precise, suppose that for each ω and each $\epsilon > 0$ there is a neighbourhood $\mathcal{N}(\omega)$ such that when $\omega' \in \mathcal{N}(\omega)$ $\Gamma_{\omega'}$ contains no nonzero element of norm $\leq \epsilon$. Then for any compact subset K of Ham(M, ω) there would be a neighbourhood $\mathcal{N}(\omega)$ such that K isotops into Ham(M, ω') for each $\omega' \in \mathcal{N}(\omega)$. For example if $K = \{\phi_t\}$ is a loop (image of a circle) in Ham(M, ω) and $\omega_s, 0 \leq s \leq 1$, is any path, this would mean that any smooth extension $\{\phi_t^s\}, s \leq 0$, of $\{\phi_t\}$ to a family of loops in Symp (M, ω_s) would be homotopic through ω_s-symplectic loops to a loop in Ham(M, ω_s).

Even if this hypothesis on Γ_ω held, it would not rule out the possibility of global instability: a loop in Ham(M, ω) could be isotopic through (nonsymplectic) loops in Diff(M) to a nonHamiltonian loop in some other far away symplectomorphism group Symp(M, ω'). One of the main results in Lalonde–McDuff–Polterovich [18] is that this global instability never occurs; any ω'-symplectic loop that is isotopic in Diff(M) to an ω-Hamiltonian loop must be homotopic in Symp(M, ω') to an ω'-Hamiltonian loop regardless of the relation between ω and ω' and no matter whether any of the groups Γ_ω are discrete. This is known as **Hamiltonian rigidity** and is a consequence of a vanishing theorem for the Flux homomorphism: see Corollary 2.3 below. As we now explain this extends to general results about the action of Ham(M) on M.

2.4 Action of Ham(M) on M

There are some suggestive but still incomplete results about the action of Ham(M) on M. The first result below is folklore. It is a consequence of the proof of the Arnold conjecture, but as we show below (see Lemma 3.2) also follows from a geometric argument. The second part is due to Lalonde–McDuff [15]. Although the statements are topological in nature, both proofs are based on the existence of the Seidel representation, a deep fact that uses the properties of J-holomorphic curves.

Proposition 2.2.

(i) *The evaluation map π_1 (Ham$(M) \to \pi_1(M)$ is zero.*

(ii) *The natural action of $H_*(\text{Ham}(M), \mathbb{Q})$ on $H_*(M, \mathbb{Q})$ is trivial.*

Here the action tr$_\phi : H_*(M) \to H_{*+k}(M)$ of an element $\phi \in H_k(\text{Ham}(M))$ is defined as follows:

if ϕ is represented by the cycle $t \mapsto \phi_t$ for $t \in V^k$ and $c \in H_*(M)$ is represented by $x \mapsto c(x)$ for $x \in C$ then tr$_\phi(c)$ is represented by the cycle

$$V^k \times C \mapsto M : (t, x) \mapsto \phi_t(c(x)).$$

It is just the action on homology induced by the map Ham$(M) \times M \to M$.

It extends to the group $(M^M)_{id}$ of self-maps of M that are homotopic to the identity, and hence depends only on the image of ϕ in $H_k(M^M)_{id}$. To say it is trivial means that

$$\text{tr}_\phi(c) = 0 \quad \text{whenever } c \in H_i(M), i > 0.$$

Note that this does *not* hold for the action of $H_1(\text{Symp}_0(M))$. Indeed by (1) the image under the Flux homomorphism of a loop $\lambda \in \pi_1(\text{Symp}_0(M))$ is simply

$$\text{Flux}(\lambda)(\gamma) = \langle \omega, \text{tr}_\lambda(\gamma) \rangle. \tag{3}$$

The rigidity of Hamiltonian loops is an immediate consequence of Proposition 2.2.

Corollary 2.3. *Suppose that* $\phi \in \pi_1(\text{Symp}(M, \omega))$ *and* $\phi' \in \pi_1(\text{Symp}(M, \omega'))$ *represent the same element of* $\pi_1((M^M)_{id})$. *Then*

$$\text{Flux}_\omega(\phi) = 0 \quad \Longleftrightarrow \quad \text{Flux}_{\omega'}(\phi') = 0$$

Proof. If $\text{Flux}_\omega(\phi) = 0$ than ϕ is an ω-Hamiltonian loop and Proposition 2.2(ii) implies that $\text{tr}_\phi : H_1(M) \to H_2(M)$ is the zero map. But, for each $\gamma \in H_1(M)$, (3) implies that

$$\text{Flux}_{\omega'}(\phi')(\gamma) = \text{Flux}_{\omega'}(\phi)(\gamma) = \langle \omega', \text{tr}_\phi(\gamma) \rangle = 0.$$

This corollary is elementary when the loops are circle subgroups since then one can distinguish between Hamiltonian and nonHamiltonian loops by looking at the weights of the action at the fixed points: a circle action is Hamiltonian if and only if there is a point whose weights all have the same sign. One can also consider maps $K \to \text{Ham}(M, \omega)$ with arbitrary compact domain K. But their stability follows from the above result because $\pi_k(\text{Ham}(M)) = \pi_k(\text{Symp}_0(M))$ when $k > 1$ by diagram (2). For more details see [16].

Thus one can compare the homotopy types of the groups $\text{Ham}(M, \omega)$ (or of $\text{Symp}(M, \omega)$) as $[\omega]$ varies in $H^2(M, \mathbb{R})$. More precisely, as Buse points out in [6], any element α in $\pi_*(\text{Ham}(M, \omega))$ has a smooth extension to a family $\alpha_t \in \pi_*(\text{Ham}(M, \omega_t))$ where $[\omega_t]$ fills out a neighborhood of $[\omega_0] = [\omega]$ in $H^2(M, \mathbb{R})$. Moreover the germ of this extension at $\omega = \omega_0$ is unique. Thus one can distinguish between **robust** elements in the homology or homotopy of the spaces $\text{Ham}(M, \omega_0)$ and $B\text{Ham}(M, \omega_0)$ whose extensions are nonzero for all t near 0 and **fragile** elements whose extensions vanish as $[\omega_t]$ moves in certain directions. For example, any class in $H^*(B\text{Ham}(M, \omega_0))$ that is detected by Gromov–Witten invariants (i.e. does not vanish on a suitable space of J-holomorphic curves as in Le–Ono [20]) is robust, while the classes w_ℓ of Proposition 1.3 are fragile. For some interesting examples in this connection, see Kronheimer [13] and Buse [6].

2.5 c-splitting for Hamiltonian bundles

From now on, we assume that (co)homology has rational coefficients. Since the rational cohomology $H^*(G)$ of any H-space (or group) is freely generated by the dual of its rational homotopy, it is easy to see that part (ii) of Proposition 2.2 holds if and only if it holds for all spherical classes $\phi \in H_k(\text{Ham}(M))$. Each such ϕ gives rise to a locally trivial fiber bundle $M \to P_\phi \to S^{k+1}$ with structural group $\text{Ham}(M)$. Moreover, the differential in the corresponding Wang sequence is precisely tr_ϕ. In other words, there is an exact sequence

$$\ldots H_i(M) \xrightarrow{\text{tr}_\phi} H_{i+k}(M) \to H_{i+k}(P_\phi) \xrightarrow{\cap[M]} H_{i-1}(M) \to \ldots \tag{4}$$

Hence $\text{tr}_\phi = 0$ for $k > 0$ if and only if this long exact sequence breaks up into short exact sequences

$$0 \to H_{i+k}(M) \to H_{i+k}(P_\phi) \xrightarrow{\cap[M]} H_{i-1}(M) \to 0.$$

Thus Proposition 2.2(ii) is equivalent to the following statement.

Proposition 2.4. *For every Hamiltonian bundle $P \to S^{k+1}$, with fiber (M, ω) the rational homology $H_*(P)$ is isomorphic as a vector space to the tensor product $H_*(M) \otimes H_*(S^{k+1})$.*

Observe that the corresponding isomorphism in cohomology need not preserve the ring structure. We say that a bundle $M \to P \to B$ is **c-split** if the rational cohomology $H^*(P)$ is isomorphic as a vector space to $H^*(M) \otimes H^*(B)$.

Question 2.5. *Is every fiber bundle $M \to P \to B$ with structural group $\text{Ham}(M)$ c-split?*

It is shown in [15] that the answer is affirmative if B has dimension ≤ 3 or is a product of spheres and projective spaces with fundamental group of rank ≤ 3. By an old result of Blanchard, it is also affirmative if (M, ω) satisfies the hard Lefschetz condition, i.e. if

$$\wedge[\omega]^k : \quad H^{n-k}(M, \mathbb{R}) \to H^{n+k}(M, \mathbb{R})$$

is an isomorphism for all $0 < k < n$. (This argument has now been somewhat extended by Haller [9] using ideas of Mathieu about the harmonic cohomology of a symplectic manifold.) If the structural group of $P \to B$ reduces to a finite dimensional Lie group G, then c-splitting is equivalent to a result of Atiyah–Bott [4] about the structure of the equivariant cohomology ring $H^*_G(M)$. This is the cohomology of the universal Hamiltonian G-bundle with fiber M

$$M \to M_G = EG \times_G M \to BG$$

and was shown in [4] to be isomorphic to $H^*(M) \otimes H^*(BG)$ as a $H^*(BG)$-module. Hence a positive answer to Question 2.5 in general would imply that this aspect of the homotopy theroy of Hamiltonian actions is similar to the more rigid cases, when the group is finite dimensional or when the manifold is Kähler. For further discussion see [15] [16] and Kedra [12].

Note finally that all results on the action of $\mathrm{Ham}(M)$ on M can be phrased in terms of the universal Hamiltonian bundle

$$M \to M_{\mathrm{Ham}} = E\mathrm{Ham} \times_{\mathrm{Ham}} M \to B\mathrm{Ham}(M).$$

For example, Proposition 2.2 part (i) states that this bundle has a section over its 2-skeleton. Such a formulation has the advantage that it immediately suggests further questions. For example, one might wonder if the bundle $M_{\mathrm{Ham}} \to B\mathrm{Ham}$ always has a global section. However this fails when $M = S^2$ since the map $\pi_3(\mathrm{Ham}(S^2)) = \pi_3(SO(3)) \to \pi_3(S^2)$ is nonzero.

3 Symplectic geometry of bundles over S^2

The proofs of Propositions 2.2 and 2.4 above rely on properties of Hamiltonian bundles over S^2. We now show how the Seidel representation

$$\pi_1(\mathrm{Ham}(M, \omega)) \to (\mathrm{QH}_{\mathrm{ev}}(M))^{\times}$$

of $\pi_1(\mathrm{Ham}(M, \omega))$ into the group of even units in quantum homology gives information on the homotopy properties of Hamiltonian bundles. As preparation, we first discuss quantum homology.

3.1 The small quantum homology ring $QH_*(M)$

There are several slightly different ways of defining the small quantum homology ring. We adopt the conventions of [18] [21].

Set $c_1 = c_1(TM) \in H^2(M, \mathbb{Z})$. Let Λ be the Novikov ring of the group $\mathcal{H} = H_2^S(M, \mathbb{R})/\sim$ with valuation I_ω where $B \sim B'$ if $\omega(B-B') = c_1(B-B') = 0$. Then Λ is the completion of the rational group ring of $\mathcal{H}$ with elements of the form

$$\sum_{B \in \mathcal{H}} q_B \, e^B$$

where for each κ there are only finitely many nonzero $q_B \in \mathbb{Q}$ with $\omega(B) > -\kappa$. Set

$$QH_*(M) = QH_*(M, \Lambda) = H_*(M) \otimes \Lambda.$$

We may define an $\mathbb{R}$ grading on $QH_*(M, \Lambda)$ by setting

$$\deg(a \otimes e^B) = \deg(a) + 2c_1(B)$$

and can also think of $QH_*(M, \Lambda)$ as $\mathbb{Z}/2\mathbb{Z}$-graded with

$$QH_{\mathrm{ev}} = H_{\mathrm{ev}}(M) \otimes \Lambda, \qquad QH_{\mathrm{odd}} = H_{\mathrm{odd}}(M) \otimes \Lambda.$$

Recall that the quantum intersection product

$$a * b \in QH_{i+j-2n}(M), \quad \text{for} \quad a \in H_i(M), b \in H_j(M)$$

is defined as follows

$$a * b = \sum_{B \in \mathcal{H}} (a * b)_B \otimes e^{-B} \tag{5}$$

where $(a * b)_B \in H_{i+j+2n+2c_1(B)}(M)$ is defined by the requirement that

$$(a * b)_B \cdot c = \mathrm{GW}_M(a, b, c; B) \quad \text{for all } c \in H_*(M). \tag{6}$$

Here $\mathrm{GW}_M(a, b, c; B)$ denotes the Gromov–Witten invariant that counts the number of B-spheres in M meeting the cycles $a, b, c \in H_*(M)$, and we have written $\cdot$ for the usual intersection pairing on $H_*(M) = H_*(M, \mathbb{Q})$. Thus $a \cdot b = 0$ unless $\dim(a) + \dim(b) = 2n$ in which case it is the algebraic number of intersection points of the cycles.

Alternatively, one can define $a * b$ as follows: if $\{e_i\}$ is a basis for $H_*(M)$ with dual basis $\{e_i^*\}$, then

$$a * b = \sum_i \mathrm{GW}_M(a, b, e_i; B) e_i^* \otimes e^{-B}.$$

The product $*$ is extended to $QH_*(M)$ by linearity over Λ, and is associative. Moreover, it preserves the $\mathbb{R}$-grading in the homological sense, i.e. it obeys the same grading rules as does the intersection product.

This product $*$ gives $QH_*(M)$ the structure of a graded commutative ring with unit $1 = [M]$. Further, the invertible elements in $QH_{\mathrm{ev}}(M)$ form a commutative group $(QH_{\mathrm{ev}}(M, \Lambda))^\times$ that acts on $QH_*(M)$ by quantum multiplication. By Poincaré duality one can transfer this product to cohomology. Although this is very frequently done, it is often easier to work with homology when one wants to understand the relation to geometry.

3.2 The Seidel representation Ψ

Consider a smooth bundle $\pi : P \to S^2$ with fiber M. Here we consider S^2 to be the union $D_+ \cup D_-$ of two copies of D, with the orientation of D_+. We denote the equator $D_+ \cap D_-$ by ∂, oriented as the boundary of D_+, and choose some point $*$ on ∂ as the base point of S^2. We assume also that the fiber M_* over $*$ has a chosen identification with M.

Since every smooth bundle over a disc can be trivialized, we can build any smooth bundle $P \to S^2$ by taking two product manifolds $D_\pm \times M$ and gluing them along the boundary $\partial \times M$ by a based loop $\lambda = \{\lambda_t\}$ in $\text{Diff}(M)$. Thus

$$P = (D_+ \times M) \cup (D_- \times M)/\sim, \qquad (e^{2\pi i t}, x)_- \equiv (e^{2\pi i t}, \lambda_t(x))_+.$$

A **symplectic bundle** is built from a based loop in $\text{Symp}(M)$ and a **Hamiltonian bundle** from one in $\text{Ham}(M)$. Thus the smooth bundle $P \to S^2$ is symplectic if and only if there is a smooth family of cohomologous symplectic forms ω_b on the fibers M_b. It is shown in [29] [24] [15] that a symplectic bundle $P \to S^2$ is Hamiltonian if and only if the fiberwise forms ω_b have a closed extension Ω. (Such forms Ω are called ω-**compatible**.) Note that in any of these categories two bundles are equivalent if and only if their defining loops are homotopic.

From now on, we restrict to Hamiltonian bundles, and denote by $P_\lambda \to S^2$ the bundle constructed from a loop $\lambda \in \pi_1(\text{Ham}(M))$. By adding the pullback of a suitable area form on the base we can choose the closed extension Ω to be symplectic. The manifold P_λ carries two canonical cohomology classes, the first Chern class of the vertical tangent bundle

$$c_{\text{vert}} = c_1\left(TP_\lambda^{\text{vert}}\right) \in H^2(P_\lambda, \mathbb{Z})$$

and the coupling class u_λ i.e. the unique class in $H^2(P_\lambda, \mathbb{R})$ such that

$$i^*(u_\lambda) = [\omega], \qquad u_\lambda^{n+1} = 0$$

where $i : M \to P_\lambda$ is the inclusion of a fiber.

The next step is to choose a canonical (generalized) section class in $\sigma_\lambda \in H_2(P_\lambda, \mathbb{R})/\sim$. By definition this should project on to the positive generator of $H^2(S^2, \mathbb{Z})$. In the general case, when c_1 and $[\omega]$ induce linearly independent homomorphisms $H_2^S(M) \to \mathbb{R}$, σ_λ is defined by the requirement that

$$c_{\text{vert}}(\sigma_\lambda) = u_\lambda(\sigma_\lambda) = 0 \tag{7}$$

which has a unique solution modulo the given equivalence. If either $[\omega]$ or c_1 vanishes on $H_2^S(M)$ then such a class σ_λ still exists.[2] In the remaining case (the

---

[2] See [21, Remark 3.1] for the case when $[\omega] = 0$ on $H_2^S(M)$. If $c_1 = 0$ on $H_2^S(M)$ but $[\omega] \neq 0$ then we can choose σ_λ so that $u_\lambda(\sigma_\lambda) = 0$. Since c_{vert} is constant on section classes, we must show that it always vanishes. But the existence of the Seidel representation implies every Hamiltonian fibration $P \to S^2$ has some section σ_P with $n \leq c_{\text{vert}}(\sigma_P) \leq 0$ (since it only counts such sections), and the value must be 0 because $c_{\text{vert}}(\sigma_{P\lambda \# P - \lambda}) = c_{\text{vert}}(\sigma_{P\lambda}) + c_{\text{vert}}(\sigma_{P-\lambda})$: see [21, Lemma 2.2].

monotone case), when c_1 is some nonzero multiple of $[\omega] \neq 0$ on $H_2^S(M)$, we choose σ_λ so that $c_{\text{vert}}(\sigma_\lambda) = 0$.

We then set

$$\Psi(\lambda) = \sum_{B \in \mathcal{H}} a_B \otimes e^B \qquad (8)$$

where, for all $c \in H_*(M)$

$$a_{B \cdot M} c = \text{GW}_{P\lambda}([M], [M], c; \sigma_\lambda - B). \qquad (9)$$

Note that $\Psi(\lambda)$ belongs to the strictly commutative part QH_{ev} of $QH_*(M)$. Moreover $\deg(\Psi(\lambda)) = 2n$ because $c_{\text{vert}}(\sigma_\lambda) = 0$. Since all ω-compatible forms are deformation equivalent, Ψ is independent of the choice of Ω.

Here is the main result.

Proposition 3.1. *For all* $\lambda_1, \lambda_2 \in \pi_1(\text{Ham}(M))$

$$\Psi(\lambda_1 + \lambda_2) = \Psi(\lambda_1) * \Psi(\lambda_2), \qquad \Psi(0) = 1$$

where 0 denotes the constant loop. Hence $\Psi(\lambda)$ *is invertible for all* λ *and* Ψ *defines a group homomorphism*

$$\Psi : \pi_1(\text{Ham}(M, \omega)) \to (QH_{\text{ev}}(M, \Lambda))^\times.$$

In the case when (M, ω) satisfies a suitable positivity condition, this is a variant of the main result in Seidel [29]. The general proof is due to McDuff [21] using ideas from Lalonde–McDuff–Polterovich [18]. It uses a refined version of the ideas in the proof of Lemma 3.2 below.

3.3 Homotopy theoretic consequences of the existence of Ψ

First of all, note that because $\Psi(\lambda) \neq 0$ there must always be J-holomorphic sections of $P_\lambda \to S^2$ to count. Thus every Hamiltonian bundle $\pi : P \to S^2$ must have a section $S^2 \to P$. If we trivialize P over the two hemispheres $D_\pm$ of S^2 and homotop the section to be constant over one of the discs, it becomes clear that there is a section if and only if the defining loop λ of P has trivial image under the evaluation map $\pi_1(\text{Ham}(M)) \to \pi_1(M)$. This proves part (i) of Proposition 2.2.

In fact one does not need the full force of Proposition 3.1 in order to arrive at this conclusion, since we only have to produce one section.

Lemma 3.2. *Every Hamiltonian bundle* $P \to S^2$ *has a section.*

Sketch of Proof. Let $\lambda = \{\lambda_t\}$ be a Hamiltonian loop and consider the family of trivial bundles $P_{\lambda,R} \to S^2$ given by

$$P_{\lambda,R} = (D_+ \times M) \cup (S^1 \times [-R, R] \times M) \cup (D_- \times M)$$

with attaching maps

$$(e^{2\pi i t}, \lambda_t(x))_+ \equiv (e^{2\pi i t}, -R, x), \qquad (e^{2\pi i t}, R, x) \equiv (e^{2\pi i t}, \lambda_t(x))_-.$$

Thus, $P_{\lambda,R}$ can be thought of as the fiberwise union (or Gompf sum) of P_λ with $P_{-\lambda}$ over a neck of length R. It is possible to define a family Ω_R of ω-compatible symplectic forms on $P_{\lambda,R}$ in such a way that the manifolds $(P_{\lambda,R}, \Omega_R)$ converge in a well-defined sense as $R \to \infty$. The limit is a singular manifold $P_\lambda \cup P_{-\lambda} \to S_\infty$ that is a locally trivial fiber bundle over the nodal curve consisting of the one point union of two 2-spheres. To do this, one first models the convergence of the 2-spheres in the base by a 1-parameter family S_R of disjoint holomorphic spheres in the one point blow up of $S^2 \times S^2$ that converge to the pair $S_\infty = \Sigma_+ \cup \Sigma_-$ of exceptional divisors at the blow up point. Then one builds a suitable smooth Hamiltonian bundle

$$\pi_{\mathcal{X}} : (\mathcal{X}, \tilde{\Omega}) \to \mathcal{S}$$

with fiber (M, ω) where $\mathcal{S}$ is a Leighbourhood of $\Sigma_+ \cup \Sigma_-$ in the blow up that contains the union of the spheres $S_R, R \geq R_0$: see [21] §2.3.2. The almost complex structures $\tilde{J}$ that one puts on $\mathcal{X}$ should be chosen so that the projection to $\mathcal{S}$ is holomorphic. Then each submanifold $P_{\lambda,R} : \pi_{\mathcal{X}}^{-1}(S_R)$ is $\tilde{J}$-holomorphic.

The bundles $(P_{\lambda,R}, \Omega_R) \to S^2$ are all trivial, and hence there is one $\tilde{J}$-holomorphic curve in the class $\sigma_0 = [S^2 \times pt]$ through each point $q_R \in P_{\lambda,R}$. (It is more correct to say that the corresponding Gromov–Witten invariant $GW_{P_{\lambda,R}}([M], [M], pt; \sigma_0)$ is one; i.e. one counts the curves with appropriate multiplicities.) Just as in gauge theory, these curves do not disappear when one stretches the neck, i.e. lets $R \to \infty$. Therefore as one moves the point q_R to the singular fiber the family of $\tilde{J}$-holomorphic curves through q_R converges to some cusp-curve (stable map) C_∞ in the limit. Moreover, C_∞ must lie entirely in the singular fiber $P_\lambda \cup P_{-\lambda}$ and projects to a holomorphic curve in $\mathcal{S}$ in the class $[\Sigma_+] + [\Sigma_-]$. Hence it must have at least two components, one a section of $P_\lambda \to \Sigma_+$ and the other a section of $P_{-\lambda} \to \Sigma_+$. There might also be some bubbles in the M-fibers, but this is irrelevant. $\square$

The above argument is relatively easy, in that it only uses the compactness theorem for J-holomorphic curves and not the more subtle gluing arguments needed to prove things like the associativity of quantum multiplication. However the proof of the rest of Proposition 2.2 is based on the fact that each element $\Psi(\lambda)$ is a multiplicative unit in quantum homology. The only known way to prove this is via some sort of gluing argument. Hence in this case it seems that one does need the full force of the gluing arguments, whether one works as here with J-holomorphic spheres or as in Seidel [29] with Floer homology.

We now show how to deduce part (ii) of Proposition 2.2 from Proposition 3.1. So far, we have described $\Psi(\lambda)$ as a unit in $QH_*(M)$. This unit induces an automorphism of $QH_*(M)$ by quantum multiplication on the left

$$b \mapsto \Psi(\lambda) * b, \quad b \in QH_*(M).$$

The next lemma shows that when $b \in H_*(M)$ then the element $\Psi(\lambda) * b$ can also be described by counting curves in P_λ rather than in the fiber M.

Lemma 3.3. *If* $\{e_i\}$ *is a basis for* $H_*(M)$ *with dual basis* $\{e_i^*\}$, *then*

$$\Psi(\lambda) * b = \sum_i \mathrm{GW}_{P_\lambda}([M], b, e_i; \sigma_\lambda - B)\, e_i^* \otimes e^B.$$

Sketch of Proof: To see this, one first shows that for any section class σ the invariant $\mathrm{GW}_{P_\lambda}([M], b, c; \sigma)$ may be calculated using a fibered J (i.e. one for which the projection $\pi : P \to S^2$ is holomorphic) and with representing cycles for b, c that each lie in a fiber. Then one is counting sections of $P \to S^2$. If the representing cycles for b, c are moved into the same fiber, then the curves must degenerate. Generically the limiting stable map will have two components, a section in some class $\sigma - C$ together with a C curve that meets b and c. Thus, using much the same arguments that prove the usual 4-point decomposition rule, one shows that

$$\mathrm{GW}_{P_\lambda}([M], b, c; \sigma)$$
$$= \sum_{A,i} \mathrm{GW}_{P\lambda}([M], [M], e_i; \sigma - A) \cdot \mathrm{GW}_M(e_i^*, b, c; A). \qquad (10)$$

But $\Psi(\lambda) = \Sigma q_i e_j^* \otimes e^B$ where

$$q_j = \mathrm{GW}_{P\lambda}([M], [M], e_j; \sigma_\lambda - B) \in \mathbb{Q}.$$

Therefore

$$\Psi(\lambda) * b$$

$$= \sum_{C,k} \mathrm{GW}_M(\Psi(\lambda), b, e_k; C) \, e_k^* \otimes e^{-C}$$

$$= \sum_{B,C,j,k} \mathrm{GW}_{P_\lambda}([M], [M], e_j; \sigma_\lambda - B) \cdot \mathrm{GW}_M(e_j^*, b, e_k; C) \, e_k^* \otimes e^{B-C}$$

$$= \sum_{A,k} \mathrm{GW}_{P_\lambda}([M], b, e_k; \sigma_\lambda - A) \, e_k^* \otimes e^A$$

where the first equality uses the definition of $*$, the second uses the definition of $\Psi(\lambda)$ and the third uses (10) with $\sigma = \sigma_\lambda - (B - C)$. For more details, see [21, Prop 1.2].

Since $\Psi(\lambda)$ is a unit, the map $b \mapsto \Psi(\lambda) * b$ is injective. Hence for every $b \in H_*(M)$ there has to be some nonzero invariant, $\mathrm{GW}_{P_\lambda}([M], b, c; \sigma_\lambda - B)$ in P_λ. In particular, the image $i_*(b)$ of the class b in $H_*(P_\lambda)$ cannot vanish. Thus the map

$$i_* : H_*(M) \to H_*(P_\lambda)$$

of rational homology groups is injective. By (4), this implies that the homology of P_λ is isomorphic to the tensor product $H_*(S^2) \otimes H_*(M)$. Equivalently, the map

$$\mathrm{tr}_\lambda : H_*(M) \to H_{*+1}(M)$$

is identically zero. This proves Proposition 2.2 (ii) in the case of loops. The proof for the higher homology $H_*(\mathrm{Ham})$ with $* > 1$ is purely topological. Since $H^*(\mathrm{Ham})$ is generated multiplicatively by elements dual to the homotopy, one first reduces to the case when $\phi \in \pi_k(\mathrm{Ham})$. Thus we need only see that all Hamiltonian bundles $M \to P \to B$ with base $B = S^{k+1}$ are c-split, i.e. that Proposition 2.4 holds. Now observe:

Lemma 3.4.

(i) *Let $M \to P' \to B'$ be the pullback of $M \to P \to B$ by a map $B' \to B$ that induces a surjection on rational homology. Then if $M \to P' \to B'$ is c-split, so is $M \to P \to B$.*

(ii) *Let $F \to X \to B$ be a Hamiltonian bundle in which B is simply connected. Then if all Hamiltonian bundles over F and over B are c-split, the same is true for Hamiltonian bundles over X.*

(The proof is easy and is given in [15].) This lemma implies that in order to establish c-splitting when B is an arbitrary sphere it suffices to consider the cases $B = \mathbb{C}P^n$, $B =$, the 1-point blow up X_n of $\mathbb{C}P^n$, and $B = T^2 \times \mathbb{C}P^n$. But the first two cases can be proved by induction using the lemma above and the Hamiltonian bundle

$$\mathbb{C}P^1 \to X_n \to \mathbb{C}P^{n-1}$$

and the third follows by considering the trivial bundle

$$T^2 \to T^2 \times \mathbb{C}P^n \to \mathbb{C}P^n.$$

This completes the proof of Proposition 2.4. Though these arguments can be somewhat extended, they do not seem powerful enough to deal with all Hamiltonian bundles. For some further work in this direction, see Kedra [12].

References

[1] Abreu, M. Topology of symplectomorphism groups of $S^2 \times S^2$, *Inv. Math.*, **131** (1998), 1–23.

[2] Anjos, S. The homotopy type of symplectomorphism groups of $S^2 \times S^2$, *Geometry and Topology*, **6** (2002), 195–218.

[3] Anjos, S. and G. Granja, Homotopy decomposition of a group of symplectomorphisms of $S^2 \times S^2$, AT/0303091.

[4] Atiyah, M. F. and R. Bott, The moment map and equivariant cohomology. *Topology*, **23** (1984), 1–28.

[5] Abreu, M. and D. McDuff, Topology of symplectomorphism groups of rational ruled surfaces, SG/9910057, *Journ. of Amer. Math. Soc.*, **13** (2000) 971–1009.

[6] Buse, O. Relative family Gromov–Witten invariants and symplectomorphisms, SG/01110313.

[7] Entov, M. and L. Polterovich, Calabi quasimorphism and quantum homology, SG/0205247, *International Mathematics Research Notes* (2003).

[8] Gromov, M. Pseudo holomorphic curves in symplectic manifolds, *Inventiones Mathematicae*, **82** (1985), 307–47.

[9] Haller, S. Harmonic cohomology of symplectic manifolds, preprint (2003).

[10] Hofer, H. On the topological properties of symplectic maps. *Proceedings of the Royal Society of Edinburgh*, **115** (1990), 25–38.

[11] Kedra, J. Remarks on the flux groups, *Mathematical Research Letters* (2000).

[12] Kedra, J. Restrictions on symplectic fibrations, SG/0203232.

[13] Kronheimer, P. Some nontrivial families of symplectic structures, preprint (1998).

[14] Lalonde, F. and D. McDuff, The classification of ruled symplectic 4-manifolds, *Math. Research Letters* **3**, (1996), 769–778.

[15] Lalonde, F. and D. McDuff, Symplectic structures on fiber bundles, SG/0010275, *Topology* **42** (2003), 309–347.

[16] Lalonde, F. and D. McDuff, Cohomological properties of ruled symplectic structures, SG/0010277, in *Mirror symmetry and string geometry*, ed Hoker, Phong, Yau, *CRM Proceedings and Lecture Notes*, Amer Math Soc. (2001).

[17] Lalonde, F., D. McDuff and L. Polterovich, On the Flux conjectures, *CRM Proceedings and Lecture Notes* **15** Amer Math Soc., (1998), 69–85.

[18] Lalonde, F., D. McDuff and L. Polterovich, Topological rigidity of Hamiltonian loops and quantum homology, *Invent. Math* **135**, 369–385 (1999).

[19] Lalonde, F. and M. Pinsonnault, The topology of the space of symplectic balls in rational 4-manifolds, preprint 2002, SG/0207096.

[20] Le, H. V. and K. Ono, Parameterized Gromov–Witten invariants and topology of symplectomorphism groups, preprint #28, MPIM Leipzig (2001).

[21] McDuff, D. Quantum homology of Fibrations over S^2, *International Journal of Mathematics*, **11**, (2000), 665–721.

[22] McDuff, D. Symplectomorphism Groups and Almost Complex Structures, SG/0010274, *Enseignment Math.* **38** (2001), 1–30.

[23] McDuff, D. and D. A. Salamon, *J-holomorphic curves and Symplectic Topology*, American Mathematical Society, Providence, RI, to appear.

[24] McDuff, D. and D. Salamon, *Introduction to Symplectic Topology*, 2nd edition (1998) OUP, Oxford, UK.

[25] Oh, Yong-Geun Minimax theory, spectral invariants and geometry of the Hamiltonian diffeomorphism group, SG/0206092.

[26] Pinsonnault, M. Remarques sur la Topologie du Groupe des Automorphismes Symplectiques de l'Éclatment de $S^2 \times S^2$, Ph. D. Thesis, UQAM Montreal (2001).

[27] Polterovich, L. *The Geometry of the Group of Symplectic Diffemorphisms*, Lectures in Math, ETH, Birkhauser (2001).

[28] Schwarz, M. On the action spectrum for closed symplectially aspherical manifolds, *Pacific Journ. Math* **193** (2000), 419–461.

[29] Seidel, P. π_1 of symplectic automorphism groups and invertibles in quantum cohomology rings, *Geom. and Funct. Anal.* **7** (1997), 1046–1095.

[30] Seidel, P. On the group of symplectic automorphisms of $\mathbb{C}P^m \times \mathbb{C}P^n$, *Amer. Math. Soc. Transl.* (2) **196** (1999), 237–250.

[31] Seidel, P. Graded Lagrangian submanifolds, *Bull Math. Soc. France* **128** (2000), 103–149.

E-mail: dusa@math.sunysb.edu

8

K-theory from a physical perspective

GREGORY MOORE

Rutgers University

Abstract

This is an expository paper which aims at explaining a physical point of view on the K-theoretic classification of D-branes. We combine ideas of renormalization group flows between boundary conformal field theories, together with spacetime notions such as anomaly cancellation and D-brane instanton effects. We illustrate this point of view by describing the twisted K-theory of the special unitary groups $SU(N)$.

1 Introduction

This is an expository paper devoted to explaining some aspects of the K-theoretic classification of D-branes. Our aim is to address the topic in ways complementary to the discussions of [1], [2]. Reviews of the latter approaches include [3], [4], [5], [6]. Our intended audience is the mathematician who is well-versed in conformal field theory and K-theory, and has some interest in the wider universe of (nonconformal) quantum field theories.

Our plan for the paper is to begin in Section 2 by reviewing the relation of D-branes and K-theory at the level of topological field theory. Then in Section 3 we will move on to discuss D-branes in conformal field theory. We will advocate a point of view emphasizing 2-dimensional conformal field theories as elements of a larger space of 2-dimensional quantum field theories. 'D-branes' are identified with conformal quantum field theories on 2-dimensional manifolds with boundary. From this vantage, the topological classification of D-branes is the classification of the connected components of the space of 2-dimensional theories on manifolds with boundary which only break conformal invariance through their boundary conditions.

In Section 4 we will turn to conformal field theories which are used to build string theories. In this case, there is a spacetime viewpoint on the classification

194

of D-branes. We will present a viewpoint on D-brane classification, based on anomaly cancellation and 'instanton effects', that turn out to be closely related to the Atiyah–Hirzebruch spectral sequence.

In Section 5 we examine a detailed example, that of branes in WZW models, and show how, using the approach explained in Sections 3 and 4, we can gain an intuitive understanding of the twisted K-theory of $SU(N)$. The picture is in beautiful harmony with a rigorous computation of M. Hopkins.

Let us warn the reader at the outset that in this modest review we are only attempting to give a broad brush overview of some ideas. We are not attempting to give a detailed and rigorous mathematical theory, nor are we attempting to give a comprehensive review of the subject.

2 Branes in 2-dimensional topological field theory

The relation of D-branes and K-theory can be illustrated very clearly in the extremely simple case of 2-dimensional (2D) topological field theory. This discussion was developed in collaboration with Graeme Segal [7].

We will regard a 'field theory' along the lines of Segal's contribution to this volume. It is a functor from a geometric category to some linear category. In the simple case of 2D topological field theory the geometric category has as objects disjoint collections of circles and as morphisms diffeomorphism classes of oriented cobordisms between the objects. The target category is the category of vector spaces and linear transformations. Recall that to give a 2D topological field theory of closed strings is to give a commutative, finite dimensional Frobenius algebra $\mathcal{C}$. For example, the algebra structure follows from Figure 1.

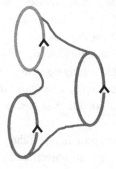

Figure 1. The 3-holed surface corresponds to the basic multiplication of the Frobenius algebra

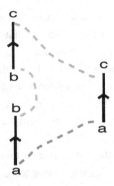

Figure 2. Multiplication defining the nonabelian Frobenius algebra of open strings

Let us now enlarge our geometric category to include open as well as closed strings. Now there are ingoing/outgoing circles and intervals, while the morphisms are surfaces with two kinds of boundaries: ingoing/outgoing boundaries as well as 'free-boundaries', traced out by the endpoints of the in/outgoing intervals. These free boundaries must be labelled by 'boundary conditions' which, for the moment, are merely labels $a, b, \ldots$.

Because we have a functor, to any pair of boundary conditions we associate a vector space ('a statespace') $\mathcal{H}_{ab}$. Moreover, there is a coherent system of bilinear products

$$\mathcal{H}_{ab} \otimes \mathcal{H}_{bc} \to \mathcal{H}_{ac} \qquad (2.1)$$

defined by Figure 2. This leads us to ask the key question: *What boundary conditions are compatible with* $\mathcal{C}$? 'Compatibility' means coherence with 'sewing' or 'gluing' of surfaces; more precisely, we wish to have a well-defined functor. Thus, just the way Figure 1 defines an associative commutative algebra structure on $\mathcal{C}$, Figure 2, in the case $a = b = c$, defines a (not necessarily commutative) algebra structure on $\mathcal{H}_{aa}$. Moreover $\mathcal{H}_{aa}$ is a Frobenius algebra. Next there are further sewing conditions relating the open and closed string structures. Thus, for example, we require that the operators defined by Figures 3 and 4 be equal, a condition sometimes referred to as the 'Cardy condition.'

As observed by Segal some time ago [8], the proper interpretation of (2.1) is that the boundary conditions are objects $a, b, \ldots$ in an additive category with

$$\mathcal{H}_{ab} = \mathrm{Mor}(a, b). \qquad (2.2)$$

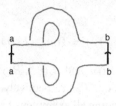

Figure 3. In the open string channel this surface defines a natural operator
$\pi : \mathcal{H}_{aa} \to \mathcal{H}_{bb}$ on noncommutative Frobenius algebras

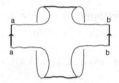

Figure 4. In the closed string channel this surface defines a composition of
open–closed and closed–open transitions $\iota_{c-o}\iota_{o-c} : \mathcal{H}_{aa} \to \mathcal{H}_{bb}$ that factors through
the center

Therefore, we should ask what the sewing constraints imply for the category
of boundary conditions. This question really consists of two parts: First, coherence of sewing is equivalent to a certain algebraic structure on the target
category. Once we have identified that structure we can ask for a classification of the examples of such structures. The first part of this question has been
completely answered: The open–closed sewing conditions were first analyzed
by Cardy and Lewellen [9], [10], and the resulting algebraic structure was described in [7], [11]. The result is the following:

Proposition. *To give an open and closed 2D oriented topological field theory
is to give*

(1) *A commutative Frobenius algebra* $\mathcal{C}$.
(2) *Frobenius algebras* $\mathcal{H}_{aa}$ *for each boundary condition a.*
(3) *A homomorphism* $\iota_a : \mathcal{C} \to Z(\mathcal{H}_{aa})$, *where* $Z(\mathcal{H}_{aa})$ *is the center, such
that* $\iota_a(1) = 1$, *and such that, if* ι^a *is the adjoint of* ι_a *then*

$$\pi_b{}^a = \iota_b \iota^a. \tag{2.3}$$

Here $\pi_b{}^a : \mathcal{H}_{aa} \to \mathcal{H}_{bb}$ is the morphism, determined purely in terms
of open string data, described by Figure 3. When $\mathcal{H}_{ab}$ is the nonzero vector space it is a Morita equivalence bimodule and $\pi_b{}^a$ can be written as
$\pi_b{}^a(\psi) = \sum \psi^\mu \psi \psi_\mu$ where ψ_μ is a basis for $\mathcal{H}_{ab}$ and ψ^μ is a dual basis

for $\mathcal{H}_{ba}$. More invariantly

$$\theta_b(\pi_b{}^a(\psi)\chi) = \mathrm{Tr}_{\mathcal{H}_{ab}}(L(\psi)R(\chi)) \tag{2.4}$$

where θ_b is the trace on $\mathcal{H}_{bb}$ and $L(\psi)$, $R(\chi)$ are the left- and right-representations of $\mathcal{H}_{aa}$, $\mathcal{H}_{bb}$ on $\mathcal{H}_{ab}$, respectively.

The second step, that of finding all examples of such structures was analyzed in [7] in the case where $\mathcal{C}$ is a semisimple Frobenius algebra. The answer turns out to be very crisp:

Theorem 2.1. *Let $\mathcal{C}$ be semisimple. Then the set of isomorphism classes of objects in the category of boundary conditions is*

$$K^0(\mathrm{Spec}(\mathcal{C})) = K_0(\mathcal{C}). \tag{2.5}$$

There are important examples of the above structure when $\mathcal{C}$ is *not* semisimple, such as the topological A- and B-twisted $\mathcal{N} = 2$ supersymmetric sigma models. As far as we are aware, the classification of examples for non-semisimple $\mathcal{C}$ is an open problem.

Even in this elementary setting, there are interesting and nontrivial generalizations. When a 2D closed topological field theory has a symmetry G it is possible to 'gauge it'. The cobordism category is enhanced by considering cobordisms of principal G-bundles. In this case the closed topological field theory corresponds to a choice of 'Turaev algebra', [12], [7], a G-equivariant extension of a Frobenius algebra which, in the semisimple case, is characterized by a 'spacetime' consisting of a discrete sets of points (corresponding to the idempotents of the algebra), a 'dilaton', encoding the trace of the Frobenius algebra on the various idempotents, and a 'B-field'. In this case we have

Theorem 2.2. *The isomorphism classes of objects in the category of boundary conditions for a G-equivariant open and closed theory with spacetime X and 'B-field' $[b] \in H_G^2(X; C^*)$ are in 1–1 correspondence with the K-group of G-equivariant, b-twisted K-theory classes: $K_{G,b}(X)$.*

These results are, of course, very elementary. What I find charming about them is precisely the fact that they *are* so primitive: they rely on nothing but topological sewing conditions and a little algebra, and yet K-theory emerges ineluctably.

A more sophisticated category-theoretic approach to the classification of branes in rational conformal field theories has been described in [13], [14].

3 K-theory and the renormalization group

3.1 Breaking conformal invariance on the boundary

Let us now consider the much more difficult question of the topological classification of D-branes in a full conformal field theory (CFT). This immediately raises the question of what we even mean by a 'D-brane'. Perhaps the most fruitful point of view is that D-branes are *local* boundary conditions in a 2D CFT $\mathcal{C}$ which preserve conformal symmetry. While there is an enormous literature on the subject of D-branes, the specific branes which have been studied are really a very small subset of what is possible.

One way to approach the classification of D-branes is to consider the space of 2D quantum field theories (QFTs), defined on surfaces with boundary, which are *not* conformal, but which only break conformal invariance via their boundary conditions. Formally, there is a space $\mathcal{B}$ of such boundary QFT's compatible with a fixed 'bulk' CFT, $\mathcal{C}$. The tangent space to $\mathcal{B}$ is the space of local operators on the boundary because a local operator $\mathcal{O}$ can be used to deform the action on a surface Σ by

$$S_{\text{worldsheet}} = S_{\text{bulkCFT}} + \int_{\partial \Sigma} ds \mathcal{O}. \qquad (3.1)$$

Here ds is a line element. Note that in general we have introduced explicit metric-dependence in this term, thus breaking conformal invariance on the boundary.

As a simple example of what we have in mind, consider a massless scalar field $x^\mu : \Sigma \to \mathbb{R}^n$ with action

$$S_{\text{worldsheet}} = \int_\Sigma \partial x^\mu \bar{\partial} x^\mu + \int_{\partial \Sigma} ds T\left(x^\mu(\tau)\right) \qquad (3.2)$$

where $T(x^\mu)$ is 'any function' on $\mathbb{R}^n$ and τ is a coordinate on $\partial \Sigma$. Then the boundary interaction in can be expanded

$$\mathcal{O} = T(x) + A_\mu(x)\frac{dx^\mu}{d\tau} + B_\mu(x)\frac{d^2 x^\mu}{d\tau^2} + C_{\mu\nu}(x)\frac{dx^\mu}{d\tau}\frac{dx^\nu}{d\tau} + \cdots \qquad (3.3)$$

The coefficients $T(x)$, $A_\mu(x)$, ... are viewed as spacetime fields on the target space $\mathbb{R}^n$.[1]

We expect that $\mathcal{B}$ can be given a topology such that renormalization group flow (see below) is a continuous evolution on this space. In this topology $\mathcal{B}$ is a disconnected space. The essential idea is that the *connected components* of this space are classified by some kind of K-theory. For example, if the conformal field theory is supersymmetric and has a target space interpretation in terms

---

[1] G. Segal points out to me that the proper formulation of the tangent space to a boundary CFT would naturally use the theory of jets.

of a nonlinear σ model, we expect the components of $\mathcal{B}$ to correspond to the K-theory of the target space X

$$\pi_0(\mathcal{B}) = K(X). \qquad (3.4)$$

Remarks:

(1) In equation we are being deliberately vague about the precise form of K-theory (e.g. K, vs. KO, KR, $K_{\pm}$ etc.). This depends on a discrete set of choices one makes in formulating the 2D field theory.

(2) From this point of view the importance of some kind of supersymmetry on the worldsheet is clear. As an example in the next section makes clear, the RG flow corresponding to taking $\mathcal{O}$ to be the unit operator always flows to a trivial fixed point with 'no boundary'. Therefore, unless the unit operator can be projected out, there cannot be interesting path components in $\mathcal{B}$. In spacetime terms, we must cancel the 'zero momentum tachyon'.

(3) One might ask what replaces (3.4) when the CFT does not have an obvious target space interpretation. One possible answer is that one should define some kind of algebraic K-theory for an open string vertex operator algebra. There has been much recent progress in understanding more deeply Witten's Chern–Simons open string field theory (see, e.g. [15], [16], [17], [18], [19]). This holds out some hope that the K-theory of the open string vertex algebra could be made precise. In string field theory D-branes are naturally associated to projection operators in a certain algebra, so the connection between K-theory and branes is again quite natural. See also Section 3.5 below.

(4) The space $\mathcal{B}$ appeared in a proposal of Witten's for a background-independent string field theory [20] (Witten's 'other' open string field theory).

(5) It is likely that the classification of superconformal boundary conditions in the supersymmetric Gaussian model is complete [21], [22], [23]. The classification is somewhat intricate and it would be interesting to see if it is compatible with the general proposal of this paper.

3.2 Boundary renormalization group flow

One way physicists explore the path components of $\mathcal{B}$ is via 'renormalization group (RG) flow'. Since conformal invariance is broken on the boundary, we can ask what happens as we scale up the size of the boundary. This scaling defines 1-dimensional flows on $\mathcal{B}$. These are the integral flows of a vector field β on $\mathcal{B}$ usually referred to as the 'beta function'. A D-brane, or conformal fixed point, corresponds to a zero of β. Two D-branes which are connected by RG flow are in the same path component, and therefore have the same 'K-theory charge'.

Let us recall a few facts about boundary RG flow. For a good review see [24]. For simplicity we will consider the bosonic case. A boundary condition $a \in \mathcal{B}$ is a zero of β. At such an RG fixed point the theory is conformal, and hence the Virasoro algebra acts on the tangent space $T_a\mathcal{B}$. We may choose a basis of local operators such that $L_0 \mathcal{O}_i = \Delta_i \mathcal{O}_i$. Here L_0 is the scaling operator in the Virasoro algebra. We may then choose coordinates $\mathcal{O} = \sum_i \lambda^i \mathcal{O}_i$ such that, in an open neighborhood of $a \in \mathcal{B}$

$$\beta \cong -\sum_i (1 - \Delta_i)\lambda^i \frac{d}{d\lambda^i} \tag{3.5}$$

Thus, as usual, perturbations by operators with $\Delta_i < 1$ correspond to unstable flows in the infrared (IR). It turns out there is an analog of Zamolodchikov's c-theorem. Boundary RG flow is gradient flow with respect to an 'action functional'. To construct it one introduces the natural function on $\mathcal{B}$ given by the disk partition function. Then set

$$g := (1 + \beta)Z_{\text{disk}}. \tag{3.6}$$

Next one introduces a metric on $\mathcal{B}$. Recalling that the local operators are to be identified with the tangent space we write

$$G(\mathcal{O}_1, \mathcal{O}_2) = \oint d\tau_1 d\tau_2 \sin^2(\frac{\tau_1 - \tau_2}{2})\langle \mathcal{O}_1(\tau_1)\mathcal{O}_2(\tau_2)\rangle_{\text{disk}}. \tag{3.7}$$

Then, the 'g-theorem' states that

$$\dot{g} = -\beta^i \beta^j G_{ij}. \tag{3.8}$$

The main nontrivial statement here is that $\iota(\beta)G$ is a locally exact one-form.

Remarks:

(1) The g-theorem was first proposed by Affleck and Ludwig [25], [26], who verified it in leading order in perturbation theory. An argument for the g-theorem, based on string field theory ideas, was proposed in [27], [28].

(2) In the Zamolodchikov theorem, the c-function at a conformal fixed point is the value of the Virasoro central charge of the fixed point conformal field theory. It is therefore natural to ask: 'What is the meaning of g at a conformal fixed point?' The answer is the 'boundary entropy'. For example, when the CFT $\mathcal{C}$ is an RCFT with irreps $\mathcal{H}_i$, $i \in I$ of

the chiral algebra, the boundary CFT's preserving the symmetry are labelled by $i \in I$ and the g-function for these conformal fixed points is expressed in terms of the modular S-matrix via

$$g = \frac{S_{0i}}{\sqrt{S_{00}}} \tag{3.9}$$

where 0 denotes the unit representation. It is notable that this can also be interpreted as a regularized dimension of the open string statespace $\sqrt{\dim \mathcal{H}_{ii}}$. If the CFT is part of a string theory with a target space interpretation then we can go further. In a string theory we have gravity and in this context the value of g at a conformal fixed point is the brane tension, or energy/volume of the brane [29].

(3) In the case of $\mathcal{N} = 1$ worldsheet supersymmetry we should take instead [28], [30], [31]

$$g := Z_{\text{disk}}. \tag{3.10}$$

(4) In an interesting series of papers Connes and Kreimer have reinterpreted perturbative renormalization of field theory and the renormalization group in terms of the structure of Hopf algebras [32], [33]. We believe that the case of boundary RG flow in two-dimensions might be a very interesting setting in which to apply their ideas.

3.3 Tachyon condensation from the worldsheet viewpoint

Here is a simple example of the g theorem. Consider a single scalar field on the disk $x : D \to \mathbb{R}$, where the disk D has radius r. Then

$$Z_{\text{disk}} = \int [dx] e^{-\int_D \partial x \bar{\partial} x + \oint_{\partial D} T(x)}. \tag{3.11}$$

Let us just take $T(x) = t = $ constant. Then, trivially, $Z_{\text{disk}}(t) = Z_{\text{disk}}(0)e^{-2\pi rt} = Z_{\text{disk}}(0)e^{-2\pi t(r)}$. Then

$$\beta^t := -\frac{\partial t(r)}{\partial (\log r)} = -t \quad \Rightarrow \quad \beta = -t\frac{d}{dt} \tag{3.12}$$

and an easy computation shows the metric is

$$ds^2 = e^{-t}(dt)^2. \tag{3.13}$$

The g-function, or action, in this case is

$$g(t) = (1 + \beta)Z_{\text{disk}} = (1 + 2\pi rt)e^{-2\pi rt}g(0) = (1 + 2\pi t(r))e^{-2\pi t(r)}g(0). \tag{3.14}$$

At $t = 0$, Z_{disk} is r-independent (hence conformally invariant) if we choose, say, Neumann boundary conditions for x. Thus at $t = 0$ we begin with an open/closed CFT consisting of a 'D1-brane' wrapping the target $\mathbb{R}$ direction. Under RG flow to the IR, $t \to \infty$. At $t = \infty$ all boundary amplitudes are infinitely suppressed and 'disappear'. We are left with a theory only of closed strings!

Remarks:

(1) The RG flow (3.14) is unusual in that we can give exact formulae. This is due to its rather trivial nature. Moreover, note that this boundary interaction cancels out of all normalized correlators. Nevertheless, we feel that the above example nicely captures the essential idea. A less trivial example based on the boundary perturbation $\oint u X^2$ is analyzed in [34], [35], [27].

(2) Let us return to remark (1) of section 3.1. It is precisely the zero-momentum tachyon (i.e. the unit operator) whose flow we wish to suppress in order to define a space $\mathcal{B}$ with interesting path components.

(3) The example of this section is essentially the 'boundary string field theory' (BSFT) interpretation of Sen's tachyon condensation [36]. In [20] Witten introduced an alternative formulation of open string field theory, in which (at least when ghosts decouple), the function g is the spacetime action. This theory was further developed by Witten and Shatashvili in [34], [37], [38], [39]. Interest in the theory was revived by [40], [41], [35], [27], [28]. These papers showed, essentially using the above example, that the dependence of the spacetime effective potential on the tachyon field is

$$V(T) \sim (T + 1)e^{-T} \qquad (3.15)$$

for the bosonic string and

$$V(T) \sim e^{-T^2} \qquad (3.16)$$

for the type IIA string (on an unstable D9 brane). The tachyon potential is minimized by $T \to \infty$, and at its minimum the open strings 'disappear'.

3.4 g-function for the nonlinear sigma model

Suppose the closed CFT $\mathcal{C}$ is a σ-model with spacetime X, dilaton Φ, metric $g_{\mu\nu}$ and 'gerbe connection' $B_{\mu\nu}$. A typical boundary condition involves, first of all, a choice of topological K-homology cycle [42], that is, an embedded subvariety $\iota : W \hookrightarrow X$ with Spin^c structure (providing appropriate Dirichlet boundary conditions for the open strings) together with a choice of (complex) vector bundle

$$E \to W \tag{3.17}$$

modulo some equivalence relations. We say a 'D-brane wraps W with Chan–Paton bundle E'.

In the supersymmetric case the most important boundary interaction is a choice of a (unitary) connection A_μ on E and a (non-abelian) section of the normal bundle. In this paper we will set the normal bundle scalars to zero (although they are very interesting). Thus the g-function becomes

$$g = \left\langle \text{Tr}_E P \exp\left(\oint_{\partial D} d\tau A_\mu(x(\tau)) \dot{x}^\mu(\tau) + F_{\mu\nu} \psi^\mu \psi^\nu + \cdots \right) \right\rangle \tag{3.18}$$

where ψ^μ are the susy partners of x^μ. When E is a line bundle, g can be computed for a variety of backgrounds and turns out to be the Dirac–Born–Infeld (DBI) action [43]

$$g = \int_W e^{-\Phi} \sqrt{\det_{\mu\nu}(g_{\mu\nu} + B_{\mu\nu} + F_{\mu\nu})} + \mathcal{O}((DF)^2). \tag{3.19}$$

If E has rank ≥ 1 to get a nice formula we need to add the condition $F_{\mu\nu} \ll 1$. In this case we have

$$g = \text{rank}(E) \int e^{-\Phi} \sqrt{\det(g + B)} + \int_X e^{-\Phi} \text{Tr}(F \wedge *F) + \cdots \tag{3.20}$$

Remarks:

(1) It follows from (3.20) that in the long-distance limit the gradient flows of the 'g-theorem' generalize nicely some flows which appeared in the work of Donaldson on the Hermitian–Yang–Mills equations [44].[2] Let X be a Calabi–Yau manifold. To X we associate an $\mathcal{N} = (2, 2)$ superconformal field theory $\mathcal{C}$. The boundary interaction (3.18) preserves $\mathcal{N} = 2$ supersymmetry iff F is of type $(1, 1)$, i.e., iff $F^{2,0} = 0$ [45], [46]. RG flow preserves $\mathcal{N} = 2$ susy, and hence preserves the $(1, 1)$ condition on the fieldstrength. A boundary RG fixed point is defined (in the $\alpha' \to 0$ limit) by an Hermitian Yang–Mills connection. The RG flow is precisely the flow

$$\frac{dA_\mu}{dt} = D_\nu F^\nu{}_\mu. \tag{3.21}$$

---

[2] This remark is based on discussions with M. Douglas.

Thus, one can view the flow from a perturbation of an unstable bundle to a stable one as an example of tachyon condensation. It might be interesting to think through systematically the implications for tachyon condensation of Donaldson's results on the convergence of these flows.

(2) The tachyon condensation from unstable $D9$ branes (or $D9\overline{D9}$ branes) to lower dimensional branes involves the Atiyah–Bott–Shapiro construction and Quillen's superconnection in an elegant way. This has been demonstrated in the context of BSFT advocated in this section in [28], [47], [48]. Given the boundary data in (3.18) one is naturally tempted to see a role for the 'differential K-theory' described in [49]. However, the non-abelian nature of the normal bundle scalars show that this is only part of the story. See [50], [51], [52] for some relevant discussions.

3.5 The Dirac–Ramond operator and the topology of $\mathcal{B}$

Let us now make some tentative remarks on how one might try to distinguish different components of $\mathcal{B}$. There are many indications that K-homology is a more natural framework for thinking about the relation of D-branes and K-theory [53], [4], [54], [55], [56], [57], [50], [51], [6]. It was pointed out some time ago by Atiyah that the Dirac operator defines a natural K-homology class [58]. Indeed, abstracting the crucial properties of the Dirac operator leads to the notion of a Fredholm module [59].

Now, in string theory, the Dirac operator is generalized to the Dirac–Ramond operator, that is, the supersymmetry operator Q; (often denoted G_0) acting in the Ramond sector of a superconformal field theory. Q is a kind of Dirac operator on loop space as explained in [60], [61], [63], [64].

In the case of open strings it is still possible to define Q in the Ramond sector, and Q still has an interpretation as a Dirac operator on a path space. For example, suppose the $\mathcal{N} = 1$ CFT has a sigma model interpretation with closed string background data $g_{\mu\nu} + B_{\mu\nu}$. Suppose that the open string boundary conditions are $x(0) \in \mathcal{W}_1$, $x(\pi) \in \mathcal{W}_2$, where the submanifolds $\mathcal{W}_i$ are equipped with vector bundles E_i with connections A_i. The supersymmetry operator will take the form

$$Q = \int_0^\pi d\sigma\, \psi^\mu(\sigma) \left(\frac{\delta}{\delta x^\mu(\sigma)} + g_{\mu\nu}(x(\sigma)) \frac{dx^\nu}{d\sigma} + \left(\omega_{\mu\nu\lambda} + H_{\mu\nu\lambda} \right) \psi^\nu \psi^\lambda \right)$$

$$+ \psi^\mu(0) A_{1,\mu}(x(0)) - \psi^\mu(\pi) A_{2,\mu}(x(\pi)) \tag{3.22}$$

where $\omega_{\mu\nu\lambda}$ is the Riemannian spin connection on X, and $H_{\mu\nu\lambda}$ is the field-strength of the B-field. Just as in the closed string case, Q can be understood more conceptually as a Dirac operator on a bundle over the path space

$$\mathcal{P}(\mathcal{W}_1, \mathcal{W}_2) = \{x : [0, \pi] \to X | x(0) \in \mathcal{W}_1, x(\pi) \in \mathcal{W}_2\} \qquad (3.23)$$

(Preservation of supersymmetry imposes further boundary conditions on x. See, for example, [64], [65], [66] for details.) Quantization of $\psi^\mu(\sigma)$ for fixed $x^\mu(\sigma)$ produces a Fermionic Fock space. This space is to be regarded as a spin representation of an infinite dimensional Clifford algebra. These Fock spaces fit together to give a Hilbert bundle $\mathcal{S}$ over $\mathcal{P}(\mathcal{W}_1, \mathcal{W}_2)$. The data $g_{\mu\nu} + B_{\mu\nu}$ induce a connection on this bundle, as indicated in (3.22). The effect of the boundaries is merely to change the bundle to

$$\mathcal{S} \to \mathrm{ev}_0^*(E_1) \otimes (\mathrm{ev}_\pi^*(E_2))^* \otimes \mathcal{S} \qquad (3.24)$$

where ev is the evaluation map. The connections A_1, A_2 induce connections on (3.24). In the zeromode approximation Q becomes the Dirac operator on $E_1 \otimes E_2^* \to \mathcal{W}_1 \cap \mathcal{W}_2$

$$Q \to \not{D}_{E_1 \otimes E_2^*} + \cdots \qquad (3.25)$$

Now let us consider RG flow. If RG flow connects boundary conditions a to a' then the target space interpretation of the superconformal field theories $\mathcal{H}_{ab}$ and $\mathcal{H}_{a'b'}$ can be very different. For example, tachyon annihilation can change the dimensionality of $\mathcal{W}$. Another striking example is the decay of many D0-branes to a single D2-brane discussed in Section 5.5 below. It follows that any formulation of an RG invariant involving geometrical constructions such as vector bundles over path space is somewhat unnatural. However, what *does* make sense throughout the renormalization group trajectory is the supersymmetry operator Q (so long as we restrict attention to $\mathcal{N} = 1$ supersymmetry-preserving flows). Moreover, it is physically 'obvious' that Q changes continuously under RG flow. This suggests that the components of $\mathcal{B}$ should be characterized by some kind of 'homotopy class' of Q.

The conclusion of the previous paragraph immediately raises the question of where the homotopy class of Q should take its value. We need to define a class of operators and define what is meant by continuous deformation within that class. While we do not yet have a precise proposal we can again turn to the zero-slope limit for guidance. In this limit, as we have noted, $Q \to \not{D}_{E_1 \otimes E_2^*}$, and $\not{D}$ defines, in a well-known way, a 'θ-summable K-cycle' for $\mathcal{A}$, the C^* algebra completion of $C^\infty(\mathcal{W}_1 \cap \mathcal{W}_2)$, acting on the Hilbert space of L^2 sections of $S \otimes E_1 \otimes E_2^*$ over $\mathcal{W}_1 \cap \mathcal{W}_2$. That is $[(\mathcal{H}, \not{D})] \in K^0(\mathcal{A})$ [59]. What is the generalization when we do *not* take the zeroslope limit? One possibility, in the closed string case, has been discussed in[67], [68], [69], [59]. Another possibility is that one can define a notion of Fredholm module for vertex operator algebras. This has the disadvantage that it is tied to a particular conformal boundary condition a. It is possible, however, that the open string vertex operator algebras $\mathcal{A}_{aa}$ for different boundary conditions a are 'Morita equivalent' and that the homotopy class of Q defines an element of some K-theory (which remains to be defined) '$K^0(\mathcal{A}_{aa})$'. This group should be independent of a and only depend on $\mathcal{C}$. (See Section 6.4 of [70] and [7] for some discussion of this idea.)

Remarks:

(1) For some boundary conditions a it is also possible to introduce a 'tachyon field'. In this case the connection term $\psi^\mu(0)A_\mu(x(0))$ is replaced by Quillen's superconnection. This happens, for example, if a represents a $D-\bar{D}$ pair with $\mathbb{Z}_2$-graded bundle $E^+\oplus E^-$. If the tachyon field $T \in \mathrm{End}(E^+, E^-)$ is everywhere an isomorphism then boundary conditions with a are in the same component as the trivial boundary condition, essentially by the example of Section 3.3.

(2) One strong constraint on the above considerations is that the Witten index

$$\mathrm{Tr}_{\mathcal{H}^R_{ab}}(-1)^F e^{-\beta Q^2} \tag{3.26}$$

must be a renormalization group invariant. In situations where we have the limit (3.25) we can use the index theorem to classify, in part, the components of $\mathcal{B}$. Of course, this will miss the torsion elements of the K-theory.

4 K-theory from anomalies and instantons

In this section we consider the question of understanding the connected components of $\mathcal{B}$ in the case where there is a geometrical target space interpretation of the CFT. We will be shifting emphasis from the worldsheet to the target space. We will use an approach based on a spacetime picture of branes as objects wrapping submanifolds of X to give an argument that (twisted) K-theory should classify components of $\mathcal{B}$.

For concreteness, suppose our CFT is part of a background in type II string theory in a spacetime

$$X_9 \times \mathbb{R} \tag{4.1}$$

where the $\mathbb{R}$ factor is to be thought of as time, while X_9 is compact and spin. Suppose moreover that spacetime is equipped with a B-field with fieldstrength

H. This can be used to introduce a twisted K-group $K_H(X_9)$.[3] We will show how $K_H(X_9)$, arises naturally in answering the question: *What subvarieties of X_9 can a D-brane wrap?* The answer involves *anomaly cancellation* and *instanton effects,* and leads to the slogan: 'K-theory = anomalies modulo instantons.' As an example of this viewpoint, we apply it to compute the twisted K-theory $K_H(SU(N))$ for $N = 2, 3$. We will be following the discussion of [71]. For other discussions of the relation of twisted K-theory to D-branes see [2], [72], [73], [74], [75], [49]. The point of view presented here has been further discussed in [76], [77].

4.1 What subvarieties of X_9 can a D-brane wrap?

Since we are discussing topological restrictions and classification, we will identify D-brane configurations which are obtained via continuous deformation. Traditionally, then, we would replace the cycle W wrapped by a D-brane by its homology class. This leads to the 'cohomological classification of D-branes.' In the cohomological classification of branes we follow two rules:

(A) Free branes[4] can wrap any nontrivial homology cycle, if $\iota^*(H_{DR})$ is exact.

(B) A brane wrapping a nontrivial homology cycle is absolutely stable.

In the K-theoretic classification of branes we have instead the modified rules:

(A') D-branes can wrap $W \subset X_9$ only if $W_3(W) + [H]|_W = 0$ in $H^3(W, \mathbb{Z})$.

(B') Branes wrapping homologically nontrivial W can be unstable if, for some $W' \subset X_9$, $PD(W \subset W') = W_3(W') + [H]|_{W'}$.

Here and below, $W_3(W) := W_3(NW)$ is the Stiefel–Whitney class of the *normal* bundle of W in X_9.

We will first explain the physical reason for (A') and (B') and then explain the relation of (A') and (B') to twisted K-theory.

To begin, condition (A') is a condition of anomaly cancellation. Consider a string worldsheet D with boundary on a D-brane wrapping $W \times \mathbb{R}$ as in Figure 5. The g-function is, schematically

---

[3] In fact, H should be refined to a 3-cocycle for integral cohomology. In the examples considered in detail below this refinement is not relevant.

[4] i.e., branes considered in isolation, with no other branes ending on them.

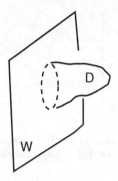

Figure 5. A disk string worldsheet ends on a D-brane worldvolume $\mathcal{W}$

$$g = \int [Dx][D\psi] \, e^{-\int_D \partial x \bar{\partial} x + \psi \partial \psi + \cdots} \, e^{i \int_D B} \, \text{Pfaff}(\not{D}_D) \, \text{Tr}_E P e^{i \oint_{\partial D} A}. \quad (4.2)$$

The measure of the path integral must be well-defined on the space of all maps

$$\{x : D \to X_9 \quad : \quad x(\partial D) \subset \mathcal{W}\}. \quad (4.3)$$

By considering a loop of paths such that ∂D sweeps out a surface in $\mathcal{W}$ it is easy to see that, at the level of the DeRham complex

$$\iota^*(H_{DR}) = d\mathcal{F} \quad (4.4)$$

must be trivialized. Heuristically $\mathcal{F} := F + \iota^*(B)$, although neither F nor $\iota^*(B)$ is separately well-defined. Note that it is the combination $\mathcal{F}$ which appears in the g-function (3.19), and hence *must* be globally well-defined on the brane worldvolume $\mathcal{W} \times \mathbb{R}$. The equation $d\mathcal{F} = \iota^*(H_{DR})$ means H_{DR} is a magnetic source for $\mathcal{F}$ on the brane worldvolume $\mathcal{W} \times \mathbb{R}$.

A more subtle analysis of global anomaly cancellation by Freed and Witten [78] shows that

$$\iota^*[H] + W_3(T\mathcal{W}) = 0 \quad (4.5)$$

at the level of integral cohomology. (See also the discussion of [72].)

Let us now turn to the stability condition (B'). Suppose there is a cycle $\mathcal{W}' \subset X_9$ on which

$$W_3(\mathcal{W}') + [H]|_{\mathcal{W}'} \neq 0. \quad (4.6)$$

As we have just seen, anomaly cancellation implies that we cannot wrap a D-brane on $\mathcal{W}'$. However, while a free brane wrapping $\mathcal{W}'$ is anomalous, we can cancel the anomaly by adding a magnetic source for $\mathcal{F}$. A D-brane ending

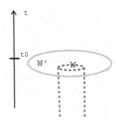

Figure 6. A D-brane wrapping spatial cycle $\mathcal{W}$ propagates in time and terminates on a configuration $\mathcal{W}'$ localized in time. This configuration of D-branes is anomaly-free

on a codimension 3-cycle $\mathcal{W} \subset \mathcal{W}'$ provides such a magentic source. Hence, we can construct an anomaly free configuration by adding a D-brane wrapping a cycle $\mathbb{R}^- \times \mathcal{W}$ that *ends* on $\mathcal{W} \subset \mathcal{W}'$, where $\mathcal{W}$ is such that

$$PD(\mathcal{W} \subset \mathcal{W}') = W_3(\mathcal{W}') + [H]|_{\mathcal{W}'}. \qquad (4.7)$$

Here $\mathbb{R}^-$ should be regarded as a semiinfinite interval in the time-direction as in Figure 6.

Figure 6 suggests a clear physical interpretation. A brane wraps a spatial cycle $\mathcal{W}$, propagates in time, and terminates on a D-*'instanton'* wrapping $\mathcal{W}'$. This means the brane wrapping a spatial cycle $\mathcal{W}$ can be unstable, and decays due to the configuration wrapping $\mathcal{W}'$.[5] The basic mechanism is closely related to the 'baryon vertex' discussed by Witten in the AdS/CFT correspondence [79].

4.2 Relation to K-theory via the Atiyah–Hirzebruch spectral sequence

(A') and (B') are in fact conditions of K-theory. In order to understand this, let us recall the Atiyah–Hirzebruch spectral sequence (AHSS). Let X be a manifold. A K-theory class $x \in K^0(X)$ determines a system of integral cohomology classes: $c_i(x) \in H^{2i}(X, \mathbb{Z})$, while $x \in K^1(X)$ determines $\omega_{2i+1}(x) \in H^{2i+1}(X, \mathbb{Z})$. Let us ask the converse. Given a system of cohomology classes, $(\omega_1, \omega_3, \dots)$ does there exist an $x \in K^1(X)$? The AHSS is a successive approximation scheme: $E_1^*, E_3^*, E_5^*, \dots$ for describing when such a system of cohomology classes $(\omega_1, \omega_3, \dots)$ arises from a K-theory class.

In order to relate the AHSS to D-branes we regard $PD(\omega_k)$ in X as the spatial cycle of a (potentially unstable) brane of spatial dimension $\dim X - k$.

---

[5] While we use the term 'instanton' for brevity, the process illustrated in Figure 6 need not be non-perturbative in string theory. Indeed, the example of Section 5.5 below is a process in *classical* string theory. The decay process is simply localized in the time direction.

In this way, a system $(\omega_1, \omega_3, \dots)$ determines a collection of branes, and hence the AHSS helps us decide which subvarieties of X can be wrapped. Now let us look at the AHSS in more detail.

The first approximation is the cohomological classification of D-branes.

$$K^0(X) \sim E_1^0(X) := H^{\mathrm{even}}(X, \mathbb{Z})$$
$$K^1(X) \sim E_1^1(X) := H^{\mathrm{odd}}(X, \mathbb{Z}). \tag{4.8}$$

The first nontrivial approximation is

$$K^0(X) \sim E_3^0(X) := \left(\mathrm{Ker}\, d_3|_{H^{\mathrm{even}}}\right) / \left(\mathrm{Im}\, d_3|_{H^{\mathrm{odd}}}\right)$$
$$K^1(X) \sim E_3^1(X) := \left(\mathrm{Ker}\, d_3|_{H^{\mathrm{odd}}}\right) / \left(\mathrm{Im}\, d_3|_{H^{\mathrm{even}}}\right) \tag{4.9}$$

with

$$d_3(a) := Sq^3(a) + [H] \smile a. \tag{4.10}$$

Let us pause to define $Sq^3(a)$. Let us suppose, for simplicity, that the Poincaré dual $PD(a)$ can be represented by a manifold W and let $\iota : W \hookrightarrow X_9$ be the inclusion. Then we let

$$Sq^3(a) = \iota_*(W_3(W)) \tag{4.11}$$

where ι_* is a composition of three operations: first take the Poincaré dual of $W_3(W)$ within W, then push forward the homology cycle, and then take the Poincaré dual in X_9. Equivalently, regard a as a class compactly supported in a tubular neighborhood of W and consider the class $W_3(W) \smile a$ where $W_3(W)$ is pulled back to the tubular neighborhood.

Returning to the AHSS, in general one must continue the approximation scheme. This is true, for example, when computing the twisted K-theory of $SU(N)$ for $N \geq 3$.

Now, let us interpret the procedure of taking d_3 cohomology in physical terms. To interpret $\mathrm{Ker}\, d_3$ note that from (4.11) it follows that

$$d_3(a) = 0 \quad \Leftrightarrow \quad \left(W_3(W) + [H]\right) \smile a = 0. \tag{4.12}$$

Recall that global anomaly cancellation for a D-brane wrapping W implies

$$W_3(W) + [H]|_W = 0 \tag{4.13}$$

and this in turn implies $d_3(a) = 0$. Thus, the physical condition (A') implies $PD(W) \in \mathrm{Ker}\, d_3$.

Next, let us interpret the quotient by the image of d_3 in (4.9). Suppose $a = d_3(a') = (Sq^3 + [H])(a')$. Then, choose representatives

$$PD(a) = \mathcal{W} \qquad PD(a') = \mathcal{W}' \tag{4.14}$$

where $\mathcal{W}$ is codimension 3 in $\mathcal{W}'$. A D-brane terminating on $\mathcal{W}$ can be the magnetic source for the D-brane gauge field on $\mathcal{W}'$ and

$$PD\big(\mathcal{W} \hookrightarrow \mathcal{W}'\big) = W_3(\mathcal{W}') + [H]|_{\mathcal{W}'} \;\Rightarrow\; a = d_3(a') \tag{4.15}$$

Therefore, the physical process of D-instanton induced brane instability implies one should take the quotient by the image of d_3 [80], [71]. (In fact, conditions (A'), (B') contain more information than d_3.)

4.3 Examples: Twisted K-groups of $SU(N)$

As an illustration of the above point of view let us consider the twisted K-groups of $SU(2)$ and $SU(3)$.

Consider first $K_H(SU(2))$. Then $H = k\omega$ where ω generates $H^3(SU(2); \mathbb{Z})$. In the cohomological model of branes we have $H^{\text{even}}(SU(2)) = H^0 \cong H_3 = \mathbb{Z}$, corresponding to 'D3-branes' (or D2-instantons) while $H^{\text{odd}}(SU(2)) = H^3 \cong H_0 = \mathbb{Z}$ corresponding to 'D0-branes'. Now condition (A') shows that we can only have D0-branes. Indeed, D2-instantons wrapping $SU(2) = S^3$ violate D0-brane charge by k units as in Figure 7. For this reason if we take $SU(2)$ as the cycle $\mathcal{W}'$ in condition (B') then it follows that a system with k D0-branes is in the same connected component of $\mathcal{B}$ as a system with no D0-branes at all. In Section 5.5 we will explain in more detail how this can be.

In this way we conclude that

$$K_H^0(SU(2)) = 0$$
$$K_H^1(SU(2)) = \mathbb{Z}/k\mathbb{Z} \tag{4.16}$$

as is indeed easily confirmed by rigorous mathematical arguments.

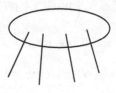

Figure 7. k D0-branes terminate on a wrapped D2-brane instanton in the $SU(2)$ level k theory

Let us now consider $K_H(SU(3))$. Here the AHSS is not powerful enough to determine the K-group. However, it is important to bear in mind that the physical conditions A', B' contain more information, and are stronger, than the d_3-cohomology. Once again we take $H = k\omega$, where ω generates $H^3(SU(3); \mathbb{Z})$.

In the cohomological model we have $H^{\text{odd}} \cong H_3 \oplus H_5$. Now, 3-branes cannot wrap $SU(2) \subset SU(3)$ since $[H]_{DR} \neq 0$. But 5-branes can wrap the cycle $M_5 \subset SU(3)$, where M_5 is Poincaré dual to ω. Now

$$\int_{SU(3)} \omega Sq^2 \omega = 1 \tag{4.17}$$

and hence

$$\iota^*(\omega) = W_3(M_5) \tag{4.18}$$

is nonzero. (In fact, it turns out that the cycle M_5 can be represented by the space of symmetric $SU(3)$ matrices. This space is diffeomorphic to $SU(3)/SO(3)$ and is a simple example of a non-Spin^c manifold.)

It follows from (4.18) that if M_5 is wrapped r times, anomaly cancellation implies

$$r(k+1)W_3 = 0. \tag{4.19}$$

The D-brane instantons relevant to condition (B') are just the $D2$-branes wrapping $SU(2)$. We thus conclude that

$$K^1_{H=k\omega}(SU(3)) = \begin{cases} \mathbb{Z}/k\mathbb{Z} & k \text{ odd} \\ 2\mathbb{Z}/k\mathbb{Z} & k \text{ even.} \end{cases} \tag{4.20}$$

Let us now turn to the even-dimensional branes, $H^{\text{even}} \cong H_0 \oplus H_8$. 8-branes are anomalous because $\lfloor H \rfloor_{DR} \neq 0$, but 0-branes are anomaly-free.

Now, D0-brane charge is not conserved because of the standard process of Figure 8. There is, however, a more subtle instanton, illustrated in Figure 9, in which a 3-chain ends on a nontrivial element in $H_2(M_5; \mathbb{Z}) \cong \mathbb{Z}_2$. This instanton violates D0-charge by $\frac{1}{2}k$ units, when k is even. In this way we conclude that

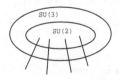

Figure 8. k D0-branes terminate on a wrapped D2-brane instanton in an $SU(2)$ subgroup of $SU(3)$

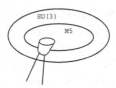

Figure 9. When k is even $\frac{1}{2}k$ D0-branes can terminate on a hemisphere of $SU(2)$ which terminates on a generator of $H_5(SU(3), \mathbb{Z})$

$$K_H^0(SU(3)) = \begin{cases} \mathbb{Z}/k\mathbb{Z} & k \text{ odd} \\ \mathbb{Z}/\frac{k}{2}\mathbb{Z} & k \text{ even} \end{cases} \qquad (4.21)$$

One could probably extend the above procedure to compute the twisted K-theory of higher rank groups using (at least for $SU(N)$), Steenrod's cell-decomposition, but this has not been done. In part inspired by the above results (and the result for D0-charge quantization explained in the next section) M. Hopkins computed the twisted K-homology of $SU(N)$ rigorously. He finds that, for $H = k\omega$

$$K_{H,*}(SU(N)) = (Z/d_{k,N}Z) \otimes \Lambda_Z[w_5, \ldots, w_{2N-1}] \qquad (4.22)$$

where

$$d_{k,N} = \gcd\left[\binom{k}{1}, \binom{k}{2}, \ldots, \binom{k}{N-1}\right]. \qquad (4.23)$$

We find perfect agreement for $G = SU(2)$, $SU(3)$ above.

It is interesting to compare (4.22) with

$$H_*(SU(N)) = \Lambda_Z[w_3, w_5, \ldots, w_{2N-1}]. \qquad (4.24)$$

Recall that $SU(N) \sim S^3 \times S^5 \times \cdots S^{2N-1}$, rationally. Evidently, the topologically distinct D-branes can be pictured as wrapping different cycles in $SU(N)$, subject to certain decay processes. In the next section we will return to the worldsheet RG point of view to explain the most important of these decay processes. We will also give a simple physical argument (which in fact predated Hopkins' computation) for why the group of charges should be torsion of order $d_{k,N}$.

5 The example of branes in $SU(N)$ WZW models

In this section we will use the theory of 'symmetry-preserving branes' to determine the order $d_{k,N}$ of the D0-charge group for $SU(N)$ level k WZW model.

Different versions of the argument are given in [81], [82], [71], [83]. For reviews with further details on the material of this section see [84], [85], [86] and references therein.

Let us summarize the strategy of the argument here:

(1) We define the 'elementary' or 'singly-wrapped' symmetry-preserving boundary conditions algebraically using the formalism of boundary conformal field theory. These boundary conditions are labelled by the unitary irreps $\lambda \in P_k^+$ of the centrally extended loop group.

(2) We give a semiclassical picture of these boundary conditions as branes wrapping special regular conjugacy classes with a nontrivial Chan–Paton line bundle. See equations (5.14) and (5.24) below.

(3) We then discuss how it is that 'multiply-wrapped' symmetry preserving branes can lie in components of $\mathcal{B}$ corresponding to certain singly wrapped branes. For example, a 'stack of L D0-branes' can be continuously connected by RG flow to a symmetry-preserving brane labelled by λ, provided the number of D0-branes L is equal to the dimension $d(\lambda)$ of the representation λ of the group G.

(4) This implies that the symmetry-preserving brane λ has, in some sense, D0-charge L. On the other hand, as we have seen in the previous section, the D0-charge must be finite and cyclic. Thus the D0-charge is $d(\lambda) \bmod d_{k,N}$, for some integer $d_{k,N}$.

(5) Finally we note that symmetry-preserving branes for different values of λ can sometimes be related by a rigid rotation continuously connected to 1. Such branes are obviously in the same component of $\mathcal{B}$, and this suffices to determine the order $d_{k,N}$ of the torsion group.

5.1 WZW Model for $G = SU(N)$

Let us set our notation. The WZW field $g : \Sigma \to G$ has action

$$S = \frac{k}{8\pi} \int_\Sigma \mathrm{Tr}_N[(g^{-1}\partial g)(g^{-1}\bar\partial g)] + 2\pi k \int \omega \qquad (5.1)$$

where the trace is in the fundamental representation. The target space $G = SU(N)$ has a metric

$$ds^2 = -\frac{k}{2}\mathrm{Tr}_N(g^{-1}dg \otimes g^{-1}dg) \qquad (5.2)$$

and a 'B-field' with fieldstrength

$$H = k\omega, \qquad \omega := -\frac{1}{24\pi^2}\mathrm{Tr}(g^{-1}dg)^3 \qquad (5.3)$$

where $[\omega]$ generates $H^3(G; Z) \cong Z$. The CFT state space is

$$\mathcal{H}^{\text{closed}} \cong \oplus_{P_k^+} \mathcal{H}_\lambda \otimes \tilde{\mathcal{H}}_{\lambda^*} \tag{5.4}$$

where $\mathcal{H}_\lambda, \tilde{\mathcal{H}}_{\lambda^*}$ are the left- and right-moving unitary irreps of the loop group $\widetilde{LG}_k$, as described in [87].

Amongst the set of conformal boundary conditions (i.e. branes) there is a distinguished set of 'symmetry-preserving boundary conditions' leaving the diagonal sum of left-and right-moving currents $J + \tilde{J}$ unbroken (see [84], [85] for more details). Since there is an unbroken affine symmetry the open string morphism spaces $\mathcal{H}_{ab}^{\text{open}}$ are themselves representations of $\widetilde{LG}_k$. Accordingly, these are objects in the category of boundary conditions labelled by $\lambda \in P_k^+$. The decomposition of the morphism spaces as irreps of $\widetilde{LG}_k$ is given by

$$\mathcal{H}_{\lambda_1,\lambda_2}^{\text{open}} = \oplus_{\lambda_3 \in P_k^+} N_{\lambda_1,\lambda_2}^{\lambda_3} \mathcal{H}_{\lambda_3} \tag{5.5}$$

where $N_{\lambda_1,\lambda_2}^{\lambda_3}$ are the fusion coefficients.

The most efficient way to establish (5.5) is via the 'boundary state formalism'. In the 2D topological field theory of Section 2, the boundary state associated to boundary condition a is defined to be $\iota^a(1_a)$ where 1_a is the unit in the open string algebra $\mathcal{H}_{aa}$. This is an element of the closed string algebra $\mathcal{C}$ which 'creates' a free boundary with boundary condition a. Similarly, in boundary CFT, to every conformal boundary condition a one associates a corresponding 'boundary state'

$$|B(a)\rangle\rangle \in \mathcal{H}^{\text{closed}}. \tag{5.6}$$

For the symmetry-preserving WZW boundary conditions the corresponding boundary state is given by the Cardy formula

$$|B(\lambda)\rangle\rangle = \sum_{\lambda' \in P_k^+} \frac{S_\lambda^{\lambda'}}{\sqrt{S_0^{\lambda'}}} 1_{\mathcal{H}_{\lambda'}} \in \mathcal{H}^{\text{closed}} \tag{5.7}$$

where $S_\lambda^{\lambda'}$ is the modular S-matrix, 0 denotes the basic representation, and we think of the closed string statespace as

$$\mathcal{H}^{\text{closed}} \cong \oplus_{P_k^+} \mathcal{H}_\lambda \otimes \tilde{\mathcal{H}}_{\lambda^*} \cong \oplus_{P_k^+} \text{Hom}(\mathcal{H}_\lambda, \mathcal{H}_\lambda) \tag{5.8}$$

Applying the Cardy condition to (5.7) we get (5.5).

Since the disk partition function is the overlap of the ground state with the boundary state, $Z_{\text{disk}} = \langle 0|B(\lambda)\rangle\rangle$, the g-function for these conformal fixed points follows immediately from (5.7)

$$g(\lambda) = S_{\lambda,0}/\sqrt{S_{00}}. \tag{5.9}$$

Finally, as we noted before, it is important to introduce worldsheet super-symmetry in order to have any stable branes at all. It suffices to introduce $\mathcal{N} = 1$ supersymmetry, although when embedded in a type II string back-ground the full background can have $\mathcal{N} = 2$ supersymmetry. It is also impor-tant to have a well-defined action by $(-1)^F$ on the conformal field theory. This distinguishes the cases where the rank of G is odd and even. When the rank is odd we can always add an $\mathcal{N} = 1$ Feigin–Fuks superfield, as indeed is quite natural when building a type II string background.

5.2 Geometrical interpretation of the symmetry-preserving branes

We would like to discuss the *geometrical* interpretation of the symmetry-preserving boundary condition labelled by λ. That is, we would like some semiclassical picture of the brane as an extended object in the group manifold. In this section we explain how that is derived.

Let us first recall how the geometry of the compact target space is recovered in the WZW model. In the WZW model the metric is proportional to k, so the path integral measure has weight factor

$$\sim e^{-kS}. \tag{5.10}$$

Thus, we expect semiclassical pictures to emerge in the limit $k \to \infty$. In this limit the vertex operator algebra 'degenerates' to become the algebra of functions on the group G. For example, CFT correlators become integrals over the group manifold

$$\langle \hat{F}_1(g(z_1, \bar{z}_1)) \cdots \hat{F}_n(g(z_n, \bar{z}_n)) \rangle \to \int_G d\mu(g) F_1(g) \cdots F_n(g) \tag{5.11}$$

On the left-hand side, $\hat{F}_i$ are suitable vertex operators of dimension $\sim 1/k$. On the right-hand side, F_i are corresponding L^2 functions on G. Roughly speak-ing, the CFT statespace degenerates as

$$\mathcal{H}^{\text{closed}} \cong \oplus_{\mathrm{P}_k^+} \mathcal{H}_\lambda \otimes \tilde{\mathcal{H}}_{\lambda^*} \to L^2(G) \otimes \mathcal{H}^{\text{string}} \tag{5.12}$$

where $L^2(G)$ is the limit of the primary fields and $\mathcal{H}^{\text{string}}$ contains the 'oscil-lator excitations'. In this limit the boundary state degenerates

$$|B(\lambda)\rangle\rangle \to B_\lambda + \cdots \tag{5.13}$$

where $B_\lambda \in L^2(G)$ and becomes a distribution in the $k \to \infty$ limit. While (5.12) is clearly heuristic, (5.13) has a well-defined meaning because the over-laps of $|B(\lambda)\rangle\rangle$ with primary fields of dimension $\sim 1/k$ have well-defined limits.

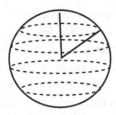

Figure 10. Distinguished conjugacy classes in $SU(2)$. These are the semiclassical worldvolumes of the symmetry-preserving branes

Using equation (5.7), the formulae for the modular S-matrix, and the Peter–Weyl theorem one finds that the function B_λ is concentrated on the regular conjugacy class

$$\mathcal{O}_{\lambda,k} := \left[\exp\left(2\pi i \frac{\lambda + \rho}{k + h}\right) \right] \tag{5.14}$$

leading to the semiclassical picture of branes in Figure 10. Here ρ is the Weyl vector and h is the dual Coxeter number. (As usual, replace $k \to k - h$, $\lambda \in P_{k-h}^+$ for the supersymmetric case.) See [88], [89], [71] for more details.

Remarks:

(1) Since k is the semiclassical expansion parameter we only expect to be able to localize the branes to within a length-scale $\ell_{\text{string}} \sim 1/\sqrt{k}$ when using closed string vertex operators [90]. Let $\underline{t}$ be the the Lie algebra of the maximal torus, and let $\chi \in \underline{t}$ parametrize conjugacy classes. Then the metric $ds^2 \sim k(d\chi)^2$ and hence vertex operators can only 'resolve' angles $\delta\chi \geq \frac{1}{\sqrt{k}}$. This uncertainty encompasses many different conjugacy classes (5.14). Nevertheless, the semiclassical geometrical pictures give exact results for many important physical quantities. The reason for this is that the relevant exact CFT results are polynomials in $1/k$, and hence can be exactly computed in a semiclassical expansion.

(2) The basic representation $\lambda = 0$ gives the 'smallest' brane. We will refer to this as a 'D0-brane'. In a IIA string compactification built with the WZW model this state is used to construct a D0-brane. Note, however, that in this description it is *not* pointlike, but rather has a size of order the string length $\sim 1/\sqrt{k}$.

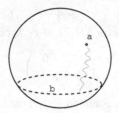

Figure 11. Using a D0-brane as boundary condition a we can probe for the location of brane b by studying the lowest mass of the stretched strings

5.3 Using CFT to measure the distance between branes

To lend further support to the geometrical picture advocated above, let us show that D-branes can be rotated in the group, and that the distance between them can then be measured using CFT techniques.

First, we explain how to 'rotate' D-branes. $G_L \times G_R$ acts on $\mathcal{H}^{\text{closed}}$, and therefore we can consider the boundary state

$$g_L g_R |B(\lambda)\rangle\rangle. \tag{5.15}$$

In the $k \to \infty$ limit this state has a limit similar to (5.13). In particular, it is supported on the subset

$$g_L \mathcal{O}_{\lambda,k} g_R \subset G. \tag{5.16}$$

Now, let us consider the open string statespace $\mathcal{H}^{\text{open}}_{ba}$ with a corresponding to a rotated D0-brane, $a = g_L \cdot |B(\lambda = 0)\rangle\rangle$ and $b = |B(\lambda)\rangle\rangle$. A typical string in this space may be pictured as in Figure 11. This picture suggests a way to 'measure the distance' between the two branes, and thereby to define the positions of branes in the spirit of [91], [92], [93]. The picture suggests that the open string channel partition function has an expansion for small q_o

$$\text{Tr}_{\mathcal{H}^{\text{open}}_{b,a}} q_o^{L_0} \overset{?}{=} q_o^{(T_f D)^2} + \cdots \tag{5.17}$$

where D is the geodesic distance between the center of the $D0$-brane at g_L and the brane b. T_f is the fundamental string tension (we set $\alpha' = 1$ here, so $T_f = 1/(2\pi)$).

We can actually compute the q_o expansion of (5.17) using the expression for the boundary state together with the Cardy condition

$$\text{Tr}_{\mathcal{H}_{b,a}} q_o^{L_0 - c/24} = \langle\langle B(\lambda)| q_c^{\frac{1}{2}(L_0 + \tilde{L}_0 - c/12)} \rho_L(g) |B(0)\rangle\rangle \tag{5.18}$$

where $q_c = e^{2\pi i \tau_c}$, $q_o = e^{-2\pi i/\tau_c} = e^{2\pi i \tau_o}$. The computation is straightforward. Let us quote the result for $SU(2)$. If g_L is conjugate to

$$\begin{pmatrix} e^{i\chi} & 0 \\ 0 & e^{-i\chi} \end{pmatrix} \tag{5.19}$$

with $0 \le \chi \le \pi$, then, in the Ramond sector, the leading power of q_o in (5.17) is

$$k \left(\frac{\hat{\chi}_j - \chi}{2\pi} \right)^2 . \tag{5.20}$$

Here $\hat{\chi}_j = \pi(2j+1)/k$, and the brane b is labelled by $j \in \{0, \frac{1}{2}, \ldots, \frac{k-2}{2}\}$. The formula (5.20) is *precisely* $(T_f D)^2$, as naïvely expected. Again, we see that the geometrical picture of the branes is beautifully reproduced from the conformal field theory.[6] Very similar remarks hold for the D-branes in coset models [89].

5.4 Why are the branes stable?

The geometrical picture advocated in the previous sections raises an interesting puzzle. We will now describe this puzzle, and its beautiful resolution in [94], [95].

Consider a D-brane wrapping $\mathcal{O}_{\lambda,k} \subset G$ once, as in Figure 10. In the context of the type II string theory, the brane has a nonzero tension T, with units of energy/volume. Hence, wrapping a submanifold $\mathcal{W}$ with a brane costs energy $E \sim T \text{vol}(\mathcal{W})$. However, the regular conjugacy classes $\mathcal{O}_{\lambda,k} \subset G$ are homologically trivial. For example, for $SU(2)$, $\mathcal{O} = S^2 \subset S^3$. We therefore expect the brane to be unstable and to contract to a point.

This leads to a paradox: We know from conformal field theory that the brane is absolutely stable. From the expression for the boundary state we can compute the spectrum of operators in the open string statestate from

$$\text{Tr}_{\mathcal{H}_{\lambda,\lambda}} q^{L_0 - c/24} \tag{5.21}$$

and we find all $\Delta_i \ge 1$. According to (3.5) it follows that there are no unstable flows under β away from this point!

The resolution of the paradox lies in the fact that D-branes also have gauge theory degrees of freedom on them. The brane carries a $U(1)$ line bundle

---

[6] In the bosonic case (5.20) turns out to be $\frac{k+2}{4} \left(\frac{\hat{\chi}_j - \chi}{\pi} \right)^2 + \frac{1}{2} \frac{\chi}{\pi}(1 - \frac{\chi}{\pi}) - \frac{1}{4(k+2)}$ so it is only for $k(\delta \chi)^2 \gg 1$ that the conclusion holds.

$\mathcal{L} \to \mathcal{O}$ with connection. If this bundle is twisted then there is a stabilizing force opposing the tension.

To illustrate the resolution in the simplest terms, consider the example of $SU(2)$ with conjugacy class $\mathcal{O} = S^2$, of radius $R = \sqrt{k} \sin \chi$. If the Chan–Paton line bundle of the brane has Chern class $n \in \mathbb{Z}$, then $\int_{\mathcal{O}} F = 2\pi n$. It follows that the Yang–Mills action is

$$\int_{S^2} F \wedge *F \sim \frac{n^2}{R^2} \tag{5.22}$$

and hence we can evaluate the g-function (3.20)

$$g(\chi) \sim R^2 + \frac{n^2}{R^2}. \tag{5.23}$$

This has a minimum at $\chi \sim \pi n/k$, and hence we expect an RG flow in the sector of $\mathcal{B}$ determined by n to evolve to this configuration.

The above arguments have been generalized from $SU(2)$ to higher rank groups in [71], [96]. The result that emerges is that $|B(\lambda)\rangle\rangle$ can be pictured, semiclassically, as wrapping the conjugacy class $\mathcal{O}_{\lambda,k}$. The brane is singly wrapped, and its Chan–Paton line bundle $\mathcal{L}_\lambda \to \mathcal{O}_{\lambda,k}$ has first Chern class

$$c_1(\mathcal{L}_\lambda) = \lambda + \rho \in H^2(G/T; Z) \cong \Lambda_{\text{weight}} \tag{5.24}$$

(for further details see [71]).

It is interesting to study the g-function and its approximation by the DBI action in this problem. We restrict attention to the bosonic WZW model. Let χ parametrize the conjugacy classes in G. For the Chan–Paton line bundle (5.24) the DBI action

$$g_{DBI}(\chi) := \int_{\mathcal{O}_\chi} \sqrt{\det(g + F + B)} \tag{5.25}$$

as a function of χ is minimized at $\chi_* = 2\pi(\lambda + \rho)/(k+h)$, where it takes the value

$$g_{DBI}(\chi_*)/g_{DBI}(0) = \prod_{\alpha>0} \left(\frac{k \sin \frac{1}{2}\alpha \cdot \chi}{\pi \alpha \cdot \rho} \right). \tag{5.26}$$

Here the product is over positive roots. This compares remarkably well with the exact CFT answer

$$g(\lambda)/g(0) = \prod_{\alpha>0} \left(\frac{\sin \pi\alpha \cdot (\lambda + \rho)/(k+h)}{\sin \pi\alpha \cdot \rho/(k+h)} \right). \tag{5.27}$$

Note that the right-hand side is the quantum dimension $d_q(\lambda)$, in harmony with (5.9) above.

5.5 *How collections of D0-branes evolve to symmetry-preserving branes*

The semiclassical picture of the symmetry-preserving branes we have just described raises an important new point. In type II string compactification, if a brane carries a topologically nontrivial Chan–Paton bundle then it carries a nontrivial induced D-brane charge. In the present case since the Chan–Paton line bundle has $\int_{\mathcal{W}} e^{c_1(\mathcal{L})} \neq 0$ it carries D0-charge. This suggests that the conformal fixed point characterized by $|B(\lambda)\rangle\rangle$ is in the same component of $\mathcal{B}$ as the fixed point corresponding to a 'collection of D0-branes'. In this section we review why that is true.[7]

By a 'stack of D0-branes' physicists mean the boundary state $L|B(0)\rangle\rangle$ for some positive integer L. By definition, the open string sectors for such a stack of D0-branes have state spaces

$$\mathcal{H}^{\text{open}}_{LB(0),b} = \mathbf{C}^L \otimes \mathcal{H}^{\text{open}}_{B(0),b} \tag{5.28}$$

for any boundary condition b.

Claim: If $\lambda \in P_k^+$ and $L = d(\lambda)$, then $L|B(0)\rangle\rangle$ is in the same component of $\mathcal{B}$ as $|B(\lambda)\rangle\rangle$.

Note that

$$\frac{g(B(\lambda))}{g(LB(0))} = \frac{S_{\lambda,0}}{LS_{00}} = \frac{d_q(\lambda)}{d(\lambda)} < 1 \tag{5.29}$$

so the claim is nicely consistent with the g-theorem. In particular, if these fixed points can be connected by RG flow then L D0s are unstable to λ, and not vice versa. We sometimes refer to this instability as the 'blowing up effect'.

The RG flow in question arises in the theory of the Kondo effect and was studied by Affleck and Ludwig [25], [26]. Their results were applied in the present context by Schomerus and collaborators. See [84], and references therein. Kondo model trajectories are obtained by perturbing a conformal fixed point by the holonomy of the unbroken current algebra in some representation. The flow, which should take $L|B(0)\rangle\rangle$ to $|B(\lambda)\rangle\rangle$, is given by considering the disk partition function

$$Z(u) = \left\langle \text{Tr}_\lambda \left(P \exp \oint d\tau u J(\tau) \right) \right\rangle. \tag{5.30}$$

As explained in [84], the results of Affleck and Ludwig lend credence to the main claim.

---

[7] Actually, the most obvious embedding of the $SU(2)$ WZW model into a type IIA background using a Feigin–Fuks superfield produces a background for which the definition of RR D0-charge is in fact subtle. The relevant $U(1)$ RR gauge group is spontaneously broken to $\mathbb{Z}_k$ due to the condensation of a spacetime scalar field of charge k. See [97] for more discussion.

Actually, it is important to take into account $\mathcal{N} = 1$ supersymmetry in this problem.[8] In the supersymmetric WZW model we have a superfield

$$\mathbf{J}_a(z) = \psi_a(z) + \theta I_a(z) \tag{5.31}$$

where we have chosen an orthonormal basis for the Lie algebra, labelled by $a = 1, \ldots, \dim G$. The OPE's are [99], [100], [101]

$$I_a(z)I_b(w) \sim \frac{k\delta_{ab}}{(z-w)^2} + f_{ab}{}^c \frac{I_c(w)}{z-w} + \cdots$$

$$I_a(z)\psi_b(w) \sim \frac{f_{ab}{}^c\psi_c(w)}{z-w} + \tag{5.32}$$

$$\psi_a(z)\psi_b(w) \sim \frac{k\delta_{ab}}{(z-w)} + \cdots$$

By a standard argument the currents $J_a = I_a + \frac{1}{2k}f_{abc}\psi_b\psi_c$ decouple from the fermions and satisfy a current algebra with level $k - h$. The Hamiltonian and supersymmetry charge are given (in the Ramond sector) by

$$Q = \oint dz \left(\frac{1}{k} J_a \psi_a - \frac{1}{6k^2} f_{abc} \psi_a \psi_b \psi_c \right)$$

$$H = \frac{1}{2k} \oint dz \, (: J_a J_a : + \partial \psi_a \psi_a). \tag{5.33}$$

The supersymmetry transformations are $[Q, \psi_a] = I_a$ and $[Q, I_a] = \partial \psi_a$.

Now, let us add a Kondo-like boundary perturbation preserving $\mathcal{N} = 1$ supersymmetry. This is given by choosing a representation λ of G and taking

$$g(u) = \left\langle \mathrm{Tr}_\lambda \left(P \exp \oint d\tau u I(\tau) \right) \right\rangle. \tag{5.34}$$

Using $[Q, I_a] = \partial \psi_a$ to vary the perturbed action in (5.34) we may compute the perturbed supercharge Q_u. This operator acts on $\mathbf{C}^L \otimes \mathcal{H}^{\mathrm{open}}$ as

$$Q_u = Q + u\psi^a(0)S^a$$

$$= Q + u \sum_{n \in \mathbb{Z}} \psi_n^a S^a \tag{5.35}$$

In the first line we have passed to a Hamiltonian formalism for the open string on a space $[0, \pi]$ (and we are only modifying the boundary condition at $\sigma = 0$), and we have introduced explicit generators S^a for the finite-dimensional representation $\mathbf{C}^L$ of the group G. In the second line we have used the doubling

---

[8] The following argument combines elements from [84], [98], [71].

trick to express the result in terms of modes of a single-valued chiral vertex operator on the plane $\mathbf{C}$. We can now compute the perturbed Hamiltonian

$$H_u = Q_u^2 = H + uI^a(0)S^a + u^2(\psi^a(0)S^a)^2. \tag{5.36}$$

The third term in (5.36) is singular, but the renormalization of this term is fixed by the requirement of supersymmetry. The Hamiltonian can be written as

$$H_u = \frac{1}{2k} \sum_n \left(: (J_n^a + ukS^a)(J_{-n}^a + ukS^a) : +n\psi_n^a\psi_{-n}^a \right)$$

$$+ \frac{1}{2}u\left(u - \frac{1}{k}\right) \sum_{n,m} f^{abc}\psi_n^a\psi_m^b S^c \tag{5.37}$$

so that the vacuum of the theory evolves in a complicated way as a function of u. Note that, exactly for $u = u_* = 1/k$, the Hamiltonian simplifies into

$$H_* = \frac{1}{2k} \sum_n \left(: \mathcal{J}_n^a \mathcal{J}_{-n}^a : +n\psi_n^a\psi_{-n}^a \right) \tag{5.38}$$

where

$$\mathcal{J}_n^a := J_n^a + S^a \tag{5.39}$$

also satisfy a current algebra with level $k - h$. Thus, at $u = u_*$, we can build a new superconformal algebra with these currents.

The previous paragraph strongly suggests that $u_* = 1/k$ is a second critical point for the boundary conformal field theory. Now, we can use an observation of Affleck and Ludwig. If $\mathcal{H}_{\lambda'}$ is a representation of J_n^a, then with respect to a new current algebra $\mathcal{J}_n^a$ we can decompose

$$\mathbf{C}^L \otimes \mathcal{H}_{\lambda'} \cong \oplus_{\lambda'' \in P_k^+} N_{\lambda,\lambda'}^{\lambda''} \mathcal{H}_{\lambda''}. \tag{5.40}$$

(An easy way to prove (5.40) is to consider the cabling of Wilson lines in 3D Chern–Simons theory, and use the Verlinde algebra.) Therefore it follows that

$$\mathrm{Tr}_{\lambda_1}\left(P\exp\oint u_*I\right)|B(\lambda_2)\rangle\rangle = \sum_{\lambda_3} N_{\lambda_1,\lambda_2}^{\lambda_3}|B(\lambda_3)\rangle\rangle \tag{5.41}$$

where the boundary states on the RHS are constructed using $\mathcal{J}_n^a$. It would be worthwhile to give a direct proof of (5.41). The identity has been verified at large k in [98].

Let us close this subsection with a number of remarks.

(1) Note that when there is more than one term on the right-hand side of (5.40) a local boundary condition has evolved into a (mildly) nonlocal boundary condition. Regrettably, this muddies the proposed definition

of D-branes as local boundary conditions preserving conformal invariance.

(2) The instability of a stack of L D0-branes to decay to a symmetry-preserving brane has been much discussed in the literature in the framework of noncommutative gauge theory (see [84] and references therein). The arguments show that the 'D-brane instantons' of the previous section should be viewed as real-time processes taking place in classical string theory.

(3) The Kondo flows are integrable flows. The g function has been studied in [102], [103], [104], [105], [106], [107] in several examples and for certain boundary conditions related to the free fermion construction of current algebras. It is possible that the techniques of [102], [103], [104] can be used to give exact results for how the boundary state evolves along the RG trajectory. This could be very interesting indeed.

(4) The 'blowing up effect' is closely related to some work of [108], [109]. These authors study families of Fredholm operators over the space of gauge fields on S^1. The perturbed supersymmetry operator along the RG flow is related to the family of Fredholm operators studied in [108], [109].

(5) One of the most remarkable aspects of the blowing-up effect is the disappearance of k D0-branes 'into nothing'. Let us stress that this is an effect studied in the laboratory! One studies electrons coupled to a magnetic 'impurity'. Translating this system into conformal field theory terms [110], [105] reveals the boundary $SU(2)$ model with $k = 1$; the presence of the magnetic impurity, in the high temperature regime, translates into the presence of a single D0-brane. The RG flow parameter is the temperature, and, as $T \to 0$, the magnetic impurity is screened and 'disappears'. The absence of the magnetic impurity corresponds to the disappearance of the D0-brane.

(6) The effect we are discussing can be related, by U-duality, to the Myers effect [111]. (Apply S-duality to a IIB solitonic 5-brane.)

(7) Actual evaluation of the standard D0-brane charge formula $\int_{O_{\lambda,k}} e^{F+B}$ yields a quantum dimension for the group. As far as we know, this curious fact has not yet been properly understood.

5.6 The D0-charge group

At this point we have *two* notions of D0-brane charge. On the one hand, we have naïve D0-charge $L = d(\lambda)$. On the other hand, as we explained in Section 4, due to D-brane instantons, the true D0-charge, which we denote by $q(\lambda)$,

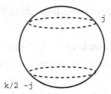

Figure 12. Two symmetry-preserving branes related by a rigid rotation

must be a torsion. Indeed, we know that D-brane instanton effects will impose a relation

$$q(\lambda) = d(\lambda) \bmod d_{k,N} \tag{5.42}$$

for some integer $d_{k,N}$, but (without Hopkins) it is hard to account for all possible D-brane instantons. So, we will now determine the order, $d_{k,N}$, of the torsion group using the blowing up effect and a simple observation regarding rotated branes.

Recall from Section 5.3 that we can rotate our branes by $G_L \times G_R$. Sometimes it can happen that the special conjugacy classes can be rotated into one another

$$g_L \mathcal{O}_{\lambda,k} g_R = \mathcal{O}_{\lambda',k}. \tag{5.43}$$

For example, if $G = SU(2)$ the conjugacy classes $\mathcal{O}_{j,k}$ and $\mathcal{O}_{k/2-j,k}$ can be rotated into each other

$$(-1) \cdot \mathcal{O}_{j,k} = \mathcal{O}_{\frac{1}{2}k-j,k} \tag{5.44}$$

as in Figure 12. Let us ask which representations are related in this way.

In order to answer this question we use the well-known relation between the center of a compact connected, simply-connected Lie group G and the automorphisms of the extended Dynkin diagram. For example, if $G = SU(N)$, $Z(G) \cong Z_N$, and Z_N acts on the extended Dynkin diagram by rotation. Next, the automorphisms of the extended Dynkin diagram act on the space of level k integrable representations P_k^+. For example, for $SU(2)$

$$j \to j' = \frac{1}{2}k - j \tag{5.45}$$

while for $\widehat{SU(N)}_k$ the generator of $Z(G)$ acts on the Dynkin labels by

$$\lambda = (a_1, \ldots, a_{N-1}) \to \lambda' = (k - \sum a_i, a_1, \ldots, a_{N-2}). \tag{5.46}$$

A beautiful result of group theory is that if $z \in Z(G)$ then

$$z\mathcal{O}_{\lambda,k} = \mathcal{O}_{z \cdot \lambda,k} \tag{5.47}$$

Now we can use (5.47) to determine the order of the D0-charge group. To see this note first that two branes related by a rigid rotation must have the same D0-charge! On the other hand, λ and $z \cdot \lambda$ have *different* dimensions, and hence have different naïve D0-charge. Therefore, we seek an integer $d_{k,N}$ such that

$$d(z \cdot \lambda) = \pm d(\lambda) \bmod d_{k,N} \qquad \forall z \in Z(G), \lambda \in P_k^+ \qquad (5.48)$$

where the sign $\pm$ depends on z and the rank of G, and accounts for orientation (see [71] for more details). It turns out that this condition determines $d_{k,N}$

$$d_{k,N} = \gcd\left[\binom{k}{1}, \binom{k}{2}, \ldots, \binom{k}{N-1} \right]. \qquad (5.49)$$

in perfect agreement with Hopkins' result! (See [82], [71] for details of some of the arithmetic involved.)

Remarks:

(1) The generalization of $d_{k,N}$ to other compact simple Lie groups has been discussed in [112].

(2) After my talk at the conference, M. Hopkins made a curious remark which I would like to record here. There is a simple mathematical relation between twisted *equivariant* K-theory and twisted K-theory of G which has several of the same ingredients as the physical discussion we have just given. If $\pi_1(G)$ is torsion free the Kunneth formula of [113], [114] suggests that the two twisted K-theories are related by

$$\mathbb{Z} \otimes_{R(G)} K_{G,H}(G) = K_H(G). \qquad (5.50)$$

Here the representation ring $R(G)$ is to be thought of as the ring of functions on the representation variety G/T with T acting by conjugation. $R(G)$ acts on $1 \in \mathbb{Z}$ by the dimension of the representation, while $K_{G,H}(G)$ is the Verlinde algebra, thanks to the theorem of Freed, Hopkins, and Teleman [115], [116], [117]. Curiously, from the point of view of algebraic geometry this means that the special conjugacy classes have an intersection with the identity element, when considered as varieties over $\mathbb{Z}$.

5.7 *Comment on cosets*

The point of view explained above has potentially interesting applications to branes in coset models. Roughly speaking, if $L \subset G$ is a subgroup then the branes in the $\mathcal{N} = 1$ supersymmetric coset model G/L should be classified by the twisted equivariant K-theory $K_{L,H}(G)$, where the twisting comes from the WZW G-theory. The branes in such coset models have been studied in many papers. See [118], [89], [86], [119] [121], [122] for a sampling. For the $SU(2)/U(1)$ model the stable A-banes described in [89] are in perfect accord with the twisted equivariant K-theory. The higher rank situation is somewhat more subtle and is currently under study by S. Schafer-Nameki [123].

6 Conclusion

Our goal in this talk was not to establish rigorous mathematical theorems but to explain how physics can suggest some intuitions for K-theory which are complementary to the more traditional (and rigorous!) approaches to the subject. Such alternative viewpoints and heuristics can sometimes suggest new and surprising directions for enquiry, or can suggest simple heuristics for already known results. The above 'derivation' of the twisted K-theory of $SU(N)$ is just one example, but there are others. For example, the symmetry-preserving branes are precisely the branes which descend to branes in the G/G gauged WZW model. The reason is that the gauge group acts on G by conjugation, and only the symmetry-preserving boundary conditions preserve this gauge symmetry. Now, the G/G WZW model is a topological field theory whose Frobenius algebra is the Verlinde algebra. This provides a simple perspective on the physics underlying the result of Freed, Hopkins, and Teleman [115], [116], [117]. (This remark is also related to the discussion of [115].)

Let us conclude by mentioning some future directions which might prove to be interesting to the mathematics community, and which are suggested by the more physical approach to K-theory advocated in this paper.

First, in the context of spacetime supersymmetric models a special class of boundary conditions, the so-called 'BPS states' might have an interesting *product* structure [124], [125]. Thus, perhaps the category of boundary conditions (or an appropriate subcategory) can also be given the structure of a tensor category.

Second, the RG approach to D-branes suggests an interesting generalization of the McKay correspondence to *non-crepant* toric resolutions of orbifold singularities [126].

Finally, the K-theoretic classification of D-branes in type II string theory must somehow be compatible with the U-duality symmetries these theories enjoy, and must somehow be compatible with 11-dimensional M-theory. Only bits and pieces of this story are at present understood. It is possible that the full resolution will be deep and will have interesting mathematical applications.

Acknowledgements

I would like to thank my collaborators on the work which was reviewed above, E. Diaconescu, D. Kutasov, J. Maldacena, M. Mariño, N. Saulina, G. Segal, N. Seiberg, and E. Witten. I have learned much from them. I am also indebted to N. Andrei, I. Brunner, M. Douglas, D. Freed, D. Friedan, E. Getzler, M. Hopkins, S. Lukyanov, H. Saleur, S. Schafer-Nameki, V. Schomerus, S. Shatashvili, C. Teleman, and A. Zamolodchikov for useful discussions and correspondence on this material. I would like to thank D. Freed for many useful comments on the draft. Finally, I would also like to thank the Isaac Newton Institute for hospitality while this talk was written, and U. Tillmann for the invitation to speak at the conference. This work is supported in part by DOE grant DE-FG02-96ER40949.

References

[1] Minasian, R. and G.W. Moore, *K*-theory and Ramond–Ramond charge, *JHEP* **9711**, 002 (1997) [arXiv:hep-th/9710230].

[2] Witten, E., D-branes and *K*-theory, *JHEP* **9812**, 019 (1998) [arXiv:hep-th/9810188].

[3] Olsen, K. and R.J. Szabo, Constructing D-branes from *K*-theory, *Adv. Theor. Math. Phys.* **3**, 889 (1999) [arXiv:hep-th/9907140].

[4] Witten, E., Overview of *K*-theory applied to strings, *Int. J. Mod. Phys.* A **16**, 693 (2001) [arXiv:hep-th/0007175].

[5] Freed, D.S., *K*-theory in quantum field theory, [arXiv:math-ph/0206031].

[6] Szabo, R.J., D-branes, tachyons and *K*-homology, *Mod. Phys. Lett.* A **17**, 2297 (2002) [arXiv:hep-th/0209210].

[7] Moore, G. and G. Segal, unpublished. The material is available at http://www.physics.rutgers.edu/~gmoore/clay.html, or at http://online.kitp.ucsb.edu/online/mp01/moore1/ and /moore2.

[8] See lectures by G. Segal at http:// online.kitp.ucsb.edu/ online/geom99, and lecture notes at http:// www.cgtp.duke.edu/ITP99/segal/

[9] Cardy, J.L. and D. C. Lewellen, Bulk and boundary operators in conformal Field theory, *Phys. Lett.* B **259**, 274 (1991).

[10] Lewellen, D.C. Sewing constraints for conformal field theories on surfaces with boundaries, *Nucl. Phys.* B **372**, 654 (1992).

[11] Lazaroiu, C.I., On the structure of open-closed topological field theory in two dimensions, *Nucl. Phys.* B **603**, 497 (2001) [arXiv:hep-th/0010269].

[12] Turaev, V. Homotopy field theory in dimension 2 and group-algebras, [arXiv: math.qa/ 9910010].

[13] Fuchs, J. and C. Schweigert, 'Category theory for conformal boundary conditions,' [arXiv:math.ct/0106050].

[14] Fuchs, J. I. Runkel and C. Schweigert, Boundaries, defects and Frobenius algebras, [arXiv:hep-th/0302200].

[15] Bars, I., Map of Witten's * to Moyal's *, *Phys. Lett.* B **517**, 436 (2001) [arXiv:hep-th/0106157].

[16] Douglas, M.R., H. Liu, G. Moore and B. Zwiebach, Open string star as a continuous Moyal product, *JHEP* **0204**, 022 (2002) [arXiv:hep-th/0202087].

[17] Bars, I. and Y. Matsuo, Computing in string field theory using the Moyal star product, *Phys. Rev.* D **66**, 066003 (2002) [arXiv:hep-th/0204260].

[18] Bars, I., MSFT: Moyal star formulation of string field theory [arXiv:hep-th/0211238].

[19] Belov, D. and C. Lovelace, to appear.

[20] Witten, E., On background independent open string field theory, *Phys. Rev.* D **46**, 5467 (1992) [arXiv:hep-th/9208027].

[21] Friedan, D. The space of conformal boundary conditions for the $c = 1$ Gaussian Model, 1993, unpublished.

[22] Gaberdiel, M.R. and A. Recknagel, Conformal boundary states for free bosons and fermions, *JHEP* **0111**, 016 (2001) [arXiv:hep-th/0108238].

[23] Janik, R.A., Exceptional boundary states at c = 1, *Nucl. Phys.* B **618**, 675 (2001) [arXiv:hep-th/0109021].

[24] Martinec, E.J. Defects, decay, and dissipated states, [arXiv:hep-th/0210231].

[25] Affleck, I. and A.W. Ludwig, Universal noninteger 'Ground State Degeneracy' In Critical Quantum Systems, *Phys. Rev. Lett.* **67**, 161 (1991).

[26] Affleck, I. and A.W. Ludwig, Exact conformal field theory results on the nultichannel Kondo effect: single Fermion Green's function, selfenergy and resistivity, UBCTP-92-029.

[27] Kutasov, D., M. Marino and G.W. Moore, Some exact results on tachyon condensation in string field theory, *JHEP* **0010**, 045 (2000) [arXiv:hep-th/0009148].

[28] Kutasov, D., M. Marino and G.W. Moore, Remarks on tachyon condensation in superstring field theory [arXiv:hep-th/0010108].

[29] Harvey, J.A., S. Kachru, G.W. Moore and E. Silverstein, Tension is dimension, *JHEP* **0003**, 001 (2000) [arXiv:hep-th/9909072].

[30] Marino, M. On the BV formulation of boundary superstring field theory, *JHEP* **0106**, 059 (2001) [arXiv:hep-th/0103089].

[31] Niarchos, V. and N. Prezas, Boundary superstring field theory, *Nucl. Phys.* B **619**, 51 (2001) [arXiv:hep-th/0103102].

[32] Connes, A. and D. Kreimer, Renormalization in quantum field theory and the Riemann–Hilbert problem. II: the beta-function, diffeomorphisms and the renormalization group, *Commun. Math. Phys.* **216**, 215 (2001) [arXiv:hep-th/0003188].

[33] Connes, A. and D. Kreimer, Renormalization in quantum field theory and the Riemann–Hilbert problem. I: the Hopf algebra structure of graphs and the main theorem, *Commun. Math. Phys.* **210**, 249 (2000) [arXiv:hep-th/9912092].

[34] Witten, E., Some computations in background independent off-shell string theory, *Phys. Rev.* D **47**, 3405 (1993) [arXiv:hep-th/9210065].

[35] Gerasimov, A.A. and S.L. Shatashvili, On exact tachyon potential in open string field theory, *JHEP* **0010**, 034 (2000) [arXiv:hep-th/0009103].

[36] Sen, A., Tachyon condensation on the brane antibrane system, *JHEP* **9808**, 012 (1998) [arXiv:hep-th/9805170].

[37] Shatashvili, S.L. Comment on the background independent open string theory, *Phys. Lett.* B **311**, 83 (1993) [arXiv:hep-th/9303143].

[38] Li, K. and E. Witten, Role of short distance behavior in off-shell open string field theory, *Phys. Rev.* D **48**, 853 (1993) [arXiv:hep-th/9303067].

[39] Shatashvili, S.L. On the problems with background independence in string theory [arXiv:hep-th/9311177].

[40] Harvey, J.A., D. Kutasov and E.J. Martinec, On the relevance of tachyons [arXiv:hep-th/0003101].

[41] Gerasimov, A.A. and S.L. Shatashvili, Stringy Higgs mechanism and the fate of open strings, *JHEP* **0101**, 019 (2001) [arXiv:hep-th/0011009].

[42] Baum, P. and R.G. Douglas, K homology and index theory, *Proc. Symp. Pure Math.* **38**(1982), 117.

[43] Tseytlin, A.A., Born-Infeld action, supersymmetry and string theory [arXiv:hep-th/9908105].

[44] Donaldson, S.K. and P.B. Kronheimer, *The Geometry of Four-Manifolds*, Oxford, 1990.

[45] Hori, K., A. Iqbal and C. Vafa, D-branes and mirror symmetry [arXiv:hep-th/0005247].

[46] Hori, K., Linear models of supersymmetric D-branes, [arXiv:hep-th/0012179].

[47] Kraus, P. and F. Larsen, Boundary string field theory of the DD-bar system, *Phys. Rev.* D **63**, 106004 (2001) [arXiv:hep-th/0012198].

[48] Takayanagi, T., S. Terashima and T. Uesugi, Brane-antibrane action from boundary string field theory, *JHEP* **0103**, 019 (2001) [arXiv:hep-th/0012210].

[49] Freed, D.S. Dirac charge quantization and generalized differential cohomology, [arXiv:hep-th/0011220].

[50] Asakawa, T., S. Sugimoto and S. Terashima, D-branes, matrix theory and *K*-homology, *JHEP* **0203**, 034 (2002) [arXiv:hep-th/0108085].

[51] Szabo, R.J., Superconnections, anomalies and non-BPS brane charges, *J. Geom. Phys.* **43**, 241 (2002) [arXiv:hep-th/0108043].

[52] Asakawa, T., S. Sugimoto and S. Terashima, Exact description of D-branes via tachyon condensation [arXiv:hep-th/0212188].

[53] Periwal, V., D-brane charges and *K*-homology. (Z), *JHEP* **0007**, 041 (2000) [arXiv:hep-th/0006223].

[54] Witten, E., Noncommutative tachyons and string field theory [arXiv:hep-th/0006071].

[55] Matsuo, Y., Topological charges of noncommutative soliton, *Phys. Lett.* B **499**, 223 (2001) [arXiv:hep-th/0009002].

[56] Harvey, J.A. and G.W. Moore, Noncommutative tachyons and *K*-theory, *J. Math. Phys.* **42**, 2765 (2001) [arXiv:hep-th/0009030].

[57] Mathai, V. and I.M. Singer, Twisted *K*-homology theory, twisted Ext-theory [arXiv:hep-th/0012046].

[58] Atiyah, M.F., Global theory of elliptic operators, *Proc. Int. Conf. Func. Anal. and Related Topics*, pp. 21–30, Univ. of Tokyo Press, Tokyo, 1970.

[59] Connes, A., *Noncommutative Geometry*, Academic Press, 1994.

[60] Witten, E., Supersymmetry and Morse theory, *J. Diff. Geom.* **17**, 661 (1982).

[61] Witten, E., Global anomalies in string theory, Print-85-0620 (PRINCETON) *To appear in Proc. of Argonne Symp. on Geometry, Anomalies and Topology, Argonne, IL, 28–30 March, 1985.*

[62] Segal, G., Elliptic cohomology, *Asterisque* **161–162**(1988) exp. no. 695, 187–201.

[63] Stolz, S., Contribution to this volume.

[64] Albertsson, C., U. Lindstrom and M. Zabzine, N = 1 supersymmetric sigma model with boundaries. I [arXiv:hep-th/0111161].

[65] Albertsson, C., U. Lindstrom and M. Zabzine, N = 1 supersymmetric sigma model with boundaries. II [arXiv:hep-th/0202069].

[66] Lindstrom, U., M. Rocek and P. van Nieuwenhuizen, Consistent boundary conditions for open strings [arXiv:hep-th/0211266].

[67] Jaffe, A., A. Lesniewski and J. Weitsman, Index of a family of Dirac operators on loop space, *Commun. Math. Phys.* **112**, 75 (1987).

[68] Jaffe, A., A. Lesniewski and K. Osterwalder, Quantum K-theory. 1. The Chern character, *Commun. Math. Phys.* **118**, 1 (1988).

[69] Ernst, K., P. Feng, A. Jaffe and A. Lesniewski, Quantum K-theory. 2. Homotopy invariance of the Chern character, HUTMP-88/B-228.

[70] Seiberg, N. and E. Witten, String theory and noncommutative geometry, *JHEP* **9909**, 032 (1999) [arXiv:hep-th/9908142].

[71] Maldacena, J.M., G.W. Moore and N. Seiberg, D-brane instantons and K-theory charges, *JHEP* **0111**, 062 (2001) [arXiv:hep-th/0108100].

[72] Kapustin, A., D-branes in a topologically nontrivial B-field, *Adv. Theor. Math. Phys.* **4**, 127 (2000) [arXiv:hep-th/9909089].

[73] Bouwknegt, P. and V. Mathai, D-branes, B-fields and twisted K-theory, *JHEP* **0003**, 007 (2000) [arXiv:hep-th/0002023].

[74] Bouwknegt, P., A.L. Carey, V. Mathai, M.K. Murray and D. Stevenson, Twisted K-theory and K-theory of bundle gerbes, *Commun. Math. Phys.* **228**, 17 (2002) [arXiv:hep-th/0106194].

[75] Gawedzki, K. and N. Reis, WZW branes and gerbes [arXiv:hep-th/0205233].

[76] Evslin, J. and U. Varadarajan, K-theory and S-duality: starting over from square 3 [arXiv:hep-th/0112084].

[77] Evslin, J., Twisted K-theory from monodromies [arXiv:hep-th/0302081].

[78] Freed, D.S. and E. Witten, Anomalies in string theory with D-branes [arXiv:hep-th/9907189].

[79] Witten, E., Baryons and branes in anti de Sitter space, *JHEP* **9807**, 006 (1998) [arXiv:hep-th/9805112].

[80] Diaconescu, D.E., G.W. Moore and E. Witten, E(8) gauge theory, and a derivation of K-theory from M-theory [arXiv:hep-th/0005090].

[81] Alekseev, A. and V. Schomerus, RR charges of D2-branes in the WZW model [arXiv:hep-th/0007096].

[82] Fredenhagen, S. and V. Schomerus, Branes on group manifolds, gluon condensates, and twisted K-theory, *JHEP* **0104**, 007 (2001) [arXiv:hep-th/0012164].

[83] Stanciu, S., An illustrated guide to D-branes in SU(3) [arXiv:hep-th/0111221].

[84] Schomerus, V., Lectures on branes in curved backgrounds, *Class. Quant. Grav.* **19**, 5781 (2002) [arXiv:hep-th/0209241].

[85] Schweigert, C., J. Fuchs and J. Walcher, Conformal field theory, boundary conditions and applications to string theory [arXiv:hep-th/0011109].

[86] Gawedzki, K., Boundary WZW, G/H, G/G and CS theories, *Annales Henri Poincaré* **3**, 847 (2002) [arXiv:hep-th/0108044].

[87] Pressley, A. and G. Segal, *Loop Groups*, Oxford, 1986.

[88] Felder, G., J. Frohlich, J. Fuchs and C. Schweigert, The geometry of WZW branes, *J. Geom. Phys.* **34**, 162 (2000) [arXiv:hep-th/9909030].

[89] Maldacena, J.M., G.W. Moore and N. Seiberg, Geometrical interpretation of D-branes in gauged WZW models, *JHEP* **0107**, 046 (2001) [arXiv:hep-th/0105038].

[90] Klebanov, I.R. and L. Thorlacius, The size of p-branes, *Phys. Lett.* B **371**, 51 (1996) [arXiv:hep-th/9510200].

[91] Douglas, M.R., Two lectures on D-geometry and noncommutative geometry [arXiv:hep-th/9901146].

[92] Douglas, M.R., D-branes on Calabi–Yau manifolds [arXiv:math.ag/0009209].

[93] Douglas, M.R., D-branes in curved space, *Adv. Theor. Math. Phys.* **1**, 198 (1998) [arXiv:hep-th/9703056].

[94] Bachas, C., M.R. Douglas and C. Schweigert, Flux stabilization of D-branes, *JHEP* **0005**, 048 (2000) [arXiv:hep-th/0003037].

[95] Pawelczyk, J., SU(2) WZW D-branes and their noncommutative geometry from DBI action, *JHEP* **0008**, 006 (2000) [arXiv:hep-th/0003057].

[96] Bordalo, P., S. Ribault and C. Schweigert, Flux stabilization in compact groups, *JHEP* **0110**, 036 (2001) [arXiv:hep-th/0108201].

[97] Maldacena, J.M., G.W. Moore and N. Seiberg, D-brane charges in five-brane backgrounds, *JHEP* **0110**, 005 (2001) [arXiv:hep-th/0108152].

[98] Hikida, Y., M. Nozaki and Y. Sugawara, Formation of spherical D2-brane from multiple D0-branes, *Nucl. Phys.* B **617**, 117 (2001) [arXiv:hep-th/0101211].

[99] Friedan, D., Z. Qiu, and S.H. Shenker, *Phys. Lett.* **151B**(1985) 37.

[100] di Vecchia, P., V.G. Knizhnik, J.L. Petersen and P. Rossi, *Nucl. Phys.* **B253**(1985) 701.

[101] Kac, V.G. and I.T. Todorov, *Comm. Math. Phys.* **102**(1985) 337.

[102] Bazhanov, V.V., S.L. Lukyanov and A.B. Zamolodchikov, Integrable structure of conformal field theory, quantum KdV theory and thermodynamic Bethe ansatz, *Commun. Math. Phys.* **177**, 381 (1996) [arXiv:hep-th/9412229].

[103] Bazhanov, V.V., S.L. Lukyanov and A.B. Zamolodchikov, Integrable Structure of Conformal Field Theory II. Q-operator and DDV equation, *Commun. Math. Phys.* **190**, 247 (1997) [arXiv:hep-th/9604044].

[104] Bazhanov, V.V., S.L. Lukyanov and A.B. Zamolodchikov, Integrable structure of conformal field theory. III: The Yang–Baxter relation, *Commun. Math. Phys.* **200**, 297 (1999) [arXiv:hep-th/9805008].

[105] Andrei, N., Integrable models in condensed matter physics, cond-mat/9408101.

[106] Fendley, P., F. Lesage and H. Saleur, A unified framework for the Kondo problem and for an impurity in a Luttinger liquid [arXiv:cond-mat/9510055].

[107] Lesage, F., H. Saleur and P. Simonetti, Boundary flows in minimal models, *Phys. Lett.* B **427**, 85 (1998) [arXiv:hep-th/9802061].

[108] Mickelsson, J., Gerbes, (twisted) K -theory, and the supersymmetric WZW model [arXiv:hep-th/0206139].

[109] Freed, D., M. Hopkins and C. Teleman, presented by D. Freed at Elliptic Cohomology and Chromatic Phenomena, December 2002, Isaac Newton Institute.

[110] Affleck, I., Conformal field theory approach to the Kondo effect, *Acta Phys. Polon.* B **26**, 1869 (1995) [arXiv:cond-mat/9512099].

[111] Myers, R.C., Dielectric-branes, *JHEP* **9912**, 022 (1999) [arXiv:hep-th/9910053].

[112] Bouwknegt, P., P. Dawson and D. Ridout, D-branes on group manifolds and fusion rings, *JHEP* **0212**, 065 (2002) [arXiv:hep-th/0210302].

[113] Snaith, V.P., On the Kunneth formula spectral sequence in equivariant K -theory, *Proc. Camb. Phil. Soc.* **72**(1972) 167.

[114] Hodgkin, L., The equivariant Kunneth theorem in K -theory, *Springer Lect. Notes in Math*, **496**, pp. 1–101, 1975.

[115] Freed, D.S., The Verlinde algebra is twisted equivariant K -theory [arXiv:math.rt/01010381].

[116] Freed, D.S., Twisted K -theory and loop groups [arXiv:math.at/0206237].

[117] Freed, D.S., M.J. Hopkins and C. Teleman, Twisted equivariant K -theory with complex coefficients, math.AT/0206257.

[118] Lerche, W. and J. Walcher, Boundary rings and $N = 2$ coset models, *Nucl. Phys.* B **625**, 97 (2002) [arXiv:hep-th/0011107].

[119] Falceto, F. and K. Gawedzki, Boundary G/G theory and topological Poisson–Lie sigma model, *Lett. Math. Phys.* **59**, 61 (2002) [arXiv:hep-th/0108206].

[120] Fredenhagen, S. and V. Schomerus, D-branes in coset models, *JHEP* **0202**, 005 (2002) [arXiv:hep-th/0111189].

[121] Fredenhagen, S. and V. Schomerus, On boundary RG-flows in coset conformal field theories [arXiv:hep-th/0205011].

[122] Fredenhagen, S., Organizing boundary RG flows [arXiv:hep-th/0301229].

[123] Schafer-Nameki, S., D-branes in $N = 2$ coset models and twisted equivariant [arXiv:hep-th/0308058].

[124] Harvey, J.A., and G.W. Moore, On the algebras of BPS states, *Commun. Math. Phys.* **197**, 489 (1998) [arXiv:hep-th/9609017].

[125] Moore, G.W., String duality, automorphic forms, and generalized Kac–Moody algebras, *Nucl. Phys. Proc. Suppl.* **67**, 56 (1998) [arXiv:hep-th/9710198].

[126] Martinec, E.J. and G. Moore, On decay of K-theory [arXiv:hep-th/0212059].

Email: gmoore@physics.rutgers.edu.

9

Heisenberg groups and algebraic topology

JACK MORAVA

Abstract

We study the Madsen–Tillmann spectrum $\mathbb{C}P^{\infty}_{-1}$ as a quotient of the Mahowald pro-object $\mathbb{C}P^{\infty}_{-\infty}$, which is closely related to the Tate cohomology of circle actions. That theory has an associated symplectic structure, whose symmetries define the Virasoro operations on the cohomology of moduli space constructed by Kontsevich and Witten.

1 Introduction

A sphere S^n maps essentially to a sphere S^k only if $n \geq k$, and since we usually think of spaces as constructed by attaching cells, it follows that algebraic topology is in some natural sense upper-triangular, and thus not very self-dual: as in the category of modules over the mod p group ring of a p-group, its objects are built by iterated extensions from a small list of simple ones.

Representation theorists find semi-simple categories more congenial, and for related reasons, physicists are happiest in Hilbert space. This paper is concerned with some remarkable properties of the cohomology of the moduli space of Riemann surfaces, discovered by physicists studying two-dimensional topological gravity (an enormous elaboration of conformal field theory), which appear at first sight quite unfamiliar. Our argument is that these new phenomena are forced by the physicists' interest in self-dual constructions, which leads to objects which are (from the point of view of classical algebraic topology) very large [1, §2].

Fortunately, equivariant homotopy theory provides us with tools to manage these constructions. The first section below is a geometric introduction to the

I owe thanks to many people for help with the ideas in this paper, but it is essentially a collage of a lifetime's conversations with Graeme Segal, who more or less adopted me when we were both very young.

235

Tate cohomology of the circle group; the conclusion is that it possesses an intrinsic symplectic module structure, which pairs positive and negative dimensions in a way very useful for applications. Section 2 studies operations on this (not quite cohomology) functor, and exhibits the action of an algebraic analog of the Virasoro group on it. The third section relates rational Tate cohomology of the circle to that of the infinite loopspace $Q\mathbb{C}P_+^\infty$ considered by Madsen and Tillmann in recent work on Mumford's conjecture.

2 Geometric Tate cohomology

2.1. Let G be a compact Lie group of dimension d. We will be concerned with a cobordism category of smooth compact G-manifolds, with the action free on the boundary: this can be regarded as a categorical cofiber for the forgetful functor from manifolds with free G-action to manifolds with unrestricted action. Under reasonable assumptions this cofiber category is closed under Cartesian products (given the diagonal action).

If E is a geometric cycle theory (e.g. stable homotopy, or classical homology) then the graded E-bordism group of free G-manifolds is isomorphic to $E_{*+d}(BG_+)$. On the other hand, the homotopy quotient of a G-manifold is a bundle of manifolds over the classifying space BG, and Quillen's conventions [22] associate to such a thing, a class in the graded cobordism group $E^{-*}(BG_+)$. The forgetful functor from free to unrestricted G-manifolds defines a long exact sequence

$$\cdots \to E_{*-d}(BG_+) \to E^{-*}(BG_+) \to t_G^{-*}(E) \to E_{*-d-1}(BG_+) \to \cdots$$

which interprets the relative groups as the (Tate–Swan [28]) E-bordism of manifolds with G-action free on the boundary. The geometric boundary homomorphism

$$\partial_E : t_G^{-*}(E) \to E_{*-d-1}(BG_+) \to E_{*-d-1}$$

sends a manifold with G-free boundary to the quotient of that action on the boundary; it will be useful later.

Remarks:

(1) $t_G(E)$ is a ring-spectrum if E is; in fact it is an E-algebra, at least in some naïve sense;

(2) Tom Dieck stabilization [4] extends this geometric bordism theory to an equivariant theory;

(3) the functor t_G sends cofibrations to cofibrations, but it lacks good limit properties: it is defined by a kind of hybrid of homology and cohomology,

and Milnor's limit fails. In more modern terms [11], the construction sends a G-spectrum E to the equivariant function spectrum $[EG_+, E] \wedge \tilde{E}G_+$.

(4) The eventual focus of §9.2 is the case when G is the circle group $\mathbb{T}$, and E is ordinary cohomology: this is closely related to cyclic cohomology [2], but I do not know enough about that subject to say anything useful.

2.2. Suppose now that E is a general complex-oriented ring-spectrum; then $E^*(B\mathbb{T}_+)$ is a formal power series ring generated by the Euler (or first Chern) class $\mathbf{e}$. If E_* is concentrated in even degrees, then the cofiber sequence above reduces for dimensional reasons to a short exact sequence

$$0 \to E^*(B\mathbb{T}_+) = E^*[[\mathbf{e}]] \to t_{\mathbb{T}}^* E = E^*((\mathbf{e})) \to E_{-*-2}(B\mathbb{T}_+) \to 0$$

with middle group the ring of formal Laurent series in $\mathbf{e}$. By a lemma of [12, §2.4] we can think of $t_{\mathbb{T}}^* E$ as the homotopy groups of a pro-object

$$S^2\mathbb{C}P_{-\infty}^{\infty} \wedge E := \{S^2\mathrm{Th}(-k\eta) \wedge E \mid k > -\infty\}$$

in the category of spectra, constructed from the filtered vector bundle

$$\cdots (k-1)\eta \subset k\eta \subset (k+1)\eta \subset \cdots$$

defined by sums of copies of the tautological line bundle over $\mathbb{C}P^{\infty} \cong B\mathbb{T}$, as discussed in the appendix to [6] (see also [18]). More precisely: the Thom spectrum can be taken to be

$$\mathrm{Th}(-k\eta) := \lim_n S^{-2(n+1)k}\mathbb{C}P_n^{k\eta^{\perp}}$$

where $\eta^{\perp}$ is the orthogonal complement to the canonical line η in $\mathbb{C}^{n+1}$.

When E is **not** complex-orientable, $t_{\mathbb{T}} E$ can behave very differently: the Segal conjecture for Lie groups implies that, up to a profinite completion [10]

$$t_{\mathbb{T}} S^0 \sim S^0 \vee S[\prod B\mathbb{T}/C]$$

where the product runs through proper subgroups C of $\mathbb{T}$ (and S denotes suspension).

In the universal complex-oriented case, the class $\mathbf{e}^{-1} \in t_{\mathbb{T}}^{-2}\mathrm{MU}$ is represented geometrically by the unit disk in $\mathbb{C}$ with the standard action of $\mathbb{T}$ as unit complex numbers; more generally, the unit ball in $\mathbb{C}^k$ represents $\mathbf{e}^{-k}$. The geometric boundary homomorphism sends that $\mathbb{T}$-manifold to $\mathbb{C}P_{k-1}$; this observation can be restated, using Mishchenko's logarithm, as the formula

$$\partial_E(f) = \mathrm{res}_{\mathbf{e}=0} f(\mathbf{e})\, d\log_{\mathrm{MU}}(\mathbf{e}) : t_{\mathbb{T}}^* \mathrm{MU} \to \mathrm{MU}_{-*-2}$$

where the algebraic residue homomorphism

$$\mathrm{res}_{\mathbf{e}=0} : \mathrm{MU}^{*-2}((\mathbf{e})) \to \mathrm{MU}^*$$

is defined by $\mathrm{res}_{\mathbf{e}=0} \mathbf{e}^k\, d\mathbf{e} = \delta_{k+1,0}$, cf. [19] [21] [29].

2.3. The relative theory of manifolds with free group action on the boundary alone defines bordism groups $\tau_G^*(E)$ analogous to a truncation of Tate cohomology, with useful geometric applications. In place of the long exact sequence above, we have

$$\cdots \to E^{-*}(S^0) \to \tau_G^{-*}(E) \to E_{*-d-1}(BG_+) \to \cdots$$

compatible with a natural transformation $t_G^*(E) \to \tau_G^*(E)$ which forgets the interior G-action. In our case (when E is complex-oriented), this is just the E-homology of the collapse map

$$\mathbb{C}P_{-\infty}^\infty \to \mathbb{C}P_{-1}^\infty := \mathrm{Th}(-\eta)$$

defined by the pro-spectrum in the previous paragraph.

A Riemann surface with geodesic boundary is in a natural way an orientable manifold with a free $\mathbb{T}$-action on its boundary, and a family of such things, parametrized by a space X, defines an element of

$$\tau_\mathbb{T}^{-2}\mathrm{MU}(X_+) \cong [X, \mathbb{C}P_{-1}^\infty \wedge \mathrm{MU}] .$$

The Hurewicz image of this element in ordinary cohomology is the homomorphism

$$H^*(\mathbb{C}P_{-1}^\infty, \mathbb{Z}) \to H^*(X_+, \mathbb{Z})$$

defined by the classifying map of Madsen and Tillmann, which will be considered in more detail below.

3 Automorphisms of classical Tate cohomology

3.1. There are profound analogies – and differences – among the Tate cohomology rings of the groups $\mathbb{Z}/2\mathbb{Z}$, $\mathbb{T}$, and $\mathrm{SU}(2)$ [3]. A property unique to the circle is the existence of the nontrivial involution $I : z \to z^{-1}$.

When E is complex oriented, the symmetric bilinear form

$$f, g \mapsto (f, g) = \partial_E(fg)$$

on the Laurent series ring $t_\mathbb{T}^* E$ is nondegenerate, and the involution on $\mathbb{T}$ defines a symplectic form

$$\{f, g\} = (I(f), g)$$

which restricts to zero on the subspace of elements of positive (or negative) degree. This Tate cohomology thus has an intrinsic inner product, with canonical polarization and involution.

This bilinear form extends to a generalized Kronecker pairing

$$t_\mathbb{T} E^*(X) \otimes_E t_\mathbb{T} E_*(X) \to E_{*-2}$$

which can be interpreted as a kind of Spanier–Whitehead duality between $t_\mathbb{T} E_*$, viewed as a pro-object as in §2.1, and the direct system $\{E^*(\mathrm{Th}(-k\eta)) \mid k > -\infty\}$ defined by the cohomology of that system. This colimit again defines a Laurent series ring, but this object is not quite its own dual: a shift of degree two intervenes, and it is most natural to think of the (non-existent) functional dual of $t_\mathbb{T} E$ as $S^{-2} t_\mathbb{T} E$. The residue map $t_\mathbb{T} E \to S^2 E$ can thus be understood as dual to the unit ring-morphism $E \to t_\mathbb{T} E$.

3.2. The Tate construction is too large to be conveniently represented, so the usual Hopf-algebraic approach to the study of its automorphisms is technically difficult. Fortunately, methods from the theory of Tannakian categories can be applied: we consider automorphisms of $t_\mathbb{T} E$ as E varies, and approximate the resulting group-valued functor by representable ones. There is no difficulty in carrying this out for a general complex-oriented theory E, but the result is a straightforward extension of the case of ordinary cohomology.

To start, it is clear that the group(scheme, representing the functor

$$A \mapsto \mathbb{G}_0(A) = \{g(x) = \sum_{k \geq 0} g_k x^{k+1} \in A[[x]] \mid g_0 = 1\}$$

on commutative rings A) of automorphisms of the formal line acts as multiplicative natural transformations of the cohomology-theory-valued functor $A \mapsto t_\mathbb{T}^* H A$, with $g \in \mathbb{G}_0(A)$ sending the Euler class $\mathbf{e}$ to $g(\mathbf{e})$. [I am treating these theories as graded by $\mathbb{Z}/2\mathbb{Z}$, with A concentrated in degree zero; but one can be more careful.]

Clearly $\mathbb{G}_0$ is represented by a polynomial Hopf algebra on generators g_k, with diagonal

$$(\Delta g)(x) = (g \otimes 1)((1 \otimes g)(x)).$$

However, $\mathbb{G}_0$ is a subgroup of a larger system $\mathbb{G}$ of natural automorphisms, which is a colimit of representable functors (though not itself representable): following [16], let

$$A \mapsto \mathbb{G}(A) = \{g(x) = \sum_{k \gg -\infty} g_k x^{k+1} \in A((x)) \mid g_0 \in A^\times, g_k \in \sqrt{A} \text{ if } 0 > k\}$$

be the group of invertible **nil-Laurent** series, i.e. Laurent series with g_0 a unit, and g_k nilpotent for negative k. It is clear that $\mathbb{G}$ is a monoid, but in fact [20] it possesses inverses. The Lie algebra of $\mathbb{G}$ is spanned by the derivations $x^{k+1} \partial_x$, $k \in \mathbb{Z}$: it is the algebra of vector fields on the circle.

3.3. A related group-valued functor preserves the symplectic structure defined above: to describe it, I will specialize even further, and work over a field in which two is invertible: $\mathbb{R}$, for convenience. Thus let $\check{\mathbb{G}}$ be the (ind-pro)-algebraic groupscheme defined by invertible nil-Laurent series over the field

$\mathbb{R}((\sqrt{x}))$ obtained from $\mathbb{R}((x))$ by adjoining a formal square root of x, and let $\check{\mathbb{G}}_{odd}$ denote the subgroup of odd invertible series $\check{g}(\sqrt{x}) = -\check{g}(-\sqrt{x})$. The homomorphism

$$\check{g} \mapsto g(x) := \check{g}(\sqrt{x})^2 : \check{\mathbb{G}}_{odd} \to \mathbb{G}$$

is then a kind of double cover.

The functor $\check{\mathbb{G}}$ acts by symplectic automorphisms of the module $\mathbb{R}((\sqrt{x}))$, given the bilinear form

$$\langle u, v \rangle := \pi \, \mathrm{res}_{x=0} \, u(x) \, dv(x)$$

[27]; it is in fact a group of restricted symplectic automorphisms of this module. The Galois group of $\mathbb{R}((\sqrt{x}))/\mathbb{R}((x))$ defines a $\mathbb{Z}/2\mathbb{Z}$-action, and the subgroup $\check{\mathbb{G}}_{odd}$ preserves the subspace $\mathbb{R}((\sqrt{x}))_{odd}$ of odd power series.

Proposition. *The linear transformation*

$$t_{\mathbb{T}}^* H\mathbb{R} \to \mathbb{R}((\sqrt{x}))_{odd}$$

defined on normalized basis elements by

$$\mathbf{e}^k \mapsto \gamma_{-k-\frac{1}{2}}(x)$$

(where $\gamma_s(x) = \Gamma(1+s)^{-1} x^s$ denotes a divided power), is a dense symplectic embedding.

Proof: We have

$$\{\mathbf{e}^k, \mathbf{e}^l\} = (-1)^k \, \mathrm{res}_{\mathbf{e}=0} \, \mathbf{e}^{k+l} \, d\mathbf{e} = (-1)^k \delta_{k+l+1,0}$$

while

$$\langle \gamma_s, \gamma_t \rangle = \mathrm{res}_{x=0} \, \gamma_s(x) \gamma_{t-1}(x) \, dx = \frac{\pi}{\Gamma(t)\Gamma(1+s)} \delta_{s+t,0} \, .$$

The assertion then follows from the duplication formula for the Gamma function.

The half-integral divided powers lie in $\mathbb{Q}((\sqrt{x}))$, aside from distracting powers of π. The remaining rational coefficients involve the characteristic 'odd' factorials of 2D topological gravity [8, 15], e.g. when k is positive

$$\Gamma(k + \tfrac{1}{2}) = (2k - 1)!! \, 2^{-k} \sqrt{\pi}.$$

4 Symmetries of the stable cohomology of the Riemann moduli space

The preceding sections describe the construction of a polarized symplectic structure on the Tate cohomology of the circle group. The algebra of symmetric functions on the Lagrangian submodule

$$H^*(\mathbb{C}P_+^\infty, \mathbb{Q}) \subset t_{\mathbb{T}} H\mathbb{Q}$$

of that cohomology can be identified with the homology of the infinite loopspace

$$Q\mathbb{C}P_+^\infty = \lim_n \Omega^n S^n \, \mathbb{C}P_+^\infty.$$

On the other hand, this module of functions admits a canonical action of the Heisenberg group associated to its defining symplectic module [24, §9.5].

The point of this paper is that the homology of this infinite loopspace, considered in this way as a Fock representation, manifests the Virasoro representation constructed by Witten and Kontsevich on the stable cohomology of the moduli space of Riemann surfaces, identified with $H_*(Q\mathbb{C}P_+^\infty, \mathbb{Q})$ through the work of Madsen, Tillman, and Weiss. Some of those results are summarized in the next two subsections; a more thorough account can be found in Michael Weiss's survey in these Proceedings. The third subsection below discusses their connection with representation theory.

4.1. Here is a very condensed account of one component of [17]: if $F \subset \mathbb{R}^n$ is a closed connected two-manifold embedded smoothly in a high-dimensional Euclidean space, its Pontrjagin–Thom construction $\mathbb{R}^n_+ \to F^\nu$ maps compactified Euclidean space to the Thom space of the normal bundle of the embedding. The tangent plane to F is classified by a map $\tau : F \to \mathrm{Grass}_{2,n}$ to the Grassmannian of oriented two-planes in $\mathbb{R}^n$, and the canonical two-plane bundle η over this space has a complementary $(n-2)$-plane bundle, which I will call $(n-\eta)$. The normal bundle ν is the pullback along τ of $(n-\eta)$; composing the map induced on Thom spaces with the collapse defines the map

$$\mathbb{R}^n_+ \to F^\nu \to \mathrm{Grass}_{2,n}^{(n-\eta)}.$$

The space $\mathrm{Emb}(F)$ of embeddings of F in $\mathbb{R}^n$ becomes highly connected as n increases, and the group $\mathrm{Diff}(F)$ of orientation-preserving diffeomorphisms of F acts freely on it, defining a compatible family

$$\mathbb{R}^n_+ \wedge_{\mathrm{Diff}} \mathrm{Emb}(F) \to \mathrm{Grass}_{2,n}^{(n-\eta)}$$

which can be interpreted as a morphism

$$B\mathrm{Diff}(F) \to \lim \Omega^n \mathrm{Grass}_{2,n}^{(n-\eta)} := \Omega^\infty \mathbb{C}P_{-1}^\infty.$$

This construction factors through a map

$$\coprod_{g \geq 0} B\mathrm{Diff}(F_g) \to \mathbb{Z} \times B\Gamma_\infty^+ \to \Omega^\infty \mathbb{C}P_{-1}^\infty$$

of infinite loopspaces. Collapsing the bottom cell defines a cofibration

$$S^{-2} \to \mathbb{C}P^{\infty}_{-1} \to \mathbb{C}P^{\infty}_{+}$$

of spectra; the fiber of the corresponding map

$$\Omega^2 QS^0 \to \Omega^{\infty}\mathbb{C}P^{\infty}_{-1} \to Q\mathbb{C}P^{\infty}_{+}$$

of spaces has torsion homology, and the resulting composition

$$\mathbb{Z} \times B\Gamma^{+}_{\infty} \to Q\mathbb{C}P^{\infty}_{+} \sim QS^0 \times Q\mathbb{C}P^{\infty}$$

is a rational homology isomorphism which identifies Mumford's polynomial algebra on classes κ_i, $i \geq 1$, with the symmetric algebra on positive powers of **e**. The rational cohomology of QS^0 adds a copy of the group ring of $\mathbb{Z}$, which can be interpreted as a ring of Laurent series in a zeroth Mumford class κ_0.

The standard convention is to write b_k for the generators of $H_*\mathbb{C}P^{\infty}_{+}$ dual to $\mathbf{e}^k$, and to use the same symbols for their images in the symmetric algebra $H_*(Q\mathbb{C}P^{\infty}_{+}, \mathbb{Q})$. The Thom construction defines a map

$$\mathbb{C}P^{\infty} \to \mathrm{MU}$$

which extends to a ring isomorphism

$$H_*(Q\mathbb{C}P^{\infty}, \mathbb{Q}) \to H_*(\mathrm{MU}, \mathbb{Q})$$

sending the b_k to classes usually denoted t_k, with $k \geq 1$; but it is convenient to extend this to allow $k = 0$.

4.2. The homomorphism

$$\lim \mathrm{MU}^{*+n-2}(\mathrm{Th}(n - \eta)) \to \mathrm{MU}^{*-2}(B\mathrm{Diff}(F))$$

defined on cobordism by the Madsen–Tillmann construction sends the Thom class to a kind of Euler class: according to Quillen, the Thom class of $n - \eta$ is its zero-section, regarded as a map between manifolds. Its image is the class defined by the fiber product. This is the space of equivalence classes, under the action of $\mathrm{Diff}(F)$, of pairs (x, ϕ), with $x \in \phi(F) \subset \mathbb{R}^n$ a point of the surface (i.e., in the zero-section of ν), and ϕ an embedding. Up to suspension, this image is thus the element

$$[Z_n \to \mathbb{R}^n_{+} \wedge_{\mathrm{Diff}} \mathrm{Emb}] \mapsto \mathrm{MU}^{n-2}(S^n B\mathrm{Diff}(F))$$

defined by the tautological family $F \times_{\mathrm{Diff}} E\mathrm{Diff}(F)$ of surfaces over the classifying space of the diffeomorphism group. It is primitive in the Hopf-like structure defined by gluing: in fact it is the image of

$$\sum_{k \geq 1} \kappa_k t_{k+1} \in \mathrm{MU}^{-2}(B\Gamma^{+}_{\infty}) \otimes \mathbb{Q}.$$

If v is a formal indeterminate of cohomological degree two, then the class

$$\Phi = \exp(v\text{Th}(-\eta)) \in \text{MU}^0_{\mathbb{Q}}(\Omega^\infty \mathbb{C}P^\infty_{-1})[[v]]$$

defined by finite unordered configurations of points on the universal surface (with v a book-keeping indeterminate of cohomological degree two) is a kind of exponential transformation

$$\tilde{\Phi}_* : H_*(Q\mathbb{C}P^\infty_+, \mathbb{Q}) \to H_*(\text{MU}, \mathbb{Q}[[v]]).$$

From this perspective it is natural to interpret the Thom class in $\text{MU}^{-2}(\mathbb{C}P^\infty_{-1}) \otimes \mathbb{Q}$ as the sum $\sum_{k \geq -1} t_{k+1} \mathbf{e}^k$, with $t_0 = v^{-1}\mathbf{e}$.

4.3. A class in the cohomology group

$$H^2_{\text{Lie}}(V, \mathbb{R}) \cong \Lambda^2(V^*)$$

of a real vector space V defines a Heisenberg extension

$$0 \to \mathbb{T} \to H \to V \to 0.$$

The representation theory of such groups, and in particular the construction of their Fock representations, is classical [5]. What is important to us is that these are **projective** representations of V, with positive energy; such representations have very special properties.

The loop group of a circle is a key example; it possesses an intrinsic symplectic form, defined by formulae much like those of §2 [23 §5, §7b]. Diffeomorphisms of the circle act on any such loop group, and it is a deep property of positive-energy representations, that they extend to representations of the resulting semidirect product of the loop group by $\text{Diff}S^1$. Therefore by restriction a positive-energy representation of a loop group automatically provides a representation of $\text{Diff}S^1$. This [Segal–Sugawara [25 §13.4]] construction yields the action of Witten's Virasoro algebra on the Fock space

$$\text{Symm}(H^*(\mathbb{C}P^\infty_+)) \cong \mathbb{Q}[t_k \mid k \geq 0] .$$

In Kontsevich's model, the classes t_k are identified with the symmetric functions

$$\text{Trace } \gamma_{k-\frac{1}{2}}(\Lambda^2) \sim -(2k-1)!! \text{ Trace } \Lambda^{-2k-1}$$

of a positive-definite Hermitian matrix Λ.

Note, however, that the deeper results of Kontsevich and Witten [31] are inaccessible in this toy model: that theory is formulated in terms of compactified moduli spaces $\overline{\mathcal{M}}_g$ of algebraic curves. The rational homology of $Q(\bigsqcup \overline{\mathcal{M}}_g)$ (suitably interpreted, for small g) contains a fundamental class

$$\exp\left(\sum_{g\geq 0}[\overline{\mathcal{M}}_g]v^{3(g-1)}\right)$$

for the moduli space of not necessarily connected curves. Witten's tau-function is the image of this 'highest-weight' vector under the analog of $\tilde{\Phi}$; it is killed by the subalgebra of Virasoro generated by the operators L_k with $k \geq -1$.

5 Concluding remarks

5.1. Witten has proposed a generalization of 2D topological gravity which encompasses surfaces with higher spin structures: for a closed smooth surface F an r-spin structure is roughly a complex line bundle L together with a fixed isomorphism $L^{\otimes r} \cong T_F$ of two-plane bundles, but for surfaces with nodes or marked points the necessary technicalities are formidable [14]. The group of automorphisms of such a structure is an extension of its group of diffeomorphisms by the group of rth roots of unity, and there is a natural analog of the group completion of the category defined by such surfaces. The generalized Madsen–Tillmann construction maps this loopspace to the Thom spectrum $\mathrm{Th}(-\eta^r)$, and it is reasonable to expect that this map is equivariant with respect to automorphisms of the group of roots of unity. This fits with some classical homotopy theory: if (for simplicity) $r = p$ is prime, multiplication by an integer u relatively prime to p in the H-space structure of $\mathbb{C}P^\infty$ defines a morphism

$$\mathrm{Th}(-\eta^p) \to \mathrm{Th}(-\eta^{up})$$

of spectra, and the classification of fiber-homotopy equivalences of vector bundles yields an equivalence of $\mathrm{Th}(-\eta^{up})$ with $\mathrm{Th}(-\eta^p)$ after p-completion. There is an analogous decomposition of $t_\mathbb{T} H\mathbb{Z}_p$ and a corresponding decomposition of the associated Fock representations [20 §2.4].

5.2. Tillmann has also studied categories of surfaces above a parameter space X; the resulting group completions have interesting connections with both Tate and quantum cohomology. When X is a compact smooth almost-complex manifold, its Hodge-deRham cohomology admits a natural action of the Lie algebra generated by the Hodge dimension operator H together with multiplication by the first Chern class (E) and its adjoint $(F = *E*)$ [26]. Recently Givental [9 §8.1] has shown that earlier work of (the schools of) Eguchi, Dubrovin, and others can be reformulated in terms of structures on $t_\mathbb{T}^{*,*} H_{\mathrm{dg}}(X)$, given a symplectic structure generalizing that of §3. In this work, the relevant involution is

$$I_{\text{Giv}} = \exp(\tfrac{1}{2}H)\exp(-E)\ I\ \exp(E)\exp(-\tfrac{1}{2}H).$$

it would be very interesting if this involution could be understood in terms of the equivariant geometry of the free loopspace of X [7].

5.3. Nothing forces us to restrict the construction of Madsen and Tillmann to two-manifolds, and I want to close with a remark about the cobordism category of smooth spin four-manifolds bounded by ordinary three-spheres. A parametrized family of such objects defines, as in §2.3, an element of the truncated equivariant cobordism group

$$\tau_{\text{SU}(2)}^{-4}\text{MSpin}(X_+)\ .$$

On the other hand, it is a basic fact of four-dimensional life that

$$\text{Spin}(4) = \text{SU}(2) \times \text{SU}(2)$$

so the Madsen–Tillmann spectrum for the cobordism category of such spin four-folds is the twisted desuspension

$$B\text{Spin}(4)^{-\rho} = (\mathbb{H}P_\infty \times \mathbb{H}P_\infty)^{-V^* \otimes_{\mathbb{H}} V}$$

of the classifying space of the spinor group by the representation ρ defined by the tensor product of two standard rank one quaternionic modules over $\text{SU}(2)$ [13 §1.4]. Composition with the Dirac operator defines an interesting rational homology isomorphism

$$(\mathbb{H}P_\infty \times \mathbb{H}P_\infty)^{-V^* \otimes_{\mathbb{H}} V} \to \mathbb{H}P_\infty^{-V} \wedge \text{MSpin} \to \mathbb{H}P_\infty^{-V} \wedge k\text{O}$$

related in low dimensions to the classification of unimodular even indefinite lattices [27, 30]. This suggests that the Tate cohomology $t_{\text{SU}(2)}^* k\text{O}$ may have an interesting role to play in the study of topological gravity in dimension four.

References

[1] Adams, J.F., ... what we don't know about $\mathbb{R}P^\infty$, in *New Developments in Topology* (ed.), G. Segal, *LMS Lecture Notes* 11 (1972) 1–9.

[2] Adem, A., R.L. Cohen, and W. Dwyer, Generalized Tate homology, homotopy fixed points and the transfer, in *Algebraic topology (Evanston 88)* 1–13, *Contemp. Math.* 96 (1989).

[3] Arnol'd, V.I., Symplectization, complexification and mathematical trinities, in *The Arnoldfest* 23–37, *Fields Inst. Commun.* 24 (1999).

[4] Th. Bröcker, E.C. Hook, Stable equivariant bordism, *Math. Zeits.* 129 (1972) 269–277.

[5] Cartier, P., Quantum mechanical commutation relations and theta functions, in *Algebraic Groups and Discontinuous Subgroups, Proc. Sympos. Pure Math.* 9 (1966) 361–383.

[6] Cohen, R.L., J.D.S. Jones, and G.B. Segal, Floer's infinite dimensional Morse theory and homotopy theory, in the *Floer Memorial Volume*, Birkhäuser, Progress in Mathematics 133 (1995) 297–326.

[7] Cohen, R.L. and A. Stacey, Fourier decompositions of loop bundles, Proc. Northwestern Conf (2001), to appear.

[8] DiFrancesco, P., C. Itzykson, and J.-B. Zuber, Polynomial averages in the Kontsevich model, *CMP* 151 (1993) 193–219.

[9] Givental, A., Gromov–Witten invariants and quantization of quadratic Hamiltonians, available at math.AG/0108100

[10] Greenlees, J.P.C., A rational splitting theorem for the universal space for almost free actions, *Bull. London Math. Soc.* 28 (1996) 183–189.

[11] Greenlees, J.P.C. and J.P. May, **Generalized Tate cohomology**, Mem. AMS 113 (1995).

[12] Greenlees, J.P.C. and H. Sadofsky, The Tate spectrum of v_n-periodic complex-oriented theories, *Math. Zeits.* 222 (1996) 391–405.

[13] Gompf, R. and A. Stipsicz **Four-manifolds and Kirby calculus**, *AMS Grad Texts 20* (1999).

[14] Jarvis, T., T. Kimura, and A. Vaintrob, Moduli spaces of higher spin curves and integrable hierarchies, available at math.AG/9905034

[15] Józefiak, T., Symmetric functions in the Kontsevich-Witten intersection theory of the moduli space of curves, *Lett. Math. Phys.* 33 (1995) 347–351.

[16] Kapranov, M. and E. Vasserot, Vertex algebras and the formal loop space, available at math.AG/0107143

[17] Madsen, I. and U. Tillmann, The stable mapping-class group and $Q(\mathbb{C}P_+^\infty)$, *Invent. Math.* 145 (2001) 509–544.

[18] Mahowald, M., **On the metastable homotopy of S^n**, Mem. AMS 72 (1967).

[19] Morava, J., Cobordism of involutions revisited, revisited, in *The Boardman Festschrift*, *Contemporary Math.* 239 (1999).

[20] Morava, J., An algebraic analog of the Virasoro group, *Czech. J. Phys.* 51 (2001), available at math.QA/0109084

[21] Quillen, D., On the formal group laws of unoriented and complex cobordism theory, *BAMS* 75 (1969) 1293–1298.

[22] Quillen, D., Elementary proofs of some results of cobordism theory using Steenrod operations, Adv. in Math 7 (1971) 29–56.

[23] Segal, G., Unitary representations of some infinite-dimensional groups, *Comm. Math. Phys.* 80 (1981) 301–342.

[24] Segal, G. and A. Pressley, *Loop groups*, Oxford (1986).

[25] Segal, G. and A. Pressley, *Algebres de Lie semisimples complexes*, Benjamin (1966).

[26] Serre, J.P., *Corps Locaux*, Hermann (1968).

[27] Serre, J.P., *A Course in Arithmetic*, Springer (1973).

[28] Swan, R., Periodic resolutions for finite groups, Annals of Math. 72 (1960) 267–291.

[29] Tate, J., Residues of differentials on curves, Ann. Sci. Ecole Norm. Sup. 1 (1968) 149–159.

[30] Wall, C.T.C., On simply-connected four-manifolds, J. London Math. Soc. 39 (1964) 141–149.

[31] Witten, E., Two-dimensional gravity and intersection theory on moduli space, *Surveys in Differential Geometry* 1 (1991) 243–310.

DEPARTMENT OF MATHEMATICS, JOHNS HOPKINS UNIVERSITY, BALTIMORE, MARYLAND 21218
Email address: jack@math.jhu.edu

10

What is an elliptic object?

STEPHAN STOLZ and PETER TEICHNER

University of California at San Diego and University of Notre Dame

Dedicated to Graeme Segal on the occasion of his 60th birthday

Contents

1 Introduction

In these notes we propose an approach towards *enriched elliptic objects* over a manifold X. We hope that once made precise, these new objects will become cocycles in the generalized cohomology theory $tmf^*(X)$ introduced by Hopkins and Miller [Ho], in a similar way as vector bundles over X represent elements in $K^*(X)$. We recall that one important role of $K^*(X)$ is as the home of the index of a family of Fredholm operators parametrized by X, e.g. the family of Dirac operators of a fiber bundle $E \to X$ with spin fibers. This is the parametrized version of the $\widehat{A}$-genus in the sense that the family index in $K^*(X)$ reduces to the $\widehat{A}$-genus of the fiber if X is a point. Similarly, one important role of $tmf^*(X)$ is that it is the home of the parametrized version of the Witten genus in the sense that a fiber bundle $E \to X$ whose fibers are *string manifolds* (cf. section 5) gives rise to an element in $tmf^*(X)$ [HBJ]. If X is a point this reduces to the Witten genus [Wi1] of the fiber (modulo torsion).

It would be very desirable to have a geometric/analytic interpretation of this parametrized Witten genus along the lines described above for the parametrized $\widehat{A}$-genus. For $X = $ pt, a heuristic interpretation was given by Witten who described the Witten genus of a string manifold M as the S^1-equivariant index of the 'Dirac operator on the free loop space' of M [Wi1] or as the 'partition function of the super symmetric non-linear σ-model' with target M [Wi2]; alas, neither of these have been rigorously constructed yet. The construction of the parametrized Witten genus in $tmf^*(X)$ is instead purely homotopy theoretic; the main ingredient is a Thom-isomorphism in tmf-cohomology for vector bundles with string structures. This is completely analogous to a description of the $\widehat{A}$-genus based on the Thom isomorphism in K-theory for spin vector bundles.

The cohomology theory $tmf^*(X)$ derives from cohomology theories of the 'elliptic' flavor. The first such theory was constructed by Landweber and Stong [La] using Landweber's Exact Functor Theorem and the elliptic genus introduced by Ochanine [Och]. Ochanine's genus can be interpreted as coming from the formal group law associated to a particular elliptic curve; varying the elliptic curve used in the Landweber–Stong construction leads to a plethora of *elliptic cohomology theories*. The cohomology theory $tmf^*(X)$ is not strictly speaking one of these, but essentially the 'inverse limit' (over the category of elliptic curves) of all these cohomology theories. (There are considerable technical difficulties with making this precise, in fact so far no complete written account is available.) Since *integral modular forms* can be defined as an inverse limit of an abelian group valued functor over the same category of elliptic curves, the elements of tmf*(pt) are called *topological modular forms*. There is a ring homomorphism from tmf*(pt) to the ring of integral modular forms which is rationally an isomorphism.

Unfortunately, the current *geometric* understanding of elliptic cohomology still is very much in its infancy despite the efforts of various people; see [Se1], [KS], [HK], [BDR]. The starting point of our new approach are the elliptic objects suggested by Graeme Segal in [Se1], which we call *Segal elliptic objects*. Segal's idea was to view a vector bundle $E \to X$ with connection as a *1-dimensional field theory over X* in the following sense: To each point $x \in X$, the bundle E associates a vector space E_x, and to each path in X the connection on E associates a linear map between these vector spaces. Segal suggested that a *2-dimensional conformal* field theory over X could be used as a cocycle for some elliptic cohomology theory. It would associate Hilbert spaces to loops in X, and Hilbert–Schmidt operators to conformal surfaces (with boundary) in X.

The main problem with Segal elliptic objects is that excision does not seem to hold. One of our contributions is to suggest a modification of the definition in order to get around this problem. This is where von Neumann algebras (associated to points in X) and their bimodules (associated to arcs in X) enter the picture. We will explain our modification in detail in the coming sections of this introduction. In the case $X =$ pt, we obtain in particular a modification of the notion of a *vertex operator algebra* (which was shown to be equivalent to a Segal elliptic object in [Hu], at least for genus zero surfaces; the super symmetric analogue appeared in [Ba]).

Another, more technical, problem in Segal's definition is that he had to introduce 'riggings' of 1- and 2-manifolds. These are certain additional structures (like parametrizations of the boundary circles) which we shall recall after Definition 4.1. Our first observation is that one can avoid these extra structures alltogether by enriching the conformal surfaces with fermions, and that these fermions give rise naturally to the degree of an elliptic object. This degree coincides for closed surfaces with the correct power of the determinant line as explained in [Se2] and in fact the space of fermions is a natural extension of the determinant line to surfaces with boundary (in the absence of parametrizations).

In Definition 4.3 we explain the resulting *Clifford elliptic objects of degree n* which includes a 'super symmetric' aspect. The reader should be warned that there remains an issue with how to make this super symmetric aspect precise; we formulate what we need as Hypothesis 3.29 and use it in the proof of Theorem 1.2 below.

We motivate these Clifford elliptic objects by first explaining carefully the K-theoretic analogues in Section 3. It turns out that the idea of a connection has to be modified because one needs the result of 'parallel transport' to depend on the length of the parametrizing interval. In other words, we explain how a K-cocycle is given by a super symmetric 1-dimensional *Euclidean* field theory, see Definition 3.5. In this simpler case, we do formulate the super symmetric aspect in detail and we discuss why it is essential for K-theory. As is well known, the best way to define K-theory in degree n, $K^n(X)$, is to introduce the finite dimensional Clifford algebras C_n. We shall explain how these algebras arise naturally when enriching intervals with fermions. We conclude in Section 10.3.2 the following new description of the K-theory spectrum:

Theorem 1.1. *For any $n \in \mathbb{Z}$, the space of super symmetric 1-dimensional Euclidean field theories of degree n has the homotopy type of K_{-n}, the $(-n)$th space in the Ω-spectrum representing periodic K-theory.*

There is an analogous statement for periodic KO-theory, using real field theories.

Roughly speaking, Segal's idea, which we are trying to implement here, was to replace 1-dimensional by 2-dimensional field theories in the above theorem in order to obtain the spectrum of an elliptic cohomology theory. The following result is our first point of contact with modular forms and hence with $\mathrm{tmf}^n(X)$.

Theorem 1.2. *Given a degree n Clifford elliptic object E over X, one gets canonically a Laurent series*

$$MF(E) \in K^{-n}(X)[[q]][q^{-1}].$$

Moreover, if n is even and $X = \mathrm{pt}$, then $MF(E) \in \mathbb{Z}[[q]][q^{-1}]$ is the q-expansion of a 'weak' modular form of weight $n/2$. This means that the product of $MF(E)$ with a sufficiently large power of the discriminant Δ is a modular form.

In terms of our new definition of K-theory, the map $E \mapsto MF(E)$ is given by crossing with the standard circle S^1, and hence is totally geometric. As we shall explain, the length of an interval is very important in K-theory, and by crossing with S^1 it is turned into the conformal modulus of an annulus. The above result shows that the modularity aspects of an elliptic object are satisfied with only minor modifications of Segal's original definition. This is related to the fact that for $X = \mathrm{pt}$ the deficiency regarding excision is not present.

In Section 4 we make a major modification of Segal's elliptic objects and explain our *enriched elliptic objects* which are defined so that excision can be satisfied in the theory. Each enriched elliptic object gives in particular a Clifford elliptic object (which is closely related to a Segal elliptic object) but there are also data assigned to points and arcs in X, see Definition 1.4. Roughly speaking, in addition to Hilbert spaces associated to loops in X, we assign von Neumann algebras $\mathcal{A}(x)$ to points $x \in X$ and bimodules to arcs in X, in a way that Segal's Hilbert space can be decomposed as a Connes fusion of bimodules whenever the loop decomposes into arcs, see Section 1.2. One purpose of the paper is to make these statements precise. We shall not, however, give the ultimate definition of elliptic cocycles because various aspects of the theory have not been completely worked out yet.

Our main result, which to our mind justifies all definitions, is the following analogue of the tmf-orientation for string vector bundles [Ho, §6], [AHS]. As the underlying Segal elliptic object, we in particular recover in the case $E = TX$ the 'spinor bundle' over the loop space LX. Our enrichment expresses the locality (in X) of this spinor bundle. We expect that this enriched elliptic object will play the role of an elliptic Euler class and, in a relative version, of the elliptic Thom class.

Theorem 1.3. *Let E be an n-dimensional vector bundle over a manifold X. Assume that E comes equipped with a string structure and a string connection. Then there is a canonical degree n enriched elliptic object over X such that for all $x \in X$ the algebras $\mathcal{A}(x)$ are hyperfinite type III_1 factors. Moreover, if one varies the string connection then the resulting enriched elliptic objects are isomorphic.*

A vector bundle over X has a *string structure* if and only if the characteristic classes w_1, w_2 and $p_1/2$ vanish. In Section 5 we define a string structure on an n-dimensional spin bundle as a lift of the structure group in the following extension of topological groups

$$1 \longrightarrow PU(A) \longrightarrow \text{String}(n) \longrightarrow \text{Spin}(n) \longrightarrow 1.$$

Here A is an explicit hyperfinite type III_1 factor, the 'local fermions on the circle', cf. Example 4.5. Its unitary group is contractible (in the strong operator topology) and therefore the resulting projective unitary group $PU(A) = U(A)/\mathbb{T}$ is a $K(\mathbb{Z}, 2)$. The extension is constructed so that $\pi_3 \text{String}(n) = 0$, which explains the condition on the characteristic class $p_1/2$. This interpretation of string structures is crucial for our construction of the enriched elliptic object in Theorem 1.3, the relation being given by a monomorphism

$$\mathrm{String}(n) \longrightarrow \mathrm{Aut}(A)$$

which arises naturally in the definition of the group extension above. It should be viewed as the 'fundamental representation' of the group $\mathrm{String}(n)$. The notion of a *string connection*, used in the above theorem, will be explained before Corollary 5.20.

1.1 Segal elliptic objects and excision

A *Segal elliptic object* over X [Se1, p. 199] associates to a map γ of a closed rigged 1-manifold to a target manifold X a topological vector space $H(\gamma)$, and to any conformal rigged surface Σ with map $\Gamma : \Sigma \to X$ a vector $\Psi(\Gamma)$ in the vector space associated to the restriction of Γ to $\partial\Sigma$ (we will define *riggings* in Definition 4.2 below). This is subject to the axiom

$$H(\gamma_1 \amalg \gamma_2) \cong H(\gamma_1) \otimes H(\gamma_2)$$

and further axioms for Ψ which express the fact that the gluing of surfaces (along closed submanifolds of the boundary) corresponds to the composition of linear operators. Thus an elliptic object over a point is a *conformal field theory* as axiomatized by Atiyah and Segal: it is a functor from a category $\mathcal{C}(X)$ to the category of topological vector spaces. Here the objects in $\mathcal{C}(X)$ are maps of closed rigged 1-manifolds into X, and morphisms are maps of conformal rigged surfaces into X.

Originally, the hope was that these elliptic objects would lead to a geometric description of elliptic cohomology. Unfortunately, excision for the geometric theory defined via elliptic objects did not seem to work out. More precisely, consider the Mayer–Vietoris sequence

$$\cdots \longrightarrow E^n(X) \longrightarrow E^n(U) \oplus E^n(V) \longrightarrow E^n(U \cap V) \longrightarrow \cdots$$

associated to a decomposition $X = U \cup V$ of X into two open subsets $U, V \subset X$. This is an *exact sequence* for any cohomology theory $X \mapsto E^n(X)$. For K-theory the exactness of the above sequence at $E^n(U) \oplus E^n(V)$ comes down to the fact that a vector bundle $E \to X$ can be reconstructed from its restrictions to U and V.

Similarly, we expect that the proof of exactness for a cohomology theory built from Clifford elliptic objects of degree n would involve being able to reconstruct an elliptic object over $U \cup V$ from its restriction to U and V. This does not seem to be the case: suppose (H, Ψ) is an elliptic object over $U \cup V$ and consider two paths γ_1, γ_2 between the points x and y. Assume that the path

γ_1 lies in U, that γ_2 lies in V, and denote by $\bar{\gamma}_2$ the path γ_2 run backwards. Then the restriction of (H, Ψ) to U (resp. V) contains not enough information on how to reconstruct the Hilbert space $H(\gamma_1 \cup \bar{\gamma}_2)$ associated to the loop $\gamma_1 \cup \bar{\gamma}_2$ in $U \cup V$.

1.2 Decomposing the Hilbert space

Our basic idea on how to overcome the difficulty with excision is to notice that in the basic geometric example coming from a vector bundle with string connection (see Theorem 1.3), there is the following additional structure: To a point $x \in X$ the string structure associates a graded type III_1-factor $\mathcal{A}(x)$ and to a finite number of points x_i it assigns the spatial tensor product of the $\mathcal{A}(x_i)$. Moreover, to a path γ from x to y, the string connection gives a graded right module $\mathcal{B}(\gamma)$ over $\mathcal{A}(\partial\gamma) = \mathcal{A}(x)^{\text{op}} \bar{\otimes} \mathcal{A}(y)$. There are canonical isomorphisms over $\mathcal{A}(\partial\bar{\gamma}) = \mathcal{A}(y)^{\text{op}} \bar{\otimes} \mathcal{A}(x)$ (using the 'conjugate' module from Section 4.3)

$$\mathcal{B}(\bar{\gamma}) \cong \overline{\mathcal{B}(\gamma)}.$$

The punchline is that Hilbert spaces like $H(\gamma_1 \cup \bar{\gamma}_2)$ discussed above can be decomposed as

$$H(\gamma_1 \cup \bar{\gamma}_2) \cong \mathcal{B}(\gamma_1) \boxtimes_{\mathcal{A}(\partial\gamma_i)} \mathcal{B}(\bar{\gamma}_2)$$

where we used the *fusion product* of modules over von Neumann algebras introduced by Connes [Co1, V.B.δ]. Following Wassermann [Wa], we will refer to this operation as *Connes fusion*. Connes' definition was motivated by the fact that a homomorphism $A \rightarrow B$ of von Neumann algebras leads in a natural way to an $B - A$-bimodule such that composition of homomorphisms corresponds to his fusion operation [Co1, Prop. 17 in V.B.δ]. In [Wa], Wassermann used Connes fusion to define the correct product on the category of positive energy representations of a loop group *at a fixed level*.

We abstract the data we found in the basic geometric example from Theorem 1.3 by giving the following preliminary

Definition 1.4 (Preliminary!). A *degree n enriched elliptic object* over X is a tuple $(H, \Psi, \mathcal{A}, \mathcal{B}, \phi_{\mathcal{B}}, \phi_H)$, where:

(1) (H, Ψ) is a degree n Clifford elliptic object over X. In particular, it gives a Hilbert space bundle over the free loop space LX.
(2) $\mathcal{A}$ is a von Neumann algebra bundle over X.
(3) $\mathcal{B}$ is a module bundle over the free path space PX. Here the end point map $PX \rightarrow X \times X$ is used to pull back two copies of the algebra bundle $\mathcal{A}$

to PX, and these are the algebras acting on $\mathcal{B}$. The modules $\mathcal{B}(\gamma)$ come equipped with gluing isomorphisms (of $\mathcal{A}(x_1)^{\mathrm{op}}\bar{\otimes}\mathcal{A}(x_3)$-modules)

$$\phi_{\mathcal{B}}(\gamma, \gamma') : \mathcal{B}(\gamma \cup_{x_2} \gamma') \xrightarrow{\cong} \mathcal{B}(\gamma) \boxtimes_{\mathcal{A}(x_2)} \mathcal{B}(\gamma')$$

if γ is a path from x_1 to x_2, and γ' is a path from x_2 to x_3.

(4) ϕ_H is an isomorphism of Hilbert spaces associated to each pair of paths γ_1 and γ_2 with $\partial\gamma_1 = \partial\gamma_2$

$$\phi_H(\gamma_1, \bar{\gamma}_2): H(\gamma_1 \cup_{\partial\gamma_i} \gamma_2) \xrightarrow{\cong} \mathcal{B}(\gamma_1) \boxtimes_{\mathcal{A}(\partial\gamma_i)} \mathcal{B}(\bar{\gamma}_2).$$

All algebras and modules are $\mathbb{Z}/2$-graded and there are several axioms that we require but have not spelled out above.

Remark 1.5. This is only a preliminary definition for several reasons. Among others:

- We left out the conditions for surfaces glued along non-closed parts of their boundary. The vectors $\Psi(\Gamma)$ of a Clifford elliptic objects compose nicely when two surfaces are glued along *closed* submanifolds of the boundary, compare Lemma 2.24. Our enriched elliptic objects compose in addition nicely when two surfaces are glued along arcs in the boundary, see Proposition 4.12.

- We left out the super symmetric part of the story. We will explain in Section 3.2 why super symmetric data are essential even in the definition of K-theory.

- We left out the fermions from the discussions. These will be used to define the degree n of an elliptic object, and there are extra data needed so that a conformal spin surface actually gives a vector in the relevant Hilbert space. If the surface Σ is closed, then a fermion is a point in the nth power of the Pfaffian line of Σ. Since the Pfaffian line is a square-root of the determinant line, this is consistent with the fact that a degree n elliptic object should give a modular form of weight $n/2$ when evaluated on tori, see Sections 3.3 and 4.1.

- Segal's Hilbert space associated to a circle will actually be *defined* by 4 above, rather than introducing the isomorphisms ϕ_H. So it will not play a central role in the theory, but can be reconstructed from it. At this point, we wanted to emphasize the *additional* data needed to resolve the problem with excision, namely a decomposition of Segal's Hilbert space.

Most of these deficiencies will be fixed in Section 4 by defining a degree n enriched elliptic object as a certain functor from a bicategory $\mathcal{D}_n(X)$ made from

d-manifolds (with n fermions) mapping into X, $d = 0, 1, 2$, to the bicategory vN of von Neumann algebras, their bimodules and intertwiners.

We end this introduction by making a brief attempt to express the meaning of an enriched elliptic object over X in physics lingo: it is a conformal field theory with $(0, 1)$ super symmetry (and target X), whose fermionic part has been quantized but whose bosonic part is classical. This comes from the fact that the Fock spaces are a well-established method of fermionic quantization, whereas there is up to date no mathematical way of averaging the maps of a surface to a curved target X. Moreover, the enriched (0-dimensional) aspect of the theory is some kind of an open string theory. It would be very interesting to relate it to Cardy's boundary conformal field theories.

1.3 Disclaimer and acknowledgments

This paper is a survey of our current understanding of the geometry of elliptic objects. Only ideas of proofs are given, and some proofs are skipped all together. We still believe that it is of service to the research community to make such work in progress accessible.

It is a pleasure to thank Dan Freed, Graeme Segal and Antony Wassermann for many discussions about conformal field theory. Graeme's deep influence is obvious, and Dan's approach to Chern–Simons theory [Fr1] motivated many of the considerations in Section 5. Antony's groundbreaking work [Wa] on Connes fusion for positive energy representations was our starting point for the central definitions in Section 4. He also proof-read the operator algebraic parts of this paper, all remaining mistakes were produced later in time.

Many thanks go to Vincente Cortez, Mike Hopkins, Justin Roberts, Markus Rosellen and Hans Wenzl for discussions about various aspects of this paper. Part of this project was developed during our stay at the Max-Planck Institute in Bonn, and we are very grateful for the wonderful research environment it provided. In March 2002, we held a preliminary workshop at the Sonderforschungsbereich in Münster, and we thank all the participants for their support, and in particular Wolfgang Lück for initiating that workshop.

2 Field theories

Following Graeme Segal [Se2], we explain in this section the axiomatic approach to field theories, leading up to a definition of 'Clifford linear field theories of degree n' (cf. Definitions 2.26 and 2.29) after introducing the necessary background on Fock spaces, spin structures and Dirac operators.

2.1 d-dimensional field theories

Roughly speaking, a d-dimensional field theory associates to a closed manifold Y of dimension $d - 1$ a Hilbert space $E(Y)$ and to a bordism Σ from Y_1 to Y_2 a Hilbert Schmidt operator $E(Y_1) \to F(Y_2)$ (a bounded operator T is *Hilbert–Schmidt* if the sum of the norm squares of its matrix elements is finite). The main requirement is that gluing bordisms should correspond to composing the associated operators. As is well-known, this can be made precise by defining a *d-dimensional field theory* to be a functor

$$E : \mathcal{B}^d \longrightarrow \text{Hilb}$$

from the d-dimensional bordism category $\mathcal{B}^d$ to the category Hilb of complex Hilbert spaces which are compatible with additional structures on these categories spelled out below. The precise definition of the categories $\mathcal{B}^d$ and Hilb is the following:

- The objects of the *d-dimensional bordism category* $\mathcal{B}^d$ are closed oriented manifolds of dimension $d - 1$, equipped with geometric structures which characterize the flavor of the field theory involved (see remarks below). If Y_1, Y_2 are objects of $\mathcal{B}^d$, the orientation preserving geometric diffeomorphisms from Y_1 to Y_2 are morphisms from Y_1 to Y_2 which form a subcategory of $\mathcal{B}^d$. There are other morphisms, namely oriented geometric bordisms from Y_1 to Y_2; i.e., d-dimensional oriented manifolds Σ equipped with a geometric structure, together with an orientation preserving geometric diffeomorphism $\partial \Sigma \cong \bar{Y}_1 \sqcup Y_2$, where $\bar{Y}_1$ is Y_1 equipped with the opposite orientation. More precisely, two bordisms Σ and Σ' are considered the *same* morphism if they are orientation preserving geometric diffeomorphic relative boundary. Composition of bordisms is given by gluing; the composition of a bordism Σ from Y_1 to Y_2 and a diffeomorphism $Y_2 \to Y_3$ is again the bordism Σ, but with the identification $\partial \Sigma \cong \bar{Y}_1 \sqcup Y_2$ modified by composition with the diffeomorphism $Y_2 \to Y_3$.
- The objects of Hilb are separable Hilbert spaces (over the complex numbers). The morphisms from H_1 to H_2 are the bounded operators $T : H_1 \to H_2$; the strong topology on the space of bounded operators makes Hilb a topological category.

Without additional geometric structures on the objects and the bordisms, such a field theory would be referred to as a *topological* field theory. If the geometric structures are conformal structures on bordisms and objects, the associated field theory is called *conformal* (for short CFT). If the conformal structure is replaced by a Riemannian metric, one obtains what is usually referred to as a *Euclidean* field theory (EFT) to distinguish it from the Lorentz case. We sometimes use the term *field theory* (FT) if the geometric structures are not specified.

The main examples of field theories in these notes will have at least a conformal structure on the manifolds, and in addition all manifolds under consideration will be equipped with a spin structure (see Definition 2.18 for a careful explanation of spin structures on conformal manifolds). It is important to point out that every spin manifold has a canonical involution associated to it (which does not move the points of the manifold but flips the two sheets of the spin bundle). This has the effect that all algebraic objects associated to spin manifolds will be $\mathbb{Z}/2$-graded. This is the first step towards super symmetry and our reason for introducing spin structures in the main Definitions 2.26 and 2.29. We should point out that those definitions (where the categories of geometric manifolds are denoted by $\mathcal{CB}_n^2$ respectively $\mathcal{EB}_n^1$) introduce the spin structures (and the degree n) *for the first time*. The following warm-up discussions, in particular Definition 2.3, only use an orientation, not a spin structure (even though the notation $\mathcal{CB}^2$ respectively $\mathcal{EB}^1$ is very similar).

Summarizing, the reader should expect spin structures whenever there is a degree n in the discussion. Indeed, we will see that the degree makes sense only in the presence of spin structures.

Definition 2.1 (Additional structures on the categories $\mathcal{B}^d$ and Hilb).

- **Symmetric monoidal structures.** The disjoint union of manifolds (respectively the tensor product of Hilbert spaces) gives $\mathcal{B}^d$ (resp. Hilb) the structure of symmetric monoidal categories. The unit is given by the empty set and $\mathbb{C}$, respectively.

- **Involutions and anti-involutions.** There are involutions $\mathcal{B}^d \to \mathcal{B}^d$ and Hilb $\to$ Hilb. On the category $\mathcal{B}^d$ this involution is given by reversing the orientation on the d-manifold (objects) as well as the bordisms (morphisms); this operation will be explained in detail in Definition 2.15. We note that if Σ is a bordism from Y_1 to Y_2, then Σ with the opposite orientation can be interpreted as a bordism from $\bar{Y}_1$ to $\bar{Y}_2$. For an object $H \in$ Hilb, $\bar{H}$ is the space H with the opposite complex structure; for a morphism $f : H_1 \to H_2$, the morphism $\bar{f} : \bar{H}_1 \to \bar{H}_2$ is equal to f as a map of sets.

  There are also anti-involutions (i.e., contravariant functors) $* : \mathcal{B}^d \to \mathcal{B}^d$ and $* :$ Hilb $\to$ Hilb. These are the identity on objects. If $T : H_1 \to H_2$ is a bounded operator, then $T^* : H_2 \to H_1$ is its adjoint; similarly, if Σ is a bordism from Y_1 to Y_2, then Σ^* is Σ with the opposite orientation, considered as a morphism from Y_2 to Y_1. Finally, if ϕ is a diffeomorphism from Y_1 to Y_2 then $\phi^* \overset{\text{def}}{=} \phi^{-1}$.

- **Adjunction transformations.** There are natural transformations

$$\mathcal{B}^d(\emptyset, Y_1 \sqcup Y_2) \longrightarrow \mathcal{B}^d(\bar{Y}_1, Y_2) \qquad \text{Hilb}(\mathbb{C}, H_1 \otimes H_2) \longrightarrow \text{Hilb}(\bar{H}_1, H_2)$$

On $\mathcal{B}^d$ this is given by reinterpreting a bordism Σ from $\emptyset$ to $Y_1 \sqcup Y_2$ as a bordism from $\bar{Y}_1$ to Y_2. On Hilb we can identify $\text{Hilb}(\mathbb{C}, H_1 \otimes H_2)$ with the space of Hilbert–Schmidt operators from H_1 to H_2 and the transformation is the inclusion from Hilbert–Schmidt operators into all bounded operators. It should be stressed that neither transformation is in general surjective: in the category $\mathcal{B}^d$, a diffeomorphism from Y_1 to Y_2 is not in the image; in Hilb, not every bounded operator is a Hilbert–Schmidt operator. For example, a diffeomorphism ϕ gives a unitary operator $E(\phi)$ (if the functor E preserves the anti-involution $*$ above). In infinite dimensions, a unitary operator is never Hilbert–Schmidt.

Remark 2.2. In the literature on field theory, the functor E always respects the above involution, whereas E is called a *unitary* field theory if it also respects the anti-involution. It is interesting that in our honest example in Section 4.3 there are actually 3 (anti) involutions which the field theory has to respect.

Definition 2.3. The main examples of field theories we will be interested in are 2-dimensional conformal field theories and 1-dimensional Euclidean field theories. We will use the following terminology: A *conformal field theory* or *CFT* is a functor

$$E: \mathcal{CB}^2 \to \text{Hilb}$$

compatible with the additional structures on the categories detailed by Definition 2.1 where $\mathcal{CB}^2$ is the 'conformal' version of $\mathcal{B}^2$, i.e., the bordisms in this category are 2-dimensional and equipped with a conformal structure. A *Euclidean field theory* or *EFT* is a functor

$$E: \mathcal{EB}^1 \to \text{Hilb}$$

compatible with the additional structures of Definition 2.1 where $\mathcal{EB}^1$ is the 'Euclidean' version of $\mathcal{B}^1$, i.e., the bordisms in this category are 1-dimensional and equipped with a Riemannian metric.

Example 2.4. Let M be a closed Riemannian manifold, let $H = L^2(M)$ be the Hilbert space of square integrable functions on M and let $\Delta: H \to H$ be the Laplace operator. Then we can construct a 1-dimensional EFT $E: \mathcal{EB}^1 \to$

Hilb by defining

$$E(\text{pt}) = H \qquad E(I_t) = e^{-t\Delta} \qquad E(S_t) = \text{tr}(e^{-t\Delta}).$$

Here pt is the one-point object in $\mathcal{E}\mathcal{B}^1$, I_t is the interval of length t, considered as a morphism from pt to pt, and $S_t \in \mathcal{E}\mathcal{B}^1(\emptyset, \emptyset)$ is the circle of length t. We note that unlike the Laplace operator Δ the heat operator $e^{-t\Delta}$ is a bounded operator, even a trace class operator and hence it is meaningful to take the trace of $e^{-t\Delta}$.

It is not hard to show that the properties in Definition 2.1 allow us to extend E uniquely to a real EFT. More interestingly, the operator $E(\Sigma)$: $E(Y_1) \to E(Y_2)$ associated to a bordism Σ from Y_1 to Y_2 can be described in terms of a path integral over the space of maps from Σ to M. This is the Feynman–Kac formula, which for $\Sigma = I_t$ gives $e^{-t\Delta}$.

Definition 2.5. More generally, if X is a manifold, we may replace the category $\mathcal{B}^d$ above by the category $\mathcal{B}^d(X)$, whose objects are closed oriented $(d - 1)$-manifolds equipped with a piecewise smooth map to X; similarly the morphisms of $\mathcal{B}^d(X)$ are oriented bordisms equipped with maps to X and orientation preserving diffeomorphisms compatible with the given maps to X. We note that $\mathcal{B}^d(X)$ can be identified with $\mathcal{B}^d$ if X is a point. The four structures described above on the bordism category $\mathcal{B}^d$ can be extended in an obvious way to the category $\mathcal{B}^d(X)$. We define a *d-dimensional field theory over* X to be a functor $E: \mathcal{B}^d(X) \to$ Hilb which is compatible with the monoidal structure, (anti)-involutions, and adjunction transformations mentioned above. Analogously, we can form the categories $\mathcal{C}\mathcal{B}^2(X)$ (resp. $\mathcal{E}\mathcal{B}^1(X)$) of 2-dimensional conformal bordisms over X (resp. 1-dimensional Euclidean bordisms over X).

Example 2.6. This is a 'parametrized' version of Example 2.4 (which in the notation below is the case $X = $ pt and $Z = M$). Suppose that $\pi: Z \to X$ is a Riemannian submersion. Then we can construct a 1-dimensional EFT $E: \mathcal{E}\mathcal{B}^1(X) \to$ Hilb over X as follows. On objects, E associates to a map γ from a 0-manifold Y to X the Hilbert space of L^2-functions on the space of lifts $\{\tilde{\gamma}: Y \to Z \mid \pi \circ \tilde{\gamma} = \gamma\}$ of γ; in particular if $Y = $ pt and $\gamma(\text{pt}) = x$, then $E(Y, \gamma)$ is just the space of L^2-functions on the fiber over x. We can associate an operator $E(\Sigma, \Gamma)$: $E(Y_1, \gamma_1) \to E(Y_2, \gamma_2)$ to a bordism (Σ, Γ) from (Y_1, γ_1) to (Y_2, γ_2) by integrating over the space of maps $\tilde{\Gamma}: \Sigma \to Z$ which are lifts of $\Gamma: \Sigma \to X$. For $\Sigma = I_t$ and if Γ maps all of Σ to the point x, then the operator constructed this way is via the Feynman–Kac formula just $e^{-t\Delta_x}$, where Δ_x is the Laplace operator on the fiber over x.

2.2 *Clifford algebras and Fock modules*

Definition 2.7 (Clifford algebras). Let V be a real or complex Hilbert space equipped with an isometric involution $\alpha \colon V \to V$, $v \mapsto \bar{v} = \alpha(v)$ ($\mathbb{C}$-*anti-linear* in the complex case). This implies that

$$b(v, w) \overset{\text{def}}{=} \langle \bar{v}, w \rangle$$

is a symmetric bilinear form (here $\langle \ , \ \rangle$ is the inner product on V, which is $\mathbb{C}$-anti-linear in the first and linear in the second slot in the complex case).

The *Clifford algebra* is the quotient of the real resp. complex tensor algebra generated by V by imposing the Clifford relations

$$v \cdot v = -b(v, v) \cdot 1 \qquad v \in V.$$

Suppressing the dependence on the involution in the notation, we'll just write $C(V)$ for this algebra. It is a $\mathbb{Z}/2$-graded algebra with grading involution $\epsilon \colon C(V) \to C(V)$ induced by $v \mapsto -v$ for $v \in V$; the inner product on V extends to an inner product on the Clifford algebra $C(V)$.

We will write $-V$ for the Hilbert space furnished with the involution $-\alpha$. We will adopt the convention that if an involution α on V has not been explicitly specified, then it is assumed to be the identity. For example:

- $C_n \overset{\text{def}}{=} C(\mathbb{R}^n)$ is the Clifford algebra generated by vectors $v \in \mathbb{R}^n$ subject to the relation $v \cdot v = -|v|^2 \cdot 1$;
- $C_{-n} \overset{\text{def}}{=} C(-\mathbb{R}^n)$ is the Clifford algebra generated by vectors $v \in \mathbb{R}^n$ subject to the relation $v \cdot v = |v|^2 \cdot 1$; and
- $C_{n,m} \overset{\text{def}}{=} C(\mathbb{R}^n \oplus -\mathbb{R}^m)$ is the Clifford algebra generated by vectors $v \in \mathbb{R}^n$, $w \in \mathbb{R}^m$ subject to the relations $v \cdot v = -|v|^2 \cdot 1$, $w \cdot w = |w|^2 \cdot 1$, $v \cdot w + w \cdot v = 0$. We will use repeatedly that $C_{n,n}$ is of real type for all n

$$C_{n,n} \cong M_{2^n}(\mathbb{R}).$$

The reader should be warned that conventions in the literature concerning Clifford algebras vary greatly; our conventions with regards to C_n and $C_{n,m}$ agree for example with [Ka, Ch. III, 3.13] and [LM, Ch. I, §3].

Remark 2.8 (Properties of Clifford algebras). Useful properties of the construction $V \mapsto C(V)$ include natural isomorphisms

$$C(V \oplus W) \cong C(V) \otimes C(W) \qquad \text{and} \qquad C(-V) \cong C(V)^{\text{op}}. \qquad (2.1)$$

Here as throughout the paper $\otimes$ stands for the *graded tensor product*; here the adjective 'graded' stipulates that the product of elements $a \otimes b$, $a' \otimes b' \in A \otimes B$

is defined by

$$(a \otimes b) \cdot (a' \otimes b') \overset{\text{def}}{=} (-1)^{|b||a'|} aa' \otimes bb'$$

where $|b|$, $|a'|$ are the degrees of b and a' respectively. The *opposite* B^{op} of a graded algebra B is B as graded vector space but with new a multiplication $*$ defined by $a * b \overset{\text{def}}{=} (-1)^{|a||b|} b \cdot a$ for homogeneous elements $a, b \in B$ of degree $|a|, |b| \in \mathbb{Z}/2$, respectively. Any graded left $A \otimes B$-module M can be interpreted as a bimodule over $A - B^{\text{op}}$ via

$$a \cdot m \cdot b \overset{\text{def}}{=} (-1)^{|m||b|} (a \otimes b)m$$

for homogeneous elements $a \in A$, $b \in B$, $m \in M$) and vice-versa. In particular, a left module M over $C(V \oplus -W)$ may be interpreted as a left module over $C(V) \otimes C(W)^{\text{op}}$; or equivalently, as a $C(V) - C(W)$-bimodule, and we will frequently appeal to this move. We note that this construction is compatible with 'passing to the opposite module', where we define the *opposite* of a graded $A - B$-module M, denoted $\overline{M}$, to be M with the grading involution ϵ replaced by $-\epsilon$ and the right B-action modified by the grading automorphism of B. This is consistent (by the above formula) with changing the grading and keeping *the same* $A \otimes B^{\text{op}}$-module structure.

Definition 2.9 (Fermionic–Fock spaces). Let V be a Hilbert space with an isometric involution as in Definition 2.7. There is a standard construction of modules over the resulting Clifford algebra $C(V)$ (cf. [PS, Ch. 12], [A]); the input datum for this construction is a *Lagrangian* $L \subset V$. By *definition*, this means that L is closed, b vanishes identically on L and that $V = L \oplus \bar{L}$. Note that the existence of L is a serious condition on our data, for example a Lagrangian cannot exist if the involution on V is trivial.

Given a Lagrangian L, the exterior algebra

$$\Lambda(\bar{L}) = \Lambda^{ev}(\bar{L}) \oplus \Lambda^{\text{odd}}(\bar{L}) = \bigoplus_{p \text{ even}} \Lambda^p(\bar{L}) \oplus \bigoplus_{p \text{ odd}} \Lambda^p(\bar{L})$$

is a $\mathbb{Z}/2$-graded module over the Clifford algebra $C(V)$:

- for $\bar{v} \in \bar{L} \subset V \subset C(V)$, the corresponding operator $c(\bar{v}) \colon \Lambda(\bar{L}) \to \Lambda(\bar{L})$ is given by exterior multiplication by $\bar{v}$ ('creation operator'),
- for $v \in L$, the operator $c(v)$ is given by interior multiplication by v ('annihilation operator'); i.e., $c(v)$ acts as a graded derivation on $\Lambda(\bar{L})$, and for $\bar{w} \in \bar{L} = \Lambda^1(\bar{L})$ we have $c(v)\bar{w} = b(v, \bar{w}) = \langle w, v \rangle$.

We define the *fermionic Fock space* $F(L)$ to be the completion of $\Lambda(\bar{L})$ with respect to the inner product induced by the inner product on $\bar{L} \subset V$. We will refer to $F_{alg}(L) \stackrel{\text{def}}{=} \Lambda(\bar{L})$ as the *algebraic Fock space*; both of these $C(V)$-modules will play an important role for us.

We note that the adjoint $c(v)^*$ of the operator $c(v) \colon F(L) \to F(L)$ is given by $c(v)^* = -c(\bar{v})$ for any $v \in V$. It is customary to call $1 \in \Lambda^0\bar{L} \subset F(L)$ the *vacuum vector* and to write $\Omega \in F(L)$ for it. It is easy to see that Ω is a cyclic vector and hence $F(L)$ is a graded irreducible module over $C(V)$. The classification of these modules is given by the following well-known result (cf. [A]).

Theorem 2.10. *[I. Segal–Shale equivalence criterion] Two Fock representations $F(L)$ and $F(L')$ of $C(V, b)$ are isomorphic if and only if the composition of orthogonal inclusion and projection maps*

$$L' \hookrightarrow V \twoheadrightarrow \bar{L}$$

is a Hilbert–Schmidt operator. Moreover, this isomorphism preserves the grading if and only if $\dim(\bar{L} \cap L')$ is even.

We recall that an operator $T \colon V \to W$ between Hilbert spaces is a *Hilbert–Schmidt operator* if and only if the sum $\sum_{i=1}^{\infty} ||Te_i||^2$ converges, where $\{e_i\}$ is a Hilbert space basis for V. We note that the space $\bar{L} \cap L'$ is finite dimensional if the map $L' \to \bar{L}$ is a Hilbert–Schmidt operator.

Remark 2.11 (Orientations and bimodules). Let V be a real inner product space of dimension $n < \infty$. Then there is a homeomorphism

$$\{\text{isometries } f \colon \mathbb{R}^n \to V\} \longrightarrow \mathcal{L} \stackrel{\text{def}}{=} \{\text{Lagrangian subspaces } L \subset V \oplus -\mathbb{R}^n\}$$

given by sending an isometry f to its graph. By passing to connected components, we obtain a bijection between orientations on V and $\pi_0\mathcal{L}$. According to the Segal–Shale Theorem (plus the fact that in finite dimensions *any* irreducible module is isomorphic to some Fock space), sending a Lagrangian L to the Fock space $F(L)$ induces a bijection between $\pi_0\mathcal{L}$ and the set of isomorphism classes of irreducible graded (left) $C(V \oplus -\mathbb{R}^n)$-modules; as explained in (2.1), these may in turn be interpreted as $C(V) - C_n$-bimodules. Summarizing, we can identify orientations on V with isomorphism classes of irreducible $C(V) - C_n$-bimodules $S(V)$. We observe that the opposite bimodule $\overline{S(V)}$, defined above 2.9, corresponds to the opposite orientation.

Remark 2.12 (Functorial aspects of the Fock space construction). Let V_1, V_2 be Hilbert spaces with involutions as in Definition 2.7 and let $L_1 \subset V_2 \oplus -V_1$ be a Lagrangian. The associated algebraic Fock space $F_{alg}(L_1)$ (cf. Definition 2.9) is then a graded module over the Clifford algebra $C(V_2 \oplus -V_1)$;

alternatively we can view it as a bimodule over $C(V_2) - C(V_1)$. We wish to discuss in which sense the constructions $V \mapsto C(V)$ and $L \mapsto F_{\text{alg}}(L)$ give a *functor* (cf. [Se2, §8]). Here the objects of the 'domain category' are Hilbert spaces V with involutions, and morphisms from V_1 to V_2 are Lagrangian subspaces of $V_2 \oplus -V_1$. Given morphisms $L_1 \subset V_2 \oplus -V_1$ and $L_2 \subset V_3 \oplus -V_2$, their composition is given by the Lagrangian $L_3 \subset V_3 \oplus -V_1$ obtained by 'symplectic' reduction from the Lagrangian $L \overset{\text{def}}{=} L_2 \oplus L_1 \subset V \overset{\text{def}}{=} V_3 \oplus -V_2 \oplus V_2 \oplus -V_1$, namely

$$L^{\text{red}} \overset{\text{def}}{=} L \cap U^{\perp b} / L \cap U \subset V^{\text{red}} \overset{\text{def}}{=} V \cap U^{\perp b} / U.$$

Here U is the isotropic subspace $U = \{(0, v_2, v_2, 0) \mid v_2 \in V_2\} \subset V$ and $U^{\perp b}$ is its annihilator with respect to the bilinear form b. We note that the reduced space V^{red} can be identified with $V_3 \oplus -V_1$.

The objects of the 'range category' are graded algebras; the morphisms from A to B are pointed, graded $B - A$-bimodules; the composition of a pointed $B - A$-bimodule (M, m_0) and a pointed $C - B$-bimodule (N, n_0) is given by the $C - A$-bimodule $(N \otimes_B M, n_0 \otimes m_0)$. The following lemma shows that in the type I case composition of Lagrangians is compatible with the tensor product of pointed bimodules, i.e., the construction $V \mapsto C(V)$, $L \mapsto (F_{\text{alg}}(L), \Omega)$ is a (lax) functor. Here 'type' refers to the type of the von Neumann algebra generated by $C(V)$ in $B(F(L))$ as explained in Section 4.3. Type I is the easiest case where the von Neumann algebra is just the bounded operators on some Hilbert space. This corresponds geometrically to gluing along closed parts of the boundary. Gluing along, say, arcs in the boundary corresponds to type III for which a more difficult gluing lemma is needed: Connes fusion appears, see Proposition 4.12. It actually covers all types, so we restrict in the arguments below to the finite-dimensional case. That is all one needs for 1-dimensional EFT's, i.e. for K-theory.

Gluing Lemma 2.13. *If the von Neumann algebra generated by $C(V_2)$ has type I, there is a unique isomorphism of pointed, graded $C(V_3) - C(V_1)$ bimodules*

$$(F_{\text{alg}}(L_2) \otimes_{C(V_2)} F_{\text{alg}}(L_1), \Omega_2 \otimes \Omega_1) \cong (F_{\text{alg}}(L_3), \Omega_3).$$

Here we assume that L_i intersect V_j trivially (which is satisfied in the geometric applications if there are no closed components, cf. Definition 2.23).

Proof. We note that $F_{\text{alg}}(L_2) \otimes_{C(V_2)} F_{\text{alg}}(L_1)$ is the quotient of $F_{\text{alg}}(L_2) \otimes F_{\text{alg}}(L_1) = F_{\text{alg}}(L_2 \oplus L_1) = F_{\text{alg}}(L)$ modulo the subspace $\bar{U} F_{\text{alg}}(L)$. Here $\bar{U} \subset V$ is the subspace obtained from U defined above by applying the involution $v \mapsto \bar{v}$; explicitly, $\bar{U} = \{(0, -v_2, v_2, 0) \mid v_2 \in V_2\}$; we observe that for

$\bar{u} = (0, -v_2, v_2, 0) \in \bar{U}$, and $\psi_i \in F_{\text{alg}}(L_i)$ we have

$$c(\bar{u})(\psi_2 \otimes \psi_1) = (-1)^{|\psi_2|}(-\psi_2 c(v_2) \otimes \psi_1 + \psi_2 \otimes c(v_2)\psi_1).$$

We recall that an element $\bar{u} \in \bar{U} \subset V$, which decomposes as $\bar{u} = u_1 + \bar{u}_2 \in V = L \oplus \bar{L}$ with $u_i \in L$ acts on $F_{\text{alg}}(L) = \Lambda(\bar{L})$ as the sum $c(u_1) + c(\bar{u}_2)$ of the 'creation' operator $c(u_1)$ and the 'annihilation' operator $c(\bar{u}_2)$. We observe that by assumption the map $L^{\text{red}} \oplus \bar{U} \to L$ given by $(v, \bar{u}) \mapsto v + u_1$ is an isomorphism. In finite dimensions, a filtration argument shows that the $C(V^{\text{red}})$-linear map

$$\Lambda(\bar{L}^{\text{red}}) \longrightarrow \Lambda(\bar{L}^{\text{red}} \oplus U)/c(\bar{U})\Lambda(\bar{L}^{\text{red}} \oplus \bar{U})$$

is in fact an isomorphism. $\square$

Definition 2.14 (Generalized Lagrangian). For our applications to geometry, we will need a slightly more general definition of a Lagrangian. This will also avoid the assumption in the gluing lemma above. A *generalized Lagrangian* of a Hilbert space V with involution is a homomorphism $L : W \to V$ with finite-dimensional kernel so that the closure $L_W \subset V$ of the image of L is a Lagrangian. In the geometric situation we are interested in, W will be the space of harmonic spinors on a manifold Σ, V will be the space of all spinors on the boundary $\partial \Sigma$, and L is the restriction map. Then we define the *algebraic Fock space*

$$F_{alg}(L) \overset{\text{def}}{=} \Lambda^{\text{top}}(\ker L)^* \otimes \Lambda(\bar{L}_W)$$

where $\Lambda^{\text{top}}(\ker L)^* = \Lambda^{\dim(\ker L)}(\ker L)^*$ is the top exterior power of the dual space of the kernel of L. The algebraic Fock space is a module over the Clifford algebra $C(V)$ via its action on $\Lambda(\bar{L}_W)$.

Unlike the case discussed previously, this Fock space has only a canonical vacuum element $\Omega = 1 \otimes 1$ if $\ker L = 0$. Otherwise the vacuum vector is zero which is consistent with the geometric setting where it corresponds to the Pfaffian element of the Dirac operator: it vanishes, if there is a nontrivial kernel. Therefore, the gluing Lemma 2.24 in the following section has to be formulated more carefully than the gluing lemma above.

2.3 Clifford linear field theories

We recall that a d-dimensional field theory is a functor $E : \mathcal{B}^d \to \text{Hilb}$; in particular on objects, it assigns to a closed oriented $(d-1)$-manifold Y a Hilbert space $E(Y)$. It is the purpose of this section to define *Clifford linear field theories E of degree n* (for $d = 1, 2$). Such a theory assigns to Y as above a Hilbert space $E(Y)$ which is a right module over $C(Y)^{\otimes n}$, where $C(Y)$ is a Clifford algebra associated to Y. The formal definition (see Definition 2.26 for $d = 2$ and Definition 2.29 for $d = 1$) is quite involved. The reader might find it helpful to look first at Example 2.17, which will be our basic example of a Clifford linear field theory (for $d = 1$) and which motivates our definition. This example is a variation of Example 2.4 with the Laplace operator replaced by the square of the Dirac operator.

Definition 2.15 (Spin structures on Riemannian vector bundles). Let V be an inner product space of dimension d. Motivated by Remark 2.11 we define a *spin structure* on V to be an irreducible graded $C(V) - C_d$-bimodule $S(V)$ (equipped with a compatible inner product as in the case of Fock spaces). If W is another inner product space with spin structure, a *spin isometry* from V to W is an isometry $f: V \to W$ together with an isomorphism $\hat{f}: S(V) \overset{\cong}{\to} f^*S(W)$ of graded $C(V) - C_d$-bimodules with inner products. We note that $f^*S(W)$ is isomorphic to $S(V)$ if and only if f is orientation preserving; in that case there are *two* choices for $\hat{f}$. In other words, the space of spin isometries $\mathrm{Spin}(V, W)$ is a double covering of the space $SO(V, W)$ of orientation preserving isometries. It is clear that spin isometries can be composed and so they can be regarded as the morphisms in a category of inner product spaces with spin structures.

Now we can use a 'parametrized version' of the above to define spin structures on vector bundles as follows. Let $E \to X$ be a real vector bundle of dimension d with Riemannian metric, i.e., a fiberwise positive definite inner product. Let $C(E) \to X$ be the Clifford algebra bundle, whose fiber over x is the Clifford algebra $C(E_x)$. A *spin structure* on E is a bundle $S(E) \to X$ of graded irreducible $C(E) - C_d$-bimodules. It is tempting (at least for topologists) to think of two isomorphic bimodule bundles as giving the *same* spin structure. However, it is better to think of the 'category of spin structures' (with the obvious morphisms), since below we want to consider the space of sections of $S(E)$ and that is a functor from this category to the category of vector spaces. Then the usual object topologists are interested in are the *isomorphism classes* of spin structures. The group $H^1(X; \mathbb{Z}/2)$ acts freely and transitively on the set of isomorphism classes.

To relate this to the usual definition of spin structure expressed in terms of a principal $\mathrm{Spin}(d)$-bundle $\mathrm{Spin}(E) \to X$ (cf. [LM, Ch. II, §1]), we note that we obtain a $C(E) - C_d$-bimodule bundle if we define

$$S(E) \overset{\mathrm{def}}{=} \mathrm{Spin}(E) \times_{\mathrm{Spin}(d)} C_d.$$

Moreover, we note that $S(E)$ determines an orientation of E by Remark 2.11.

We define the *opposite spin structure* on E to be $\overline{S(E)}$ (whose fiber over $x \in X$ is the bimodule opposite to $S(E_x)$ in the sense of Remark 2.11); this induces the opposite orientation on E.

Remark 2.16. We note that there is a functor F from the category of spin structures on $E \oplus \mathbb{R}$ to the category of spin structures on E. Given a spin structure on $E \oplus \mathbb{R}$, i.e., a $C(E \oplus \mathbb{R}) - C_{d+1}$-bimodule bundle $S \to X$ over, we define $F(S) \overset{\text{def}}{=} S^+(E \oplus \mathbb{R})$, the even part of $S(E \oplus \mathbb{R})$. This is a graded $C(E) - C_d$-bimodule, if we define the grading involution on $S^+(E \oplus \mathbb{R})$ by $\psi \mapsto e_1 \psi e_1$ ($e_1 \in \mathbb{R}$ is the standard unit vector), the left action of $v \in E \subset C(E)$ by $\psi \mapsto v e_1 \psi$ and the right action of $w \in \mathbb{R}^d \subset C_d$ by $\psi \mapsto \psi e_1 w$. The functor F is compatible with 'passing to the opposite spin structure' in the sense that there is an isomorphism of spin structures $\overline{F(S)} \cong F(\overline{S})$, which is natural in S.

Example 2.17 (EFT associated to a Riemannian spin manifold). Let M be a closed manifold of dimension n with a spin structure; i.e., a spin structure on its cotangent bundle T^*M. In other words, M comes equipped with a graded irreducible $C(T^*M) - C_n$-bimodule bundle $S \to M$. A Riemannian metric on M induces the Levi-Civita connection on the tangent bundle TM which in turn induces a connection ∇ on S. The *Dirac operator* $D = D_M$ is the composition

$$D : C^\infty(M; S) \overset{\nabla}{\longrightarrow} C^\infty(M; T^*M \otimes S) \overset{c}{\longrightarrow} C^\infty(M; S)$$

where c is Clifford multiplication (given by the left action of $T^*M \subset C(T^*M)$ on S). The Dirac operator D is an (unbounded) Fredholm operator on the real Hilbert space $L^2(M; S)$ of square integrable sections of S. As in Example 2.4 we can construct a 1-dimensional EFT $E : \mathcal{EB}^1 \to \text{Hilb}$ by defining

$$E(\text{pt}) = L^2(M; S) \otimes_{\mathbb{R}} \mathbb{C} \qquad E(I_t) = e^{-tD^2}.$$

However, there is more structure in this example: the fibers of S and hence the Hilbert space $L^2(M; S)$ is a $\mathbb{Z}/2$-*graded right module* over C_n (or equivalently by Remark 2.8, a left module over $C_n^{\text{op}} = C_{-n}$). Moreover, D and hence $E(I_t)$ *commute* with this action. It should be emphasized that we are working in the *graded* world; in particular, saying that the odd operator D commutes with the left C_{-n}-action means $D(c \cdot x) = (-1)^{|x|} c \cdot D(x)$ for a homogeneous element $c \in C_{-n}$ of degree $|c|$ and $x \in E(\text{pt})$.

Definition 2.18 (Spin structures on conformal manifolds). Let Σ be a manifold of dimension d and for $k \in \mathbb{R}$ let $L^k \to \Sigma$ be the oriented real line bundle (and hence trivializable) whose fiber over $x \in \Sigma$ consists of all maps $\rho \colon \Lambda^d(T_x \Sigma) \to \mathbb{R}$ such that $\rho(\lambda \omega) = |\lambda|^{k/d} \rho(\omega)$ for all $\lambda \in \mathbb{R}$. Sections of L^d are referred to as *densities*; they can be integrated over Σ resulting in a real number.

Now assume that Σ is equipped with a conformal structure (i.e., an equivalence class of Riemannian metrics where we identify a metric obtained by multiplication by a function with the original metric). We remark that for any $k \neq 0$ the choice of a metric in the conformal class corresponds to the choice of a positive section of L^k. Moreover, the conformal structure on Σ induces a *canonical* Riemannian metric on the *weightless cotangent bundle*

$$T_0^* \Sigma \overset{\text{def}}{=} L^{-1} \otimes T^* \Sigma.$$

A *spin structure* on a conformal d-manifold Σ is by definition a spin structure on the Riemannian vector bundle $T_0^* \Sigma$. The *opposite spin structure on* Σ is the opposite spin structure on the vector bundle $T_0^* \Sigma$. We will use the notation $\bar{\Sigma}$ for Σ equipped with the opposite spin structure.

If Σ' is another conformal spin d-manifold, a *conformal spin diffeomorphism* from Σ to Σ' is a conformal diffeomorphism $f \colon \Sigma \to \Sigma'$ together with an isometry between the $C(T_0^* \Sigma) - C_d$-bimodule bundles $S(\Sigma)$ and $f^* S(\Sigma')$. We observe that every conformal spin manifold Σ has a canonical spin involution $\epsilon = \epsilon_\Sigma$, namely the identity on Σ together with the bimodule isometry $S(\Sigma) \to S(\Sigma)$ given by multiplication by -1.

Example 2.18 (Examples of spin structures). The manifold $\Sigma = \mathbb{R}^d$ has the following 'standard' spin structure: identifying $T_0^* \Sigma$ with the trivial bundle $\underline{\mathbb{R}^d}$, the bundle $S \overset{\text{def}}{=} \mathbb{R}^d \times C_d \to \mathbb{R}^d$ becomes an irreducible graded $C(T_0^* \Sigma) - C_d$-bimodule bundle. Restricting S we then obtain spin structures on codimension zero submanifolds like the disc $D^d \subset \mathbb{R}^d$ or the interval $I_t = [0, t] \subset \mathbb{R}$.

The above spin structure on $\mathbb{R}^d$ makes sense even for $d = 0$; here $\mathbb{R}^0$ consists of one point and $S = \mathbb{R}$ is a graded bimodule over $C_d = \mathbb{R}$ (i.e., a graded real line). We will write pt for the point equipped with this spin structure, and $\overline{\text{pt}}$ for the point equipped with its opposite spin stucture (the bimodule for pt is an 'even' real line, while the bimodule for $\overline{\text{pt}}$ is an 'odd' real line).

If Σ has a boundary $\partial \Sigma$, we note that the restriction $T_0^* \Sigma_{|\partial \Sigma}$ is *canonically* isometric to $T_0^* \partial \Sigma \oplus \underline{\mathbb{R}}$. It follows by Remark 2.16 that a spin structure on Σ, i.e., a $C(T_0^* \Sigma) - C(\mathbb{R}^d)$-bimodule bundle $S \to \Sigma$ restricts to a spin structure $S^+ \to \partial \Sigma$ on the boundary $\partial \Sigma$. In particular, the standard spin structure on D^2 restricts to a spin structure on $S^1 = \partial D^2$, which we refer to as the *zero-bordant* or *anti-periodic* spin structure; we'll use the notation S^{ap}.

Definition 2.19 (The C_{d-1}-Hilbert space $V(Y)$). If Y^{d-1} is a conformal spin manifold with spinor bundle $S \to Y$, we define

$$V(Y) \overset{\text{def}}{=} L^2(Y, L^{\frac{d-1}{2}} \otimes S)$$

the space of square-integrable sections of the real vector bundle $E = L^{\frac{d-1}{2}} \otimes S$. We note that using the fiberwise inner product of the spinor bundle S, we can pair sections φ, ψ of E to obtain a section of L^{d-1} which in turn may be integrated over Y to obtain a real valued inner product $\langle \varphi, \psi \rangle$ on the space of smooth sections of E; completion then gives the real Hilbert space $V(Y)$. We note that each fiber of E is a graded right C_{d-1}-module, which induces the same structure on $V(Y)$.

Definition 2.20 (The Clifford algebra $C(Y)$, $d = 1, 2$). Let Y^{d-1} be a conformal spin manifold and let $V(Y)$ be as above. In particular, for $d = 1$, $V(Y)$ is just a graded real Hilbert space; for $d = 2$, the Clifford algebra C_{d-1} is isomorphic to $\mathbb{C}$ and hence $V(Y)$ is a complex vector space on which the grading involution acts by a $\mathbb{C}$-anti-linear involution. After extending the $\mathbb{R}$-valued inner product to a $\mathbb{C}$-valued hermitian product, we can regard $V(Y)$ as a graded complex Hilbert space. So for $d = 1, 2$, $V(Y)$ has the structures needed to form the Clifford algebra $C(Y) \overset{\text{def}}{=} C(V(Y))$ as described in Definition 2.7. Here the involution α is given by the grading involution (which for $d = 2$ anticommutes with the action of $C_1 = \mathbb{C}$).

Example 2.21 (Examples of Clifford algebras $C(Y)$). If pt, $\overline{\text{pt}}$ are the point equipped with its standard resp. its opposite spin structure as defined in Definition 2.18, then $C(\text{pt}) = C_1$ and $C(\overline{\text{pt}}) = C_{-1}$.

If $Y = \emptyset$, then $V(Y)$ is zero-dimensional and consequently, $C(\emptyset) = \mathbb{R}$ (for $d = 1$) resp. $C(\emptyset) = \mathbb{C}$ (for $d = 2$).

Definition 2.22 (The generalized Lagrangian $L(\Sigma) : W(\Sigma) \to V(\partial\Sigma)$). Let Σ^d be a conformal spin manifold. Picking a Riemannian metric in the given conformal class determines the Levi–Civita connection on the tangent bundle of Σ, which in turn determines connections on the spinor bundle $S = S(T_0^*\Sigma)$, the line bundles L^k and hence $L^k \otimes S$ for all $k \in \mathbb{R}$. The corresponding *Dirac operator* $D = D_\Sigma$ is the composition

$$D \colon C^\infty(\Sigma; L^k \otimes S) \overset{\nabla}{\longrightarrow} C^\infty(\Sigma; T^*\Sigma \otimes L^k \otimes S)$$

$$= C^\infty(\Sigma; L^{k+1} \otimes T_0^*\Sigma \otimes S) \overset{c}{\longrightarrow} C^\infty(\Sigma; L^{k+1} \otimes S) \quad (2.2)$$

where c is Clifford multiplication (given by the left action of $T_0^* \Sigma \subset C(T_0^* \Sigma)$ on S). It turns out that for $k = \frac{d-1}{2}$ the Dirac operator is in fact *independent* of the choice of the Riemannian metric.

According to Green's formula, we have

$$\langle D\psi, \phi \rangle - \langle \psi, D\phi \rangle = \langle c(\nu)\psi_|, \phi_| \rangle \qquad \psi, \phi \in C^\infty(\Sigma, L^{\frac{d-1}{2}} \otimes S)$$

where $\psi_|, \phi_|$ is the restriction of ψ resp. ϕ to $\partial \Sigma$ and ν is the unit conormal vector field (the section of $T_0^* \Sigma_{|\partial \Sigma}$ corresponding to $1 \in \mathbb{R}$ under the natural isomorphism $T_0^* \Sigma_{|\partial \Sigma} \cong T_0^* \partial \Sigma \oplus \mathbb{R}$). Replacing ψ by ψe_1 in the formula above and using the fact that multiplication by e_1 is skew-adjoint, we obtain

$$\langle D\psi e_1, \phi \rangle + \langle \psi, D\phi e_1 \rangle = \langle c(\nu)\psi_| e_1, \phi_| \rangle. \tag{2.3}$$

Let $W(\Sigma) \stackrel{\text{def}}{=} \ker D^+$ where D^+ has domain $C^\infty(\Sigma, L^{\frac{d-1}{2}} \otimes S^+)$ and consider the restriction map to the boundary

$$L(\Sigma) : W(\Sigma) \longrightarrow L^2(\partial \Sigma, L^{\frac{d-1}{2}} \otimes S) = V(\partial \Sigma).$$

The closure L_Σ of the image of $L(\Sigma)$ is the *Hardy space* of boundary values of harmonic sections of $L^{\frac{d-1}{2}} \otimes S^+$. The kernel of $L(\Sigma)$ is the space of harmonic spinors on Σ which vanish on the boundary. If $\Sigma_0 \subseteq \Sigma$ denotes the subspace of closed components of Σ then $\ker L(\Sigma) = \ker D^+_{\Sigma_0}$ is the (finite-dimensional) subspace of harmonic spinors on Σ_0.

The Green formula shows that L_Σ is isotropic with respect to the bilinear form $b(v, w) = \langle \epsilon(v), w \rangle$, where the involution ϵ is given by $\epsilon(v) = c(\nu)\nu e_1$. Comparison with Remark 2.16 shows that ϵ is precisely the grading involution on S^+ defining the spin structure on $\partial \Sigma$ and it agrees with the grading involution on $V(\partial \Sigma)$. Analytically, much more involved arguments show that L_Σ is in fact a *Lagrangian* subspace [BW]. This implies that $L(\Sigma)$ is a generalized Lagrangian in the sense of Definition 2.14.

Moreover, the map $L(\Sigma) : W(\Sigma) \to V(\partial \Sigma)$ is linear with respect to $C_d^{ev} = C_{d-1}$, since the Dirac operator D commutes with the right C_d-action.

We give the following definition only for dimensions $d = 1, 2$ because these are the cases where C_{d-1} is commutative and hence one has a good definition of the 'exterior algebra' over C_{d-1}. For higher dimensions, one could ignore the C_{d-1}-action, but we will not discuss this case as it is not important for our applications.

Definition 2.23 (The $C(\partial \Sigma)$-modules $F_{\text{alg}}(\Sigma)$ and $F(\Sigma)$). Using the generalized Lagrangian from the previous definition, we define $F_{\text{alg}}(\Sigma) \stackrel{\text{def}}{=} F_{\text{alg}}(L(\Sigma))$, the algebraic Fock module over $C(\partial \Sigma)$ from Definition 2.14. This is a real vector space for $d = 1$ and a complex vector space for $d = 2$. Recall that

$$F_{\mathrm{alg}}(L(\Sigma)) = \Lambda^{\mathrm{top}}(\ker L(\Sigma))^* \otimes \Lambda(\bar{L}_\Sigma) \qquad (2.4)$$

and that $\bar{L}_\Sigma$ (and hence the exterior algebra) is equipped with a natural inner product. If $\Sigma_0 \subseteq \Sigma$ denotes again the subspace of closed components of Σ then $\ker L(\Sigma) = \ker D_{\Sigma_0}^+$. We note that $D_{\Sigma_0}^+$ is is skew-adjoint by equation (2.3) with respect to the natural hermitian pairing between the domain and range of this operator: for $\psi \in C^\infty(\Sigma; L^{(d-1)/2} \otimes S^+)$ and $\phi \in C^\infty(\Sigma; L^{(d+1)/2} \otimes S^-)$ the point-wise inner product of $\psi \cdot e_1$ and ϕ gives a section of L^2 which may be integrated over Σ to give a complex number; this allows us to identify $L^2(\Sigma, L^{(d+1)/2} \otimes S^-)$ with the dual of $L^2(\Sigma, L^{(d-1)/2} \otimes S^+)$. In particular, $\Lambda^{\mathrm{top}}(\ker L(\Sigma))^* = \Lambda^{\mathrm{top}}(\ker D_{\Sigma_0}^+)^*$ is the *Pfaffian line* $\mathrm{Pf}(\Sigma)$ of the skew-adjoint operator $D_{\Sigma_0}^+$, which comes equipped with the Quillen metric [BF] (this is a *real* line for $d = 1$ and a *complex* line for $d = 2$). Hence both factors on the right-hand side of equation (2.4) are equipped with natural inner products and we obtain a Hilbert space $F(\Sigma)$ as the completion of $F_{\mathrm{alg}}(\Sigma)$, which is still a module over $C(\partial\Sigma)$. We note that the Fock space $F(\Sigma)$ can be regarded as a generalization of the Pfaffian line, since for a *closed* Σ the Fock space $F(\Sigma)$ is *equal* to $\mathrm{Pf}(\Sigma)$. For $d = 1$ we have $F_{\mathrm{alg}}(\Sigma) = F(\Sigma)$ because both are finite-dimensional.

If Σ is a conformal spin bordism from Y_1 to Y_2, then $F(\Sigma)$ is a left module over $C(\partial\Sigma) = C(Y_1)^{\mathrm{op}} \otimes C(Y_2)$; in other words, a $C(Y_2) - C(Y_1)$-bimodule.

We need to understand how these Fock modules behave under gluing surfaces together (we shall not discuss the 1-dimensional analogue explicitly but the reader will easily fill this gap). So let Σ_i be conformal spin surfaces with decompositions

$$\partial\Sigma_1 = Y_1 \cup Y_2, \quad \partial\Sigma_2 = Y_2 \cup Y_3$$

where $Y_i \cap Y_{i+1}$ could be nonempty (but always consists of the points ∂Y_i). Let $\Sigma_3 \overset{\mathrm{def}}{=} \Sigma_1 \cup_{Y_2} \Sigma_2$, then this geometric setting leads to the algebraic setting in Remark 2.12. We have $V_i \overset{\mathrm{def}}{=} V(Y_i)$ and $L_i \overset{\mathrm{def}}{=} L_{\Sigma_i}$, so that we can derive a gluing isomorphism. Note that there are two cases, depending on the type of the von Neumann algebra generated by $C(V_2) = C(Y_2)$. If Y is closed then we are in type I, and if Y has boundary we are in type III where a more sophisticated gluing lemma is needed. Note also that we really have generalized Lagrangians $L(\Sigma_i)$ which are used in the gluing lemma below. It follows from our algebraic gluing lemma (for type I) together with the canonical isomorphisms of Pfaffian lines for disjoint unions of closed surfaces.

Gluing Lemma 2.24. *If Y_2 is a closed 1-manifold, there are natural isomorphisms of graded $C(Y_3) - C(Y_1)$ bimodules*

$$F_{\text{alg}}(\Sigma_2) \otimes_{C(Y_2)} F_{\text{alg}}(\Sigma_1) \cong F_{\text{alg}}(\Sigma_3).$$

Again there is a refined version of this lemma for all types of von Neumann algebras which uses Connes fusion, see Proposition 4.12. It will actually imply that the above isomorphism are isometries and hence carry over to the completions $F(\Sigma_i)$.

Remark 2.25. A different way to see the isometry for completions is to observe that our assumption on Y_2 being closed (i.e. that the von Neumann algebra $A(Y_2)$ is of type I) implies that $A(Y_i) \cong B(H_i)$ for some Hilbert spaces H_i and also that

$$F(\Sigma_1) \cong HS(H_2, H_1), \quad F(\Sigma_2) \cong HS(H_3, H_2), \quad F(\Sigma_3) \cong HS(H_3, H_1).$$

Then the isomorphism for Lemma 2.24 is just given by composing these Hilbert–Schmidt operators. Note that if Y_i bound conformal spin surfaces S_i then we may choose $H_i = F(S_i)$ in which case everything becomes canonical. It is important to note that in the case relevant for string vector bundles, this last assumption will be satisfied because we will be working in a relative situation where Y_i consists of two copies of the same manifold, one with a trivial bundle, and with a nontrivial bundle over it.

After these preliminaries, we are now ready to define Clifford linear field theories of degree n. To motivate the following definition, we recall Example 2.17 of a 1-dimensional EFT: here the Hilbert space $E(\text{pt})$ associatated to the point pt has additional structure: $E(\text{pt})$ is a $\mathbb{Z}/2$-graded left module over the Clifford algebra $C_{-n} = (C(\text{pt})^{\text{op}})^{\otimes n}$ (see Example 2.21). Roughly speaking, a *Clifford field theory of degree n* is a field theory (of dimension $d = 1$ or 2) with extra structure ensuring that the Hilbert space $E(Y)$ associated to a manifold Y of dimension $d - 1$ is a graded left module over the Clifford algebra $C(Y)^{-n} \overset{\text{def}}{=} (C(Y)^{\text{op}})^{\otimes n}$. To make this precise, we define Clifford linear field theories of degree n as functors from $\mathcal{CB}_n^2$ (resp. $\mathcal{EB}_n^1$) to the category of Hilbert spaces; here $\mathcal{CB}_n^2$ (resp. $\mathcal{EB}_n^1$) are 'larger' versions of the categories $\mathcal{CB}^2$ (resp. $\mathcal{EB}^1$) such that the endomorphisms of the object given by Y contains the Clifford algebra $C(Y)^{-n}$. This implies that for such a functor E the Hilbert space $E(Y)$ is left module over $C(Y)^{-n}$ (or equivalently, a right module over $C(Y)^{\otimes n}$).

Definition 2.26 (CFT of degree n). A *Clifford linear conformal field theory of degree $n \in \mathbb{Z}$* is a continuous functor

$$E \colon \mathcal{CB}_n^2 \longrightarrow \text{Hilb}$$

compatible with the additional structures in Definition 2.1 on both categories. We recall that these are the monodial structures, involutions and anti-involutions, and adjunction transformations on both categories. In addition we require that the functor E is compatible with the linear structure on morphisms in the sense that the equations (2.5) below hold. For brevity's sake, we will refer to such a theory also just as *CFT of degree n* (we note that we have defined the notion of 'degree' only for these Clifford linear theories).

The objects of $\mathcal{CB}_n^2$ are closed conformal spin 1-manifolds Y. If Y_1, Y_2 are objects of $\mathcal{CB}_n^2$, there are two types of morphisms from Y_1 to Y_2, namely:

- pairs (f, c) consisting of a spin diffeomorphism $f \colon Y_1 \to Y_2$ and an element $c \in C(Y_1)^{-n}$; here $C(Y_1)^k$ stands for the graded tensor product of $|k|$ copies of $C(Y_1)$ if $k \geq 0$ resp. $C(Y_1)^{\mathrm{op}}$ if $k < 0$. In particular, there are morphisms

$$f \overset{\text{def}}{=} (f, 1 \in C(Y_1)^{-n}) \in \mathcal{CB}_n^2(Y_1, Y_2) \quad \text{and} \quad c \overset{\text{def}}{=} (1_{Y_1}, c) \in \mathcal{CB}_n^2(Y_1, Y_1)$$

- pairs (Σ, Ψ), where Σ is a conformal spin bordism from Y_1 to Y_2, and $\Psi \in F_{\mathrm{alg}}(\Sigma)^{-n}$. Here $F = F_{\mathrm{alg}}(\Sigma)$ is the algebraic Fock space, and F^k stands for the graded tensor product of $|k|$ copies of F if $k \geq 0$ resp. of $\bar{F}$ if $k \leq 0$. A conformal spin bordism Σ from Y_1 to Y_2 is a conformal spin manifold together with a spin diffeomorphism $\partial \Sigma \cong \bar{Y}_1 \amalg Y_2$. More precisely, we identify the morphisms (Σ, Ψ) and (Σ', Ψ') if there is a conformal spin diffeomorphism $\Sigma \to \Sigma'$ compatible with the boundary identification with $\bar{Y}_1 \amalg Y_2$ such that Ψ is sent to Ψ' under the induced isomorphism on Fock spaces. We recall from Definition 2.23 that if Σ has no closed components, then $F_{\mathrm{alg}}(\Sigma)$ is a Fock space which by definition 2.9 has a *canonical* cyclic vector Ω. Then $\Omega^{-n} \in F_{\mathrm{alg}}(\Sigma)^{-n}$ and we will write

$$\Sigma \overset{\text{def}}{=} (\Sigma, \Omega^{-n}) \in \mathcal{CB}_n^2(Y_1, Y_2).$$

We note that every 1-manifold has a unique conformal structure; hence our definition of spin structure and the construction of the Clifford algebra $C(Y)$ applies to every oriented 1-manifold Y. Composition of morphisms is given as follows:

- If (f_1, c_1) is a morphism from Y_1 to Y_2, and (f_2, c_2) is a morphism from Y_2 to Y_3, then $(f_2, c_2) \circ (f_1, c_1) = (f_2 \circ f_1, f_1^* c_2 \cdot c_1)$. In particular, interpreting as above a spin diffeomorphism $f \colon Y_1 \to Y_2$ as a morphism from Y_1 to Y_2, and an element $c \in C(Y_1)^{-n}$ as an endomorphism of Y_1 we have $(f, c) = f \circ c$.

- If (Σ_1, Ψ_1) is a morphism from Y_1 to Y_2, and (Σ_2, Ψ_2) is a morphism from Y_2 to Y_3, their composition is given by $(\Sigma_2 \cup_{Y_2} \Sigma_1, \Psi_2 \cup_{Y_2} \Psi_1)$, where

$\Sigma_3 = \Sigma_2 \cup_{Y_2} \Sigma_1$ is obtained by 'gluing' along the common boundary component Y_2, and the fermion $\Psi_3 = \Psi_2 \cup_{Y_2} \Psi_1$ on Σ_3 is obtained by 'gluing' the fermions Ψ_2 and Ψ_1, i.e., it is the image of $\Psi_2 \otimes \Psi_1$ under the $((-n)$th power of the) 'fermionic gluing homomorphism' from Lemma 2.24

$$F_{\mathrm{alg}}(\Sigma_2) \otimes_{C(Y_2)} F_{\mathrm{alg}}(\Sigma_1) \longrightarrow F_{\mathrm{alg}}(\Sigma_3).$$

In the present context where Y_2 is closed, the assumptions of Lemma 2.24 are indeed satisfied.

- Composing a morphism (Σ, Ψ) from Y_1 to Y_2 with a diffeomorphism $f \colon Y_2 \to Y_3$ is again (Σ, Ψ), but now regarding Σ as a bordism from Y_1 to Y_3, and $F_{\mathrm{alg}}(\Sigma)$ as a bimodule over $C(Y_3) - C(Y_1)$ by means of f. Precomposition of (Σ, Ψ) by a diffeomorphism is defined analogously.
- For $c_i \in C(Y_i)^{-n} \subset \mathcal{CB}_n^2(Y_i, Y_i)$ we have

$$c_2 \circ (\Sigma, \Psi) = (\Sigma, c_2 \cdot \Psi) \qquad \text{and} \qquad (\Sigma, \Psi) \circ c_1 = (\Sigma, \Psi \cdot c_1).$$

We note that $F_{\mathrm{alg}}(\Sigma)$ is a $C(Y_2) - C(Y_1)$-bimodule and hence $F_{\mathrm{alg}}(\Sigma)^{-n}$ is a $C(Y_2)^{-n} - C(Y_1)^{-n}$-bimodule, which explains the products $c_2 \cdot \Psi$ and $\Psi \cdot c_1$.

We require that a CFT $E \colon \mathcal{CB}_n^2 \to \mathrm{Hilb}$ is compatible with the linear structure on morphisms in the sense that given a spin diffeomorphism $f \colon Y_1 \to Y_2$ or a conformal spin bordism Σ from Y_1 to Y_2 the maps

$$C(Y_1)^{-n} \longrightarrow \mathrm{Hilb}(E(Y_1), E(Y_1)), \ F_{\mathrm{alg}}(\Sigma)^{-n} \longrightarrow \mathrm{Hilb}(E(Y_1), E(Y_2)) \ (2.5)$$

given by $c \mapsto E(f, c)$ (resp. $\Psi \mapsto E(\Sigma, \Psi)$) are linear maps.

Remark 2.27 (Basic properties of Clifford conformal field theories). Let Σ be a bordism from Y_1 to Y_2 with no closed components. Then Ω^{-n} is a cyclic vector in the $C(Y_2)^{-n} - C(Y_1)^{-n}$-bimodule $F_{\mathrm{alg}}(\Sigma)^{-n}$ and hence every morphism (Σ, Ψ) can be written as $(\Sigma, c_2 \Omega^{-n} c_1) = c_2 \circ (\Sigma, \Omega^{-n}) \circ c_1$. This shows that the morphisms in the category $\mathcal{CB}_n^2$ are generated by diffeomorphisms f, Clifford elements c, and conformal bordisms $\Sigma = (\Sigma, \Omega^{-n})$ (with no closed components).

We note that the spin involution $\epsilon = \epsilon_Y$ (see Definition 2.18) on a conformal spin 1-manifold Y induces the grading involution on the associatated Clifford algebra $C(Y)$. This implies that as morphisms in the category $\mathcal{CB}_n^2$, it commutes with the even elements of the Clifford algebra $C(Y)^{-n}$, while it anticommutes with the odd elements. In particular, if $E \colon \mathcal{CB}_n^2 \to \mathrm{Hilb}$ is a CFT of degree n, then the Hilbert space $E(Y)$ is a *graded* left $C(Y)^{-n}$-module (or equivalently, a right $C(Y)^n$-module). If Σ is a conformal spin bordism from Y_1

to Y_2, then the 'spin involution' ϵ_Σ restricts to ϵ_{Y_i} on the boundary and hence we have the relation

$$\epsilon_{Y_2} \circ \Sigma = \Sigma \circ \epsilon_{Y_1}$$

in $\mathcal{CB}_n^2$. In particular, the corresponding bounded operator $E(\Sigma): E(Y_1) \to E(Y_2)$ is *even*.

We claim that $E(\Sigma)$ is in fact a *Hilbert–Schmidt operator* from $E(Y_1)$ to $E(Y_2)$. To see this, observe that $\Sigma \in \mathcal{CB}_n^2(Y_1, Y_2)$ is in the image of the natural transformation

$$\mathcal{CB}_n^2(\emptyset, \bar{Y}_1 \amalg Y_2) \longrightarrow \mathcal{CB}_n^2(Y_1, Y_2)$$

by regarding Σ as a bordism from $\emptyset$ to $\bar{Y}_1 \amalg Y_2$. This implies that $E(\Sigma)$ is in the image of the corresponding natural transformation in Hilb

$$\mathrm{Hilb}(\mathbb{C}, \overline{E(Y_1)} \otimes E(Y_2)) \longrightarrow \mathrm{Hilb}(E(Y_1), E(Y_2))$$

which consists exactly of the Hilbert–Schmidt operators from $E(Y_1)$ to $E(Y_2)$.

Remark 2.28. We note that if Σ is a bordism from Y_1 to Y_2, and E is a Clifford linear theory of degree n then the map $F(\Sigma)^{-n} \longrightarrow \mathrm{Hom}(E(Y_1), E(Y_2))$, $\Psi \mapsto E(\Sigma, \Psi)$ in fact induces a $C(Y_2)^{-n}$-linear map

$$E(\Sigma): F(\Sigma)^{-n} \otimes_{C(Y_1)^{-n}} E(Y_1) \longrightarrow E(Y_2)$$

Definition 2.29 (EFT of degree n). A *Clifford linear 1-dimensional Euclidean field theory of degree n* is a continuous functor

$$\mathcal{EB}_n^1 \longrightarrow \mathrm{Hilb}$$

compatible with the additional structures in Definition 2.1 and the linear structure on the morphisms (equation (2.5)). Here the 1-dimensional degree n bordism category $\mathcal{EB}_n^1$ is defined as for $\mathcal{CB}_n^2$, except that the dimension of all manifolds involved is down by one: the objects of $\mathcal{EB}_n^1$ are 0-dimensional spin manifolds Y and the bordisms Σ are 1-dimensional; furthermore the geometric structure on these bordisms are *Riemannian metrics* rather than conformal structures. We want to emphasize that now the Clifford algebras $C(Y)$ and the Fock spaces $F(\Sigma)$ are *finite-dimensional real vector spaces*.

We note that we can define *real* EFT's; these are functors from $\mathcal{EB}_n^1$ to the category $\mathrm{Hilb}^{\mathbb{R}}$ of *real Hilbert spaces* with the same properties. In fact our motivating example 2.17 is the complexification of a real EFT.

It should be pointed out that there are no naïve 'real' versions of CFT's, since e.g. the map $C(Y)^{-n} \to \mathrm{Hilb}(E(Y), E(Y))$ is required to be linear, which means *complex* linear if Y is 1-dimensional (in which case $C(Y)$ is an algebra over $\mathbb{C}$). Consequently, we can not restrict the vector spaces $E(Y)$ to be real.

Definition 2.30 (Clifford linear field theories over a manifold X). As in Definition 2.5 we define Clifford linear field theories over a manifold X as follows. Let $\mathcal{CB}_n^2(X)$ resp. $\mathcal{EB}_n^1(X)$ be categories whose objects are as in the categories $\mathcal{CB}_n^2$ resp. $\mathcal{EB}_n^1$ except that all objects Y (given by manifolds of dimension 1 resp. 0) come equipped with piecewise smooth maps to X. Similarly all bordisms Σ come with piecewise smooth maps to X. The additional structures on $\mathcal{CB}_n^2$ and $\mathcal{EB}_n^1$ extend in an obvious way to $\mathcal{CB}_n^2(X)$ and $\mathcal{EB}_n^1(X)$, respectively. We define a *Clifford linear CFT of degree n over X* to be a functor $E \colon \mathcal{CB}_n^2 \to$ Hilb compatible with the additional structures.

Similarly a *Clifford linear EFT of degree n over X* is a functor $E \colon \mathcal{EB}_n^1(X) \to$ Hilb compatible with the additional structures.

Example 2.31 (Basic example of a Clifford linear EFT over X). Let $\xi \to X$ be an n-dimensional spin vector bundle with metric and compatible connection over a manifold X. Let $S(\xi) \to X$ be the associated spinor bundle (a $C(\xi) - C_n$-bimodule bundle, see Definition 2.15). Then there is a Clifford linear EFT over X of degree n: this is a functor $E \colon \mathcal{EB}_n^1(X) \to$ Hilb which maps the object of $\mathcal{EB}_n^1(X)$ given by pt $\mapsto x \in X$ to the Hilbert space $S(\xi)_x$; on morphisms, $E(c)$ for $c \in C(\text{pt})^{-n} = C_n^{\text{op}}$ is given by the right C_n-module structure on $S(\xi)$. If $\gamma \colon I_t \to X$ is a path from x to y representing a morphism in $\mathcal{EB}_n^1$, then $E(\gamma) \colon S(\xi)_x \to S(\xi)_y$ is given by parallel translation along γ. The properties of a Clifford linear field theory then determine the functor E.

2.4 Twisted Clifford algebras and Fock modules

In this section we shall generalize all the definitions given in Section 2.3 to the twisted case, i.e. where the manifolds are equipped with vector bundles and connections. This is a straightforward step, so we shall be fairly brief. At the end of the definition of the twisted Clifford algebra (respectively twisted Fock module), we will explain the relative version of the constructions, which involves the twisted and untwisted objects. It is these relative objects which will be used in Section 5.

Definition 2.32 (The C_{d-1}-Hilbert space $V(\xi)$). Let Y^{d-1} be a conformal spin manifold with spinor bundle S, and let $\xi \to Y$ be a vector bundle, equipped with a Riemannian metric. Define

$$V(\xi) \stackrel{\text{def}}{=} L^2(Y, L^{\frac{d-1}{2}} \otimes S \otimes \xi)$$

the space of square-integrable sections of the real vector bundle $E = L^{\frac{d-1}{2}} \otimes S \otimes \xi$. Each fiber of E is a graded right C_{d-1}-module, which induces the same structure on $V(\xi)$.

Definition 2.33 (The Clifford algebras $C(\xi)$ and $C(\gamma)$). The above definition gives for $d = 1$ a graded real Hilbert space $V(\xi)$; for $d = 2$, the Clifford algebra C_{d-1} is isomorphic to $\mathbb{C}$ and hence $V(\xi)$ is a complex vector space on which the grading involution acts by a $\mathbb{C}$-anti-linear involution. As in Definition 2.20, $V(\xi)$ has thus the structures needed to form the Clifford algebra $C(\xi) \overset{\text{def}}{=} C(V(\xi))$ for $d = 1, 2$.

In case that $\xi = \gamma^* E$ is the pullback of an n-dimensional vector bundle $E \to X$ via a smooth map $\gamma : Y \to X$, we define the following *relative* Clifford algebra

$$C(\gamma) \overset{\text{def}}{=} C(\gamma^* E) \otimes C(Y)^{-n}.$$

For example, if $Y = \text{pt}$ and $\gamma(\text{pt}) = x \in X$ then this gives the algebra $C(x) = C(E_x) \otimes C_{-n}$. Recall that a spin structure on E_x can then be described as a graded irreducible (left) $C(x)$-module.

Definition 2.34 (The generalized Lagrangian $L(\xi) : W(\xi) \to V(\partial \xi)$). Let Σ^d be a conformal spin manifold with boundary Y. Assume that the bundle ξ extends to a vector bundle with metric and connection on Σ. We denote it again by ξ and let $\partial \xi$ be its restriction to Y. Let S be the spinor bundle of Σ and recall from Definition 2.18 that the restriction of S^+ to Y is the spinor bundle of Y. Consider the *twisted (conformal) Dirac operator*

$$D_\xi : C^\infty(\Sigma; L^{\frac{d-1}{2}} \otimes S \otimes \xi) \overset{\nabla}{\longrightarrow} C^\infty(\Sigma; T^* \Sigma \otimes L^{\frac{d-1}{2}} \otimes S \otimes \xi)$$

$$= C^\infty(\Sigma; L^{\frac{d+1}{2}} \otimes T_0^* \Sigma \otimes S \otimes \xi) \overset{c}{\longrightarrow} C^\infty(\Sigma; L^{\frac{d+1}{2}} \otimes S \otimes \xi) \quad (2.6)$$

where ∇ is the connection on $L^{\frac{d-1}{2}} \otimes S \otimes \xi$ determined by the connection on ξ and the Levi–Civita connection on $L^{\frac{d-1}{2}} \otimes S$ for the choice of a metric in the given conformal class. Let $W(\xi) \overset{\text{def}}{=} \ker D_\xi^+$ where D_ξ^+ has domain $C^\infty(\Sigma, L^{\frac{d-1}{2}} \otimes S^+ \otimes \xi)$ and consider the restriction map to the boundary

$$L(\xi) : W(\xi) \longrightarrow L^2(\partial \Sigma, L^{\frac{d-1}{2}} \otimes S \otimes \xi) = V(\partial \xi).$$

The closure L_ξ of the image image of $L(\xi)$ is the *twisted Hardy space* of boundary values of harmonic sections of $L^{\frac{d-1}{2}} \otimes S^+ \otimes \xi$. $\ker L(\xi)$ is the space of twisted harmonic spinors which vanish on the boundary. If $\Sigma_0 \subseteq \Sigma$ denotes the subspace of closed components of Σ and ξ_0 is the restriction of ξ, then

$\ker L(\xi) = \ker D_{\xi_0}^+$ is the (finite-dimensional) subspace of twisted harmonic spinors on Σ_0. As before, one shows that $L(\xi)$ is C_{d-1}-linear and that L_ξ is a *Lagrangian* subspace of $V(\partial\xi)$.

Definition 2.35 (The $C(\partial\xi)$-modules $F_{\mathrm{alg}}(\xi)$ and $F(\xi)$). We define $F_{\mathrm{alg}}(\xi) \overset{\mathrm{def}}{=} F_{\mathrm{alg}}(L(\xi))$, the algebraic Fock module over $C(\partial\xi)$ determined by the generalized Lagrangian $L(\xi) : W(\xi) \to V(\partial\xi)$, see Definition 2.14. As before, this is a real Hilbert space for $d = 1$ and a complex Hilbert space for $d = 2$. As in Definition 2.23, $F_{\mathrm{alg}}(\xi)$ can be completed to the Hilbert space $F(\xi) = F(L(\xi))$.

In case that $\xi = \Gamma^*E$ is the pullback of an n-dimensional vector bundle $E \to X$ via a smooth map $\Gamma : \Sigma \to X$, we define the following *relative* Fock modules

$$F(\Gamma) \overset{\mathrm{def}}{=} F(\Gamma^*E) \otimes F(\Sigma)^{-n} \text{ and } F_{\mathrm{alg}}(\Gamma) \overset{\mathrm{def}}{=} F_{\mathrm{alg}}(\Gamma^*E) \otimes F_{\mathrm{alg}}(\Sigma)^{-n}.$$

These are left modules over the relative Clifford algebra $C(\gamma)$ from Definition 2.33, where $\gamma = \Gamma|_Y$. It is important to note that the vacuum vector for Γ is by definition $\Omega_\Gamma \in F_{\mathrm{alg}}(\Gamma^*E)$. If Σ is closed then $\mathrm{Pf}(\Gamma) \overset{\mathrm{def}}{=} F_{\mathrm{alg}}(\Gamma)$ is the relative Pfaffian line.

There are again gluing laws for twisted Fock spaces as in Lemma 2.24 and Proposition 4.12.

3 K-theory and 1-dimensional field theories

3.1 The space of 1-dimensional Euclidean field theories

We recall from Definition 2.29 that an EFT of degree n is a continuous functor E from the Euclidean bordism category $\mathcal{EB}_n^1$ to the category Hilb of Hilbert spaces compatible with the symmetric monoidal structure, the (anti-) involutions $*$ and $\bar{}$, the 'adjunction transformations' (see 2.1) and the linear structure on morphisms (see equation (2.5)). An important feature is that the Hilbert space $E(\mathrm{pt})$ associated to the point is a graded left C_{-n}-module, or equivalently, a graded right C_n-module. In our basic Example 2.17, $E(\mathrm{pt})$ is the space of square integrable sections of the spinor bundle $S \to M$ of a spin n-manifold, where the right C_n-action is induced by the right C_n-action on S.

It might be important to repeat the reason why the algebra C_{-n} comes up. The geometric example dictates that $E(\mathrm{pt})$ be a *right* C_n-module. (This goes back to the fact that a frame for a vector space V is an isometry $\mathbb{R}^n \to V$, and hence $O(n)$ acts on the right on these frames.) However, from a functorial point of view, the endomorphisms of an object in a category act on the left.

This is preserved under the covariant functor E. Since we built in C_{-n} as the endomorphisms of the object pt $\in \mathcal{EB}_n^1$, $E(\text{pt})$ becomes a *left* C_{-n}-module. Equivalently, this is a right C_n-module, exactly what we want.

In this subsection we consider the space of EFT's. We want to assume that the (right) C_n-module $E(\text{pt})$ is a submodule of some fixed graded complex Hilbert space H (equipped with a right C_n-action such that all irreducible modules occur infinitely often) in order to obtain a *set* of such functors.

Proposition 3.1. *There is a bijection*

$$\{EFTs\ of\ degree\ n\} \xrightarrow{R} \text{Hom}(\mathbb{R}_+, HS_{C_n}^{ev,sa}(H)).$$

Here $\mathbb{R}_+$ is the additive semi-group of positive real numbers, and $HS_{C_n}^{ev,sa}(H)$ is the semi-group of Clifford linear, even (i.e., grading preserving), self-adjoint Hilbert–Schmidt operators with respect to composition.

Definition 3.2 (Construction of R). Let $\mathbb{R}$ be equipped with the standard spin structure (see Example 2.18). We note that the translation action of $\mathbb{R}$ on itself is by spin isometries, allowing us to identify all the spin 0-manifolds $\{t\}$ with the object pt of the bordism category $\mathcal{EB}_n^1$. We recall $C(\text{pt}) = C_1$ (Example 2.21) and hence $C(\text{pt})^{-n} = C_{-n}$.

For $t > 0$ let $I_t \in \mathcal{EB}_n^1(\text{pt}, \text{pt})$ be the endomorphism given by the Riemannian spin 1-manifold $[0, t] \subset \mathbb{R}$. We note that the composition $I_t \circ I_{t'}$ is represented by gluing together the spin 1-manifolds $[0, t']$ and $[0, t]$, identifying $0 \in [0, t]$ with $t' \in [0, t']$ by means of the translation $t' \in \mathbb{R}_+$. This results in the spin 1-manifold $[0, t + t']$. We note that $I_t^* = I_t$, since reflection at the midpoint of the interval I_t is a spin structure reversing isometry.

As discussed in Remark 2.27, if $E: \mathcal{EB}_n^1 \to \text{Hilb}$ is a Clifford linear EFT of degree n, then $E(\text{pt})$ is a right C_n-module and $E(I_t): E(\text{pt}) \to E(\text{pt})$ is an even, Clifford linear Hilbert–Schmidt operator. Furthermore, due to $I_t^* = I_t$, and the required compatibility of E with the anti-involution $*$, the operator $E(I_t)$ is self-adjoint. The relation $I_t \circ I_{t'} = I_{t+t'}$ in the category $\mathcal{EB}_n^1$ implies that

$$\mathbb{R}_+ \longrightarrow HS_{C_n}^{ev,sa}(E(\text{pt})) \qquad t \mapsto E(I_t) \qquad (3.1)$$

is a semi-group homomorphism. Extending the Hilbert–Schmidt endomorphism $E(I_t)$ of $E(\text{pt}) \subset H$ to all of H by setting it zero on $E(\text{pt})^{\perp}$ defines the desired semi-group homomorphism $R(E): \mathbb{R}_+ \to HS_{C_n}^{ev,sa}(H)$.

Sketch of proof of Proposition 3.1. Concerning the injectivity of the map R, we observe that the functor $E: \mathcal{EB}_n^1 \to \text{Hilb}$ can be recovered from $E(\text{pt})$ (as graded right module over C_n) and $E(I_t)$ as follows. Every spin 0-manifold Z

is a disjoint union of copies of pt and $\overline{\text{pt}}$ and hence $E(Z)$ is determined by the Hilbert space $E(\text{pt})$, $E(\overline{\text{pt}}) = \overline{E(\text{pt})}$ and the requirement that E sends disjoint unions to tensor products. Concerning the functor E on morphisms, we note that $E(c)$ for $c \in C(\text{pt})^{-n} = C_{-n} \subset \mathcal{EB}_n^1(\text{pt}, \text{pt})$ is determined by the (left) C_{-n}-module structure on $E(\text{pt}) \subset H$. Similarly, the image of the endomorphism $\epsilon \in B_1^n(\text{pt}, \text{pt})$ is the grading involution on $E(\text{pt})$. Now the morphisms of the category $\mathcal{EB}_n^1$ are *generated* by I_t, $c \in C(\text{pt})^{-n}$ and ϵ using the operations of composition, disjoint union, the involution $\bar{\ }$ and the adjunction transformations $\mathcal{EB}_n^1(\emptyset, Z_1 \amalg Z_2) \to \mathcal{EB}_n^1(\overline{Z}_1, Z_2)$. For example, I_t can be interpreted as an element of $\mathcal{EB}_n^1(\text{pt}, \text{pt})$ or $\mathcal{EB}_n^1(\emptyset, \overline{\text{pt}} \amalg \text{pt})$ or $\mathcal{EB}_n^1(\text{pt} \amalg \overline{\text{pt}}, \emptyset)$. The second and third interpretation correspond to each other via the involution $\bar{\ }$; the first is the image of the second under the natural transformation $\mathcal{EB}_n^1(\emptyset, \overline{\text{pt}} \amalg \text{pt}) \to \mathcal{EB}_n^1(\text{pt}, \text{pt})$. It can be shown that the composition

$$\emptyset \xrightarrow{I_t} \overline{\text{pt}} \amalg \text{pt} \xrightarrow{I_{t'}} \emptyset \tag{3.2}$$

is the circle $S_{t+t'}^{\text{ap}}$ of length $t + t'$ with the anti-periodic spin structure, while

$$\emptyset \xrightarrow{I_t} \overline{\text{pt}} \amalg \text{pt} \xrightarrow{\epsilon \amalg 1} \overline{\text{pt}} \amalg \text{pt} \xrightarrow{I_{t'}} \emptyset \tag{3.3}$$

is $S_{t+t'}^{\text{per}}$, the circle of length $t + t'$ with the periodic spin structure. The best way to remember this result is to embed I as the upper semi-circle into the complex plane. For $x \in I$, the real line $S_x^+(I)$ can be identified with the complex numbers whose square lies in $T_x I \subset \mathbb{C}$. It follows that the spinor bundle $S^+(I)$ is a band twisted by $\pi/2$ (or a 'quarter twist'). This is consistent with the fact that $S(\partial I)$ consists of one even line, and one odd line (which are orthogonal). Gluing two such quarter twisted bands together gives a half twisted band (i.e. the anti-periodic spin structure on the corresponding circle). This also follows from the fact that this circle bounds a disk in the complex plane, and is thus spin zero-bordant. Gluing together two quarter twisted bands using the half twist ϵ gives a fully twisted band (i.e. the periodic spin structure on S^1).

The fact that the morphisms in $\mathcal{EB}_n^1$ are generated by I_t, ϵ and $c \in C(\text{pt})^{-n}$ implies that the functor E is determined by the semi-group homomorphism $E(I_t)$. Hence the map $E \mapsto E(I_t)$ is injective. Surjectivity of this map is proved by similar arguments by analyzing the relations between these generators. $\qquad\square$

Remark 3.3. As in our motivating Example 2.17 for a Clifford linear field theory, let M be a Riemannian spin manifold of dimension n and consider the semi-group of Hilbert–Schmidt operators $t \mapsto e^{-tD^2}$ acting on the Hilbert space $L^2(M; S)$. Then Proposition 3.1 (or rather its version for *real* EFT's) shows that there is a real Clifford linear EFT of degree n with $E(\text{pt}) = L^2(M; S)$ and $E(I_t) = e^{-tD^2}$.

This EFT contains interesting information, namely the *Clifford index* of D, an element of $KO_n(\text{pt})$, see [LM, §II.10]. We recall (see e.g. [LM, Ch. I, Theorem 9.29]) that $KO_n(\text{pt})$ can described as $KO_n(\text{pt}) = \mathfrak{M}(C_n)/i^*\mathfrak{M}(C_{n+1})$, where $\mathfrak{M}(C_n)$ is the Grothendieck group of graded right modules over the Clifford algebra C_n, and i^* is induced by the inclusion map $C_n \to C_{n+1}$. Hence the C_n-module ker D^2 represents an element of $KO_n(\text{pt})$. The crucial point is that $[\ker D^2] \in KO_n(\text{pt})$ is *independent of the choice of Riemannian metric used in the construction of D*. The argument is this: the eigenspace E_λ of D^2 with eigenvalue λ is a C_n-module; for $\lambda > 0$ the automorphism $\lambda^{-1/2}\epsilon D$ of E_λ has square -1 and anti-commutes with right multiplication by $v \in \mathbb{R}^n \subset C_n$. In other words, the graded C_n-module structure on E_λ extends to a C_{n+1}-module structure. This shows that $[\ker D^2] = [E_{<\rho}] \in KO_n(\text{pt})$, where $E_{<\rho}(D^2)$ for $\rho > 0$ is the (finite-dimensional) sum of all eigenspaces E_λ with eigenvalue $\lambda < \rho$. Choosing a ρ not in the spectrum of D^2, the C_n-module $E_{<\rho}(D^2)$ can be identified with $E_{<\rho}((D')^2)$ for any sufficiently close operator D', in particular for Dirac operators corresponding to slightly deformed metrics on M. This shows that $[E_{<\rho}(D^2)] \in KO_n(\text{pt})$ is independent of the choice of $\rho > 0$ as well as the metric on M. We note that in terms of the EFT, the Clifford index can be described as $[E_{>\rho}(E(I_t))] \in KO_n(\text{pt})$, where $E_{>\rho}(E(I_t))$ is the sum of all eigenspaces of $E(I_t)$ with eigenvalue $> \rho$ (a finite-dimensional graded C_n-module); the argument above shows that this is independent of t and $\rho > 0$.

This example suggests that the space of 1-dimensional EFT's of degree n contains interesting 'index information' and that we should analyze its homotopy type. Unfortunately, the result is that it is contractible! To see this, use Proposition 3.1 to identify this space with the space of semi-groups $t \mapsto P_t$ of even, self-adjoint, C_n-linear Hilbert–Schmidt operators. We note that if P_t is such a semi-group, then so is $t \mapsto s^t P_t$ for any $s \in [0, 1]$, which implies that the space of these semi-groups is contractible.

3.2 Super symmetric 1-dimensional field theories

After the 'bad news' expressed by the last remark, we will bring the 'good news' in this section: if we replace 1-dimensional EFTs by *super symmetric* EFTs, then we obtain a space with a very interesting homotopy type. Before stating this result and explaining what a super symmetric EFT *is*, let us motivate a little better why super symmetry is to be expected to come in here.

Remark 3.4. Let E be a real EFT of degree n. Then motivated by Remark 3.3, one is tempted to define its Clifford index in $KO_n(\text{pt})$ to be represented by the C_n-module $E_{>\rho}(E(I_t))$ (the sum of the eigenspace of $E(I_t)$ with eigenvalue $> \rho$). However, in general, this *does* depend on t and ρ; moreover, for fixed t, ρ replacing the semi-group $E(I_t)$ by the deformed operator $s^t E(I_t)$ leads to a *trivial* module for sufficiently small s! This simply comes from the fact that this operator has no Eigenvalues $> \rho$ for sufficiently small s.

What goes wrong is this: the arguments in Remark 3.14 show that there is a non-negative, self-adjoint, even operator A (not bounded!) on some subspace $H' \subset H$, which is an infinitesimal generator of the semi-group $E(I_t)$ in the sense that $E(I_t) = e^{-tA} \in HS_{C_n}^{\text{ev,sa}}(H') \subset HS_{C_n}^{\text{ev,sa}}(H)$ (this inclusion is given by extending by 0 on the orthogonal complement of H' in H). However, in general, A is *not* the square of an odd operator D, and so the argument in Remark 3.3 showing that $[E_{>\rho}(E(I_t))] \in KO_n(\text{pt})$ is independent of t, ρ fails.

The argument goes through for those semi-groups $\mathbb{R}_+ \to HS_{C_n}^{\text{ev,sa}}(H)$ whose generators are squares of odd operators; we will see that these are precisely those semi-group which extend to 'super homomorphims' $\mathbb{R}_+^{1|1} \to HS_{C_n}^{\text{sa}}(H)$.

Definition 3.5 (Susy EFT of degree n). A *super symmetric* 1-*dimensional Euclidean field theory* (or susy EFT) of degree n is a continuous functor

$$E: \mathcal{SEB}_n^1 \longrightarrow \text{Hilb}$$

satisfying the compatibility conditions 2.1. Here $\mathcal{SEB}_n^1$ is the 'super' version of the 1-dimensional bordism category $\mathcal{EB}_n^1$, where 1-dimensional Riemannian manifolds (which are morphisms in $\mathcal{EB}_n^1$) are replaced by *super manifolds* of dimension $(1|1)$ with an appropriate 'super' structure corresponding to the metric.

We refer to [DW] or [Fr2] for the definition of super manifolds. To a super manifold M of dimension $(n|m)$ we can in particular associate:

- its 'algebra of smooth functions' $C^\infty(M)$, which is a $\mathbb{Z}/2$-graded, graded commutative algebra;
- an ordinary manifold M^{red} of dimension n so that $C^\infty(M^{\text{red}})$ (the smooth functions on M^{red}) is the quotient of $C^\infty(M)$ by its nil radical.

One assumes that $C^\infty(M)$ is a locally free module over $C^\infty(M^{\text{red}})$. A basic example of a super manifold of dimension $(n|m)$ is $\mathbb{R}^{n|m}$ with

$$(\mathbb{R}^{n|m})^{\text{red}} = \mathbb{R}^n \quad \text{and} \quad C^\infty(\mathbb{R}^{n|m}) = C^\infty(\mathbb{R}^n) \otimes \Lambda^* \mathbb{R}^m.$$

More generally, if Σ is a manifold of dimension n and $E \rightarrow \Sigma$ is a real vector bundle of dimension m, then there is an associated super manifold M of dimension $(n|m)$ with

$$M^{\text{red}} = \Sigma \quad \text{and} \quad C^{\infty}(M) = C^{\infty}(\Sigma, \Lambda^* E^*)$$

where $C^{\infty}(\Sigma, \Lambda^* E^*)$ is the algebra of smooth sections of the exterior algebra bundle $\Lambda^* E^*$ generated by the dual vector bundle E^*.

In particular, if Σ is a spin bordism between 0-manifolds Y_1 and Y_2, then we can interpret Σ as a super manifold of dimension $(1|1)$ (using the even part $S^+ \rightarrow \Sigma$ of the spinor bundle) and Y_1, Y_2 as super manifolds of dimension $(0|1)$; in fact then Σ is a 'super bordisms' between Y_1 and Y_2 with Σ^{red} being the original bordism between Y_1^{red} and Y_2^{red}. The question is what is the relevant *geometric structure* on Σ, which reduces to the Riemannian metric on the underlying 1-manifold Σ^{red}? We note that a Riemannian metric on an oriented 1-manifold determines a unique 1-form which evaluates to 1 on each unit vector representing the orientation. Conversely, a nowhere vanishing 1-form determines a Riemannian metric. We generalize this point of view by defining a *metric structure* on a $(1|1)$-manifold Σ to be an even 1-form ω (see [DW, §2.6] for the theory of differential forms on super manifolds) such that

- ω and $d\omega$ are both nowhere vanishing (interpreted as sections of vector bundles over Σ^{red}) and
- the Berezin integral of ω over $(0|1)$-dimensional submanifolds is *positive*.

On $\Sigma^{\text{red}} \subset \Sigma$ such a form ω restricts to a nowhere vanishing 1-form which in turn determines a Riemannian metric on Σ^{red}.

Example 3.6. For example, on the $(1|1)$-dimensional super manifold $\mathbb{R}^{1|1}$ with even coordinate t and odd coordinate θ, the form $\omega = dz + \eta d\eta$ is a metric structure (we note that $d\omega = d\eta \wedge d\eta \neq 0$; the form $d\eta$ is an odd 1-form and hence *commutes* with itself according to equation (2.6.3) in [DW]). The Berezin integral of ω over $\{t\} \times \mathbb{R}^{0|1}$ gives the value 1 for every t (the form $dt - \eta d\eta$ gives the value -1 and hence is *not* a metric structure). The form ω restricts to the standard form dz on $(\mathbb{R}^{1|1})^{\text{red}} = \mathbb{R}$ by setting $\eta = 0$. In particular, the metric structure ω induces the *standard* Riemannian metric on $(\mathbb{R}^{1|1})^{\text{red}} = \mathbb{R}$.

With this terminology in place we can define $\mathcal{SEB}_n^1$. It is a category (enriched over super manifolds!) and its morphisms consist of 'super bordisms' as in $\mathcal{EB}_n^1$, except that 1-dimensional spin bordisms Σ equipped with Riemannian metrics are replaced by (1|1)-dimensional super bordisms equipped with a metric structure. In particular, the endomorphism spaces of each object are now super semi-groups; compare Definition 3.9.

1-EFTs and the K-theory spectrum. We can now give a precise formulation of Theorem 1.1 from the introduction. It says that the space $\mathcal{EFT}_n$ of susy EFTs of degree n has the homotopy type of K_{-n}, the $(-n)$th space in the Ω-spectrum K representing periodic complex K-theory. Here $K_0 = \Omega^\infty K$ is the 0th space in the spectrum K, and all the spaces K_n, $n \in \mathbb{Z}$ are related to each other by $\Omega K_n \simeq K_{n-1}$. Note that this implies that the connected components of the space of susy EFTs of degree n are the homotopy groups of the spectrum.

$$\pi_0(\mathcal{EFT}_n) = \pi_0(K_{-n}) = K^{-n}(\text{pt}) = K_n(\text{pt}). \tag{3.4}$$

Remark 3.7. There is an $\mathbb{R}$-version of the above result (with the same proof), namely that the space $\mathcal{EFT}_n^{\mathbb{R}}$ of *real* susy EFT's of degree n is homotopy equivalent to KO_{-n}, the $(-n)$th space in the real K-theory spectrum KO.

To convince the reader that one can really do explicit constructions in terms of these spaces of field theories, we will describe the Thom class of a spin vector bundle and the family Dirac index of bundle with spin fibers in terms of maps into these spaces (see Remarks 3.23 and 3.22).

The 'super' analog of Proposition 3.1 is then the following result.

Proposition 3.8. *There is a bijection*

$$\mathcal{EFT}_n \xrightarrow{R} \text{Hom}(\mathbb{R}_+^{1|1}, HS_{C_n}^{\text{sa}}(H)).$$

Here $HS_{C_n}^{\text{sa}}(H)$ is the super ($\mathbb{Z}/2$-graded) algebra of self-adjoint C_n-linear Hilbert–Schmidt operators on a Hilbert space H which is a graded right module over C_n (containing all irreducible C_n-modules infinitely often). The 'super semi-group' $\mathbb{R}_+^{1|1}$ and homomorphisms between super semi-groups are defined as follows.

Definition 3.9 (The super group $\mathbb{R}^{1|1}$). We give $\mathbb{R}^{1|1}$ the structure of a 'super group' by defining a multiplication

$$\mathbb{R}^{1|1} \times \mathbb{R}^{1|1} \longrightarrow \mathbb{R}^{1|1} \qquad (t_1, \theta_1), (t_2, \theta_2) \mapsto (t_1 + t_2 + \theta_1\theta_2, \theta_1 + \theta_2).$$

Here the points of $\mathbb{R}^{1|1}$ are parametrized by pairs (t, θ), where t is an even and θ is an oddvariable.

What do we mean by 'odd' and 'even' variables, and how do we make sense of the above formula? A convenient way to interpret these formulas is to extend scalars by some exterior algebra Λ and form $\mathbb{R}^{1|1}(\Lambda) \stackrel{\text{def}}{=} (\mathbb{R}^{1|1} \otimes \Lambda)^{\text{ev}}$, called the Λ-*points* of $\mathbb{R}^{1|1}$. Here $\mathbb{R}^{1|1} = \mathbb{R} \oplus \mathbb{R}$ is just considered as a graded vector space, with one copy of $\mathbb{R}$ in even, the other copy of $\mathbb{R}$ of odd degree, so that $(\mathbb{R}^{1|1} \otimes \Lambda)^{\text{ev}} = \Lambda^{\text{ev}} \oplus \Lambda^{\text{odd}}$. Now considering (t, θ) as an element of $\mathbb{R}^{1|1}(\Lambda)$ the formula in Definition 3.9 makes sense: for $t_1, t_2 \in \Lambda^{\text{ev}}$ and $\theta_1, \theta_2 \in \Lambda^{\text{odd}}$, we have $t_1 + t_2 + \theta_1\theta_2 \in \Lambda^{\text{ev}}$ and $\theta_1 + \theta_2 \in \Lambda^{\text{odd}}$ and it is easy to check that in this fashion we have given $\mathbb{R}^{1|1}(\Lambda)$ the structure of a group. But how about a 'super group structure' on $\mathbb{R}^{1|1}$ itself? Well, in one approach to super groups putting a super group structure on $\mathbb{R}^{1|1}$ is *by definition* the same as putting a group structure on $\mathbb{R}^{1|1}(\Lambda)$ for every Λ, depending functorially on Λ. In particular, the formula in Definition 3.9 gives $\mathbb{R}^{1|1}$ the structure of a super group.

Hopefully, the reader can now guess what a homomorphism $A \rightarrow B$ between super groups is: it is a family of ordinary group homomorphisms $A(\Lambda) \rightarrow B(\Lambda)$ depending functorially on Λ. A particularly interesting example of a super homomorphism is given in Example 3.11.

Remark 3.10. It is well-known that the group of orientation preserving isometries of $\mathbb{R}$ equipped with its standard orientation and metric can be identified with $\mathbb{R}$ acting on itself by translations. Similarly the group of automorphisms of the super manifold $\mathbb{R}^{1|1}$ preserving the metric structure $\omega = dz + \eta d\eta$ can be identified with the super group $\mathbb{R}^{1|1}$ acting on itself by translations. Let us check that for $(t, \theta) \in \mathbb{R}^{1|1}$ the translation

$$T_{t,\theta} : \mathbb{R}^{1|1} \longrightarrow \mathbb{R}^{1|1} \qquad (z, \eta) \mapsto (t + z + \theta\eta, \theta + \eta)$$

preserves the form ω

$$T_{t,\theta}^* dz = d(t + z + \theta\eta) = dz - \theta d\eta \qquad T_{t,\theta}^* d\eta = d(\theta + \eta) = d\eta$$

and hence

$$T_{t,\theta}^* \omega = (dz - \theta d\eta) + (\theta + \eta)d\eta = dz + \eta d\eta = \omega.$$

The Λ-points $\mathbb{R}_+^{1|1}(\Lambda)$ of the super space $\mathbb{R}_+^{1|1}$ consist of all $(t, \theta) \in \Lambda_+^{\text{ev}} \oplus \Lambda^{\text{odd}}$, where the Λ^0-component of t is positive. We note that the multiplication on $\mathbb{R}^{1|1}$ restricts to a multiplication on $\mathbb{R}_+^{1|1}$, but there are no inverses; i.e., $\mathbb{R}_+^{1|1}$ is a 'super semi-group'. It is the analog of $\mathbb{R}_+$ (where 'multiplication' is given by addition): we can interpret $\mathbb{R}^+$ as the moduli space of of intervals equipped with Riemannian metrics; similarly, $\mathbb{R}_+^{1|1}$ can be interpreted as the moduli space of 'super intervals with metric structures'. The multiplication on $\mathbb{R}_+$ (resp. $\mathbb{R}_+^{1|1}$) corresponds to the gluing of intervals (resp. super intervals).

Example 3.11 (A super homomorphism). Let H be the $\mathbb{Z}/2$-graded Hilbert space of L^2-sections of the spinor bundle on a compact spin manifold, let D be the Dirac operator acting on H, and let $HS^{\text{sa}}(H)$ be the space of self-adjoint Hilbert–Schmidt operators on H. Then we obtain a map of super spaces $\mathbb{R}^{1|1}_+ \to HS^{\text{sa}}_{C_{-n}}(H)$ by defining it on Λ-points in the following way

$$\mathbb{R}^{1|1}_+(\Lambda) = \Lambda^{\text{ev}}_+ \times \Lambda^{\text{odd}} \longrightarrow HS^{\text{sa}}(H)(\Lambda) = (HS^{\text{sa}}(H) \otimes \Lambda)^{\text{ev}}$$

$$(t, \theta) \mapsto e^{-tD^2} + \theta D e^{-tD^2}.$$

Here e^{-tD^2} is defined for real-valued $t > 0$ via functional calculus; for a general t, we decompose t in the form $t = t_B + t_S$ with $t_B \in \mathbb{R}_+ = \Lambda^0_+$ and $t_S \in \bigoplus_{p=1}^{\infty} \Lambda^{2p}$ (physics terminology: t_B is the 'body' of t, while t_S is the 'soul' of t). Then we use Taylor expansion to define $e^{-tD^2} = e^{-(t_B+t_S)D^2}$ as an element of $HS^{\text{sa}}(H) \otimes \Lambda$ (we note that the Taylor expansion gives a *finite* sum since t_S is nilpotent). We note that θ and D are both odd, so that $e^{-tD^2} + \theta D e^{-tD^2}$ is indeed in the *even* part of the algebra $HS^{\text{sa}}(H) \otimes \Lambda$. We will check in the proof of Lemma 3.15 that the map defined above is in fact a super homomorphism.

Definition 3.12 (Construction of R). We recall from Remark 3.10 that for $(t, \theta) \in \mathbb{R}^{1|1}$ the translation $T_{t,\theta} \colon \mathbb{R}^{1|1} \longrightarrow \mathbb{R}^{1|1}$ preserves the metric structure given by the even 1-form $\omega = dz + \eta d\eta$ from example 3.6. For $t > 0$, we will write $I_{t,\theta}$ for the $(1|1)$-dimensional super manifold $[0, t] \times \mathbb{R}^{0|1} \subset \mathbb{R}^{1|1}$ equipped with the metric structure given by ω (the Λ-points of $[0, t] \times \mathbb{R}^{0|1}$ consist of all $(z, \eta) \in \mathbb{R}^{1|1}$ such that z_B, the 'body' of z is in the interval $[0, t]$). We consider $I_{t,\theta}$ as a super bordism between pt $\overset{\text{def}}{=} 0 \times \mathbb{R}^{0|1}$ and itself by identifying $0 \times \mathbb{R}^{0|1}$ with $\{t\} \times \mathbb{R}^{0|1}$ by means of the translation $T_{(t,\theta)}$. The composition of I_{t_1,θ_1} and I_{t_2,θ_2} in the category $\mathcal{SEB}^1_n$ (given by gluing of these 'super bordisms') is then given by

$$I_{t_2,\theta_2} \circ I_{t_1,\theta_1} = I_{t_1+t_2+\theta_1\theta_2, \theta_1+\theta_2}$$

since we use the translation T_{t_1,θ_1} to identify $[0, t_2] \times \mathbb{R}^{0|1}$ with $[t_1, t_1 + t_2] \times \mathbb{R}^{0|1}$. This then 'fits' together with $[0, t_1] \times \mathbb{R}^{0|1}$ to form the bigger domain $[0, t_1 + t_2] \times \mathbb{R}^{0|1}$, the relevant identification between $\{0\} \times \mathbb{R}^{0|1}$ and the right-hand boundary $\{t_1 + t_2\} \times \mathbb{R}^{0|1}$ of this domain is given by $T_{t_2,\theta_2} \circ T_{t_1,\theta_1} = T_{t_1+t_2+\theta_1\theta_2, \theta_1+\theta_2}$.

This implies that

$$R(E) \colon \mathbb{R}^{1|1}_+ \longrightarrow HS(E(\text{pt})) \subset HS(H) \qquad (t, \theta) \mapsto E(I_{t,\theta})$$

is a super homomorphism. As in Definition 3.2 it follows that $E(I_{t,\theta})$ is a self-adjoint operator. However, in general it will not be *even*; unlike the situation in the category $\mathcal{E}\mathcal{B}_n^1$ in the super bordism category $\mathcal{S}\mathcal{E}\mathcal{B}_n^1$ the spin involution $\epsilon = \epsilon_{\text{pt}}$ (see Definition 2.18) does *not* commute with $I_{t,\theta}$; rather we have

$$\epsilon \circ I_{t,\theta} \circ \epsilon = I_{t,-\theta}. \tag{3.5}$$

To see this, we recall that the spin involution ϵ_Σ of a conformal spin manifold Σ is the identity on Σ and multiplication by -1 on the fibers of the spinor bundle $S(\Sigma) \to \Sigma$. If Σ is d-dimensional, this is an involution on the $(d|2^d)$-dimensional super manifold $S(\Sigma)$. In particular for $\Sigma = \{0\} \subset \mathbb{R}$, $S(\Sigma)$ can be identified with $\{0\} \times \mathbb{R}^{0|1} \subset \mathbb{R}^{1|1}$ on which ϵ acts by $(z, \eta) \mapsto (z, -\eta)$. It is easy to check that the translation $T_{t,\theta}$ and the involution ϵ (considered as automorphism of $\mathbb{R}^{1|1}$) satisfy the relation $\epsilon \circ T_{t,\theta} \circ \epsilon = T_{t,-\theta}$. This in turn implies the relation (3.5) between endomorphism of pt $= \{0\} \times \mathbb{R}^{0|1}$.

The equation (3.5) shows that restricting a super symmetric EFT $E \colon \mathcal{S}\mathcal{E}\mathcal{B}_n^1 \to \text{Hilb}$ to the semi-group of morphisms $I_{t,\theta} \in \mathcal{S}\mathcal{E}\mathcal{B}_n^1(\text{pt}, \text{pt})$ gives a $\mathbb{Z}/2$-equivariant semi-group homomorphism $R(E)$ as desired.

The proof of Proposition 3.8 is analogous to the proof of Proposition 3.1, so we skip it. The proof of Theorem 1.1 is based on a description of K-theory in terms of homomorphisms of C^*-algebras. We recall that a C^*-algebra A is a subalgebra of the algebra of bounded operators on some Hilbert space, which is closed under the operation $a \mapsto a^*$ of taking adjoints and which is a closed subset of all bounded operators with respect to the operator norm. Equivalently, A is an algebra (over $\mathbb{R}$ or $\mathbb{C}$) equipped with a norm and an anti-involution $*$ satisfying some natural axioms [BR], [Co1], [HR].

The examples of C^*-algebras relevant to us are:

• The C^*-algebra $C_0(\mathbb{R})$ of continuous real valued functions on $\mathbb{R}$ which vanish at ∞ with the supremum norm and trivial $*$-operation. This is a $\mathbb{Z}/2$-graded algebra with grading involution $\epsilon \colon C_0(\mathbb{R}) \to C_0(\mathbb{R})$ induced by $t \mapsto -t$ for $t \in \mathbb{R}$.

• The C^*-algebra $\mathcal{K}(H)$ of compact operators on a Hilbert space H; if H is graded, $\mathcal{K}(H)$ is a graded C^*-algebra. More generally, if H is a graded module over the Clifford algebra C_n, then the algebra $\mathcal{K}_{C_n}(H)$ of C_n-linear compact operators is a graded C^*-algebra.

If A, B are graded C^*-algebras, let $C^*(A, B)$ be the space of grading preserving $*$-homomorphisms $f \colon A \to B$ (i.e., $f(a^*) = f(a)^*$; such maps are automatically continuous) equipped with the topology of pointwise convergence, i.e., a sequence f_n converges to f if and only if for all $a \in A$ the sequence $f_n(a)$ converges to $f(a)$.

Theorem 3.13. *[Higson–Guentner [HG]] Let H be a real Hilbert space which is a graded right module over the Clifford algebra C_n (containing all irreducible C_n-modules infinitely often). Then the space $C^*(C_0(\mathbb{R}), \mathcal{K}_{C_n}(H))$ is homotopy equivalent to the $(-n)$th space in (the Ω-spectrum equivalent to) the real K-theory spectrum KO.*

Remark 3.14. This picture of K-theory is derived from Kasparov's KK-theory, see e.g. [HR]. It is also closely related to a geometric picture of KO-homology due to Graeme Segal [Se3]. We note that if $\varphi\colon C_0(\mathbb{R}) \to \mathcal{K}(H)$ is a $*$-homomorphism (not necessarily grading preserving), then $\varphi(f)$ for $f \in C_0(\mathbb{R})$ is a family of commuting self-adjoint compact operators. By the spectral theorem, there is a decomposition of H into mutually perpendicular simultaneous eigenspaces of this family. On a particular eigenspace the corresponding eigenvalue $\lambda(f)$ of $\varphi(f)$ determines a real number $t \in \mathbb{R} \cup \infty$ such that $\lambda(f) = f(t)$ (any algebra homomorphism $C_0(\mathbb{R}) \to \mathbb{C}$ is given by evaluation at some point $t \in \mathbb{R} \cup \infty$). The eigenspaces are necessarily finite-dimensional (except possibly for $t = \infty$), and the only accumulation point of points $t \in \mathbb{R}$ corresponding to a non-trivial eigenspace is ∞. Hence a C^*-homomorphism φ determines a configuration of points on the real line with labels which are mutually perpendicular subspaces of H (given by the corresponding eigenspaces); conversely, such a configuration determines a C^*-homomorphism φ. The conjugation involution on the space of all $*$-homomorphisms $C_0(\mathbb{R}) \to \mathcal{K}(H)$ (whose fixed point set is $C^*(C_0(\mathbb{R}), \mathcal{K}(H))$) corresponds to the involution on the configuration space induced by $t \mapsto -t$ and the grading involution on H. We observe that this implies that every grading preserving $*$-homomorphism $\varphi\colon C_0(\mathbb{R}) \to \mathcal{K}(H)$ is of the form $f \mapsto f(D) \in \mathcal{K}(H') \subset \mathcal{K}(H)$, where $H' \subset H$ is the subspace given by the direct sum of all subspaces E_t of H which occur as 'labels' of points $t \in \mathbb{R}$ of the configuration corresponding to φ. The operator $D\colon H' \to H'$ has E_t as its eigenspace with eigenvalue t; the equivariance condition implies that D is an *odd* operator.

A geometric model for $C^*(C_0(\mathbb{R}), \mathcal{K}_{C_n}(H))$ is obtained by requiring that the points $t \in \mathbb{R}$ are labeled by C_n-linear subspaces of H.

Theorem 1.1 follows from Theorem 3.13 and the following result.

Lemma 3.15. *The inclusion map $HS_{C_{-n}}^{\mathrm{sa}}(H) \to \mathcal{K}_{C_{-n}}^{\mathrm{sa}}(H)$ induces a homotopy equivalence of the corresponding spaces of homomorphisms of super groups from $\mathbb{R}_+^{1|1}$ to $HS_{C_{-n}}^{\mathrm{sa}}(H)$ resp. $\mathcal{K}_{C_{-n}}^{\mathrm{sa}}(H)$. Moreover, there is a homeomorphism*

$$C^*(C_0(\mathbb{R}), \mathcal{K}_{C_n}(H)) \xrightarrow{\approx} \mathrm{Hom}(\mathbb{R}_+^{1|1}, \mathcal{K}_{C_n}^{\mathrm{sa}}(H))$$

Sketch of proof. Let us outline the proof of the second part. The home-omorphism is given by sending a grading preserving $*$-homomorphism $\varphi\colon C_0(\mathbb{R}) \to \mathcal{K}_{C_n}(H)$ (which may be considered a super homomorphism!) to the composition $\mathbb{R}_+^{1|1} \xrightarrow{\chi} C_0(\mathbb{R}) \xrightarrow{\varphi} \mathcal{K}_{C_n}(H)$, where χ is the map of super spaces given on Λ-points by

$$\mathbb{R}_+^{1|1}(\Lambda) = \Lambda^{\mathrm{ev}} \times \Lambda^{\mathrm{odd}} \xrightarrow{\chi(\Lambda)} C_0(\mathbb{R})(\Lambda) = (C_0(\mathbb{R}, \Lambda))^{\mathrm{ev}}$$

$$(t, \theta) \mapsto e^{-tx^2} + \theta x e^{-tx^2}.$$

Here the expression $e^{-tx^2} + \theta x e^{-tx^2}$ is interpreted the same way as in Example 3.11: for $t \in \mathbb{R}_+ \subset \Lambda_+^{\mathrm{ev}}$, e^{-tx^2}, xe^{-tx^2} are obviously elements of $C_0(\mathbb{R})$; for general $t = t_B + t_S$ we use Taylor expansion around t_B.

Let us check that $\chi(\Lambda)$ is in fact a homomorphism

$$(e^{-t_1 x^2} + \theta_1 x e^{-t_1 x^2})(e^{-t_2 x^2} + \theta_2 x e^{-t_1 x^2})$$

$$= e^{-(t_1+t_2)x^2} - \theta_1 \theta_2 x^2 e^{-(t_1+t_2)x^2} + (\theta_1 + \theta_2) x e^{-(t_1+t_2)x^2}$$

$$= e^{-(t_1+t_2+\theta_1\theta_2)x^2} + (\theta_1 + \theta_2) x e^{-(t_1+t_2)x^2}$$

$$= e^{-(t_1+t_2+\theta_1\theta_2)x^2} + (\theta_1 + \theta_2) x e^{-(t_1+t_2+\theta_1\theta_2)x^2}.$$

Here the minus sign in the second line comes from permuting the odd element x past the odd element θ_2; the second equality follows by taking the Taylor expansion of $e^{-(t_1+t_2+\theta_1\theta_2)x^2}$ around the point $t_1 + t_2$; the third equality follows from the observation that the higher terms of that expansion are annihilated by multiplication by $\theta_1 + \theta_2$.

To finish the proof, one needs to show that the above map χ induces an isomorphism of graded C^*-algebras

$$C^*(\mathbb{R}_+^{1|1}) \cong C_0(\mathbb{R})$$

where the left-hand side is the C^*-algebra generated by the super semigroup $\mathbb{R}_+^{1|1}$.   □

Theorem 1.1 and its real analog identify in particular the components of the space $\mathcal{EFT}_n$ (resp. $\mathcal{EFT}_n^\mathbb{R}$) of complex (resp. real) super symmetric 1-field theories of degree n; there is a commutative diagram

$$
\begin{array}{ccc}
\pi_0 \mathcal{EFT}_n^\mathbb{R} & \xrightarrow[\cong]{\Theta} & KO_n(\mathrm{pt}) \overset{\mathrm{def}}{=} \mathfrak{M}(C_n)/i^*\mathfrak{M}(C_{n+1}) \\
\downarrow & & \downarrow \\
\pi_0 \mathcal{EFT}_n & \xrightarrow[\cong]{\Theta} & K_n(\mathrm{pt}) \overset{\mathrm{def}}{=} \mathfrak{M}^\mathbb{C}(C_n)/i^*\mathfrak{M}^\mathbb{C}(C_{n+1})
\end{array}
$$

where the horizontal isomorphisms are given by the theorem and the vertical maps come from complexification; $\mathfrak{M}(C_n)$ (resp. $\mathfrak{M}^{\mathbb{C}}(C_n)$) is the Grothendieck group of real (resp. complex) graded modules over the Clifford algebra C_n, and i^* is induced by the inclusion map $C_n \to C_{n+1}$ (this way of relating K-theory and Clifford algebras is well-known; see for example [LM, Ch. I, Theorem 9.29]). Explicitly, the map Θ is given by associating to a field theory E (real or complex) the graded C_n-module $E_{>\rho}(E(I_t))$ (the finite-dimensional sum of the eigenspaces of $E(I_t)$ with eigenvalue $> \rho$); the argument in Remark 3.3 shows that its class in K-theory is independent of t, ρ and depends only on the path component of E in $\mathcal{EFT}_n$.

By Bott-periodicity, $K_{2k+1}(\mathrm{pt}) = 0$ and $K_{2k} \cong \mathbb{Z}$. This isomorphism can be described explicitly as follows.

Lemma 3.16. *For n even, the map*

$$\mathfrak{M}^{\mathbb{C}}(C_n)/i^*\mathfrak{M}^{\mathbb{C}}(C_{n+1}) \longrightarrow \mathbb{Z} \qquad [M] \mapsto \mathrm{str}(\gamma^{\otimes n} : M \to M)$$

is an isomorphism. Here $\gamma \overset{\mathrm{def}}{=} 2^{-1/2}i^{1/2}e_1 \in C_1 \otimes \mathbb{C}$ *and* $\gamma^{\otimes n} = \gamma \otimes \cdots \otimes \gamma \in \mathbb{Cl}_n = \mathbb{Cl}_1 \otimes \cdots \otimes \mathbb{Cl}_1$, *where* $\mathbb{Cl}_n = C_n \otimes_{\mathbb{R}} \mathbb{C}$ *is the complexified Clifford algebra.*

Proof. The complex Clifford algebra $\mathbb{Cl}_2$ is isomorphic to the algebra of complex 2×2-matrices. Let $\Delta = \mathbb{C}^2$ be the irreducible module over $\mathbb{Cl}_2$; make Δ a graded module by declaring the grading involution ϵ to be multiplication by the 'complex volume element' $\omega_{\mathbb{C}} = ie_1e_2 \in \mathbb{Cl}_2$ [LM, p. 34]. It is well known that $\Delta^{\otimes k}$ (the graded tensor product of k copies of Δ) represents a generator of $\mathfrak{M}^{\mathbb{C}}(C_{2k})/i^*\mathfrak{M}^{\mathbb{C}}(C_{2k+1})$ [LM, Ch. I, Remark 9.28]. We have

$$\gamma^{\otimes 2} = \gamma \otimes \gamma = \frac{i}{2}e_1e_2 = \frac{1}{2}\omega_{\mathbb{C}} \in \mathbb{Cl}_1 \otimes \mathbb{Cl}_1 = \mathbb{Cl}_2.$$

We note that for any homomorphism $f : M \to M$ on a graded vector space with grading involution ϵ we have $\mathrm{str}(f) = \mathrm{tr}(\epsilon f)$. In particular, for $M = \Delta$ with grading involution $\epsilon = \omega_{\mathbb{C}}$ we obtain

$$\mathrm{str}(\omega_{\mathbb{C}} : \Delta \to \Delta) = \mathrm{tr}(\omega_{\mathbb{C}}^2) = \mathrm{tr}(1_\Delta) = \dim \Delta = 2.$$

This implies $\mathrm{str}(\gamma^{\otimes 2} : \Delta \to \Delta) = 1$ and hence $\mathrm{str}(\gamma^{\otimes 2k} : \Delta^{\otimes k} \to \Delta^{\otimes k}) = 1$. $\qquad\square$

The above lemma motivates the following

Definition 3.17. *If M is a finite-dimensional graded C_n module we define its Clifford super dimension as*

$$\mathrm{sdim}_{C_n}(M) \overset{\mathrm{def}}{=} \mathrm{str}(\gamma^{\otimes n} : M \to M).$$

More generally, if $f : M \to M$ is a C_n-linear map then we define its *Clifford super trace* as

$$\mathrm{str}_{C_n}(f) \overset{\text{def}}{=} \mathrm{str}(\gamma^{\otimes n} f : M \to M).$$

We note that the definition of the Clifford super trace continues to make sense for not necessarily finite-dimensional modules M, provided f is of trace class, i.e., f is the composition of two Hilbert–Schmidt operators (this guarantees that the infinite sums giving the super trace above converge).

The simplest invariants associated to a field theory E are obtained by considering a closed d-manifold Σ equipped with a fermion $\Psi \in F_{\mathrm{alg}}(\Sigma)$, and to regard (Σ, Ψ) as an endomorphism of the object $\emptyset$; then $E(\Sigma, \Psi) \in \mathrm{Hilb}(E(\emptyset), E(\emptyset)) = \mathrm{Hilb}(\mathbb{C}, \mathbb{C}) = \mathbb{C}$. For $d = 1$ we obtain the following result:

Lemma 3.18. *Let I_t the interval of length t, let S_t^{per} (resp. S_t^{ap} resp. S_t) be the circle of length t with the periodic (resp. anti-periodic resp. unspecified) spin structure. Let $\mu : F_{\mathrm{alg}}(I_t) \to F_{\mathrm{alg}}(S_t)$ be the fermionic gluing map and let $\Psi \in F_{\mathrm{alg}}(I_t)^{-n}$. If E is a EFT, then*

$$E(S_t^{\mathrm{per}}, \mu(\Psi)) = \mathrm{str}(E(I_t, \Psi)) \qquad E(S_t^{\mathrm{ap}}, \mu(\Psi)) = \mathrm{tr}(E(I_t, \Psi)).$$

This lemma follows from decomposing $S_{t+t'}^{\mathrm{per}}$ resp. $S_{t+t'}^{\mathrm{ap}}$ as in equation (3.2) (resp. (3.3)) and noting that the algebraic analog of this chain of morphisms is just the trace (resp. super trace) of $E(I_{t_1} \circ I_{t_2}) = E(I_{t_1 + t_2})$.

Remark 3.19. We remark that unlike $\mathrm{tr}(E(I_t, \Psi))$ the function $\mathrm{str}(E(I_t, \Psi))$ is *independent of* t: super symmetry implies that the generator A of the semigroup $E(I_t) = e^{-tA} : E(\mathrm{pt}) \to E(\mathrm{pt})$ is the square of an odd operator. This implies that for any $c \in C(\mathrm{pt})^{-n} = C_{-n}$ the contributions to the super trace of $E(I_t, c\Omega^{-n}) = E(c)E(I_t) = E(c)e^{-tA}$ coming from eigenspaces of A with non-zero eigenvalues vanish and hence $\mathrm{str}(E(c)e^{-tA}) = \mathrm{str}(E(c) : \ker A \to \ker A)$ is independent of t (cf. Remark 3.4).

Definition 3.20. If E is a EFT of degree n, we will call the function

$$Z_E(t) \overset{\text{def}}{=} \mathrm{str}_{C_n}(E(I_t)) = \mathrm{str}(\gamma^{\otimes n} E(I_t)) = E(S_t^{\mathrm{per}}, \mu(\gamma\Omega)^{\otimes n})$$

the *partition function* of E. The previous remark shows that this function is constant if E is super symmetric. This terminology is motivated by the fact that physicists refer to the analogous function for higher-dimensional field theories as *partition function* (see Definition 3.26 for the case of 2-dimensional conformal field theories).

Putting Theorem 1.1, Lemma 3.16 and Lemma 3.18 together, we obtain:

Corollary 3.21. *There is a bijection*

$$\pi_0 \mathcal{EFT}_{2k} \longrightarrow \mathbb{Z}$$

which sends the EFT E to its (constant) partition function $Z_E(t) \in \mathbb{Z}$.

It is desirable to describe certain important K-theory classes (like the *families* index or the *Thom class*) as maps to the space $\mathcal{EFT}_n$ of field theories of degree n. Below we do something a little less: we describe maps to the space $\mathrm{Hom}(\mathbb{R}_+^{1|1}, HS_{C_n}^{\mathrm{sa}}(H))$, which is homeomorphic to $\mathcal{EFT}_n$ by Proposition 3.8; in other words, we describe the associated EFT only on the standard super interval $I_{t,\theta}$.

Remark 3.22 (The index of a family of spin manifolds). Let $\pi : Z \to X$ be a fiber bundle with fibers of dimension n. Assume that the tangent bundle τ along the fibers has a spin structure; this implies that π induces a map $\pi_* : KO(Z) \to KO^{-n}(X)$ called 'Umkehr map' or 'integration over the fiber'. If $\xi \to Z$ is a real vector bundle, the element

$$\pi_*(\xi) \in KO^{-n}(X) = [X, KO_{-n}] = [X, \mathcal{EFT}_n^{\mathbb{R}}]$$

can be described as follows. Let $S \to Z$ be the $C(\tau) - C_n$-bimodule bundle representing the spin structure on τ. Then we obtain a C_n-bundle over X whose fiber over $x \in X$ is $L^2(Z_x, (S \otimes \xi)_{|Z_x})$, the Hilbert space H_x of L^2-sections of $S \otimes \xi$ restricted to the fiber Z_x. The Dirac operator $D_x \otimes \xi$ on Z_x 'twisted by ξ' acts on H_x and commutes with the (right) C_n-action on H_x induced by the action on S. Since the space of C_n-linear grading preserving isometries is contractible, we may identify H_x with a *fixed* real Hilbert space $H_{\mathbb{R}}$ with C_n-action. Then the map

$$X \longrightarrow \mathrm{Hom}(\mathbb{R}_+^{1|1}, HS_{C_n}^{\mathrm{sa}}(H_{\mathbb{R}})) \cong \mathcal{EFT}_n^{\mathbb{R}} \qquad x \mapsto ((t, \theta) \mapsto f_{t,\theta}(D_x \otimes \xi))$$

represents the element $\pi_*(\xi)$; here $f_{t,\theta} = e^{-tx^2 + \theta x} = e^{-tx^2} + \theta x e^{-tx^2} \in C_0(\mathbb{R})$ for $(t, \theta) \in \mathbb{R}^{1|1}$; functional calculus can then be applied to the self-adjoint operator $D_x \otimes \xi$ to produce the super semi-group $f_{t,\theta}(D_x \otimes \xi)$ of even self-adjoint Hilbert–Schmidt operators.

Remark 3.23 (The KO-theory Thom class). Let $\pi : \xi \to X$ be an n-dimensional vector bundle with spin structure given by the $C(\xi) - C_n$-bimodule bundle $S \to X$ (see Remark 2.15). We may assume that there is a real Hilbert space $H_{\mathbb{R}}$ which is a graded C_n-module, and that S is a C_n-linear subbundle of the trivial bundle $X \times H_{\mathbb{R}}$. We note that for $v \in \xi$ the Clifford multiplication operator $c(v) : S_x \to S_x$ is skew-adjoint, and $\epsilon c(v)$ is self-adjoint (ϵ is the grading involution); moreover, $\epsilon c(v)$ *commutes* with the right action of the Clifford algebra $C_{-n} = C(-\mathbb{R}^n)$ if we let $w \in \mathbb{R}^n$ act via $\epsilon c(w)$. Then the map

$$\xi \longrightarrow \operatorname{Hom}(\mathbb{R}^{1|1}_+, HS^{sa}_{C_{-n}}(H_\mathbb{R})) \cong \mathcal{EFT}^\mathbb{R}_{-n}$$

$$v \mapsto \left((t, \theta) \mapsto f_{t,\theta}(\epsilon c(v))\right) \in HS^{sa}_{C_{-n}}(S_{\pi(v)}) \subset HS^{sa}_{C_{-n}}(H_\mathbb{R})$$

extends to the Thom space X^ξ and represents the KO-theory Thom class of ξ in $KO^n(X^\xi) = [X^\xi, \mathcal{EFT}^\mathbb{R}_{-n}]$.

3.3 Conformal field theories and modular forms

In this section we will show that CFTs of degree n (see Definition 2.26) are closely related to modular forms of weight $n/2$. For a precise statement see Theorem 3.25 below. Let us first recall the definition of modular forms (cf. [HBJ, Appendix]).

Definition 3.24. A *modular form of weight k* is a function $f : \mathfrak{h} \to \mathbb{C}$ which is holomorphic (also at $i\infty$) and which has the following transformation property

$$f(\tfrac{a\tau+b}{c\tau+d}) = (c\tau + d)^k f(\tau) \qquad \text{for all } \left(\begin{smallmatrix} a & b \\ c & d \end{smallmatrix}\right) \in SL_2(\mathbb{Z}). \tag{3.6}$$

Let us recall what 'holomorphic at $i\infty$' means. The transformation property for the matrix $\left(\begin{smallmatrix} 1 & 1 \\ 0 & 1 \end{smallmatrix}\right)$ implies the translation invariance

$$f(\tau + 1) = f(\tau).$$

It follows that $f : \mathfrak{h} \to \mathbb{C}$ factors through $\mathfrak{h}/\mathbb{Z}$, which is conformally equivalent to the punctured open unit disc D^2_0 by means of the map

$$\mathfrak{h} \longrightarrow D^2_0 \qquad \tau \mapsto q = e^{2\pi i\tau}.$$

Then $f(\tau)$ is *holomorphic at $i\infty$* if the resulting function $f(q)$ on D^2_0 extends over the origin (note that $\tau \to i\infty$ corresponds to $q \to 0$). Equivalently, in the expansion

$$f(\tau) = \sum_{n \in \mathbb{Z}} a_n q^n \tag{3.7}$$

of $f(q)$ as a Laurent series around 0 (this is called the *q-expansion of f*), we require that $a_n = 0$ for $n < 0$.

Theorem 3.25. *If E is a CFT of degree n, then its partition function $Z_E : \mathfrak{h} \to \mathbb{C}$ (see Definition 3.26 below) has the transformation Property (3.6) of a modular form of weight $n/2$.*

Definition 3.26 (Partition function). We recall from Definition 2.23 that for a *closed* conformal spin surface Σ the fermionic Fock space $F_{\mathrm{alg}}(\Sigma)$ is the Pfaffian line $\mathrm{Pf}(\Sigma)$. Given a CFT E of degree n and an element $\Psi \in \mathrm{Pf}^{-n}(\Sigma) \overset{\mathrm{def}}{=} F_{\mathrm{alg}}(\Sigma)^{-n}$, the pair (Σ, Ψ) represents a morphisms in the category $\mathcal{CB}_n^2$ from $\emptyset$ to $\emptyset$; hence we can apply the functor E to (Σ, Ψ) to obtain an element $E(\Sigma, \Psi) \in \mathrm{Hilb}(E(\emptyset), E(\emptyset)) = \mathrm{Hilb}(\mathbb{C}, \mathbb{C}) = \mathbb{C}$. Since $E(\Sigma, \Psi)$ depends linearly on Ψ, we obtain an element

$$Z_E(\Sigma) \in \mathrm{Pf}^n(\Sigma) = \mathrm{Hom}(\mathrm{Pf}^{-n}(\Sigma), \mathbb{C}) \qquad \text{given by} \qquad \Psi \mapsto E(\Sigma, \Psi).$$

We recall that if $g \colon \Sigma \to \Sigma'$ is a conformal spin diffeomorphism and $g \colon \mathrm{Pf}^{-n}(\Sigma) \to \mathrm{Pf}^{-n}(\Sigma')$ is the induced isomorphism of Pfaffian lines, then for any $\Psi \in \mathrm{Pf}^{-n}(\Sigma)$ the pairs (Σ, Ψ) and $(\Sigma', g\Psi)$ represent the *same* morphism in $\mathcal{CB}_n^2(\emptyset, \emptyset)$. In particular we have

$$E(\Sigma, \Psi) = E(\Sigma', g\Psi) \in \mathbb{C} \quad \text{and} \quad g(Z_E(\Sigma)) = Z_E(\Sigma') \in \mathrm{Pf}^n(\Sigma'). \quad (3.8)$$

This property can be used to interpret Z_E as a section of a complex line bundle Pf^n over the Teichmüller spaces of conformal spin surfaces, which is equivariant under the action of the mapping class groups. The *partition function* of E is obtained by restricting attention to conformal surfaces of genus one with the non-bounding spin structure. This spin structure is preserved up to isomorphism by any orientation preserving diffeomorphism, i.e. by $SL_2(\mathbb{Z})$, whereas the other 3 spin structure is permuted. This will ultimately have the effect of obtaining a modular form for the full modular group $SL_2(\mathbb{Z})$ rather than for an (index 3) subgroup.

Let us describe the Teichmüller space of this spin torus explicitly. Given a point τ in the upper half plane $\mathfrak{h} \subset \mathbb{C}$, let $\Sigma_\tau \overset{\mathrm{def}}{=} \mathbb{C}/(\mathbb{Z} + \mathbb{Z}\tau)$ be the *conformal torus* obtained as the quotient of the complex plane (with its standard conformal structure) by the free action of the group $\mathbb{Z} + \mathbb{Z}\tau \subset \mathbb{C}$ acting by translations. Let A_τ be the *conformal annulus* obtained as a quotient of the strip $\{z \in \mathbb{C} \mid 0 \le \mathrm{im}(z) \le \mathrm{im}(\tau)\}$ by the translation group $\mathbb{Z}$. The annulus is a bordism from $S^1 = \mathbb{R}/\mathbb{Z}$ to itself if we identify $\mathbb{R}$ with the horizontal line $\{z \in \mathbb{C} \mid \mathrm{im}(z) = \mathrm{im}(\tau)\}$ via $s \mapsto s + \tau$. So while A_τ as a manifold depends only on the imaginary part of τ, the identification between ∂A_τ and the disjoint union of S^{per} (circle with periodic spin structure) and $\bar{S}^{\mathrm{per}}$ depends on the real part of τ. Note that if we equip $\mathbb{C}$ with the standard spin structure given by the bimodule bundle $S = \mathbb{C} \times C_2$, then the translation action of $\mathbb{C}$ on itself lifts to an action on S (trivial on the second factor). This implies that the spin structure on $\mathbb{C}$ induces a spin structure on Σ_τ (which is the non-bounding spin structure) and A_τ.

For $Y = S^{\text{per}}$ the space Hilbert $V(Y)$ from Definition 2.19 can be identified with complex valued functions on the circle. In particular, the constant real functions give us an isometric embedding $\mathbb{R} \subset V(S^{\text{per}})$ and hence an embedding of Clifford algebras $C_1 = C(\mathbb{R}) \to C(S^{\text{per}})$. Let $\Omega \in F_{\text{alg}}(A_\tau)$ be the vacuum vector, let $\gamma \in C_1 \otimes \mathbb{C} \subset C(S^{\text{per}}) \otimes \mathbb{C}$ be the element constructed in Lemma 3.16, and let $\xi_\tau \in F_{\text{alg}}(\Sigma_\tau)$ be the image of $\gamma\Omega$ under the fermionic gluing map $\mu \colon F_{\text{alg}}(A_\tau) \to F_{\text{alg}}(\Sigma_\tau)$. If E is a CFT of degree n (not necessarily super symmetric), we define its *partition function* to be the function

$$Z_E \colon \mathfrak{h} \longrightarrow \mathbb{C} \qquad \tau \mapsto E(\Sigma_\tau, \xi_\tau^{-n}).$$

We note that for $g = \left(\begin{smallmatrix} a & b \\ c & d \end{smallmatrix}\right) \in SL_2(\mathbb{Z})$ and $\tau \in \mathfrak{h}$ the map $\mathbb{C} \to \mathbb{C}$, $z \mapsto (c\tau+d)^{-1}z$ sends the lattice $\mathbb{Z}+\mathbb{Z}\tau$ to $\mathbb{Z}+\mathbb{Z}g\tau$, $g\tau = \frac{a\tau+b}{c\tau+d}$. This conformal diffeomorphism lifts to a conformal spin diffeomorphism $g \colon \Sigma_\tau \to \Sigma_{g\tau}$, which induces an isomorphism of Pfaffian lines $g_* \colon \text{Pf}(\Sigma_\tau) \to \text{Pf}(\Sigma_{g\tau})$.

Lemma 3.27. *The induced map* $\text{Pf}(\Sigma_\tau)^{\otimes 2} \to \text{Pf}(\Sigma_{g\tau})^{\otimes 2}$ *sends* $\xi_\tau^{\otimes 2}$ *to* $(c\tau + d)\xi_{g\tau}^{\otimes 2}$.

Proof of Theorem 3.25. We note that the previous lemma and equation (3.8) implies that if E is a CFT of degree $2k$, then

$$Z_E(\tau) = E(\Sigma_\tau, \xi_\tau^{-2k}) = E(\Sigma_{g\tau}, g_*(\xi_\tau^{-2k})) = E(\Sigma_{g\tau}, (c\tau+d)^{-k}\xi_{g\tau}^{-2k}))$$
$$= (c\tau+d)^{-k} E(\Sigma_{g\tau}, \xi_{g\tau}^{-2k}) = (c\tau+d)^{-k}Z_E(g\tau). \quad (3.9)$$

This shows that the partition function of E has the transformation property (3.6) of a modular form of weight k as claimed by Theorem 3.25. $\qquad \square$

Now we would like to discuss whether the partition function $Z_E(\tau)$ of a conformal field theory E of degree $2k$ *is* a modular form of weight k; in other words, whether $Z_E(\tau)$ is holomorphic and holomorphic at $i\infty$. The key to this discussion is the following result whose proof is analogous to that of the corresponding result Lemma 3.18 for the partition function of a 1-dimensional EFT.

Lemma 3.28. $Z_E(\tau) = \text{str}(E(A_\tau, (\gamma\Omega)^{-n}))$.

We note that $E(S^{\text{per}})$ is a graded module over the Clifford algebra $C(S^{\text{per}})^{-n}$ by letting $c \in C(S^{\text{per}})^{-n}$ act on $E(S^{\text{per}})$ via the operator $E(c)$. We note that the operator $E(A_\tau)$ does not commute with the action of the algebra $C(S^{\text{per}})^{-n}$, but it *does* commute with the subalgebra $C_{-n} = C_1^{-n} \subset C(S^{\text{per}})^{-n}$ generated by constant functions. Expressing $(A_\tau, (\gamma\Omega)^{-n})) \in \mathcal{CB}_n^2(S^{\text{per}}, S^{\text{per}})$ as the composition $\gamma^{-n} \circ A_\tau$ we obtain the following alternative expression for the partition function (cf. Definition 3.17)

$$Z_E(\tau) = \text{str}(E(\gamma^{\otimes n})E(A_\tau)) = \text{str}_{C_n}(E(A_\tau)). \tag{3.10}$$

The compatibility of $E \colon \mathcal{CB}_n^2 \to \text{Hilb}$ with the involution $*$ on both categories implies that the homomorphism

$$\mathfrak{h} \to HS_{C_n}^{\text{ev}}(E(S^{\text{per}})) \qquad \tau \mapsto E(A_\tau) \tag{3.11}$$

is $\mathbb{Z}/2$-equivariant, where $\mathbb{Z}/2$ acts on $\mathfrak{h}$ by $\tau \mapsto -\bar{\tau}$ and on $HS_{C_n}(H)$ by taking adjoints. *Any* homomorphism $\rho \colon \mathfrak{h} \to HS_{C_n}^{\text{ev}}(H)$ has the form

$$\rho(\tau) = q^{L_0}\bar{q}^{\bar{L}_0} \qquad q = e^{2\pi i \tau}$$

where $L_0, \bar{L}_0$ are two commuting, even, C_n-linear operators (in general unbounded), such that the eigenvalues of $L_0 - \bar{L}_0$ are integral. Moreover, the homomorphism is $\mathbb{Z}/2$-equivariant if and only if the operators $L_0, \bar{L}_0$ are self-adjoint.

We want to emphasize that the homomorphism (3.11) is completely analogous to the homomorphism

$$\mathbb{R}_+ \longrightarrow HS_{C_n}^{\text{ev,sa}}(E(\text{pt})) \qquad t \mapsto E(I_t)$$

associated to a 1-dimensional EFT (see equation (3.1)). We've shown that $E(I_t)$ is always of the form $E(I_t) = e^{-tA}$ for an (unbounded) self-adjoint, even operator A. Moreover, if E is the restriction of a *super symmetric* EFT, i.e., if E extends from the bordism category $\mathcal{EB}_n^1$ to the 'super bordism category' $\mathcal{SEB}_n^1$, then A is the square of an odd operator. This had the wonderful consequence that for any $c \in C_n$ the super trace $\text{str}(cE(I_t))$ is in fact *independent* of t.

Similarly, we would like to argue that if the Clifford linear CFT E is the restriction of a 'super conformal' field theory of degree n, then the infinitesimal generator $\bar{L}_0$ of

$$E(A_\tau) = q^{L_0}\bar{q}^{\bar{L}_0} \tag{3.12}$$

is the square of an odd self-adjoint operator $\bar{G}_0$. Here by 'super conformal' field theory of degree n we mean a functor $E \colon \mathcal{SCB}_n^2 \to \text{Hilb}$ (satisfying the usual requirements), where $\mathcal{SCB}_n^2$ is the 'super version' of the category $\mathcal{CB}_n^2$, in which the conformal spin bordisms are replaced by super manifolds equipped with an appropriate 'geometric super structure', which induces a conformal structure on the underlying 2-dimensional spin manifold. Unfortunately our ignorance about super geometry has kept us from identifying the correct version of this 'geometric super structure', but we are confident that this can be done (or has been done already). In this situation it seems reasonable to proceed assuming this. In other words, from now on the results in this section will all be subject to the following.

Hypothesis 3.29. *There is an appropriate notion of 'super conformal structure' with the following properties:*

(1) *on the underlying 2-dimensional spin manifold it amounts to a conformal structure;*

(2) *if $E\colon \mathcal{CB}_n^2 \to$ Hilb is a CFT of degree n which extends to a (yet undefined) 'super symmetric CFT of degree n', then $\bar{L}_0$ is the square of an odd operator $\bar{G}_0$ (where $\bar{L}_0$ is as in equation (3.12)).*

We want to emphasize that the usual notion of 'super conformal structure' is *not* what is needed here; we will comment further in Remark 3.32.

Theorem 3.30. *Assuming Hypothesis 3.29, the partition function Z_E of a susy CFT of degree n is a weak integral modular form of weight $\frac{n}{2}$.*

Definition 3.31 (Weak integral modular forms). A *weak modular form* is a holomorphic function $f\colon \mathfrak{h} \to \mathbb{C}$ with the transformation property (3.6), whose q-expansion (3.7) has only finitely many terms with negative powers of q; equivalently, the function $f(q)$ on the disc has a pole, not an essential singularity at $q = 0$.

A (weak) modular form is *integral* if all coefficients a_n in its q-expansion are integers (this low-brow definition is equivalent to more sophisticated definitions). An example of an integral modular form is the *discriminant* Δ whose q-expansion has the form

$$\Delta = q \prod_{n=1}^{\infty} (1 - q^n)^{24}.$$

Other examples of integral modular forms are the the *Eisenstein series*

$$c_4 = 1 + 240 \sum_{k>0} \sigma_3(k) q^k \qquad c_4 = 1 - 504 \sum_{k>0} \sigma_5(k) q^k$$

(modular forms of weight 4 respectively 6) where $\sigma_r(k) = \Sigma_{d|k} d^r$. The ring of integral modular forms is equal to the quotient of the polynomial ring $\mathbb{Z}[c_4, c_6, \Delta]$ by the ideal generated by $c_4^3 - c_6^2 - (12)^3 \Delta$.

Let us denote by MF_* the graded ring of weak integral modular forms; it is graded by the *degree of a modular form* to be *twice* its weight; the motivation here being that with this definition the degree of the partition function of a susy CFT E agrees with the degree of E. Recall that the discriminant Δ has a simple zero at $q = 0$. It follows that if f is a weak modular form, then $f\Delta^N$ is a modular form for N sufficiently large. As a consequence

$$MF_* = \mathbb{Z}[c_4, c_6, \Delta, \Delta^{-1}]/(c_4^3 - c_6^2 - (12)^3\Delta).$$

Proof of Theorem 3.30. By equations (3.10) and (3.12) we have

$$Z_E(\tau) = \mathrm{str}_{C_n}(q^{L_0}\bar{q}^{\bar{L}_0}).$$

We recall that the eigenvalues of $L_0 - \bar{L}_0$ are integral; let $H_k \subset E(S^{\mathrm{per}})$ be the subspace corresponding to the eigenvalue $k \in \mathbb{Z}$. According to our Hypothesis 3.29 we have $\bar{L}_0 = \bar{G}_0^2$. This allows us to calculate the partition function $Z_E(\tau)$ as follows

$$Z_E(\tau) = \mathrm{str}_{C_n}(q^{L_0}\bar{q}^{\bar{L}_0}) = \mathrm{str}_{C_n}(q^{L_0}_{|\ker \bar{L}_0}) \tag{3.13}$$

$$= \sum_{k\in\mathbb{Z}} \mathrm{str}_{C_n}(q^{L_0}_{|\ker \bar{L}_0 \cap H_k}) = \sum_{k\in\mathbb{Z}} q^k \, \mathrm{sdim}_{C_n}(\ker \bar{L}_0 \cap H_k). \tag{3.14}$$

Here the second equality follows from the fact that the eigenspace of $\bar{L}_0$ with non-zero eigenvalue does not contribute to the supertrace (see Remark 3.4); the last equality follows from the fact that restricted to the kernel of $\bar{L}_0$ the operator $L_0 = L_0 - \bar{L}_0$ which in turn is just multiplication by k on H_k. This implies that $Z_E(\tau)$ is a holomorphic function with integral coefficients in its q-expansion.

To see that all but finitely many coefficients a_k with negative k must be zero, we note that if $\ker \bar{L}_0 \cap H_k$ were $\neq 0$ for an infinite sequence of negative values of k, we would run into a contradiction with the fact that $q^{L_0}\bar{q}^{\bar{L}_0}$ is a Hilbert–Schmidt operator. $\square$

The above proof suggests to associate to a susy CFT E of degree n the following homomorphism of super groups

$$\psi_k \colon \mathbb{R}^{1|1} \longrightarrow HS_{C_n}(H_k) \qquad (t, \theta) \mapsto e^{2\pi i(it\bar{L}_0 + i^{1/2}\theta\bar{G}_0)}.$$

By the result of the previous section, the super homomorphism ψ_k represents an element $\Psi_k(E) \in \pi_0(\mathcal{EFT}_n) \cong K_n(\mathrm{pt})$. Let us calculate the image of this element under the isomorphism $K_n(\mathrm{pt}) \cong \mathbb{Z}$ (for n even), which is given by associating to $\Psi_k(E)$ its partition function. By the arguments leading to Corollary 3.21, it is given by

$$\mathrm{str}_{C_n}(e^{2\pi i(it\bar{L}_0)} \text{ acting on } H_k) = \mathrm{sdim}_{C_n}(\ker \bar{L}_0 \cap H_k)$$

which is the coefficient a_k in the q-expansion of $Z_E(\tau)$ by comparison with the proof of Theorem 3.25.

Sketch of proof of Theorem 1.2. This is a 'parametrized' version of the above argument: if C is a Clifford elliptic object over X, then given a point $x \in X$, we obtain a susy Clifford linear CFT as the composition $E_x \colon \mathcal{SCB}_n^2 \to \mathcal{SCB}_n^2(X) \overset{C}{\longrightarrow}$ Hilb; here the first map is given by using the constant map to $x \in X$. From E_x we manufacture a collection of homomorphisms $\psi_k(x) \colon \mathbb{R}^{1|1} \to HSC_n(H_k)$ as above; these depend continuously on x. By the results of the last subsection, the map ψ_k from X to the space of homomorphisms represents an element of $K^{-n}(X)$, giving the coefficient of q^k in the Laurent series $MF(C) \in K^{-n}(X)[[q]][q^{-1}]$. $\qquad\square$

Remark 3.32. We would like to conclude this section with some comments on our Hypothesis 3.29. The function $\rho(\tau) = q^{L_0}\bar{q}^{\bar{L}_0}$ can be rewritten in the form

$$\rho(\tau) = q^{L_0}\bar{q}^{\bar{L}_0} = e^{2\pi i[uL_0 + v\bar{L}_0]}$$

where $u = \tau$ and $v = -\bar{\tau}$. Geometrically, the new coordinates u, v can be interpreted as follows: If we think of $\tau = x + iy$ with coordinates x, y of $\mathfrak{h} \subset \mathbb{R}^2$ and with Euclidean metric $ds^2 = dx^2 + dy^2$, then the *Wick rotation* $y \mapsto t = iy$ gives a new coordinate t with respect to which the metric becomes the Minkowski metric $dx^2 - dt^2$. We note that u, v are the light cone coordinates with respect to the Minkowski metric (i.e., the light cone consists of the points with $uv = 0$). In other words, if we write $\mathbb{R}_E^2$ for $\mathbb{R}^2$ with coordinates x, y and the usual Euclidean metric and $\mathbb{R}_M^2$ for $\mathbb{R}^2$ with x, t coordinates and the Minkowski metric, then the Wick rotation gives us an identification $\mathbb{R}_E^2 \otimes \mathbb{C} = \mathbb{R}_M^2 \otimes \mathbb{C}$ between the complexifications. Let $\rho_* \colon \mathrm{Lie}(\mathbb{R}_E^2) \subset \mathrm{Lie}(\mathbb{R}_E^2)_\mathbb{C} \to \mathrm{End}_{C_n}(H)$ be the Lie algebra homomorphism induced by ρ, extended to the complexification $\mathrm{Lie}(\mathbb{R}_E^2)_\mathbb{C}$. Interpreting the translation invariant vectorfields $\frac{\partial}{\partial x}, \frac{\partial}{\partial y}$ on $\mathbb{R}_E^2$ as elements of $\mathrm{Lie}(\mathbb{R}_E^2)$ and $\frac{\partial}{\partial u}, \frac{\partial}{\partial v}$ as elements of $\mathrm{Lie}(\mathbb{R}_M^2)_\mathbb{C}$, the above equation shows that

$$\rho_*\left(\frac{\partial}{\partial u}\right) = 2\pi i L_0 \qquad \rho_*\left(\frac{\partial}{\partial v}\right) = 2\pi i \bar{L}_0.$$

Let $\mathbb{R}_M^{2|1}$ be the super Minkowski space (super space time) of dimension $(2|1)$ with even coordinates u, v and one odd coordinate θ (see [Wi2, §2.8] or [Fr2, Lecture 3]). This comes equipped with a natural geometric 'super structure' extending the Minkowski metric on the underlying $\mathbb{R}_M^2$. It has a group structure such that the translation action on itself preserves that geometric structure. The corresponding super Lie algebra is given by the space of invariant vector fields; a basis is provided by the two even vector fields $\frac{\partial}{\partial u}, \frac{\partial}{\partial v}$ and the odd vector field $Q = \partial_\theta + \theta \frac{\partial}{\partial v}$. The odd element Q commutes (in the graded sense) with the even elements; the crucial relation in this super Lie algebra $\text{Lie}(\mathbb{R}_M^{2|1})$ is (see [Wi2, p. 498])

$$\frac{1}{2}[Q, Q] = Q^2 = \frac{\partial}{\partial v}.$$

This shows that if the representation $\rho \colon \mathbb{R}_M^2 \to \text{End}_{C_n}(H)$ extends to a representation of the super Poincaré group $\mathbb{R}_M^{2|1}$, then the operator $\bar{L}_0$ is the *square of an odd operator*.

We note that the usual 'super conformal structure' on super manifolds of dimension $(2|2)$ [CR], [Ba] is *not* the geometric structure we are looking for. This is *too much* super symmetry in the sense that if a Clifford linear conformal field theory $E \colon \mathcal{CB}_n^2 \to \text{Hilb}$ would extend over these super manifolds, then *both generators* L_0 and $\bar{L}_0$ of the semi-group $E(A_\tau)$ would be squares of odd operators, thus making the partition function $Z_E(\tau)$ *constant*.

4 Elliptic objects

In this section we describe various types of elliptic objects (over a manifold X). We start by recalling Segal's original definition, then we modify it by introducing fermions. After adding a super symmetric aspect (as in Section 3.2) we arrive at so-called *Clifford elliptic objects*. These are still not good enough for the purposes of excision, as explained in the introduction. Therefore, we add data associated to points, rather than just circles and conformal surfaces. These are our *enriched elliptic objects*, defined as certain functors from a geometric bicategory $\mathcal{D}_n(X)$ to the bicategory of von Neumann algebras vN.

4.1 Segal and Clifford elliptic objects

We first remind the reader of a definition due to Segal [Se1, p. 199].

Definition 4.1. A *Segal elliptic object* over X is a projective functor $\mathcal{C}(X) \to \mathcal{V}$ satisfying certain axioms. Here $\mathcal{V}$ is the category of topological vector spaces and trace class operators; the objects of $\mathcal{C}(X)$ are closed oriented 1-manifolds equipped with maps to X and the morphisms are 2-dimensional oriented bordisms equipped with a conformal structure and a map to X. In other words, $\mathcal{C}(X)$ is the subcategory of the bordism category $\mathcal{CB}^2(X)$ (see Definition 2.5) with the same objects, but excluding those morphisms which are given by *diffeomorphisms*.

The adjective 'projective' basically means that the vector space (resp. operator) associated to map of a closed 1-manifold (resp. a conformal 2-manifold) to X is only defined up to a scalar. As explained by Segal in §4 of his paper [Se2] (after Definition 4.4), a projective functor from $\mathcal{C}(X)$ to $\mathcal{V}$ can equivalently be described as a functor

$$\widehat{E}: \mathcal{C}_n(X) \longrightarrow \mathcal{V}$$

where $n \in \mathbb{Z}$ is the *central charge* of the elliptic object. Here $\mathcal{C}_n(X)$ is some 'extension' of the category $\mathcal{C}(X)$, whose objects and morphisms are like those of $\mathcal{C}(X)$, but the 1-manifolds and 2-manifolds (giving the objects resp. morphisms) are equipped with an extra structure that we will refer to as *n-riggings*. The functor $\widehat{E}$ is required to satisfy a linearity condition explained below. We will use the notation $\widehat{E}$ for Segal elliptic objects and E for Clifford elliptic objects (Definition 4.3).

The following definition of n-riggings is not in Segal's papers, but it is an obvious adaptation of Segal's definitions if we work with manifolds with spin structures as Segal proposes to do at the end of §6 in [Se1].

Definition 4.2 (Riggings). Let Y be a closed spin 1-manifold which is zero bordant. We recall that associated to Y is a Clifford algebra $C(Y)$ (see Definitions 2.19 and 2.20), and that a conformal spin bordism Σ' from Y to the empty set determines an irreducible (right) $C(Y)$-module $F(\Sigma')$ (the 'Fock space' of Σ'; see Definition 2.23). The isomorphism type of $F(\Sigma')$ is *independent* of Σ'.

Given an integer n, we define an *n-rigging* of Y to be a right $C(Y)^{-n}$-module R isomorphic to $F(\Sigma')^{-n}$ for some Σ'. In particular, such a conformal spin bordism Σ' from Y to $\emptyset$ determines an n-rigging for Y, namely $R = F(\Sigma')^{-n}$. This applies in particular to the case Segal originally considered: if Y is parametrized by a disjoint union of circles then the same number of disks can be used as Σ'.

Let Σ be a conformal spin bordism from Y_1 to Y_2 and assume that Y_i is equipped with an n-rigging R_i. An *n-rigging* for Σ is an element λ in the complex line

$$\mathrm{Pf}^n(\Sigma, R_1, R_2) \stackrel{\text{def}}{=} \mathrm{Hom}_{C(Y_1)^{\otimes n}}(R_1, R_2 \boxtimes_{C(Y_2)^{-n}} F(\Sigma)^{-n})$$

which is well defined since by Definition 2.23 $F(\Sigma)$ is a left module over $C(Y_1)^{\mathrm{op}} \otimes C(Y_2)$. If Σ is closed, this is just the $(-n)$th power of the Pfaffian line $\mathrm{Pf}(\Sigma) = F(\Sigma)$. More generally, if the riggings R_i come from conformal spin bordisms Σ_i' from Y_i to $\emptyset$, then the line $\mathrm{Pf}^n(\Sigma, R_1, R_2)$ can be identified with the $(-n)$th power of the Pfaffian line of the closed conformal spin surface $\Sigma_2' \cup_{Y_2} \Sigma \cup_{Y_1} \overline{\Sigma_1'}$.

We want to point out that our definition of a 1-rigging for a closed spin 1-manifold Y is basically the 'spin version' of Segal's definition of a rigging (as defined in Section 4 of [Se2], after Definition 4.4). As Segal mentions in a footnote in §6 of [Se2] a rigging (in his sense) is the datum needed on the boundary of a conformal surface Σ in order to define the determinant line $\mathrm{Det}(\Sigma)$ (which is the dual of the top exterior power of the space of holomophic 1-forms on Σ for a closed Σ). Similarly, an n-rigging on the boundary of a conformal spin surface Σ makes it possible to construct the nth power of the Pfaffian line of Σ.

Our notion of rigging for a conformal surface, however, is different from Segal's (see Definition 5.10 in [Se2]); his is designed to give the datum needed to define non-integral powers of the determinant line $\mathrm{Det}(\Sigma)$ (corresponding to a non-integral central charge). This is then used to resolve the phase indeterminancy and get a non-projective functor. In our setting, only integral powers of the Pfaffian line arise, so that our definition has the same effect.

A Segal elliptic object $\widehat{E} \colon \mathcal{C}_n(X) \to \mathcal{V}$ is required to be linear on morphisms in the sense that the operator $\widehat{E}(\Sigma, \Gamma, \lambda)$ associated to a bordism Σ equipped with an n-rigging λ depends complex linearly on λ.

If Σ is a torus, $\mathrm{Pf}^{-2}(\Sigma)$ is canonically isomorphic to the *determinant line* $\mathrm{Det}(\Sigma)$. In particular, an n-rigging on a closed conformal spin torus amounts to the choice of an element $\lambda \in \mathrm{Det}^{n/2}(\Sigma)$. Evaluating a Segal elliptic object $\widehat{E}$ over $X = \mathrm{pt}$ of central charge n on the family of tori $\Sigma_\tau = \mathbb{C}/(\mathbb{Z} + \mathbb{Z}\tau)$ parametrized by points $\tau \in \mathfrak{h}$ of the upper half plane (equipped with the non-bounding spin structure), we obtain a section of the complex line bundle $\mathrm{Pf}^n \to \mathfrak{h}$; it is given by $\tau \mapsto (\lambda \mapsto \widehat{E}(\Sigma_\tau, \lambda)) \in \mathrm{Hom}(\mathrm{Pf}^{-n}(\Sigma_\tau), \mathbb{C})$ (compare Definition 3.26). By construction, this section is $SL_2(\mathbb{Z})$-equivariant; it is holomorphic by Segal's requirement that the operator $\widehat{E}(\Sigma, \lambda)$ associated to a conformal spin bordism Σ equipped with a rigging λ depends *holomorphically* on the conformal structure (the Teichmüller space of conformal structures on Σ is a complex manifold); such CFTs are referred to as *chiral*. Moreover, the section is holomorphic at infinity, thanks to the 'contraction condition' on $\widehat{E}$ (see [Se1, §6]). In other words, this construction associates a *modular form of weight $n/2$* to a Segal elliptic object over $X = \mathrm{pt}$.

Chiral CFTs are 'rigid' in a certain sense so that they are not general enough to obtain an interesting space of such objects (this is *not* to say that elliptic cohomology could not be described in terms of chiral CFT's; in fact it might well be possible to obtain the elliptic cohomology spectrum from a suitable symmetric monoidal category of chiral CFTs in the same way that the K-theory spectrum is obtained from the symmetric monoidal category of finite-dimensional vector spaces). We propose to study 'super symmetric' CFTs which are non-chiral, but whose partition functions are holomorphic as a consequence of the built-in super symmetry, see the proof of Theorem 1.2 in Section 3.3.

Definition 4.3. A *Clifford elliptic object* over X is a Clifford linear 2-dimensional CFT in the sense of Definition 2.30, together with a super symmetric refinement. The latter is given just like in the K-theoretic context described in Section 3.2 by replacing conformal surfaces by their (complex) $(1|1)$-dimensional partners. This is explained in more detail in Section 3.3.

Roughly speaking, the relationship between a Segal elliptic object of central charge n and a Clifford elliptic object of degree n is analogous to the relationship between the *complex* spinor bundle $S_{\mathbb{C}}(\xi)$ and the *Clifford linear* spinor bundle $S(\xi)$ associated to a spin vector bundle ξ of dimension $n = 2k$ over a manifold X (see [LM, II.5]). Given a point $x \in X$ the fiber $S_{\mathbb{C}}(\xi)_x$ is a vector space, while $S(\xi)_x$ is a graded right module over C_n, or equivalently, a graded left module over $C_{-n} = C(\text{pt})^{-n}$; we can recover $S_{\mathbb{C}}(\xi)$ from $S(\xi)$ as $S_{\mathbb{C}}(\xi) = \Delta^{\otimes k} \otimes_{C_{-n}} S(\xi)$.

Similarly, given a Clifford elliptic object E over X of degree n, we can produce an associated functor

$$\widehat{E} \colon \mathcal{C}_n(X) \longrightarrow \mathcal{V}$$

as follows:

- given an object of $\mathcal{C}_n(X)$, i.e. a closed spin manifold Y with a rigging R and a map $\Gamma \colon Y \to X$, we define $\widehat{E}(Y, R, \Gamma)$ to be the Hilbert space $R \boxtimes_{C(Y)^{-n}} E(\gamma)$;
- Given a morphism $(\Sigma, \lambda, \Gamma)$ in $\mathcal{C}_n(X)$ from (Y_1, γ_1, R_1) to (Y_2, γ_2, R_2), i.e. a conformal spin bordism Σ from Y_1 to Y_2 equipped with an n-rigging $\lambda \in \text{Pf}^n(\Sigma, R_1, R_2)$, we define the operator $E(\Sigma, \Gamma, \lambda)$ to be the composition

$$\widehat{E}(\gamma_1) = R_1 \boxtimes_{C(Y_1)^{-n}} E(\gamma_1) \xrightarrow{\lambda \boxtimes 1} R_2 \boxtimes_{C(Y_2)^{-n}} F(\Sigma)^{-n} \boxtimes_{C(Y_1)^{-n}} E(\gamma_1)$$

$$\xrightarrow{1 \boxtimes E(\Sigma)} R_2 \boxtimes_{C(Y_2)^{-n}} E(\gamma_2) = \widehat{E}(\gamma_2) \quad (4.1)$$

(see Remark 2.28 for the meaning of $E(\Sigma)$).

It should be emphasised that the resulting functor $\widehat{E}$ is *not* a Segal elliptic object: in general the operators $\widehat{E}(\Sigma, \lambda)$ will not depend holomorphically on the complex structure on Σ, since there is no such requirement for $E(\Sigma)$ in the definition of Clifford elliptic objects. However, as mentioned above, the built-in super symmetry for E will imply that the partition function (for $X = \text{pt}$) of $\widehat{E}$ is holomorphic and so we obtain a (weak) modular form of weight $n/2$.

4.2 The bicategory $\mathcal{D}_n(X)$ of conformal 0-,1-, and 2-manifolds

A *bicategory* $\mathcal{D}$ consists of objects (represented by points), 1-morphisms (*horizontal* arrows) and 2-morphisms (*vertical* double arrows). There are composition maps of 1-morphisms which are associative only up to a natural transformation between functors, and an identity 1-morphism exists (but it is only an identity up to natural transformations). There are also compositions of 2-morphisms (which are strictly associative) and strict identity objects. In particular, given objects a, b there is a category $\mathcal{D}(a, b)$ whose objects are the 1-morphisms from a to b and whose morphisms are the 2-morphisms between two such 1-morphisms; the composition in $\mathcal{D}(a, b)$ is given by vertical composition of 2-morphisms in $\mathcal{D}$. Given another object c, horizontal composition gives a functor

$$\mathcal{D}(b, c) \times \mathcal{D}(a, b) \longrightarrow \mathcal{D}(a, c)$$

which is associative only up to a natural transformation. We refer to [Be] for more details.

We will first describe the geometric bicategory $\mathcal{D}_n(X)$. The objects, morphisms and 2-morphisms will be manifolds of dimension 0, 1 and 2, respectively, equipped with conformal and spin structures, and maps to X as well as the fermions from Definition 2.26. Note that the conformal structure is only relevant for surfaces.

Following is a list of data necessary to define a bicategory. We only spell out the case $X = \text{pt}$; in the general case one just has to add piecewise smooth maps to X, for all the 0-,1-, and 2-manifolds below. So it will be easy for the reader to fill in those definitions.

Objects: The objects of $\mathcal{D}_n = \mathcal{D}_n(\text{pt})$ are 0-dimensional spin manifolds Z, i.e. a finite number of points with a graded real line attached to each of them.

Morphisms: A morphism in $\mathcal{D}_n(Z_1, Z_2)$ is either a spin diffeomorphism $Z_1 \to Z_2$, or a 1-dimensional spin manifold Y, together with a spin diffeomorphism $\partial Y \to \bar{Z}_1 \amalg Z_2$.

Composition of morphisms: For two diffeomorphisms, one uses the usual composition, and for 2 bordisms, composition is given by gluing 1-manifolds, and pictorially by

$$Z_3 \xleftarrow{\quad Y_2 \quad} Z_2 \xleftarrow{\quad Y_1 \quad} Z_1 = Z_3 \xleftarrow{\quad Y_2 \cup_{Z_2} Y_1 \quad} Z_1 \, .$$

The composition of a diffeomorphism and a bordism is given as for the category $\mathcal{B}_d$, namely by using the diffeomorphism to change the parametrization of one of the boundary pieces of the bordism.

2-morphisms: Given two bordism type morphisms Y_1, Y_2 from the object Z_1 to the object Z_2, then a 2-morphism in $\mathcal{D}_n(Y_1, Y_2)$ consists either of a spin diffeomorphism $Y_1 \to Y_2$ (rel. boundary), or it is given by a conformal spin surface Σ together with a diffeomorphism $\partial\Sigma \cong Y_1 \cup_{Z_1 \cup Z_2} Y_2$; this is schematically represented by the following picture

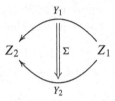

As in the category $\mathcal{CB}_2^n$, we need in addition the following datum for a 2-morphism from Y_1 to Y_2: In the case of a diffeomorphism, we have an element $c \in C(Y_1)^{\otimes n}$; in the case of a bordism, we need a fermion Ψ in the nth power of the algebraic Fock space $F_{\text{alg}}(\Sigma)$ from Definition 2.23. Moreover, we define two such pairs (Σ, Ψ) and (Σ', Ψ') to give *the same* 2-morphism from Y_1 to Y_2, if there is a conformal spin diffeomorphism $\alpha : \Sigma \to \Sigma'$ sending Ψ to Ψ'.

Given one spin diffeomorphism $\phi : Z_1 \to Z_2$ and one bordism Y, then one can form a closed spin 1-manifold Y_ϕ by gluing the ends of Y together along ϕ. Then a 2-morphism from ϕ to Y is a conformal spin surface Σ together with a fermion in $F_{\text{alg}}(\Sigma)$ and a diffeomorphism $\partial\Sigma \cong Y_\phi$. Again, two such 2-morphisms are considered equal if they are related by a conformal spin diffeomorphism.

Composition of 2-morphisms: Depending on the case at hand (horizontal and vertical) composition in the 2-category $\mathcal{D}_n$ is either given by gluing surfaces or composing diffeomorphisms. This is very similar to the category $\mathcal{CB}_2^n$, so details are omitted.

As for K-cocycles, it will be important that our enriched elliptic objects preserve a symmetric monoidal structure, certain involutions on the bicategories, as well as certain adjunction transformations. The monoidal structure on $\mathcal{D}_n$ is simply given by a disjoint union, which has a unit given by the empty set.

The involutions in $\mathcal{D}_n$. In the case of $\mathcal{B}_d$ we mentioned two involutions, called $^-$ and $*$. The first reversed the spin structure on $(d-1)$-manifolds, the second on d-manifolds. So in the case of $\mathcal{D}_n$ it is natural to have 3 involutions in the game, each of which reverses the spin structure of manifolds exactly in dimension $d = 0, 1$ respectively 2. We call these involutions (in that order), op, $\bar{}$ and $*$ even though it might seem funny to distinguish these names. Note however, that there will be analogous involutions on the von Neumann bicategory vN and we wish to be able to say which involutions are taken to which by our enriched elliptic object.

The adjunction transformation in $\mathcal{D}_n$. Given objects Z_1, Z_2 of $\mathcal{D}_n$, there is a functor

$$\mathcal{D}_n(\emptyset, Z_1 \sqcup Z_2) \longrightarrow \mathcal{D}_n(Z_1^{\mathrm{op}}, Z_2). \qquad (4.2)$$

On objects, it reinterprets a bordism Y from $\emptyset$ to $Z_1 \sqcup Z_2$ as a bordism from Z_1^{op} to Z_2. Similarly, if Y_1, Y_2 are two such bordisms, and Σ is a morphism from Y_1 to Y_2 in the category $\mathcal{D}_n(\emptyset, Z_1 \sqcup Z_2)$, then it can be reinterpreted as a morphism between Y_1 and Y_2 considered as morphisms in $\mathcal{D}_n(Z_1^{\mathrm{op}}, Z_2)$. This is natural in Z_1, Z_2; expressed in technical terms, it is a natural transformation between the two functors from $\mathcal{D}_n \times \mathcal{D}_n$ to the category of topological categories given by the domain resp. range of the functor (4.2). It is clear that the functor (4.2) is not surjective on objects or morphisms, since no diffeomorphisms can lie in the image.

4.3 Von Neumann algebras and their bimodules

References for this section are [vN], [Co1], [BR] and [Ta] for the general theory of von Neumann algebras. For the fusion aspects we recommend in addition [J2], [J4] and [Wa]. We thank Antony Wassermann for his help in writing this survey.

General facts on von Neumann algebras. A *von Neumann algebra* A is a unital $*$-subalgebra of the bounded operators $B(H)$, closed in the weak (or equivalently strong) operator topology. We assume here that H is a complex separable Hilbert space. For example, if S is any $*$-closed subset of $B(H)$, then the *commutant* (or symmetry algebra)

$$S' \overset{\text{def}}{=} \{a \in B(H) \mid as = sa \quad \forall s \in S\}$$

is a von Neumann algebra. By von Neumann's double commutant theorem, any von Neumann algebra arises in this way. In fact, the double commutant S'' is exactly the von Neumann algebra *generated by* S. For example, given two von Neumann algebras $A_i \subseteq B(H_i)$ one defines the *spatial tensor product*

$A_1 \bar{\otimes} A_2 \subseteq B(H_1 \otimes H_2)$ to be the von Neumann algebra generated by A_1 and A_2.

Just like a commutative C^*-algebra is nothing but the continuous functions on a topological space, one can show that a commutative von Neumann algebra is isomorphic to the algebra of bounded measurable functions on a measure space. The corresponding Hilbert space consists of the L^2-functions which are acted upon by multiplication.

On the opposite side of the story, one needs to understand *factors* which are von Neumann algebras with center $\mathbb{C}$. By a direct integral construction (which reduces to a direct sum if the measure space corresponding to the center is discrete), one can then combine the commutative theory with the theory of factors to understand all von Neumann algebras.

The factors come in 3 types, depending on the range of the Murray-von Neumann dimension function $d(p)$ on projections $p \in A$. This function actually characterizes equivalence classes of projections p, or equivalently, isomorphism classes of A'-modules pH. Type I factors are those von Neumann algebras isomorphic to $B(H)$ where the range of the dimension is just $\{0, 1, 2, \ldots, \dim_{\mathbb{C}}(H)\}$ (where $\dim_{\mathbb{C}}(H) = \infty$ is not excluded). For type II_1 factors, $d(p)$ can take any real value in $[0, 1]$ and for type II_∞ any value in $[0, \infty]$ ('continuous dimension'). Finally, there are type III factors for which the dimension function can only take the values 0 and ∞. Thus all nontrivial projections are equivalent. It is an empirical fact that most von Neumann algebras arising in quantum field theory are of this type.

Example 4.4 (Group von Neumann algebras). For a discrete countable group Γ one defines the *group von Neumann algebra* as the weak operator closure of the group ring $\mathbb{C}\Gamma$ in the bounded operators on $\ell^2(\Gamma)$. It is always of type II_1 and a factor if and only if each conjugacy class (of a nontrivial group element) is infinite. There are many deep connections between such factors and topology described for example in [Lu]. An application to knot concordance is given in [COT].

Example 4.5 (Local Fermions). Consider the Fock space $H = F(\Sigma)$ of a conformal spin surface Σ as in Definition 2.23. If Y is a compact submanifold of the boundary of Σ we can consider the Clifford algebra $C(Y)$ inside $B(H)$. The weak operator closure is a factor $A(Y)$ which is of type I if Y itself has no boundary. Otherwise $A(Y)$ is a type III factor known as the *local fermions* [Wa].

We remark that by taking an increasing union of finite-dimensional subspaces of the Hilbert space of spinors $V(Y)$, it follows that $A(Y)$ is *hyperfinite*, i.e. it is (the weak operator closure of) an increasing union of finite-dimensional von Neumann algebras. It is a much deeper fact that a group von Neumann algebra as in Example 4.4 is hyperfinite if and only if the group is amenable.

There is a classification of all hyperfinite factors due to Connes [Co1, p. 45] (and Haagerup [H] in the III_1 case). The complete list is very short:

I_n: $A = B(H)$ where $n = \dim_{\mathbb{C}}(H)$ is finite or countably infinite.

II_1: Group von Neumann algebras (of amenable groups with infinite conjugacy classes). All of these turn out to be isomorphic!

II_∞: The tensor product of types I_∞ and II_1.

III_0: The Krieger factor associated to a non-transitive ergodic flow.

III_λ: The Powers factors, where $\lambda \in (0, 1)$ is a real parameter coming from the 'flow of weights'.

III_1: The local fermions explained in Example 4.5. Again, these are all isomorphic.

This classification is obtained via the modular theory to which we turn in the next section. For example, a factor is of type III_1 if and only if there is a vacuum vector Ω for which the modular flow Δ^{it} on the vacuum representation only fixes multiples of Ω.

Tomita–Takesaki theory. We start with a factor $A \subseteq B(H_0)$ and assume that there is a cyclic and separating vector $\Omega \in H_0$. (Recall that this just means that $A\Omega$ and $A'\Omega$ are both dense in H_0.) Then H_0 is called a *vacuum representation* (or standard form) for A, and the vector Ω is the *vacuum vector*. It has the following extra structure: Consider the (unbounded) operator $a\Omega \mapsto a^*\Omega$ and let S be its closure. Then S has a polar decomposition $S = J\Delta^{1/2}$, where J is a conjugate linear isometry with $J^2 = \mathrm{id}$ and Δ is a positive operator (usually unbounded). By functional calculus one gets a unitary flow Δ^{it}, referred to as the *modular flow* corresponding to Ω, and the main fact about this theory is that

$$JAJ = A' \quad \text{and} \quad \Delta^{it}A\Delta^{-it} = A.$$

Note that in particular H_0 becomes a bimodule over A by defining a right action of A on H_0 by $\pi^0(a) \overset{\text{def}}{=} J\pi(a)^*J$ in terms of the original left action $\pi(a)$. This structure encodes the 'flow of weights' which classifies all hyperfinite factors as explained in the previous section.

It turns out that up to unitary isomorphism, there is a unique pair (H_0, J) consisting of a (left) A-module H_0 and a conjugate linear isometry $J : H_0 \to H_0$ with $JAJ = A'$; such a pair is referred to as *vacuum representation of*

A. For a given von Neumann algebra A there is a more sophisticated construction of such a pair, even in the absence of a cyclic and separating vector. In this invariant definition (see [Co1, p. 527]), the vacuum representation is denoted by $L^2(A)$, in analogy to the commutative case where $A = L^\infty(X)$ and $L^2(A) = L^2(X)$ for some measure space X. Similarly, if A is a group von Neumann algebra corresponding to Γ then $L^2(A) = \ell^2(\Gamma)$. If one chooses Ω to be the δ-function concentrated at the unit element of Γ, then $J(\sum_i a_i g_i) = \sum_i \bar{a}_i g_i^{-1}$ and $\Delta = \mathrm{id}$.

Remark 4.6. A vacuum vector Ω defines a faithful normal state on A via

$$\varphi_\Omega(a) \stackrel{\text{def}}{=} \langle a\Omega, \Omega \rangle_{H_0}.$$

Defining $\sigma(a) \stackrel{\text{def}}{=} \Delta^{1/2} a \Delta^{-1/2} \in A$ for *entire* elements $a \in A$ (this is a dense subset of A for which σ is defined, see [BR, I 2.5.3]) one can then verify the relation [BR, I p. 96]

$$\varphi_\Omega(ba) = \varphi_\Omega(\sigma^{-1}(a)\sigma(b))$$

for all entire elements a, b in A. It follows that φ_Ω is a trace if and only if $\Delta = \mathrm{id}$. Such vacuum vectors can be found for types I and II.

Remark 4.7. The independence from Ω implies that the image of the modular flow (given by conjugation with Δ^{it} on A) defines a canonical central subgroup of $\mathrm{Out}(A) \stackrel{\text{def}}{=} \mathrm{Aut}(A)/\mathrm{Inn}(A)$. As discussed in the previous remark, this quotient flow is nontrivial exactly for type III. Alain Connes sometimes refers to it as an 'intrinsic time', defined only in the most noncommutative setting of the theory.

Example 4.8. Consider the example of local fermions in the special case that $\Sigma = D^2$ and $\partial\Sigma = Y \cup Y_c$ is the decomposition into the upper and lower semi-circle. It was shown in [Wa] that in this case the operators J and Δ^{it} can be described geometrically: J acts on the Fock space $F(\Sigma)$ by reflection in the real axis, which clearly is of order two and interchanges $A(Y)$ and $A(Y_c) = A(Y)'$. Moreover, the modular flow Δ^{it} on $A(Y)$ is induced by the Möbius flow on D^2 (which fixes $\pm 1 = \partial Y$). This implies in particular that the Fock space is a vacuum representation (with Ω_Σ as the vacuum vector)

$$F(\Sigma) \cong L^2(A(Y)).$$

The last statement is actually true for any surface Σ and any $Y \subset \partial\Sigma$ which is not the full boundary. In the latter case, $A(\partial\Sigma) = B(F(\Sigma))$, so $F(\Sigma)$ is not the vacuum representation of $A(\partial\Sigma)$. In fact, the vacuum representation $L^2(B(H))$ of $B(H)$ is given by the ideal of all Hilbert–Schmidt operators on

H (with the operator J given by taking adjoints, and $\Delta = \mathrm{id}$). This is a good example of a construction of the vacuum representation without any canonical vacuum vector in sight.

Bimodules and Connes fusion. Given two von Neumann algebras A_i, an $A_2 - A_1$-*bimodule* is a Hilbert space F together with two normal (i.e. weak operator continuous) $*$-homomorphisms $A_2 \to \mathcal{B}(F)$ and $A_1^{\mathrm{op}} \to \mathcal{B}(F)$ with commuting images. Here A^{op} denotes the opposite von Neumann algebra which is the same underlying vector space (and same $*$ operator) as A but with the order of the multiplication reversed. One can imagine A_2 acting on the left on F and A_1 acting on the right.

Given an $A_3 - A_2$ bimodule F_2 and an $A_2 - A_1$ bimodule F_1, on can construct an $A_3 - A_1$ bimodule $F_2 \boxtimes_{A_2} F_1$ known as the *Connes fusion* of F_2 and F_1 over A_2. This construction is *not* the algebraic tensor product but it introduces a certain twist (by the modular operator Δ) in order to stay in the category of Hilbert spaces.

Definition 4.9 (Connes fusion). It is the completion of the pre-Hilbert space given by the algebraic tensor product $\mathfrak{F}_2 \odot F_1$, where

$$\mathfrak{F}_2 \overset{\text{def}}{=} B_{A_2^{\mathrm{op}}}(H_0, F_2)$$

are the bounded intertwiners from the vacuum $H_0 = L^2(A_2)$ to F_2. An inner product is obtained by the formula

$$\langle x \otimes \xi, y \otimes \eta \rangle \overset{\text{def}}{=} \langle \xi, (x, y) \cdot \eta \rangle_{F_1} \quad \xi, \eta \in F_1, \ x, y \in \mathfrak{F}_2$$

where we have used the following A_2-valued inner product on $\mathfrak{F}_2$

$$(x, y) \overset{\text{def}}{=} x^* y \in B_{A_2^{\mathrm{op}}}(H_0, H_0) = A_2.$$

Note that this makes $\mathfrak{F}_2$ into a right Hilbert module over A_2 and that the Connes fusion is nothing but the Hilbert module tensor product with F_1 and its A_2-action. Since A_1^{op} and A_3 still act in the obvious way, it follows that the Hilbert space $F = F_2 \boxtimes_{A_2} F_1$ is an $A_3 - A_1$-bimodule.

This definition looks tantalizingly simple, for example one can easily check that the relations

$$xa \otimes \xi - x \otimes a\xi = 0, \quad a \in A_2, \xi \in F_1, x \in \mathfrak{F}_2$$

are satisfied in $F_2 \boxtimes_{A_2} F_1$ (because this vector is perpendicular to all elements of $\mathfrak{F}_2 \odot F_1$ with respect to the above inner product). This assertion is true using the obvious A_2^{op}-action on $\mathfrak{F}_2$ for which $xa(v) = x(av)$, $v \in H_0$. However, if one wants to write elements in the Connes fusion in terms of *vectors* in the original bimodules F_1 and F_2 (rather than using the intertwiner space $\mathfrak{F}_2$), then it turns out that this action has to be twisted by Δ. This can be seen more precisely as follows: First of all, one has to pick a vacuum vector $\Omega \in H_0$ (or at least a normal, faithful semifinite weight) and the construction below will depend on this choice. There is an obvious embedding

$$i_\Omega : \mathfrak{F}_2 \hookrightarrow F_2, \quad x \mapsto x(\Omega)$$

and the crucial point is that this map is *not* A_2^{op}-linear. To do this calculation carefully, write $\pi(a)$ for the left A_2 action on H_0 and $\pi^0(a)$ for the right action. Recall from the previous section that $\pi^0(a) = J\pi(a)^*J$ which implies the following formulas, using $J\Omega = \Omega = \Delta\Omega$ and that $S = J\Delta^{1/2}$ has the defining property $S(\pi(a)\Omega) = \pi(a)^*\Omega$.

$$\begin{aligned}
x(\Omega) \cdot a &= x(\pi^0(a)\Omega) = x(J\pi(a)^* J\Omega) \\
&= x(J\Delta^{1/2}\Delta^{-1/2}\pi(a)^*\Delta^{1/2}\Omega) \\
&= x(S(\Delta^{1/2}\pi(a)\Delta^{-1/2})^*\Omega) \\
&= x(\Delta^{1/2}\pi(a)\Delta^{-1/2})\Omega)
\end{aligned}$$

Recall from Remark 4.6 that $\sigma(a) = \Delta^{1/2}a\Delta^{-1/2} \in A_2$ is defined for entire elements $a \in A_2$. Then we see that i_Ω has the intertwining property

$$i_\Omega(x\sigma(a)) = i_\Omega(x)a \quad \text{for all entire } a \in A_2. \tag{4.3}$$

This explains the connection between the Connes fusion defined above and the one given in [Co1, p. 533] as follows. Consider the A_2^{op}-invariant subset $\mathfrak{F}_2\Omega = \mathrm{im}(i_\Omega)$ of the bimodule F_2. These are exactly the 'ν-bounded vectors' in [Co1, Prop.6, p. 531] where in our case the weight ν is simply given by $\nu(a) = \langle a\Omega, \Omega \rangle_{H_0}$. One can then start with algebraic tensors

$$\xi_2 \otimes \xi_1 \text{ with } \xi_1 \in F_1, \xi_2 \in \mathfrak{F}_2\Omega \subset F_2$$

instead of the space $F_1 \odot \mathfrak{F}_2$ used above. This is perfectly equivalent except that the σ-twisting of the map i_Ω translates the usual algebraic tensor product relations into the following 'Connes' relations which hold for all entire $a \in A_2$

$$\xi_2 a \otimes \xi_1 = \xi_2 \otimes \sigma(a)\xi_1 = \xi_2 \otimes \Delta^{1/2}a\Delta^{-1/2}\xi_1, \quad \xi_1 \in F_1, \xi_2 \in \mathfrak{F}_2\Omega \subset F_2.$$

Remark 4.10 (Symmetric form of Connes fusion). There is the following more symmetric way of defining the Connes fusion which was introduced in [Wa] in order to actually *calculate* the fusion ring of positive energy representations of the loop group of $SU(n)$. One starts with the algebraic tensor product

$\mathfrak{F}_2 \odot \mathfrak{F}_1$ and defines the inner product by

$$\langle x_2 \otimes x_1, y_2 \otimes y_1 \rangle \overset{\text{def}}{=} \langle x_1^* y_1 \Omega, x_2^* y_2 \Omega \rangle_{H_0}$$
$$= \langle y_2^* x_2 x_1^* y_1 \Omega, \Omega \rangle_{H_0} \quad \text{for } x_i, y_i \in \mathfrak{F}_i.$$

One can translate this '4-point formula' to the definition given above by substituting $\xi = x_1(\Omega), \eta = y_1(\Omega)$. It uses again that $x_i^* y_i \in A_2$ and also the choice of a vacuum vector. After translating this definition into the subspaces $\mathfrak{F}_i \Omega$ of F_i, one can also write the Connes relations (for entire a in A_2) in the following symmetric form

$$\xi_2 \Delta^{-1/4} a \Delta^{1/4} \otimes \xi_1 = \xi_2 \otimes \Delta^{1/4} a \Delta^{-1/4} \xi_1, \quad \xi_i \in \mathfrak{F}_i \Omega \subset F_i.$$

Remark 4.11 (Subfactors). We should mention that the fusion of bimodules has had a tremendous impact on low-dimensional topology through the work of Jones, Witten and many others, see [J3] for a survey. In the context of the Jones polynomial for knots, only the hyperfinite II_1 factor was needed, so the subtlety in the Connes fusion disappears (because $\Delta = $ id if one uses the trace to define the vacuum). However, the interesting data came from *subfactors* $A \subset B$, i.e. inclusions of one factor into another. They give rise to the $A - B$ bimodule $L^2(B)$. Iterated fusion leads to very interesting bicategories and tensor categories, compare Remark 4.15.

The main reason Connes fusion arises in our context, is that we want to glue two conformal spin surfaces along *parts* of their boundary. As explained in Section 2.2, the surfaces lead naturally to Fock modules over Clifford algebras. Using the notation from the Gluing Lemma 2.24, the question arises how to express $F(\Sigma_3)$ as a $C(Y_3) - C(Y_1)$ bimodule in terms of $F(\Sigma_2)$ and $F(\Sigma_1)$. In Lemma 2.24 we explained the case of type I factors, where the modular operator $\Delta = $ id so there is no difference between the algebraic tensor product and Connes fusion. This case includes the finite-dimensional setting (and hence K-theory) as well as the gluing formulas for Segal and Clifford elliptic objects (as explained in Example 4.5, type I corresponds exactly to the case where the manifold Y along which one glues is a *closed* 1-manifold).

After all the preparation, the following answer might not come as a surprise. We only formulate it in the absence of closed components in Σ_i, otherwise the vacuum vectors might be zero. There is a simply modification in the general case which uses isomorphisms of Pfaffian lines of disjoint unions of closed surfaces. Similarly, there is a twisted version of this result which we leave to the reader. We note that in the following the definition of fusion has to be adjusted to take the grading on the Fock modules into account. This can be done by the usual trick of Klein transformations.

Proposition 4.12. *There is a unique unitary isometry of $C(Y_3) - C(Y_1)$ bimodules*

$$F(\Sigma_2) \boxtimes_{A(Y_2)} F(\Sigma_1) \xrightarrow{\cong} F(\Sigma_3)$$

sending $\Omega_2 \otimes \Omega_1$ to Ω_3.

Recall that $A(Y_2)$ is the von Neumann algebra generated by $C(Y_2)$ in the bounded operators on $F(\Sigma_2)$. One knows that $F(\Sigma_2)$ is a vacuum representation for $A(Y_2)$ with vacuum vector Ω_2 [Wa]. Therefore, the above expression $\Omega_2 \otimes \Omega_1$ is well defined in the Connes fusion. The uniqueness of the isomorphism follows from the fact that both sides are irreducible $C(Y_3) - C(Y_1)$ bimodules.

The bicategory vN of von Neumann algebras. The objects of vN are von Neumann algebras and a morphism from an object A_1 to an object A_2 is a an $A_2 - A_1$ bimodule.

Composition of morphisms. Is given by Connes fusion which will be denoted pictorially by

$$A_3 \xleftarrow{\ F_2\ } A_2 \xleftarrow{\ F_1\ } A_1 = A_3 \xleftarrow{\ F_2 \boxtimes_{A_2} F_1\ } A_1.$$

Recall that this operation is associative up to higher coherence (which is fine in a bicategory). Moreover, the identity morphism from A to A is the vacuum representation $H_0 = L^2(A)$ which therefore plays the role of the 'trivial' bimodule. This is in analogy to the trivial 1-dimensional representation of a group.

2-morphisms. Given two morphisms F_1, F_2 from the object A_1 to the object A_2, then a 2-morphism from F_1 to F_2 is a bounded intertwining operator $T \in \mathcal{B}_{A_2-A_1}(F_1, F_2)$, i.e. a bounded operator which commutes with the actions of A_1 and A_2. Pictorially

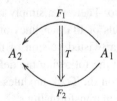

Vertical composition. Let F_i, $i = 1, 2, 3$ be three morphisms from A_1 to A_2, let T_1 be a 2-morphism from F_1 to F_2 and let T_2 be a 2-morphism from F_2 to F_3. Then their vertical composition is the 2-morphism $T_2 \circ T_1$, which is just the composition of the bounded operators T_1 and T_2. Pictorially

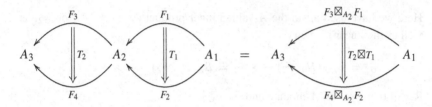

Horizontal composition. The following picture should be self explaining

Additional structures on vN. The bicategory vN has a symmetric monoidal structure given on objects by the spatial tensor product of von Neumann algebras. There are also monodial structures on the categories of bimodules by considering the Hilbert tensor product of the underlying Hilbert spaces.

Involutions on vN. There are also 3 involutions

$$A \mapsto A^{\mathrm{op}}, \qquad F \mapsto \bar{F}, \qquad T \mapsto T^*$$

on the bicategory vN, where the first was explained above and the third is the usual adjoint map. The *conjugate* $A_1 - A_2$ bimodule $\bar{F}$ (for a $A_2 - A_1$ bimodule F) is given by the formula

$$a_1 \cdot \bar{v} \cdot a_2 \overset{\mathrm{def}}{=} \overline{(a_2^* \cdot v \cdot a_1^*)}, \quad v \in F.$$

We leave it to the reader to extend the above definitions so that they really define involutions on the bicategory vN. This should be done so that the functoriality agrees with the 3 involutions in the bicategory $\mathcal{D}_n(X)$ because our enriched elliptic object will have to preserve these involutions.

Adjunction transformations on vN. Just like in $\mathcal{D}_n(X)$, we are looking for adjunction transformations of 1- respectively 2-morphisms

$$\text{vN}(\mathbb{C}, A_1 \bar{\otimes} A_2) \longrightarrow \text{vN}(A_1^{\text{op}}, A_2) \quad \text{and} \quad \text{vN}(\mathbb{C}, F_2 \boxtimes_A F_1) \longrightarrow \text{vN}(\bar{F}_2, F_1).$$

where the left-hand side is defined by considering the inclusion of the algebraic tensor product in $A_1 \bar{\otimes} A_2$. The resulting bimodule is still referred to as a $A_1 \bar{\otimes} A_2$-module because the bimodule structure is in some sense boring.

To address the right-hand side, let $A = A_1 \bar{\otimes} A_2$ and consider an A-module F_1 and an A^{op}-module F_2 (both thought of as lying in the image of the left hand side transformation). Then we may form the Connes fusion $F_2 \boxtimes_A F_1$ as the completion of $\mathfrak{F}_2 \odot F_1$, see Definition 4.9. There is a natural map

$$\Theta : \mathfrak{F}_2 \odot F_1 \longrightarrow B_A(\bar{\mathfrak{F}}_2, F_1), \quad x \otimes \eta \mapsto \theta_{x,\eta} \text{ where } \theta_{x,\eta}(\bar{y}) \overset{\text{def}}{=} (y, x)\eta.$$

Here we have used again the A-valued inner product $(y, x) = y^* x$ on $\mathfrak{F}_2$, as well as the linear isometry

$$\mathfrak{F}_2 = B_{A^{\text{op}}}(H_0, F_2) \longrightarrow \bar{\mathfrak{F}}_2 \overset{\text{def}}{=} B_A(H_0, \bar{F}_2) \quad x \mapsto \bar{x} \overset{\text{def}}{=} xJ.$$

Recall that $\bar{F}_2$ is an A-module and so is $\bar{\mathfrak{F}}_2$.

Lemma 4.13. *In the above setting, the mapping $\theta_{x,\eta}$ is indeed A-linear.*

Proof. We use the careful notation used to derive equation (4.3), where $\pi(a)$ denotes the A-action on H_0 and $\pi^0(a) = J\pi(a)^* J$ the A^{op}-action. Then we get that for $a \in A$ and $y \in \mathfrak{F}_2$

$$a\bar{y} = \bar{y}\pi^0(a) = yJ(J\pi(a)^* J) = (y\pi(a)^*)J = \overline{ya^*}.$$

This implies

$$\theta_{x,\eta}(a\bar{y}) = (ya^*, x)\eta = (ya^*)^* x\eta$$
$$= ay^* x\eta = a(y, x)\eta$$
$$= a\theta_{x,\eta}(\bar{y})$$

which is exactly the statement of our lemma. $\square$

Note that Θ takes values in the Banach space of A-intertwiners with the operator norm. It actually turns out that it is an isometry with respect to the fusion inner product. To check this statement, we assume for simplicity that A is of type III. Then there is a unitary A-intertwiner $U : H_0 = L^2(A) \to F_2$ and hence $y^* x = (y^* U)(U^* x)$ is a product of two elements in A and

$||y^*U|| = ||\bar{y}||$. Now let $f = \sum_i x_i \otimes \eta_i$ be and arbitrary element in $\mathfrak{F}_2 \odot F_1$. Then the norm squared of $\Theta(f)$ is calculated as follows

$$||\Theta(f)||^2 = \sup_{0 \neq \bar{y} \in \bar{\mathfrak{F}}_2} \frac{||\sum_i (y, x_i)\eta_i||^2}{||\bar{y}||^2}$$

$$= \sup_{0 \neq \bar{y} \in \bar{\mathfrak{F}}_2} \frac{||\sum_i (y^*U)(U^*x_i)\eta_i||^2}{||\bar{y}||^2}$$

$$= ||\sum_i (U^*x_i)\eta_i||^2 = \sum_{i,j} \langle \eta_i, x_i^* U U^* x_j \eta_j \rangle_{F_1}$$

$$= \sum_{i,j} \langle \eta_i, (x_i, x_j)\eta_j \rangle_{F_1} = \langle \sum_i x_i \otimes \eta_i, \sum_j x_j \otimes \eta_j \rangle_{\mathfrak{F}_2 \odot F_1}$$

$$= ||f||^2_{\mathfrak{F}_2 \odot F_1}.$$

This implies the following result because we have a functorial isometry which for $F_2 = L^2(A)$ clearly is an isomorphism. Note that the same result holds for bimodules, if there are two algebras acting on the left of F_2 respectively the right of F_1.

Proposition 4.14. *The above map* Θ *extends to an isometry*

$$\Theta : F_2 \boxtimes_A F_1 \xrightarrow{\cong} B_A(\bar{\mathfrak{F}}_2, F_1).$$

In order to define our adjunction transformation announced above, we now have to compare the right-hand side of the isometry to $vN(\bar{F}_2, F_1)$. If one thinks of the latter as all A-intertwiners then there is a serious problem in relating the two, because of the twisting property (4.3) of the inclusion $i_\Omega : \mathfrak{F}_2 \hookrightarrow F_2$. This is where the modular operator Δ has to come in. At this moment in time, we do not quite know how to resolve the issue, but it seems very likely that one has to change the definition of $vN(\bar{F}_2, F_1)$ slightly. Note that one can not use the right-hand side of the above isometry because these intertwiners cannot be composed, certainly not in an obvious way. This problem is related to the fact that in the example of a string vector bundle, we can only associate vectors in the fusion product to conformal spin surfaces.

Remark 4.15. It is interesting to point out the following special subcategories of the bicategory above. In the Jones example for a subfactor $A \subset B$, there are two objects (namely A and B) and the morphisms are all bimodules obtained by iterated fusion from $_A L^2(B)_B$. The crucial finite index property of Jones guarantees that for all irreducible bimodules $_A F_B$ that arise the vacuum representation H_0 is contained exactly once in $F \boxtimes_B \bar{F}$ and $\bar{F} \boxtimes_A F$. This condition expresses the fact that F has *finite* 'quantum dimension'. In [Oc], these bicategories were further developed and applied to obtain 3-manifold invariants.

If one fixes a single von Neumann algebra A, then one can consider the bicategory 'restricted to A'. This means that one has only $A - A$ bimodules and their intertwiners, together with the fusion operation. This is an example of a *tensor category*. Borrowing some notation from Section 5.4 we get the following interesting subcategory: Fix a compact simply connected Lie group G and a level $\ell \in H^4(BG)$. Then there is a canonical III_1-factor A and an embedding

$$\Phi : G \hookrightarrow \text{Out}(A) = \text{Aut}(A)/\text{Inn}(A).$$

We can thus consider those bimodules which are obtained from twisting $L^2(A)$ by an element in $\text{Aut}(A)$ which projects to $\Phi(g)$ for some $g \in G$. We believe that a certain 'quantization' of this tensor category gives the category of positive energy representation of the loop group LG at level ℓ.

4.4 Enriched elliptic objects and the elliptic Euler class

Definition 4.16. An *enriched elliptic object of degree n* over X is a continuous functor $\mathcal{D}_n(X) \to \text{vN}$ to the bicategory of von Neumann algebras. It is assumed to preserve the monodial structures (disjoint union gets taken to tensor product), the 3 involutions op, $\bar{\ }$ and $*$, as well as the adjunction transformations explained above. Finally, it has to be $\mathbb{C}$-linear in an obvious sense on Clifford algebra elements and fermions.

Again, this is only a preliminary definition because some of the categorical notions have not been defined yet, and it does not contain super symmetry. The main example of an enriched elliptic object comes from a string vector bundle, hopefully leading to an Euler class and a Thom class in elliptic cohomology. In fact, we hope that it will ultimately lead to a map of spectra

$$M \text{ String} \longrightarrow \text{tmf}.$$

We explain the construction of the Euler class momentarily class but we shall use several notions which are only developed in the coming sections. Thus the following outline can be thought of as a motivation for the reader to read on.

We next outline the construction of a degree n enriched elliptic object corresponding to an n-dimensional vector bundle $E \to X$ with string connection. This is our proposed 'elliptic Euler class' of E and it is the main example that guided many of our definitions. In Remark 5.6 we explain briefly how the analogous K-theory Euler class is defined for a vector bundle with spin connection. As usual, this will be our guiding principle.

Recall from Definition 4.16 that an enriched elliptic object of degree n in the manifold X is a certain functor between bicategories

$$\mathcal{E}_E : \mathcal{D}_n(X) \longrightarrow vN.$$

So we have to explain the values of $\mathcal{E}_E$ in dimensions $d = 0, 1, 2$. Let $\mathcal{S}$ be the string connection on E as explained in Definition 5.8. This definition is crucial for the understanding of our functor $\mathcal{E}_E$ and we expect the reader to come back to this section once she is familiar with the notion of a string connection.

The functor $\mathcal{E}_E$ in dimension 0. This is the easiest case because we can just set $\mathcal{E}_E(x) = \mathcal{S}(x)$ for a map $x : Z \to X$ of a 0-dimensional spin manifold Z. Recall that $\mathcal{S}(x)$ is a von Neumann algebra which is completely determined by the string structure on E (no connection is needed). By construction, the monoidal structures on the objects of our bicategories are preserved.

The functor $\mathcal{E}_E$ in dimension 1. For each piecewise smooth map $\gamma : Y \to X$ of a spin 1-manifold Y, the string connection $\mathcal{S}(\gamma)$ is a graded irreducible $C(\gamma) - \mathcal{S}(\partial\gamma)$ bimodule. Here $C(\gamma)$ is the relative Clifford algebra from Definition 2.33. To define $\mathcal{E}_E(\gamma)$ we use the same Hilbert space but considered only as a $C(Y)^{-n} - \mathcal{S}(\partial\gamma)$-bimodule. Note that this means that the module is far from being irreducible. If one takes orientations into account, one gets a bimodule over the incoming–outgoing parts of $\partial\gamma$. The gluing law of the string connection $\mathcal{S}$ translates exactly into the fact that our functor $\mathcal{E}_E$ preserves composition of 1-morphisms, i.e. it preserves Connes fusion.

The functor $\mathcal{E}_E$ in dimension 2. Consider a conformal spin surface Σ and a piecewise smooth map $\Gamma : \Sigma \to X$, and let $Y = \partial\Sigma$ and $\gamma = \Gamma|_Y$. Then the string connection on Γ is a unitary isometry of left $C(\gamma)$-modules

$$\mathcal{S}(\Gamma) : F(\Gamma) \cong \mathcal{S}(\gamma).$$

Here we used the relative Fock module $F(\Gamma) = F(\Gamma^* E) \otimes F(\Sigma)^{-n}$ from Definition 2.35. Recall from the same definition that the vacuum vector Ω_Γ of Γ lies in $F(\Gamma^* E)$. Given a fermion $\Psi \in F_{\mathrm{alg}}(\Sigma)^{-n}$, we may thus define

$$\mathcal{E}_E(\Gamma, \Psi) \overset{\text{def}}{=} \mathcal{S}(\Gamma)(\Omega_\Gamma \otimes \Psi) \in \mathcal{S}(\gamma) = \mathcal{E}_E(\gamma).$$

This is exactly the datum we need on 2-morphisms (Γ, Ψ) in $\mathcal{D}(X)_n$. The behavior of the string connection with respect to a conformal spin diffeomorphism $\phi : \Sigma \to \Sigma'$ implies the following important condition on an enriched elliptic object. Assuming that ϕ restricts to the identity on the boundary and noting that conformality implies $\Omega_{\Gamma'} = F(\phi)(\Omega_\Gamma)$, we may conclude that

$$\mathcal{E}_E(\Gamma', F(\phi)(\Psi)) = \mathcal{S}(\Gamma')(F(\phi)(\Omega_\Gamma \otimes \Psi)) = \mathcal{S}(\Gamma)(\Omega_\Gamma \otimes \Psi) = \mathcal{E}_E(\Gamma, \Psi).$$

Finally, $\mathcal{E}_E$ preserves horizontal and vertical composition by the gluing laws of the string connection $\mathcal{S}$ as well as those of the vacuum vectors.

5 String structures and connections

Given an n-dimensional vector bundle $E \to X$, we want to introduce a topological notion of a *string structure* and then the geometric notion of a *string connection* on E. As usual we start with the analogy of a spin structure. It is the choice of a principal Spin(n)-bundle $P \to X$ together with an isomorphism of the underlying principal $GL(n)$-bundle with the frame bundle of E. In particular, one gets an inner product and an orientation on E because one can use the sequence of group homomorphisms

$$\text{Spin}(n) \xrightarrow{2} \text{SO}(n) \le O(n) \le GL(n).$$

Recall that the last inclusion is a homotopy equivalence and that, for $n > 8$, the first few homotopy groups of the orthogonal groups $O(n)$ are given by the following table:

k	0	1	2	3	4	5	6	7
$\pi_k O(n)$	$\mathbb{Z}/2$	$\mathbb{Z}/2$	0	$\mathbb{Z}$	0	0	0	$\mathbb{Z}$

It is well known that there are topological groups and homomorphisms

$$S(n) \to \text{Spin}(n) \to \text{SO}(n) \to O(n)$$

which kill exactly the first few homotopy groups. More precisely, SO(n) is connected, Spin(n) is 2-connected, $S(n)$ is 6-connected, and the above maps induce isomorphisms on all higher homotopy groups. This homotopy theoretical description of k-connected covers actually works for any topological group in place of $O(n)$ but it only determines the groups up to homotopy equivalence. For the 0th and 1st homotopy groups, it is also well known how to construct the groups explicitly, giving the smallest possible models: one just takes the connected component of the identity, and then the universal covering. In our case this gives SO(n), an index 2 subgroup of $O(n)$, and Spin(n), the universal double covering of SO(n). In particular, both of these groups are Lie groups. However, a group $S(n)$ cannot have the homotopy type of a Lie group since π_3 vanishes. To our best knowledge, there has yet not been found a canonical construction for $S(n)$ which has reasonable 'size' and a geometric interpretation.

The groups String(n). In Section 5.4 we construct such a concrete model for $S(n)$ as a subgroup of the automorphism group of 'local fermions' on the circle. These are certain very explicit von Neumann algebras, the easiest examples of hyperfinite type III_1 factors. We denote by String(n) our particular models of the groups of homotopy type $S(n)$, and we hope that the choice of this name will become apparent in the coming sections. In fact, Section 5.4 deals with the case of compact Lie groups rather than just Spin(n), and we thank Antony Wassermann for pointing out to us this generalization. It is also his result that the unitary group $U(A)$ of a hyperfinite III_1-factor is contractible (see Theorem 5.17). This is essential for the theorem below because it implies that the corresponding projective unitary group $PU(A) \overset{\text{def}}{=} U(A)/\mathbb{T}$ is a $K(\mathbb{Z}, 2)$.

Theorem 5.1. *Consider a compact, simply connected Lie group G and a level $\ell \in H^4(BG)$. Then one can associate to it a canonical von Neumann algebra $A_{G,\ell}$, which is a hyperfinite factor of type III_1. There is an extension of topological groups*

$$1 \longrightarrow PU(A_{G,\ell}) \overset{i}{\longrightarrow} G_\ell \longrightarrow G \longrightarrow 1$$

such that the boundary map $\pi_3 G \to \pi_2 PU(A_{G,\ell}) \cong \mathbb{Z}$ is given by $\ell \in H^4(BG) \cong \mathrm{Hom}(\pi_3 G, \mathbb{Z})$. Moreover, there is a monomorphism

$$\Phi : G_\ell \hookrightarrow \mathrm{Aut}(A_{G,\ell})$$

such that the composition $\Phi \circ i$ is given by the inclusion of inner automorphisms into all of $\mathrm{Aut}(A_{G,\ell})$.

Applied to $G = \mathrm{Spin}(n)$ and $\ell = p_1/2 \in H^4(B \, \mathrm{Spin}(n))$ (or 'level 1') this gives type III_1 factors A_n ($\cong A_1^{\bar{\otimes} n}$) and groups String($n$) as discussed above.

Definition 5.2. A G_ℓ-structure on a principal G-bundle is a lift of the structure group through the above extension. In particular, a *string structure* on a vector bundle is a lift of the structure group from $SO(n)$ to String(n) using the homomorphisms explained above.

Corollary 5.3. *A G_ℓ-structure on a principal G-bundle $E \to X$ gives a bundle of von Neumann algebras over X.*

This bundle is simply induced by the monomorphism Φ above, and hence over each $x \in X$ the fiber $A(x)$ comes equipped with a G-equivariant map

$$\alpha_x : \mathrm{Iso}_G(G, E_x) \longrightarrow \mathrm{Out}(A_{G,\ell}, A(x))$$

where $\mathrm{Out}(A, B) \overset{\text{def}}{=} \mathrm{Iso}(A, B)/\mathrm{Inn}(A)$ are the outer isomorphisms. G-equivariance is defined using the homomorphism $\tilde{\Phi} : G \to \mathrm{Out}(A)$. It is not hard to see that the pair $(A(x), \alpha_x)$ contains exactly the same information as the string structure on E_x. We shall use this observation in Definition 5.18 and hence introduce the following notation (abstracting the case $V = E_x$ above):

Definition 5.4. Given (G, ℓ) and a G-torsor V, define a $G_\ell - V$-*pointed factor* to be a factor A together with a G-equivariant map

$$\alpha : \mathrm{Iso}_G(G, V) \longrightarrow \mathrm{Out}(A_{G,\ell}, A)$$

where G-equivariance is defined using the homomorphism $\tilde{\Phi} : G \to \mathrm{Out}(A)$. The choice of (A, α) is a G_ℓ-*structure* on V.

For the purposes of our application, it is actually important that all the von Neumann algebras are graded. It is possible to improve the construction for $G = \mathrm{Spin}(n)$ so that the resulting algebra is indeed graded by using local fermions rather than local loops, see Section 5.4. The above algebra A_n is then just the even part of this graded algebra.

Characteristic classes. The homotopy theoretical description given at the beginning of Section 5 implies the following facts about existence and uniqueness of additional structures on a vector bundle E in terms of characteristic classes. We point out that we are more careful about spin (and string) structure as is customary in topology: A spin structure is really the choice of a principal $\mathrm{Spin}(n)$-bundle, and not only up to isomorphism. In our language, we obtain the usual notion of a spin structure by taking isomorphism classes in the category of spin structures. Similar remarks apply to string structures. The purpose of this refinement can be seen quite clearly in Proposition 5.5.

Let E be a vector bundle over X. Then

- E is *orientable* if and only if the Stiefel–Whitney class $w_1 E \in H^1(X; \mathbb{Z}/2)$ vanishes. Orientations of E are in 1–1 correspondence with $H^0(X; \mathbb{Z}/2)$.
- In addition, E has a *spin structure* if and only if the Stiefel–Whitney class $w_2 E \in H^1(X; \mathbb{Z}/2)$ vanishes. Isomorphism classes of spin structures on E are in 1–1 correspondence with $H^1(X; \mathbb{Z}/2)$.
- In addition, E has a *string structure* if and only if the characteristic class $p_1/2(E) \in H^4(X; \mathbb{Z})$ vanishes. Isomorphism classes of string structures on E are in 1–1 correspondence with $H^3(X; \mathbb{Z})$.

More generally, a principal G-bundle E (classified by $c : X \to BG$) has a G_ℓ-structure if and only if the characteristic class $c^*(\ell) \in H^4(X; \mathbb{Z})$ vanishes. Isomorphism classes of G_ℓ-structures on E are in 1–1 correspondence with $H^3(X; \mathbb{Z})$.

In the next two sections we will enhance these topological data by geometric ones, namely with the notion of a *string connection*. These are needed to construct our enriched elliptic object for a string vector bundle, just like a spin connection was needed to define the K-cocycles in Section 3.

Since $\mathrm{String}(n)$ is not a Lie group, it is necessary to come up with a new notion of a connection on a principal $\mathrm{String}(n)$-bundle. We first present such a new notion in the spin case, assuming the presence of a metric connection on the bundle.

Spin connections. By Definition 2.15, a spin structure on an n-dimensional vector bundle $E \to X$ with Riemannian metric is a graded irreducible bimodule bundle $S(E)$ over the Clifford algebra bundle $C(E) - C_n$. For a point $x \in X$, we denote the resulting bimodule $S(E_x)$ by $S(x)$. It is a left module over the algebra $C(x) = C(E_x) \otimes C_{-n}$ from Definition 2.33. We now assume in addition that X is a manifold and that E is equipped with a metric connection.

Proposition 5.5. *A spin connection S on E gives for each piecewise smooth path γ from x_1 to x_2, an isomorphism between the following two $C(\partial\gamma) \overset{\mathrm{def}}{=} C(x_1)^{\mathrm{op}} \otimes C(x_2)$ (left) modules*

$$S(\gamma) : F(\gamma) \overset{\cong}{\longrightarrow} \mathrm{Hom}_{\mathbb{R}}(S(x_1), S(x_2))$$

where $F(\gamma)$ is the relative Fock module from Definition 2.35 (defined using the connection on E). We assume that S varies continuously with γ and is independent of the parametrization of I. Moreover, S satisfies the following gluing condition: Given another path γ' from x_2 to x_3, there is a commutative diagram

$$
\begin{array}{ccc}
F(\gamma' \cup_{x_2} \gamma) & \xrightarrow{\;\;S(\gamma' \cup_{x_2}\gamma)\;\;} & \mathrm{Hom}(S(x_1), S(x_3)) \\[4pt]
\Big\downarrow{\scriptstyle\cong} & & {\scriptstyle\circ}\Big\uparrow{\scriptstyle\cong} \\[4pt]
F(\gamma') \otimes_{C(x_2)} F(\gamma) & \xrightarrow[\cong]{\;\;S(\gamma')\otimes S(\gamma)\;\;} & \mathrm{Hom}(S(x_2), S(x_3)) \otimes_{C(x_2)} \mathrm{Hom}(S(x_1), S(x_2))
\end{array}
$$

where the left vertical isomorphism is the gluing isomorphisms from Lemma 2.24.

Remark 5.6. The vacuum vectors in the Fock modules $\Omega_\gamma \in F(\gamma^* E)$ define a parallel transport in $S(E)$ as follows: Recall that $F(\gamma) = F(\gamma^* E) \otimes F(I)^{-n}$ and that $F(I) = C_1$. Thus the vector $\Omega_\gamma \in F(\gamma^* E)$ together with the identity $\mathrm{id} \in C_{-n} = F(I)^{-n}$ gives a homomorphism from $S(x_1)$ to $S(x_2)$ via $S(\gamma)(\Omega_\gamma \otimes \mathrm{id})$. One checks that this homomorphism is in fact C_n-linear and coincides with the usual parallel transport in the spinor bundle $S(E)$.

It is interesting to observe that these vacuum vectors Ω_γ exist for any vector bundle E with metric and connection but it is the spin connection in the sense above which makes it possible to view them as a parallel transport.

Remark 5.7. There is a *unique* spin connection in the setting of the above proposition. In the usual language, this is well known and follows from the fact that the fiber of the projection $\mathrm{Spin}(n) \to SO(n)$ is discrete. For our definitions, existence and uniqueness follows from the fact that all the bimodules are irreducible (and of real type) and hence the isometries $\mathcal{S}(\gamma)$ are determined up to sign. Since they vary continuously and satisfy the gluing condition above, it is possible to see this indeterminacy in the limit where γ is the constant map with image $x \in X$. Then the right-hand side contains a canonical element, namely $\mathrm{id}_{\mathcal{S}(x)}$ and our gluing condition implies that it is the image under $\mathcal{S}(\gamma)$ of $\Omega_\gamma \otimes \mathrm{id}$. Hence the indeterminacy disappears.

These are the data a spin structure associates to points in X and a spin connection associates to paths in X. It is easy to extend the spin connection to give data associated to arbitrary 0- and 1-dimensional spin manifolds mapping to X, just like in Proposition 3.1.

In the next section, and in particular Lemma 5.12, we shall explain how all these data are really derived from 'trivializing' a 2-dimensional field theory (called Stiefel–Whitney theory in this paper). This derivation is necessary to motivate our definition of a string connection as a 'trivialization' of the Chern–Simons (3-dimensional) field theory. Because of the shift of dimension from 2 to 3, a string connection will necessarily have 0-, 1- and 2-dimensional data. As above, it is enough to formulate the top-dimensional data for manifolds with boundary (intervals in the case of spin, conformal surfaces in the case of string), since the usual gluing formulas determine the data on closed manifolds. Also as above, the 0-dimensional data are purely topological, and in the case of a string structure are given by the von Neumann algebra bundle from Corollary 5.3.

String connections. We recall from Definition 2.33 that there is a (relative) complex Clifford algebra $C(\gamma)$ defined for every piecewise smooth map $\gamma : Y \to X$, where Y is a spin 1-manifold and X comes equipped with a metric vector bundle E. If Y is closed then a connection on $\gamma^* E$ gives a preferred isomorphism class of graded irreducible (left) $C(\gamma)$-modules as follows: Consider the conformal spin surface $Y \times I$ and extend the bundle

$$\gamma^* E \cup \underline{\mathbb{R}}^{\dim(E)} \overset{\mathrm{def}}{=} (\gamma^* E \times 0) \amalg (Y \times 1 \times \mathbb{R}^{\dim(E)})$$

over $Y \times \{0, 1\}$ to a bundle E' (with connection) on $Y \times I$ (this uses the fact that E is orientable, and hence trivial over 1-manifolds). In Definition 2.35 we explained how to construct a Fock module $F(E')$ from boundary values of harmonic sections on $Y \times I$. It is a graded irreducible $C(\gamma)$ module and the isomorphism class of $F(E')$ is independent of the extension of the bundle with connection. It will be denoted by $[F(\gamma)]$. If $\Gamma : \Sigma \to X$ is a piecewise smooth map of a conformal spin surface with boundary $\gamma : Y \to X$, then a connection on $\Gamma^* E$ gives a particular representative $F(\Gamma)$ in this isomorphism class as explained in Definition 2.35.

Definition 5.8. Let $E \to X$ be an n-dimensional vector bundle with spin connection. Assume further that a string structure on E has been chosen and denote by $A(x)$ the fiber of the corresponding von Neumann algebra bundle. A *string connection* $\mathcal{S}$ on E consists of the following data.

dim 0: For each map $x : Z \to X$ of a 0-dimensional spin manifold Z, $\mathcal{S}(x)$ is a von Neumann algebra given by the von Neumann tensor product $A(x_1) \bar{\otimes} \ldots \bar{\otimes} A(x_n)$ if $x(Z)$ consists of the spin points $x_1, \ldots, x_n$. By definition, $A(\bar{x}) = A(x)^{op}$ and $\mathcal{S}(\emptyset) = \mathbb{C}$. All these data are completely determined by the string structure alone.

dim 1: For each piecewise smooth map $\gamma : Y \to X$ of a spin 1-manifold Y, $\mathcal{S}(\gamma)$ is a graded irreducible $C(\gamma) - \mathcal{S}(\partial \gamma)$ bimodule. These fit together to bimodule bundles over $\mathrm{Maps}(Y, X)$ and we assume that on these bundles there are lifted actions $\mathcal{S}(\phi)$ of the spin diffeomorphisms $\phi \in \mathrm{Diff}(Y, Y')$ which are the identity on the boundary. It is clear that these are bimodule maps only if one takes the action of ϕ on $C(\gamma)$ into account, as well as the action of $\phi|_{\partial\gamma}$ on $\mathcal{S}(\partial \gamma)$.

   Given another such $\gamma' : Y' \to X$ with 0-dimensional intersection on the boundary $x \stackrel{\mathrm{def}}{=} \partial\gamma \cap \partial\gamma' = \partial_{\mathrm{in}}\gamma = \partial_{\mathrm{out}}\gamma'$, there are $C(\gamma \cup_x \gamma') - \mathcal{S}(\partial(\gamma \cup_x \gamma'))$ bimodule isomorphisms

$$\mathcal{S}(\gamma, \gamma') : \mathcal{S}(\gamma \cup_x \gamma') \stackrel{\cong}{\longrightarrow} \mathcal{S}(\gamma) \boxtimes_{\mathcal{S}(x)} \mathcal{S}(\gamma')$$

where we used Connes fusion of bimodules on the right-hand side, and also the identifications

$$C(\gamma \cup_x \gamma') \cong C(\gamma) \otimes C(\gamma') \text{ and } \mathcal{S}(\partial(\gamma \cup_x \gamma')) \subset \mathcal{S}(\partial\gamma)\bar{\otimes}\mathcal{S}(\partial\gamma').$$

The isomorphisms $\mathcal{S}(\gamma, \gamma')$ must satisfy the obvious associativity constraints. Note that for closed Y, we just get an irreducible $C(\gamma)$-module $\mathcal{S}(\gamma)$, multiplicative under disjoint union. We assume that $\mathcal{S}(\gamma)$ is a (left) module in the preferred isomorphism class $[F(\gamma)]$ explained above.

dim 2: Consider a conformal spin surface Σ and a piecewise smooth map Γ :
$\Sigma \to X$, and let $Y = \partial\Sigma$ and $\gamma = \Gamma|_Y$. Then there are two irreducible
(left) $C(\gamma)$-modules in the same isomorphism class, namely $F(\Gamma)$ and
$\mathcal{S}(\gamma)$. The string connection on Γ is a unitary isometry of left $C(\gamma)$-
modules

$$\mathcal{S}(\Gamma) : F(\Gamma) \cong \mathcal{S}(\gamma)$$

such that for each conformal spin diffeomorphism $\phi : (\Sigma, \Gamma) \to$
(Σ', Γ') the following diagram commutes

$$
\begin{array}{ccc}
F(\Gamma) & \xrightarrow[\cong]{\mathcal{S}(\Gamma)} & \mathcal{S}(\gamma) \\
F(\phi) \downarrow & & \downarrow \mathcal{S}(\phi|_{\partial\gamma}) \\
F(\Gamma') & \xrightarrow[\cong]{\mathcal{S}(\Gamma')} & \mathcal{S}(\gamma')
\end{array}
$$

The module maps $\mathcal{S}(\Gamma)$ fit together to continuous sections of the resulting bundles over the relevant moduli spaces. The irreducibility of
the modules in question implies that there is only a circle worth of
possibilities for each $\mathcal{S}(\Gamma)$. This is the *conformal anomaly*.

Finally, there are gluing laws for surfaces which meet along a part Y of their
boundary. If Y is closed this can be expressed as the composition of Hilbert–
Schmidt operators. If Y has itself boundary one uses Connes fusion, see Proposition 4.12.

Note that the irreducible $C(\gamma)$-module $\mathcal{S}(\gamma)$ for $\gamma \in LM$ plays the role of
the spinor bundle on loop space LM. We explain in Section 4.4 how the vac-
uum vectors for conformal surfaces lead to a 'conformal connection' of this
spinor bundle. All of Section 5.2 is devoted to discuss the motivation behind
our above definition of a string connection. This definition can also be given in
the language of gerbes with 1- and 2-connection, see e.g. [Bry]. But the gerbe
in question needs to be defined on the total space of the principal Spin(n)-
bundle, restricting to the Chern–Simons gerbe on each fiber. We feel that such
a definition is at least as complicated as ours, and it lacks the beautiful connection to von Neumann algebras and Connes fusion.

5.1 Spin connections and Stiefel–Whitney theory

We first explain a 2-dimensional field theory based on the second Stiefel–
Whitney class. We claim no originality and thus skip most proofs. Stiefel–
Whitney theory is defined on manifolds with the geometric structure (or

classical field) given by an oriented vector bundle with inner product, and hence is a functor

$$SW : \mathcal{B}_2^{SO} \longrightarrow Hilb_\mathbb{R}$$

where $\mathcal{B}_2^{SO}$ is the category explained in Section 2.1, where the geometric structure is an oriented vector bundle with inner product. In the following definitions we could used $\mathbb{Z}/2$ instead of $\mathbb{R}$ as the values, but it will be convenient for further use to stay in the language of (real) Hilbert spaces. We use the embedding $\mathbb{Z}/2 = \{\pm 1\} \subset \mathbb{R}$ and note that these are the numbers of unit length. Note also that a $\mathbb{Z}/2$-torsor is the same thing as a 1-dimensional real Hilbert space, also called a *real line* below.

Definition of Stiefel–Whitney theory. Stiefel–Whitney theory SW associates to a closed geometric 2-manifold $E \to \Sigma$ the second Stiefel–Whitney number

$$SW(E \to \Sigma) \overset{\text{def}}{=} \langle w_2(E), [\Sigma] \rangle \in \mathbb{Z}/2 = \{\pm 1\} \subset \mathbb{R}.$$

To a closed geometric 1-manifold $E \to Y$ it associates the real line

$$SW(E \to Y) \overset{\text{def}}{=} \{(F, r) \mid r \in \mathbb{R}, F \to Y \times I, F|_{Y \times \{0,1\}} = E \cup \underline{\mathbb{R}}^{\dim(E)}\}/ \sim$$

where $\underline{\mathbb{R}}^n$ denotes the trivial bundle and $(F_1, r_1) \sim (F_2, r_2)$ if and only if $w_2(F_1 \cup F_2 \to Y \times S^1) \cdot r_1 = r_2$. If $\partial \Sigma = Y$ and $E' \to \Sigma$ extends $E \to Y$, then the equivalence class of $(F, w_2(E \cup F \cup \underline{\mathbb{R}}^{\dim(E)}))$ is a well-defined element

$$SW(E' \to \Sigma) \in SW(E \to \partial \Sigma).$$

It is independent of the choice of the bundle F by additivity of w_2. This theory by itself is not very interesting but we shall make several variations, and ultimately generalize it to Chern–Simons theory. The first observation is that one can also define the value $SW(E \to Z)$ for a 0-manifold Z. According to the usual field theory formalism we expect that this is a category whose morphism spaces are real lines (which can then be used to calculated the value of the field theory on 1-manifolds). In the spirit of the above definition, we start with vector bundles $F \to Z \times I$ which extend the bundle $E \cup \underline{\mathbb{R}}^{\dim(E)}$ on $Z \times \{0, 1\}$. These are the objects in a category $SW(E \to Z)$ with morphisms defined by

$$Mor(F_1, F_2) \overset{\text{def}}{=} SW(F_1 \cup F_2 \to Z \times S^1) = Mor(F_2, F_1).$$

To complete the description of the theory, we need to associate something to a bundle $E' \to Y$ over a 1-manifold with boundary $Z = \partial Y$ (with restricted bundle $E = E'|_Z$). It should be an 'element' in the category $SW(E \to Z)$ which can then be used to formulate the appropriate gluing laws of the theory. There are various possible interpretations of such an 'element' but in the best case, it would mean an object a in the category. In order to find such an object, we slightly enlarge the above category, allowing as objects not just vector bundles over $Z \times I$ but, more generally, vector bundles over Y with $\partial Y = Z \times \{0, 1\}$. The reader will easily see that this has the desired effect.

Definition 5.9. Stiefel–Whitney theory is the *extended* 2-dimensional field theory described above, where the geometric structure on Y is given by an oriented vector bundle. Here the word *extended* refers to the fact that SW also assigns a small category to 0-manifolds, and objects of this category to 1-manifolds with boundary.

Relative, real Dirac theory. There is an interesting reformulation of the theory which uses the fact that our domain manifolds Σ are equipped with a spin structure and that the bundle E comes with a connection. Enhance for a moment the geometrical structure on Σ by a conformal structure. Then we have the Dirac operator D_Σ, as well as the twisted Dirac operator D_E. If Σ^2 is closed, we get an index in $KO_2 \cong \mathbb{Z}/2$. For a closed conformal 1-manifold Y, the Dirac operator is just covariant differentiation in the spinor bundle from Definition 2.15. Hence it comes equipped with a real Pfaffian line $\mathrm{Pf}(D_Y)$, see Definition 2.23. If $Y = \partial \Sigma$ then the relative index of Σ is an element of unit length in $\mathrm{Pf}(D_Y)$. The same holds for the twisted case. Finally, for a bundle with metric over a 0-manifold, we define the following *relative, real Dirac category*. The objects are Lagrangian subspaces L in $V \perp -\mathbb{R}^n$, where V is again the orthogonal sum of the fibers and n is the dimension of V. These Lagrangians should be thought of as boundary conditions for the Dirac operator on a bundle on $Z \times I$ which restricts to $V \cup \mathbb{R}^n$ on the boundary. In particular, the boundary values of harmonic sections of a bundle E over 1-manifold Y define an object in the category for $E|_{\partial Y} = V_0 \cup V_1$ by rewriting the spaces in question as follows

$$-(V_0 \perp -\mathbb{R}^n) \perp (V_1 \perp -\mathbb{R}^n) = (-V_0 \perp V_1) \perp (\mathbb{R}^n \perp -\mathbb{R}^n).$$

This is in total analogy to the above rewriting of the isometry groups. The morphisms in the category are given by the real lines

$$\mathrm{Mor}(L_1, L_2) \stackrel{\mathrm{def}}{=} \mathrm{Hom}_{C(V)-C_n}(F(L_1), F(L_2))$$

where $F(L_i)$ are the Fock spaces from Definition 2.9. They are irreducible graded bimodules over the Clifford algebras $C(V) - C_n$. Recall from Remark 2.11 that the orientation of V specifies a connected component of such Lagrangians L and we only work in this component.

Lemma 5.10. *There is a canonical isomorphism between the two extended 2-dimensional field theories, Stiefel–Whitney theory and relative, real Dirac theory. For $n = \dim(E)$ this means the following statements in the various dimensions:*

dim 2: $\mathrm{SW}(E \to \Sigma) = \mathrm{index}(D_\Sigma \otimes E) - n \cdot \mathrm{index}(D_\Sigma) \in \mathbb{Z}/2$,

dim 1: $\mathrm{SW}(E \to Y) \cong \mathrm{Pf}(D_Y \otimes E) \otimes \mathrm{Pf}^{-n}(D_Y)$, *such that for $Y = \partial \Sigma$ the element $\mathrm{SW}(E' \to \Sigma)$ is mapped to the relative index.*

dim 0: *For an inner product space V, the category $\mathrm{SW}(V)$ is equivalent to the above relative Dirac category, in a way that the objects defined by 1-manifolds with boundary correspond to each other.*

The extra geometric structure of bundles with connection is needed to define the right-hand side theory, as well as for the isomorphisms above.

Proof. The 2-dimensional statement follows from index theory, and for the 1-dimensional statement one uses the relative index on $Y \times I$. In dimension zero, recall from Remark 2.11 that a Lagrangian subspace L in $V \perp -\mathbb{R}^n$ is given by the graph of a unique isometry $V \to \mathbb{R}^n$. Moreover, parallel transport along a connection gives exactly the Lagrangian of boundary values of harmonic spinors along an interval. $\qquad\square$

Spin structures as trivializations of Stiefel–Whitney theory. Fix a manifold X and an n-dimensional oriented vector bundle $E \to X$ with metric connection. One may restrict the Stiefel–Whitney theory to those bundles (with connection) that are pull-backs of E via a piecewise smooth map $Y \to X$. Thus geometric structures on Y make up the set $\mathrm{Maps}(Y, X)$, and we call the resulting theory SW_E.

Lemma 5.11. *A spin structure on $E \to X$ gives a trivialization of the Stiefel–Whitney theory SW_E in the following sense:*

dim 2: $\mathrm{SW}_E(\Sigma \to X) = 0$ *if Σ is a closed 2-manifold.*

dim 1: $\mathrm{SW}_E(Y \to X)$ *is canonically isomorphic to $\mathbb{R}$ for a closed spin 1-manifold and all elements $\mathrm{SW}_E(\Sigma \to X)$ with $\partial \Sigma = Y$ are mapped to 1.*

dim 0: *The set of objects $\mathrm{ob}(\mathrm{SW}_E(E_x)) = SO(E_x, \mathbb{R}^n)$ of the category for a point $x \in X$ comes with a nontrivial real line bundle ξ and isomorphisms $\mathrm{Mor}(b_1, b_2) \cong \mathrm{Hom}(\xi_{b_1}, \xi_{b_2})$ which are compatible with composition in the category.*

Moreover, the last item is equivalent to the usual definition of a spin structure, and so all the other items follow from it.

Proof. The 2-dimensional statement follows from the fact that $w_2(E) = 0$, and the isomorphism in dimension 1 is induced by the relative second Stiefel–Whitney class. To see why the last item is the usual definition of a spin structure on E_x, recall that the real line bundle is the same information as a double covering $\mathrm{Spin}(E_x, \mathbb{R}^n)$, and that the isomorphisms between the morphism spaces follow from the group structures on $SO(n)$ and $\mathrm{Spin}(n)$. $\qquad\square$

Spin connections as trivializations of relative, real Dirac theory. For the next lemma, we recall from Remark 2.11 that for a inner product space E_x of dimension n, the isometries $O(E_x, \mathbb{R}^n)$ are homeomorphic to the space $\mathcal{L}(x)$ of Lagrangians of $E_x \perp -\mathbb{R}^n$.

Lemma 5.12. *A spin connection $\mathcal{S}$ on an oriented bundle $E \to X$ with metric and connection gives a trivialization of the relative, real Dirac theory on* Maps$(\cdot, X)$ *in the following sense:*

dim 2: index$(D_{f^*E}) = n \cdot$ index$(D_\Sigma) \in \mathbb{Z}/2$ *if Σ is a closed 2-manifold and $f : \Sigma \to X$ is used to twist the Dirac operator on Σ by E.*

dim 1: *For $f : Y \to X$, Y a closed spin 1-manifold, there is an isomorphism $\mathcal{S}(f) : \mathrm{Pf}(f^*E) \cong \mathrm{Pf}^n(Y)$, taking twisted to untwisted indices of Dirac operators of surfaces Σ with $\partial\Sigma = Y$.*

dim 0: *For each $x \in X$, there is a graded irreducible $C(E_x) - C_n$ bimodule $\mathcal{S}(x) = S(E_x)$ which gives a nontrivial line bundle over the connected component of $\mathcal{L}(x)$ (given by the orientation of E_x). Here the line over a Lagrangian $L \in \mathcal{L}(x)$ is $\mathrm{Hom}_{C(x)}(\mathcal{S}(x), F(L))$, where $C(x) = C(E_x) \otimes C_{-n}$ (and hence $\mathcal{S}(x)$ is a left $C(x)$-module). Moreover, for each path γ from x_1 to x_2, the spin structure on E induces an isomorphism between the following two left modules over $C(x_1)^{\mathrm{op}} \otimes C(x_2)$:*

$$\mathcal{S}(\gamma) : F(\gamma) \xrightarrow{\cong} \mathrm{Hom}_{\mathbb{R}}(\mathcal{S}(x_1), \mathcal{S}(x_2))$$

where $F(\gamma)$ is the relative Fock module from Definition 2.35. When two paths are composed along one point, then the gluing laws from Proposition 5.5 hold.

Note that the bimodules $\mathcal{S}(x) = S(E_x)$ fit together to give the C_n-linear spinor bundle $S(E)$, so we have finally motivated our Definition 2.15 of spin structures. The vacuum vectors in the Fock modules $F(\gamma^*E)$ define a parallel transport in $S(E)$. It is interesting to note that these vacuum vectors exist even for an oriented vector bundle E (with metric and connection) but it is the spin structure in the sense above which makes it possible to view them as a parallel transport in the spinor bundle.

Proof. The result follows directly from Lemmas 5.10 and 5.11. In dimension 0 one defines $S(E_x)$ in the following way: For a given Lagrangian L we have a Fock space $F(L)$ but also the line ξ_L from Lemma 5.11 (since L is the graph of a unique isometry). Moreover, given two Lagrangians L_i we have given isomorphisms

$$F(L_1) \otimes \xi_{L_1} \cong F(L_2) \otimes \xi_{L_2}$$

which are associative with respect to a third Lagrangian. Therefore, we may define $S(E_x)$ as the direct limit of this system of bimodules. Note that $S(E_x)$ is then canonically isomorphic to each bimodule of the form $F(L) \otimes \xi_L$ and so one can recover the line bundle ξ from $S(E_x)$. In fact, the bimodule and the line bundle ξ contain the exact same information. $\square$

5.2 String connections and Chern–Simons theory

We want to explain the steps analogous to the ones in the previous section with $w_2 \in H^2(BSO(n); \mathbb{Z}/2)$ replaced by a 'level' $\ell \in H^4(BG; \mathbb{Z})$. The most interesting case for us is the generator $p_1/2$ of $H^4(B\,\mathrm{Spin}(n); \mathbb{Z})$ which will lead to string structures. The analogue of Stiefel–Whitney theory is (classical) Chern–Simons theory which we briefly recall, following [Fr1]. We shall restrict to the case where the domain manifolds are spin as this is the only case we need for our applications.

Let G be a compact Lie group and fix a level $\ell \in H^4(BG; \mathbb{Z})$. For $d = 0, \ldots, 4$ we consider compact d-dimensional spin manifolds M^d together with connections $\mathfrak{a}$ on a G-principal bundle $E \to M$. The easiest invariant is defined for a closed 4-manifold M^4, and is given by the characteristic number $\langle c_E^*(\ell), [M] \rangle \in \mathbb{Z}$, where $c_E : M \to BG$ is a classifying map for E. It is independent of the connection and one might be tempted to view it as the analog of $\mathrm{SW}(E \to \Sigma) \in \mathbb{Z}/2$ of a closed surface Σ. However, this is not quite the right point of view. In fact, Chern–Simons theory is a 3-dimensional field theory

$$\mathrm{CS} = \mathrm{CS}_\ell : \mathcal{B}_3^G \longrightarrow \mathrm{Hilb}_\mathbb{C}$$

in the sense of Section 2.1, with geometric structure being given by G-bundles with connection. The value $\mathrm{CS}(M^3, \mathfrak{a}) \in S^1$ for a closed 3-manifold is obtained by extending the bundle and connection over a 4-manifold W with boundary M, and then integrating the Chern–Weil representative of ℓ over W. By the integrality of ℓ on closed 4-manifolds, it follows that one gets a well-defined invariant in $S^1 = \mathbb{R}/\mathbb{Z}$, viewed as the unit circle in $\mathbb{C}$ (just like $\mathbb{Z}/2 = \{\pm 1\}$ was the unit circle in $\mathbb{R}$). Thus we think of this *Chern–Simons invariant* as the analogue of $\mathrm{SW}(E \to \Sigma)$. One can then use the tautological definitions explained in the previous sections to get the following values for the invariant $CS(M^d, \mathfrak{a})$, leading to an *extended* 3-dimensional field theory.

d	M^d closed	$\partial M^d \neq \emptyset$
4	element in $\mathbb{Z}$	element in $\mathbb{R}$ reducing to invariant of ∂M
3	element in S^1	point in the hermitian line for ∂M
2	hermitian line	object in the $\mathbb{C}$-category for ∂M
1	$\mathbb{C}$-category	

By a $\mathbb{C}$-category we mean a category where all morphism spaces are hermitian lines. In Stiefel–Whitney theory we associated an $\mathbb{R}$-category to 0-manifolds. So $\mathbb{R}$ has been replaced by $\mathbb{C}$ and all dimensions have moved up by one. It will be crucial to understand the 0-dimensional case in Chern–Simons theory, where von Neumann algebras enter the picture.

Relative, complex Dirac theory. First we stick to dimensions 1 to 4 as above and explain the relation to Dirac operators.

Theorem 5.13. *For $G = \mathrm{Spin}(n)$ at level $\ell = p_1/2$, the above extended Chern–Simons theory is canonically isomorphic to relative, complex Dirac theory.*

In Dirac theory one has conformal structures on the spin manifolds M, which enables one to define the Dirac operator D_M, as well as the twisted Dirac operator $D_{\mathfrak{a}}$. Here we use the fundamental representation of $\mathrm{Spin}(n)$ to translate a principal $\mathrm{Spin}(n)$bundle into a spin vector bundle, including the connections $\mathfrak{a}$. In the various dimensions $d = 1, \ldots, 4$, relative, complex Dirac theory is given by the following table of classical actions. It is a (well-known) consequence of our theorem that the relative theory is metric independent. Let M be a closed d-manifold and E an n-dimensional vector bundle E over M with connection $\mathfrak{a}$.

d	$\mathfrak{D}(M^d, \mathfrak{a}) \overset{\text{def}}{=}$
4	$\mathrm{index}^{\mathrm{rel}}(M, \mathfrak{a}) \overset{\text{def}}{=} \frac{1}{2}\mathrm{index}(D_{\mathfrak{a}}) - \frac{n}{2}\mathrm{index}(D_M) \in \mathbb{Z}$
3	$\eta^{\mathrm{rel}}(M^3, \mathfrak{a}) \in S^1$
2	$\mathrm{Pf}^{\mathrm{rel}}(M^2, \mathfrak{a})$, a hermitian line
1	$[F^{\mathrm{rel}}(M^1, \mathfrak{a})]$, a $\mathbb{C}$-category of representations

Proof of Theorem 5.13. The statement in dimension 4 follows from the index theorem (see below) which implies that the relative index in the above table equals the characteristic class $\langle p_1(E)/2, [M^4]\rangle$ on closed 4-manifolds. In dimension 3, we first need to explain the invariant η^{rel}. It is one half of the reduced η-invariant which shows up in the Atiyah–Patodi–Singer index theorem for 4-manifolds with boundary (where we are using the Dirac operator twisted by the virtual bundle $E \oplus -\mathbb{R}^n$)

$$\text{index}(D_{M^4, \mathfrak{a}}) - n \cdot \text{index}(D_M) = \int_M \hat{A}(M)\widetilde{ch}(E \otimes \mathbb{C}, \mathfrak{a}) - \tilde{\eta}(\partial M, \mathfrak{a}) \in \mathbb{R}$$

Both indices above are even-dimensional because of a quaternion structure on the bundles (coming from the fact that the Clifford algebra C_4 is of quaternion type). Applying this observation together with the fact that the Chern character in degree 4 is given by $p_1(E \otimes \mathbb{C})/2 = p_1(E)$ one gets

$$\int_M p_1(E, \mathfrak{a}) \equiv \tilde{\eta}(\partial M, \mathfrak{a}) \qquad \text{mod } 2\mathbb{Z}.$$

Since we are assuming that E is a spin bundle, the left-hand side is an even integer for closed M. Therefore, we may divide both sides by 2 to obtain a well-defined invariant $\eta^{rel}(\partial M, \mathfrak{a})$ in $\mathbb{R}/\mathbb{Z}$ which equals $CS(\partial M, \mathfrak{a})$.

In dimension 2, one needs to understand the Pfaffian line of the skew-adjoint operator $D_{\mathfrak{a}}^+$, as well as the corresponding relative Pfaffian line

$$\text{Pf}^{rel}(M, \mathfrak{a}) = \overline{\text{Pf}(\mathfrak{a})} \otimes \text{Pf}(D_M)^{\otimes n}$$

in the above table. The main point is that the relative η-invariant above can be extended to 3-manifolds with boundary so that it takes values in this relative Pfaffian line. Therefore, one can define an isomorphism of hermitian lines $CS(M^2, \mathfrak{a}) \to \text{Pf}^{rel}(M^2, \mathfrak{a})$ by associating this relative η-invariant to a connection on $M^2 \times I$ (extending $\mathfrak{a}$ respectively the trivial connection).

Finally, for a closed 1-manifold, $[F^{rel}(M^1, \mathfrak{a})]$ is the isomorphism class of twisted Fock spaces explained in Definition 2.23. They can be defined from harmonic boundary values of twisted Dirac operators on $M \times I$. The isomorphism class of the bimodule does not depend on the extension of bundle and connection to $M \times I$. Each of these Fock spaces is a complex graded irreducible representation of the Clifford algebra

$$C^{rel}(M, \mathfrak{a}) = C(E, \mathfrak{a})^{op} \otimes C(M)^{\otimes \dim(E)}$$

the latter replacing $C_n = C(\text{pt})^{\otimes n}$ from Stiefel–Whitney theory. Given the isomorphism class of such a bimodule, there is an associated $\mathbb{C}$-category whose objects are actual representations in this isomorphism type, and whose morphisms are intertwiners. The equivalence of categories from $CS(M, \mathfrak{a})$ to the

$\mathbb{C}$-category defined by $[F^{\text{rel}}(M, \mathfrak{a})]$ is on objects given by sending a connection on $M^1 \times I$ (extending $\mathfrak{a}$ respectively the trivial connection) to the Fock space defined from harmonic boundary values of twisted Dirac operators on $M \times I$. By definition, this is an object in the correct category. To define the functor on morphisms, one uses the canonical isomorphism

$$\text{Pf}^{\text{rel}}(M^1 \times S^1, \mathfrak{a}) \cong \text{Hom}_{C^{\text{rel}}(M \times S^1, \mathfrak{a})}(F^{\text{rel}}(M \times I, \mathfrak{a}_0), F^{\text{rel}}(M \times I, \mathfrak{a}_1))$$

where $\mathfrak{a}$ is a connection on a bundle over $M \times S^1$ which is obtained by gluing together two connections $\mathfrak{a}_0, \mathfrak{a}_1$ on $M \times I$. $\qquad\square$

5.3 Extending Chern–Simons theory to points

Fix a compact, simply connected, Lie group G and a level $\ell \in H^4(BG)$. Recall from Theorem 5.1 that there is a von Neumann algebra $A = A_{G,\ell}$ and a G-kernel (see Remark 5.23)

$$\widetilde{\Phi} : G \longrightarrow \text{Out}(A) \overset{\text{def}}{=} \text{Aut}(A)/\text{Inn}(A)$$

canonically associated to (G, ℓ). Moreover, $\widetilde{\Phi}$ defines the extension G_ℓ of G by $PU(A) = \text{Inn}(A)$ and lifts to a monomorphism $\Phi : G_\ell \to \text{Aut}(A)$. We want to use these data to define the Chern–Simons invariant of a point.

On a G-bundle V over a point, we first pick a G_ℓ-structure. Recall from Definition 5.4 that this is an algebra A_V together with a G-equivariant map

$$\alpha_V : \text{Iso}_G(G, V) \longrightarrow \text{Out}(A, A_V) \overset{\text{def}}{=} \text{Iso}(A, A_V)/\text{Inn}(A).$$

It turns out that the $\text{Out}(A)$-torsor $\text{Out}(A, A_V)$ is actually defined independently of the choice of such a G_ℓ-structure.

Definition 5.14. We define $CS(V)$ to be the $\text{Out}(A)$-torsor $\text{Out}(A, A_V)$.

The above independence argument really shows that there is a map

$$\text{Iso}_G(V_1, V_2) \longrightarrow \text{Out}(A_{V_1}, A_{V_2})$$

which is well defined without choosing G_ℓ-structures. In particular, without knowing what the algebras A_{V_i} really are. When applying this map to the parallel transport of a G-connection $\mathfrak{a}$ an interval I, we get the value $CS(I, \mathfrak{a})$.

To motivate why this definition really extends Chern–Simons theory to points, we propose a whole new picture of the theory.

Chern–Simons theory revisited. We propose a rigidified picture of the Chern–Simons actions motivated by our definition in dimension 0. For a given (G, ℓ), the Chern–Simons invariant $CS(M^d, \mathfrak{a})$ for connected d-manifolds would then take values in mathematical objects listed in the table below. Note that the values for closed spin manifolds are special cases of manifolds with boundary, i.e. the entries in the middle column are subsets of the entries in the right-hand column.

d	M^d closed	$\partial M^d \neq \emptyset$
4	$\mathbb{Z}$	$\mathbb{R}$
3	S^1	$U(A)$
2	$PU(A)$	$\text{Aut}(A)$
1	$\text{Out}(A)$	$\text{Out}(A)$-equivariant maps
0	space of $\text{Out}(A)$-torsors	

Here $A = A_{G,\ell}$ is the von Neumann algebra discussed in the previous section. The guiding principle in the table above is that for closed connected d-manifolds M, the Chern–Simons invariant $CS(M^d, \mathfrak{a})$ should be a point in a particular version of an Eilenberg-MacLane space $K(\mathbb{Z}, 4 - d)$ (whereas for manifolds M with boundary one gets a point in the corresponding contractible space). More precisely, for $K(\mathbb{Z}, 4 - d)$ we used the models

$$\mathbb{Z}, \quad S^1 = \mathbb{R}/\mathbb{Z}, \quad \text{Inn}(A) \cong PU(A) = U(A)/S^1,$$

$$\text{Out}(A) = \text{Aut}(A)/\text{Inn}(A)$$

for $d = 4, 3, 2, 1$. Our model of a $K(\mathbb{Z}, 4)$ is the space of $\text{Out}(A)$-torsors. This is only a conjectural picture of classical Chern–Simons theory but it should be clear why it rigidifies the definitions in Section 5.2: Every point in $PU(A)$ defines a hermitian line via the S^1-torsor of inverse images in $U(A)$. Moreover, every point in $g \in \text{Out}(A)$ defines an isomorphism class of $[F_g]$ of $A - A$-bimodules by twisting the standard bimodule $L^2(A)$ by an automorphism in $\text{Aut}(A)$ lying above g. This defines the $\mathbb{C}$-category of $A - A$-bimodules isomorphic to $[F_g]$.

Remark 5.16. The homomorphism $\tilde{\Phi} : G \longrightarrow \text{Out}(A)$ should be viewed as follows: An element in G gives a G-bundle with connection $\mathfrak{a}_g$ on S^1 via the clutching construction. Then $\tilde{\Phi}(g) = CS(S^1, \mathfrak{a}_g)$. In the previous section we explained a similar construction which gives $CS(I, \mathfrak{a})$, and also the value of CS on points. Thus we have not explained the definition of the rigidified Chern–Simons invariant only for surfaces.

We will not seriously need the new picture of Chern–Simons theory in the following because we really only want to explain what a 'trivialization' it is. But we do spell out the basic results which are necessary to make this picture precise. Since we could not find these statements in the literature (only the analogous results for II_1-factors were known before), we originally formulated them as conjectures. We recently learned from Antony Wassermann that they are also true in the III_1-context. His argument for the contractibility of $U(A)$ is a variation of the argument for type I given in [DD], for $\text{Aut}(A)$ he reduces the problem to the type II case.

Theorem 5.17 (Wassermann). *If A is a hyperfinite type III_1-factor then the unitary group $U(A)$ is contractible in the weak (or equivalently strong) operator topology. Moreover, the automorphism group $\text{Aut}(A)$ is also contractible in the topology of pointwise norm convergence in the predual of A.*

It follows that $PU(A) = U(A)/\mathbb{T}$ is a $K(\mathbb{Z}, 2)$. One has to be more careful with the topology on $\text{Out}(A) = \text{Aut}(A)/PU(A)$ because with the quotient topology this is not a Hausdorff space (using the above topologies, $PU(A)$ is *not* closed in $\text{Aut}(A)$). A possible strategy could be to *define* a continuous map $X \to \text{Out}(A)$ to be any old map but together with local continuous sections to $\text{Aut}(A)$.

String connections as trivializations of Chern–Simons theory. Let $E \to X$ be a principal G-bundle with connection. We get a Chern–Simons theory for E by restricting to those bundles with connection on spin manifolds M^d which come from piecewise smooth maps $M \to X$ via pullback. Thus the new geometric structures on M are $\text{Maps}(M, X)$ and we get the Chern–Simons theory CS_E.

Definition 5.18. Let G_ℓ be the group extension of G at level ℓ constructed in Section 5.4. Then a *geometric G_ℓ-structure* $\mathcal{S}$ on E is a trivialization of the extended Chern–Simons theory CS_E. For a closed spin manifold M in dimension d this amounts to the following 'lifts' of the Chern–Simons action on piecewise smooth maps $f : M^d \to X$:

d	values of $\mathcal{S}(f) = \mathcal{S}(f : M^d \to X)$
4	the equation $CS(f) = 0$, no extra structure!
3	$\mathcal{S}(f) \in \mathbb{R}$ reduces to $CS(f) \in \mathbb{R}/\mathbb{Z}$
2	$\mathcal{S}(f)$ is a point in the line $CS(f)$
1	$\mathcal{S}(f)$ is an object in the $\mathbb{C}$-category $CS(f)$
0	for $x \in X$, $\mathcal{S}(x)$ is a $G_\ell - E_x$-pointed factor

The last line uses Definition 5.4. There are also data associated for manifolds M with boundary, and these data must fit together when gluing manifolds and connections. Note that for $d \leq 3$, $\mathcal{S}(f : M^d \to X)$ takes values in the same objects as $\mathrm{CS}(F)$ if $F : W^{d+1} \to X$ extends f, i.e. $\partial W = M$. By construction, they both project to the same point in the corresponding quotient given by $\mathrm{CS}(f)$. The geometric G_ℓ-structure on $F : W^{d+1} \to X$ gives by definition a point in this latter group. For example, if $d = 2$ then $\mathcal{S}(F)$ is the element of S^1 such that

$$\mathcal{S}(F) \cdot \mathcal{S}(f) = \mathrm{CS}(F) \in \mathrm{CS}(f).$$

Finally, we assume that these data fit together to give bundles (respectively sections in these bundles) over the relevant mapping spaces.

Note that the data associated to points combine exactly to a G_ℓ-structure on E as explained in Definition 5.2. Thus a geometric G_ℓ-structure has an underlying (topological) G_ℓ-structure.

Theorem 5.19. *Every principal G-bundle with G_ℓ-structure admits a geometric G_ℓ-structure, unique up to isomorphism.*

In fact, the 'space' of geometric G_ℓ-structures is probably contractible. The proof of this theorem will appear elsewhere but it is important to note that the construction uses a 'thickening' procedure at every level, i.e. one crosses all manifolds M^d with I and extends the bundle f^*E with connection over $M \times I$ in a way that it restricts to the trivial bundle on the other end. So one seriously has to use the fact that all the structures explained above are really 'relative', i.e. twisted tensor untwisted structures.

In the case $G = \mathrm{Spin}(n)$ and $\ell = p_1/2$ we need a more geometric interpretation. This is given by the following result which incorporates Definition 5.8. There, a geometric $\mathrm{String}(n)_{p_1/2}$ structure was called a *string connection* and we stick to this name.

Corollary 5.20. *Given an n-dimensional vector bundle $E \to X$ with spin connection. Then a string connection $\mathcal{S}$ on E induces the following data for closed conformal spin manifolds M^d. In the table below, $D_{M,f}$ is the conformal Dirac operator twisted by $f^*(E)$ for a piecewise smooth map $f : M \to X$ and the data fit together to give bundles (respectively sections in these bundles) over the relevant mapping spaces.*

d	values of $\mathcal{S}(f) = \mathcal{S}(f : M^d \to X)$
4	the equation index$(D_{f^*E}) = n$ index(D_M), no extra structure!
3	$\mathcal{S}(f) \in \mathbb{R}$ reduces to $\eta^{\mathrm{rel}}(M, f^*E) \in \mathbb{R}/\mathbb{Z}$
2	an isomorphism $\mathcal{S}(f) : \mathrm{Pf}(f^*E) \cong \mathrm{Pf}(M)^{\otimes n}$
1	a representation $\mathcal{S}(f)$ isomorphic to $[F(f)]$
0	for $x \in X$, $\mathcal{S}(x)$ is a String$(n) - E_x$-pointed factor

Again, the last line uses Definition 5.4 and the data in dimension 0 give exactly a string structure on E.

The precise gluing conditions in dimensions 0, 1, 2 were explained in Definition 5.8 and that is all we shall need. The main point is that the von Neumann algebras for 0-manifolds can be used to decompose the representations of closed 1-manifolds into the Connes fusion of bimodules. That is the locality condition we need for our purposes of constructing a cohomology theory in the end. Note that by Theorem 5.19 such string connections exist and are up to isomorphism determined by the topological datum of a string structure.

Remark 5.21. In Section 5.2 we have not taken care of the actions of diffeomorphisms of d-manifolds, $d = 0, 1, 2, 3, 4$. This is certainly necessary if one wants the correct theory and we have formulated the precise conditions only in Definition 5.8 (which is important for elliptic objects). However, we felt that the theory just presented is complicated enough as it stands and that the interested reader will be able to fill the gaps if necessary.

5.4 Type III$_1$-factors and compact Lie groups

In this section we discuss canonical extensions of topological groups

$$1 \longrightarrow PU(A_\rho) \longrightarrow G_\rho \longrightarrow G \longrightarrow 1 \qquad (5.6)$$

one for each projective unitary representation ρ of the loop group LG of a Lie group G. The above extensions were first found for $G = \mathrm{Spin}(n)$ and ρ the positive energy vacuum representation at level $\ell = p_1/2$. We used 'local fermions' in the construction, and arrived at the groups String$(n) = G_\rho$. Antony Wassermann explained to us the more general construction (in terms of 'local loops') which we shall discuss below.

In the extension above, A_ρ is a certain von Neumann algebra, the 'local loop algebra', and one can form the projective unitary group $PU(A_\rho) = U(A_\rho)/\mathbb{T}$. If $U(A_\rho)$ is contractible, the projective group has the homotopy type of a

$K(\mathbb{Z}, 2)$. In that case one gets a boundary map

$$\pi_3 G \longrightarrow \pi_2 PU(A_\rho) \cong \mathbb{Z}$$

which we call the *level* of ρ. In the special case where G is compact and ρ is the vacuum representation of LG at level $\ell \in H^4(BG)$, this leads to an extension $G_\ell \to G$ which was used in Theorem 5.1. By Wassermann's Theorem 5.17, the unitary group is contractible in this case.

Lemma 5.22. *If G is simply connected and compact, then the two notions of level above agree in the sense that*

$$\ell \in H^4(BG) \cong \text{Hom}(\pi_3 G, \mathbb{Z})$$

gives the boundary map $\pi_3 G \to \pi_2 PU(A_\rho) \cong \mathbb{Z}$ in extension (5.6) if ρ is the positive energy vacuum representation of LG at level ℓ.

The proof is given at the end of this section. It is interesting to remark that the 'local equivalence' result in [Wa, p. 502] implies that the construction leads to canonically isomorphic algebras A_ρ and groups G_ρ if one uses any other positive energy representation of LG at the same level ℓ.

Remark 5.23. The extension (5.6) is constructed as a pullback from a homomorphism $G \to \text{Out}(A_\rho)$. Such homomorphisms are also called *G-kernels* and they were first studied by Connes in [Co2]. He showed that for G a finite cyclic group, G-kernels into the hyperfinite II_1 factor are classified (up to conjugation) by an obstruction in $H^3(G; \mathbb{T}) \cong H^4(BG)$. This result was extended in Jones' thesis to arbitrary finite groups [J1]. In a sense, the above construction is an extension of this theory to compact groups (and hyperfinite III_1 factors). More precisely, Wassermann pointed out that the extensions 5.6 are extensions of Polish groups and by a general theorem have therefore Borel sections. There is then an obstruction cocycle in C. Moore's [Mo] third *Borel* cohomology of G which measures the nontriviality of the extension. By a result of D. Wigner [Wig], one in fact has $H^4(BG) \cong H^3_{\text{Borel}}(G; \mathbb{T})$. In the simply connected case (and for tori), Wassermann has checked that the obstruction cocycle in Borel cohomology actually agrees with the level $\ell \in H^4(BG)$. This lead Wassermann to a similar classification as for finite groups, using the unique minimal action (cf. [PW]) of the constant loops on A_ρ.

For our applications to homotopy theory, this Borel cocycle is not as important as the boundary map on homotopy groups in Lemma 5.22. However, it might be an important tool in the understanding of non-simply connected groups because the isomorphism $H^4(BG) \cong H^3_{\text{Borel}}(G; \mathbb{T})$ continues to hold for all compact Lie groups (even non-connected).

Remark 5.24. One drawback with this more general construction is that the von Neumann algebras A_ρ are not graded, whereas our original construction in terms of local fermions gives graded algebras via the usual grading of Clifford algebras. Whenever such a grading is needed, we shall revert freely to this other construction.

Remark 5.25. The 'free loop group' LG is the group consisting of all piecewise smooth (and continuous) loops. The important fact is that the theory of positive energy representations of smooth loop groups extends to these larger groups (cf. [PS] and [J4]).

Let ρ be a projective unitary representation of LG, i.e., a continuous homomorphism $\rho: LG \to PU(H)$ from LG to the projective unitary group $PU(H) \stackrel{\text{def}}{=} U(H)/\mathbb{T}$ of some complex Hilbert space H. This group carries the quotient topology of the weak (or equivalently strong) operator topology on $U(H)$. Note that by definition, we are assuming that ρ is defined for all piecewise smooth loops in G. Pulling back the canonical circle group extension

$$1 \longrightarrow \mathbb{T} \longrightarrow U(H) \longrightarrow PU(H) \longrightarrow 1$$

via ρ, we obtain an extension $\mathbb{T} \longrightarrow \tilde{L}G \longrightarrow LG$, and a unitary representation $\tilde{\rho}: \tilde{L}G \to U(H)$.

Let $I \subset S^1$ be the upper semi-circle consisting of all $z \in S^1$ with nonnegative imaginary part. Let $L_I G \subset LG$ be the subgroup consisting of those loops $\gamma: S^1 \to G$ with support in I (i.e., $\gamma(z)$ is the identity element of G for $z \notin I$). Let $\tilde{L}_I G < \tilde{L}G$ be the preimage of $L_I G$. Define

$$A_\rho \stackrel{\text{def}}{=} \tilde{\rho}(\tilde{L}_I G)'' \subset B(H)$$

to be the von Neumann algebra generated by the operators $\tilde{\rho}(\gamma)$ with $\gamma \in \tilde{L}_I G$. Recall that von Neumann's double commutant theorem implies that this is precisely the weak operator closure (in the algebra $B(H)$ of all bounded operators on H) of linear combinations of group elements $\tilde{L}_I G$.

To construct the group extension (5.6) we start with the group extension

$$1 \longrightarrow L_I G \longrightarrow P_{\mathbb{1}}^I G \longrightarrow G \longrightarrow 1 \tag{5.7}$$

where $P_{\mathbb{1}}^I G = \{\gamma: I \to G \mid \gamma(1) = \mathbb{1}\}$, the left map is given by restriction to $I \subset S^1$ (alternatively we can think of $L_I G$ as maps $\gamma: I \to G$ with

$\gamma(1) = \gamma(-1) = \mathbb{1}$), and the right map is given by evaluation at $z = -1$. The idea is to modify this extension by replacing the normal subgroup $L_I G$ by the projective unitary group $PU(A_\rho)$ of the von Neumann algebra A_ρ (the unitary group $U(A_\rho) \subset A_\rho$ consists of all $a \in A_\rho$ with $aa^* = a^*a = 1$), using the homomorphism

$$\rho : L_I G \longrightarrow PU(A_\rho)$$

given by restricting the representation ρ to $L_I G \subset LG$. We note that by definition of $A_\rho \subset B(H)$, we have $\rho(L_I G) \subset PU(A_\rho) \subset PU(H)$.

We next observe that $P_{\mathbb{1}}^I G$ acts on $L_I G$ by conjugation and that this action extends to a left action on $PU(A_\rho)$. In fact, this action exists for the group $P^I G$ of all piecewise smooth path $I \to G$ (of which $P_{\mathbb{1}}^I G$ is a subgroup): to describe how $\delta \in P^I G$ acts on $PU(A_\rho)$, extend $\delta : I \to G$ to a piecewise smooth loop $\gamma : S^1 \to G$ and pick a lift $\tilde{\gamma} \in \tilde{L}G$ of $\gamma \in LG$. We decree that $\delta \in P^I G$ acts on $PU(A_\rho)$ via

$$[a] \mapsto [\tilde{\rho}(\tilde{\gamma}) a \tilde{\rho}(\tilde{\gamma}^{-1})].$$

Here $a \in U(A_\rho) \subset B(H)$ is a representative for $[a] \in PU(A_\rho)$. It is clear that $\tilde{\rho}(\tilde{\gamma}) a \tilde{\rho}(\tilde{\gamma}^{-1})$ is a unitary element in $B(H)$; to see that it is in fact in A_ρ, we may assume that a is of the form $a = \tilde{\rho}(\tilde{\gamma}_0)$ for some $\tilde{\gamma}_0 \in \tilde{L}_I G$ (these elements generate A_ρ as von Neumann algebra). Then $\tilde{\rho}(\tilde{\gamma}) a \tilde{\rho}(\tilde{\gamma}^{-1}) = \tilde{\rho}(\tilde{\gamma} \tilde{\gamma}_0 \tilde{\gamma}^{-1})$, which shows that this element is in fact in A_ρ and that it is independent of how we extend the path $\delta : I \to G$ to a loop $\gamma : S^1 \to G$, since $\gamma_0(z) = 1$ for $z \notin I$.

Lemma 5.26. *With the above left action of $P^I G$ on $PU(A_\rho)$, the representation $\rho : L_I G \to PU(A_\rho)$ is $P^I G$-equivariant. Therefore, there is a well-defined monomorphism*

$$r : L_I G \longrightarrow PU(A_\rho) \rtimes P^I G, \quad r(\gamma) \stackrel{\text{def}}{=} (\rho(\gamma^{-1}), \gamma)$$

into the semi-direct product, whose image is a normal subgroup.

Before giving the proof of this Lemma, we note that writing the semi-direct product in the order given, one indeed needs a *left* action of the right-hand group on the left-hand group. This follows from the equality

$$(u_1 g_1)(u_2 g_2) = u_1(g_1 u_2 g_1^{-1}) g_1 g_2$$

because $u \mapsto g u g^{-1}$ is a left action on $u \in U$.

Proof. The first statement is obvious from our definition of the action on $PU(A_\rho)$. To check that r is a homomorphism, we compute

$$
\begin{aligned}
r(\gamma_1)r(\gamma_2) &= (\rho(\gamma_1^{-1}), \gamma_1)(\rho(\gamma_2^{-1}), \gamma_2) \\
&= (\rho(\gamma_1^{-1})[\rho(\gamma_1)\rho(\gamma_2^{-1})\rho(\gamma_1^{-1})], \gamma_1\gamma_2) \\
&= (\rho(\gamma_2^{-1})\rho(\gamma_1^{-1}), \gamma_1\gamma_2) = (\rho(\gamma_1\gamma_2)^{-1}, \gamma_1\gamma_2) \\
&= r(\gamma_1\gamma_2).
\end{aligned}
$$

To check that the image of r is normal, it suffices to check invariance under the two subgroups $PU(A_\rho)$ and $P^I G$. For the latter, invariance follows directly from the $P^I G$-equivariance of ρ. For the former, we check

$$
\begin{aligned}
(u^{-1}, 1)(\rho(\gamma^{-1}), \gamma)(u, 1) &= (u^{-1}\rho(\gamma^{-1}), \gamma)(u, 1) \\
&= (u^{-1}\rho(\gamma^{-1})\rho(\gamma)u\rho(\gamma)^{-1}, \gamma) \\
&= (r(\gamma^{-1}), \gamma)
\end{aligned}
$$

This actually shows that the two subgroups $r(L_I G)$ and $PU(A_\rho)$ commute in the semi-direct product group. Finally, projecting to the second factor $P^I G$ one sees that r is injective. $\qquad\square$

Definition 5.27. We define the group G_ρ to be the quotient of $PU(A_\rho) \rtimes P^I_{\mathbb{1}} G$ by the normal subgroup $r(L_I G)$, in short

$$
G_\rho \overset{\text{def}}{=} PU(A_\rho) \rtimes_{L_I G} P^I_{\mathbb{1}} G.
$$

Then there is a projection on to G by sending $[u, \gamma]$ to $\gamma(-1)$ which has kernel $PU(A_\rho)$. This gives the extension in 5.6.

The representation of G_ρ into $\mathrm{Aut}(A_\rho)$. We observe that there is a group extension

$$
G_\rho \longrightarrow PU(A_\rho) \rtimes_{L_I G} P^I G \longrightarrow G
$$

where the right-hand map sends $[u, \gamma]$ to $\gamma(1)$. This extension splits because we can map g to $[\mathbb{1}, \gamma(g)]$, where $\gamma(g)$ is the constant path with value g. This implies the isomorphism

$$
G_\rho \rtimes G \cong PU(A_\rho) \rtimes_{L_I G} P^I G
$$

with the action of G on G_ρ defined by the previous split extension. Note that after projecting G_ρ to G this action becomes the conjugation action of G on G because the splitting used constant paths.

Lemma 5.28. *There is a homomorphism*

$$\Phi : PU(A_\rho) \rtimes_{L_I G} P^I G \longrightarrow \text{Aut}(A_\rho) \quad \Phi([u], \gamma) \overset{\text{def}}{=} c_u \circ \phi(\gamma)$$

where c_u is conjugation by $u \in U(A_\rho)$ and $\phi(\gamma)$ is the previously defined action of $P^I G$ on A_ρ (which was so far only used for its induced action on $PU(A_\rho)$).

Proof. The statement follows (by calculations very similar to the ones given above) from the fact that

$$\phi(\gamma) \circ c_u = c_{\rho(\gamma) u \rho(\gamma)^{-1}} \circ \phi(\gamma).$$

$\square$

We summarize the above results as follows.

Proposition 5.29. *There is a homomorphism $\Phi : G_\rho \rtimes G \longrightarrow \text{Aut}(A_\rho)$ which reduces to the conjugation action $PU(A_\rho) \twoheadrightarrow \text{Inn}(A_\rho) \subset \text{Aut}(A_\rho)$ on*

$$PU(A_\rho) = \ker(G_\rho \longrightarrow G) = \ker(G_\rho \rtimes G \longrightarrow G \rtimes G).$$

The action of G on G in the right-hand semi-direct product is given by conjugation. This implies that the correct way to think about the homomorphism Φ is as follows: It is a homomorphism $\Phi_0 : G_\rho \to \text{Aut}(A_\rho)$, together with a lift to $\text{Aut}(A_\rho)$ of the conjugation action of G on $\text{Out}(A_\rho)$ (which is given via $\widetilde{\Phi}_0 : G \to \text{Out}(A_\rho)$).

Proof of Lemma 5.22. Since $P_{\parallel}^I G$ is contractible, the boundary maps in extension 5.7 are isomorphisms. Therefore, we need to show that $\rho_* : \pi_2 L_I G \to \pi_2 PU(A_\rho)$ is the same map as the level $\ell \in H^4(BG)$. If G is simply connected the latter can be expressed as the induced map $\rho_* : \pi_2 LG \to \pi_2 PU(H)$. Note that we use the same letter ρ for the original representation $\rho : LG \to PU(H)$ as well as for its restriction to $L_I G$. Now the inclusion $L_I G \hookrightarrow LG$ induces an isomorphism on π_2 and so does the inclusion $PU(A_\rho) \hookrightarrow PU(H)$. For the latter one has to know that $U(A_\rho)$ is contractible by Theorem 5.17 (which is well known for $U(H)$). Putting this information together, one gets the claim of our lemma. $\square$

References

[AHS]   Ando, M., M. Hopkins and N. Strickland, *Elliptic spectra, the Witten genus and the theorem of the cube.* Invent. Math. 146 (2001) 595–687.

[A]   Araki, H., *Bogoliubov automorphisms and Fock representations of canonical anticommutation relations.* Contemp. Math., 62, AMS, 1987.

[Ba]   Barron, K., *The moduli space of N = 1 supersphere with tubes and the sewing operation. Memoirs of the AMS* 772, 2003.

[BDR]   Baas, N., B. Dundas and J. Rognes, *Two-vector bundles and forms of elliptic cohomology.* In these proceedings.

[Be]   Bénabou, J., *Introduction to bicategories. LNM* 47, Springer, 1967, 1–77.

[BF]   Bismut, J.-M. and D. Freed, *The analysis of elliptic families I. Metrics and connections on determinant bundles. Comm. Math. Phys.* 106 (1986), 159–176.

[BR]   Bratteli, O. and D. Robinson, *Operator Algebras and Quantum Statistical Mechanics I and II.* Texts and Monographs in Physics. Springer, 1979.

[BW]   Booß-Bavnbek, B. and K. Wojciechowski, *Elliptic Boundary Problems for Dirac Operators.* Birkhäuser, 1993.

[Bry]   Brylinski, J.-L., *Loop spaces, Characteristic classes and Geometric quantization. Progress in Math.* 107, Birkhäuser, 1993.

[Ca]   Calderbank, D., *Clifford analysis for Diac operators on manifolds with boundary.* Max-Planck Institute Preprint No. 13, 1996.

[COT]   Cochran, T., K. Orr and P. Teichner, *Knot concordance, Whitney towers and von Neumann signatures. Annals of Math.* 157 (2003) 433–519.

[Co1]   Connes, A., *Noncommutative Geometry.* Academic Press, 1994.

[Co2]   Connes, A., *Periodic automorphisms of the hyperfinite factor of type* II$_1$. *Acta Sci. Math.* (Szeged) 39 (1977) 39–66.

[CR]   Crane, L. and J. Rabin, *Super Riemann Surfaces: Uniformization and Teichmüller theory. Comm. in Math. Phys.* 113 (1988) 601–623.

[DD]   Dixmier, J. and A. Douady, *Champs continus d'espace hilbertiens et de C*-algèbres. Bull. Soc. Math. Fr.* 91 (1963) 227–284.

[DW]   DeWitt, Bryce *Supermanifolds.* Second edition. Cambridge Monographs on Mathematical Physics. Cambridge University Press, 1992.

[Fr1]   Freed, D., *Classical Chern–Simons theory, Part 1. Adv. Math.* 113 (1995) 237–303.

[Fr2]   Freed, D., *Five lectures on supersymmetry.* AMS, 1999.

[Ha]   Haag, R., *Local Quantum Physics. Fields, Particles, Algebras.* Second edition. Texts and Monographs in Physics. Springer, 1996.

[H]   Haagerup, U., *Connes' bicentralizer problem and the uniqueness of the injective factor of type* III$_1$. *Acta Math.* 158 (1987) 95–148.

[HG]   Higson, N. and E. Guentner, *Group C*-algebras and K-theory.* Preprint.

[HK]   Hu, P. and I. Kriz, *Conformal field theory and elliptic cohomology.* Preprint.

[HR]   Higson, N. and J. Roe, *Analytic K-homology.* Oxford Math. Monographs, Oxford Science Publications, 2000.

[HBJ]   Hirzebruch, F., T. Berger and R. Jung, *Manifolds and Modular Forms.* Publication of the Max-Planck Institut für Mathematik, Bonn. Aspects of Math., Vieweg 1992.

[Ho]   Hopkins, M., *Algebraic Topology and Modular Forms.* Plenary Lecture, ICM Beijing 2002.

[Hu]   Huang, Y.-Z., *Two-dimensional conformal geometry and vertex operator algebras. Progress in Math.* 148, Birkhäuser 1997.

[J1]   Jones, V., *Actions of finite groups on the hyperfinite type* II$_1$ *factor. Memoirs of the AMS* 237 (1980).

[J2]   Jones, V., *Index for subfactors. Invent. math.* 72 (1983) 1–25.

[J3]   Jones, V. *Von Neumann Algebras in Mathematics and Physics.* ICM talk, Kyoto Proceedings, Vol.1 (1990) 127–139.

[J4]   Jones, V., *Fusion en algèbres de von Neumann et groupes de lacets (d'après A. Wassermann)*. Sém. Bourbaki, Vol. 1994/95. Astérisque No. 237, (1996) 251–273.

[Ka]   Karoubi, M., *K-Theory, An Introduction. Grundlehren der Math.* Wissenschaften, Springer, 1978.

[KS]   Kreck, M. and S. Stolz, $\mathbb{HP}^2$-*bundles and elliptic homology*. Acta Math. 171 (1993) 231–261.

[La]   Landweber, P., *Elliptic cohomology and modular forms. Elliptic Curves and Modular Forms in Alg. Top.*, Princeton Proc. 1986, LNM 1326, Springer, 55–68.

[LM]   Lawson, H.-B. and M.L. Michelsohn, *Spin Geometry*. Princeton University Press, 1989.

[Lu]   Lück, W., L^2-*invariants: theory and applications to geometry and K-theory*. Ergebnisse der Math. Series of Modern Surveys in Math. 44. Springer, 2002.

[Mo]   Moore, C., *Group extensions and cohomology for locally compact groups III + IV.* Transactions of the AMS 221 (1976) 1–33 and 35–58.

[Oc]   Ocneanu, A., *Quantized groups, string algebras and Galois theory for algebras. Operator algebras and applications*, Vol. 2, London Math. Soc. Lecture Note Ser. 136, Cambridge Univ. Press (1988) 119–172.

[Och]   Ochanine, S., *Genres elliptiques equivariants*. Elliptic curves and modular forms in alg. top., Princeton Proc. 1986, LNM 1326, Springer, 107–122.

[PW]   Popa, S. and A. Wassermann, *Actions of compact Lie groups on von Neumann algebras*. C. R. Acad. Sci. Paris Sér. I Math. 315 (1992) 421–426.

[PS]   Pressley, A. and G. Segal, *Loop groups*. Oxford University Press, 1986.

[Se1]   Segal, G. *Elliptic Cohomology*. Séminaire Bourbaki 695 (1988) 187–201.

[Se2]   Segal, G. *The definition of conformal field theory*. In these proceedings.

[Se3]   Segal, G. *K-homology theory and algebraic K-theory*. K-theory and operator algebras, Georgia Proc. 1975, LNM 575, Springer, 113–127.

[Ta]   Takesaki, M., *Theory of operator algebras I–III. Encyclopaedia of Mathematical Sciences*, 124, 125, 127. *Operator Algebras and Non-commutative Geometry*, 5, 6, 8. Springer, 2003.

[vN]   von Neumann, J., *Rings of Operators*. Collected Works, Volume 3, Pergamon Press 1961.

[Wa]   Wassermann, A., *Operator algebras and conformal field theory. Inventiones Math.* 133 (1998) 467–538.

[Wig]   Wigner, D., *Algebraic cohomology of topological groups. Transactions of the AMS* 178 (1973) 83–93.

[Wi1]   Witten, E., *The index of the Dirac operator on loop space*. Elliptic curves and modular forms in alg. top., Princeton Proc. 1986, LNM 1326, Springer, 161–181.

[Wi2]   Witten, E., *Index of Dirac operators. Quantum fields and strings: a course for mathematicians*, Vol. 1, 2 (Princeton, NJ, 1996/1997), AMS (1999) 475–511.

E-mail: stolz.1@nd.edu and teichner@math.ucsd.edu.

11

Open and closed string field theory interpreted in classical algebraic topology

DENNIS SULLIVAN
SUNY at Stony Brook and CUNY Graduate Center

Abstract

There is an interpretation of open–closed string field theory in algebraic topology. The interpretation seems to have much of the expected structure but notably lacks the vacuum expectations. All the operations are defined by classical transversal intersection of ordinary cycles and homologies (derived from chains in path spaces) inside finite-dimensional smooth manifolds. The closed string theory can be deduced from the open string theory by the known equivariant chain or homology construction. One obtains the interpretation of open and closed string field theory combined. The algebraic structures derived from the first layer of open string interactions realize algebraic models discussed in work of Segal and collaborators. For example Corollary 1 of §11.1 says that the homology of the space of paths in any manifold beginning and ending on any submanifold has the structure of an associative dialgebra satisfying the module compatibility (equals Frobenius compatibility). See the appendix for the definition of six kinds of dialgebras. Corollary 2 gives another dialgebra structure which is less known. Corollary 3 gives yet another, the Lie bialgebra of [3].

1 Open string states in M

The open string theory interpretation in topology takes place on the homology or on the chain level–referred to respectively as 'on-shell' and 'off-shell'. On-shell there will be a linear category $[\partial M]$ for each ambient space M, a finite dimensional oriented smooth manifold possibly with general singularities. A morphism in this category is called an (on-shell) open string state. In Greg Moore's paper in this volume the idea to connect formal properties of open string states in physics to morphisms of a category is credited to Graeme Segal. Eventually the categories here become dicategories generalizing the dialgebras of the appendix.

344

The objects in the category $[\vartheta M]$ include the smooth oriented submanifolds (without singularities) $L_a, L_b, L_c, \ldots$ of M. The set of morphisms $[\vartheta_{ab}]$ between two such objects L_a and L_b is the graded homology (with coefficients in $\mathbb{Z}/n$, $\mathbb{Z}$, or $\mathbb{Q}$) of $P(a, b)$, the space of smooth paths starting in L_a and ending in L_b. The composition of morphisms $[\vartheta_{ab}] \otimes [\vartheta_{bc}] \xrightarrow{\ \cdot\ } [\vartheta_{ac}]$ is defined as follows. Choose representative cycles x in $P(a, b)$ and y in $P(b, c)$. The endpoints of x and beginning points of y define respectively two cycles (of points) in L_b. These cycles can be intersected transversally in L_b after small perturbation in L_b to obtain a cycle z of dimension equal $\dim x + \dim y - \dim L_b$.

Now z parametrizes a set of paths from L_a to L_b and a set of paths from L_b to L_c which are composable along z. These are made out of the original paths plus small pieces from the perturbation. After composition (joining and parametrizing) z defines a cycle in the space of paths from L_a to L_c defining an element in $[\vartheta_{ac}]$. The composition is well defined and associative on the level of homology, namely on-shell, using familiar arguments (see the discussion immediately following).

When M is an oriented manifold without singularities each of the submanifolds $L_a, L_b, \ldots$ has an oriented normal bundle and the additional structure of dicategory can be defined. By this we mean for each triple of objects L_a, L_b and L_c there is a cocomposition or cutting operation $[\vartheta_{ac}] \xrightarrow{\vee_t} [\vartheta_{ab}] \otimes [\vartheta_{bc}]$ defined as follows. Choose a $t \epsilon [0, 1]$ and a representative cycle z for an element in $[\vartheta_{ac}]$. Evaluating each path labelled by z at time t yields a cycle (of points) in M. After small perturbation one can transversally intersect in M this cycle (of points in M) with L_b to obtain a cycle w in L_b of dimension equal to (dimension z–codimension L_b). The cycle w labels pairs of paths, one from L_a to L_b (the part of z's path from 0 to t) and one from L_b to L_c (the part of z's path from t to 1). Passing to homology classes and applying the Kunneth property yields an element in $[\vartheta_{ab}] \otimes [\vartheta_{ac}]$ (at least with coefficients in $\mathbb{Z}/n$ or $\mathbb{Q}$).

If t' is a different time in $[0,1]$ we can evaluate z between t and t' to obtain a chain W in M of dimension equal to dimension $(z) + 1$. Assuming z at t and at t' is transversal to L_b, W will also be transversal to L_b near its boundary (at t or t'). A small relative perturbation can make W transversal to L_b without changing it near the boundary. (This kind of argument is used often to show that these transversally defined operations are well defined in homology.)

This provides a homology between the cycle w defining $\vee_t(z)$ and the cycle w' defining $\vee_{t'}(z)$. Thus $\vee_t$ on homology is independent of t in $[0,1]$. A similar argument shows $\vee_t$ is coassociative on the level of homology.

The independence of t allows two different computations of $\vee_t(x \wedge y)$. By choosing t in x's time we get on the cycle level $\vee_t(x) \wedge y$ (composing in the right factor of the pair of paths). By choosing t in y's time we get on the cycle level $x \wedge \vee_t(y)$ (composing on the left factor of the pair of paths). Summarizing, we have:

Theorem 1 (on shell Frobenius compatibility). *For each oriented manifold the sets of homology classes $[\vartheta_{ab}]$ of the path spaces between arbitrary oriented submanifolds L_a and L_b are the sets of morphisms of a dicategory satisfying module compatibility (equals Frobenius compatibility). By this phrase we mean we have objects and morphisms and there are associative compositions of morphisms $[\vartheta_{ab}] \otimes [\vartheta_{bc}] \xrightarrow{\wedge} [\vartheta_{ac}]$ and coassociative cocompositions of morphisms $[\vartheta_{ac}] \xrightarrow{\vee_t} [\vartheta_{ab}] \otimes [\vartheta_{bc}]$ satisfying $x \wedge \vee_t(y) = \vee_t(x \wedge y) = \vee_t(x) \wedge y$. (In this formulation the coefficients are $\mathbb{Z}/n, \mathbb{Z},$ or $\mathbb{Q}$ for the composition and $\mathbb{Z}/n$ or $\mathbb{Q}$ for the cocomposition.)*

Corollary 1. *For each object $L_a \subset M$, the on shell self morphisms of the object L_a of the dicategory $[\vartheta M]$, the homology of paths in M beginning and ending on L_a with coefficients in $\mathbb{Z}/n$ or $\mathbb{Q}$, forms an associative dialgebra satisfying the module compatibility (equals Frobenius compatibility). See the appendix for a discussion of dialgebras.*

Example 1 (manifolds). If L_a is taken to be all of M, the space of paths from L_a to L_a is homotopy equivalent to M itself. Then $\wedge$ is identified on shell with ordinary intersection of homology in M. Also $\vee_t$ is identified on shell with the diagonal map on homology of M. We recover the classical fact that the homology of an oriented manifold has the structure of an associative dialgebra with module compatibility (equals Frobenius compatibility). If M is also closed the intersection multiplication and the diagonal comultiplication are in hom duality via the Poincaré duality inner product and we have the special case of a (graded) commutative Frobenius algebra.

Example 2 (free loop space). If the ambient space is $M \times M$ and the submanifold is the diagonal M in $M \times M$, then the space of paths in $M \times M$ beginning and ending in M is homotopy equivalent to the free loop space of M defined by smooth maps of the circle into M. The homology of the free loop space receives a product $\wedge$ and a coproduct $\vee_t$. The product agrees (on shell) with the loop product from 'String Topology' [2]. The coproduct is only non-zero if M is a closed manifold with non-zero Euler characteristic. (Otherwise $M = L_a$ can be deformed in $M \times M$ off of itself to $L_{a'}$. We can compute $[\vartheta_{aa}] \xrightarrow{\vee_t} [\vartheta_{aa'}] \otimes [\vartheta_{a'a}]$ for $t = 0$ (or $t = 1$) at the cycle level to see we get zero. Then we identify $[\vartheta_{aa}]$ with $[\vartheta_{aa'}]$ and with $[\vartheta_{a'a}]$ using the deformation of L_a to $L_{a'}$.)

Example 3 (based loop space). If L_a is a point in M the space of paths beginning and ending on L_a is just ΩM the based loop space of M. The product on the homology of ΩM defined above agrees (by definition) with the usual Pontryagin product on the homology of the based loop space. The coproduct defined above (when M is a manifold near L_a) is zero because L_a can be deformed in M off of itself (see example 2). The fact that the Pontryagin algebra is a Hopf algebra (for the diagonal coproduct) and the fact that a (finite dimensional) Hopf algebra is a Frobenius algebra is only suggestive at this point *vis-à-vis* the above theory.

Remark. There is an algebra homomorphism (on shell) of the free loop space algebra to the base loop space algebra of degree − dim (ambient space). It is defined by transversally intersecting the cycle of marked points from a cycle of loops with the base point (compare [2]).

Remark. The readers comfortable with homology intersection defined by geometric cycles and transversality will be able to add details they feel are needed above and below to the proofs of Theorems 1 and 2 and Corollaries 1 and 2 except possibly for orientations. Orientations are discussed in [3] along with specific information about the proof of Theorem 3 and Corollary 3 to be expanded elsewhere. Otherwise the text should be regarded as an outline or sketch of the proof of these results.

2 On-shell and off-shell

In the above constructions at the cycle or chain level the conclusions were stated at the homology level. We refer to these two levels respectively as 'off-shell' and 'on-shell'. A remark about these expressions in terms of familiar topology may be useful. If one tries to lift a homological structure to the geometric level discrepancies often show up. For example one can associate harmonic forms to cohomology classes. Harmonic forms are 'on-shell' for physicists because they satisfy the critical point equations associated to the energy action. The cohomological product is represented by the wedge product of these harmonic forms which is (almost always) not harmonic and therefore 'off-shell'.

One knows that putting in chain homotopies resolving this discrepancy of the product (and continuing) constructs algebraic models of the real (or rational) homotopy type [13] [15]. Recently a remarkable result [10] of Michael Mandell shows similar chain homotopies for integral cochains and their cup product when suitably organized determines the entire homotopy type for simply connected spaces.

Bearing this in mind it seems worthwhile to also study the above string theory at the cycle and chain level–namely off-shell. The idea is that an algebraic structure on-shell will be reflected in a more elaborate structure off-shell made out of a hierarchy of chain homotopies. A further idea is that the off-shell structure may be easier to work with in certain respects than the on-shell structure. For example Quillen's model of rational homotopy theory is a *differential* on a *free Lie algebra* which (we now know) is organizing the off-shell strong homotopy commutative associative cup product structure. In some sense the on-shell structure 'graded commutative algebra' is harder to classify and understand than the off-shell structure, which is 'free differential graded Lie algebra', because of the freeness.

Another idea comes up here. The notion of these hierarchical homotopies or strong homotopy structures due to Stasheff is very intuitive but combinatorially complicated. However in a number of cases this complexity is absorbed in a single operator of square zero on a free object for a dual structure (see Ginzburg–Kapranov [7] for the definition and explanation of this property of Koszul dual pairs of structures over operads and see [6] for algebras over dioperads).

Let us return to the rational homotopy example and the graded Lie algebra of homotopy groups of a space. The off-shell version of the Lie algebra of homotopy groups would be a strong homotopy Lie algebra which can be described by a differential on a *free* graded commutative algebra (by Koszul duality between commutative algebra and Lie algebra). The latter differential may be computed [15] inside the differential forms starting from harmonic forms (or any other lift) and iteratively correcting the off-shell wedge products by chain homotopies.

An early example of this on-shell–off-shell discussion in topology (and the first exactly in this vein) was Stasheff's notion of a strong homotopy associative algebra, or A_∞ algebra. The latter may be described by a differential on the free tensor algebra (of the dual space). An analogous notion of A_∞ category is also defined where composition is only associative up to homotopy etc. (see [1]).

3 Open strings off shell

Now we work off-shell with the set of cycles and chains ϑ_{ab} in the path spaces $P(a, b)$. For example, we can take ϑ_{ab} to be linear combinations (over $\mathbb{Z}/n$, $\mathbb{Z}$ or $\mathbb{Q}$) of smooth maps of standard simplices into $P(a, b)$ (namely, (simplex) x $[0,1] \to M$ is smooth).

The discussion in §1 of operations defined by transversality can now be considered off-shell at the chain level. Intersection of ordinary chains in M was developed as an A_∞ structure [1]. A similar discussion should show composition or joining of off-shell open string states (chains in $P(a, b), \ldots$) will generate an A_∞ category. Going further the off-shell analogue of Theorem 1 becomes.

Conjecture 1. *The off-shell open string states, the chains ϑ_{ab} on path spaces $(P(a, b), \ldots)$ form the morphisms of a strong homotopy dicategory, satisfying the module compatibility (equals Frobenius compatibility) between composition and cocomposition.*

A special case of Conjecture 1 is that the chains on paths in M from L_a back to L_a has the structure of a strong homotopy dialgebra satisfying the module compatibility (equals Frobenius compatibility).

There is another structure beyond this we could mention now. Consider cutting paths at any time and then use Eilenberg Zilber relating chains in a Cartesian product to tensor product of chains to define a new cocomposition $\vartheta_{ac} \overset{\vee}{\to} \vartheta_{ab} \otimes \vartheta_{bc}$ of one degree higher than $\vee_t$. This operator does not commute with the ∂ operator in general. In fact (as proved above) $\partial \vee + \vee \partial = \vee_1 - \vee_0$.

This operator satisfies a new compatibility with $\wedge$ called derivation compatibility.

Theorem 2 (off-shell derivation compatibility). *For appropriately transversal chains $\wedge$ and $\vee$ are defined and satisfy $\vee(x \wedge y) = \vee(x) \wedge y \pm x \wedge \vee(y)$.*

Corollary 2. *When L_a is deformable in M off of itself the homology of the space of paths in M beginning and ending on L_a has the structure of an associative dialgebra satisfying derivation compatibility.*

Remark. In the papers of Aguilar dialgebras with derivation compatibility are called 'infinitesimal bialgebras'. Aguilar attributes the concept to Gian Carlo-Rota *et al.* who introduced it in the 1960s to study certain combinatorial problems.

Proof of Theorem 2. $x \wedge y$ is represented by paths of x joined to paths of y (where the appropriate endpoints transversally intersect). Cutting along some L transversally we get two terms where the cut belongs to the x part or to the y part. This is the right-hand side of compatibility.

Proof of Corollary 2. As mentioned above when L_a is deformable off of itself to $L_{a'}$ then with regard to cutting paths from L_a back to itself along $L_{a'}$ $\vee_0$ and $\vee_1$ are zero at the chain level. Then $\vee$ defined transversally commutes with ∂ and passes to homology. The required identity on-shell follows from Theorem 2.

Remark. The structure of Corollary 2 may depend on the isotopy class of the push off.

Corresponding to Theorem 2 there is:

Conjecture 2. *The off-shell open strings states have a strong homotopy structure involving* $\wedge$, $\vee_0$, $\vee_1$, *and* $\vee$, *the various associativities, the two compatibilities Frobenius and derivation, and* $\partial \vee + \vee \partial = \vee_1 - \vee_0$.

Adding to the fun of formulating conjecture 2 exactly we note that in [6] it is asserted that the algebraic structures associated to $(\wedge, \vee_t)$ and to $(\wedge, \vee)$ on-shell in Corollary 1 and Corollary 2 are Koszul dual (for more see Appendix).

4 Closed strings

We have seen in example 2 of §1 above that open strings beginning and ending on the diagonal in $M \times M$ gives the free loop space of M, namely smooth maps of the circle into M. Now the free loop space also has a circle action by rotating the domain. *The closed string states in M on-shell or off-shell will be defined as the equivariant homology or chains relative to this circle action.* There are several models for the equivariant theories. We will employ here a geometric one called 'closed string space'.

A point in closed string space $S(M)$ is a pair (L, f) where L is a complex line in $\mathbb{C}^\infty = \{$finite sequences of complex numbers$\}$ and f is a smooth map of the unit circle in L into M.

Remark. Note that:

(1) $S(M)$ fibres over $\mathbb{C}P^\infty$ with fibre the free loop space of M.
(2) $S(M)$ is the base of a circle fibration with total space equivariantly homotopy equivalent to the free loop space of M.

Proof. Projection onto the first factor of the pair (L, f) proves 1. For 2 let the total space be triples (v, L, f) where v is a unit vector in a complex line L in $\mathbb{C}^\infty$ and f is a smooth map of the unit circle in L into M. Note the set of (v, L) is contractible. $\quad\square$

Definition. The homology classes of closed string space $S(M)$ are the *on-shell closed string states*. The chains on $S(M)$ are the *off-shell closed string states*.

Remark. The projection of the circle bundle (or the inclusion of the fibre of S(M) over $\mathbb{C}P^\infty$) defines E a degree zero chain or homology mapping from the free loop space of M to the closed string space (E for erase the isometric parametrization (or mark) of the circle). Taking the pre-image of the circle bundle projection defines a degree one chain or homology mapping M in the opposite direction (M for add a mark or isometric parametrization to a closed string in all ways to get a circle of loops.) We hope the double use of 'M' here does not cause a problem.

The composition EM produces a degenerate chain and may be regarded as zero by working in the quotient by degenerate chains. The composition ME is usually denoted $\triangle$. It is the operator of degree $+1$ on chains or homology of the free loop space associated to the circle action. Since $\triangle \cdot \triangle = (ME)(ME) = M(EM)E$, we have $\triangle^2 = 0$ on-shell and even off-shell mod degenerate chains.

In [2] it was shown the operator $\triangle$ on the homology of the free loops space with the open string product defined a BV or Batalin Vilkovisky algebra. Namely, the deviation of $\triangle$ from being a derivation of the open string product is a Lie bracket (of degree $+ 1$) compatible with the open string product via the Leibniz identity.

Remark. (1) This bracket was also defined [2] from an off-shell operation $*$ by skew symmetrization just as Gerstenhaber did in the Hochshild complex of an associative algebra. This fits with the idea that the Hochshild complex $\oplus_k$ Hom $(A^{\otimes k}, A)$ of the intersection algebra A of chains models the free loop space of a simply connected closed manifold (cf [5][16]).

(2) We will discuss below a Lie product or bracket on the closed string states which is compatible via the mapping M (adding a mark) with the BV or Gerstenhaber Lie bracket mentioned in 1.

(3) This closed string product or bracket generalizes to all manifolds the Goldman bracket on the vector space generated by conjugacy classes in the fundamental group of oriented surfaces. The Goldman bracket is a universal version of the Poisson structure on the moduli space of flat bundles over a surface. We suppose the off-shell string bracket for $S(M)$ bears a similar relation to general bundles with general connections over M (compare Cattaneo Frohlich et al.).

The string product on closed string states satisfying Jacobi (on the transversal chain level [3]) may be defined by the formula $[\alpha, \beta] = E(M\alpha \wedge M\beta)$ where $\wedge$ is the open string product. Other closed string operations c_n can be defined by $c_n(\alpha_1, \alpha_2, \ldots, \alpha_n) = E(M\alpha_1 \wedge M\alpha_2 \wedge \ldots \wedge M\alpha_n)$. These all commute with the ∂ operator and satisfy commutation identities transversally [2].

The collision operators c_n pass to the reduced equivariant complex or *reduced closed string* states which is defined to be the equivariant homology for the S^1 pair (free loop space, constant loops). This passage follows from the formulae for c_n because the marking operator M takes a chain of constant loops to a degenerate chain of constant loops.

We can define a closed string cobracket s_2 by the formula $s_2(\alpha) = (E \otimes E)(\vee(M\alpha))$. In the reduced complex s_2 commutes with ∂ and passes to homology (but not so in the unreduced complex [3]).

Theorem 3 (closed string bracket and cobracket). *The closed string bracket* $c_2(\alpha, \beta) = E(M\alpha \wedge M\beta)$ *where* $x \wedge y = \wedge(x \otimes y)$ *and the closed string cobracket* $s_2(\alpha) = (E \otimes E)(\vee M\alpha)$ *satisfy respectively jacobi, cojacobi, and derivation compatibility (equals Drinfeld compatibility). The term satisfy means either on the level of $\mathbb{Z}/n$ or $\mathbb{Q}$ homology or for transversal chains on the chain level (see appendix for discussion of compatibilities).*

Proof. These formulae in terms of open strings are reinterpretations as in [2] of the definitions given in 'Closed string operators in topology leading to Lie bialgebras and higher string algebra' [3]. In [3] the identities at the transversal chain level were considered. □

Corollary 3. *Homology of reduced closed string states forms a Lie bialgebra,* [3].

Remark. The corollary generalized Turaev's discovery [17] of a Lie bialgebra for surfaces to all manifolds. Questions in [17] motivated this work. See [4] for some answers and further developments.

Conjecture 3. *The off-shell closed string states (reduced) have the structure of a strong homotopy Lie bialgebra.*

Remark. Other cobracket or splitting operations $s_3, s_4, \ldots$ can be defined similarly by iterations of $\vee$, $s_n(\alpha) = E \otimes \ldots \otimes E(\ldots \vee \otimes 1 \cdot \vee(M\alpha))$. These also commute with ∂ and pass to homology in the reduced equivariant theory. A conjecture about $c_2, c_3, \ldots$; $s_2, s_3, \ldots$ generating genus zero closed string operators and the algebraic form of this structure was proposed in [3] and relates to [9].

5 Interplay between open and closed string states

Let $\mathcal{C}$ denote the closed string states in M, a manifold of dimension d, and let ϑ denote any of the complexes ϑ_{ab} of open string states. Transversality yields an action of closed strings on open strings

$$\mathcal{C} \otimes \vartheta \to \vartheta \qquad \text{degree} = (-d + 2)$$

and a coaction of closed strings on open strings

$$\vartheta \to \mathcal{C} \otimes \vartheta \qquad \text{degree} = (-d + 2).$$

The operations are defined off-shell for transversal chains. In the coaction we let the open string hit itself transversally inside M of dimension d at any two times and split the event into a closed string and an open string. In the action we let a closed string combine with an open string to yield an open string. We lose d dimensions by the intersection in M and gain two from the possible positions on each string of the attaching points.

The action is a Lie action of the Lie algebra of closed strings by derivations on the algebra of open strings. This is seen by looking directly at the construction at the transversal chain level. Both the action and the coaction have a non-trivial commutator with the boundary operator on chains. These boundary terms are expressed by interactions between the closed string and the open string at the endpoints of the open string.

For the action the individual boundary terms commute with the boundary operator and pass to homology. For the coaction the individual boundary terms have themselves additional boundary terms to be elucidated.

Problem and Conjecture 4. *The action and coaction between open strings and closed strings and their boundary interaction terms are described by a strong homotopy structure to be elucidated.*

6 Connection to work of Segal and collaborators

Dialgebras satisfying the module or Frobenius compatibility give examples of $1 + 1$ TQFT's without vacuum expectations. In the commutative case we associate the underlying vector space to a directed circle, its tensor products to a disjoint union of directed circles and to a connected 2D oriented bordism between two non-empty collections the morphism obtained by decomposing the bordism into pants and composing accordingly the algebra or coalgebra map. The module compatibility (equals Frobenius compatibility) is just what is required for the result to be independent of the choice of pants decomposition.

N.B. this description differs from the usual one because we do not have disks to close up either end of the bordism. One knows these discs at both ends would force the algebra to be finite dimensional and the algebra and coalgebra to be related by a non-degenerate inner product. We refer to these generalizations of the Atiyah–Segal concepts as the positive boundary version of TQFT (a name due to Ralph Cohen). The editor notes that Segal refers to the underlying algebras of positive boundary 2D TQFT as non compact Frobenius algebras.

An exactly similar discussion with non-commutative associative dialgebras satisfying the Frobenius compatibility leads to a positive boundary version of a TQFT using open intervals. Now the algebra and coalgebra are associated to 1/2 pants (a disc with ∂ divided into six intervals–three (1/2 seams) alternating with two (1/2 cuffs) and one (1/2 waist)). Any planar connected bordism between two non-empty collections of intervals determines a mapping between inputs and outputs.

The structures we have found (including ∂ labels $L_a, L_b, \ldots$) for open strings using the composition $\wedge$ and fixed time cutting $\vee_t$ satisfies this Frobenius compatiblity up to a chain homotopy and we can apply it at the homology level in the relative TQFT scheme just mentioned.

Remark. One can show the on-shell structure of open and closed strings gives an example of the structure described in Moore's article of these proceedings. This follows by showing cycles and homologies on the moduli space of open closed string Riemann surfaces acts on open closed string states. For example, there is an operation on pairs of open strings which combine and reconnect at arbitrary interior points. A general operation is essentially a composition of the latter with all the above.

7 Summary

We have described the part of the interpretation of open and closed string field theory in topology associated to the basic product and coproduct (and in the equivariant setting certain implied n-variable splitting and collision operators as in [3]). The coproduct discussion has two levels involving a coproduct $\vee_t$ and an associated chain homotopy coproduct $\vee$.

We found the open string product and the coproduct $\vee_t$ satisfied the module compatibility (equals Frobenius compatibility) on the level of homology namely on-shell. In a setting where $\vee_0$ and $\vee_1$ were zero or even deformable to zero, $\vee$ emerges as or can be deformed to a coproduct commuting with ∂ and thus a coproduct $\vee$ on homology of one higher degree. Then a new compatibility with the product is observed – the derivation or infinitesimal bialgebra compatibility (also true at the transversal chain level and therefore suggesting a corresponding strong homotopy structure which was Conjecture 2).

Remark. The submanifolds which are the objects of the open string categories here are called D-branes in the math physics literature. We are currently considering more general boundary conditions forced on us by 3D computations which lead us to flat bundles along submanifolds and more general sheaves.

For closed strings in M we considered the equivariant theory associated to open strings on the diagonal in $M \times M$.

The higher genus interpretation of open closed string field theory in topology involves full families of arbitrary cutting and reconnecting operations of a string in an ambient space M. For closed curves some full families of these operators were labelled combinatorially by decorated even valence ribbon graphs obtained by collapsing chords in general chord diagrams in [3]. There is a compactness issue for the full families discussed there for realizing these in algebraic topology. The issue is a correct computation of the boundary. The problem has a parallel with renormalization in Feynman graphs (see the Bott–Taubes [18] treatment of configuration space integrals).

In both cases algebraic topology transversality and Feynman graphs the loops in collapsing subgraphs cause the problems. We hope to address this issue using Penner's intriguing paper [12].

Appendix: (dialgebras and compatibilities)

Let us call a linear space V with two maps $V \otimes V \xrightarrow{\wedge} V$ and $V \xrightarrow{\vee} V \otimes V$ a *dialgebra*. *Associative dialgebra* means $\wedge$ is associative and $\vee$ is coassociative. *Commutative dialgebra* means besides being associative $\wedge$ and $\vee$ are symmetric. *Lie dialgebra* means both maps are skew symmetric and that jacobi and cojacobi hold.

In all these cases V and $V \otimes V$ have module structures over V and there are two kinds of compatibilities between $\wedge$ and $\vee$ relative to these.

The compatibilities we consider here are:

derivation compatibility $\quad \vee(a \cdot b) = (\vee a) \cdot b + a \cdot \vee(b)$ and
module compatibility $\quad\quad \vee(a \cdot b) = \vee(a) \cdot b = a \cdot \vee(b)$

Where the $\cdot$ refers to the algebra structure or the module structure (which means in the associative case $a \cdot (b \otimes c) = (a \cdot b) \otimes c, (a \otimes b) \cdot c = a \otimes (b \cdot c)$

and in the Lie case $a \cdot (b \otimes c) = -(b \otimes c) \cdot a = [a, b] \otimes c + b \otimes [a, c]$ where $[x, y] = \wedge(x \otimes y)$).

We get six kinds of structures (five appear in this paper, see table below) which are examples of definitions of algebras over dioperads [6]. Algebras over dioperads are structures whose generators and relations are described diagrammatically by trees.

The familiar example of a compatibility studied by Hopf that $\vee$ is a map of algebras (associative or commutative case but not Lie) is described by a non-tree diagram and is not an algebra over a dioperad.

Table with names of compatibility and/or structure and/or examples.

	Module compatibility	Derivation compatibility
Associative dialgebra	Frobenius compatibility Special case = Frobenius algebra = associative algebra with non-degenerate invariant inner product	These are called infinitesimal bialgebras by Aguilar
Commutative dialgebra	Frobenius compatibility Special case = Commutative Frobenius algebra	commutative cocommutative infinitesimal bialgebra
Lie dialgebra	Frobenius compatibility Special case = Lie algebra with non-degenerate invariant inner product	Drinfeld compatibility These are called Lie bialgebras in the literature

In [6] Koszul dual pairs are defined and there it is proved that upper left and upper right are Koszul dual pairs and that middle left and lower right are Koszul dual pairs. We suppose that the lower left and middle right are also Koszul dual pairs.

We emphasize these Koszul relations because in several important situations a strong homotopy algebraic structure of one kind is very naturally expressed by freely generated diagrams decorated with tensors labeled by the Koszul dual structure. Our main conjecture in the above discussion is that *all the structures that are true transversally will lead to strong homotopy versions on the entire space of states*. These might be usefully expressed in this graphical Koszul dual way.

References

[1] Fukaya, K., Oh, and Ohta, Ono 'Lagrangin intersection Floer theory-anomaly and obstruction', (2000) See Fukaya website.

[2] Chas, M. and D. Sullivan, 'String Topology' GT/ 9911159. *Annals of Mathematics* (to appear).

[3] Chas, M. and D. Sullivan, Closed string operators in topology leading to Lie bialgebras and higher string algebra, GT/ 0212358. Abel Bicentennial Proceedings (to appear).

[4] Chas, Moira, Combinatorial Lie bialgebras of curves on surfaces, to appear in *Topology*. Also arXiv GT/0105178.

[5] Cohen-Jones, A homotopy theoretic realization of string topology, math GT/0107187.

[6] Wee Liang Gan, Koszul duality for dioperads, preprint University of Chicago 2002 QA/0201074.

[7] Ginzburg, V. and M.M. Kapranov, Koszul duality for operads, *Duke Mathematical Journal* 76 (1994).

[8] Goldman, William M. Invariant functions on Lie group and Hamiltonian flows of surface group representations, *Invent. Math.* 85(1986), 263–302.

[9] Manin, Yuri, Frobenius Manifolds, Quantum cohomology and moduli spaces, *AMS Colloquium Publications*, Vol. 47.

[10] Mandell, Michael, Cochains and Homotopy Type, Preprint, September 2001, January 2002.

[11] Moore, Greg, Some comments on Branes, G-flux, and K-theory, Part II and references to Segal notes therein. *International Journal of Modern Physics* A. [arXiv:hep-th/0012007] v1. 1 Dec. 2000.

[12] Penner, R.C., The decorated Teichmuller space of punctured surface, *Communications in Mathematical Physics* 113 (1987) 299–339.

[13] Quillen, Daniel, Rational homotopy theory, *Ann. of Math.*, 90 (1969), 205–295.

[14] Stasheff, James, H-spaces from a homotopy point of view, *Lecture Notes in Mathematics* 161, Springer-Verlag, Berlin (1970), ii-95.

[15] Sullivan, Dennis, Infinitesimal Computations in Topology, Publ. Math. *IHES* 47 (1977), 269 332.

[16] Tradler, Thomas, The BV algebra on Hochschild cohomology induced by infinity inner products, GT/0210150.

[17] Turaev, V., Skein quantization of Poisson algebra of loops on surfaces, *Ann. Sci. Ecole Norm. Sup.* (1) 24 (1991), no.6, 635–704.

[18] Taubes, C. and R. Bott, Configuration space integrals and knot invariants.

Email: dsullivan@gc.cuny.edu

12

K-theory of the moduli space of bundles on a surface and deformations of the Verlinde algebra

CONSTANTIN TELEMAN

Cambridge University

Abstract

We conjecture that index formulas for K-theory classes on the moduli of holomorphic G-bundles over a compact Riemann surface Σ are controlled, in a precise way, by Frobenius algebra deformations of the Verlinde algebra of G. The Frobenius algebras in question are twisted K-theories of G, equivariant under the conjugation action, and the controlling device is the equivariant Gysin map along the 'product of commutators' from G^{2g} to G. The conjecture is compatible with naïve virtual localization of holomorphic bundles, from G to its maximal torus; this follows by localization in twisted K-theory.

1 Introduction

Let G be a compact Lie group and let M be the moduli space of flat G-bundles on a closed Riemann surface Σ of genus g. By well-known results of Narasimhan, Seshadri and Ramanathan [NS], [R], this is also the moduli space of stable holomorphic principal bundles over Σ for the complexified group $G_{\mathbb{C}}$; as a complex variety, it carries a fundamental class in complex K-homology. This paper is concerned with index formulas for vector bundles over M. The analogous problem in cohomology – integration formulas over M for top degree polynomials in the tautological generators – has been extensively studied [N], [K], [D], [Th], [W], and, for the smooth versions of M, the moduli of vector bundles of fixed degree co-prime to the rank, it was completely solved in [JK]. In that situation, the tautological classes generate the rational cohomology ring $H^*(M; \mathbb{Q})$. Knowledge of the integration formula leads to the intersection pairing, and from here, Poincaré duality determines this ring as the quotient of the polynomial ring in the tautological generators by the null ideal of the pairing.

358

For smooth M, index formulas result directly from the Riemann–Roch theorem and the integration formula. This breaks down in the singular case, and no index formula can be so obtained, for groups other than $SU(n)$. Follow-up work [Kie] has extended the results of [JK] to the study of the duality pairing in intersection cohomology, for some of the singular moduli spaces. Whether a useful connection to K-theory can be made is not known; so, to that extent, the formulas I propose here are new. But I should point out the novel features of the new approach, even in the smooth case.

More than merely giving numbers, the conjecture posits a structure to these indices; to wit, they are controlled by (finite-dimensional) Frobenius algebras, in the way the Verlinde algebras control the index of powers of the *determinant line bundle D*. The Frobenius algebras in question are formal deformations of Verlinde algebras. This is best explained by the twisted K-theory point of view, [FHT1], [FHT2], which identifies the Verlinde algebra with a twisted equivariant K-theory ${}^\tau K_G^*(G)$. The determinantal twistings appearing in that theorem (cf. §3.2 below) correspond to powers of D. Other K-theory classes, involving index bundles over Σ (see §2.iii), relate to higher twistings in K-theory, and these effect infinitesimal deformations of the Verlinde algebra. If ordinary (=determinantal) twistings can be represented by gerbes [BCMMS], higher twistings are realized by what could be called virtual gerbes, which generalize gerbes in the way one-dimensional virtual bundles generalize line bundles. In this picture, K-theory classes over M of virtual dimension one are automorphisms of virtual gerbes, and arise by comparing two trivializations of a twisting τ for $K_G(G)$ (§4.14). Relevant examples are the transgressions over Σ of delooped twistings for BG (§3.9).[1] All these K-classes share with D the property of being 'multiplicative in a piece of surface'.

The first formulation, §4.11 of the conjecture expresses the index of such a K-class over the moduli of G-bundles as the *partition function* for the surface Σ, in the 2D topological field theory defined by ${}^\tau K_G^*(G)$. This is a sum of powers of the *structure constants* of the Frobenius algebra, for which explicit formulas can be given (Theorem 4.7). There is a restriction on the allowed twistings, but they are general enough to give a satisfactory set of K-classes.

We can reformulate the conjectures in §4.15, §4.17 by encoding part of the Frobenius algebra structure into the *product of commutators map* $\Pi : G^{2g} \to G$. This map has the virtue of lifting the transgressed twistings for $K_G(G)$

---

[1] For the expert, we mean G-equivariant $B^2 BU_\otimes$-classes of a point.

(§3.9) to trivializable ones, which allows one to identify $K^*_G(G^{2g})$ with its τ-twisted version. In particular, we get a class $^\tau 1$ in $^\tau K^*_G(G^{2g})$. The conjecture then asserts that the index of the K-class associated to τ over the moduli of G-bundles equals the Frobenius algebra trace of the Gysin push-forward of $^\tau 1$ along Π.

Having (conjecturally) reduced this index to a map of compact manifolds, ordinary localization methods allows us to express the answer in terms of the maximal torus T and Weyl group of G. This reduction to T, it turns out, can be interpreted as a virtual localization theorem from the moduli of holomorphic $G_{\mathbb{C}}$-bundles to that of holomorphic $T_{\mathbb{C}}$-bundles. (The word 'virtual' reflects the use of the virtual normal bundle, defined by infinitesimal deformations). For this interpretation, however, it turns out that we must employ the moduli $\mathfrak{M} = \mathfrak{M}_G$, $\mathfrak{M}_T$ of *all* $G_{\mathbb{C}}$-and $T_{\mathbb{C}}$-bundles, not merely the semi-stable ones. These moduli have the structure of *smooth stacks*, with an infinite descending stratification by smooth algebraic substacks. Even the simplest case, the Verlinde formula, cannot be reduced to a single integral over the variety of topologically trivial $T_{\mathbb{C}}$-bundles; the correct expression arises only upon summing over all topological T-types. (Recall that the non-trivial T-types define unstable $G_{\mathbb{C}}$-bundles.)

Now is the right time to qualify the advertised statements. The fact that the Verlinde formula, the simplest instance of our conjecture, expresses the indices of positive powers D over M is a fortunate accident. It is an instance of the 'quantization commutes with reduction' conjecture of Guillemin and Sternberg [GS], which in this case [T2] equates the indices of positive powers of D over M and over the stack $\mathfrak{M}$ of all holomorphic $G_{\mathbb{C}}$-bundles. This does not hold for more general K-theory classes, for which there will be contributions from the unstable Atiyah–Bott strata, and our deformed Verlinde algebras really control not the index over M, but that over $\mathfrak{M}$. This incorporates information about the moduli of flat G-bundles, and the moduli of flat principal bundles of various subgroups of G. In other words, the index information which assembles to a nice structure refers to the stack $\mathfrak{M}$ and not to the space M.

Hence, the third formulation of the conjecture, §5.3, expresses the index of any admissible K-class (Definition 2.7) over the moduli stack $\mathfrak{M}$ of all holomorphic $G_{\mathbb{C}}$-bundles over Σ by virtual localization to the stack of holomorphic $T_{\mathbb{C}}$-bundles. This involves integration over the Jacobians, summation over all degrees, leading to a distribution on T, and, finally, integration over T (to extract the invariant part). In §5, these steps are carried out explicitly for the group $SU(2)$.

However, even if our interest lies in M (which, our approach suggests, it should not), all is not lost, because a generalization of 'quantization commutes with reduction' (first proved in [TZ], for compact symplectic manifolds) asserts, in this case, the equality of indices over M and $\mathfrak{M}$, after a large D-power twist, for the class of bundles we are considering.[2] This follows easily from the methods of [T2], but there is at present no written account. Because M is projective algebraic, the index of $\mathcal{E} \otimes D^{\otimes n}$ over M, for any coherent sheaf $\mathcal{E}$, is a polynomial in n; so its knowledge for large n determines it for all n, including, by extrapolation, $n = 0$. Thus, the information contained in $\mathfrak{M}$, which combines index information for the moduli of bundles of subgroups of G, can be disassembled into its constituent parts; the leading contribution, as $n \to \infty$, comes from M itself. When M is smooth, the '$\mathcal{E}$-derivatives' of the $n \to \infty$ asymptotics of the index of $\mathcal{E} \otimes D^{\otimes n}$ give integration formulas for Chern polynomials of $\mathcal{E}$ over M; and the author suggests that the Jeffrey–Kirwan residue formulas for the integrals can be recovered in this manner. What one definitely recovers in the large level limit are Witten's conjectural formulas [W]. Indeed, there is evidence that the relevant field theories are topological limits of Yang–Mills theory coupled to the WZW model (in other words, the G/G coset model with a Yang–Mills term); this would fit in nicely with the physical argument of [W].

2 The moduli space M, the moduli stack $\mathfrak{M}$ and admissible K-classes

In this section, we recall some background material; some of it is logically needed for the main conjecture, but mostly, it sets the stage for my approach to the question. This is anchored in Theorem 2.8.

2.1. Recall the set-up of [AB]: let $\mathcal{A}$ be the affine space of smooth connections, and $\mathcal{G}$ the group of smooth gauge transformations on a fixed smooth principal G-bundle P over Σ. The $(0, 1)$-component of such a connection defines a $\bar{\partial}$-operator, hence a complex structure on the principal $G_{\mathbb{C}}$-bundle $P_{\mathbb{C}}$ associated to P. We can identify $\mathcal{A}$ with the space of smooth connections of type $(0, 1)$ on $P_{\mathbb{C}}$; the latter carries an action of the complexification $\mathcal{G}_{\mathbb{C}}$ of $\mathcal{G}$, and the quotient $\mathcal{A}/\mathcal{G}_{\mathbb{C}}$ is the set of isomorphism classes of holomorphic principal $G_{\mathbb{C}}$-bundles on Σ with underlying topological bundle $P_{\mathbb{C}}$.

2.2. The space $\mathcal{A}$ carries a $\mathcal{G}_{\mathbb{C}}$-equivariant stratification, according to the instability type of the holomorphic bundle. The semi-stable bundles define the open subset $\mathcal{A}^0$, whose universal Hausdorff quotient by $\mathcal{G}_{\mathbb{C}}$ is a projective algebraic variety M, the moduli space of semi-stable holomorphic $G_{\mathbb{C}}$-bundles over Σ. The complex structure is descended from that of $\mathcal{A}^0$, in the sense that

---

[2] An explicit bound for the power can be given, linear in the highest weight.

a function on M is holomorphic in an open subset if and only if its lift to $\mathcal{A}$
is so. We can restate this, to avoid troubles relating to holomorphy in infinite
dimensions. The gauge transformations that are based at one point $* \in \Sigma$
act freely on $\mathcal{A}^0$, with quotient a smooth, finite-dimensional algebraic variety M_*. This is the moduli of semi-stable bundles with a trivialization of the
fibre over $*$. Its algebro-geometric quotient under the residual gauge group
$G_{\mathbb{C}}$ is M. The other strata $\mathcal{A}^\xi$ are smooth, locally closed complex submanifolds, of finite codimension; they are labeled by the non-zero dominant co-
weights ξ of G, which give the destabilizing type of the underlying holomorphic $G_{\mathbb{C}}$-bundle. The universal Hausdorff quotient of $\mathcal{A}^\xi$ can be identified with
a moduli space of semi-stable principal bundles under the centralizer $G_{(\xi)}$ of ξ
in $G_{\mathbb{C}}$.

2.3. The stack $\mathfrak{M}$ of all holomorphic $G_{\mathbb{C}}$-bundles over Σ is the homotopy
quotient $\mathcal{A}/\mathcal{G}_{\mathbb{C}}$. As such, it seems that we are using new words for an old
object, and so it would be, were our interest confined to ordinary cohomology,
$H^*(\mathfrak{M}; \mathbb{Z})$. However, we will need to discuss its K-theory, and the index map
to $\mathbb{Z}$. The abstract setting for this type of question is a homotopy category
of analytic spaces or algebraic varieties (see e.g. Simpsons work [Si], and,
with reference to $\mathfrak{M}$, [T1]). Fortunately, little of that general abstraction is
necessary here. It turns out that $\mathfrak{M}$ is homotopy equivalent to the quotient,
by G-conjugation, of a principal ΩG-bundle over G^{2g}. (This is the homotopy
fibre of $\Pi : G^{2g} \rightarrow G$; cf. §4.13). This follows from Segal's *double coset
presentation* of $\mathfrak{M}$ ([PS], Ch. 8). As a result, there is a sensible topological
definition of $K^*(M)$, which makes it into an inverse limit of finite modules for
the representation ring R_G of G.

2.4. As in the cohomological setting of Atiyah–Bott, this $K^*(M)$ can be shown
to surject on to the $\mathcal{G}_{\mathbb{C}}$-equivariant K-theory of $\mathcal{A}^0$. The latter can be defined
as the G-equivariant version of $K^*(M_*)$, for the variety M_* of §2.2. This is
as close as we can get to $K^*(M)$. When the action of G on M_* is free, the
two groups coincide, but this only happens when $G = PU(n)$ and the degree of our bundle P is prime to n. However, some relation between $K_G^0(M_*)$
and M always exists. Namely, every holomorphic, $G_{\mathbb{C}}$-equivariant bundle $\mathcal{E}$
over M_* has an invariant direct image $q_*^G(\mathcal{E})$, which is a coherent analytic
sheaf over M. (This is the sheaf of G-invariant holomorphic sections along
the fibres of the projection $q : M_* \rightarrow M$.) The coherent sheaf cohomology groups of $q_*^G(\mathcal{E})$ are finite-dimensional vector spaces, and the alternating sum of their dimensions is our definition of the G-invariant index of $\mathcal{E}$
over M_*.

2.5. A similar construction, applied to a stratum $\mathcal{A}^\xi$, allows us to define the $\mathcal{G}_{\mathbb{C}}$-invariant index of holomorphic vector bundles over it. Because $\mathcal{A}$ is stratified by the $\mathcal{A}^\xi$, there is an obvious candidate for the $\mathcal{G}_{\mathbb{C}}$-invariant index of holomorphic vector bundles over $\mathcal{A}$, as a sum over all ξ. (Contributions from the normal bundle must be taken into account; see [T2], §9). This sum may well be infinite. The task, therefore, is to identify a set of admissible K-theory classes, for which the sum is finite; we then define that sum to be the index over $\mathfrak{M}$. For a good class of bundles, this can be done, and shown to agree with a more abstract global definition, as the *coherent-sheaf cohomology Euler characteristic over the algebraic site of the stack* $\mathfrak{M}$. I shall not prove any of the assertions above, instead will take a low-brow approach and define directly the set of admissible K-theory classes. When G is simply connected, they turn out to generate a dense subring of $K^*(\mathfrak{M}; \mathbb{Q})$, rather in the way that a polynomial ring is dense in its power series completion; and their restrictions to the semi-stable part generate $K_G^*(M_*) \otimes \mathbb{Q}$. (This can be deduced from the cohomological result of [AB], by using equivariant Chern characters.)

2.6. Note, first, that the pull-back of the bundle $P_{\mathbb{C}}$ to $\Sigma \times \mathcal{A}$ carries a natural, $\mathcal{G}_{\mathbb{C}}$-equivariant, holomorphic structure, as defined by the $(0, 1)$-part of the universal connection along Σ. We might call this the universal bundle on $\Sigma \times \mathfrak{M}$. A representation V of G defines an associated holomorphic, $\mathcal{G}_{\mathbb{C}}$-equivariant vector bundle E^*V on $\Sigma \times \mathcal{A}$. (We think of E as the classifying map of the universal bundle to BG.) Let $\pi : \Sigma \times \mathcal{A} \to \mathcal{A}$ be the projection, and fix a square root $K^{1/2}$ of the canonical bundle on Σ. We now associate the following objects to V, which we shall call the *tautological classes* in $K^*(M)$.

(i) For a point $x \in \Sigma$, the restriction E_x^*V of E^*V to $\{x\} \times \mathfrak{M}$;

(ii) The *index bundle* $\alpha(V) := R^*\pi_*(E^*V \otimes K^{1/2})$ along Σ over $\mathcal{A}$;

(iii) For any class $C \in K_1(\Sigma)$, its slant product with E^*V (the index of E^*V along a 1-cycle).

Object (i) is an equivariant holomorphic vector bundle, objects (ii) and (iii) are equivariant K^0 and K^1 classes over $\mathcal{A}$, the misnomer 'bundle' in (ii) notwithstanding. For example, we can represent (ii) by a $\mathcal{G}_{\mathbb{C}}$-equivariant Fredholm complex based on the relative $\bar{\partial}$-operator. (The square root of K leads to the Dirac, rather than $\bar{\partial}$ index, and the notation $\alpha(V)$ stems from the Atiyah index map, which we get when Σ is the sphere.) We shall not consider type (iii) objects in this paper, so we refrain from analyzing them further. Note that the topological type of the bundles in (i) is independent of x; we shall indeed see that their index is so as well.

Any reasonable definition of $K^*(M)$ should include the tautological classes, but another distinguished object plays a crucial role:

(iv) The determinant line bundle $D(V) := \det R^* \pi_*(E^* V)$ over $\mathcal{A}$.

This is a holomorphic, $\mathcal{G}_{\mathbb{C}}$-equivariant line bundle, or a holomorphic line bundle over $\mathfrak{M}$. When G is semi-simple, such line bundles turn out to be classified by their Chern classes in $H^2(M; \mathbb{Z})$ [T1]; in the case of $D(V)$, this is the transgression along Σ of $c_2(V) \in H^4(BG)$. Not all line bundles are determinants, but they are fractional powers thereof. The convex hull of the $D(V)$ define the semi-positive cone in the group of line bundles; its interior is the positive cone. In particular, when G is simple and simply connected, $H^2(\mathfrak{M}; \mathbb{Z}) \cong \mathbb{Z}$, and the positive cone consists of positive powers of a single D; for $G = SU(n)$, this is $D(\mathbb{C}^n)$, for the standard representation.

Definition 2.7. An *admissible class* in $K^*(\mathfrak{M})$ is a polynomial in the tautological classes and the semi-positive line bundles.

For simply connected G, an admissible K-class is a finite sum of terms $p_n \otimes D^{\otimes n}$, where $n \geq 0$ and p_n is a polynomial in the objects (i)–(iii) above. We can actually allow some small[3] negative values of n, but the index of such classes turns out to vanish, so little is gained. The following theorem allows our approach to $K^*(M)$ to get off the ground. To simplify the statements, we assume that G is semi-simple.

Theorem 2.8. (i) *The coherent sheaf cohomology groups, over the algebraic site of the stack $\mathfrak{M}$, of any admissible class $\mathcal{E} \in K^*(\mathfrak{M})$, are finite-dimensional, and vanish in high degrees.*
(ii) *The index* $\mathrm{Ind}(\mathfrak{M}; \mathcal{E})$ *over $\mathfrak{M}$, defined as the alternating sum of cohomology dimensions, is also expressible as a sum of index contributions over the Atiyah–Bott strata $\mathcal{A}^\xi$. Each contribution is the index of a coherent sheaf over a moduli space of semi-stable $G_{(\xi)}$-bundles. These contributions vanish for large ξ.*
(iii) *For sufficiently positive D, all $\xi \neq 0$-contributions of $\mathcal{E} \otimes D$ to the index vanish.*
(iv) *Hence, for sufficiently positive D,* $\mathrm{Ind}(\mathfrak{M}; \mathcal{E} \otimes D) = \mathrm{Ind}\left(M; q_*^G(\mathcal{E} \otimes D)\right)$.

Proof. (Sketch) For a product of 'evaluation bundles' §2.6.i, the results were proved in [T1] and [T2]. This generalizes immediately to a family of bundles,

---

[3] Larger than the negative of the dual Coxeter number

parametrized by a product of copies of Σ, and integration along the curves leads to index bundles §2.6.ii. Slant products with odd K-homology classes on that product of Riemann surfaces lead to the same conclusion for arbitrary admissible $\mathcal{E}$. $\qquad\qquad\square$

Remark 2.9. As noted in the introduction, the theorem allows us to determine the index of $q_*^G(\mathcal{E})$ over M, for any admissible $\mathcal{E}$, if index formulas over $\mathfrak{M}$ are known. Indeed, suitable positive line bundles D descend to M, and so $\mathrm{Ind}\left(M; q_*^G(\mathcal{E} \otimes D^{\otimes n})\right)$ is a polynomial in n. Its value at $n = 0$ can then be determined from large n, where it agrees with $\mathrm{Ind}\left(\mathfrak{M}; \mathcal{E} \otimes D^{\otimes n}\right)$.

3 Twistings and higher twistings in K-theory

We start with some background on twisted K-theory and its equivariant versions. The statements are of the 'known to experts' kind, but, unfortunately, references do not always exist. They will be proved elsewhere.

3.1. Let X be a compact, connected space. Units in the ring $K^*(X)$, under tensor product, are represented by the virtual vector bundles of dimension ± 1. A distinguished set of units in the 1-dimensional part GL_1^+ is the Picard group $\mathrm{Pic}(X)$ of topological line bundles; it is isomorphic to $H^2(X; \mathbb{Z})$, by the Chern class. In the other direction, the determinant defines a splitting

$$GL_1^+\left(K^*(X)\right) \cong \mathrm{Pic}(X) \times SL_1\left(K^*(X)\right) \qquad (3.1)$$

where the last factor denotes 1-dimensional virtual bundles with trivialized determinant line.

We shall ignore here the twistings coming from the group $\{\pm 1\}$. The splitting (3.1) refines to a decomposition of the spectrum $BU_\otimes$ of 1-dimensional units in the classifying spectrum for complex K-theory [MST]; in self-explanatory notation, we have a factorization

$$BU_\otimes \cong K\left(\mathbb{Z}; 2\right) \times BSU_\otimes. \qquad (3.2)$$

3.2. A twisting of complex K-theory over X is a principal $BU_\otimes$-bundle over that space. By (3.2), this is a pair $\tau = (\delta, \chi)$ consisting of a *determinantal twisting* δ, which is a $K(\mathbb{Z}; 2)$-principal bundle over X, and a *higher twisting* χ, which is a $BSU_\otimes$-torsor. Twistings are classified, up to isomorphism, by a pair of classes $[\delta] \in H^3(X; \mathbb{Z})$ and $[\chi]$ in the generalized cohomology group $H^1(X; BSU_\otimes)$. This last group has some subtle features over $\mathbb{Z}$; rationally, however, $BSU_\otimes$ is a topological abelian group, isomorphic to

$\prod_{n\geq 2} K(\mathbb{Q}; 2n)$ via the logarithm of the Chern character ch. We obtain the following.

Proposition 3.3. *Twistings of rational K-theory over X are classified, up to isomorphism, by the group* $\prod_{n>1} H^{2n+1}(X; \mathbb{Q})$.

Remark 3.4. The usual caveat applies: if X is not a finite complex, we rationalize the coefficients before computing cohomologies.

The twistings in Proposition 3.3, of course, are also the twistings for rational cohomology with coefficients in formal Laurent series $\mathbb{Q}((\beta))$ in the Bott element β of degree (-2). This is not surprising, as the classifying spectra $BU \otimes \mathbb{Q}$ and $K(\mathbb{Q}((\beta)), 0)$ for the two theories are equivalent under ch. The isomorphism extends naturally to the twisted theories, by a twisted version of the Chern character, as in §3 of [FHT1], where determinantal twistings were considered:

Proposition 3.5. *There is a natural isomorphism* $^{\tau}ch : {}^{\tau}K^* (X; \mathbb{Q}) \to {}^{\tau}H^* (X; \mathbb{Q}((\beta)))$.

Remark 3.6. The strength of the proposition stems from computability of the right-hand side. Let $(A^{\bullet}, d)$ be a DGA model for the rational homotopy of X, $\eta = \eta^{(3)} + \eta^{(5)} + \cdots$ a cocycle representing the twisting, decomposed into graded parts. If we define $\eta' := \beta\eta^{(3)} + \beta^2\eta^{(5)} + \cdots$, then it turns out that $^{\tau}H^* (X; \mathbb{Q}((\beta)))$ is the cohomology of $A^{\bullet}((\beta))$ with modified differential $d + \eta'\wedge$. The latter can be computed by a spectral sequence, commencing at E_2 with the ordinary $H^* (X; \mathbb{Q}((\beta)))$, and with third differential $\beta\eta^{(3)}$ (cf. [FHT1]).

3.7. Thus far, the splitting (3.2) has not played a conspicuous role: rationally, all twistings can be treated uniformly, with $\log ch$ playing the role that c_1 plays for determinantal twistings. Things stand differently in the equivariant world, when a compact group G acts on X. There are equivariant counterparts to §3.1 and (3.2), namely, a spectrum $BU_{\otimes}^{G}$ of equivariant K-theory units, factoring into the equivariant versions of $K(\mathbb{Z}; 2)$ and $BSU_{\otimes}$. However, the group, $\prod_{n>1} H_{G}^{2n+1} (X; \mathbb{Q})$, analogous to that in Proposition 3.3, no longer classifies twistings for rational equivariant K-theory, but only for its augmentation completion. The reason is the comparative dearth of units in the representation ring R_G of G, versus its completion.

The easily salvaged part in (3.1) is the equivariant Picard group. Realizing twistings in $H_{G}^{3}(X; \mathbb{Z})$ by equivariant projective Hilbert bundles allows a construction of the associated twisted K-theory by Fredholm operators; see [F] for a relevant example.[4] One method to include more units (hence more twistings) is to localize in the representation ring. This is not an option here; one of our operations for the index formula, the trace map, will be integration over G. Instead, we introduce the extra units by adjoining formal variables to R_G. Thus, we consider the formal power series ring $R_G[[t]]$, and the associated K-theory $K_G^*(X)[[t]]$ of formal power series in t, with equivariant vector bundle coefficients. Any series $\sum t^n V_n$, where V_0 is a 1-dimensional representation of G, is now a unit. This realizes the G-spectrum

$$K^G(\mathbb{Z}; 2) \times \left(1 + t BU^G[[t]]\right)_\otimes \tag{3.3}$$

within the G-spectrum of units in $BU^G[[t]]$.

Definition 3.8. An *admissible twisting* for $K_G(X)[[t]]$-theory is a torsor under the spectrum (3.3) over the G-space X.

Admissible twistings $\tau(t) = (\delta, \chi(t))$ are classified, up to isomorphism, by a pair of classes $[\delta] \in H_G^3(X; \mathbb{Z})$ and a 'higher' class $[\chi(t)]$ in the generalized equivariant cohomology group $\left[X; B\left(1 + t BU[[t]]\right)_\otimes\right]^G$.

3.9. One way to define a higher admissible twisting over G, equivariant for the conjugation action, uses a (twice deloopable) *exponential morphism* from $BU^G[[t]]$ to $1 + t BU^G[[t]]$ (taking sums to tensor products). Segal's theory of Γ-spaces [S] shows that sufficiently natural exponential operations on the coefficient ring $R_G[[t]]$ define such morphisms. An example is the *total symmetric power*

$$V(t) \mapsto S_t[t \cdot V(t)] = \sum_{n \geq 0} t^n \cdot S^n[V(t)]. \tag{3.4}$$

Allowing rational coefficients, the naïve exponential

$$V(t) \mapsto \exp[t \cdot V(t)] = \sum_{n \geq 0} t^n/n! \cdot V(t)^{\otimes n} \tag{3.5}$$

is more closely related to the previous discussion, in the sense that completing at the augmentation ideal takes us to the ordinary K-theory of BG, and applying ch to the right leads to the earlier identification §§3.2–3.3 of $BSU_\otimes \otimes \mathbb{Q}$ with $\prod_{n \geq 2} K(\mathbb{Q}; 2n)$. Whichever exponential morphism we choose, Bott periodicity permits us to regard $BU^G[[t]]_\oplus$-classes of a point as

---

[4] Even here, we meet a new phenomenon, in that *integral* twistings are required to define the rational equivariant K-theory; for instance, the torsion part of a twisting in $H_G^3(X; \mathbb{Z})$ affects the rational answer.

$B^2BU^G[[t]]_{\oplus}$-classes, and delooping our morphism produces classes in $B^2(1 + tBU^G[[t]])$ of a point. Transgressing once gives higher twistings for $K_G(G)$. In this paper, we shall pursue the exponential (3.5) in more detail.

4 The Index from Verlinde algebras

The Verlinde algebra and its deformations 4.1. Call a twisting $\tau(t) = (\tau_0, \chi(t))$ *non-degenerate* if the invariant bilinear form h it defines on $\mathfrak{g}$, via the restriction of $[\tau_0] \in H_G^3(G)$ to $H_T^2 \otimes H^1(T)$, is so; call it *positive* if this same form is symmetric and positive definite. Integrality of τ_0 implies that h defines an isogeny from T to the Langlands dual torus, with kernel a finite, Weyl-invariant subgroup $F \subset T$. Recall the following result from [FHT1], [FHT2], referring to determinantal twisting $\tau = (\tau_0, 0)$. For simplicity, we restrict to the simply connected case. Let σ be the twisting coming from the projective cocycle of the Spin representation of the loop group; it restricts to the dual Coxeter number on H^3 for each simple factor.

Theorem 4.2. *Let G be simply connected. For a positive determinantal twisting τ, the twisted K-theory $^\tau K_G^{\dim G}(G)$ is isomorphic to the Verlinde algebra $V_G(\tau - \sigma)$ of G, at a shifted level $\tau - \sigma$. It has the structure of an integral Frobenius algebra; as a ring, it is the quotient of R_G by the ideal of representations whose characters vanish at the regular points of F. The trace form $^\tau\mathrm{Tr} : R_G \to \mathbb{Z}$ sends $V \in R_G$ to*

$$^\tau\mathrm{Tr}(V) = \sum_{f \in F^{\mathrm{reg}}/W} \mathrm{ch}_V(f) \cdot \frac{\Delta(f)^2}{|F|} \tag{4.1}$$

where ch_V is the character of V, $\Delta(f)$ is the Weyl denominator, normalized to have positive square on T_g and $|F|$ is the order of F. □

Remark 4.3. (i) The $R_G \otimes \mathbb{C}$-algebra $^\tau K_G^{\dim G}(G) \otimes \mathbb{C}$ is supported at the regular conjugacy classes of G which meet F, and has one-dimensional fibres. (ii) The trace form (4.1) determines the Frobenius algebra: by non-degeneracy, the kernel of the homomorphism from R_G is the null subspace under the bilinear form $(V, W) \mapsto {}^\tau Tr(V \otimes W)$.
(iii) After complexifying R_G, we can represent Tr by integration against an invariant distribution on G. The latter is the sum of δ-functions on the conjugacy classes in (ii), divided by the order of F; the factor $|\Delta(f)|^2$ is the volume of the conjugacy class.
(iv) The result holds for connected groups with torsion-free π_1, although some care must be taken with the ring structure when the adjoint representation does

not spin [FHT1]. When π_1 has torsion or G is disconnected, ${}^\tau K_G^{\dim G}(G)$ is still the Verlinde algebra, but it is larger than the quotient of R_G described in Theorem 4.2; so the trace form on R_G no longer determines V_G.

The complexified form of Theorem 4.2 was given a direct proof in [FHT1], using the Chern character to compute the twisted K-theory. The trace form was introduced *ad hoc*, using our knowledge of the Verlinde algebra (§7 of loc. cit.). There is in fact no choice on the matter, and the entire Frobenius structure is determined topologically (see the proof of Theorem 4.7 below).

We now incorporate higher twistings in Theorem 4.2. We tensor with $\mathbb{C}$ for convenience.

Theorem 4.4. *Let G be simply connected, $\tau(t) = (\tau_0, \chi(t))$ an admissible twisting with positive determinantal part τ_0. The twisted K-theory ${}^\tau K_G^{\dim G}(G) \otimes [[t]]$ is a Frobenius algebra, which is a quotient of $R_G[[t]] \otimes \mathbb{C}$, and a flat deformation of the Verlinde algebra at level σ.*

Remark 4.5. The use of complex higher twistings forces us to tensor with $\mathbb{C}$. The use of integral twistings of the type (3.4) would lead to a similar result over $\mathbb{Z}$, but as our goal here is an index formula, nothing is lost over $\mathbb{C}$.

Proof. (Idea) All statements follow by computing the Chern character, exactly as in [FHT1]; the completions of ${}^\tau K_G^{\dim G}(G) \otimes \mathbb{C}[[t]]$ at conjugacy classes in $G_\mathbb{C}$ are calculable by spectral sequences as in (3.6). Away from F, the E_2 term of this sequence is nil. At singular points of F, this is not so, but the third differential, which stems from the determinant of the twisting, is exact; so the limit is null again. At regular points of F, the same third differential resolves one copy of $\mathbb{C}[[t]]$ in degrees of the parity $\dim G$, and zero otherwise; so the sequence collapses there, and the abutment is a free $\mathbb{C}[[t]]$-module of rank 1. $\qquad\square$

4.6. As explained in Remark 4.3.ii, the Frobenius algebra is completely determined by the trace form ${}^\tau \mathrm{Tr} : R_G[[t]] \to \mathbb{C}[[t]]$. At $t = 0$, this is given by an invariant distribution φ_0 on G, specifically, $1/|F|$ times the sum of δ-functions on the regular conjugacy classes meeting F. We must describe how this varies with t. Recall that the determinantal part $\tau(0) = (\tau_0, 0)$ of the twisting defines an invariant metric h on $\mathfrak{g}$. We will now associate to the unit $\exp(t \cdot V)$ a one-parameter family of conjugation-invariant coordinate changes on the group G. More precisely, this is a (formal) path in the complexified (formal) group of automorphisms of the variety G/G; or, even more precisely, a formal

1-parameter family of automorphisms of the representation ring $R_G \otimes \mathbb{C}$. Because $G/G = T/W$, it suffices to describe this on the maximal torus T. In flat coordinates $\exp(\xi)$, where $\xi \in \mathfrak{t}$, this is

$$\xi \mapsto \xi + t \cdot \nabla \left[\mathrm{ch}_V \left(\exp(\xi) \right) \right] \tag{4.2}$$

the gradient being computed with respect to the metric h on $\mathfrak{t}$.

Theorem 4.7. *The trace form* $^\tau \mathrm{Tr} : R_G[[t]] \mapsto \mathbb{C}[[t]]$ *is integration against the invariant distribution* φ_t *on* $G_\mathbb{C}$, *obtained from* φ_0, *the distribution associated to* $\tau_0 = (\tau_0, 0)$, *by the formal family of coordinate changes* (4.2) *on* $G_\mathbb{C}$.

Remark 4.8. It is no more difficult to give the formula for more general exponential morphisms $V \mapsto \Phi_t(V)$. We assume Φ_t compatible with the splitting principle, via restriction to the maximal torus: in other words, its value on any line L is a formal power series in t, with coefficients Laurent polynomials in L. Then, $\log \Phi_t$ extends by linearity to an additive map $\mathfrak{t} \otimes R_T \to R_T[[t]]$, and the required change of coordinates is $\xi \mapsto \xi + \log \Phi_t [\nabla \mathrm{ch}_V]$, the metric being used to define the gradient. For instance, the symmetric power twisting (3.4) arises from $\Phi_t(L) = (1 - tL)^{-1}$; when $G = S^1$, $\mathrm{ch}_V(u) = \sum c_n u^n$, and we take level h for the determinantal part, we get the change of variable $u \mapsto u \cdot \prod (1 - tu^n)^{-nc_n/h}$.

Proof. (Sketch) The 2D field theory structure of $^\tau K_G(G)$ requires the trace form to be the inverse of the bilinear form which is the image of $1 \in K_G(G)$ in $^\tau K_G(G)^{\otimes 2}$, under the anti-diagonal morphism of spaces $G \to G \times G$ (and the diagonal inclusion of the acting groups). By localization to the maximal torus T, it suffices to check the proposition for tori. (The Euler class of the inclusion $T \subset G$ is responsible for the factor Δ^2.) Now, the twisting τ enters the computation of this direct image only via the holonomy representation $\pi_1(T) \to GL_1(R_T[[t]])$ it defines. Via the metric h, the determinantal twisting τ_0 assigns to any $p \in \pi_1(T)$ a weight of T, which gives a unit in R_T, and this defines the holonomy representation for τ_0. The change of coordinates (4.2) has the precise effect of converting this holonomy representation to the one associated to τ. $\qquad\square$

Index formulas 4.9. In the simply connected case, the class $[\tau_0] \in H_G^3(G; \mathbb{Z})$ determines a unique holomorphic line bundle over $\mathfrak{M}$ (cf. §2.6), which we call $\mathcal{O}(\tau_0)$. The simplest relation between the Verlinde algebras and indices of bundles over M is that $\mathrm{Ind}(M; \mathcal{O}(\tau_0 - \sigma))$ is the partition function of the surface

Σ, in the 2D topological field theory defined by $V_G(\tau_0 - \sigma)$. Recall (4.2) that $V_G \otimes \mathbb{C}$ is isomorphic, as an algebra, to a direct sum of copies of $\mathbb{C}$, supported on the regular Weyl orbits in F. The traces of the associated projectors are the *structure constants* θ_f ($f \in F^{\mathrm{reg}}/W$) of the Frobenius algebra; their values here are $|\Delta(f)|^2/|F|$. The partition function of a genus g surface is the sum $\sum \theta_f^{1-g}$, which leads to one version of the Verlinde formula

$$\mathrm{Ind}\,(M; \mathcal{O}(\tau_0 - \sigma)) = \sum_{f \in F^{\mathrm{reg}}/W} \Delta(f)^{2-2g} \cdot |F|^{g-1}. \qquad (4.3)$$

Remark 4.10. The downshift by σ stems from our use of the $\bar{\partial}$-index; the Dirac index would refer to $\mathcal{O}(\tau_0)$. However, there is no definition of the Dirac index in the singular case, and even less for the stack $\mathfrak{M}$.

The generalization of (4.3) to higher twistings is one form of the main conjecture. Recall the index bundle $\alpha(V)$ over $\mathfrak{M}$ associated to a representation V of G, and call $\theta_f(t)$, $f \in F^{\mathrm{reg}}/W$, the structure constants of the Frobenius algebra ${}^{\tau}K_G^{\dim G}(G) \otimes \mathbb{C}[[t]]$ over $\mathbb{C}[[t]]$.

Conjecture 4.11. $\mathrm{Ind}\,(\mathfrak{M}; \mathcal{O}(\tau_0 - \sigma) \otimes \exp[t\alpha(V)]) = \sum_{f \in F^{\mathrm{reg}}/W} \theta_f(t)^{1-g}$.

Remark 4.12. (i) Expansion in t allows the computation of indices of $\mathcal{O}(\tau_0 - \sigma) \otimes \alpha(V)^{\otimes n}$ from (4.2). More general expressions in the index bundles $\alpha(V)$, for various V, are easily obtained by the use of several formal parameters. One can also extend the discussion to include odd tautological generators §2.6.iii, but we shall not do so here.

(ii) The change in $\theta_f(t)$ is due both to the movement of the point f under the flow, and to the change in the volume form, under the change of coordinates (4.2).

Transgressed twistings and the product of commutators 4.13. Let us move to a more sophisticated version of the conjecture, which incorporates the evaluation bundles, §2.6.i. We refer to [FHT1], §7 for more motivation, in connection to loop group representations. Recall the 'product of commutators' map $\Pi : G^{2g} \to G$. If we remove a disk Δ from Σ, this map is realized by the restriction to the boundary of flat connections on $\Sigma \setminus \Delta$, based at some boundary point; the conjugation G-action forgets the base-point. The homotopy fibre of Π is the ΩG-bundle over G^{2g} mentioned in §2.3; the actual fibre over $1 \in G$ is the variety of based flat G-bundles on Σ, and its quotient by G-conjugation is M.

4.14. The twistings τ of interest to us are transgressed from G-equivariant $B^2BU[[t]]_\otimes$ classes of a point. For determinantal twistings, we are looking at the transgression from $H^4(BG)$ to $H_G^3(G)$; and, if G is simply connected, all determinantal twistings are so transgressed. In general, transgression is the integration along S^1 of the $B^2BU_\otimes$ class on the universal flat G-bundle over S^1, which is pulled back by the classifying map of the bundle. The relevant feature of a transgressed twisting τ is that its (equivariant) pull-back to G^{2g}, via Π, is trivialized by the transgression over $\Sigma - \Delta$. This trivialization gives an isomorphism

$$K_G^*(G^{2g}) \cong {}^{\Pi^*\tau}K_G^*(G^{2g}) \qquad (4.4)$$

which allows us to define a class ${}^\tau 1 \in {}^{\Pi^*\tau}K_G^0(G^{2g})$ without ambiguity.

Recall from §2.3 that the homotopy fibre of Π over $1 \in G$, when viewed G-equivariantly, is represented by the stack $\mathfrak{M}$. Thereon, we have two trivializations of the equivariant twisting $\Pi^*\tau$: one lifted from the base $\{1\}$, and one coming from transgression over $\Sigma \setminus \Delta$. The difference of the two is an element of $K^0(\mathfrak{M})[[t]]_\otimes$. For the twisting τ_0, this is the line bundle $\mathcal{O}(\tau_0)$; for admissible twistings, it will be a formal power series in t, with admissible K-class coefficients. For transgressed twistings based on the exponential morphism (3.5), we obtain the exponential $\exp[t\alpha(V)]$ of the index bundle $\alpha(V)$.

Conjecture 4.15. $\mathrm{Ind}\,(\mathfrak{M};\,\mathcal{O}(\tau_0-\sigma)\otimes\exp[t\alpha(V)]) = {}^\tau Tr\,(\Pi_!{}^\tau 1) \in \mathbb{C}[[t]].$

Remark 4.16. Equality of the right-hand sides of the conjectured formulae in §§4.11 and 4.15 is part of the definition of the 2D field theory (Frobenius algebra) structure on ${}^\tau K_G^{\dim G}(G)$. The only check there is has been incorporated into Theorem 4.7, which describes the trace map.

The last formulation has the advantage of allowing us to incorporate the evaluation bundles §2.6.i. Let W be another representation of G, and call $[W]$ the image class in ${}^\tau K_G^{\dim G}(G)$.

Amplification 4.17. $\mathrm{Ind}\,\big(\mathfrak{M};\,\mathcal{O}(\tau_0-\sigma)\otimes\exp[t\alpha(V)]\otimes E_x^*W\big) = {}^\tau Tr\,([W]\cdot\Pi_!{}^\tau 1).$

5 The index formula by virtual localization

In this section, we explain how the most naïve localization procedure, from G to its maximal torus, gives rise to an index formula for admissible K-classes, which agrees with Conjecture 4.17. There is an intriguing similarity here with localization methods used by Blau and Thomas [BT] in their path-integral calculations. We emphasize, however, that, in twisted K-theory, the localization formula from G to its maximal torus is completely rigorous, and can be applied to the Gysin map $\Pi_!$ of §4.13 to prove the equivalence of Conjectures 4.17 and 5.3 below. The role of the δ-functions which appear in this section is played, on the K-theory side, by the skyscrapers of the sheaf $^\tau K_G^{\dim G}(G)$, when localized over the conjugacy classes. For clarity of the formulas we shall confine the calculation to $SU(2)$-bundles; no new issues appear for other simply connected groups.

5.1. The maximal torus T of $G = SU(2)$ is S^1, with coordinate u, and the moduli stack $\mathfrak{M}_T$ of holomorphic $T_\mathbb{C}$-bundles is $J(\Sigma) \times \mathbb{Z} \times BT$, where $J(\Sigma)$ denotes the Jacobian variety J_0 of degree 0 line bundles and BT denotes the classifying stack of $T_\mathbb{C} \cong \mathbb{C}^\times$. A vector bundle over BT is a T-representation, and its index is the invariant subspace; the allowable representations, for which the index is well-defined, are finite-multiplicity sums of irreducibles. We have a natural isomorphism $H^1(J_0; \mathbb{Z}) \cong H_1(\Sigma; \mathbb{Z})$, and the class $\psi := \mathrm{Id} \in H^1(\Sigma) \otimes H^1(J_0)$ is the mixed $\Sigma \times J_0$ part of the first Chern class of the universal (Poincaré) bundle $\mathcal{P}$. More precisely, denoting by ω the volume class in $H^2(\Sigma)$, by η the restriction of $c_1(\mathcal{P})$ to J_0 and by $\lambda \in H^2(BT)$ the Chern class of the standard representation of T, we have, for the universal bundle on $J(\Sigma) \times \{d\} \times BT$

$$c_1(\mathcal{P}_d) = \eta + d \cdot \omega + \psi + \lambda. \tag{5.5}$$

Note that $e^\omega = 1 + \omega$, while $e^\psi = 1 + \psi - \eta \wedge \omega$, whence we get for the Chern character

$$\mathrm{ch}(\mathcal{P}_d) = (1 + \psi + d \cdot \omega - \eta \wedge \omega) \wedge e^\eta \cdot u \tag{5.6}$$

having identified the Chern character of the standard representation with its character u.

5.2. The 'virtual normal bundle' ν for the morphism $\mathfrak{M}_T \to \mathfrak{M}$ is the complex $R^*\pi_*(\mathrm{ad}\, \mathfrak{g}/\mathfrak{t})[1]$. Since $\mathrm{ad}\, \mathfrak{g}/\mathfrak{t} \cong \mathcal{P}^2 + \mathcal{P}^{-2}$, a small calculation from (5.5) gives, on the dth component

$$\mathrm{ch}\mathcal{P}_d^2 = (1 + 2\psi + 2d \cdot \omega - 4\eta \wedge \omega) \cdot e^{2\eta} u^2,$$
$$\mathrm{ch}\mathcal{P}_d^{-2} = (1 - 2\psi - 2d \cdot \omega - 4\eta \wedge \omega) \cdot e^{-2\eta} u^{-2} \tag{5.7}$$

and integrating over Σ, while remembering the shift by 1, gives

$$\begin{aligned}
\mathrm{ch}[\nu_d] &= u^2 e^{2\eta} \cdot (g - 1 - 2d) + u^{-2} e^{-2\eta} \cdot (g - 1 + 2d) \\
&\quad + 4\eta \wedge (u^2 e^{2\eta} + u^{-2} e^{-2\eta}) \\
\mathrm{ch}[\nu_d^*] &= u^{-2} e^{-2\eta} \cdot (g - 1 - 2d) + u^2 e^{2\eta} \cdot (g - 1 + 2d) \\
&\quad - 4\eta \wedge (u^2 e^{2\eta} + u^{-2} e^{-2\eta}).
\end{aligned} \tag{5.8}$$

In the same vein, note that the Chern character of the basic line bundle D, the determinant of cohomology $\det H^1 \otimes \det^{-1} H^0$ of $\mathcal{P} + \mathcal{P}^{-1}$, is

$$\mathrm{ch}(D) = e^{(2-2d)\eta} \cdot u^{-2d}. \tag{5.9}$$

The following conjecture describes the naive localization formula for the index, from $\mathfrak{M}$ to $\mathfrak{M}_T$.

Conjecture 5.3. *The index of an admissible class over $\mathfrak{M}$ is one-half the index of its restriction to $\mathfrak{M}_T$, divided by the equivariant K-theory Euler class of the conormal bundle ν^*.*

Remark 5.4. (i) The index over $\mathfrak{M}_T$ is defined as integration over each J_d, summation over degrees $d \in \mathbb{Z}$, and, finally, selection of the T-invariant part. At the third step, we shall see that the character of the T-representation obtained from the first two steps is a distribution over T, supported at the regular points of F (see §4.2). Miraculously, this corrects the problem which makes Conjecture 5.3 impossible at first sight: the equivariant Euler class of ν^* is singular at the singular conjugacy classes of G, so there is no well-defined index contribution over an individual component J_d. The sum over d acquires a meaning by extending the resulting distribution by zero, on the singular conjugacy classes.

(ii) The 'one-half' corrects for the double-counting of components in $\mathfrak{M}_T$, since opposite T-bundles induce isomorphic G-bundles. In general, we divide by the order of the Weyl group.

5.5. We need the Chern character of the equivariant K-Euler class of (5.8). The first two terms are sums of line bundles, and they contribute a multiplicative factor of

$$\begin{aligned}
&\left(1 - u^2 e^{2\eta}\right)^{g-1+2d} \left(1 - u^{-2} e^{-2\eta}\right)^{g-1-2d} \\
&\quad = (-1)^{g-1} \left(u e^\eta - u^{-1} e^{-\eta}\right)^{2g-2} \cdot \left(u e^\eta\right)^{4d}.
\end{aligned} \tag{5.10}$$

Now the log of the Euler class is additive, and we have

$$4\eta \cdot u^2 e^{2\eta} = 4\frac{d}{dx}\left(e^{x\eta} \cdot u^2 e^{2\eta}\right)\Big|_{x=0} \tag{5.11}$$

whence the Chern character of the Euler class of the remaining term is the exponential of

$$-4\frac{d}{dx}\left[\log\left(1 - e^{x\eta} \cdot u^2 e^{2\eta}\right) + \log\left(1 - e^{x\eta} \cdot u^{-2} e^{-2\eta}\right)\right]\Big|_{x=0}$$

$$= 4\frac{\eta \cdot u^2 e^{2\eta}}{1 - u^2 e^{2\eta}} + 4\frac{\eta \cdot u^{-2} e^{-2\eta}}{1 - u^{-2} e^{-2\eta}} = -4\eta \tag{5.12}$$

and we conclude that the Chern character of the K-theory Euler class of v^* is

$$(-1)^{g-1}\left(ue^{\eta} - u^{-1}e^{-\eta}\right)^{2g-2} \cdot \left(ue^{\eta}\right)^{4d} \cdot e^{-4\eta}. \tag{5.13}$$

5.6. We can now write the index formula asserted by Conjecture 5.3. For an admissible K-class of the form $D^{\otimes h} \otimes \mathcal{E}$, where $\mathcal{E}$ is a polynomial in the classes §2.6.i–iii, this is predicted to be the u-invariant part in the sum

$$\frac{1}{2}\sum_{d\in\mathbb{Z}}(-1)^{g-1} \cdot \int_{J_d} \mathrm{ch}(\mathcal{E}) \cdot \frac{u^{-2(h+2)d} \cdot e^{-2(h+2)(d-1)\eta}}{(ue^{\eta} - u^{-1}e^{-\eta})^{2g-2}}. \tag{5.14}$$

Noting, from (5.6), that $\mathrm{ch}(\mathcal{E})$ has a linear d-term for each factor $\alpha(V)$, it is now clear, as was promised in Remark (5.4.i), that (5.14) sums, away from the singular points $u = \pm 1$, to a finite linear combination of δ-functions and their derivatives, supported at the roots of unity of order $2(h + 2)$. Integration over the torus is somewhat simplified by the formal change of variables $u \mapsto ue^{\eta}$, and the identification $(u - u^{-1})^2 = -|\Delta(u)|^2$ in terms of the Weyl denominator leads to our definitive answer

$$\mathrm{Ind}(\mathfrak{M}; D^{\otimes h} \otimes \mathcal{E}) = \frac{1}{4\pi i}\oint\left[\sum_{d\in\mathbb{Z}}\frac{u^{2(h+2)d}}{|\Delta(u)|^{2g-2}} \cdot \int_{J_d}\mathrm{ch}(\mathcal{E}) \cdot e^{2(h+2)\eta}\right]\frac{du}{u} \tag{5.15}$$

where the distribution in brackets is declared to be null at $u = \pm 1$.

Example 1: Evaluation bundles 5.7. Let $\mathcal{E}$ be an evaluation bundle (2.6.i), $\mathcal{E} = E_x^* V$. Then

$$\int_{J_d} \mathrm{ch}(\mathcal{E}) \cdot e^{-2(h+2)\zeta} = (2h+4)^g \cdot \mathrm{ch}_V(u), \tag{5.16}$$

$$\sum_{d \in \mathbb{Z}} u^{2(h+2)d} = \frac{1}{2h+4} \sum_{\zeta^{2h+4}=1} \delta_\zeta(u) \tag{5.17}$$

and the index formula (5.15) reduces to the Verlinde formula for an evaluation bundle

$$\mathrm{Ind}(\mathfrak{M}; D^{\otimes h} \otimes E_x^* V) = \sum_{\substack{\zeta^{2h+4}=1 \\ \Im\zeta > 0}} \frac{(2h+4)^{g-1}}{|\Delta(\zeta)|^{2g-2}} \cdot \mathrm{ch}_V(\zeta). \tag{5.18}$$

Example 2: Exponentials of index bundles 5.8. Let V be any representation of G, with character a Laurent polynomial $f(u) = \sum f_n u^n$. We define $\dot{f}(u) := \sum n \cdot f_n u^n$, $\ddot{f}(u) := \sum n^2 \cdot f_n u^n$. For the index bundle $\alpha(V)$, we have

$$\mathrm{ch}\,\alpha(V) = \mathrm{ch}(R^* \pi_* V) - (g-1) \cdot \mathrm{ch}(E_x^* V) = d \cdot \dot{f}(ue^\eta) - \eta \cdot \ddot{f}(ue^\eta). \tag{5.19}$$

We compute the integral over J_d for insertion in (5.15) (after changing variables $u \mapsto ue^\eta$)

$$\int_{J_d} \exp\left[td \cdot \dot{f}(u)\right] \cdot \exp\left[-\left(2h+4+t\ddot{f}(u)\right)\eta\right]$$
$$= \left(2h+4+t\ddot{f}(u)\right)^g \cdot \exp\left[td \cdot \dot{f}(u)\right] \tag{5.20}$$

whereupon the sum in (5.15) becomes again a sum of δ-functions

$$\sum_{d \in \mathbb{Z}} \left[u \cdot \exp\left(\frac{t\dot{f}(u)}{2h+4}\right)\right]^{(2h+4)d} = \sum_{\zeta^{2h+4}=1} \delta_\zeta\left(u \cdot \exp\left(t\dot{f}(u)/(2h+4)\right)\right) \tag{5.21}$$

The index formula becomes now a sum over the solutions ζ_t, with positive imaginary part, of $\zeta_t^{2h+4} \cdot \exp\left(t\dot{f}(\zeta_t)\right) = 1$:

$$\mathrm{Ind}\left(\mathfrak{M}; D^{\otimes h} \otimes \exp\left[t\alpha(V)\right]\right) = \sum_{\zeta_t} \left[\frac{2h+4+t\ddot{f}(\zeta_t)}{|\Delta(\zeta_t)|^2}\right]^{g-1}. \tag{5.22}$$

One power of the numerator was swallowed up by the change of variables in the δ-function. This has precisely the form predicted by Conjecture 4.11; note that the numerator differs from (5.18) by the volume scaling factor in the co-ordinate change

$$u \mapsto u_t := u \cdot \exp\left(t \dot{f}(u)/(2h+4)\right).$$

References

[AB]   Atiyah, M.F. and R. Bott, The Yang-Mills equation over Riemann surfaces. *Philos. Trans. Roy. Soc. London* **308A** (1983), 523–615

[BCMMS] Bouwknegt, P., A.L. Carey, V. Mathai, M.K. Murray and D. Stevenson: Twisted *K*-theory and *K*-theory of bundle gerbes. *Comm. Math. Phys.* **228** (2002), 17–45

[BT]   Blau, M. and G. Thompson, Localization and diagonalization: a review of functional integral techniques for low-dimensional gauge theories and topological field theories. *J. Math. Phys.* **36** (1995), 2192–2236.

[D]   Donaldson, S.K., Gluing techniques in the cohomology of moduli spaces. In: *Topological Methods in Modern Mathematics*, Publish or Perish, (1993), 137–170

[F]   Freed, D.S., Loop Groups and twisted *K*-theory. *Proceedings of the ICM 2002*

[FHT1]   Freed, D.S. and M.J. Hopkins, C. Teleman: *Twisted K-theory with complex coefficients.* math.AT/0206257

[FHT2]   Freed, D.S. and M.J. Hopkins, C. Teleman: *Loop Group Representations and twisted K-theory.* In preparation.

[GS]   Guillemin, V. and S. Sternberg: Geometric quantization and multiplicities of group representations. *Invent. Math.* **67** (1982), 515–538

[JK]   Jeffrey, L.C. and F.C. Kirwan: Intersection theory on moduli spaces of holomorphic bundles of arbitrary rank on a Riemann surface. *Ann. of Math. (2)* **148** (1998), 109–196

[K]   Kirwan, F.C. On the homology of compactifications of moduli spaces of vector bundles over a Riemann surface. *Proc. London Math. Soc. (3)*, **53** (1986), 237–266

[Kie]   Kiem, Y.H., *Intersection cohomology of representation spaces of surface groups.* math.AG/0101256

[MST]   Madsen, I. and V. Snaith, J. Tornehave: Infinite loop maps in geometric topology. *Math. Proc. Cambridge Philos. Soc.* **81** (1977) 399–430

[NS]   Narasimhan, M.S. and C.S. Seshadri: Stable and unitary vector bundles on a compact Riemann surface. *Ann. of Math. (2)* **82** (1965), 540–567

[N]   Newstead, P.E., Topological properties of some spaces of stable bundles. *Topology* **6** (1967), 241–262

[PS]   Pressley, A. and G.B. Segal: *Loop Groups.* Oxford, 1986

[R]   Ramanathan, M.S., Stable principal bundles on a compact Riemann surface. *Math. Ann.* **213** (1975), 129–152

[S]   Segal, G.B., Categories and cohomology theories. *Topology* **13** (1974), 293–312

[Si]   Simpson, C.T., Homotopy over the complex numbers and generalized de Rham cohomology. Moduli of vector bundles (Sanda, 1994; Kyoto, 1994), 229–263, *Lecture Notes in Pure and Appl. Math.* **179**, Dekker, New York, 1996.

[T1]   Teleman, C., Borel-Weil-Bott theory on the moduli stack of G-bundles over a curve. *Invent. Math.* **134** (1998), 1–57

[T2]   Teleman, C., The quantization conjecture revisited. *Ann. of Math. (2)* **152** (2000), 1–43

[Th]   Thaddeus, M., Conformal field theory and the cohomology of the moduli space of stable bundles. *J. Differential Geom.* **35** (1992), 131–149

[TZ]   Tian, Y. and W. Zhang: An analytic proof of the geometric quantization conjecture of Guillemin-Sternberg. *Invent. Math.* **132** (1998), 229–259

[W]   Witten, E., Two-Dimensional gauge theories revisited. *J. Geom. Phys.* **9** (1992), 303–368

Email: teleman@dpmms.cam.ac.uk

13

Cohomology of the stable mapping class group

MICHAEL S. WEISS

University of Aberdeen

Abstract

The stable mapping class group is the group of isotopy classes of auto-morphisms of a connected oriented surface of 'large' genus. The Mumford conjecture postulates that its rational cohomology is a polynomial ring generated by certain classes κ_i of dimension $2i$, for $i > 0$. Tillmann's insight [38] that the plus construction makes the classifying space of the stable mapping class group into an infinite loop space led to a stable homotopy theory version of Mumford's conjecture, stronger than the original [25]. This stronger form of the conjecture was recently proved by Ib Madsen and myself [26]. I will describe some of the ideas which led to the proof, and some retrospective thoughts, rather than trying to condense large portions of [26].

1 The stable mapping class group and stable homotopy theory

Let $F_{g,b}$ be a connected, compact, oriented smooth surface of genus g with b boundary circles (and no 'punctures'). The topological group of smooth orientation preserving automorphisms of $F_{g,b}$ which restrict to the identity on $\partial F_{g,b}$ will be denoted by $\mathrm{Diff}(F_{g,b}; \partial)$. The *mapping class group* of $F_{g,b}$ is

$$\Gamma_{g,b} = \pi_0 \mathrm{Diff}(F_{g,b}; \partial) \,.$$

A fundamental result of Earle, Eells and Schatz [8][9] states that the discrete group $\Gamma_{g,b}$ is homotopy equivalent to $\mathrm{Diff}(F_{g,b}; \partial)$ in most cases. More precisely:

Theorem 1.1. *If $g > 1$ or $b > 0$, then the identity component of* $\mathrm{Diff}(F_{g,b}; \partial)$ *is contractible.*

It is often convenient to assume that each boundary circle of $F_{g,b}$ comes equipped with a diffeomorphism to the standard circle S^1. Where this is orientation preserving, the boundary circle is considered to be *outgoing*, otherwise *incoming*. It is customary to write

$$b_1 + b_2$$

instead of b, to indicate that there are b_1 incoming and b_2 outgoing boundary circles. A particularly important case is $F_{g,1+1}$. By gluing outgoing to incoming boundary circles, we obtain homomorphisms

$$\Gamma_{g,1+1} \times \Gamma_{h,1+1} \longrightarrow \Gamma_{g+h,1+1} . \tag{1.1}$$

They determine a multiplication on the disjoint union of the classifying spaces $B\Gamma_{g,1+1}$ for $g \geq 0$, so that the group completion

$$\Omega B\Big(\coprod_g B\Gamma_{g,1+1}\Big)$$

is defined. As is often the case, the group completion process can be replaced by a plus construction [1]. Namely, taking $h = 1$ in display (1.1) and using only the neutral element of $\Gamma_{h,1+1}$ leads to stabilization homomorphisms $\Gamma_{g,1+1} \to \Gamma_{g+1,1+1}$. We write $\Gamma_{\infty,1+1} = \mathrm{colim}_g\, \Gamma_{g,1+1}$. This is the stable mapping class group of the title. It is a perfect group; in fact $\Gamma_{g,b}$ is perfect for $g \geq 3$. Let $B\Gamma_\infty^+$ be the result of a plus construction on $B\Gamma_{\infty,1+1}$.

Proposition 1.2. $\Omega B \Big(\coprod_g B\Gamma_{g,1+1}\Big) \simeq \mathbb{Z} \times B\Gamma_\infty^+.$

The proof uses the group completion theorem, see [1], which concerns the effect of a group completion on homology. As the referee pointed out to me, the verification of the hypotheses in the group completion theorem is not a trivial matter in the present case. It relies on the homological stability theorem of Harer [17] which we state next, with the improvements due to Ivanov [20][21].

Theorem 1.3. *Let S be an oriented surface, $S = S_1 \cup S_2$ where $S_1 \cap S_2$ is a union of finitely many smooth circles in the interior of S. If $S_1 \cong F_{g,b}$ and $S \cong F_{h,c}$, then the inclusion-induced homomorphism $H_*(B\Gamma_{g,b}; \mathbb{Z}) \to H_*(B\Gamma_{h,c}; \mathbb{Z})$ is an isomorphism for $* < g/2 - 1$.*

The homological stability theorem is a very deep theorem with impressive applications, some of them much more surprising than proposition 1.2. A particularly surprising application is Tillmann's theorem [38]:

Theorem 1.4. $\mathbb{Z} \times B\Gamma_\infty^+$ *is an infinite loop space.*

Theorems 1.1 and 1.3 imply that the cohomology of $B\Gamma_\infty^+$ is a receptacle for characteristic classes of surface bundles, with fibers of 'large' genus. Following

Mumford, Miller and Morita we now use this point of view to construct elements in the cohomology of $B\Gamma_\infty^+$.

With the hypotheses of theorem 1.1, let $E \to B$ be any $F_{g,b}$–bundle with oriented fibers and trivialized boundary bundle $\partial E \to B$, that is, each fiber of $\partial E \to B$ is identified with a disjoint union of b standard circles. The vertical tangent bundle $T_B E$ of E is a two-dimensional oriented vector bundle, trivialized near ∂E, with Euler class $e = e(T_B E) \in H^2(E, \partial E; \mathbb{Z})$. Let

$$\kappa_i \in H^{2i}(B; \mathbb{Z})$$

be the image of $e^{i+1} \in H^{2i+2}(E, \partial E; \mathbb{Z})$ under the Gysin transfer map, also known as integration along the fibers

$$H^{2i+2}(E, \partial E; \mathbb{Z}) \longrightarrow H^{2i}(B; \mathbb{Z}). \tag{1.2}$$

The κ_i are, up to a sign, Mumford's characteristic classes [31] in the description of Miller [27] and Morita [28][29]. By theorem 1.1, the universal choice of B is $B\Gamma_{g,b}$ and we may therefore regard the κ_i as classes in the cohomology of $B\Gamma_{g,b}$. For $i > 0$, they are compatible with respect to homomorphisms $\Gamma_{g,b} \to \Gamma_{h,c}$ of the type considered in theorem 1.3 and we may therefore write

$$\kappa_i \in H^{2i}(B\Gamma_{\infty,1+1}; \mathbb{Z}).$$

1.5. Mumford's conjecture [31] (now a theorem)

$$H^*(B\Gamma_{\infty,1+1}; \mathbb{Q}) = \mathbb{Q}[\kappa_1, \kappa_2, \dots]$$

i.e., the classes $\kappa_i \in H^{2i}(B\Gamma_{\infty,1+1}; \mathbb{Q})$ are algebraically independent and generate $H^*(B\Gamma_{\infty,1+1}; \mathbb{Q})$ as a $\mathbb{Q}$-algebra.

The algebraic independence part was very soon established by Miller [27] and Morita [28][29]. About fifteen years later, after Tillmann had proved theorem 1.4, it was noticed by Madsen and Tillmann [25] that the Miller–Morita construction of the Mumford classes κ_i provides an important clue as to 'which' infinite loop space $\mathbb{Z} \times B\Gamma_\infty^+$ might be. Assume for simplicity that $b = 0$ in the above and that B is finite-dimensional. A choice of a fiberwise smooth embedding $E \to B \times \mathbb{R}^{2+n}$ over B, with $n \gg 0$, leads to a Thom–Pontryagin collapse map of Thom spaces

$$\text{Th}\,(B \times \mathbb{R}^{2+n}) \longrightarrow \text{Th}\,(T_B^\perp E) \tag{1.3}$$

where $T_B^\perp E$ is the fiberwise normal bundle of E in $B \times \mathbb{R}^{2+n}$. It is well known that (1.3) induces the Gysin transfer (1.2), modulo the appropriate Thom isomorphisms. Let now $\text{Gr}_2(\mathbb{R}^{2+n})$ be the Grassmannian of oriented 2-planes in $\mathbb{R}^{2+n}$ and let L_n, $L_n^\perp$ be the canonical vector bundles of dimension 2 and n on $\text{Gr}_2(\mathbb{R}^{2+n})$, respectively. Composing (1.3) with the tautological map $\text{Th}(T_B^{\perp}E) \to \text{Th}(L_n^{\perp})$ gives $\text{Th}(B \times \mathbb{R}^{2+n}) \longrightarrow \text{Th}(L_n^{\perp})$ and hence by adjunction $B \to \Omega^{2+n}\text{Th}(L_n^{\perp})$, and finally in the limit

$$B \longrightarrow \Omega^{2+\infty}\text{Th}(L_\infty^{\perp}) \qquad (1.4)$$

where $\Omega^{2+\infty}\text{Th}(L_\infty^{\perp}) = \text{colim}_n \, \Omega^{2+n}\text{Th}(L_n^{\perp})$. At this stage we can also allow an infinite-dimensional B, in particular $B = B\Gamma_g$ with the universal F_g-bundle. The case $B\Gamma_{g,b}$ can be dealt with by using a homomorphism $\Gamma_{g,b} \to \Gamma_g$ of the type considered in theorem 1.3. In this way, (1.4) leads to a map

$$\alpha_\infty : \mathbb{Z} \times B\Gamma_\infty^+ \longrightarrow \Omega^{2+\infty}\text{Th}(L_\infty^{\perp}). \qquad (1.5)$$

It is easy to recover the MMM characteristic classes κ_i by applying (1.5) to certain classes $\bar{\kappa}_i$ in the cohomology of $\Omega^{2+\infty}\text{Th}(L_\infty^{\perp})$. Namely, choose $n \gg i$ and let $\bar{\kappa}_i$ be the image of $(e(L_n))^{i+1}$ under the composition

$$H^{2i+2}(\text{Gr}_2(\mathbb{R}^{2+n}); \mathbb{Z}) \xrightarrow{\ u\ } H^{2i+2+n}(\text{Th}(L_n^{\perp}); \mathbb{Z})$$

$$\xrightarrow{\ \Omega^{2+n}\ } H^{2i}(\Omega^{2+n}\text{Th}(L_n^{\perp}); \mathbb{Z})$$

where u is the Thom isomorphism. Since $n \gg i$, we have

$$H^{2i}(\Omega^{2+n}\text{Th}(L_n^{\perp}); \mathbb{Z}) \cong H^{2i}(\Omega^{2+\infty}\text{Th}(L_\infty^{\perp}); \mathbb{Z}).$$

1.6. Madsen's integral Mumford conjecture [25], now a theorem: *The map α_∞ is a homotopy equivalence.*

Tillmann and Madsen noted in [25] that this would imply statement 1.5. They showed that α_∞ is a map of infinite loop spaces, with the Ω^∞ structure on $\mathbb{Z} \times B\Gamma_\infty^+$ from theorem 1.4, and used this fact to prove a p-local refinement of the Miller–Morita result on the rational independence of the classes κ_i, for any prime p. In the meantime Galatius [11] made a very elegant calculation of $H^*(\Omega^{2+\infty}\text{Th}(L_\infty^{\perp}); \mathbb{Z}/p)$.

2 Submersion theory and the first desingularization procedure

Let X be any smooth manifold. By Thom–Pontryagin theory, homotopy classes of maps $X \to \Omega^{2+\infty}\text{Th}(L_\infty^{\perp})$ are in bijective correspondence with bordism classes of triples $(M, q, \hat{q})$ where M is smooth, $q : M \to X$ is proper, and

$$\hat{q} : TM \times \mathbb{R}^n \to q^*TX \times \mathbb{R}^n$$

is a vector bundle surjection with a 2-dimensional oriented kernel bundle, for some n. (The correspondence is obtained by making pointed maps from Th $(X \times \mathbb{R}^n)$ to Th $(L_n^\perp)$ transverse to the zero section of $L_n^\perp$; the inverse image of the zero section is a smooth M equipped with data q and $\hat{q}$ as above.) The triples $(M, q, \hat{q})$ are best memorized as commutative squares

$$
\begin{array}{ccc}
TM \times \mathbb{R}^n & \xrightarrow{\hat{q}} & TX \times \mathbb{R}^n \\
\downarrow & & \downarrow \\
M & \xrightarrow{q} & X
\end{array}
\qquad (2.1)
$$

with $\hat{q}$ written in adjoint form.

In particular, any bundle of closed oriented surfaces $q : M \to X$ determines a triple $(M, q, \hat{q})$ with $\hat{q}$ equal to the differential of q, hence a homotopy class of maps from X to $\Omega^{2+\infty}\mathrm{Th}\,(L_\infty^\perp)$. This is the fundamental idea behind (1.5). From this angle, statement 1.6 is a 'desingularization' statement. More precisely, it is equivalent to the following:

For fixed i and sufficiently large g, every oriented i-dimensional bordism class of the degree g component of $\Omega^{2+\infty}\mathrm{Th}\,(L_\infty^\perp)$ can be represented by an F_g-bundle on a closed smooth oriented i-manifold; such a representative is unique up to an oriented bordism of F_g-bundles.

The translation uses theorem 1.3 and the fact that a map between simply connected spaces is a homotopy equivalence if and only if it induces an isomorphism in the generalized homology theory 'oriented bordism'.

Let $(M, q, \hat{q})$ be a triple as above, so that $q : M \to X$ is a proper smooth map and $\hat{q} : TM \to q^*TX$ is a stable vector bundle surjection with 2-dimensional oriented kernel. If $\hat{q}$ happens to agree with the differential dq of q, then q is a proper submersion, hence a surface bundle by Ehresmann's fibration theorem [4]. In general it is not possible to arrange this by deforming the pair $(q, \hat{q})$. One must settle for less. The approach taken in [26] is as follows.

Suppose for simplicity that X is closed. Let $E = M \times \mathbb{R}$ and let $\pi_E : E \to X$ be the composition $E \to M \to X$. By obstruction theory, $\hat{q}$ deforms to an honest surjection

$$
\hat{\pi}_E : TE \longrightarrow \pi_E^*TX
$$

of vector bundles on E, with kernel of the form $V \times \mathbb{R}$, where V is a 2-dimensional oriented vector bundle on E. Writing $\hat{\pi}_E$ in adjoint form, we can describe the situation by a commutative square

$$
\begin{array}{ccc}
TE & \xrightarrow{\hat{\pi}_E} & TX \\
\downarrow & & \downarrow \\
E & \longrightarrow & X.
\end{array}
\tag{2.2}
$$

By submersion theory [32], which is applicable here because E is an open manifold, the pair $(\pi_E, \hat{\pi}_E)$ deforms to a pair $(\pi, \hat{\pi})$ where $\pi : E \to X$ is a smooth submersion with differential $d\pi = \hat{\pi}$. See also section 3 below. The kernel of $d\pi : TE \to \pi^*TX$ is still of the form $V \times \mathbb{R}$ with 2-dimensional oriented V. In addition, we have a proper map $f : E \to \mathbb{R}$, the projection.

The 'first desingularization' procedure $(M, q, \hat{q}) \rightsquigarrow (E, \pi, f)$ is an important conceptual step. If we forget or ignore the product structure $E \cong M \times \mathbb{R}$, we can still recover $(M, q, \hat{q})$ from (E, π, f) up to bordism by forming $(N, \pi|N, d\pi|\dots)$, where $N = f^{-1}(c)$ for a regular value c of f. Let us now see how this reverse procedure reconstitutes the singularities.

Lemma 2.1. *For $z \in N$ with $\pi(z) = x$, the following are equivalent:*

- *$\pi|N$ is nonsingular at z;*
- *$f|E_x$ is nonsingular at z, where $E_x = \pi^{-1}(x)$.*

The following are also equivalent:

- *$\pi|N$ has a fold singularity at z;*
- *$f|E_x$ has a Morse singularity at z.*

Proof. Let T, V and H be the (total spaces of the) tangent bundle of E, the vertical subbundle (kernel of $d\pi$) and the horizontal quotient bundle, respectively, so that $H = T/V$. Let K be the tangent bundle of N. We are assuming that $df : T_z \to \mathbb{R}$ is onto, since $f(z) = c$ is a regular value. Hence $df|V_z$ is nonzero if and only if K_z is transverse to V_z in T_z, which means that the projection $K_z \to H_z$ is onto. This proves the first equivalence.

Suppose now that $df|V_z$ is zero. By definition, $\pi|N$ has a fold singularity at z if the differential $K_z \to H_z$ has corank 1 and the 'second derivative' of $\pi|N$, as a well-defined symmetric bilinear map Q from $\ker(K_z \to H_z)$ to $\mathrm{coker}(K_z \to H_z)$, is nondegenerate. In our situation, $\ker(K_z \to H_z) = V_z$ and $\mathrm{coker}(K_z \to H_z)$ is canonically identified, via $d\pi$, with T_z/K_z and hence via df with $\mathbb{R}$. Using local coordinates near z, it is not difficult to see that the second derivative of $f|E_x$ at z, regarded as a well-defined symmetric bilinear map from V_z to $\mathbb{R}$, is equal to $-Q$. Hence z is a nondegenerate critical point for $f|E_x$ if and only if $\pi|N$ has a fold singularity at z. $\qquad\square$

These ideas also steer us away from a bordism theoretic approach and towards a description of $\Omega^{2+\infty}\mathrm{Th}\,(L_\infty^\perp)$ in terms of 'families' of 3-manifolds.

Proposition 2.2. *The space* $\Omega^{2+\infty}\mathrm{Th}\,(L_\infty^\perp)$ *is a classifying space for 'families' of oriented 3-manifolds without boundary, equipped with a proper smooth map to* $\mathbb{R}$ *and an everywhere nonzero 1-form.*

To be more precise, the families in question are parametrized by a smooth manifold without boundary, say X. They are smooth *submersions* $\pi : E \to X$ with oriented 3-dimensional fibers. The additional data are: a smooth $f : E \to \mathbb{R}$ such that $(\pi, f) : E \to X \times \mathbb{R}$ is proper, and a vector bundle surjection from $\ker(d\pi)$, the vertical tangent bundle of E, to a trivial line bundle on E.

Two such families on X are *concordant* if their disjoint union, regarded as a family on $X \times \{0, 1\}$, extends to a family of the same type on $X \times \mathbb{R}$. The content of proposition 2.2 is that the set of concordance classes is in natural bijection with the set of homotopy classes of maps from X to $\Omega^{2+\infty}\mathrm{Th}\,(L_\infty^\perp)$. Note that both sets depend contravariantly on X.

Remark 2.3. When using proposition 2.2, beware that most smooth submersions are not bundles. For example, the inclusion of $\mathbb{R} \smallsetminus \{0\}$ in $\mathbb{R}$ and the first coordinate projection from $\mathbb{R}^2 \smallsetminus \{0\}$ to $\mathbb{R}$ are smooth submersions. Proposition 2.2 is therefore still rather far from being a description of $\Omega^{2+\infty}\mathrm{Th}\,(L_\infty^\perp)$ in terms of manifold bundles. But it is a start, and we will complement in the following sections with methods for improving submersions to bundles or decomposing submersions into bundles.

Remark 2.4. There exists another formulation of proposition 2.2 in which all 3-manifolds in sight have a prescribed boundary equal to $\{0, 1\} \times \mathbb{R} \times S^1$. This is more suitable where concatenation as in (1.1) matters. But since the equivalence of the two formulations is easy to prove, there is much to be said for working with boundariless manifolds until the concatenation issues need to be addressed.

3 More h-principles and the second desingularization procedure

Let M, N be smooth manifolds without boundary, $z \in M$. A k-*jet* from M to N at z is an equivalence class of smooth maps $f : M \to N$, where two such maps are considered equivalent if they agree to kth order at z. Let $J^k(M, N)_z$ be the set of equivalence classes and let

$$J^k(M, N) = \bigcup_z J^k(M, N)_z.$$

This has the structure of a differentiable manifold. The projection $J^k(M, N) \to M$ is a smooth bundle. Every smooth function $f \colon M \to N$ determines a smooth section $j^k f$ of the jet bundle $J^k(M, N) \to M$, the *k-jet prolongation* of f. The value of $j^k f$ at $z \in M$ is the k-jet of f at z.

A smooth section of $J^k(M, N) \to M$ is *integrable* or *holonomic* if it has the form $j^k f$ for some smooth $f \colon M \to N$. Most smooth sections of $J^k(M, N) \to M$ are not integrable. Nevertheless there exists a highly developed culture of integrability theorems up to homotopy, so-called *h*-principles [14], [10]. Such a theorem typically begins with the description of an open subbundle $A \to M$ of $J^k(M, N) \to M$, and states that the inclusion of the space of integrable sections of $A \to M$ into the space of all sections of $A \to M$ is a homotopy equivalence. (For us the cases where $k = 1$ or $k = 2$ are the most important.)

The relevance of these notions to the Mumford–Madsen project is clear if we adopt the bordism-free point of view developed in section 2. Consider a single oriented smooth 3-manifold E with a proper smooth $f \colon E \to \mathbb{R}$ and an everywhere nonvanishing 1-form, as in proposition 2.2. The map f and the 1-form together define a section of the jet bundle $J^1(E, \mathbb{R})$. If this is integrable, then f is a proper submersion. Hence $f \colon E \to \mathbb{R}$ is a bundle of oriented surfaces, again by Ehresmann's fibration theorem. The argument goes through in a parametrized setting: a family as in 2.2, parametrized by X, is a surface bundle on $X \times \mathbb{R}$ provided it satisfies the additional condition of integrability. From this point of view, statement 1.6 is roughly an *h*-principle 'up to group completion'. (It is unusual in that the source manifolds are allowed to vary.)

Examples 3.1 and 3.2 below are established *h*-principles. The *h*-principle of theorem 3.4 is closely related to a special case of 3.2 and at the same time rather similar to statement 1.6.

Example 3.1. An element in $J^1(M, N)$ can be regarded as a triple (x, y, g) where $(x, y) \in M \times N$ and g is a linear map from the tangent space of M at x to the tangent space of N at y. Let $U_1 \subset J^1(M, N)$ consist of the triples (x, y, g) where g is injective and let $U_2 \subset J^1(M, N)$ consist of the triples (x, y, g) where g is surjective. Let $\Gamma(U_1)$, $\Gamma(U_2)$ be the section spaces of the bundles $U_1 \to M$ and $U_2 \to M$, respectively. Let Γ_{itg} be the space of integrable (alias holonomic) sections of $J^1(M, N) \to M$. Note that $\Gamma_{\text{itg}} \cap \Gamma(U_1)$ is identified with the space of smooth immersions from M to N, and $\Gamma_{\text{itg}} \cap \Gamma(U_2)$ is identified with the space of smooth submersions from M to N. One of the main results of immersion theory [36], [18] is the statement that the inclusion

$$\Gamma_{\text{itg}} \cap \Gamma(U_1) \longrightarrow \Gamma(U_1)$$

is a homotopy equivalence if $\dim(M) < \dim(N)$. The main result of submersion theory [32] is that the inclusion

$$\Gamma_{\text{itg}} \cap \Gamma(U_2) \longrightarrow \Gamma(U_2)$$

is a homotopy equivalence if M is an open manifold. Gromov's 1969 thesis, outlined in [15], develops a general method for proving these and related h-principles using sheaf-theoretic arguments. This has become the standard. Much of it is reproduced in [14, §2.2]. See also [16] and [10].

Example 3.2. Fix positive integers m, n, k. Let $\mathfrak{A}$ be a closed semialgebraic subset [3] of the vector space $J^k(\mathbb{R}^m, \mathbb{R}^n)$. Suppose that $\mathfrak{A}$ is invariant under the right action of the group of diffeomorphisms $\mathbb{R}^m \to \mathbb{R}^m$, and of codimension $\geq m + 2$ in $J^k(\mathbb{R}^m, \mathbb{R}^n)$. Fix a smooth m-manifold M and let $\mathfrak{A}(M) \subset J^k(M, \mathbb{R}^n)$ consist of the jets which, in local coordinates about their source, belong to $\mathfrak{A}$. Let Γ be the space of smooth sections of $J^k(M, \mathbb{R}^n) \to M$, let $\Gamma_{\text{itg}} \subset \Gamma$ consist of the integrable sections, and let $\Gamma_{\neg\mathfrak{A}} \subset \Gamma$ consist of the sections which avoid $\mathfrak{A}(M)$. Note that $\Gamma_{\text{itg}} \cap \Gamma_{\neg\mathfrak{A}}$ is identified with the space of smooth maps from M to $\mathbb{R}^n$ having no singularities of type $\mathfrak{A}$. Vassiliev's h-principle [40, Thm 0.A], [39, III,1.1] states among other things that the inclusion

$$\Gamma_{\text{itg}} \cap \Gamma_{\neg\mathfrak{A}} \longrightarrow \Gamma_{\neg\mathfrak{A}}$$

induces an isomorphism in integral cohomology. (There is also a relative version in which M is compact with boundary.) If the codimension of $\mathfrak{A}$ is at least $m + 3$, then both $\Gamma_{\text{itg}} \cap \Gamma_{\neg\mathfrak{A}}$ and $\Gamma_{\neg\mathfrak{A}}$ are simply connected; it follows that in this case the inclusion map is a homotopy equivalence.

Vassiliev's proof of this h-principle is meticulously and admirably organized, but still not easy to read. As far as I can see, it is totally different from anything described in [14] or [10]. An overview is given in Appendix.

In theorem 3.4 below, we will need an analogue of proposition 2.2. Let $\text{Gr}_{\mathcal{W}}(\mathbb{R}^{3+n})$ be the space of 3-dimensional oriented linear subspaces $V \subset \mathbb{R}^{3+n}$ equipped with a certain type of map $q + \ell \colon V \to \mathbb{R}$. Here q is a quadratic form, ℓ is a linear form, and we require that q be nondegenerate if $\ell = 0$. Denote by

$$U_{\mathcal{W},n}, \quad U_{\mathcal{W},n}^{\perp}$$

the tautological 3-dimensional vector bundle on $\text{Gr}_{\mathcal{W}}(\mathbb{R}^{3+n})$ and its n-dimensional complement, respectively, so that $U_{\mathcal{W},n} \oplus U_{\mathcal{W},n}^{\perp}$ is a trivial vector bundle with fiber $\mathbb{R}^{3+n}$. Let

$$\Omega^{2+\infty}\mathrm{Th}\,(U_{W,\infty}^{\perp}) = \mathrm{colim}_n\,\Omega^{2+n}\mathrm{Th}\,(U_{W,n}^{\perp}).$$

Proposition 3.3. *The space $\Omega^{2+\infty}\mathrm{Th}\,(U_{W,\infty}^{\perp})$ is a classifying space for 'families' of oriented smooth 3-manifolds E_x without boundary, equipped with a section of $J^2(E_x, \mathbb{R}) \to E_x$ whose values are all of Morse type, and whose underlying map $f_x \colon E_x \to \mathbb{R}$ is proper.*

Some details: An element of $J^2(E_x, \mathbb{R})$ is, in local coordinates about its source $z \in E_x$, uniquely represented by a function of the form $q + \ell + c \colon \mathbb{R}^3 \to \mathbb{R}$ where q is a quadratic form, ℓ is a linear form and c is a constant. It is *of Morse type* if either $\ell \neq 0$ or q is nondegenerate.

The families in question are smooth submersions $\pi \colon E \to X$ where each fiber E_x is a 3-manifold with the structure and properties described in proposition 3.3. The content of proposition 3.3 is that the set of concordance classes of such families on X is in natural bijection with the set of homotopy classes of maps from X to $\Omega^{2+\infty}\mathrm{Th}\,(U_{\infty}^{\perp})$. The proof mainly uses the Thom-Pontryagin construction and submersion theory, just like the proof of proposition 2.2 sketched in section 2.

Theorem 3.4. *The space $\Omega^{2+\infty}\mathrm{Th}\,(U_{W,\infty}^{\perp})$ is also a classifying space for 'families' of oriented smooth 3-manifolds without boundary, equipped with a proper smooth Morse function.*

Clearly the simultaneous validity of theorem 3.4 and proposition 3.3 implies something like an h-principle for proper Morse functions on oriented 3-manifolds without boundary – the 'second desingularization procedure' which appears in the title of this section. (It can be applied to a family of 3-manifolds E_x as in proposition 2.2; the smooth function f_x and the 1-form together form a section of $J^1(E_x, \mathbb{R}) \to E_x$, which can also be regarded as a section of $J^2(E_x, \mathbb{R}) \to E_x$ after a choice of riemannian metric on E_x.) But it must be emphasized that variability of the 3-manifolds is firmly built in. No claim is made for the space of proper Morse functions on a single oriented 3-manifolds without boundary.

Here is an indication of how theorem 3.4 can be deduced from Vassiliev's h-principle (example 3.2) and proposition 3.3. It is not hard to show that the concordance classification of the 'families' under consideration remains unchanged if we impose the Morse condition only at level 0. This means that in theorem 3.4 we may allow families of oriented smooth 3-manifolds E_x without boundary, equipped with a proper smooth function $E_x \to \mathbb{R}$ whose critical points are nondegenerate *if the critical value is* 0. In proposition 3.3 we may allow families of oriented 3-manifolds E_x without boundary, equipped with a section $\hat{f}_x$ of $J^2(E_x, \mathbb{R}) \to E_x$ whose values are of Morse type *whenever their constant term is zero*, and whose underlying map $f_x : E_x \to \mathbb{R}$ is proper. Thus the elements of $J^2(E_x, \mathbb{R})$ to be avoided are those which, in local coordinates about their source, are represented by polynomial functions $\mathbb{R}^3 \to \mathbb{R}$ of degree at most two which have constant term 0, linear term 0 and degenerate quadratic term. These polynomial functions form a subset $\mathfrak{A}$ of $J^2(\mathbb{R}^3, \mathbb{R})$ which satisfies the conditions listed in 3.2; in particular, its codimension is $3 + 2$. Unfortunately E_x is typically noncompact, and depends on x. Nevertheless, with an elaborate justification one can use Vassiliev's h-principle here, mainly on the grounds that the 'integration up to homotopy' of a section

$$\hat{f}_x : E_x \to J(E_x, \mathbb{R})$$

satisfying the above conditions is easy to achieve outside the compact subset $f_x^{-1}(0)$ of E_x. This leads to a statement saying that two abstractly defined classifying spaces, corresponding to the two types of "families" being compared, are homology equivalent. Since the two classifying spaces come with a group-like addition law, corresponding to the disjoint union of families, the homology equivalence is a homotopy equivalence.

In the next statement, a variation on proposition 3.3, we identify $\mathrm{Gr}_2(\mathbb{R}^{2+n})$ with the closed subspace of $\mathrm{Gr}_W(\mathbb{R}^{3+n})$ consisting of the oriented 3-dimensional linear subspaces $V \subset \mathbb{R}^{3+n}$ which contain the 'first' factor $\mathbb{R} \cong \{(t, 0, 0, 0, \dots)\}$, with $q + \ell : V \to \mathbb{R}$ equal to the corresponding projection. The restriction of $U_n^{\perp}$ to $\mathrm{Gr}_2(\mathbb{R}^{2+n})$ is identified with $L_n^{\perp}$. This leads to a cofibration

$$\mathrm{Th}\,(L_n^{\perp}) \longrightarrow \mathrm{Th}\,(U_n^{\perp}).$$

In this way $\Omega^{2+n}(\mathrm{Th}\,(U_n^{\perp})/\mathrm{Th}\,(L_n^{\perp}))$ acquires a meaning.

For a smooth E_x and a section of $J^2(E_x, \mathbb{R}) \to E_x$, let the *formal singularity set* consist of the elements in E_x where the associated 2-jet is singular.

Proposition 3.5. *The space $\Omega^{2+\infty}(\mathrm{Th}\,(U_{\infty}^{\perp})/\mathrm{Th}\,(L_{\infty}^{\perp}))$ is a classifying space for 'families' of oriented smooth 3-manifolds E_x without boundary, equipped with a section of $J^2(E_x, \mathbb{R}) \to E_x$ whose values are all of Morse type, and whose underlying map $f_x : E_x \to \mathbb{R}$ is proper on the formal singularity set.*

The proof is similar to the proofs of propositions 2.2 and 3.3.

Theorem 3.6. *The space $\Omega^{2+\infty}(\mathrm{Th}\,(U_{\infty}^{\perp})/\mathrm{Th}\,(L_{\infty}^{\perp}))$ is also a classifying space for 'families' of oriented smooth 3-manifolds without boundary, equipped with a smooth Morse function which is proper on the singularity set.*

Again, the simultaneous validity of theorem 3.6 and proposition 3.5 implies something like an h-principle for Morse functions which are proper on their singularity set, and defined on oriented 3-manifolds without boundary. But this is much easier than the h-principle implicit in theorem 3.4.

Namely, let $\pi: E \to X$ with $f: E \to \mathbb{R}$ be a family of the type described in theorem 3.6. Thus π is a smooth submersion, $f|E_x$ is Morse for each $x \in X$ and $(\pi, f): E \to X \times \mathbb{R}$ is proper on Σ, where $\Sigma \subset E$ is the union of the singularity sets of all $f|E_x$. The stability of Morse singularities implies that Σ is a codimension 3 smooth submanifold of E, transverse to each fiber E_x of π. Hence $\pi|\Sigma$ is an *étale* map from Σ to X, that is, a codimension zero immersion. Choose a normal bundle N of Σ in E, in such a way that each fiber of $N \to \Sigma$ is contained in a fiber of π. It is easy to show that the family given by π and f is concordant to the family given by $\pi|N$ and $f|N$. This fact leads to a very neat concordance classification for such families, and so leads directly to theorem 3.6.

4 Strategic thoughts

For each $k \geq 0$, the functor $\Omega^{k+\infty}$ converts homotopy cofiber sequences of spectra into homotopy fiber sequences of infinite loop spaces. Applied to our situation, this gives a homotopy fiber sequence

$$\Omega^{2+\infty}\mathrm{Th}\,(L_\infty^\perp) \longrightarrow \Omega^{2+\infty}\mathrm{Th}\,(U_\infty^\perp) \longrightarrow \Omega^{2+\infty}(\mathrm{Th}\,(U_\infty^\perp)/\mathrm{Th}\,(L_\infty^\perp))$$

leading to a long exact sequence of homotopy groups for the three spaces. Combining this with the main results of the previous section, we obtain a homotopy fiber sequence

$$\Omega^{2+\infty}\mathrm{Th}\,(L_\infty^\perp) \longrightarrow |\mathcal{W}| \hookrightarrow |\mathcal{W}_{\mathrm{loc}}| \qquad (4.1)$$

where $|\mathcal{W}|$ and $|\mathcal{W}_{\mathrm{loc}}|$ classify (up to concordance) certain families of oriented smooth 3-manifolds without boundary, equipped with Morse functions. In the case of $|\mathcal{W}|$, we insist on proper Morse functions; in the case of $|\mathcal{W}_{\mathrm{loc}}|$, Morse functions whose restriction to the singularity set is proper. The details are as in theorems 3.4 and 3.6. The spaces $|\mathcal{W}|$ and $|\mathcal{W}_{\mathrm{loc}}|$ can, incidentally, be constructed directly in terms of the contravariant functors $\mathcal{W}$ and $\mathcal{W}_{\mathrm{loc}}$ which to a smooth X associate the appropriate set of 'families' parametrized by X.

There is an entirely different approach to $|\mathcal{W}|$ and $|\mathcal{W}_{\mathrm{loc}}|$, which eventually leads to a homotopy fiber sequence

$$\mathbb{Z} \times B\Gamma_\infty^+ \longrightarrow |\mathcal{W}| \hookrightarrow |\mathcal{W}_{\mathrm{loc}}| \qquad (4.2)$$

and so, in combination with (4.1), to a proof of (1.6). In this approach, $|\mathcal{W}|$ and $|\mathcal{W}_{\mathrm{loc}}|$ are seen as *stratified* spaces. The reasons for taking such a point of view are as follows.

Let a family of 3-manifolds E_x and proper Morse functions $f_x \colon E_x \to \mathbb{R}$ as in theorem 3.4 be given, where $x \in X$. For $x \in X$ let S_x be the finite set of critical points of f_x with critical value 0. It comes with a map $S_x \to \{0, 1, 2, 3\}$, the Morse index map. We therefore obtain a partition of the parameter manifold X into locally closed subsets $X_{\langle S \rangle}$, indexed by the isomorphism classes of finite sets S over $\{0, 1, 2, 3\}$. Namely, $X_{\langle S \rangle}$ consists of the $x \in X$ with $S_x \cong S$. If the family is sufficiently *generic*, the partition is a stratification (definition 5.1 below) and $X_{\langle S \rangle}$ is a smooth submanifold of X, of codimension $|S|$. At the other extreme we have the case where $X_{\langle S \rangle} = X$ for some $\langle S \rangle$; then the family is *pure of class* $\langle S \rangle$.

A careful elaboration of these matters results in a stratified model of $|\mathcal{W}|$, with strata $|\mathcal{W}_{\langle S \rangle}|$ indexed by the isomorphism classes $\langle S \rangle$ of finite sets over $\{0, 1, 2, 3\}$, where $|\mathcal{W}_{\langle S \rangle}|$ classifies families (as above) which are pure of class $\langle S \rangle$. There is a compatibly stratified model of $|\mathcal{W}_{\mathrm{loc}}|$. It turns out, and it is not all that hard to understand, that the strata $|\mathcal{W}_{\langle S \rangle}|$ and $|\mathcal{W}_{\mathrm{loc}, \langle S \rangle}|$ are also classifying spaces for certain genuine *bundle* types. More importantly, the homotopy fibers of the forgetful map $|\mathcal{W}_{\langle S \rangle}| \to |\mathcal{W}_{\mathrm{loc}, \langle S \rangle}|$ are classifying spaces for bundles of compact oriented smooth surfaces with a prescribed boundary which depends on the reference point in $|\mathcal{W}_{\mathrm{loc}, \langle S \rangle}|$. It is this information, coupled with the Harer stability result, which then leads to a description of the homotopy fiber of $|\mathcal{W}| \to |\mathcal{W}|_{\mathrm{loc}}$ in bundle-theoretic terms, i.e., to the homotopy fiber sequence (4.2).

5 Stratified spaces and homotopy colimit decompositions

This section is about a general method for extracting homotopy theoretic information from a stratification. In retrospective, the homotopy fiber sequence (4.2) can be regarded as an application of that general method.

Definition 5.1. A *stratification* of a space X is a locally finite partition of X into locally closed subsets, the *strata*, such that the closure of each stratum in X is a union of strata.

Example 5.2. Let Y be a nonempty Hausdorff space, S a finite set and $X = Y^S$. Then X is canonically stratified, with one stratum X_η for each equivalence relation η on S. Namely, $u \in X$ belongs to the stratum X_η if $s\eta t \Leftrightarrow (u_s = u_t)$ for $(s, t) \in S \times S$. The closure of X_η is the union of all X_ω with $\omega \supset \eta$.

Example 5.3. Let X be the space of Fredholm operators $\mathbb{H} \to \mathbb{H}$ of index 0, where $\mathbb{H}$ is a separable Hilbert space. See e.g. [2]. Then X is stratified, with one stratum X_n for each integer $n \geq 0$. Namely, X_n consists of the Fredholm operators f having $\dim(\ker(f)) = \dim(\operatorname{coker}(f)) = n$. Here the closure of X_n is the union of all X_m with $m \geq n$.

Definition 5.4. Let X be a stratified space. The set of strata of X becomes a poset, with $X_i \leq X_j$ if and only if the closure of X_i in X contains X_j. (*Warning*: This is the opposite of the obvious ordering.)

The main theme of this section is that stratifications often lead to homotopy colimit decompositions. I am therefore obliged to explain what a homotopy colimit is. Let $\mathcal{C}$ be a small category and let

$$F : \mathcal{C} \to Spaces$$

be a functor. The *colimit* of F is the quotient of the disjoint union $\coprod_c F(c)$ obtained by identifying $x \in F(c)$ with $g_*(x) \in F(d)$, for any morphism $g : c \to d$ in $\mathcal{C}$ and $x \in F(c)$. In general, the homotopy type of $\operatorname{colim} F$ is somewhat unpredictable. As a protection against that one may impose a condition on F.

Definition 5.5. A functor $F : \mathcal{C} \to Spaces$ is *cofibrant* if, given functors G, G' from $\mathcal{C}$ to spaces and natural transformations

$$F \xrightarrow{\;u\;} G \xleftarrow{\;e\;} G'$$

where $e : G'(c) \to G(c)$ is a homotopy equivalence for all c in $\mathcal{C}$, there exists a natural transformation $u' : F \to G'$ and a natural homotopy h from eu' to u.

If $F, G : \mathcal{C} \to Spaces$ are both cofibrant and $u : F \to G$ is a natural transformation such that $u : F(c) \to G(c)$ is a homotopy equivalence for each c in $\mathcal{C}$, then the induced map $\operatorname{colim} F \to \operatorname{colim} G$ is a homotopy equivalence. This follows immediately from definition 5.5. In this sense, colimits are well behaved on cofibrant functors. With standard resolution techniques, one can show that an arbitrary F from $\mathcal{C}$ to spaces admits a *cofibrant resolution*; i.e., there exist a cofibrant F' from $\mathcal{C}$ to spaces and a natural transformation $F' \to F$ such that $F'(c) \to F(c)$ is a homotopy equivalence for every c.

Definition 5.6. For $F : \mathcal{C} \to Spaces$ with a cofibrant resolution $F' \to F$, the *homotopy colimit* of F is the colimit of F'.

Definition 5.6 is unambiguous in the following sense: if $F' \to F$ and $F'' \to F$ are two cofibrant resolutions, then F' and F'' can be related by natural transformations $v: F' \to F''$ and $w: F'' \to F'$ such that vw and wv are *naturally* homotopic to the appropriate identity transformations. Hence colim $F' \simeq$ colim F''. Of course, there is always a standard choice of a cofibrant resolution $F' \to F$, and this depends naturally on F. With the standard choice, the following holds:

Proposition 5.7. *The homotopy colimit of F is naturally homeomorphic to the classifying space of the transport category $\mathcal{C} \int F$ of F. This has object space $\coprod_c F(c)$ and morphism space*

$$\coprod_{c,d} F(c) \times \mathrm{mor}_{\mathcal{C}}(c,d)$$

so that a morphism from $x \in F(c)$ to $y \in F(d)$ is an element $g \in \mathrm{mor}_{\mathcal{C}}(c,d)$ for which $g_*(x) = y$.

In particular, if $F(c)$ is a singleton for every c in $\mathcal{C}$, then the transport category determined by F is identified with $\mathcal{C}$ itself, and so the homotopy colimit of F is identified with the classifying space of $\mathcal{C}$.

Another special case worth mentioning, because it is well known, is the Borel construction. Let Y be a space with an action of a group G. The group is a category with one object, and the group action determines a functor from that category to spaces. In this case the homotopy colimit is the Borel construction alias *homotopy orbit space*, $EG \times_G Y$.

In [34], where Segal introduced classifying spaces of arbitrary (topological) categories, homotopy colimits also made their first appearance, namely as classifying spaces of transport categories. The derived functor approach in definition 5.6 was developed more thoroughly in [5], now the standard reference for homotopy colimits and homotopy limits, and later in [7].

Our theme is that most stratifications lead to homotopy colimit decompositions. Let us first note that many homotopy colimits are stratified. Compare [35].

Example 5.8. Let $\mathcal{C}$ be a small EI-category (all Endomorphisms in $\mathcal{C}$ are Isomorphisms). For each isomorphism class $[C]$ of objects in $\mathcal{C}$, we define a locally closed subset $B\mathcal{C}_{[C]}$ of the classifying space $B\mathcal{C}$, as follows. A point $x \in B\mathcal{C}$ is in $B\mathcal{C}_{[C]}$ if the unique cell of $B\mathcal{C}$ containing x corresponds to a diagram

$$C_0 \leftarrow C_1 \leftarrow \cdots \leftarrow C_k$$

without identity arrows, where C_0 is isomorphic to C. (Remember that $B\mathcal{C}$ is a CW-space, with one cell for each diagram $C_0 \leftarrow C_1 \leftarrow \cdots \leftarrow C_k$ as above.) Then $B\mathcal{C}$ is stratified, with one stratum $B\mathcal{C}_{[C]}$ for each isomorphism class $[C]$.

Example 5.9. Let $F: \mathcal{C} \to$ *Spaces* be a functor, where $\mathcal{C}$ is a small EI-category. Then $\mathcal{C} \int F$ is a topological EI-category; hence

$$B(\mathcal{C} \int F) = \text{hocolim } F$$

is stratified as in example 5.8, with one stratum for each isomorphism class $[C]$ of objects in $\mathcal{C}$. (This stratification can also be pulled back from the stratification of $B\mathcal{C}$ defined above, by means of the projection $B(\mathcal{C} \int F) \to B\mathcal{C}$.)

In order to show that 'most' stratified spaces can be obtained by the procedure described in example 5.8, we now associate to each stratified space X a topological category.

Definition 5.10. Let X be a stratified space with strata X_i. A path $\gamma: [0, c] \to X$, with $c \geq 0$, is *nonincreasing* if the induced map from $[0, c]$ to the poset of strata of X is nonincreasing. The *nonincreasing path category* $\mathcal{C}_X$ of a stratified space X has object set X (made discrete). The space of morphisms from $x \in X$ to $y \in X$ is the space of nonincreasing paths starting at x and ending at y. Composition of morphisms is Moore composition of paths.

Each diagram of the form $x_0 \leftarrow x_1 \leftarrow \cdots \leftarrow x_k$ in $\mathcal{C}_X$ determines real numbers $c_1, c_2, \ldots, c_k \geq 0$ and a nonincreasing path $\gamma: [0, c_1 + \cdots c_k] \to X$ with $\gamma(0) = x_k$ and $\gamma(c_1 + \cdots c_k) = x_0$. Composing γ with the linear map $\Delta^k \to [0, c_1 + \cdots c_k]$ taking the ith vertex to $c_{i+1} + \cdots + c_k$, we obtain a map $\Delta^k \to X$; and by 'integrating' over all such diagrams, we have a canonical map

$$B\mathcal{C}_X \longrightarrow X. \tag{5.1}$$

Definition 5.11. The stratified space X is *decomposable in the large* if (5.1) is a weak homotopy equivalence. It is *everywhere decomposable* if each open subset of X, with the stratification inherited from X, is decomposable in the large.

If X is decomposable in the large, we can think of (5.1) as a homotopy colimit decomposition of X, since $B\mathcal{C}_X = \text{hocolim } F$ for the functor F given by $F(x) = *$, for all objects x. Note that $\mathcal{C}_X$ is an EI-category 'up to homotopy'.

That is, for any object y of $\mathcal{C}_X$, the space of endomorphisms of y is a grouplike topological monoid

$$\text{mor}_{\mathcal{C}_X}(y, y) = \Omega(X_i, y)$$

where X_i is the stratum containing y. More to the point, the category $\pi_0(\mathcal{C}_X)$, with the same object set as $\mathcal{C}_X$ and morphism sets

$$\text{mor}_{\pi_0(\mathcal{C}_X)}(y, z) = \pi_0 \text{mor}_{\mathcal{C}_X}(y, z)$$

is an EI-category.

More useful homotopy colimit decompositions of X can often be constructed from the one above by choosing a continuous functor $p \colon \mathcal{C}_X \to \mathcal{D}$, where $\mathcal{D}$ is a discrete EI-category, and using

$$\underset{\mathcal{C}_X}{\text{hocolim}} \, F \simeq \underset{\mathcal{D}}{\text{hocolim}} \, p_* F. \tag{5.2}$$

Here $p_* F$ is the 'pushforward' of F along p, also known as the left homotopy Kan extension. It associates to an object d of $\mathcal{D}$ the homotopy colimit of $F \circ \varphi_d$, where φ_d is the forgetful functor from the 'over' category p/d to $\mathcal{C}_X$. An object of p/d consists of an object x in $\mathcal{C}_X$ and a morphism $p(x) \to d$ in $\mathcal{D}$. Consult the last pages of [6] for formula (5.2) and other useful tricks with homotopy colimits and homotopy limits.

Keeping the notation in (5.2), let $x \in X$ and let $g : p(x) \to d$ be a morphism in $\mathcal{D}$. A *lift* of (x, g) consists of a morphism $\gamma : x \to y$ in $\mathcal{C}_X$ and an isomorphism $u : p(y) \to d$ in $\mathcal{D}$ such that $u \circ p(\gamma) = g$. If X is locally 1-connected, then the set of lifts of (x, g) has a canonical topology which makes it into a covering space of a subspace of the space of all (Moore) paths in X.

Definition 5.12. Assume that X is locally 1-connected. We will say that p has *contractible chambers* if the space of lifts of (x, g) is weakly homotopy equivalent to a point, for each (x, g) as above.

Proposition 5.13. Suppose that X is locally 1-connected and p has contractible chambers. Then for d in $\mathcal{D}$, the value $(p_* F)(d)$ is weakly homotopy equivalent to a covering space of a union of strata of X; namely, the space of pairs (y, u) where $y \in X$ and $u : p(y) \to d$ is an isomorphism.

Example 5.14. Let $X = |\mathcal{W}|$ with the stratification discussed in the previous section. This is decomposable in the large. We can think of a point $x \in |\mathcal{W}|$ as a smooth oriented 3-manifold E_x without boundary, with a proper map

$f_x : E_x \to \mathbb{R}$. A path from x to y in $|W|$ amounts to a family of smooth oriented 3-manifolds $E_{\gamma(t)}$, each without boundary and with a proper map $f_{\gamma(t)} : E_{\gamma(t)} \to \mathbb{R}$; the parameter t runs through an interval $[0, c]$ and $\gamma(0) = x$, $\gamma(c) = y$. For each $t \in [0, c]$ let $S_{\gamma(t)}$ be the set of critical points of $f_{\gamma(t)}$ with critical value 0; this comes with a map to $\{0, 1, 2, 3\}$, the Morse index map. If the path is nonincreasing, then it is easy to identify each $S_{\gamma(t)}$ with a subset of $S_{\gamma(0)} = S_x$, in such a way that we have a nonincreasing family of subsets $S_{\gamma(t)}$ of the finite set S_x, parametrized by $t \in [0, c]$. With every $z \in S_x = S_{\gamma(0)}$ which is not in the image of $S_y = S_{\gamma(c)}$, we can associate an element $\varepsilon(z) \in \{-1, +1\}$, as follows. There is a largest $t \in [0, c]$ such that $z \in S_{\gamma(t)}$; call it $t(z)$. The stability property of nondegenerate critical points ensures that for t just slightly larger than $t(z)$, the element z viewed as a point in $S_{\gamma(t(z))} \subset E_{\gamma(t(z))}$ is close to a unique critical point of $f_{\gamma(t)} : E_{\gamma(t)} \to \mathbb{R}$. The latter has critical value either greater than 0, in which case $\varepsilon(z) = +1$, or less than 0, in which case $\varepsilon(z) = -1$. Summarizing, a morphism $\gamma : x \to y$ in $\mathcal{C}_X$ determines an injective map $\gamma^* : S_y \to S_x$ over $\{0, 1, 2, 3\}$ and a function from $S_x \smallsetminus u(S_y)$ to the set $\{+1, -1\}$.

These considerations lead us to a certain category $\mathcal{K}$. Its objects are the finite sets over $\{0, 1, 2, 3\}$; a morphism from S to T in $\mathcal{K}$ is an injective map u from S to T over $\{0, 1, 2, 3\}$, together with a function ε from $T \smallsetminus u(S)$ to $\{-1, +1\}$. The composition of two composable morphisms $(u_1, \varepsilon_1) : R \to S$ and $(u_2, \varepsilon_2) : S \to T$ in $\mathcal{K}$ is $(u_2 u_1, \varepsilon_3)$, where ε_3 agrees with ε_2 on $T \smallsetminus u_2(S)$ and with $\varepsilon_1 u_2^{-1}$ on $u_2(S \smallsetminus u_1(R))$. The rule $x \mapsto S_x$ described above is a functor p from $\mathcal{C}_X$ to $\mathcal{K}^{\mathrm{op}}$. This functor has contractible chambers. For S in $\mathcal{K}$, the space $(p_* F)(S)$ is therefore, by proposition 5.13, a classifying space for families of 3-manifolds E_x equipped with $f_x : E_x \to \mathbb{R}$ as in theorem 3.4 and with a *specified* isomorphism $S_x \to S$ in $\mathcal{K}$. Consequently, it can be identified with a finite-sheeted covering space of the stratum $|W_{(S)}|$ of $|W|$; see section 4. It is convenient to write $|W_S|$ for $(p_* F)(S)$.

Proposition 5.13 does not say anything very explicit about the map $|W_S| \to |W_R|$ induced by a morphism $(u, \varepsilon) : R \to S$ in $\mathcal{K}$, but this is easily described up to homotopy. Namely, suppose given a family of smooth 3-manifolds E_x with Morse functions f_x and isomorphisms $a_x : S_x \to S$, as above. Now perturb each f_x slightly by adding a small smooth function $g_x : E_x \to \mathbb{R}$ with support in a small neighborhood of S_x, locally constant in a smaller neighborhood of S_x, and such that for $z \in S_x$ we have

$$g_x(z) \begin{cases} = 0 & \text{if} \quad a_x(z) \in u(R) \\ > 0 & \text{if} \quad \varepsilon(a_x(z)) = +1 \\ < 0 & \text{if} \quad \varepsilon(a_x(z)) = -1. \end{cases}$$

(The g_x should also depend smoothly on the parameter x, like the f_x.) The set of critical points of $f_x + g_x$ having critical value 0 is then identified with $u(R) \cong R$. Therefore, by keeping the E_x and substituting the $f_x + g_x$ for the f_x, we obtain a family of the type which is classified by maps to $|\mathcal{W}_R|$. Letting the parameter manifold approximate $|\mathcal{W}_S|$, we obtain a well-defined homotopy class of maps $|\mathcal{W}_S| \to |\mathcal{W}_R|$.

For fixed S, the characterization of $|\mathcal{W}_S|$ as a classifying space for families of oriented 3-manifolds E_x with proper Morse functions $f_x : E_x \to \mathbb{R}$ and isomorphisms $S_x \cong S$ can be simplified. It turns out that we need only allow Morse functions f_x having no other critical points than those in S_x; that is, no critical values other than, possibly, 0. When this extra condition is imposed, the families considered are automatically *bundles* of 3-manifolds over the parameter space – not just submersions with 3-dimensional fibers. This is an easy consequence of Ehresmann's fibration theorem. Equally important is the fact that each of the 3-manifolds E_x in such a bundle can be reconstructed from the closed oriented surface $f_x^{-1}(-1)$ and certain surgery data. These data are instructions for disjoint oriented surgeries [41, §1] on the surface, one for each element of $S_x \cong S$.

In this way, we end up with a description of $|\mathcal{W}_S|$ as a classifying space for bundles of closed oriented surfaces, where each surface comes with data for disjoint oriented surgeries labelled by elements of S.

The stratification of $|\mathcal{W}_{\mathrm{loc}}|$ sketched in the previous section can be taken to pieces in a similar fashion. The result is a homotopy colimit decomposition

$$|\mathcal{W}_{\mathrm{loc}}| \simeq \operatorname*{hocolim}_{S} |\mathcal{W}_{\mathrm{loc},S}|$$

where S runs through $\mathcal{K}$. Here $|\mathcal{W}_{\mathrm{loc},S}|$ should be thought of as the space of S-tuples of oriented surgery instructions on an oriented surface – but without a specified surface! See [26] for details. The homotopy colimit decompositions for $|\mathcal{W}|$ and $|\mathcal{W}_{\mathrm{loc}}|$ are related via obvious forgetful maps.

Now, in order to obtain information about the homotopy fibers of $|\mathcal{W}| \to |\mathcal{W}_{\mathrm{loc}}|$, one can ask what the homotopy fibers of

$$|\mathcal{W}_S| \to |\mathcal{W}_{\mathrm{loc},S}|$$

are, for each S in $\mathcal{K}$, and then how they vary with S. The first question is easy to answer: the homotopy fibers of $|\mathcal{W}_S| \to |\mathcal{W}_{\mathrm{loc},S}|$ are classifying spaces for bundles of compact oriented surfaces with a prescribed boundary depending on the chosen base point in $|\mathcal{W}_{\mathrm{loc},S}|$. The dependence on S can be seen in commutative squares of the form

$$\begin{array}{ccc} |W_S| & \longrightarrow & |W_{\mathrm{loc},S}| \\ \downarrow & & \downarrow \\ |W_R| & \longrightarrow & |W_{\mathrm{loc},R}| \end{array}$$

where the vertical arrows are induced by a morphism $R \to S$ in $\mathcal{K}$. With the above geometric description, the induced maps from the homotopy fibers of the top horizontal arrow to the homotopy fibers of the bottom horizontal arrow are maps of the kind considered in Harer's stability theorem 1.3. Unfortunately the stability theorem cannot be used here without further preparation: there is no reason to suppose that all surfaces in sight are connected and of large genus. Fortunately, however, the homotopy colimit decomposition of $|W|$ described in 5.14 can be rearranged and modified in such a way that this objection can no longer be made. (At this point, concatenation matters must be taken seriously and consequently some of the main results obtained so far must be reworded, as explained in remark 2.4.) The stability theorem 1.3 can then be applied and the homotopy fiber sequence (4.2) is a formal consequence.

To conclude, it seems worthwhile to stress that the Harer stability theorem 1.3 is an enormously important ingredient in the proof of Madsen's conjecture 1.6. But in contrast to Vassiliev's h-principle (example 3.2), which is an equally important ingredient, the stability theorem only makes a very brief and decisive appearance at the end of the proof. There it is used almost exactly as in Tillmann's proof of 1.4.

Appendix Vassiliev's h-principle: An outline of the proof

This outline covers only the case where the manifold M is closed. It follows [40] in all essentials. I have made some minor rearrangements in the overall presentation, emphasizing the way in which transversality theory and interpolation theory shape the proof. I am indebted to Thomas Huettemann for suggesting this change in emphasis. Any errors and exaggerations which may have resulted from it should nevertheless be blamed on me. Besides, it is not a big change: *plus ça change, plus c'est la même chose*.

Let Z be the topological vector space of all smooth maps $M \to \mathbb{R}^n$, with the Whitney C^∞ topology [13]. Let $Z_{\mathfrak{A}} \subset Z$ be the closed subset consisting of those $f : M \to \mathbb{R}^n$ which have at least one singularity of type $\mathfrak{A}$. Then $Z \smallsetminus Z_{\mathfrak{A}}$ is identified with $\Gamma_{\mathrm{itg}} \cap \Gamma_{\neg \mathfrak{A}}$. As our starting point, we take the idea to approximate $Z \smallsetminus Z_{\mathfrak{A}}$ by subspaces of the form $D \smallsetminus Z_{\mathfrak{A}}$ where D can be any finite dimensional *affine* subspace of Z; in other words, D is a translate of a finite-dimensional linear subspace. To be more precise, let r be a positive integer; we will look for finite-dimensional affine subspaces D of Z such that the inclusion-induced map in integer cohomology

$$H^*(Z \smallsetminus Z_{\mathfrak{A}}) \longrightarrow H^*(D \smallsetminus Z_{\mathfrak{A}}) \tag{A.1}$$

is an isomorphism for $* < r$.

Vassiliev's method for solving this important approximation problem is to impose a general position condition ($\mathbf{c_1}$) and an interpolation condition ($\mathbf{c_{2,r}}$) on D. The two conditions are described just below. Further down there is a sketch of Vassiliev's argument showing that (A.1) is indeed an isomorphism for $* < r$ if D satisfies both ($\mathbf{c_1}$) and ($\mathbf{c_{2,r}}$). The h-principle then 'falls out' as a corollary.

Condition ($\mathbf{c_1}$) requires, roughly, that the finite-dimensional affine subspace $D \subset Z$ be in general position relative to $Z_{\mathfrak{A}}$.

Vassiliev is not very precise on this point, but I understand from [23] that every semialgebraic subset S of a finite-dimensional real vector space V has a preferred *regular* stratification. This is a partition of S into smooth submanifolds of V which satisfies the conditions for a stratification and, in addition, Whitney's regularity conditions [42]. In particular, the portion of $\mathfrak{A}$ lying over $0 \in \mathbb{R}^m$ has a preferred regular stratification; it follows that $\mathfrak{A}(M)$ has a preferred regular stratification as a subset of the smooth manifold $J^k(M, \mathbb{R}^n)$.

Definition A.1. Let D be a finite-dimensional affine subspace of Z. We say that D satisfies condition ($\mathbf{c_1}$) if:

- the map $u : D \times M \longrightarrow J^k(M, \mathbb{R}^n)$ given by $u(f, x) = j^k f(x)$ is transverse to each stratum of $\mathfrak{A}(M)$, so that $u^{-1}(\mathfrak{A}(M))$ is regularly stratified in $D \times M$;
- the projection from $u^{-1}(\mathfrak{A}(M))$ to D is *generic*.

The second item in definition A.1 amounts to a condition on the multijets [13] of the evaluation map $D \times M \to \mathbb{R}^n$ at finite subsets of $u^{-1}(\mathfrak{A}(M))$. The condition implies local injectivity of the projection from $u^{-1}(\mathfrak{A}(M))$ to D, and self-transversality in a stratified setting. More precision would take us too far.

The content of the much more striking condition ($\mathbf{c_{2,r}}$) is that D must contain at least one solution for each interpolation problem on M of a certain type depending on r.

Definition A.2. Let d_{kmn} be the dimension of the real vector space of degree $\leq k$ polynomial maps from $\mathbb{R}^m$ to $\mathbb{R}^n$. Let D be a finite-dimensional affine subspace of Z. We say that D satisfies condition ($\mathbf{c_{2,r}}$) if, for every

$C^\infty(M, \mathbb{R})$-submodule Y of Z with $\dim_\mathbb{R}(Z/Y) \leq r \cdot d_{kmn}$, the projection $D \to Z/Y$ is onto.

To see what this has to do with interpolation, fix distinct points $z_1, z_2, \ldots, z_s$ in M, with $s \leq r$, and k-jets $u_1, u_2, \ldots, u_s \in J^k(M, \mathbb{R}^n)$ so that z_i is the source of u_i. Let Y consist of the $f \in Z$ whose k-jets at $z_1, z_2, \ldots, z_s$ vanish. If D satisfies $(c_{2,r})$, then $D \to Z/Y$ must be onto and so there exists $g \in D$ with $j^k g(z_i) = u_i$ for $i = 1, 2, \ldots, s$.

Let $\mathcal{L}_r$ be the collection of all finite-dimensional affine subspaces D of Z satisfying (c_1) and $(c_{2,r})$. Then we have $\mathcal{L}_{r+1} \subset \mathcal{L}_r$ for all $r \geq 0$.

Lemma A.3. *There exists an increasing sequence of finite-dimensional affine subspaces $D_1, D_2, D_3, D_4, \ldots$ of Z such that $D_i \in \mathcal{L}_i$ and the union $\bigcup_i D_i$ is dense in the space Z.*

The proof of this is an application of transversality theory as in [13] and interpolation theory as in [12]. The fact that the set of all $C^\infty(M, \mathbb{R})$-submodules Y of Z satisfying the conditions in definition A.2 has a canonical topology making it into a *compact* Hausdorff space is an essential ingredient.

Theorem A.4. *The map (A.1) is an isomorphism if $D \in \mathcal{L}_r$ and $* < r$.*

Sketch proof. Write $D_\mathfrak{A} = D \cap Z_\mathfrak{A}$. In the notation of definition A.1, this is the image of $u^{-1}(\mathfrak{A}(M))$ under the projection $D \times M \to D$. Condition (c_1) on D ensures that $D_\mathfrak{A}$ is a well-behaved subset of D, so that there is an Alexander duality isomorphism

$$H^*(D \smallsetminus Z_\mathfrak{A}) \xrightarrow{\ \cong\ } H^{lf}_{\dim(D)-*-1}(D_\mathfrak{A}) \tag{A.2}$$

where the superscript *lf* indicates that locally finite chains are used. To investigate $D_\mathfrak{A}$, Vassiliev introduces a *resolution* $RD_\mathfrak{A}$ of $D_\mathfrak{A}$, as follows. Let $\Delta(M)$ be the simplex spanned by M, in other words, the set of all functions w from M to $[0, 1]$ such that $\{x \in M \mid w(x) > 0\}$ is finite and $\sum_{x \in M} w(x) = 1$. The standard topology of $\Delta(M)$ as a simplicial complex is not of interest here, since it does not reflect the topology of M. Instead, we endow $\Delta(M)$ with the smallest topology such that, for each continuous $g : M \to \mathbb{R}$, the map $w \mapsto \sum_x w(x)g(x)$ is continuous on $\Delta(M)$. We write $\Delta(M)_t$ to indicate this topology. Now $RD_\mathfrak{A}$ can be defined as the subspace of $D_\mathfrak{A} \times \Delta(M)_t$ consisting of all (f, w) such that the support of w is contained in the set of $\mathfrak{A}$-singularities of f. Because D satisfies condition (c_1), the projection

$$RD_\mathfrak{A} \longrightarrow D_\mathfrak{A}$$

is a proper map between locally compact spaces. Each of its fibers is a simplex, and it is not difficult to deduce that it induces an isomorphism in locally finite homology

$$H_*^{lf}(RD_\mathfrak{A}) \xrightarrow{\cong} H_*^{lf}(D_\mathfrak{A}). \tag{A.3}$$

For an integer p, let $RD_\mathfrak{A}^p \subset RD_\mathfrak{A}$ consist of the pairs (f, w) where the support of w has at most p elements. The filtration of $RD_\mathfrak{A}$ by the closed subspaces $RD_\mathfrak{A}^p$ leads in the usual manner to a homology spectral sequence of the form

$$E_{p,q}^1 = H_{p+q+\dim(D)}^{lf}(RD_\mathfrak{A}^p, RD_\mathfrak{A}^{p-1}) \Longrightarrow H_{p+q+\dim(D)}^{lf}(RD_\mathfrak{A}) \tag{A.4}$$

where $p, q \in \mathbb{Z}$. There are three vanishing lines: $E_{p,q}^1 = 0$ for $p < 0$ and for $p + q < -\dim(D)$ by construction, but also

$$E_{p,q}^1 = 0 \text{ when } 2p + q > 0. \tag{A.5}$$

To understand (A.5), note that by the general position condition (c_1) on D, the codimension of the image of $RD_\mathfrak{A}^p$ in D is at least $p(\operatorname{codim}(\mathfrak{A}) - m)$; here $\operatorname{codim}(\mathfrak{A})$ denotes the codimension of $\mathfrak{A}$ in $J^k(\mathbb{R}^m, \mathbb{R}^n)$. Since the fibers of the projection $RD_\mathfrak{A}^p \to D$ are at most p-dimensional, it follows that the *dimension* of $RD_\mathfrak{A}^p$ is not greater than $p + \dim(D) - p(\operatorname{codim}(\mathfrak{A}) - m)$. With our hypothesis $\operatorname{codim}(\mathfrak{A}) \geq m + 2$ this implies

$$\dim(RD_\mathfrak{A}^p) \leq \dim(D) - p$$

and (A.5) follows.

Now comes the crucial observation that $E_{p,q}^1$ does not depend on our choice of $D \in \mathcal{L}_r$, as long as $p \leq r$. To see this, we use the multi-jet prolongation map

$$RD_\mathfrak{A} \longrightarrow \Delta(\mathfrak{A}(M))_t \tag{A.6}$$

which takes $(f, w) \in RD_\mathfrak{A}$ to $\bar{w}$ with $\bar{w}(u) = w(x)$ if $u = j^k f(x)$ and $\bar{w}(u) = 0$ otherwise. Here $\Delta(\mathfrak{A}(M))_t$ is the simplex spanned by the set $\mathfrak{A}(M) \subset J^k(M, \mathbb{R}^n)$, but again topologized so that the topology of $\mathfrak{A}(M)$ is reflected; cf. the definition of $\Delta(M)_t$. Note that each fiber of (A.6) is identified with an affine subspace of D; but the fiber dimensions can vary and some fibers may even be empty. But restricting (A.6), we have

$$RD_\mathfrak{A}^p \smallsetminus RD_\mathfrak{A}^{p-1} \longrightarrow \Delta(\mathfrak{A}(M))_t^p \smallsetminus \Delta(\mathfrak{A}(M))_t^{p-1} \tag{A.7}$$

where $\Delta(\mathfrak{A}(M))_t^p$ consists of the $w \in \Delta(\mathfrak{A}(M))_t$ whose support has at most p elements, with distinct tabels in M. Now the interpolation condition $(c_{2,r})$ on D and our assumption $p \leq r$ imply that the fibers of (A.7) are *nonempty* affine spaces, and all of the same dimension; in other words, (A.7) is a bundle of affine spaces. Its base space obviously does not depend on D, and it can be shown that its first Stiefel–Whitney class, too, is independent of D. Consequently the locally finite homology of the total space

$$H_*^{lf}(RD_{\mathfrak{A}}^p \setminus RD_{\mathfrak{A}}^{p-1}) \cong H_*^{lf}(RD_{\mathfrak{A}}^p, RD_{\mathfrak{A}}^{p-1}) = E_{p,*-p-\dim(D)}^1$$

is identified with the locally finite homology of the base space, with twisted integer coefficients, and so is independent of D except for the obvious dimension shift. (Note the strong excision property of locally finite homology groups.) To state the independence result more precisely, the spectral sequence (A.4) depends contravariantly on D, and for $C, D \in \mathcal{L}_r$ with $C \subset D$, the induced map from the D-version of $E_{p,q}^1$ to the C-version of $E_{p,q}^1$ is an isomorphism whenever $p \leq r$.

Remembering (A.5) now, we can immediately deduce that $E_{p,q}^m$ is also independent of $D \in \mathcal{L}_r$, in the same sense, for any $m \geq 0$ and p, q with $p + q \geq -r$. Remembering the isomorphisms (A.3) and (A.2) also, we then conclude that for $C, D \in \mathcal{L}_r$ with $C \subset D$, the inclusion $C \setminus Z_{\mathfrak{A}} \to D \setminus Z_{\mathfrak{A}}$ induces an isomorphism

$$H^*(D \setminus Z_{\mathfrak{A}}) \longrightarrow H^*(C \setminus Z_{\mathfrak{A}})$$

for $* < r$. With lemma A.3, this leads us finally to the statement that (A.1) is an isomorphism for $* < r$ and $D \in \mathcal{L}_r$. (First suppose that D is one of the D_i in lemma A.3; then for the general case, approximate D by affine subspaces of the D_i for $i \gg 0$.) $\qquad \square$

But we have achieved much more. Letting r tend to ∞, we have a well defined spectral sequence converging to

$$H^*(Z \setminus Z_{\mathfrak{A}}) = H^*(\Gamma_{\text{itg}} \cap \Gamma_{\neg \mathfrak{A}})$$

independent of r. (Convergence is a consequence of (A.5), and again lemma A.3 is needed to show that the spectral sequence is independent of all choices.) Similar but easier reasoning leads to an analogous spectral sequence converging to the cohomology of $\Gamma_{\neg \mathfrak{A}}$. By a straightforward inspection, the inclusion of $\Gamma_{\text{itg}} \cap \Gamma_{\neg \mathfrak{A}}$ in $\Gamma_{\neg \mathfrak{A}}$ induces an isomorphism of the E^1-pages. This establishes Vassiliev's h-principle in the case where M is closed.

References

[1] Adams, J. F. Infinite loop spaces, *Annals of Math.* Studies 90, Princeton Univ. Press (1978).

[2] Atiyah, M. F. Algebraic Topology and Operators in Hilbert Space, *Lectures in Modern Analysis and Applications I, Springer Lect. Notes in Math.* 103 (1969), Springer, 101–122.

[3] Benedetti, R. and J.-J. Risler Real algebraic and semi-algebraic sets, *Actualités mathématiques*, Hermann, Paris (1990).

[4] Bröcker, T. and K. Jänich *Introduction to Differential Topology*, English edition Cambridge Univ. Press (1982); German edition Springer-Verlag (1973).

[5] Bousfield, A. K. and D. Kan *Homotopy limits, completions and localizations*, Springer Lecture Notes in Math. 304, Springer (1972).

[6] Dwyer, W. and D. M. Kan A classification theorem for diagrams of simplicial sets, *Topology* 23 (1984) 139–155.

[7] Dror, E. Homotopy and homology of diagrams of spaces, in *Algebraic Topology, Seattle Wash. 1985, Springer Lec. Notes in Math.* 1286 (1987) 93–134.

[8] Earle, C. J. and J. Eells A fibre bundle description of Teichmueller theory, *J. Differential Geom.* 3 (1969) 19–43.

[9] Earle, C. J. and A. Schatz Teichmueller theory for surfaces with boundary, *J. Differential Geom.* 4 (1970) 169–185.

[10] Eliashberg, Y. and N. Mishachev *Introduction to the h-principle*, Graduate Studies in Math., Amer Math. Soc. (2002).

[11] Galatius, S. Mod p homology of the stable mapping class group, to appear.

[12] Glaeser, G. L'interpolation des fonctions différentiables de plusieurs variables, Proceedings of Liverpool Singularities Symposium, II, 1969/1970, *Springer Lect. Notes in Math.* 209 (1971), Springer, 1–33.

[13] Golubitsky, M. and V. Guillemin *Stable mappings and their singularities*, Graduate Texts in Math. (1973), Springer-Verlag.

[14] Gromov, M. *Partial Differential Relations*, Springer-Verlag, Ergebnisse series (1984).

[15] Gromov, M. *A topological technique for the construction of solutions of differential equations and inequalities*, Actes du Congrès International des Mathematiciens, Nice 1970, Gauthier–Villars (1971).

[16] Haefliger, A. *Lectures on the theorem of Gromov*, in Proc. of 1969/70 Liverpool Singularities Symp., Lecture Notes in Math. vol. 209, Springer (1971) 128-141.

[17] Harer, J. L. *Stability of the homology of the mapping class groups of oriented surfaces*, Ann. of Math. 121 (1985) 215–249.

[18] Hirsch, M. *Immersions of manifolds*, Trans. Amer. Math. Soc. 93 (1959) 242–276.

[19] Hirsch, M. *Differential Topology*, Springer–Verlag (1976).

[20] Ivanov, N. V. *Stabilization of the homology of the Teichmueller modular groups*, Algebra i Analiz 1 (1989) 120–126; translation in: Leningrad Math. J. 1 (1990) 675–691.

[21] Ivanov, N. V. *On the homology stability for Teichmüller modular groups: closed surfaces and twisted coefficients*, in: Mapping Class groups and moduli spaces of Riemann surfaces (Göttingen 1991, Seattle 1991), Contemp. Math. 150, Amer. Math. Soc. (1993), 149–194.

[22] Lang, S. *Differential manifolds*, Addison-Wesley (1972).

[23] Lojasiewicz, S. *Stratification des ensembles analytiques avec les propriétés (A) et (B) de Whitney*, Colloq. Internat. C.N.R.S. No. 208, Paris 1972, "Agora Mathematica", no.1, Gauthier–Villars (1974) 116–130.

[24] MacLane, S. *Categories for the working mathematician*, Grad. texts in Math., Springer-Verlag (1971).

[25] Madsen, I. and U. Tillmann *The stable mapping class group and $Q(\mathbb{C}P^{\infty})$*, Invent. Math. 145 (2001) 509–544.

[26] Madsen, I. and M. Weiss *The stable moduli space of Riemann surfaces: Mumford's conjecture*, preprint, arXiv:math.AT/0212321.

[27] Miller, E. *The homology of the mapping class group*, J. Diff. Geometry 24 (1986) 1–14.

[28] Morita, S. *Characteristic classes of surface bundles*, Bull. Amer. Math. Soc. 11 (1984) 386–388.

[29] Morita, S. *Characteristic classes of surface bundles*, Invent. Math. 90 (1987) 551–557.

[30] Morita, S. *Geometry of characteristic classes*, Transl. of Math. Monographs, Amer. Math. Soc. (2001); Japanese original published by Iwanami–Shoten (1999).

[31] Mumford, D. *Towards an enumerative geometry of the moduli space of curves*, Arithmetic and geometry, Vol. II, Progr. Math. 36, Birkhäuser (1983) 271–328.

[32] Phillips, A. *Submersions of open manifolds*, Topology 6 (1967) 170–206.

[33] Quillen, D. *Elementary proofs of some results of cobordism theory using Steenrod operations*, Advances in Math. 7 (1971) 29–56.

[34] Segal, G. *Classifying spaces and spectral sequences*, Inst. Hautes Etudes Sci. Publ Math. 34 (1968) 105–112.

[35] Slominska, J. *Homotopy colimits on E-I-categories*, in Alg. topology Poznan 1989, Springer Lect. Notes in Math. vol. 1474 (1991) 273–294.

[36] Smale, S. *The classification of immersions of spheres in Euclidean spaces*, Ann. of Math. 69 (1959) 327–344.

[37] Stong, R. E. *Notes on cobordism theory*, Mathematical Notes, Princeton University Press (1968).

[38] Tillmann, U. *On the homotopy of the stable mapping class group*, Invent. Math. 130 (1997) 257–275.

[39] Vassiliev, V. *Complements of Discriminants of Smooth Maps: Topology and Applications*, Transl. of Math. Monographs Vol. 98, revised edition, Amer. Math. Soc. (1994).

[40] Vassiliev, V. *Topology of spaces of functions without complicated singularities*, Funktsional Anal. i Prilozhen 93 no. 4 (1989) 24–36; Engl. translation in Funct. Analysis Appl. 23 (1989) 266–286.

[41] Wall, C. T. C. *Surgery on compact manifolds*, Lond. Math. Soc. Monographs, Acad. Press (1970); second ed. (ed. A.A.Ranicki), Amer. Math. Soc. Surveys and Monographs 69 (1999).

[42] Whitney, H. *Tangents to an analytic variety*, Ann. of Math. 81 (1965) 496–549.

Email: mweiss@maths.abdn.ac.uk.

14

Conformal field theory in four and six dimensions

EDWARD WITTEN

Institute for Advanced Study
Princeton

1 Introduction

In this paper, I will be considering conformal field theory (CFT) mainly in four
and six dimensions, occasionally recalling facts about two dimensions. The
notion of conformal field theory is familiar to physicists. From a mathematical
point of view, we can keep in mind Graeme Segal's definition [1] of conformal
field theory. Instead of just summarizing the definition here, I will review how
physicists actually study examples of quantum field theory, as this will make
clear the motivation for the definition.

When possible (and we will later consider examples in which this is not
possible), physicists make models of quantum field theory using path integrals.
This means first of all that, for any n-manifold M_n, we are given a space of
fields on M_n; let us call the fields Φ. The fields might be, for example, real-
valued functions, or gauge fields (connections on a G-bundle over M_n for some
fixed Lie group G), or p-forms on M_n for some fixed p, or they might be maps
$\Phi : M_n \to W$ for some fixed manifold W. Then we are given a local action
functional $I(\Phi)$. 'Local' means that the Euler–Lagrange equations for a critical
point of I are partial differential equations. If we are constructing a quantum
field theory that is not required to be conformally invariant, I may be defined
using a metric on M_n. For conformal field theory, I should be defined using
only a conformal structure. For a closed M_n, the partition function $Z(M_n)$ is
defined, formally, as the integral over all Φ of $e^{-I(\Phi)}$

$$Z(M_n) = \int D\Phi \, \exp(-I(\Phi)). \tag{1.1}$$

If M_n has a boundary M_{n-1}, the integral depends on the boundary conditions.
If we let φ denote the restriction of Φ to M_{n-1}, then it formally makes sense
to consider a path integral on a manifold with boundary in which we integrate

Supported in part by NSF Grant PHY-0070928.

over all Φ for some fixed φ. This defines a function

$$\Psi(\varphi) = \int_{\Phi|_{M_{n-1}}=\varphi} D\Phi \, \exp(-I(\Phi)). \qquad (1.2)$$

We interpret the function $\Psi(\varphi)$ as a vector in a Hilbert space $\mathcal{H}(M_{n-1})$ of $\mathbf{L}^2$ functions of φ. From this starting point, one can motivate the sort of axioms for quantum field theory that Segal considered. I will not go into detail, as we will not need them here. In fact, to keep things simple, we will mainly consider closed manifolds M_n and the partition function $Z(M_n)$.

Before getting to the specific examples that we will consider, I will start with a general survey of conformal field theory in various dimensions. Two-dimensional conformal field theory plays an important role in string theory and statistical mechanics and is also relatively familiar mathematically.[1] For example, rational conformal field theory is studied in detail using complex geometry. More general conformal field theories underlie, for example, mirror symmetry.

Three- and four-dimensional conformal field theory is also important for physics. Three-dimensional conformal field theory is used to describe second-order phase transitions in equilibrium statistical mechanics, and a four-dimensional conformal field theory could conceivably play a role in models of elementary particle physics.

Physicists used to think that four was the maximum dimension for non-trivial (or non-Gaussian) unitary conformal field theory. Initially, therefore, little note was taken of a result by Nahm [2] which implies that *six* is the maximum possible dimension in the supersymmetric case. (A different result proved in the same paper – eleven is the maximal possible dimension for super-gravity – had a large impact right away.) Nahm's result follows from an algebraic argument and I will explain what it says in section 3. String theorists have been quite surprised in the last few years to learn that the higher-dimensional superconformal field theories, whose existence is suggested by Nahm's theorem, apparently do exist. Explaining this, or at least giving a few hints, is the goal of this article.

One of the surprises is that the new theories suggested by Nahm's theorem are theories for which there is apparently no Lagrangian – at least none that can be constructed using classical variables of any known sort. Yet these new theories are intimately connected with fascinating mathematics and physics of more conventional theories in four dimensions.

---

[1] In counting dimensions, we include time, so a two-dimensional theory, if formulated in Lorentz signature, is a theory in a world of one space and one time dimension. Here, we will mostly work with Euclidean signature.

In section 2, we warm up with some conventional and less conventional linear theories. Starting with the example of abelian gauge theory in four dimensions, I will describe some free or in a sense linear conformal field theories that can be constructed in arbitrary even dimensions. The cases of dimension $4k$ and $4k + 2$ are rather different, as we will see. The most interesting linear theory in $4k + 2$ dimensions is a self-dual theory that does not have a Lagrangian, yet it exists quantum mechanically and its existence is related to subtle modular behavior of the linear theories in $4k$ dimensions.

In section 3, I will focus on certain nonlinear examples in four and six dimensions and the relations between them. These examples will be supersymmetric. The importance for us of supersymmetry is that it gives severe constraints that have made it possible to get some insight about highly nonlinear theories. After reviewing Nahm's theorem, I will say a word or two about supersymmetric gauge theories in four dimensions that are conformally invariant at the quantum level, and then about how some of them are apparently related to nonlinear superconformal field theories in six dimensions.

2 Gauge theory and its higher cousins

First let us review abelian gauge theory, with gauge group $U(1)$. (For general references on some of the following discussion of abelian gauge fields and self-dual p-forms, see [3].) The connection A is locally a one-form. Under a gauge transformation, it transforms by $A \to A + d\epsilon$, with ϵ a zero-form. The curvature $F = dA$ is invariant.

For the action, we take

$$I(A) = \frac{1}{2e^2} \int_M F \wedge *F + \frac{i\theta}{2} \int_M \frac{F}{2\pi} \wedge \frac{F}{2\pi}. \tag{2.3}$$

Precisely in four dimensions, the Hodge $*$ operator on two-forms is conformally invariant and so $I(A)$ is conformally invariant. If M is a closed manifold without boundary the second term in $I(A)$ is a topological invariant, $i(\theta/2) \int_M c_1(\mathcal{L})^2$. In general, $c_1(\mathcal{L})^2$ is integral, and on a spin manifold it is actually even. So the integrand $\exp(-I(A))$ of the partition function is always invariant to $\theta \to \theta + 4\pi$, while on a spin manifold it is invariant to $\theta \to \theta + 2\pi$. In general, even when M is not closed, this is a symmetry of the theory (but in case M has a boundary, the discussion becomes a little more elaborate).

Now let us look at the partition function $Z(M) = \sum_{\mathcal{L}} \int DA \exp(-I(A))$, where we understand the sum over all possible connections A as including a sum over the line bundle $\mathcal{L}$ on which A is a connection. We can describe the path integral rather explicitly, using the decomposition $A = A' + A_h^{\mathcal{L}}$, where

A' is a connection on a trivial line bundle $\mathcal{O}$, and $A_h^{\mathcal{L}}$ is (any) connection on $\mathcal{L}$ of harmonic curvature $F_h^{\mathcal{L}}$. The action is $I(A) = I(A') + I(A_h^{\mathcal{L}})$, and the path integral is

$$\sum_{\mathcal{L}} \int DA \, \exp(-I(A)) = \int DA' \, \exp(-I(A')) \sum_{\mathcal{L}} \exp(-I(A_h^{\mathcal{L}})). \quad (2.4)$$

Here, note that $I(A_h^{\mathcal{L}})$ depends on $\mathcal{L}$, but $I(A')$ does not.

Let us look first at the second factor in eqn. (2.4), the sum over $\mathcal{L}$. On the lattice $H^2(M; \mathbf{Z})$, there is a natural, generally indefinite quadratic form given, for x an integral harmonic two-form, by $(x, x) = \int_M x \wedge x$. There is also a positive-definite but metric-dependent form $\langle x, x \rangle = \int_M x \wedge *x$, with $*$ being the Hodge star operator. The indefinite form (x, x) has signature $(b_{2,+}, b_{2,-})$, where $b_{2,\pm}$ are the dimensions of the spaces of self-dual and anti-self-dual harmonic two-forms.

Setting $x = F_h^{\mathcal{L}}/2\pi$, the sum over line bundles becomes

$$\sum_{x \in H^2(M;\mathbf{Z})} \exp\left(-\tfrac{4\pi^2}{e^2}\langle x, x \rangle + i\tfrac{\theta}{2}(x, x)\right). \quad (2.5)$$

If I set $\tau = \frac{\theta}{2\pi} + \frac{4\pi i}{e^2}$, then this function has modular properties with respect to τ. It is the non-holomorphic theta function of C.L. Siegel, which in the mid-1980s was introduced in string theory by K.S. Narain to understand toroidal compactification of the heterotic string. The Siegel–Narain function has a simple transformation law under the full modular group $SL(2, \mathbf{Z})$ if M is spin, in which case $(x, x)/2$ is integer-valued. In general, it has modular properties for a subgroup $\Gamma_0(2)$ of $SL(2, \mathbf{Z})$. In any case, it transforms as a modular function with holomorphic and anti-holomorphic weights $(b_{2,+}, b_{2,-})$.

The other factor in eqn. (2.4), namely the integral over A', $\int DA' \exp(-I(A'))$, is essentially a Gaussian integral that can be defined by zeta functions. Its dependence on the metric of M is very complicated, but its dependence on τ is very simple – just a power of $\operatorname{Im} \tau$. Including this factor, the full path-integral transforms as a modular function of weights $(1 - b_1 + b_{2,+}/2, 1 - b_1 + b_{2,-}/2) = ((\chi + \sigma)/2, (\chi - \sigma)/2)$, where b_1, χ, and σ are respectively the first Betti number, the Euler characteristic, and the signature of M.

The fact that the modular weights are linear combinations of χ and σ has an important consequence, which I will not be able to explain fully here. Because χ and σ can be written as integrals over M of quadratic polynomials in the Riemann curvature (using for example the Gauss–Bonnet–Chern formula for χ), it is possible to add to the action I a 'c-number' term – the integral of a local expression that depends on τ and on the metric of M but not on the integration variable A of the path integral – that cancels the modular weight and makes the partition function completely invariant under $SL(2, \mathbf{Z})$ or $\Gamma_0(2)$. The appropriate c-number terms arise naturally when, as we discuss later, one derives the four-dimensional abelian gauge theory from a six-dimensional self-dual theory.

2.1 *p-form analog*

Now I want to move on to the p-form analog, for $p > 2$. For our purposes, we will be informal in describing p-form fields. A 'p-form field' A_p is an object that locally is a p-form, with gauge invariance $A_p \to A_p + d\epsilon_{p-1}$ (with ϵ_{p-1} a $(p-1)$-form) and curvature $H = dA_p$. But globally there can be non-trivial periods $\int_D \frac{H}{2\pi} \in \mathbf{Z}$ for every $(p+1)$-cycle D. More precisely, H is the de Rham representative of a characteristic class x of A_p; this class takes values in $H^{p+1}(M; \mathbf{Z})$ and can be an arbitrary element of that group. The Lagrangian, for a p-form field on an n-manifold M_n, is

$$I(H) = \frac{1}{2\pi t} \int_{M_n} H \wedge *H \qquad (2.6)$$

with t a positive constant. In a more complete and rigorous description, the A_p are 'differential characters', for example A_0 is a map to $\mathbf{S}^1$, A_1 an abelian gauge field, etc. There is also a mathematical theory, not yet much used by physicists, in which a two-form field is understood as a connection on a gerbe, and the higher p-forms are then related to more sophisticated objects.

We can compute the partition function as before. We write $A_p = A'_p + A_{p,h}$, where A'_p is a globally defined p-form and $A_{p,h}$ is a p-form field with harmonic curvature. The curvature of $A_{p,h}$ is determined by the characteristic class x of A_p. This leads to a description of the partition function in which the interesting factor (for our purposes) come from the sum over x. It is[2]

$$\Theta = \sum_{x \in H^{p+1}(M_n; \mathbf{Z})} \exp\left(-\frac{\pi}{t}\langle x, x \rangle\right). \qquad (2.7)$$

As before $\langle x, x \rangle = \int_{M_n} x \wedge *x$. The $*$ operator that is used in this definition is only conformally invariant in the middle dimension, so conformal invariance only holds if n is even and $p + 1 = n/2$. Let us focus on this case.

---

[2] Θ is a function of the metric on M_n, which enters through the induced metric $\langle x, x \rangle$ on the middle-dimensional cohomology.

If $n = 4k$, then, as we have already observed for $k = 1$, another term $\frac{\theta}{(2\pi)^2} \int_{M_n} H \wedge H$ can be added to the action. This leads to a modular function, similar to what we have already described for $k = 1$.

If $n = 4k + 2$, then (H being a $(2k + 1)$-form) $\int H \wedge H = 0$, so we cannot add a θ-term to the action. But something else happens instead.

To understand this properly, we should at least temporarily return to the case that M_n is an n-manifold with *Lorentz* signature, $- + + + \cdots +$, which is the real home of physics. (In Lorentz signature we normally restrict M_n to have a global Cauchy hypersurface, and no closed timelike curves; normally, in Lorentz signature, we take M_n to have the topology $\mathbf{R} \times M_{n-1}$, where $\mathbf{R}$ parametrizes the 'time' and M_{n-1} is 'space'.) In $4k + 2$ dimensions with Lorentz signature, a self-duality condition $H = *H$ is possible for *real* H. In $4k$ dimensions, self-duality requires that H be complex. (In Euclidean signature, the conditions are reversed: a self-duality condition for a real middle-dimensional form is possible only in dimension $4k$ rather than $4k + 2$. This result may be more familiar than the corresponding Lorentzian statement.)

At any rate, in $4k + 2$ dimensions with Lorentz signature, a middle-dimensional classical H-field, obeying the Bianchi identity $dH = 0$ and the Euler–Lagrange equation $d * H = 0$, can be decomposed as $H = H_+ + H_-$, where $H_\pm$ are real and

$$*H_\pm = \pm H_\pm$$
$$dH_\pm = 0. \qquad (2.8)$$

Since classically it is consistent to set $H_- = 0$, one may suspect that there exists a quantum theory with $H_- = 0$ and only H_+. It turns out that this is true if we choose the constant t in the action eqn. (2.6) properly.

The lowest dimension of the form $4k + 2$, to which this discussion is pertinent, is of course dimension two. The self-dual quantum theory in dimension two has been extensively studied; it is important in the Segal–Frenkel–Kac vertex construction of representations of affine Lie algebras, in bosonization of fermions and its applications to statistical mechanics and representation theory, and in string theory. In these applications, it is important to consider generalizations of the theory we have considered to higher rank (by introducing several H fields). The generalization of picking a positive number t is to pick a lattice with suitable properties. After dimension two, the next possibility (of the form $4k + 2$) is dimension six, and very interesting things, which we will indicate in section 3 below, do occur in dimension six. To understand these phenomena, it is simplest and most useful to set $t = 1$. However, theories with interesting (and in general more complicated) properties can also be constructed for other rational values of t.

There is no way to write a Lagrangian for the theory with H_+ only – since for example $\int_{M_{4k+2}} H_+ \wedge H_+ = 0$. This makes the quantum theory subtle, but nevertheless it does exist, if we slightly relax our axioms. From the viewpoint that we have been developing, this can be seen by writing the non-holomorphic Siegel–Narain theta function of the lattice $\Lambda = H^{n/2}(M; \mathbf{Z})$, which appears in eqn. (2.7), in terms of holomorphic theta functions. For dimension $n = 4k + 2$, the lattice Λ has a *skew form* $(x, y) = \int x \wedge y$. It also, of course, just as in any other dimension, has the metric $\langle x, x \rangle = \int_{M_{4n+2}} x \wedge *x$. The skew form plus metric determine a complex structure on the torus $T = H^{n/2}(M; U(1))/$torsion.

Another important ingredient is a choice of 'quadratic refinement' of the skew form. A quadratic refinement of an integer-valued skew form (x, y) is a $\mathbf{Z}_2$-valued function $\phi : \Lambda \to \mathbf{Z}_2$ such that $\phi(x + y) = \phi(x) + \phi(y) + (x, y)$ mod 2. There are $2^{b_{n/2}(M)}$ choices of such a ϕ. Given a choice of ϕ, by classical formulas one can construct a unitary line bundle with connection $\mathcal{L}_\phi \to T$ whose curvature is the two-form determined by the skew form (x, y). This turns T into a 'principally polarized abelian variety', which has an associated holomorphic theta function ϑ_ϕ.

It can be shown (for a detailed discussion, see [4]) that the non-holomorphic theta function Θ of eqn. (2.7) which determines the partition function of the original theory without self-duality can be expressed in terms of the holomorphic theta functions ϑ_ϕ

$$\Theta = \sum_\phi \vartheta_\phi \overline{\vartheta}_\phi. \tag{2.9}$$

The sum runs over all choices of ϕ. If we could pick a ϕ in a natural way, we would interpret ϑ_ϕ as the difficult part, the 'numerator', of the partition function of the self-dual theory. In fact, roughly speaking, a choice of a spin structure on M determines a ϕ (for more detail, see the last two papers in [3], as well as [5] for an interpretation in terms of the Kervaire invariant). So we modify the definition of conformal field theory to allow a choice of spin structure and set the partition function Z_{sd} of the self-dual theory to be $Z_{sd} = \frac{\vartheta_\phi}{\det_+}$. Here $\det_+$ is the result of projecting the determinant that comes from the integral over topologically trivial fields on to the self-dual part. (Even in the absence of a self-dual projection, we did not discuss in any detail this determinant, which comes from the Gaussian integral over the topologically trivial field A'_p. For a discussion of it and an explanation of its decomposition in self-dual and anti-self-dual factors to get $\det_+$, see [4].)

Many assertions we have made depend on having set $t = 1$. For other values of t, to factorize Θ in terms of holomorphic objects, we would need to use theta functions at higher level; they would not be classified simply by a choice of quadratic refinement; and the structure needed to pick a particular holomorphic theta function would be more than a spin structure.

2.2 Relation between 4k and 4k + 2 dimensions

My last goal in discussing these linear theories is to indicate, following [6], how the existence of a self-dual theory in $4k + 2$ dimensions implies $SL(2, \mathbf{Z})$ (or $\Gamma_0(2)$) symmetry in $4k$ dimensions.

First let us look at the situation classically. We formulate the $(4k + 2)$-dimensional self-dual theory on the manifold $M_{4k+2} = M_{4k} \times \mathbf{T}^2$, where M_{4k} is a $(4k)$-manifold, and $\mathbf{T}^2$ a two-torus. We take $\mathbf{T}^2 = \mathbf{R}^2/L$, where L is a lattice in the $u - v$ plane $\mathbf{R}^2$. On $\mathbf{R}^2$ we take the metric $ds^2 = du^2 + dv^2$. So $E = \mathbf{T}^2$ is an elliptic curve with a τ parameter τ_E, which depends in the usual way on L.

Keeping the metric fixed on $\mathbf{T}^2$, we scale up the metric g on M_{4k} by $g \to \lambda g$, where we take λ to become very large. Any middle-dimensional form H on $M_{4k} \times \mathbf{T}^2$ can be expanded in Fourier modes on $\mathbf{T}^2$. In our limit with $\mathbf{T}^2$ much smaller than any characteristic radius of M_{4k}, the important modes (which, for example, give the main contribution to the theta function) are constant, that is, invariant under translations on the torus. So we can write $H = F \wedge du + \widetilde{F} \wedge dv + G + K \wedge du \wedge dv$, where $F, \widetilde{F}, G$, and K are pullbacks from M_{4k}.

Self-duality of H implies that $K = *G$ and that $\widetilde{F} = *F$ (where here $*$ is the duality operator on M_{4k}). The $SL(2, \mathbf{Z})$ symmetry of $\mathbf{T}^2$ acts trivially on G and K; for that reason we have not much of interest to say about them. Instead, we will concentrate on F and $\widetilde{F}$.

The fact that H is closed, $dH = 0$, implies that $dF = d\widetilde{F} = 0$. As $\widetilde{F} = *F$, it follows that $dF = d * F = 0$. These are the usual conditions (along with integrality of periods) for F to be the curvature of a $(2k - 1)$-form field in $4k$ dimensions. So, for example, if $k = 1$, then F is simply the curvature of an abelian gauge field.

So in the limit that the elliptic curve E is small compared to M_{4k}, the self-dual theory on $M_{4k} \times E$, which I will call (a), is equivalent to the theory of a $(2k - 1)$-form on M_{4k} (plus less interesting contributions from G and K), which I will call (b).

Suppose that this is true quantum mechanically. The theory (a) depends on the elliptic curve E, while (b) depends on $\tau = \theta/2\pi + 4\pi i/e^2$, which modulo $SL(2, \mathbf{Z})$ determines an elliptic curve E'.

A natural guess is that $E \cong E'$, and if so (since theory (a) manifestly depends only on E and not on a contruction of E using a specific basis of the lattice L or a specific τ-parameter) this makes obvious the $SL(2, \mathbf{Z})$ symmetry of theory (b).

The relation $E = E'$ can be established by comparing the theta functions. But instead, I will motivate this relation in a way that will be helpful when we study nonlinear theories in the next section.

Instead of reducing from $4k+2$ to $4k$ dimensions, let us first compare $4k+2$ to $4k + 1$ dimensions, and then take a further step down to $4k$ dimensions. So we formulate the self-dual theory on $M_{4k+2} = M_{4k+1} \times \mathbf{S}^1$, with $\mathbf{S}^1$ described by an angular variable v, $0 \le v \le R$. We fix the metric dv^2 on $\mathbf{S}^1$, and scale up the metric on M_{4k+1} by a large factor. In the limit, just as in the previous case, we can assume $H = F \wedge dv + G$, where F and G are pullbacks from M_{4k+1}. Moreover, $G = *F$ and $dG = dF = 0$, so F obeys the conditions $0 = dF = *dF$ to be the curvature of an 'ordinary' $(2k - 1)$-form theory on M_{4k+1}.[3]

Unlike the self-dual theory on M_{4k+2}, the 'ordinary' theory on M_{4k+1} does have a Lagrangian. This Lagrangian depends on a free parameter (called t in eqn. (2.6)). Conformal invariance on $M_{4k+1} \times \mathbf{S}^1$ implies that t must be a constant multiple of R, so that the action (apart from a constant that can be fixed by comparing the theta functions) is

$$I = \frac{1}{4\pi R} \int_{M_{4k+1}} F \wedge *F. \tag{2.10}$$

The point of this formula is that if we rescale the metric of both factors of $M_{4k+2} = M_{4k+1} \times \mathbf{S}^1$ by the same factor, then R (the circumference of $\mathbf{S}^1$) and $*$ (the Hodge $*$ operator of M_{4k+1} acting from $(2k)$-forms to $(2k + 1)$-forms) scale in the same way, so the action in eqn. (2.10) is invariant.

The formula of eqn. (2.10) has the very unusual feature that R is in the denominator. If we had a Lagrangian in $4k + 2$ dimensions, then after specializing to $M_{4k+2} = M_{4k+1} \times \mathbf{S}^1$, we would deduce what the action must be in $4k + 1$ dimensions by simply 'integrating over the fiber' of the projection $M_{4k+2} \to M_{4k+1}$. For fields that are pullbacks from M_{4k+1}, this would inevitably give an action on M_{4k+1} that is proportional to R – the volume of the fiber – and not to R^{-1}, as in eqn. (2.10). But there is no classical action in $4k + 2$ dimensions, and the 'integration over the fiber' is a quantum operation that gives a factor of R^{-1} instead of R.

Now let us return to the problem of comparing $4k+2$ to $4k$ dimensions, and arguing that E' is isomorphic to E. We specialize to the case that the lattice L is 'rectangular', generated by the points $(S, 0)$ and $(0, R)$ in the $u - v$ plane.

---

[3] $2k - 1$ is the degree of the potential, while the curvature F is of degree $2k$.

Accordingly, the torus $E \cong \mathbf{T}^2$ has a decomposition as $\mathbf{S} \times \mathbf{S}'$, where $\mathbf{S}$ and $\mathbf{S}'$ are circles of circumference, respectively, S and R.

We apply the previous reasoning to the decomposition $M_{4k+2} = M_{4k+1} \times \mathbf{S}'$, with $M_{4k+1} = M_{4k} \times \mathbf{S}$. Since $\mathbf{S}'$ has circumference R, the induced theory on M_{4k+1} has action given by eqn. (2.10). Now, let us look at the decomposition $M_{4k+1} = M_{4k} \times \mathbf{S}$. Taking the length scale of M_{4k} to be large compared to that of $\mathbf{S}$, we would like to reduce to a theory on M_{4k}. For this step, since we do have a classical action on M_{4k+1}, the reduction to a classical action on M_{4k} is made simply by integrating over the fibers of the projection $M_{4k+1} \to M_{4k}$. As the fibers have volume S, the result is the following action on M_{4k}

$$I = \frac{1}{4\pi} \frac{S}{R} \int_{M_{4k}} F \wedge *F. \tag{2.11}$$

We see from eqn. (2.11) that the τ parameter of the theory on M_{4k} is $\tau' = iS/R$. But this in fact is the same as the τ parameter of the elliptic curve $E = \mathbf{S} \times \mathbf{S}'$, so we have demonstrated, for this example, that $E \cong E'$.

In our two-step procedure of reducing from $M_{4k} \times \mathbf{S} \times \mathbf{S}'$, we made an arbitrary choice of reducing on $\mathbf{S}'$ first. Had we proceeded in the opposite order, we would have arrived at $\tau' = iR/S$ instead of iS/R; the two results differ by the expected modular transformation $\tau \to -1/\tau$.

One can extend the above arguments to arbitrary E with more work; it is not necessary in this two-step reduction for $\mathbf{S}$ and $\mathbf{S}'$ to be orthogonal. Of course, one can also make the arguments more precise by study of the theta function of the self-dual theory on M_{4k+2}.

3 Superconformal field theories in four and six dimensions

In n dimensions, the conformal group of (conformally compactified) Minkowski spacetime is $SO(2, n)$. A superconformal field theory, that is a conformal field theory that is also supersymmetric, should have a supergroup G of symmetries whose bosonic part is $SO(2, n) \times K$, with K a compact Lie group. The fermionic part of the Lie algebra of G should transform as a sum of spin representations of $SO(2, n)$. *A priori*, the spinors may appear in the Lie algebra with any multiplicity, and for n even, where $SO(2, n)$ has two distinct spinor representations, these may appear with unequal multiplicities.

Nahm considered the problem of classifying supergroups G with these properties. The result is that such groups exist only for $n \geq 6$. For $n = 6$, the algebraic solution can be described as follows. The group G is $OSp(2, 6|r)$ for some positive integer r. Thus $K = Sp(r)$. To describe the fermionic generators of G, first consider $G' = OSp(2, n|r)$ for general n. The fermionic generators of this group transform not as spinors but as the vector representation of $O(2, n)$ (tensored with the fundamental representation of $Sp(r)$). Thus for general n, the group G' does not solve our algebraic problem. However, precisely for $n = 6$, we can use the *triality* symmetry of $O(2, 6)$; by an outer automorphism of this group, its vector representation is equivalent to one of the two spinor representations. So modulo this automorphism, the group $G = OSp(2, 6|r)$ does obey the right algebraic conditions and is a possible supergroup of symmetries for a superconformal field theory in six dimensions.

The algebraic solutions of Nahm's problem for $n < 6$ are similarly related to exceptional isomorphisms of Lie groups and supergroups of low rank. (We give the example of $n = 4$ presently.) Triality is in some sense the last of the exceptional isomorphisms, and the role of triality for $n = 6$ thus makes it plausible that $n = 6$ is the maximum dimension for superconformal symmetry, though I will not give a proof here.

As I remarked in the introduction, this particular result by Nahm had little immediate impact, since it was believed at the time that the correct bound was really $n \leq 4$. But in the mid-1990s, examples were found with $n = 5, 6$. The known examples in dimension 6 have $r = 1$ and $r = 2$. My goal in what follows will be to convey a few hints about the $r = 2$ examples. A reference for some of what I will explain is [7].

3.1 Superconformal gauge theories in four dimensions

We will need to know a few more facts about gauge theories in four dimensions. The basic gauge theory with the standard Yang–Mills action $I(A) = \frac{1}{4e^2} \int \mathrm{Tr} F \wedge *F$ is conformally invariant at the classical level, but not quantum mechanically. There are many ways to introduce additional fields and achieve quantum conformal invariance.

We will focus on superconformal field theories. The superconformal symmetries predicted by Nahm's analysis are $SU(2, 2|\mathcal{N})$ for arbitrary positive integer $\mathcal{N}$, as well as an exceptional possibility $PSU(2, 2|4)$. Note that $SU(2, 2)$ is isomorphic to $SO(2, 4)$, and that the fermionic part of the super Lie algebra of $SU(2, 2|\mathcal{N})$ (or of $PSU(2, 2|4)$) transforms as $\mathcal{N}$ copies of $V \oplus \overline{V}$, where V is the defining four-dimensional representation of $SU(2, 2)$. V and $\overline{V}$ are isomorphic to the two spinor representations of $SO(2, 4)$, so $SU(2, 2|\mathcal{N})$ and $PSU(2, 2|4)$ do solve the algebraic problem posed by Nahm. The supergroups $SU(p, q|\mathcal{N})$ exist for all positive integers $p, q, \mathcal{N}$, but it takes the exceptional isomorphism $SU(2, 2) \cong SO(2, 4)$ to get a solution of the problem considered by Nahm.

Examples of superconformal field theories in four dimensions exist for $\mathcal{N} = 1, 2$, and 4. For $\mathcal{N} = 1$, there are myriads of possibilities – though much more constrained than in the absence of supersymmetry – while the examples with $\mathcal{N} = 2$ and $\mathcal{N} = 4$ are so highly constrained that a complete classification is possible. In particular, for $\mathcal{N} = 4$, the fields that must be included are completely determined by the choice of the gauge group G. For $\mathcal{N} = 2$, one also picks a representation of G that obeys a certain condition on the trace of the quadratic Casimir operator (there are finitely many choices for each given G). We will concentrate on the examples with $\mathcal{N} = 4$; they have the exceptional $PSU(2, 2|4)$ symmetry.

3.2 $\mathcal{N} = 4$ *super Yang–Mills theory*

The fields of $\mathcal{N} = 4$ super Yang–Mills theory are the gauge field A plus fermion and scalar fields required by the supersymmetry. The Lagrangian is

$$I(A, \dots) = \int_{M_4} \mathrm{Tr} \left(\frac{1}{4e^2} F \wedge *F + \frac{i\theta}{8\pi^2} F \wedge F + \dots \right) \tag{3.1}$$

where the ellipses refer to terms involving the additional fields.

If we set $\tau = \frac{\theta}{2\pi} + \frac{4\pi i}{e^2}$, then the Montonen–Olive duality conjecture [8] asserts an $SL(2, \mathbf{Z})$ symmetry acting on τ. Actually, the element

$$\begin{pmatrix} 0 & 1 \\ -1 & 0 \end{pmatrix} \tag{3.2}$$

of $SL(2, \mathbf{Z})$ is conjectured to map the $\mathcal{N} = 4$ theory with gauge group G to the same theory with the Langlands dual group, while also mapping τ to $-1/\tau$. So in general the precise modular properties are a little involved, somewhat analogous to the fact that in section 2, we found in general $\Gamma_0(2)$ rather than full $SL(2, \mathbf{Z})$ symmetry. By around 1995, many developments in the study of supersymmetric gauge theories and string theories gave strong support for the Montonen–Olive conjecture.

If we formulate the $\mathcal{N} = 4$ theory on a compact four-manifold M, endowed with some metric tensor g, the partition function $Z(M, g; \tau)$ is, according to the Montonen–Olive conjecture, a modular function of τ. It is not in general holomorphic or anti-holomorphic in τ, and it depends non-trivially on g, so it is not a topological invariant of M.

However [9], there is a 'twisted' version of the theory that is a topological field theory and still $SL(2, \mathbf{Z})$-invariant. For a four-manifold M with $b_{2,+}(M) > 1$, the partition function is holomorphic (with a pole at the 'cusp') and a topological invariant of M. In fact, setting $q = \exp(2\pi i \tau)$, the partition function can be written

$$Z(M; \tau) = q^{-c} \sum_{n=0}^{\infty} a_n q^n \qquad (3.3)$$

where, assuming a certain vanishing theorem holds, a_n is the Euler character-istic of the moduli space of G-instantons of instanton number n. In general, a_n is the 'number' of solutions, weighted by sign, for a certain coupled system of equations for the connection plus certain additional fields. These more elabo-rate equations, which are somewhat analogous to the Seiberg–Witten equations and have similarly nice Bochner formulas (related in both cases to supersym-metry), were described in [9].

3.3 Explanation from six dimensions

So if the $SL(2, \mathbf{Z})$ conjecture of Montonen and Olive holds, the functions de-fined in eqn. (3.3) are modular. But why should the $\mathcal{N} = 4$ supersymetric gauge theory in four dimensions have $SL(2, \mathbf{Z})$ symmetry?

Several explanations emerged from string theory work in the mid-1990s. Of these, one [7] is in the spirit of what we discussed for linear theories in section 2. In its original form, this explanation only works for simply laced G, that is for G of type A, D, or E. I will limit the following discussion to this case. (For simply laced G, G is locally isomorphic to its Langlands dual, and the statement of Montonen–Olive duality becomes simpler.)

The surprise which leads to an insight about Montonen–Olive duality is that in dimension $n = 6$, there is for each choice of simply laced group G a superconformal field theory that is a sort of nonlinear (and supersymmet-ric) version of the self-dual theory that we discussed in section 2. This exotic six-dimensional theory was found originally [7] by considering Type IIB su-perstring theory at an $A - D - E$ singularity.

The superconformal symmetry of this theory is the supergroup $OSp(2, 6|2)$. When it is formulated on a six-manifold $M_6 = M_4 \times E$, with E an elliptic curve, the resulting behavior is quite similar to what we have discussed in section 2 for the linear self-dual theory. Taking a product metric on $M_4 \times E$, in the limit that M_4 is much larger than E, the six-dimensional theory reduces to the four-dimensional $\mathcal{N} = 4$ theory with gauge group G and τ parameter determined by E. Just as in section 2, this makes manifest the Montonen–Olive symmetry of the $\mathcal{N} = 4$ theory. From this point of view, Montonen–Olive symmetry reflects the fact that the six-dimensional theory on $M_4 \times E$ depends only on E and not on a specific way of constructing E using a τ parameter.

Further extending the analogy with what we discussed in section 2 for linear theories, if we formulate this theory on $M_5 \times S$, where S is a circle of circumference R, we get at distances large compared to R a five-dimensional gauge theory, with gauge group G, and action proportional to R^{-1} rather than R. As in section 2, this shows that the five-dimensional action cannot be obtained by a classical process of 'integrating over the fiber'; it gives an obstruction to deriving the six-dimensional theory from a Lagrangian.

The six-dimensional theory that comes from Type IIB superstring theory at the $A - D - E$ singularity might be called a 'nonabelian gerbe theory', as it is an analog for $A - D - E$ groups of the linear theory discussed in section 2 with a two-form field and a self-dual three-form curvature. Under a certain perturbation (to a vacuum with spontaneous symmetry breaking in six dimensions), the six-dimensional $A - D - E$ theory reduces at low energies to a theory that can be described more explicitly; this theory is a more elaborate version of the theory with self-dual curvature that we considered in section 3. In this theory, the gerbe-like field has a characteristic class that takes values not in $H^3(M; \mathbf{Z})$, but in $H^3(M; \mathbf{Z}) \otimes \Lambda$, where Λ is the root lattice of G. Physicists describe this roughly by saying that, if r denotes the rank of G, there are r self-dual two-form fields (i.e., two-form fields whose curvature is a self-dual three-form).

The basic hallmark of the six-dimensional theory is that on the one hand it can be perturbed to give something that we recognize as a gerbe theory of rank r; on the other hand, it can be perturbed to give nonabelian gauge theory with gauge group G. Combining the two facts, this six-dimensional theory is a sort of quantum nonabelian gerbe theory. I doubt very much that this structure is accessible in the world of classical geometry; it belongs to the realm of quantum field theory. But it has manifestations in the classical world, such as the modular nature of the generating function (eqn. (3.3)) of Euler characteristics of instanton moduli spaces.

References

[1] Segal, G., The definition of conformal field theory, in *Differential Geometric Methods in Theoretical Physics* (proceedings, Como, 1987); see also this volume.

[2] Nahm, W., Supersymmetries and their representations, *Nucl. Phys.* **B135** (1978) 149.

[3] Witten, E., On *S*-duality in abelian gauge theory, hep-th/9505186, *Selecta Mathematica* **1** (1995) 383; Five-brane effective action in *M*-theory, hep-th/9610234, *J. Geom. Phys.* **22** (1997) 103; Duality relations among topological effects in string theory, hep-th/9912086, *JHEP* **0005:031,2000**.

[4] Henningson, M., The quantum Hilbert space of a chiral two form in $D = (5 + 1)$ dimensions, hep-th/0111150, *JHEP* **0203:021,2002**.

[5] Hopkins M. and I. M. Singer, Quadratic functions in geometry, topology, and *M*-theory, math.at/0211216.

[6] Verlinde, E., Global aspects of electric-magnetic duality, hep-th/9506011, *Nucl. Phys.* **B455** (1995) 211.

[7] Witten, E., Some comments on string dynamics, hep-th/9507121, in *Future Perspectives in String Theory* (World Scientific, 1996), ed. I. Bars et al., 501.

[8] Montonen C. and D. Olive, Magnetic monopoles as gauge particles?, *Phys. Lett.* **B72** (1977) 117; P. Goddard, J. Nuyts, and D. Olive, Gauge theories and magnetic charge, *Nucl. Phys.* **B125** (1977) 1; H. Osborn, Topological charges for $\mathcal{N} = 4$ supersymmetric gauge theories and monopoles of spin 1, *Phys. Lett.* **B83** (1979) 321.

[9] Vafa C. and E. Witten, A strong coupling test of S-duality, hep-th/9408074, *Nucl. Phys.* **B431** (1994) 3.

E-mail: witten@ias.edu

Part II
The definition of conformal field theory

Graeme Segal

Foreword and postscript

The manuscript that follows was written fifteen years ago. On balance, though, conformal field theory has evolved less quickly than I expected, and to my mind the difficulties which kept me from finishing the paper are still not altogether elucidated.

My aim when I began the work was fairly narrow. I was not trying to motivate the study of conformal field theory: I simply wanted to justify my proposed definition, on the one hand by showing that it did encode the usual structure of local field operators and their vacuum expectation values, and on the other hand by checking that all the known examples of conformal theories did fit the definition. As far as the first task is concerned, the crucial part of the paper is §9, where local fields are defined and studied. It was the second task that held me up. The known theories are

1. the σ-model of a torus, or 'free bosons compactified on a torus',
2. free fermions,
3. the Wess–Zumino–Witten theory for a compact Lie group,
4. theories obtained from WZW theories by the 'coset' construction of Goddard, Kent, and Olive,
5. theories obtained from the preceding ones by the 'orbifold' construction.

(I should stress that this is a list of explicit constructions, not a classification of theories. It ignores supersymmetry, and also what I would now call 'noncompact' theories.) The crucial case is the WZW theory, which reduces to the representation theory of loop groups. In my formulation, one must construct a 'modular functor', and prove that it is unitary. I was unable to do this. The task has since been carried out in the book [BK]* of Bakalov and Kirillov, but

---

* The references are to the list at the end of the original manuscript (page 576).

their method is very long and indirect – the book is highly non-selfcontained – resting heavily on the equivalence of categories between representations of loop groups and representations of quantum groups at roots of unity. I still hope that a more direct treatment of this beautiful subject will be found.

Another deficiency of my approach is discussed in §9: the axioms do not seem to imply that the infinitesimal deformations of a theory are given by local fields, and so I could not say anything rigorous about the moduli spaces of theories. One way to deal with this problem is by extending the two-tier structure of my definition to a three-tier structure which includes axioms about cutting a circle into intervals. The paper by Stolz and Teichner in this volume goes some way towards carrying out this programme.

I shall make some more detailed comments section by section.

Section 4

I no longer like the emphasis I placed on the operation of sewing an outgoing to an incoming boundary circle of a cobordism. The associated 'trace axiom' follows readily from the other properties of a conformal field theory. The definition I would give now is as follows.

A (not necessarily unitary) conformal field theory (H, U) consists of two pieces of data:

1. A projective functor $S \mapsto H_S$ from the category of closed oriented smooth 1-manifolds to locally convex complete topological vector spaces, which takes disjoint unions to tensor products, and
2. For each oriented cobordism X, with conformal structure, from S_0 to S_1 a ray U_X in the space of trace-class linear maps $H_{S_0} \to H_{S_1}$, subject to
   (a) $U_{X' \circ X} = U_{X'} \circ U_X$ when cobordisms are composed, and
   (b) $U_{X \sqcup X'} = U_X \otimes U_{X'}$.

Furthermore, U_X must depend smoothly on the conformal structure of X.

Given the data (H, U), it follows from the representation theory of the semi-group $\mathcal{A}$ that the vector space $H_S = H_{S,L}$ is honestly – not just projectively – associated to a *rigged* 1-manifold (S, L) (see page 30), and that a specific operator $U_{X,\xi} : H_{S_0,L_0} \to H_{S_1,L_1}$ is associated to a cobordism X together with a point ξ in the determinant line Det_X (which in turn depends on the rigging of ∂X).

It also follows from the definition that $H_{\bar{S}}$ is canonically dual to H_S (and that $U_X : H_{\bar{S}_1} \to H_{\bar{S}_0}$ is the transpose of U_X). More precisely, because we have a trace-class semigroup $\{U_T = U_{S \times [0,T]}\}_{T>0}$ acting on each vector space H_S we can define complete topological vector spaces $\check{H}_S$ and $\hat{H}_S$ with maps

$$\check{H}_S \to H_S \to \hat{H}_S$$

which are injective with dense images and make H_S into a rigged vector space in the sense of Gelfand [GV]. In particular, $\check{H}_S$ and $\hat{H}_S$ are nuclear. It is easy to see that $\check{H}_S$ and $\hat{H}_{\bar{S}}$ are canonically dual (for details, see page 15 of Lecture 1 of [S4]). In the light of these remarks the missing Appendix A is not needed.

A theory (H, U) is *unitary* if there is given a natural isomorphism $\bar{H}_S \to H_{\bar{S}}$ which makes $\check{H}_S$ into a pre-Hilbert space with H_S as its completion. ('Natural' here means that $\bar{U}_X = U_{\bar{X}}$.) In the manuscript the 'positive' part of the reflection-positivity condition was accidentally omitted.

I would now put the remarks about Minkowski space at the end of §4 in a different context. The theories axiomatized in the paper are *compact* ones: they correspond to loops moving in a compact target manifold, and the Hamiltonian operator H such that $U_T = e^{-HT}$ has discrete spectrum. One can also define *non-compact* theories, for which H has continuous spectrum and the operator U_T is not of trace-class. Such a theory is a vector-space-valued functor on a subcategory $\mathcal{C}^+$ of the basic cobordism category $\mathcal{C}$, where $\mathcal{C}^+$ consists of cobordisms every connected component of which has a non-empty outgoing boundary. If one thinks of a conformal field theory as a generalized commutative Frobenius algebra, then a theory based on $\mathcal{C}^+$ is a generalized 'non-compact commutative Frobenius algebra'. The basic example of such a structure is the cohomology algebra of a non-compact manifold. (The category $\mathcal{C}^+$ was introduced by Tillmann in [T], and also occurs in the papers of Cohen and Sullivan in this volume.)

I now feel more confident than I did that the framework of the manuscript is appropriate for describing quantum field theories which are not conformally invariant, and not necessarily 2-dimensional. But the remarks on pages 27–28 should be modified. For a d-dimensional theory $\mathcal{C}_{\text{metric}}$ should be the category whose objects are *germs* $\hat{S}$ of oriented Riemannian d-manifolds along compact $(d - 1)$-manifolds S (i.e. equivalence classes of neighbourhoods of S in a d-manifold), and whose morphisms are oriented Riemannian cobordisms. When one has a theory based on this category the vector space $H_{\hat{S}}$ will depend only on a finite jet of the Riemannian structure normal to S: e.g. in [S4] page 33 it is shown that the jet of order $[(d - 1)/2]$ is needed for free fermions.

Section 5

I would now (roughly following suggestions of Deligne) think of a modular functor as a *category-valued topological field theory*. Such a theory associates a $\mathbb{C}$-linear category $\mathcal{R}_S$ to each closed oriented 1-manifold S, and an additive functor

$$E_X : \mathcal{R}_{S_0} \to \mathcal{R}_{S_1}$$

to each cobordism X from S_0 to S_1. (No conformal structure is involved here, but we should require the manifolds to be *rigged* in the sense of (5.10) and page 30.) The functor $S \mapsto \mathcal{R}_S$ must take disjoint unions to tensor products of categories: the axioms ensure that each $\mathcal{R}_S$ is a semisimple category with only finitely many irreducible objects. More details of this approach can be found in Lecture 3 of [S4]. The main example is the representation theory of loop groups, when $\mathcal{R}_{S^1}$ is the category of positive-energy representations of a loop group $\mathcal{L}G$ at a definite level $k \in H^4(BG; \mathbb{Z})$.

This perspective is, nevertheless, just a reformulation of what is in the present manuscript. The set $\Phi_S = \Phi^k$ of labels for a 1-manifold $S = S_1 \amalg \cdots \amalg S_k$ is the set of irreducible objects of $\mathcal{R}_S$, while the functor $E_X : \mathcal{R}_{S_0} \to \mathcal{R}_{S_1}$ associated to a cobordism X is given by

$$E_X(\varphi) = \bigoplus_\psi E(X_{\varphi\psi}) \otimes \psi$$

in the notation of the manuscript, where $\varphi \in \Phi_{S_0}$ and $\psi \in \Phi_{S_1}$, and on the right-hand side the object ψ is tensored with the finite-dimensional vector space $E(X_{\varphi\psi})$.

In terms of the category-valued theory $(\mathcal{R}, E)$ what I called a 'weakly conformal' field theory assigns to each 1-manifold S an additive functor H_S from $\mathcal{R}_S$ to topological vector spaces, and to each cobordism X from S_0 to S_1 a transformation of functors

$$U_X : H_{S_0} \to H_{S_1} \circ E_X.$$

(The functors H_S must have coherent equivalences $H_S \otimes H_{S'} \to H_{S \amalg S'}$, and the transformations U_X must be compatible, as usual, with concatenation and disjoint union of cobordisms.)

Kontsevich's argument from [K] shows that a category-valued theory extends to a '3-tier' theory in which a 3-dimensional cobordism W between two cobordisms X, X' from S_0 to S_1 defines a transformation of functors

$U_S : E_X \to E_{X'}$. Restricting this structure to closed surfaces X and their cobordisms gives a 3-dimensional topological field theory, which in the loop group example is Chern–Simons theory.

Unfortunately there is a gap in the proof of the crucial Proposition (5.4) – the existence of a projectively flat connection in the modular functor – for I referred to a non-existent appendix for a proof that the Lie algebra $\mathrm{Vect}(X)$ of holomorphic vector fields on a non-closed surface X has no finite-dimensional projective representations. To fill the gap, notice that if X has genus zero then $\mathrm{Vect}(X)$ contains the Virasoro algebra, which certainly has no finite-dimensional representations. So in that case $\mathrm{Vect}(X)$ acts trivially on $E(X)$. But in general we can cut X into pieces of genus zero – say $X = X_1 \cup \cdots \cup X_k$ – and write

$$E(X) = \bigoplus E(X_{1,\varphi_1}) \otimes \cdots \otimes E(X_{k,\varphi_k}),$$

where the sum is over appropriate labellings φ_i. This decomposition is compatible with the action of $\mathrm{Vect}(X)$. But $\mathrm{Vect}(X)$ acts on $E(X_{i,\varphi_i})$ via $\mathrm{Vect}(X_i)$, which must act trivially. So $\mathrm{Vect}(X)$ acts trivially on $E(X)$.

A more illuminating account of the flat connection, though from a quite different point of view, has been given by Hitchin [H1]. Unfortunately his method, like mine, is not helpful in establishing unitarity.

Section 7

In retrospect, this section does not seem properly motivated. The part concerned with finite groups was put in to lead up to a discussion – unfortunately never written – in §12 of the orbifold construction of theories.

The material fits into the general framework of gauge theories. If a compact Lie group G acts on a quantum field theory (H, U) – i.e. G acts on each vector space H_S and the maps U_X are G-equivariant – we say 'the symmetry can be gauged' if the functor (H, U) can be extended from the usual cobordism category $\mathcal{C}$ to the category $\mathcal{C}^G$ whose objects (S, P) are 1-manifolds S equipped with a principal G-bundle P with a connection, and whose morphisms (X, Q) are conformal cobordisms X also equipped with principal G-bundles with connection. If P_0 is the trivial G-bundle on S then H_{S,P_0} should be the original H_S, and the action of G as a group of automorphisms of (S, P_0) should induce the given G-action on H_S.

As a generic example, one can think of a sigma-model whose target space M has a G-action. Then H_S is the space of L^2 functions on the mapping space

Map(S; M), while $H_{S,P}$ is the space of L^2 functions on the mapping space Map$_G(P; M)$ of G-equivariant maps, or equivalently, on the space of sections of the bundle on S with fibre M associated to the principal bundle P.

When the action can be gauged one can hope to construct a 'quotient' theory (H^G, U^G). An element of H^G_S is a function ψ which associates to each G-bundle P on S an element $\psi_P \in H_{S,P}$, and is gauge-invariant in the sense that an isomorphism $P \to P'$ takes ψ_P to $\psi_{P'}$. The operator U^G_X should in principle be an integral operator whose kernel $U^G_X(P_0; P_1)$ is the integral of $U_{X,Q}$ over the isomorphism classes of bundles Q on X which restrict to P_0, P_1 on ∂X.

If G is a finite group any bundle has a unique connection, and there are only finitely many G-bundles on any manifold. Thus we find

$$H^G_{S^1} = \bigoplus_{[g]} (H_{S^1, P_g})^{Z_g}$$

where P_g is the bundle on S^1 with holonomy $g \in G$, and Z_g is the centralizer of g in G, while the sum is over the conjugacy classes of elements g. In this case $U^G_X(P_0, P_1)$ is simply a finite sum.

In the case of conformal field theories, however, we must be careful that each operator $U_{X,Q} = U_{X,Q,\xi}$ will depend on a choice of a point ξ of a line $L_{X,Q}$ associated to (X, Q), and we can make the quotient construction only if we can identify the lines $L_{X,Q}$ for different bundles Q on X.

The passage from $\mathcal{C}^G$ to the modular functor does not look interesting as it is presented. Its significance is that a chiral theory with a group action, arising, say, from an even unimodular lattice with a finite symmetry group G, gives us a theory based on an extension of $\mathcal{C}^G$ by determinant lines $L_{X,Q}$ which do depend on Q as well as X. If one shows that the associated modular functor is unitary one can tensor the chiral theory with its conjugate to obtain a genuine conformal theory by the method of §5. In practice it is easier to show the modular functor is unitary than to deal with the individual lines $\bar{L}_{X,Q} \otimes L_{X,Q}$.

The same remarks apply to the discussion of spin structures. In (8.16) I describe the theory $\mathcal{F}$ of a free chiral fermion, which is based on the category $\mathcal{C}^{\text{spin}}$. The standard theory of free fermions, also based on $\mathcal{C}^{\text{spin}}$, is $\bar{\mathcal{F}} \otimes \mathcal{F}$. This has a quotient theory, formed by summing over spin structures, which has

$$H_{S^1} = (\bar{\mathcal{F}}_A \otimes \mathcal{F}_A)^{\text{even}} \otimes (\bar{\mathcal{F}}_P \otimes \mathcal{F}_P)^{\text{even}}$$

(for the mod 2 grading of the fermionic Fock space expresses the action of the automorphisms of the spin bundles on S^1). The quotient theory turns out to be equivalent to the sigma-model whose target is a circle of a specific length (which is $\sqrt{2}$ in the notation of this paper): this is the basic boson-fermion correspondence of 2-dimensional field theory. But the relevant point at the moment is that the line

$$\overline{\mathrm{Det}}_{\bar{\partial}_L} \otimes \mathrm{Det}_{\bar{\partial}_L}$$

where $\mathrm{Det}_{\bar{\partial}_L}$ is the determinant line of the $\bar{\partial}$-operator of a spin bundle L on X, is independent of L and is a square-root of $\overline{\mathrm{Det}}_X \otimes \mathrm{Det}_X$. The only way I know of proving this is by showing that the modular functor

$$E_X = \bigoplus_L \mathrm{Det}_{\bar{\partial}_L}$$

is unitary.

Finally, I should like to make a few remarks concerning Propositions 7.7 and 7.8. As it stands, Proposition 7.7 is almost trivial, for the set $\mathcal{S}(X, \partial_0 X)$ of spin structures on X trivialized at the base-points $\partial_0 X$ is an affine space of $H^1(X, \partial_0 X; \mathbb{F}_2)$, and so the vector space $\tilde{H}_X$ of affine-linear functions from $\mathcal{S}(X, \partial_0 X)$ to $\mathbb{F}_2$ is an extension of $H_X = H_1(X, \partial_0 X; \mathbb{F}_2)$ by $\mathbb{F}_2$ whose set of splittings is $\mathcal{S}(X, \partial_0 X)$. The point is that $\tilde{H}_X$ – which depends of course on the choice of $\partial_0 X$ – can be constructed from H_X by means of an intersection form. This is straightforward when ∂X has one component, but more complicated otherwise. The closed surface X^* is canonically associated to X, but both the isomorphisms $H_1(X, \partial_0 X) \cong H_1(X^*)$ and $\mathcal{S}(X, \partial_0 X) \cong \mathcal{S}(X^*)$ depend on the choice of the tree y, which can be fixed by choosing a cyclic order of the components of ∂X.

Turning to 7.8, a more conceptual proof of the existence of the extension of $\mathbb{C}_X^\times$ by $\mathbb{F}_2$ should be as follows. The basic extension of $\mathbb{C}_X^\times$ by $\mathbb{C}^\times$ arises by lifting the action of $\mathbb{C}_X^\times$ on the restricted Grassmannian Gr of $\Omega^{\frac{1}{2}}(\partial X; L)$ to the determinant bundle on Gr. A spin structure on X gives us a fixed point of the action on Gr, and hence a splitting of the extension. But the fibre of Det at any one of these fixed points is a square-root of the determinant line of X, and so the basic extension contains a subextension consisting of the elements which act trivially on Det_X. I do not see, unfortunately, how to relate this argument to the one using cocycles which can be extracted from §12.

Section 8

Although the alternative description of the fermionic Fock space on page 92 is often useful, to call it a 'bosonic' description is misleading, as the chiral theories in question do not correspond to anything described by a recognizable bosonic Lagrangian. They are at best chiral fragments of bosonic theories.

Section 10

The parameter space $M_{p,q}$ described on page 129 is usually called the *Narain moduli space*. It is believed to be a complete component of the moduli space of conformal theories (in the sense of algebraic geometry: it intersects other components). The 2-form ω arising in the sigma-model description of the theories is called the *B-field* (cf. [S5]). The relation between different pairs (T, ω) which give rise to the same field theory is called *T-duality*.

Section 12

I intended this section to be considerably longer, though – unlike §11 – it was not held up by any mathematical difficulties. The missing material was in two parts. The first would have described the WZW model when the compact group is a torus $T = \mathfrak{t}/\Lambda$. The central extension $\tilde{\mathcal{L}}T$ of the loop group $\mathcal{L}T$ – called the *level* of the theory – is determined by an inner product on $\mathfrak{t}$ for which the lattice Λ is integral and even. The commutator pairing is given by (12.1), but with products replaced by the inner product of $\mathfrak{t}$. To construct $\tilde{\mathcal{L}}T$ one needs its cocycle, obtained by choosing a bilinear form B on $\mathfrak{t}$, integral on Λ, such that

$$B(x, y) + B(y, x) = \langle x, y \rangle.$$

This can be done because Λ is an even lattice, and, up to isomorphism, the choice of B is immaterial.

The centre of $\tilde{\mathcal{L}}T$ is $\mathbb{T} \times A$, where A is the finite subgroup Λ^0/Λ of the group T of constant loops. (Here

$$\Lambda^0 = \{\mu \in \mathfrak{t} : \langle \mu, \lambda \rangle \in \mathbb{Z} \text{ for all } \lambda \in \Lambda\}.)$$

In any irreducible representation of $\tilde{\mathcal{L}}T$ the subgroup A acts by a character, and there is precisely one irreducible positive-energy representation E_φ with the given cocycle for each character $\varphi \in \hat{A}$.

The character group $\hat{A}$ is thus the set of labels for the modular functor defined as follows. If S is an oriented 1-manifold let us write T_S for the complexification of the group of smooth maps $S \to T$, and $\tilde{T}_S$ for the central extension by $\mathbb{C}^\times$ corresponding to the chosen level. If X is a Riemann surface with (outgoing) boundary, write T_X for the group of holomorphic maps $X \to T_\mathbb{C}$, which can be identified with a subgroup of $T_{\partial X}$, and even of $\tilde{T}_{\partial X}$ as the cocycle vanishes identically on it. If $\varphi = (\varphi_1, ..., \varphi_k)$ is a labelling of the components of $\partial\Sigma$ we define the modular functor, as explained on pages 37–40, by

$$E(X_\varphi) = \{E_{\varphi_1} \otimes \cdots \otimes E_{\varphi_k}\}^{T_X}.$$

From the representation theory of Heisenberg groups we know that this space is an irreducible representation of $\tilde{T}_X^\perp / T_X$, where $\tilde{T}_X^\perp$ is the centralizer of T_X in $\tilde{T}_{\partial X}$. One easily shows that $\tilde{T}_X^\perp / T_X$ is the Heisenberg group made from the finite group $H^1(X; A)$ with its non-degenerate cup-product pairing. This gives us a very explicit description of the modular functor from which all desired properties, including unitarity, can be read off.

The other topic of §12 was to be the chiral factorization of the sigma-model of a rational torus $T = \mathfrak{t}/\Lambda$, i.e. one for which the inner product is rational on Λ, or, equivalently, such that $\Lambda_0 = \Lambda \cap \Lambda^0$ has finite index in Λ and Λ^0. Let T_0 be the torus $\mathfrak{t}/\Lambda_0$. We have an exact sequence

$$0 \to A \to \mathcal{L}_{\text{left}}T_0 \times \mathcal{L}_{\text{right}}T_0 \to \mathcal{L}T \times \mathcal{L}T^* \to A \to 0$$

where the middle map is $(f, g) \mapsto (f + g, f - g)$, and now

$$A = \left(\frac{1}{2}\Lambda + \frac{1}{2}\Lambda^0\right) \Big/ \Lambda_0.$$

The standard cocycle on $\mathcal{L}T \times \mathcal{L}T^*$ pulls back to a product cocycle on $\mathcal{L}_{\text{left}}T_0 \times \mathcal{L}_{\text{right}}T_0$, and each factor acquires a central extension with centre A. (One should think of T_0 as $\mathfrak{t}_0/\Lambda_0$, where $\mathfrak{t}_0 = \mathfrak{t}$, but with its inner product multiplied by 2.) The projective irreducible representation H_{S^1} of $\mathcal{L}T \times \mathcal{L}T^*$ which is the Hilbert space of the sigma-model of T decomposes under $\mathcal{L}_{\text{left}}T_0 \times \mathcal{L}_{\text{right}}T_0$ as $\bigoplus \bar{E}_\varphi \otimes E_\varphi$, where φ runs through the characters of A, and E_φ is the corresponding irreducible representation of $\mathcal{L}_{\text{right}}T_0$. It is easy to check that the sigma-model of T is thereby identified as the WZW model of T_0.

The best-known case is when $T = \mathbb{R}/R\mathbb{Z}$ is a circle of circumference R. This is rational if R^2 is rational. If $R^2 = p/q$, where p and q are coprime integers. Then $\Lambda_0 = (pq)^{1/2}\mathbb{Z}$, and A is a cyclic group of order $2pq$: the WZW model is that of $\mathbb{T}$ at level $2pq$.

The Definition of Conformal Field Theory

Graeme Segal

St. Catherine's College, Oxford.

The object of this work is to present a definition of a
two-dimensional conformally invariant quantum field theory in
mathematical language, and to describe the basic examples. I hope this
will be helpful to mathematicians who are interested in physics; but
apart from that there are several areas of pure mathematics where
conformal field theories seem to play a fundamental but quite
unexpected role. I shall give five examples.

(i) The "monster" group of Griess-Fischer is the group of
automorphisms of a fairly simple and natural conformal field theory.
The graded representation of the monster group whose Poincaré series is
the modular function j is the basic Hilbert space of the field theory,
and Griess's non-associative algebra is also part of its structure.

(ii) The representation theory of loop groups and of the group
$\text{Diff}(S^1)$ of diffeomorphisms of the circle is greatly illuminated by
conformal field theory. In particular the modularity properties of the
characters of the representations fall into place.

(iii) Field theory shows how the representations of $\text{Diff}(S^1)$ are
related to the geometry of the moduli spaces of Riemann surfaces. Thus
the universal central extension of $\text{Diff}(S^1)$ "is" the determinant line
of the $\bar{\partial}$-operator on Riemann surfaces; and Mumford's classification of
the holomorphic line bundles on moduli spaces can be simply proved.

4322

(iv) Some, at least, of Vaughan Jones's new representations of braid groups arise from field theories, and his classification of subfactors in von Neumann algebras is reflected in the classification of field theories.

(v) The new "elliptic" cohomology theory of Landweber-Stong and Ochanine is undoubtedly connected with conformal field theory, though the connection is still mysterious.

This work is intended to be a coherent and self-contained exposition of material which is essentially well known. It contains no new results. The different sections are fairly independent, and aimed at slightly different readers: they are not meant to be read in order. The recent wave of interest in conformal field theory began with the well-known paper [BPZ] of Belavin, Polyakov, and Zamolodchikov, but I have not attempted the difficult task of indicating the history of the subject, or the provenance of particular ideas. I should like to point out, however, that two features of my exposition which I thought original when I wrote the first versions of this work, namely the emphasis on the semigroup $\mathcal{A}$ in §2, and the algebraic model of the fermion theory in §8, had been developed independently by Neretin [N1],[N2]. Apart from that I am greatly indebted to very many people who have taught me about the subject, especially Deligne, Frenkel, Friedan, Quillen, and Witten. Quillen was originally to be a joint author.

CONTENTS

§1. Introduction

I shall begin with a schematic description of the situation we want to axiomatize. Suppose that a string in the form of a circle S^1 is moving about in a manifold M. The configuration space of the system is then the loop space $\mathcal{L}M$, and the quantum states are the rays in a Hilbert space $\mathcal{H}$ of complex-valued wave functions on $\mathcal{L}M$. The evolution of the system is described by a one-parameter group $\{e^{iHt}\}$ of unitary operators in $\mathcal{H}$. If $T > 0$ the contraction operator e^{-HT} is an integral operator in $\mathcal{H}$:

$$(e^{-HT}\psi)(\gamma) = \int_{\mathcal{L}M} K_T(\gamma,\gamma')\psi(\gamma')\,\mathcal{D}\gamma' \ , \qquad (1.1)$$

where the kernel K_T is of the form

$$K_T(\gamma,\gamma') = \int e^{-S(\sigma)}\,\mathcal{D}\sigma \ , \qquad (1.2)$$

the integral being over all paths $\sigma : [0,T] \rightarrow \mathcal{L}M$ from γ to γ', i.e. over all maps $\sigma : S^1 \times [0,T] \rightarrow M$ which restrict to γ,γ' at the ends of the cylinder. The crucial property of the functional S is that it depends only on the <u>conformal</u> structure of the surface $X = S^1 \times [0,T]$ the basic example is $S(\sigma) = \frac{1}{2} \int_\Sigma \|D\sigma\|^2$.

Needless to say, the preceding integrals have no precise sense. We extract from the discussion simply the Hilbert space $\mathcal{H}$ and the idea of an operator in $\mathcal{H}$ depending not so much on a number T as on the <u>Riemann surface</u> $X = S^1 \times [0,T]$. We could as well or as ill perform the integral (1.2) over maps defined on <u>any</u> Riemann surface X whose boundary consists of two circles, and so obtain an evolution operator $U_X : \mathcal{H} \rightarrow \mathcal{H}$. If two such surfaces X,X' are joined end-to-end to form a new one then we expect the semigroup property

$$U_{XUX'} = U_{X'}U_X . \tag{1.3}$$

It is now natural, given a Riemann surface X with a boundary consisting of m+n circles, to interpret the integral (1.2) over maps X $\to$ M as the kernel of an operator which transforms functions of m loops to functions of n loops, i.e. as the kernel of an operator

$$U_X : \mathcal{H}^{\otimes m} \to \mathcal{H}^{\otimes n} .$$

We might think of this operator as associated with a physical process in which m strings evolve into n strings. We still expect the composition rule (1.3) to hold when it makes sense.

The structure so far described is simply a functor from a certain category $\mathcal{C}$ to the category of Hilbert spaces: the objects of $\mathcal{C}$ are all compact one-dimensional manifolds (i.e. finite disjoint unions of circles), and a morphism from S_0 to S_1 is a Riemann surface X whose boundary ∂X is the disjoint union $S_0 \amalg S_1$. Composition of morphisms in $\mathcal{C}$ is defined by sewing the surfaces together along the common part of their boundaries. A functor assigns a Hilbert space $\mathcal{H}_S$ to each 1-manifold, and an operator $U_X : \mathcal{H}_{S_0} \to \mathcal{H}_{S_1}$ to each surface X with $\partial X = S_0 \amalg S_1$. A conformal field theory is no more and no less than such a functor.[1] It must satisfy a number of simple conditions motivated by the formulae (1.1) and (1.2). The most obvious is that

$$\mathcal{H}_{S_0 \amalg S_1} = \mathcal{H}_{S_0} \otimes \mathcal{H}_{S_1} .$$

----
[1]Strictly speaking, a <u>projective</u> functor: one should allow a scalar multiplier in (1.3).

Another is that if X is a surface with exactly two boundary circles
then the trace of the operator U_X depends only on the closed surface $\check{X}$
got by sewing the two ends of X together. The motivation for this is
that integrating $K_X(\gamma,\gamma)$ over $\gamma \in \mathcal{L}M$ amounts to integrating $e^{-S(\sigma)}$
over all maps $\sigma : \check{X} \to M$.) This property implies the modularity of
the <u>partition function</u> of the theory, which is defined as the trace of
the operator e^{-HT} associated to the cylinder $X_T = S^1 \times [0,T]$. Because
the tori $\check{X}_T$ and $\check{X}_{1/T}$ are conformally equivalent the partition function
satisfies

$$\text{tr}(E^{-HT}) = \text{tr}(e^{-H/T}) .$$

There is another way to approach conformal invariance. The basic
Hilbert space $\mathcal{H} = \mathcal{H}_{S^1}$ of the theory is thought of as the quantization
of a classical system whose phase space is the tangent bundle $T\mathcal{L}M$.
This can be identified with the space of solutions $\sigma : S^1 \times \mathbb{R} \to M$ of
the classical equations of motion, which are conformally invariant,
i.e. invariant under the group $\text{Conf}(S^1 \times \mathbb{R})$ of diffeomorphisms of $S^1 \times \mathbb{R}$
which preserve $d\theta^2 - dt^2$ up to multiplication by a function of (θ,t).
Thus $\text{Conf}(S^1 \times \mathbb{R})$ acts on $T\mathcal{L}M$. We shall see that it follows from our
definition of a field theory that $\text{Conf}(S^1 \times \mathbb{R})$ acts projectively on $\mathcal{H}$.
One can think of a conformal field theory as a projective unitary
representation of $\text{Conf}(S^1 \times \mathbb{R})$ equipped with some additional structure.
Speaking very roughly, the additional structure expresses the fact that
is a representation of a disconnected "group" which has $\text{Conf}(S^1 \times \mathbb{R})$
as its identity component.

[7]

The group $\text{Conf}(S^1 \times R)$ is a Z-fold covering group of $\text{Diff}(S^1) \times \text{Diff}(S^1)$. For $S^1 \times R$ possesses a circle S_R^1 of right-moving light-paths $\{\theta = t - \alpha : \alpha \in S_R^1\}$ and a circle S_L^1 of left-moving paths $\{\theta = -t + \alpha : \alpha \in S_L^1\}$; these two circles are permuted by any conformal diffeomorphism, and so we have a homomorphism

$$\text{Conf}(S^1 \times R) \rightarrow \text{Diff}(S_L^1) \times \text{Diff}(S_R^1) \ ,$$

which is clearly surjective with kernel Z. An irreducible projective representation $\mathcal{H}$ of $\text{Conf}(S^1 \times R)$ decomposes canonically as a tensor product $\mathcal{H} = \mathcal{H}_L \otimes \mathcal{H}_R$ of representations of $\text{Diff}(S_L^1)$ and $\text{Diff}(S_R^1)$. One of the interesting questions to ask about conformal field theories is how they decompose into left-handed and right-handed theories. These so-called chiral theories are to a mathematician — not to a physicist — the basic objects of study. They are rigid in the same sense as the representations of a compact group. Theories containing both chiralities, in contrast, are capable of continuous deformation. We shall consider the simplest example of this phenomenon in §10.

[8]

§2. Diff$^+$(S^1) and the semigroup of annuli

The group Diff$^+$(S^1) of orientation-preserving diffeomorphisms of
the circle is an infinite dimensional Lie group which does not possess
a complexification. In this section I shall describe a complex Lie
semigroup $\mathcal{A}$ which can reasonably be regarded as a subsemigroup of the
non-existent complexification. The relation between Diff$^+$(S^1) and $\mathcal{A}$
is exactly the same as that between the group $T = \{z \in \mathbb{C} : |z| = 1\}$ and
the semigroup $\mathbb{C}^\times_{<1} = \{z \in \mathbb{C} : 0 < |z| < 1\}$, or, better, between the
subgroup PSU$_{1,1}$ of Diff$^+$(S^1) consisting of Mobius transformations and
the sub-semigroup

$$PSL_2^<(\mathbb{C}) = \{g \in PSL_2(\mathbb{C}) : g(D) \subset \mathring{D}\}$$

of the complexification PSL$_2(\mathbb{C})$ of PSU$_{1,1}$. (Here D is the unit disc
$\{z \in \mathbb{C} : |z| \leqslant 1\}$, and $\mathring{D}$ is its interior.) Another such pair consists
of U$_n$ and the semigroup of contraction operators $\{g \in GL_n(\mathbb{C}) : \|g\| < 1\}$.

The semigroup $\mathcal{A}$ is constructed by considering Riemann surfaces
with boundaries. The surfaces we consider in this paper will always be
compact smooth (i.e. C$^\infty$) manifolds X with boundary ∂X, with a smooth
almost complex structure defined everywhere in X. We shall
usually consider surfaces with parametrized boundaries, i.e. with a
given smooth identification of each boundary circle S $\subset$ ∂X with the
standard circle S$^1 = \mathbb{R}/\mathbb{Z}$. If the parametrization of S agrees with the
orientation induced by the complex structure of X we shall call the
circle outgoing, otherwise incoming. Surfaces with parametrized
boundaries can be sewn together by identifying incoming circles with
outgoing ones. One can also sew together an incoming and an outgoing
circle of the same surface. The sewing-together process is formally

characterized as follows. If $\check{X}$ is obtained from a (possibly disconnected) surface X by sewing together some of the circles making up ∂X, and $\pi : X \to \check{X}$ is the identification map, then a function $f : U \to \mathbb{C}$ defined in an open set U of $\check{X}$ is holomorphic if and only if the composite $f\,\pi : \pi^{-1}(U) \to \mathbb{C}$ is holomorphic. It is true, though by no means obvious, that this does define a complex structure (and hence a smooth structure too) on the interior of $\check{X}$.

Let $\mathscr{A}$ denote the set of isomorphism classes of Riemann surfaces A which are topologically annuli (i.e. diffeomorphic to $\{z \in \mathbb{C} : a < |z| < b\}$) and are equipped with parametrizations of their boundary circles, one incoming and one outgoing. Such annuli form a semigroup in which the composite $A_2 \circ A_1$ is formed by sewing the outgoing end of A_1 to the incoming end of A_2.

If one forgets the parametrization of the ends then any annulus is isomorphic to $A_r = \{z \in \mathbb{C} : r < |z| < 1\}$ for a unique $r \in (0,1)$. The only holomorphic automorphisms of A_r are rigid rotations, so we have

<u>Proposition (2.1)</u>. $\mathscr{A}$ is homeomorphic to

$$(0,1) \times (\mathrm{Diff}^+(S^1) \times \mathrm{Diff}^+(S^1))/\mathbf{T} \ .$$

Thus $\mathscr{A}$ has the right size to be a complexification of $\mathrm{Diff}^+(S^1)$. On the other hand $\mathscr{A}$ is a complex manifold in view of

<u>Proposition (2.2)</u>. Any element A of $\mathscr{A}$ is uniquely representable as an annulus in $\mathbb{C}$ bounded by the circles

$$z \mapsto f_0(z) = a_1 z + a_2 z^2 + \dots$$

$$z \mapsto f_\infty(z) = \{z^{-1} + b_2 z^{-2} + b_3 z^{-3} + \dots\}^{-1} ,$$

where f_0 extends to a holomorphic embedding $f_0 : D \to \mathbb{C}$, and f_∞ extends to a holomorphic embedding of $D_\infty = \{z \in \mathbb{C} \cup \infty : |z| \geqslant 1\}$ in the Riemann sphere S^2. (We always identify $S^1 = \mathbb{R}/\mathbb{Z}$ with $T \subset \mathbb{C}$ by $t \mapsto e^{2\pi i t}$.)

Proof: Given an annulus A, let $\hat{A}$ be the closed surface got by sewing copies of D and D_∞ to its ends. Then $\hat{A}$ can be identified holomorphically with the standard S^2, and the identification is unique with the normaliztion prescribed in the proposition.

The space Hol(D) of holomorphic functions on D with smooth boundary values has a natural topology as a subspace of $C^\infty(S^1)$. Proposition (2.2) identifies $\mathcal{A}$ with an open set in the complex vector space $E = \mathbb{C} \oplus \mathrm{Hol}_1(D) \oplus \mathrm{Hol}_1(D)$ by $A \mapsto (a_1, a_1^{-1} f_0, f_\infty^{-1})$: here

$$\mathrm{Hol}_1(D) = \{f : f(0) = 0 \text{ and } f'(0) = 1\} .$$

In fact $\mathcal{A}$ is a bounded domain in E, because $|a_1| < 1$ and each coefficient $a_1^{-1} a_i$ or b_i in (2.2) is also uniformly bounded. (The area of the annulus, as a subset of $\mathbb{C}$, is $\pi\{1 - \Sigma k|a_k|^2 - \Sigma k|b_k|^2\}$.)

Proposition (2.3). The composition $\mathcal{A} \times \mathcal{A} \to \mathcal{A}$ is holomorphic.

To prove this we must consider the tangent spaces to $\mathcal{A}$. Because any annulus can be embedded holomorphically in $\mathbb{C}$ we have

<u>Proposition (2.4)</u>. The tangent space to $\mathscr{A}$ at A is the space of complex tangent vector fields to A along ∂A, modulo those which extend holomorphically over A, i.e.

$$T_A = \{ \mathrm{Vect}_{\mathbb{C}}(S^1) \oplus \mathrm{Vect}_{\mathbb{C}}(S^1) \}/\mathrm{Vect}(A).$$

<u>Remark</u>. As A shrinks to S^1, i.e. to the absent identity element of $\mathscr{A}$, the space T_A approaches $\mathrm{Vect}_{\mathbb{C}}(S^1)$, as one would expect if $\mathscr{A}$ is a complexification of $\mathrm{Diff}(S^1)$.

<u>Proof of (2.3)</u>. We must show that the map of tangent spaces induced by $\mathscr{A} \times \mathscr{A} \to \mathscr{A}$ is complex-linear. Let A_1 and A_2 be annuli, and $A_3 = A_2 \circ A_1$. Write $\partial A_1 = S_0 \amalg S_1$, and $\partial A_2 = S_1 \amalg S_2$, so that $A_1 \cap A_2 = S_1$.

If $(\xi_0, \xi_1) \in \mathrm{Vect}_{\mathbb{C}}(S_0) \oplus \mathrm{Vect}_{\mathbb{C}}(S_1)$ represents a tangent vector to $\mathscr{A}$ at A_1, and (η_1, η_2) represents a tangent vector at A_2, then the composition law takes these vectors to $(\varsigma_0, \varsigma_2)$, where

$$\varsigma_0 = \xi_0 + \alpha_1 | S_0 \qquad\qquad \varsigma_2 = \eta_2 + \alpha_2 | S_2$$

for some $\alpha_i \in \mathrm{Vect}(A_i)$ such that

$$\alpha_1 | S_1 - \alpha_2 | S_1 = \xi_1 - \eta_1 .$$

(It follows from Laurent's theorem that any vector field on S_1 is the difference of holomorphic vector fields on A_1 and A_2.) The $\mathrm{map}((\xi_0,\xi_1),(\eta_1,\eta_2)) \mapsto (\varsigma_0, \varsigma_2)$ is clearly complex-linear.

[12]

There is an important holomorphic function q : $\mathscr{A} \to \mathbb{C}^\times$ whose value at A is the modulus of the torus $\check{A}$ obtained by sewing the ends of A together. (A torus with a preferred cycle is isomorphic to $\mathbb{C}^\times/\lambda$ for a unique λ with $0 < |\lambda| < 1$: I shall call λ the modulus.) More explicitly, if $A \subset \mathbb{C}$ is bounded by the curves f_0 and f_∞, then $q(A) = \lambda$ if there is a holomorphic map $F : A \to \mathbb{C}^\times$ such that $F(f_0(z)) = \lambda F(f_\infty(z))$. I shall omit the proof that q : $\mathscr{A} \to \mathbb{C}^\times$ is holomorphic, but the following result is almost obvious.

Proposition (2.5). We have $q(A) = q(B)$ if and only if A and B are conjugate in $\mathscr{A}$, i.e. related by the equivalence relation ~ generated by

$$A \sim B \quad \text{if} \quad A = C \circ D \quad \text{and} \quad B = D \circ C \text{ for some C, D}$$

When $\mathscr{A}$ is regarded as a bounded domain in the vector space E its boundary is made up of several different pieces. One piece lies in the hyperplane $a_1 = 0$. It is of complex codimension 1, and consists of "infinitely long" annuli. If it is adjoined to $\mathscr{A}$ we still have an open set of E. Another piece of the boundary consists of the points such that the embedded discs $f_0(D)$ and $f_\infty(D_\infty)$ in S^2 touch each other, i.e. those for which the "width" of the annulus collapses to zero at some point. This piece is of real codimension 1. It contains an extremal part $\Sigma(\mathscr{A})$ where $f_0(S^1) = f_\infty(S^1)$. This is a completion of $\mathrm{Diff}^+(S^1)$, in the sense that it contains a dense open subset $\overset{\circ}{\Sigma}(\mathscr{A})$ where $f_0|S^1$ and $f_\infty|S^1$ are injective, and $\overset{\circ}{\Sigma}(\mathscr{A})$ can be identified with $\mathrm{Diff}^+(S^1)$ by $(f_0, f_\infty) \mapsto f_\infty^{-1} \circ f_0$. There are also two other parts of the boundary consisting of points where $f_0|D$ or $f_\infty|D_\infty$ fail to be embeddings.

It is natural at this point to ask a number of questions to which I do not know the answers.

(i) Is $\Sigma(\mathcal{A})$ a Shilov boundary of $\mathcal{A}$?

(ii) Does the function q extend continuously from $\mathcal{A}$ to $\Sigma(\mathcal{A})$?

(iii) If so, what is the relation between $q|\mathrm{Diff}^{+}(S^{1})$ and the rotation number in the sense of Poincaré?

I recall that a <u>Shilov boundary</u> of $\mathcal{A}$ as a subset of E is defined as a minimal closed subset $\mathcal{B}$ of the closure $\mathcal{A}^{cl}$ of $\mathcal{A}$ with the property that

$$\sup(f|\mathcal{A}) = \sup(f|\mathcal{B})$$

for every bounded holomorphic function $f : \mathcal{A} \to \mathbb{C}$ which extends continuously to $\mathcal{A}^{cl}$. If a Shilov boundary of $\mathcal{A}$ exists then it is certainly contained in $\Sigma(\mathcal{A})$, for any boundary point A of $\mathcal{A}$ which is not contained in $\Sigma(\mathcal{A})$ belongs to a holomorphic curve in $\mathcal{A}^{cl}$ got by deforming $\partial A \subset S^{2}$ by any vector field on S^{2} which is holomorphic everywhere except for an essential singularity in the interior of A.

An optimist might hope that for diffeomorphisms of the circle the function q simultaneously measures the rotation number and how far the diffeomorphism is from being conjugate to a rotation, i.e.

<u>Conjecture</u>. The function q extends continuously from $\mathcal{A}$ to $\Sigma(\mathcal{A})$, and for a diffeomorphism f of S^{1} one has $q(f) = \rho e^{i\alpha}$, where α is the rotation number of f, and $\rho = 1$ if and only if f is conjugate to a rotation.

To conclude this section I should mention that the semigroup $\mathcal{E}$ of holomorphic embeddings $f : D \to \mathring{D}$ is a sub-semigroup of $\mathcal{A}$: one identifies f with the annulus $A_f = D - f(\mathring{D})$. Heuristically, at least, $\mathcal{E}$

[14]

is "maximal parabolic" in $\mathcal{A}$, and contains the "minimal parabolic" $\mathcal{E}_0 = \{f \in \mathcal{E} : f(0) = 0\}$. In support of this terminology we notice that (cf. [BR])

$$\mathcal{A}/\mathcal{E}_0 \cong \mathrm{Diff}^+(S^1)/T ,$$

$$\mathcal{A}/\mathcal{E} \cong \mathrm{Diff}^+(S^1)/PSU_{1,1} ,$$

and also that $\bar{\mathcal{E}}_0 \mathcal{E}_0$ is an open subset of $\mathcal{A}$.

It is easy to see that if $f \in \mathcal{E}$ then $q(f) = f'(\zeta)$, where ζ is the unique fixed point of f. Thus $q|\mathcal{E}_0$ is the homomorphism $f \mapsto f'(0)$, whose kernel is the commutator subgroup of $\mathcal{E}_0$.

§3. Wick-rotation, and representations of $\mathcal{A}$

In ordinary quantum field theory there is a Hilbert space $\mathcal{H}$ of states on which the group R^4 of translations of Minkowski space-time acts unitarily. It is well-known that the positivity of energy can be expressed by saying that the unitary action of R^4 extends to an action of the semigroup

$$C_+^4 = \{ \xi \epsilon C^4 : \operatorname{Im}(\xi) \epsilon P \} ,$$

where $P \subset R^4$ is the positive light-cone. The action of C_+^4 is by contraction operators, and is holomorphic. The "boundary" R^4 is an open dense subset of the Shilov boundary of C_+^4.

Now let us consider 2-dimensional Minkowskian space-time $\Sigma = R \times S^1$, in which space is a circle. The group T of translations is $(R \times R)/2\pi Z$, where $(\xi, \eta) \epsilon T$ acts on $(t, \theta) \epsilon \Sigma$ by

$$(t, \theta) \longmapsto (t + \xi + \eta, \ \theta + \xi - \eta) .$$

The positivity of energy is now expressed by saying that the unitary action of T is the boundary value of a holomorphic contraction representation of

$$T_C^+ = \{(\xi, \eta) \ \epsilon \ (C \times C)/2\pi Z : \operatorname{Im}(\xi) > 0, \ \operatorname{Im}(\eta) > 0\} .$$

This is a covering group of $C_{<1}^\times \times C_{<1}^\times$, where $C_{<1}^\times = \{q \ \epsilon \ C^\times : |q| < 1\}$. If one has a conformal theory one expects the group $\operatorname{Conf}(\Sigma)$ to act unitarily on $\mathcal{H}$. I have already mentioned that

$$\text{Conf}(\Sigma) = (\widetilde{\text{Diff}}(S_L^1) \times \widetilde{\text{Diff}}(S_R^1)/2\pi Z ,$$

where $\widetilde{\text{Diff}}(S^1)$ is the simply connected covering group of $\text{Diff}(S^1)$, i.e.
the group of diffeomorphisms $\varphi : R \to R$ such that $\varphi(\theta + 2\pi) = \varphi(\theta) + 2\pi$.
The main idea of conformal field theory - in one interpretation - is
that the positivity of energy is expressed by the fact that the action
of $\text{Conf}(\Sigma)$ on $\mathcal{H}$ extends to a holomorphic contraction representation of
$(\widetilde{\mathcal{A}} \times \widetilde{\mathcal{A}})/2\pi Z$, where $\widetilde{\mathcal{A}}$ is the simply connected covering group of the
semigroup $\mathcal{A}$ of annuli which was introduced in §2. In fact one wants
the action to extend to a still larger semigroup (or rather category)
which allows circles to split into two: that is described in §4.

The holomorphic action of $\mathbb{C}_+^4$ on a conventional state space is of
course completely determined by its restriction to the cone iP. A
contraction representation of iP can be extended holomorphically to $\mathbb{C}_+^4$
providing it satisfies the condition called "reflection-positivity",
and then restricted to give a unitary representation of R^4. In the
two-dimensional case the sub-semigroup of $T_{\mathbb{C}}^+$ which corresponds to iP is
the upper half-plane $\mathbb{C}_+$, the covering of $\mathbb{C}_{<1}^\times$. In the case of $\text{Conf}(\Sigma)$
the corresponding semigroup is $\mathcal{A}$, embedded diagonally in
$(\widetilde{\mathcal{A}} \times \widetilde{\mathcal{A}})/2\pi Z$. We are therefore interested in two questions:

(i) when are unitary representations of $\text{Diff}(S^1)$ the boundary
values of holomorphic contraction representations of $\mathcal{A}$, and

(ii) when can contraction representations of $\mathcal{A}$ be continued
analytically to holomorphic representations of $(\widetilde{\mathcal{A}} \times \widetilde{\mathcal{A}})/2\pi Z$?

Concerning the first question I should mention that the
corresponding finite dimensional situation - where a Lie group G is
essentially the Shilov boundary of an open semigroup $G_{\mathbb{C}}^+$ contained in
the complexification $G_{\mathbb{C}}$ - occurs frequently and has been much studied

(cf. [ ]). For example when G is the subgroup $PSU_{1,1} \cong PSL_2(R)$ of $\text{Diff}(S^1)$, and $G_{\mathbb{C}}^+$ is the sub-semigroup of $PSL_2(\mathbb{C})$ described at the beginning of §2, it is well-known (and obvious) that the irreducible unitary representations of G which extend to $G_{\mathbb{C}}^+$ are precisely the discrete series representations, i.e. the representations of G on the spaces of holomorphic forms on D, and also the trivial representation.

Returning to $\text{Diff}^+(S^1)$ and $\mathscr{A}$, the representations of $\text{Diff}^+(S^1)$ which have a chance of extending to $\mathscr{A}$ are the ones of positive energy ([S2], [PS]), i.e. those for which the subgroup T of rigid rotations acts by characters $\{e^{ik\theta}\}$ for which the values of k are bounded below. These are all projective representations. In the following discussion we shall tacitly restrict our attention to representations for which the action of $\mathbb{C}_{<1}^{\times} \subset \mathscr{A}$ is diagonalizable and extends to an action of $T \subset \text{Diff}^+(S^1)$. I shall also not distinguish between representations which are "essentially equivalent" in the sense of [PS] Chapter 9.

Proposition (3.1). There is a 1-1 correspondence between positive energy projective representations of $\text{Diff}^+(S^1)$ and holomorphic projective representations of $\mathscr{A}$. Unitary representations of $\text{Diff}^+(S^1)$ correspond to representations of $\mathscr{A}$ which are reflection-positive in the sense that $U_A^* = U_{\bar{A}}$.

Proof: First suppose given a representation $A \mapsto U_A$ of $\mathscr{A}$ on a topological vector space E. Let A_q be the standard annulus with parameter $q \in \mathbb{C}_{<1}^{\times}$, and let $U_q = U_{A_q}$. The union of the subspaces $U_q.E$ for all q is a dense subspace $\check{E}$ of E. I shall prove that the group $\text{Diff}_{an}^+(S^1)$ of real-analytic diffeomorphisms of S^1 acts on $\check{E}$. It is, however, well known that all positive energy representations of Diff_{an}^+ extend to Diff^+.

If A is an annulus and φ is a diffeomorphism of S^1 I shall write φA (resp. $A\varphi^{-1}$) for the annulus obtained by changing the outgoing (resp. incoming) parametrization of A by φ. Let us call an annulus real-analytic if both its boundary parametrizations (in the sense of (2.2)) are real-analytic. If φ is a real-analytic diffeomorphism let U_φ denote the densely-defined operator $U_{\varphi A} U_A^{-1}$ in E, where A is a real-analytic annulus. This does not depend on A, for if A' is another choice then there is a standard annulus $B = A_q$ such that $A = B \circ C$ and $A' = B \circ C'$, and

$$U_{\varphi A} U_A^{-1} = U_{\varphi B} U_B^{-1} = U_{\varphi A'} U_{A'}^{-1} .$$

(We are suppressing a possible projective multiplier, which is immaterial.) Then U_φ maps $\check{E}$ to itself, and defines a representation of $\mathrm{Diff}_{an}^+(S^1)$, because

$$U_{\psi\varphi} = U_{\psi(\varphi A)} U_{\varphi A}^{-1} U_{\varphi A} U_A^{-1} = U_\psi U_\varphi .$$

Conversely, if E is a positive energy representation of $\mathrm{Diff}^+(S^1)$ then there is an obvious candidate for the operator U_q associated to A_q. But for any annulus A we can by (2.1) write $A = \varphi A_q \psi^{-1}$ in an essentially unique way, and then define $U_A = U_\varphi U_q U_\psi^{-1}$. We must show that U_A depends holomorphically on A, and that it defines a representation of $\mathcal{A}$. For the first, recall from (2.4) that the tangent space to $\mathcal{A}$ at A is $(\mathrm{Vect}_{\mathbb{C}}(S^1) \oplus \mathrm{Vect}_{\mathbb{C}}(S^1))/\mathrm{Vect}(A)$. Let $\xi \mapsto L_\xi$ be the derivative of $\varphi \mapsto U_\varphi$. Writing the derivative of $A \mapsto U_A$ as

$$\delta U_A = U_\varphi \{ (U_\varphi^{-1} \delta U_\varphi) U_q - U_q (U_\psi^{-1} \delta U_\psi) \} U_\psi^{-1}$$

we see that $A \mapsto U_A$ is holomorphic if the map

$$(\xi, \eta) \mapsto L_\xi U_q - U_q L_\eta \, ,$$

defined on $\text{Vect}_{\mathbb{C}}(S^1) \oplus \text{Vect}_{\mathbb{C}}(S^1)$, vanishes on $\text{Vect}(A_q)$. But that is obvious. (Holding q fixed in this calculation is permissible because q = constant defines a submanifold of $\mathcal{A}$ of <u>real</u> codimension one.) Finally, to show that $A \mapsto U_A$ is a homomorphism amounts to proving that two holomorphic maps $\mathcal{A} \times \mathcal{A} \to \text{End}(E)$ coincide. But they coincide by definition at points of the form $(\varphi A_q, A_q, \psi^{-1})$, and as they are holomorphic that is enough.

The correspondence between unitarity and reflection-positivity needs no comment, except perhaps to point out that if $A = \varphi A_q \psi^{-1}$ then $\bar{A} = \psi A_{\bar{q}} \varphi^{-1}$.

I have little to say about question (ii) above, when a non-holomorphic representation of $\mathcal{A}$ can be continued to a holomorphic representation of the complexification $\mathcal{A}_{\mathbb{C}} = (\tilde{\mathcal{A}}_L \times \tilde{\mathcal{A}}_R)/2\pi\mathbb{Z}$. It is certainly true in the reflection-positive case. For any representation of $\mathcal{A}$ gives us a representation of the Lie algebra of $\mathcal{A}_{\mathbb{C}}$, which is the complexification of the Lie algebra of $\text{Diff}(S_L^1) \times \text{Diff}(S_R^1)$. But it is known that any unitary positive energy representation of this Lie algebra extends to a representation of the group, and then the representation of $\text{Diff}(S_L^1) \times \text{Diff}(S_R^1)$ gives rise to a holomorphic representation of $\mathcal{A}_{\mathbb{C}}$ as in the proof above. It would be interesting, however, to have a better treatment of this question.

[20]

§4. <u>The category $\mathscr{C}$ and the definition of a field theory</u>

The category $\mathscr{C}$

The category $\mathscr{C}$ is defined as follows. There is a set of objects $\{C_n\}_{n \geqslant 0}$, where C_n is the disjoint union of a set of n parametrized circles. A morphism $C_m \to C_n$ is a Riemann surface X with boundary ∂X together with an orientation-preserving identification $C_n - C_m \to \partial X$. (Here $C_n - C_m$ means $C_n \amalg C_m$ with the orientation of C_m reversed.) We identify two surfaces if they are isomorphic by a map which respects the parametrization of the boundary. Composition of morphisms is defined by sewing surfaces together.[1]

The set $\mathscr{C}_{mn}$ of morphisms $C_m \to C_n$ is a topological space with one connected component $\mathscr{C}_\alpha$ for each topological type of surface. Thus when α is an annulus $\mathscr{C}_\alpha$ is the semigroup $\mathscr{A}$ of §2. Two other cases are worth mentioning.

(i) If α is a disc $\mathscr{C}_\alpha$ is $\mathrm{Diff}^+(S^1)/PSU_{1,1}$, for all discs are the same except for the parametrization of the boundary. This gives a description of the complex structure on $\mathrm{Diff}^+(S^1)/PSU_{1,1}$ (cf. [BR]). In terms of the semigroups of §2 we have $\mathscr{C}_\alpha \cong \mathscr{A}/\mathscr{C}$.

(ii) If α is a disc with two holes then $\mathscr{C}_\alpha$ has a Shilov boundary which consists of the space of ways in which a circle can split into two:

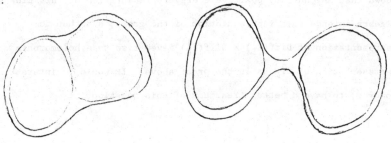

---

[1]Purists will object that the category $\mathscr{C}$ has no identity morphisms, and will have their preferred remedies.

As a space $\mathcal{C}_\alpha$ is the quotient of the contractible space of complex structures on a ∧fixed smooth surface Σ_α of type α by the group of all diffeomorphisms of Σ_α which are the identity on $\partial\Sigma_\alpha$. On the other hand it is well known that the moduli space of closed surfaces of a given topological type is a finite dimensional complex variety with some mild singularities. If α is a connected surface with $k > 0$ boundary components the complex structure of $\mathcal{C}_\alpha$ can be described by analogy with the description of $\mathcal{A}$ in §2, as follows. Let g be the genus of the closed surface got by adding k caps to α - we shall call g simply "the genus of α" - and let $\mathcal{M}'_{g,k}$ be the moduli space of closed surfaces of genus g with k marked points $\{x_i\}$ and prescribed tangent vectors $\{\xi_i\}$ at the points $\{x_i\}$. The space $\mathcal{M}'_{g,k}$ is a finite dimensional complex manifold with no singularities, and there is a tautological fibre bundle over it whose fibre at $\hat{X}$ is $\hat{X}$. The space $\mathcal{C}_\alpha$ is a fibration over $\mathcal{M}'_{g,k}$ whose fibre at $(\hat{X},\{x_i\},\{\xi_i\})$ is the space of k-tuples of disjointly embedded discs $f_i : D \to \hat{X}$ such that $f_i(0) = x_i$ and $f'_i(0) = \xi_i$. (This description needs adjustment when α is a disc: then $\mathcal{C}_\alpha$ is the space of embeddings $f: D \to S^2$ such that $f(0) = 0$, $f'(0) = 1$, and $f''(0) = 0$.

Composition of morphisms is a holomorphic map $\mathcal{C}_{km} \times \mathcal{C}_{mn} \to \mathcal{C}_{kn}$. It is enough to prove this when the composite surface has no closed components, and in that case it follows as in §2 from

Proposition (4.1). If α has no closed components the tangent space to $\mathcal{C}_\alpha$ at X is $\text{Vect}_\mathbb{C}(\partial X)/\text{Vect}(X)$, the space of tangent vector fields to X along ∂X modulo those which extend holomorphically to X.

<u>Remark</u>. The dual cotangent space is therefore the space of holomorphic quadratic differentials on $\overset{\circ}{X}$ which have distributional boundary values on ∂X.

<u>Proof of (4.1)</u>. The argument is the same as in §2, except that we need to know that if X is obtained from a closed surface $\hat{X}$ by removing discs, then any $Y \in \mathscr{C}_\alpha$ which is sufficiently close to X can be embedded holomorphically in $\hat{X}$. That is true because X is a Stein manifold.

<u>Remark</u>. We could put a finer topology on $\mathscr{C}_{mn}$ - without changing the topology on each $\mathscr{C}_\alpha$ - so that $\mathscr{C}_{mn}$ had just one connected component for each genus. For if β is disconnected $\mathscr{C}_\beta$ can be stuck on to the boundary of $\mathscr{C}_\alpha$ for an appropriate connected α. Thus the space $\{\mathrm{Diff}^+(S^1)/PSU_{1,1}\}^{\times 2}$ of pairs of discs can be attached to the boundary of $\mathscr{A}$ by collapsing the divisor $\{\mathrm{Diff}^+(S^1)/T\}^{\times 2}$ consisting of infinitely long cylinders. The resulting connected $\mathscr{C}_{mn}$ would be a complex variety with bad singularities. We shall not pursue this, however.

The definition: first version

We shall define a conformal field theory as a functor from $\mathscr{C}$ to complex topological vector spaces. We assume the vector spaces H are locally convex and complete, and equipped with a continuous hermitian form $\bar{H} \times H \to \mathbb{C}$. We shall not restrict ourselves to Hilbert spaces, as we want to allow <u>indefinite</u> inner products. We shall state the definition in terms of tensor products. These should be interpreted in the sense explained in Appendix A. But if H is a Hilbert space the tensor products can equally well be taken in the Hilbert space sense.

We shall make use of a number of elementary operations which can be performed on the morphisms of $\mathscr{C}$.

(a) The symmetric groups S_m and S_n act on $\mathscr{C}_{mn}$ by permuting the numbering of the boundary circles.

(b) If $X \in \mathscr{C}_{mn}$ then the complex conjugate surface $\bar{X}$ belongs to $\mathscr{C}_{nm}$, and $X \mapsto \bar{X}$ is an antiholomorphic map.

(c) By reversing the orientation of the incoming boundary circles we obtain the "crossing" isomorphism $\mathscr{C}_{mn} \to \mathscr{C}_{0,m+n}$, which I shall write $X \mapsto |X|$.

(d) By sewing k incoming to k outgoing circles we obtain a holomorphic map $\mathscr{C}_{mn} \to \mathscr{C}_{m-k,n-k}$.

We now give the provisional definition of a conformal field theory. We should warn the reader, however, that it is unsatisfactory because it does not allow for projective multipliers.

<u>Definition (4.2)</u>. Let H be a topological vector space with a symmetric complex bilinear form and a given real structure (i.e. an anti-involution $H \to \bar{H}$). A conformal field theory based on H is a continuous functor U from $\mathscr{C}$ to topological vector spaces with the following properties.

(i) $U(C_n) = H \otimes \ldots \otimes H = H^{\otimes n}$.

(ii) The map $\mathscr{C}_{mn} \times H^{\otimes m} \to H^{\otimes n}$ is compatible with the action of the symmetric groups S_m and S_n.

(iii) "Crossing": for each $X \in \mathscr{C}_{mn}$ the operator $U(X) : H^{\otimes m} \to H^{\otimes n}$ is of trace class, and is defined by the element $U(|X|)$ of $H^{\otimes m} \otimes H^{\otimes n}$ together with the bilinear form on H.

(iv) "Sewing": the map $\mathscr{C}_{mn} \to \mathscr{C}_{m-k,n-k}$ of (d) above is compatible with the map

$$\mathrm{Hom}(H^{\otimes m};H^{\otimes n}) \to \mathrm{Hom}((H^{\otimes(m-k)};H^{\otimes(n-k)})$$

got by taking the trace over $H^{\otimes k}$. In particular, if $X \in \mathscr{C}_{11}$ and $\check{X}$ is
the associated closed surface, then

$$\mathrm{trace}\ U(X) = U(\check{X})\ .$$

(v) "Reflection positivity": U is a *-functor in the sense that
$U(\bar{X}) = U(X)*$ for all morphisms X. Here the adjoint $U(X)*$ refers to the
hermitian structure on H got by combining the real structure with the
complex bilinear form.

Notes. (i) In this definition we ought certainly to allow the space H to
have a mod 2 grading. Then the permutations of $H^{\otimes n}$ should be performed
with the usual sign conventions, and - most importantly - the trace in
property (iv) should be replaced by the supertrace. We shall for the
most part not bother to make this generalization explicit.

(ii) If we omit to give the real structure on H and the associated
axiom (v) of reflection-positivity then we have a "non-unitary" field
theory.

A conformal field theory is thus, among other things, a trace-
class representation of the semigroup $\mathscr{A}$. As we saw in §3, this gives
us a pseudo-unitary action on H of the Lie algebra of the conformal
group $\mathrm{Conf}(S^1 \times R)$, i.e. of the Lie algebra of $\mathrm{Diff}^+(S_L^1) \times \mathrm{Diff}^+(S_R^1)$.
Under the action of the rigid motions the space H breaks up as a
discrete sum of finite dimensional pieces: $H = \oplus H_{a,b}$, where
$(a,b) \in R^2$, and $a-b \in Z$.

The <u>partition function</u> Z_U of the theory is the function on the upper half-plane defined by

$$Z_U(\tau) = \text{trace } U(A_q) = \Sigma \ \bar{q}^a q^b \dim(H_{a,b}) \ ,$$

where $q = e^{2\pi i\tau}$ and A_q is the standard annulus $\{z : |q| < |z| < 1\}$ described in §2. Because the annuli A_q and $A_{\tilde{q}}$ produce isomorphic tori if $\tilde{q} = e^{-2\pi i/\tau}$ the partition function satisfies

$$Z_U(-\tau^{-1}) = Z_U(\tau) \tag{4.3}$$

(but cf. Proposition (6.11)). The partition function completely determines H as a representation of $\text{Diff}^+(S^1_L) \times \text{Diff}^+(S^1_R)$, for the characters of the representations of $\text{Diff}^+(S^1)$ are all known, and are linearly independent.

Another aspect of the structure is seen by choosing, once for all, a disc with two holes Σ, regarded as an element of $\mathcal{C}_{21}$. For any theory, Σ gives us a map $H \otimes H \to H$ which makes H into a non-associative algebra. This composition law is called the <u>operator product expansion</u>. Together with the partition function the product in H determines the theory completely, for any Riemann surface can be obtained by sewing together discs, cylinders, and copies of Σ, by suitable diffeomorphisms. In the case of the theory whose group of automorphisms is the monster group, the algebra H contains Griess's non-associative algebra as a subalgebra.

Friedan has conjectured that a field theory U is completely determined by its restriction to <u>closed</u> surfaces, i.e. by the homomorphism $U : \mathcal{C}_{00} \to \mathbb{C}^\times$ defined on the commutative semigroup $\mathcal{C}_{00}$. This seems plausible, but I do not know a proof.

A field theory is called <u>holomorphic</u> if the operators U(X) depend holomorphically on X. That is the case if and only if $\text{Diff}^{+}(S_L)$ acts trivially on H, and also if and only if the partition function is holomorphic. A theory is called <u>chiral</u> if it is either holomorphic or antiholomorphic.

The conformal anomaly

The preceding definition is too restrictive, and we must introduce slightly more general structures. In the usual terminology these are theories which have a "conformal anomaly". Mathematically this amounts to passing to <u>projective</u> representations of the category $\mathscr{C}$, i.e. the operator U(X) associated to a surface X is given only up to an indeterminate scalar multiplier. Physically one should think that U(X) is associated not to the surface X alone, but to the surface together with a chosen metric compatible with its conformal structure. The dependence on the metric is slight: if the volume element ω is multiplied by $e^{2\varphi}$, for some $\varphi : X \to \mathbb{R}$, then U(X) is multiplied by $e^{icS(\varphi)}$, where c is a constant depending on the theory (the "central charge") and $S(\varphi)$ is the Liouville action

$$S(\varphi) = \int_X \tfrac{1}{2}\{d\varphi \wedge *d\varphi + 4\varphi R\} .$$

Here R is the curvature 2-form of the metric.

To digress briefly, one can define a general notion of two dimensional field theory as a representation of a category $\mathscr{C}_{\text{metric}}$ made from circles and surfaces equipped with metrics. The metrics must be piecewise twice differentiable, and the boundary circles must be geodesics. Circles of different lengths are, of course, non-isomorphic objects of $\mathscr{C}_{\text{metric}}$. It may be that the intriguing work of Zamolodchikov [Z] can be formulated in this language, but perhaps something more subtle is needed.

Returning to mathematics, just as a projective representation of a group G is a genuine representation of an extension $\tilde{G}$ of G by $\mathbb{C}^{\times}$, so a projective representation of a category $\mathscr{C}$ is an ordinary representation of an extension category $\tilde{\mathscr{C}}$ of $\mathscr{C}$ by $\mathbb{C}$. To give such an extension category is the same as giving a rule which assigns a complex line L_X to each morphism X of $\mathscr{C}$, and a map

$$\mu_{XY} : L_X \otimes L_Y \to L_{X \circ Y}.$$

to each composable pair of morphisms. The maps μ_{XY} must be associative in the obvious sense. The objects of $\tilde{\mathscr{C}}$ are the same as the objects of $\mathscr{C}$, and a morphism in $\tilde{\mathscr{C}}$ is a pair (X, λ), where X is a morphism in $\mathscr{C}$ and $\lambda \in L_X$. In the next section we shall prove that there is essentially only one such extension of $\mathscr{C}$, got by assigning to X the determinant line Det_X of its $\bar{\partial}$-operator, in the sense of Quillen [Q] (cf. also Appendix B). More precisely, the most general extension is of the form[1] $L_X = (\text{Det}_X)^{\otimes p} \otimes (\overline{\text{Det}_X})^{\otimes q}$. If $p = q = c$ one says that the theory has <u>central charge</u> c. (The determinant bundle will be discussed in detail in §6.)

The conditions of (4.2) make sense for a projective functor providing $X \mapsto L_X$ has the properties:

(i) $L_X = L_{\tilde{X}}$ if $\tilde{X}$ is obtained from X by reversing the parametrization of some boundary components;

----

[1] See (5.18).

(ii) $L_{\overline{X}} = \overline{L}_X$;

(iii) there is a natural map $L_X \to L_{\overset{\vee}{X}}$ when $\overset{\vee}{X}$ is made from X by sewing boundary circles together.

When X is an annulus there is a preferred element $\epsilon_X \in L_X$, and so we can define the partition function as the trace of the operator $U(\epsilon_X)$. It has a modularity property analogous to (4.3), but we shall postpone discussion of that (and also the definition of ϵ_X) until §6.

An improved version of the definition

Definition (4.2) is cumbersome and unnatural, and the following reformulation is cleaner. I shall give it in the projective version.

We begin with a hermitian vector space H with a projective unitary action of $\text{Diff}(S^1)$ in which the orientation-reversing diffeomorphisms act <u>antilinearly</u>. There is a unique way to associate to H a projective functor $S \mapsto H_S$ from compact oriented 1-manifolds (and orientation-preserving diffeomorphisms) to hermitian vector spaces (and $\mathbb{C}$-linear operators given up to an arbitrary scalar multiplier) with the two properties:

   (a) $H_{\overline{S}} = \overline{H}_S$ if $\overline{S}$ is S with reversed orientation;

   (b) $H_{S_1 \amalg S_2} = H_{S_1} \otimes H_{S_2}$.

Definition (4.4). A conformal field theory based on H is a continuous natural transformation which assigns to each Riemann surface X with (unparametrized) boundary a ray H_X in $H_{\partial X}$ satisfying

   (i) $H_{\overline{X}} = \overline{H}_X$,

   (ii) $H_{X \amalg Y} = H_X \otimes H_Y$,

   (iii) $H_{\overset{\vee}{X}} = \text{trace } H_X$ if $X \to \overset{\vee}{X}$ is a sewing map.

Here a sewing map $X \to \check{X}$ is one which identifies two disjoint parts S_1 and S_2 of ∂X by an orientation-reversing diffeomorphism; and the trace map $H_{\partial X} \to H_{\partial \check{X}}$ is induced by the bilinear form

$$H(S_1) \otimes H(S_2) \to \mathbb{C}.$$

In (4.4) it is important that we do <u>not</u> use Hilbert space tensor products, for then the Hermitian form $\bar{H} \times H \to \mathbb{C}$ would not extend to $\bar{H} \otimes H$.

The idea of a "projective" functor may seem unappealingly vague. The additional structure which an oriented 1-manifold S needs in order to define a vector space H_S rather than just a projective space can be described as follows.

We define a <u>rigged</u> 1-manifold as an oriented 1-manifold S together with a specific choice L of a determinant line bundle on the restricted Grassmannian $Gr(\Omega^0(S))$ of the space of smooth functions on S (see Appendix B). For given S the bundle L is canonically defined up to isomorphism, but the isomorphism is arbitrary up to an element of $\mathbb{C}^{\times}$. (A parametrization of S is more than enough to provide a canonical choice of L.) A morphism from (S_0, L_0) to (S_1, L_1) is a diffeomorphism $f : S_0 \to S_1$ together with an isomorphism $L_0 \cong f*L_1$.

A surface X with $\partial X = S$ defines a point $Hol(X)$ in $Gr(\Omega^0(S))$. If S is rigged by L then we define as the fibre L_X of L at $Hol(X)$. the determinant line of X. To obtain a vector in $H_{S,L}$ corresponding to Σ we must choose a point of L_X.

To describe chiral theories we shall need an even more general definition than (4.4), in which a surface X defines a subspace H_X of $H_{\partial X}$ which need not be one-dimensional. That is the subject of §5.

Minkowski space, ghosts, and BRS cohomology

Apart from the question of projective multipliers there are two
other respects in which Definition (4.2) is not quite general enough
for the needs of string theory. It is usual to study strings moving in
a product $V \times M$, where V is Minkowski space of some dimension, and M is
a compact Riemannian manifold. In that case the space of states of the
string is a direct integral $H = \int H_p$, where H_p is the states of momentum
p, and p runs through the dual space V* of V. A surface $X \in \mathscr{C}_{mn}$
defines an operator $U(X) : H^{\otimes m} \to H^{\otimes n}$ which is an integral of operators

$$U(X)_{\underline{p},\underline{q}} : H_{p_1} \otimes \ldots \otimes H_{p_m} \to H_{q_1} \otimes \ldots \otimes H_{q_n},$$

where $\Sigma p_i = \Sigma q_i$. Each operator $U(X)_{\underline{p},\underline{q}}$ is of trace class, but $U(X)$
itself is not.

More importantly, strings are not supposed to be parametrized,
while the spaces H we have been discussing describe parametrized
strings. One would expect to replace H by the subspace which
is invariant under $\mathrm{Conf}(S^1 \times R)$. In fact the spaces H which arise are
projective representations of $\mathrm{Conf}(S^1 \times R)$ with a positive central
charge c, and the invariant subspace would be 0. Instead of the
invariant part of H one has recourse to its <u>BRS cohomology</u> H_{BRS}. The
essential points about this are:

(i) it is defined only for a theory with $c = 26$,

(ii) it has a bi-grading (called the "ghost number"),

(iii) in good cases, at least, it has a positive definite metric,

(iv) instead of an operator $H_{BRS}^{\otimes m} \to H_{BRS}^{\otimes n}$ for each surface $X \in \mathscr{C}_{mn}$
one has a top-dimensional differential form ω_{mn} on the finite

dimensional moduli space $\mathcal{M}_{m+n}$ of all closed surfaces with m+n marked points, with values in the space of operators $H_{BRS}^{\otimes m} \rightarrow H_{BRS}^{\otimes n}$ of bidegree $(-m, -m)$.

In particular, m elements of H_{BRS} of bidegree $(1,1)$ define a top-dimensional scalar-valued form on $\mathcal{M}_m$.

We shall say only a little about BRS cohomology in this paper. To define it one tensors the theory H with another theory H_{ghost} which has $c = -26$. The resulting theory $H \otimes H_{ghost}$ has a genuine (non-projective) action of $\mathcal{A}$. The space $H \otimes H_{ghost}$ has an operator $Q = Q_L + Q_R$ which satisfies $Q^2 = 0$, and Q_L and Q_R raise degree by $(1,0)$ and $(0, 1)$ respectively. The cohomology (ker Q)/(im Q) is the BRS cohomology. The theory H_{ghost} will be described in §8, and we shall return to Q and the property (iv) in §9.

§5. Modular functors

Definition and main properties

In studying chiral field theories and also the representations of loop groups one meets the concept of a modular functor. From one point of view this is a generalization of the idea of a central extension of $\text{Diff}^+(S^1)$. On the other hand it can also be regarded as a coherent family of projective representations of the braid groups and mapping class groups.

We start with a finite set Φ of labels. Let $\mathcal{S}_\Phi$ be the category whose objects are Riemann surfaces with each boundary circle parametrized and equipped with a label from Φ. A morphism in $\mathcal{S}_\Phi$ is a holomorphic sewing map $X \to \check{X}$, i.e. one which sews together pairs of edges in accordance with the parametrization; we allow a pair of edges to be identified only if they have the same label. A morphism is allowed to permute the boundary circles, but it must preserve their parametrization.

Definition (5.1). A modular functor is a holomorphic functor E from $\mathcal{S}_\Phi$ to finite dimensional complex vector spaces with the following properties.

(i) $E(X \amalg Y) = E(X) \otimes E(Y)$.

(ii) If X_φ is obtained from X by cutting it along a simple closed curve and giving the label φ to the two new edges then the natural map

$$\bigoplus_{\varphi \, \epsilon \, \Phi} E(X_\varphi) \to E(\check{X})$$

is an isomorphism.

[33]

(iii) For the Riemann sphere S^2 we have dim $E(S^2) = 1$.

Notes. (a) To say that E is holomorphic means that when $\{X_b\}_{b \in B}$ is a holomorphic family of surfaces parametrized by a complex manifold B the spaces $E(X_b)$ fit together to form a holomorphic vector bundle on B. In particular, E defines a holomorphic vector bundle E_α on the moduli space $\mathcal{C}_\alpha$ of surfaces of (labelled) topological type α, at least if α has no closed components. (Recall that $\mathcal{C}_\alpha$ was defined in §4. We exclude closed surfaces to avoid the singularities caused by their possible automorphisms.)

(b) The isomorphism of (i) above is supposed to be compatible with the maps interchanging the summands on each side. As in §4 we should certainly allow modular functors to be graded mod 2, and should use the graded tensor product in (i). The determinant line, for example, is a mod 2 graded modular functor for which $E(S^2)$ is in degree 1.

For any modular functor E we have a map $E(X) \otimes E(Y) \to E(X \circ Y)$ when X and Y are composable morphisms in $\mathcal{C}$ with their boundaries compatibly labelled. So E defines an extension $\mathcal{C}^E$ of the category $\mathcal{C}$. An object of $\mathcal{C}^E$ is a collection of circles each with a label from Φ, and a morphism is a pair (X, ϵ), where X is an morphism in $\mathcal{C}$ and $\epsilon \in E(X)$.

Definition (5.2). A **weakly conformal** field theory is a representation of $\mathcal{C}^E$ for some modular functor E, satisfying conditions as in (4.4).

Thus such a theory assigns a vector space H_S to each one-dimensional manifold and a vector space E_X to each surface, and there is a natural map $E_X \to H_{\partial X}$ for each X.

One may as well assume that the labelling set Φ of a modular functor contains no superfluous elements, i.e. no labels φ such that $E(X) = 0$ whenever X has an edge labelled by φ. We can then make the following elementary observations.

<u>Proposition (5.3)</u>

(i) There is a distinguished label $1 \in \Phi$ such that $\dim E(D) = 1$ when D is a disk with ∂D labelled 1, and $E(D) = 0$ if ∂D has any other label.

(ii) If $A_{\varphi\psi}$ is a annulus with ends labelled φ, ψ then $\dim E(A_{\varphi\psi}) = 1$ if $\varphi = \psi$ and $E(A_{\varphi\psi}) = 0$ otherwise. In particular, E defines a central extension $\mathcal{A}_{\varphi}$ of $\mathcal{A}$ by $\mathbb{C}^{\times}$ for each label φ.

(iii) There is an involution $\varphi \mapsto \bar{\varphi}$ of Φ such that if B is an annulus with both ends <u>outgoing</u> then $\dim E(B_{\varphi\psi}) = 1$ if $\psi = \bar{\varphi}$ and $E(B_{\varphi\psi}) = 0$ otherwise.

(iv) If $\tilde{X}$ is obtained from X by reversing the parametrization of an incoming boundary circle and changing its label from φ to $\bar{\varphi}$ then $E(\tilde{X}) = E(X) \otimes E(B_{\varphi\bar{\varphi}})$.

<u>Proof</u>: We first prove (ii) by observing that the $\Phi \times \Phi$ matrix $\dim E(A_{\varphi\psi})$ is idempotent with positive integer entries. The matrix $\dim E(B_{\varphi\psi})$ is then symmetric and invertible, so we obtain (iii). Assertion (iv) follows immediately, and finally we get (i) by considering the decomposition $S^2 = D \cup D$.

From now on we shall assume modular functors are <u>normalized</u> so that $E(D) = \mathbb{C}$ when ∂D is outgoing and labelled 1.

The sense in which a modular functor is a coherent family of projective representations of discrete groups is explained by

Proposition (5.4). For any modular functor there is a canonical flat connection in the projective bundle of the bundle E_α on $\mathcal{C}_\alpha$, for every non-closed labelled surface α. These connections are compatible with the sewing-together of surfaces.

If the modular functor has central charge 0 (see below) then there is a canonical flat connection in the bundle E_α itself.

In other words, if X and X' are surfaces of type α there is an isomorphism $P(E(X)) \to P(E(X'))$ for each homotopy class of paths from X to X' in $\mathcal{C}_\alpha$. Thus for each α a modular functor gives a projective representation $\pi_1(\mathcal{C}_\alpha) \to PGL_{n_\alpha}(\mathbb{C})$. For example if α is a disc with k holes then $\pi_1(\mathcal{C}_\alpha) = \mathbb{Z}^k \times CBr_k$, where CBr_k is the coloured braid group on k strands. If α is a surface of genus g with one hole then $\pi_1(\mathcal{C}_\alpha)$ is the mapping class group of α.

Verlinde's algebra

An attractive way of looking at modular functors has been developed by Verlinde [V], following the "fusion-rule" approach of Belavin-Polyakov-Zamolodchikov [BPZ]. Let Σ be a disc with two holes, labelled

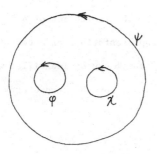

Let $n_{\varphi \chi \psi}$ = dim E(Σ). Then the free abelian group Z[Φ] is clearly a commutative ring under the multiplication

$$(\varphi, \chi) \mapsto \sum_{\psi} n_{\varphi \chi \psi} \, \psi \, .$$

The element 1 ϵ Φ is the identity element of the ring. We shall say more about this ring later on. For the moment let us notice that the ring structure of Z[Φ] is a very compact way of encoding the dimension of E(X) for all labelled surfaces X. Thus if M_φ is the operator of multiplication by φ on Z[φ] then the dimension of E(X) when X is a torus with an incoming and an outgoing hole labelled φ, ψ is

$$p_{\varphi \psi} = \text{trace}(M_\varphi M_{\overline{\psi}}) \, ,$$

and if X_g is a closed surface of genus g then

$$\dim E(X_g) = \text{trace}(P^{g-1}) \, ,$$

where P is the matrix $(p_{\varphi \psi})$.

Loop groups

The natural examples of modular functors arise from representations of loop groups in the following way. I shall suppose for simplicity that G is the complexification of a simply connected compact group. Let $\{E_\varphi\}_{\varphi \epsilon \Phi}$ be the finite set of all irreducible projective positive energy representations of a certain level of the loop group LG. The indexing set Φ can be identified with a set of irreducible representations of G, for the zero-energy subspace of E_φ is an irreducible representation φ of G. The involution $\varphi \mapsto \bar{\varphi}$ takes a

representation of G to its dual, and $1 \in \Phi$ is the trivial

representation of G. Let X be a surface with k boundary components,

all outgoing, labelled by $\varphi_1, \ldots, \varphi_k \in \Phi$. Then the group of holomorphic

maps Hol(X;G) acts on $E_{\varphi_1} \otimes \ldots \otimes E_{\varphi_k}$ via restriction to ∂X, for the

central extension of $(LG)^k$ is canonically split over Hol(X;G). We

define E(X) as the part of $E_{\varphi_1} \otimes \ldots \otimes E_{\varphi_k}$ fixed under Hol(X;G). If

some of the boundary circles are incoming we replace the corresponding

factor E_φ by $t^*E_{\bar{\varphi}}$, where $t : S^1 \to S^1$ reverses the parametrization.

Then $X \mapsto E(X)$ is a modular functor. This will be proved in §11. The

point of the definition is that a surface X with p incoming and q

outgoing circles labelled $\varphi_1, \ldots, \varphi_p$ and $\psi_1, \ldots, \psi_q$, together with an

element ϵ of E(X) - i.e. a morphism (X, ϵ) in the extended category $\mathscr{C}^E$

- defines a trace-class operator

$$U_{X, \epsilon} : E_{\varphi_1} \otimes \ldots \otimes E_{\varphi_p} \to E_{\psi_1} \otimes \ldots \otimes E_{\psi_q} .$$

This is because for each φ there is a natural duality pairing

$$E_\varphi \otimes t^*E_{\bar{\varphi}} \to \mathbb{C} .$$

The concept of a modular functor is designed, among other things,

to express the modularity properties of the characters of

representations of loop groups. A representation E_φ decomposes under

the action of the rigid rotations of S^1 as a sum $E_\varphi = \bigoplus_{k \geqslant 0} E_{\varphi, k}$ of finite dimensional pieces, where the rotation through the angle α acts as $e^{ik\alpha}$ on $E_{\varphi,k}$. Each piece $E_{\varphi,k}$ is a representation of the subgroup G of constant loops in LG. (Thus $E_{\varphi,0} = \varphi$.) The <u>partition function</u> and the <u>character</u> of E_φ are defined as the formal series

$$\chi_\varphi(q) = \Sigma \, q^k \dim(E_{\varphi,k})$$

and

$$\chi_\varphi(q,g) = \Sigma \, q^k \text{trace}(g|E_{\varphi,k})$$

respectively. In fact these series converge when $|q| < 1$, and $\chi_\varphi(q)$ is best regarded as a function of an annulus with modulus q. More precisely,

$$\chi_\varphi(q) = \text{trace}(U_{A,\epsilon} : E_\varphi \to E_\varphi) \ ,$$

where A is the standard annulus A_q, and ϵ is the standard element of the line $E(A_q)$, where the ends of A_q are labelled with φ. Then $\chi_\varphi(q)$ depends only on the image, say $\epsilon_{q,\varphi}$, of ϵ in E(X), where X is the torus got by sewing together the ends of A_q. We know from (5.1)(ii) that the elements $\epsilon_{q,\varphi}$ form a basis for E(X). On the other hand by (5.4) the modular group $SL_2(\mathbb{Z})$ acts projectively on E(X). This means that the partition function χ_φ is transformed by a modular transformation into a linear combination of characters of the same level.

The character $\chi_\varphi(q,g)$ should similarly be regarded as a function of a pair (A,P), where A is an annulus and P is a holomorphic principal G-bundle on A with a given trivialization of $P|\partial A$. Thus $\chi_\varphi(q,g) = \chi_\varphi(A_q, P_g)$, where P_g is $A_q \times G$ with the obvious trivialization over the incoming circle and g times the obvious trivialization over the outgoing end. The character depends only on $\epsilon_{q,\varphi} \in E(X)$ and the holomorphic G-bundle on X got by joining the ends of P_g. We shall explain this in detail in §11.

An example

The most basic modular functor is the determinant line, which is the subject of §6. We shall see (see (5.17)) that it and its powers are the only modular functors with only one label. A more typical example which can be described very explicitly is the following one, which corresponds to the level one representations of the loop group of U_n. (We shall meet other simple examples in §7.)

Let Φ be the set of characters of $\mathbb{Z}/n$. To a surface X with boundary we associate a Heisenberg group H_X which is an extension of $H_1(X; \mathbb{Z}/n)$ by $\mathbb{C}^\times$ with the commutator given by the intersection pairing. The centre of H_X is the image of $H_{\partial X} = \mathbb{C}^\times \oplus H_1(\partial X; \mathbb{Z}/n)$. A labelling $\varphi = (\varphi_1, \ldots, \varphi_k)$ of the boundary components defines a character χ_φ of $H_{\partial X}$ which is the identity on $\mathbb{C}^\times$. There is a unique irreducible representation E(X) of H_X in which $H_{\partial X}$ acts by χ_φ. It is zero unless $\partial \varphi = 0$ in $H^2(X, \partial X; \mathbb{C}^\times)$, i.e. unless $\Pi \varphi_i = 1$, in which case it has dimension n^g, where g is the genus of X. (It can be identified with the space of θ-functions of level n on the Jacobian of X.)

The ring $\mathbb{Z}[\Phi]$ in this case is simply the group ring of Φ.

Note. The preceding description is imprecise in two ways. First, H_X is defined only up to non-canonical isomorphism by the commutator pairing. Secondly, even when H_X is given, the representation E(X) is only uniquely defined as a projective space.

To clarify the definition of H_X we first introduce the extension H_X^Z of $H_1(X;Z)$ by C^X defined by the cocycle

$$(\xi, \eta) \mapsto e^{2\pi i <\xi, \eta>/2n} , \qquad\qquad (5.5)$$

and then we define H_X as the quotient of H_X^Z by the central subgroup $H_1(X;nZ)$.

To deal with the second point we consider the extension H_X^R of $H_1(X;R)$ by C^X defined by the same formula (5.5). The group H_X^R has a standard Heisenberg representation on the space F_X of holomorphic functions on $H_1(\hat{X};R)$, and we define $E(X)$ as the part of F_X fixed by $H_1(X;nZ)$. (Here $\hat{X}$ is X with caps added to its boundary circles, and the complex structure of $H_1(\hat{X};R)$ comes from that of X.)

Extensions of $\mathcal{A}$

Modular functors give us extensions of $\mathcal{A}$ by C^X, and we shall now explain how these are classified.

Proposition (5.6). Holomorphic extensions of $\mathcal{A}$ by C^X correspond precisely to extensions of $Diff^+(S^1)$ by C^X.

Proof: We use the argument of (3.1). If A is an annulus we write φA (resp. $A\varphi^{-1}$) for the same annulus with its outgoing (resp. incoming) edge reparametrized by a diffeomorphism φ of S^1. Suppose that we are given a line L_A for each annulus A. Then we define L_φ for a real-analytic diffeomorphism φ by $L_\varphi = L_{\varphi A} \otimes L_A^*$, where A is a real-analytic annulus. The line L_φ does not depend on A, and it defines a central extension of the real-analytic diffeomorphism group, because

$$L_{\psi\varphi} = L_{\psi\varphi A} \otimes L_A^*$$

$$\cong L_{\psi\varphi A} \otimes L_{\varphi A}^* \otimes L_{\varphi A} \otimes L_A^*$$

$$\cong L_{\psi} \otimes L_{\varphi} \ ,$$

for we can choose A so that φA is also real-analytic. On the other hand it is known that the classification of extensions of $\mathrm{Diff}^+(S^1)$ is the same as that for the real-analytic diffeomorphisms.

Conversely, if we are given an extension $\varphi \mapsto L_{\varphi}$ of $\mathrm{Diff}^+(S^1)$ we define an extension of $\mathscr{A}$ by setting $L_{A_q} = \mathbb{C}$ for the standard annulus A_q, and $L_{\varphi A_q \psi} = L_{\varphi} \otimes L_{\psi}$. It is easy to see that $L_{\varphi} \otimes L_{\psi}$ depends only on the annulus $A = \varphi A_q \psi$. The dependence is holomorphic because any central extension of the Lie algebra $\mathrm{Vect}_{\mathbb{C}}(\partial A)$ is canonically split over $\mathrm{Vect}(A)$: see (6.7). We have therefore defined a correspondence between extensions of $\mathrm{Diff}^+(S^1)$ and extensions of $\mathscr{A}$.

Central extensions of $\mathrm{Diff}^+(S^1)$ were classified in [S2]. The universal central extension has kernel $\mathbb{R} \oplus \mathbb{Z}$, so extensions by $\mathbb{C}^\times$ correspond to homomorphisms $\mathbb{R} \oplus \mathbb{Z} \to \mathbb{C}^\times$, i.e. to elements of $\mathbb{C} \times \mathbb{C}^\times$. An extension can be completely described by its Lie algebra cocycle, in the following sense. The Lie algebra $\mathrm{Vect}(S^1)$ has the traditional basis $\{L_n = e^{in\theta} d/d\theta\}$. When one has a projective representation of $\mathrm{Diff}^+(S^1)$ one can choose the representatives of the L_n so that

$$[L_{-n}, L_n] = -2inL_0 + \frac{1}{12} cn(n^2-1) .$$ (5.7)

Then the classification is given by

Proposition (5.8). A central extension of $\text{Diff}^+(S^1)$ by $\mathbb{C}^\times$ is described by $(c,h) \in \mathbb{C} \times (\mathbb{C}/\mathbb{Z})$, where the "central charge" c is defined by (5.7), and h is any eigenvalue of L_0. (Thus h is detected by the restriction of the extension to the subgroup $PSL_2\mathbb{R}$.)

A modular functor gives us an extension of $\mathcal{A}$ for each label. We shall see in (6.9) that the extensions corresponding to the different labels all have the same central charge c, which will be called the central charge of the modular functor. The extension corresponding to the label 1 necessarily[1] has h = 0. The extension defined by the determinant line has $(c,h) = (-2,0)$.

The proof of (5.4)

A modular functor gives us an extension $\tilde{\mathcal{A}}_\varphi$ of $\mathcal{A}$, and hence an extension $\tilde{V}_\varphi$ of $\text{Vect}_\mathbb{C}(S^1)$, for each label φ. Consider the bundle E_α on the moduli space $\mathcal{C}_\alpha$. There is an action of $\mathcal{A}$ on $\mathcal{C}_\alpha$ for each boundary circle, and it is covered by an action of the appropriate $\tilde{\mathcal{A}}_\varphi$, and hence of $\tilde{V}_\varphi$, on E_α. Putting these actions together gives us an action on E_α of a central extension $\tilde{V}_{\partial X}$ of $\text{Vect}_\mathbb{C}(\partial X)$. At a point $X \in \mathcal{C}_\alpha$ the tangent space to $\mathcal{C}_\alpha$ is $\text{Vect}_\mathbb{C}(\partial X)/\text{Vect}(X)$, and so an extension $\tilde{V}_X$ of $\text{Vect}(X)$ acts on the fibre $E(X)$. But the Lie algebra $\text{Vect}(X)$ has no finite dimensional projective representations (see Appendix *), so the
the Foreword and Postscript

---

[1]Because $D \circ A = D$ when A belongs to the subsemigroup $\mathcal{E}$ of $\mathcal{A}$ (see §2) the extension is split when restricted to $\mathcal{E}$, and hence when restricted to $PSL_2\mathbb{R}$.

extension $\tilde{V}_X$ is canonically split, and $\tilde{V}_X$ acts scalarly on $E(X)$. Thus at X we have a differentiation operator D_ξ on sections of E_α for each $\xi \epsilon \tilde{V}_{\partial X}/\text{Vect}(X)$, i.e. D_ξ is defined up to an additive scalar for each tangent vector ξ to $\mathscr{C}_\alpha$ at X. This is a connection in the bundle of projective spaces of E_α, and it is flat because it comes from a Lie algebra action of $\tilde{V}_{\partial X}$. The nature of the definition of the connection makes it automatically compatible with sewing surfaces together.

Modular functors from a topological viewpoint

Immediately after the first version of this work was written the study of modular functors was transformed by Witten''s realization [W] that the vector spaces in question are in fact the state spaces of 2+1 dimensional "topological" field theories. To explain this it is best to look at modular functors in a slightly different way.

The main point is that for any modular functor E we know from Theorem (5.4) that the space $E(X)$ is almost independent of the complex structure of X. For if X is a smooth surface the space $\mathscr{J}(X)$ of all complex structures on X (not identifying structures which are diffeomorphic) is contractible. The modular functor gives us a vector bundle on $\mathscr{J}(X)$: let $E_J(X)$ denote its fibre at J. By (5.4) the projective space of $E_J(X)$ is independent of J. But we can do better. There is a line bundle on $\mathscr{J}(X)$ whose fibre $\text{Det}_J(X)$ at J is the determinant line of the Riemann surface (X,J). If the functor E has central charge c the bundle with fibres

$$E_J(X) = E_J(X) \otimes \text{Det}_J(X)^{\otimes(\frac{1}{2}c)} \qquad (5.9)$$

has a flat connection, i.e. $E_J(X)$ is independent of J. To define $\mathrm{Det}^{\otimes(\frac{1}{2}c)}$, however, we must make a choice (unless $\frac{1}{2}c$ is an integer). This can be done universally for all $c \in \mathbb{C}$ by choosing a universal covering space $\widetilde{\mathcal{O}}_X$ of the principal $\mathbb{C}^\times$-bundle $\mathcal{O}_X$ of the line bundle Det on $\mathcal{J}(X)$. The space $\widetilde{\mathcal{O}}_X$ then has an action of $\mathbb{C}$, and $\mathrm{Det}^{\otimes(\frac{1}{2}c)}$ can be defined as $\widetilde{\mathcal{O}}_X \times_{\mathbb{C}} \mathbb{C}$, where $\mathbb{C}$ acts on $\mathbb{C}$ by $(\lambda,\xi) \to e^{\pi i c \lambda}\xi$.

Definition (5.10). A <u>rigged surface</u> is a smooth surface X together with a choice of a universal covering space of $\mathcal{O}_X$.

Of course any two riggings of the same surface are isomorphic, but the group of automorphisms of a rigged surface $(X, \widetilde{\mathcal{O}}_X)$ is a central extension by $\mathbb{Z}$ of the group of diffeomorphisms of X. In fact for a surface of genus >1 it is the universal central extension of the diffeomorphism group.

I have not been able to think of a less sophisticated definition of a rigged surface, although there are many possible variants. The essential idea is to associate <u>functorially</u> to a smooth surface a space - such as $\mathcal{O}_X$ - which has fundamental group $\mathbb{Z}$. Instead of $\mathcal{O}_X$ one can take $\mathcal{J}(H^1(\hat{X};\mathbb{R}))$, which is obtained by replacing the determinant line on $\mathcal{O}_X$ by the determinant line on the Siegel domain $\mathcal{J}(H^1(\hat{X};\mathbb{R}))$ of complex structures on the symplectic vector space $H^1(\hat{X};\mathbb{R})$. (Here $\hat{X}$ is X with discs attached to its boundary circles.) There is an obvious natural transformation $\mathcal{O}_X \to \mathcal{J}(H^1(X;\mathbb{R}))$. Another variant is to replace $\mathcal{O}_X$ by the Grassmannian of oriented Lagrangian subspaces of $H^1(X;\mathbb{R})$. (Let us notice that if Y is a 3-manifold with $\partial Y = X$ the image of $H^1(Y)$ in $H^1(X)$ is a Lagrangian subspace.) In [A2] Atiyah, following Witten, considers the space $\mathfrak{J}_X$ of 2-framings of X, i.e. trivializations of the sum of two copies of the tangent bundle of X. We have $\pi_1(\mathfrak{J}_X) = Z$, but the natural map $\pi_1(\mathscr{G}_X) \to \pi_1(\mathfrak{J}_X)$ is multiplication by 12. The 2-framings therefore lead to an extension of the mapping-class group by Z whose class is 12 times that of the extension considered here. (In particular, Atiyah's extension is trivial when X is a torus, whereas ours is the extension of $SL_2(Z)$ induced by the universal covering group of $SL_2(R)$, and is isomorphic to the braid group on three strings.) In any case, we can now reformulate (5.4) as follows.

Proposition (5.11). A modular functor defines a functor on the category of rigged smooth surfaces and isotopy classes of rigged diffeomorphisms.

The functor on the category of rigged smooth surfaces so obtained will be called a reduced modular functor. From what we have said so far it is defined for surfaces with parametrized boundaries (or, better, with rigged boundaries in the sense explained after Definition (4.4)) but it is clear that we could equally well regard it as a functor on the category of closed rigged smooth surfaces equipped with a finite number of labelled marked points with a preferred tangent direction at each. If the tangent directions are rotated there is a flat connection in the resulting vector bundle over the torus of tangent directions, and the holonomy of a rotation of 2π about a point labelled φ is $e^{2\pi i h}\varphi$.

The central charge of a reduced modular functor is defined only modulo 1. (It is well-defined modulo 1 because $H^2(\Gamma; \mathbb{C}^\times) = \mathbb{C}^\times$ when Γ is the mapping-class group of a surface of large genus.) The original modular functor can be recovered from the reduced one up to tensoring with an arbitrary integral power of the determinant line.

When two rigged surfaces are sewn together the result is rigged, and a reduced modular functor inherits the composition properties of Definition (5.1). This follows from the simple behaviour of the determinant line, which we shall treat in §6. (The variant definitions of rigging mentioned above are less well adapted to sewing.)

It seems appropriate at this point to mention "Verlinde's conjecture", which gives a remarkable description of the algebra $Z[\Phi]$ associated to a modular functor E, which we can assume to be reduced. (The statement and the idea of the proof are due to Verlinde [V]; a complete proof was first given by Moore and Seiberg [MS].) Let X be a torus, and let α,β be simple closed curves on X representing a basis for the homology $H_1(X)$. Because an annulus is canonically rigged we can identify E(X) with $\mathbb{C}[\Phi]$ by cutting X along α and using (5.1)(ii). The mapping-class group of X is $SL_2(\mathbb{Z})$. It acts projectively on E(X), and we transfer the action to $\mathbb{C}[\Phi]$. Let $S : \mathbb{C}[\Phi] \to \mathbb{C}[\Phi]$ be a representative of $\begin{pmatrix} 0 & 1 \\ -1 & 0 \end{pmatrix}$, and let M_φ denote as above the operation of multiplication by φ in the Verlinde algebra. Then we have

Theorem (5.12). The matrix of $SM_\varphi S^{-1}$ is diagonal with respect to the natural basis of $\mathbb{C}[\Phi]$.

The theorem implies that the structure of the Verlinde algebra is completely determined by the matrix S. In fact

Corollary (5.13). The structural constants $n_{\varphi\chi\psi}$ of $Z[\Phi]$ are given by

$$n_{\varphi\chi\psi} = \sum_\theta (S^{-1})_{\chi\theta} S_{\theta\varphi} S_{\theta\psi} / S_{\theta 1} .$$

The corollary follows from the theorem because $n_{\varphi\chi\psi}$ is the (χ,ψ) matrix element of M_φ and the (θ,θ) entry of the diagonal matrix $SM_\varphi S^{-1}$ is $S_{\theta\varphi}/S_{\theta 1}$. That is proved by equating the $(\theta,1)$ entries of the matrices $SM_\varphi = (SM_\varphi S^{-1})S$, for $(M_\varphi)_{\chi 1} = \delta_{\varphi\chi}$.

We shall not give a proof of (5.12), but shall explain how it follows from the 2+1 dimensional description of modular functors to which we now turn.

Topological field theories

A topological field theory in d+1 dimensions can be defined, by analogy with (4.4), as a system comprising

(i) a functor $X \mapsto H(X)$ from closed oriented d-dimensional smooth manifolds to finite dimensional complex vector spaces,

(ii) a non-singular pairing $H(\overline{X}) \otimes H(X) \to \mathbb{C}$ for each X, where $\overline{X}$ denotes X with reversed orientation, and

(iii) a vector $\psi_Y \in H(\partial Y)$ for each smooth oriented (d+1)-dimensional manifold Y with boundary.

These data are required to obey the following two axioms.

(a) Multiplicativity: $H(X_1 \amalg X_2) = H(X_1) \otimes H(X_2)$ and
$$\psi_{Y_1 \amalg Y_2} = \psi_{Y_1} \otimes \psi_{Y_2} .$$

(b) Sewing: if $\partial Y = X_0 \amalg X_1 \amalg X_2$, and $\check{Y}$ is formed from Y by sewing X_2 to X_1 by an orientation-reversing diffeomorphism, then $\psi_Y \mapsto \psi_{\check{Y}}$ under the map $H(\partial Y) \to H(\partial \check{Y})$ induced by the pairing $H(X_1) \otimes H(X_2) \to \mathbb{C}$.

Witten realized that the modular functors coming from representations of loop groups are the state spaces of 2+1 dimensional theories,

and subsequently Kontsevich [K] and others have given arguments - a
little sketchy - to show that the two concepts are actually equivalent.
Of course one must first widen the definition of a topological theory a
little so that it is defined on the category of rigged surfaces and
3-manifolds. An oriented 3-manifold Y whose boundary ∂Y is rigged has
itself a set of riggings which form a principal homogeneous set under
the group Z which is the centre of the central extension of $\text{Diff}(\partial Y)$.
I do not know an altogether straightforward way to define a rigging of
a 3-manifold. One approach is to introduce the contractible space $\mathcal{M}_Y$
of metrics on Y. Each metric has an "η-invariant" (see [APS]) which is
a non-zero element of the determinant line of ∂Y. (The invariant is
essentially the phase of the determinant of the signature operator.)
Thus we have a map

$$\eta : \mathcal{M}_Y \to \mathcal{P}_{\partial Y} .$$

A rigging of Y is a lift of this map to the covering space $\widetilde{\mathcal{P}}_{\partial Y}$ which
defines the rigging of ∂Y.

To relate modular functors to 2+1 dimensional theories it is
helpful to introduce the intermediate idea of a relative 2+1
dimensional theory. Like a modular functor this has a set Φ of labels,
and assigns a vector space H(X) to each rigged oriented surface with
labelled boundary circles. It has the same sewing-together property
as a modular functor. As for a field theory there is a vector
$\psi_Y \in H(\partial Y)$ for each rigged 3-manifold with boundary, but it is required
to satisfy a stronger sewing property than (b) above, for one must
allow ∂Y to be decomposed $X_0 \cup X_1 \cup X_2$, where the X_i are surfaces with
boundary which intersect along various boundary circles. An

orientation-reversing diffeomorphism $f : X_1 \to X_2$ allows one to sew

together X_1 and X_2 to form a 3-manifold $\check{Y}$ such that $\partial\check{Y} = \check{X}_0$ is obtained

by sewing from X_0. To see that there is a natural map $H(\partial Y) \to H(\partial\check{Y})$,

we write

$$H(\partial Y) \; = \; \bigoplus_{\varphi_{01},\varphi_{02},\varphi_{12}} \quad H(X_0;\varphi_{01},\varphi_{02}) \otimes H(X_1;\varphi_{01},\varphi_{12}) \otimes H(X_2;\varphi_{12},\varphi_{02}),$$

where φ_{ij} is a multi-label for $X_i \cap X_j$. We project this sum to the sum

of the terms where $\varphi_{01} = \varphi_{02}$. Then the last two factors in the tensor

product are in duality under f, so the sum maps to

$$\bigoplus_{\varphi_{01} = \varphi_{02}} H(X_0;\varphi_{01},\varphi_{02}) \quad = \quad H(\check{X}_0) \; .$$

The axiom we require is that $H(\partial Y) \to H(\partial\check{Y})$ takes ψ_Y to $\psi_{\check{Y}}$.

When a reduced modular functor E is given it is obvious that there

is at most one way to define the vectors ψ_Y corresponding to

3-manifolds Y. One begins with the standard 3-disc D and chooses ψ_D in

the line $E(S^2)$. This can be done arbitrarily, because any modular

functor has an automorphism which multiplies by $\lambda^{\chi(X)}$ on E(X). Any

other 3-manifold Y can be obtained by sewing copies of D together, and

its vector ψ_Y is determined by the sewing axiom. Kontsevich [K] has

given a simple argument to show that the vector obtained is independent

of the chosen decomposition of Y. I feel, however, that the matter is

still far from well-understood.

I shall conclude this section with the proof of Verlinde's

conjecture (5.12) for a 2+1 dimensional field theory. Let Σ be a disc with two holes, and let $Y = \Sigma \times [0,1]$. Then $\partial Y = \Sigma \cup \overline{\Sigma} \cup A \cup A \cup A$, where A is an annulus. We have

$$H(\partial Y) = \bigoplus_{\varphi, \chi, \psi} H(\Sigma_{\varphi\chi\psi}) \otimes H(\Sigma_{\varphi\chi\psi})^* \otimes H(A_{\varphi\varphi}) \otimes H(A_{\chi\chi}) \otimes H(A_{\psi\psi}) ,$$

in what I hope is obvious notation. Let

$$\psi_Y = \sum_{\varphi, \chi, \psi} t_{\varphi\chi\psi} \otimes \epsilon_\varphi \otimes \epsilon_\chi \otimes \epsilon_\psi ,$$

where $t_{\varphi\chi\psi}$ is an endomorphism of $H(\Sigma_{\varphi\chi\psi})$ and ϵ_φ is the canonical element of $A_{\varphi\varphi}$. When two copies of the cylinder Y are sewn end-to-end we have $Y \cup Y \cong Y$, and hence $\psi_Y^2 = \psi_Y$ in the algebra $H(\partial Y)$. But $\epsilon_\varphi^2 = \epsilon_\varphi$, etc., so $t_{\varphi\chi\psi}$ is the identity map. Joining the ends of Y together to get $\check{Y} = \Sigma \times S^1$ we have

$$\psi_{\check{Y}} = \sum n_{\varphi\chi\psi} \, \epsilon_\varphi \otimes \epsilon_\chi \otimes \epsilon_\psi \qquad (5.14)$$

On the other hand we can form $\check{Y}$ also from $\Delta \times S^1$, where Δ is the disc

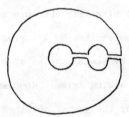

From this point of view $\partial(\Delta \times S^1)$ is the union of eight annuli, and ψ_Y has to be of the form

$$\Sigma \; \lambda_\varphi \tilde{\epsilon}_\varphi \; \otimes \; \tilde{\epsilon}_\varphi \; \otimes \; \tilde{\epsilon}_\varphi \; , \tag{5.15}$$

where $\tilde{\epsilon}_\varphi \; \epsilon \; H(S^1 \times S^1)$ is formed in the same way as ϵ_φ, but with the axes of the torus interchanged. In terms of the modular transformation $S \; : \; H(S^1 \times S^1) \rightarrow H(S^1 \times S^1)$ we know that $\tilde{\epsilon}_\varphi$ is a multiple of $S\epsilon_\varphi$, and so the equality of (5.14) and (5.15) is exactly Verlinde's assertion (5.12).

Mumford's theorem

We can now easily prove the crucial theorem of Mumford which determines all one-dimensional modular functors. (I am greatly indebted to Deligne for showing me how to correct an earlier version of the following proof.)

Proposition (5.16). If a modular functor E satisfies dim E(X) = 1 for all X then it is determined by its restriction to $\mathcal{A}$.

Corollary (5.17). The only such modular functors are integral powers of the determinant line.

The same argument will prove

Proposition (5.18). The only central extensions of the category $\mathcal{C}$ by $\mathbb{C}^\times$ are those given by $X \mapsto \text{Det}_X^{\otimes p} \otimes \overline{\text{Det}}_X^{\otimes q}$ for $p, q \; \epsilon \; \mathbb{C}$ such that $p-q \; \epsilon \; \mathbb{Z}$.

Example. Let $E_m(X)$ denote the determinant line of the $\bar{\partial}$-operator acting on differentials of order m. Thus $E_m(X) \cong \Lambda^{(2m-1)(g-1)} \Omega_{\text{hol}}^{\otimes m}(X)^*$

if m,g > 1 (cf. §6). Calculating the Lie algebra cocycles (see
(8.14)) shows that

$$E_m(X) \cong E_0(X)^{\otimes(6m^2-6m+1)} \qquad (5.19)$$

when X is an annulus, and (5.8) shows that this isomorphism holds for
all surfaces X.

Proof of (5.16). Let E_1 and E_2 be two functors with the same
restriction to $\mathcal{A}$. Then $E = E_1^* \otimes E_2$ is a modular functor which is
trivial on $\mathcal{A}$. The argument of (5.3) shows that for any α there is a
connection in E_α which is flat - not just projectively flat - and
compatible with sewing. This means that E_α is determined by a
representation of $\pi_1(\mathcal{C}_\alpha)$. But it is a classical result that $\pi_1(\mathcal{C}_\alpha)$ is
generated by "Dehn twists" along various curves γ in the surface α. In
our language, if X is a point of $\mathcal{C}_\alpha$ one can write $X = Y \cup A$, where A is
an annulus containing the curve γ. Holding Y fixed we have a
map $\mathcal{A} \to \mathcal{C}_\alpha$, and the Dehn twist is the image of $\pi_1(\mathcal{A}) = Z$. But $E|\mathcal{A}$ is
trivial by hypothesis, so the action of $\pi_1(\mathcal{C}_\alpha)$ is trivial, and all the
fibres of E_α can be canonically identified. This means that $X \mapsto E(X)$
is a functor on the category of smooth surfaces and diffeomorphisms,
and also that the group of diffeomorphisms of X acts trivially on E(X).
The isomorphisms $E(X) \otimes E(Y) \to E(X \amalg Y)$ and $E(X) \to E(\check{X})$ are still, of
course, natural.

Let us write E_g for $E(X_g)$ when X_g is an arbitrary closed surface
of genus g. If $X_g^{(k)}$ is got by removing k discs from X_g then $E(X_g^{(k)})$
can also be identified canonically with E_g. The complete data provided
by the functor are then described by the sequence of lines E_g together
with the maps

$$i : E_g \to E_{g+1}$$

$$m : E_{g_1} \otimes E_{g_2} \to E_{g_1+g_2} \;,$$

where i is defined by sewing together the boundary circles of $X_g^{(2)}$, and m by sewing together $X_{g_1}^{(1)}$ and $X_{g_2}^{(1)}$. To prove the theorem we must show that one can choose isomorphisms $\epsilon_g : \mathbb{C} \to E_g$ which are compatible with the maps i and m. That is possible because the diagrams

$$
\begin{array}{ccc}
E_{g_1} \otimes E_{g_2} & \to & E_{g_1+g_2} \\
\downarrow & & \downarrow \\
E_{g_1+1} \otimes E_{g_2} & \to & E_{g_1+g_2+1}
\end{array}
\qquad \text{and} \qquad
\begin{array}{ccc}
E_{g_1} \otimes E_{g_2} & \to & E_{g_1+g_2} \\
\downarrow & & \downarrow \\
E_{g_1} \otimes E_{g_2+1} & \to & E_{g_1+g_2+1}
\end{array}
$$

commute.

Proofs of (5.17) and (5.18). We have seen in (5.7) that a holomorphic extension of $\mathcal{A}$ is determined by a pair (c,h), and that h = 0 for a modular functor with one label. In view of (5.16) it is therefore enough to show that c must be an even integer. That is true because the $(c/2)^{th}$ power of Det — which is defined for rigged surfaces — does not descend to unrigged surfaces unless $c/2$ is an integer. One reason is that the first Chern class of Det generates $H^2(\mathcal{C}_\alpha ; \mathbb{Z}) \cong \mathbb{Z}$ when α is a surface of large genus with one hole. I do not know if there is a simpler reason.

Unitarity

All the examples known to me of modular functors are <u>unitary</u> in the following sense.

<u>Definition (5.20)</u>. A modular functor E is <u>unitary</u> if there is a positive non-degenerate transformation

$$\overline{E(X)} \otimes E(X) \to |Det_X|^c$$

for each surface X with labelled boundary, such that, in the notation of (5.1), the diagram

$$\oplus_{\varphi} \overline{E(X_\varphi)} \otimes E(X_\varphi) \quad \to \quad |Det_X|^c$$
$$\downarrow \qquad\qquad\qquad\qquad \downarrow$$
$$\overline{E(X)} \otimes E(X) \quad\quad \to \quad |Det_X|^c$$

commutes.

Thus a unitary modular functor provides unitary projective representations of the braid groups, etc. More importantly, the definition is designed to give us

<u>Proposition (5.21)</u>. A pair of weakly conformal holomorphic field theories H and H' corresponding to the same unitary modular functor E with index set Φ gives rise to a conformal field theory based on the space $\oplus_{\varphi \in \Phi} \overline{H}_\varphi \otimes H'_\varphi$ and the central extension $|Det|^c$ of $\mathscr{C}$.

§6. The determinant line

Definition and basic properties

The determinant line Det_X of a Riemann surface X with parametrized boundary[1] is the dual of the top exterior power of the space of holomorphic differentials on the closed surface $\hat{X}$ obtained by adding caps to the boundary circles of X, i.e.

$$\text{Det}_X = \Lambda^g \, \Omega^1_{\text{hol}}(\hat{X})^* . \qquad (6.1)$$

This definition, however, does not lead one to expect the canonical isomorphism

$$\text{Det}_X \otimes \text{Det}_Y \cong \text{Det}_{X \cup Y} \qquad (6.2)$$

which exists when surfaces are sewn together.

An alternative definition of Det_X is as the determinant line of the $\bar{\partial}$-operator of X in the sense of Quillen [Q]. To define this, recall that on any Riemann surface there is a $\bar{\partial}$-operator

$$\bar{\partial}_X : \Omega^0(X) \rightarrow \Omega^{01}(X)$$

mapping smooth functions to (0,1)-forms. If X has a parametrized boundary then $\bar{\partial}_X$ has a natural boundary condition which makes it a Fredholm operator: one restricts it to the subspace $\Omega^0(X, \partial X)$ of functions which on each incoming boundary circle are of the form

----

[1]Sophisticated readers should notice that to define Det_X we do not need the boundary of X to be parametrized, but only to be <u>rigged</u>, as was explained after Defn. (4.4).

$\Sigma_{n \geqslant 0} a_n e^{in\theta}$, and on each outgoing circle of the form $\Sigma_{n<0} b_n e^{in\theta}$. Any
Fredholm operator P : E → F between topological vector spaces has a

determinant line Det_P, which can be defined in various ways. A

convenient definition for our purposes is given in Appendix B. For a

single operator P we have

$$\text{Det}_P = \text{Det (ker P)}^* \otimes \text{Det (coker P)} \ ,$$

where on the right Det denotes the top exterior power. For the

operator $\bar\partial_X$ this reduces to (6.1), but the important property of the

definition is that the lines Det_P fit together to form a holomorphic

line bundle on the space of Fredholm operators E → F. More

generally, if E and F are holomorphic bundles of topological vector

spaces over some base space, and P : E → F is holomorphic and

Fredholm (cf. Appendix B), then Det_P is a holomorphic line bundle on

the base space.

It should be remembered that the determinant line of a Fredholm

operator is a vector space with a mod 2 grading. The degree is the

index of the operator. For a surface of genus g with m incoming and n

outgoing circles the degree of the determinant line is m + 1 - g. Thus

Det_{S^2} is canonically $\mathbb{C}$, but in degree 1. This means that

$\text{Det}_{S^2 \amalg S^2 \amalg \ldots \amalg S^2}$ is also $\mathbb{C}$, but that the group of permutations of

the spheres acts on it by the sign representation.

If the surface X has no closed components there is another

description of Det_X. Let Hol(X) be the vector space of holomorphic

functions on X. The space of smooth functions $\Omega^0(\partial X)$ has a splitting

$\Omega^0_+(\partial X) \oplus \Omega^0_-(\partial X)$, where Ω^0_- denotes the functions which satisfy the

boundary condition above.

Proposition (6.3). If X has no closed components then Det_X is canonically isomorphic to the determinant line of the operator $\pi_X : \text{Hol}(X) \to \Omega^0_+(\partial X)$ given by restriction to ∂X followed by projection on to Ω^0_+.[1]

Corollary (6.4). The lines Det_X form a holomorphic bundle on each moduli space $\mathcal{C}_\alpha$ of surfaces with parametrized boundaries.

We shall return to the case of closed surfaces below: see the remark after Proposition (6.5).

Proof of (6.3). We consider the diagram

$$
\begin{array}{ccccccccc}
0 & \to & \text{Hol}(X) & \to & \Omega^0(X) & \overset{\bar\delta}{\to} & \Omega^{01}(X) & \to & 0 \\
 & & \downarrow \pi_X & & \downarrow \bar\delta\oplus\text{pr} & & \downarrow \text{id} & & \\
0 & \to & \Omega^0_+(\partial X) & \to & \Omega^{01}(X) \oplus \Omega^0_+(\partial X) & \to & \Omega^{01}(X) & \to & 0 \ .
\end{array}
$$

As the rows are exact this defines an isomorphism

$$
\text{Det}_X \cong \text{Det}_{\bar\delta\oplus\text{pr}} \cong \text{Det}_{\pi_X} \otimes \text{Det}_{\text{id}} \cong \text{Det}_{\pi_X} \ .
$$

The essential property of the determinant line is (6.2), which, as it is obvious that $\text{Det}_{X\sqcup Y} \cong \text{Det}_X \otimes \text{Det}_Y$, is a particular case of

Proposition (6.4). A sewing map $X \to \check{X}$, i.e. one which sews outgoing edges of X to incoming ones, induces a canonical isomorphism $\text{Det}_X \cong \text{Det}_{\check{X}}$.

---

[1]It is more convenient for the sequel if we change the definition of π_X by composing it with the automorphism of $\Omega^0_+(\partial X)$ which multiplies by -1 on the incoming circles. That does not affect the truth of (6.3).

Proof: First assume that $\check{X}$ is not closed, and that it is formed by sewing together the parts S_1 and S_2 of ∂X to form a curve S in $\check{X}$. We have a commutative diagram

$$
\begin{array}{ccccccccc}
0 & \to & \mathrm{Hol}(\check{X}) & \to & \mathrm{Hol}(X) & \overset{\Delta}{\to} & \Omega^0(S) & \to & 0 \\
 & & {}^{\pi\check{}}_X \downarrow & & \tilde{\pi} \downarrow & & \mathrm{id} \downarrow & & \\
0 & \to & \Omega_+(\partial\check{X}) & \to & \Omega_+(\partial\check{X}) \oplus \Omega^0(S) & \to & \Omega^0(S) & \to & 0 .
\end{array}
$$

Here Δ is defined by $\Delta(f) = f|S_1 - f|S_2$, and $\tilde{\pi}$ by

$$\tilde{\pi}(f) = (\ (f|\partial\check{X})_+,\ \Delta(f)\ ) .$$

The rows are exact. (To see that Δ is surjective, let Y be the surface formed from two copies X_1 and X_2 of X by attaching $S_1 \subset X_1$ to $S_2 \subset X_2$. Because Y is a Stein manifold, any smooth function f on $S = X_1 \cap X_2$ can be written $f_1|S - f_2|S$, with $f_i \in \mathrm{Hol}(X_i)$.) Thus $\mathrm{Det}_X \cong \mathrm{Det}(\tilde{\pi})$. But Det_X is the determinant of $\pi_X : \mathrm{Hol}(X) \to \Omega^0_+(\partial X)$. We can identify $\Omega^0(S)$ with $\Omega^0_+(S_1) \oplus \Omega^0_-(S_2)$, and hence $\Omega^0_+(\partial X)$ with $\Omega^0_+(\partial\check{X}) \oplus \Omega^0(S)$. Then $\pi_X - \tilde{\pi}$ is the map $f \mapsto (f|S_1)_- - (f|S_2)_+$. This is of trace class by Lemma (6.6) below. But the determinant line does not change when the operator is changed by an operator of trace class (see Appendix B), so the result is proved.

The case when $\check{X}$ is closed can be dealt with by making a hole in $\check{X}$ so that it does have a boundary, and then using the following result.

Proposition (6.5). For any surface X the line Det_X does not change when the interiors of one or more holomorphically embedded discs are removed from X.

If X is not closed this is already implied by (6.4). It is therefore in keeping with the spirit of our approach to <u>define</u> Det_X when X is closed as $\text{Det}_{X-\mathring{D}}$, where D is any disc in X. This definition does not depend on the disc chosen, for if D_1 and D_2 are disjoint discs in X then we know that

$$\text{Det}_{X-\mathring{D}_1} \cong \text{Det}_{X-\mathring{D}_1-\mathring{D}_2} \cong \text{Det}_{X-\mathring{D}_2} .$$

We shall therefore leave the proof of (6.5) to Appendix B (B12).

In proving (6.4) we made use of

<u>Lemma (6.6)</u>. If S is a union of outgoing boundary circles of a Riemann surface X then the map $f \mapsto (f|S)_-$ from $\text{Hol}(X)$ to $\Omega^0_-(S)$ is of trace class.

<u>Proof</u>: It is enough to prove this when X is an annulus and $S = S_1$ is its outgoing end. Indeed because diffeomorphisms preserve the decomposition $\Omega^0(S) = \Omega^0_+(S) \oplus \Omega^0_-(S)$ up to trace class operators (see [PS]( )) we can assume that X is $\{z \in \mathbb{C} : r \leq |z| \leq 1\}$ with the standard parametrization. But then $\text{Hol}(X) \to \Omega^0_-(S_1)$ factorizes

$$\text{Hol}(X) \to \Omega^0_+(S_0) \to \Omega^0_-(S_1) ,$$

where the second map is the diagonal operator taking z^k to $r^k z^k$. This is clearly of trace class.

The central extension of $\text{Diff}^+(S^1)$

We can now see why - as was mentioned at the beginning of this
paper - the determinant line gives rise to the basic central extension[1]
of $\text{Diff}^+(S^1)$. One way of formulating the result of [S2] §7(b) is that
the basic central extension of $\text{Diff}^+(S^1)$ consists of pairs (φ, λ) with
$\varphi \in \text{Diff}^+(S^1)$ and $\lambda \in L_\varphi$, where L_φ is the determinant line of the
Toeplitz operator $T_\varphi : \Omega^0_+(S^1) \to \Omega^0_+(S^1)$ which is the (++) block of the
action of φ on $\Omega^0(S^1)$. But L_φ is equivalently the determinant line of

$$P_\varphi : \Omega^0(S^1) \to \Omega^0_+(S^1) \oplus \Omega^0_-(S^1)$$

$$f \mapsto ((\varphi*f)_+, \ f_-) \ .$$

If the diffeomorphism φ is regarded as the limit of a family of annuli
A then P_φ is evidently the limit of the operators

$$\pi_A : \text{Hol}(A) \to \Omega^0_+(\partial A) \ .$$

More precisely, in terms of the proof of (5.5), if A is the standard
annulus $\{r \leq |z| \leq 1\}$, we have $\pi_{\varphi A} = P_\varphi \pi_A$, and hence

$$\det(P_\varphi) = \text{Det}_{\varphi A} \otimes \text{Det}^*_A \ .$$

This makes clear the sense in which the central extension of $\text{Diff}^+(S^1)$
is the "boundary" of an extension of the semigroup $\mathscr{A}$.

To understand why the pairs $(\varphi, \lambda \in L_\varphi)$ form a group it is best to
regard L_φ, in the notation of Appendix B, as $\text{Det}(W; \varphi W)$, where W is an
element of $\text{Gr}(\Omega^0(S^1))$: this line does not depend on W.

----

[1]The cleanest statement is simply that the extension is the group of
automorphisms of a rigged circle.

An important property of the central extension of $\text{Diff}(S^1)$ which is easy to see in terms of the determinant line - or rather in terms of the splitting of $\Omega^0(S^1)$ into positive and negative frequency - is the following "reciprocity law". When we have a surface X with boundary we can put together a copy of the standard extension of $\text{Vect}_{\mathbb{C}}(S^1)$ for each boundary circle to obtain an extension $\widetilde{\text{Vect}}_{\mathbb{C}}(\partial X)$ of $\text{Vect}_{\mathbb{C}}(\partial X)$,

Proposition (6.7). The restriction of the extension $\widetilde{\text{Vect}}_{\mathbb{C}}(\partial X)$ to the subalgebra $\text{Vect}(X)$ of holomorphic vector fields on X is canonically split.

Proof: The extension of $\text{Vect}_{\mathbb{C}}(\partial X)$ measures the extent to which the vector fields fail to preserve the decomposition $\Omega^0(\partial X) = \Omega^0_+(\partial X) \oplus \Omega^0_-(\partial X)$. If we write T_ξ for the $\Omega_+ \to \Omega_+$ component of the action of a vector field ξ then we have the following explicit formula for the cocycle (cf. [PS](6.6.5)):

$$(\xi, \eta) \mapsto \text{trace}([T_\xi, T_\eta] - T_{[\xi, \eta]}) \qquad (6.8)$$
$$= \text{trace}(J[J, \xi][J, \eta])$$

Here J is the operator which defines the splitting $\Omega = \Omega_+ \oplus \Omega_-$, i.e. $J|\Omega_\pm = \pm 1$. If the decomposition is changed by replacing J by another operator J_X such that $K = J_X - J$ is of trace class (and $J_X^2 = 1$) then the cocycle (6.8) changes by the coboundary

$$(\xi, \eta) \mapsto 2\,\text{trace}([\xi, \eta]K) .$$

Let us choose J_X corresponding to the decomposition $\Omega^0(\partial X) = \text{Hol}(X) \oplus \text{Hol}'(Y)$, where Y is a collection of discs with $\partial Y = \partial X$, so that $X \cup Y$ is a closed surface, and $\text{Hol}'(Y)$ means the functions which are

[62]

holomorphic except for a pole (or zero) of an appropriate order at the
centre of one of the discs. The difference J_X - J is given by an
integral operator on ∂X with a smooth kernel, so it is certainly of
trace-class. On the other hand, the subspace Hol(X) is preserved by
holomorphic vector fields on X, so the cocycle trace$(J_X[J_X,\xi][J_X,\eta])$
vanishes on Vect(X).

Another result which fits in naturally at this point is

Proposition (6.9). For any modular functor the extensions of Vect(S^1)
corresponding to its labels all have the same central charge.

Proof: Suppose that X is a disc with two holes with boundary circles
S_0, S_1, S_2 labelled $\varphi_0, \varphi_1, \varphi_2$. Let c_i be the central charge corresponding
to φ_i. We must show that $c_1 = c_2 = c_3$. Let $\xi_i \in H^2(\text{Vect}(X); \mathbb{C})$ be the
class of the extension of Vect(X) pulled back from the determinant line
extension of Vect(S_i). Then $\xi_0 + \xi_1 + \xi_2 = 0$; but ξ_1 and ξ_2 are
linearly independent, because by filling in, say, the second hole we
embed X in an annulus A in such a way that $\xi_2 \mapsto 0$ but $\xi_1 \not\mapsto 0$ in
$H^2(\text{Vect}(A); \mathbb{C}) = \mathbb{C}$. Now the modular functor gives us an extension of
Vect(X) with class $\Sigma c_i \xi_i$. We know that it is split, and therefore
$\Sigma c_i \xi_i = 0$, and hence $c_1 = c_2 = c_3$.

Modularity and the η-function

Our final task in this section is to give a completely explicit
description of the isomorphism $\text{Det}_A \cong \text{Det}_X$ when A is an annulus and
X = $\check{A}$ is the torus got by joining its ends. We need this to find the
modularity properties of partition functions. Suppose, for example,

that we have a holomorphic field theory with central charge c, so that
an operator $U_{c,\alpha}$ in H is associated to the annulus A together with a
choice of $\alpha \in Det_A$, and $U_{A,\lambda\alpha} = \bar{\lambda}^c U_{A,\alpha}$ for $\lambda \in \mathbb{C}$. Then the trace of
$U_{A,\alpha}$ depends only on the image $\check{\alpha}$ of α in the line Det_X. There is a
canonical element $\epsilon_A \in Det_A$, for the $\bar{\partial}$-operator is an isomorphism. The
partition function Z of the theory is defined by

$$Z(\tau) = \text{trace } U_{A,\epsilon_A} \ ,$$

where A is the standard annulus determined by $q = e^{2\pi i\tau}$, with $Im(\tau) > 0$.
If τ is replaced by $\tau' = (a\tau + b)/(c\tau + d)$, for some $g = \begin{pmatrix} a & b \\ c & d \end{pmatrix}$ in the
modular group $\Gamma = SL_2(\mathbb{Z})$, then A changes to A', but the torus X does
not change, so we have

$$Z(\tau') = \rho(\tau,g)^{\frac{1}{2}c} Z(\tau) \ , \tag{6.10}$$

where $\rho(\tau,g)$ is the ratio of the images of the elements $\epsilon_{A'}$ and ϵ_A in
Det_X. (Note that $\rho(\tau,g)$ depends on g, and not just on τ and τ',
because one must choose an isomorphism between the tori $\check{A}$ and $\check{A}'$.) The
crucial result is

<u>Proposition (6.11)</u>. We have $\rho(\tau,g) = u(g)e^{2\pi i(\tau'-\tau)/12}$, where
$u : \Gamma \to \mu_{12}$ is a canonical homomorphism from Γ to the group μ_{12} of
12th roots of unity. In other words

$$(q')^{-c/12}Z(\tau') = u(g)^{\frac{1}{2}c} q^{-c/12}Z(\tau) \ .$$

The line Det_X attached to a torus X is the dual of the line of
holomorphic differentials, so it contains a lattice $\Lambda = H_1(X,\mathbb{Z})$ given
by the geometrical cycles. Let $\xi_\gamma \in Det_X$ correspond to the cycle γ.

If X is formed from an annulus A this gives us a preferred element ξ_A.
It is natural to expect that ξ_A should be related to the image of ϵ_A,
and in fact it is very easy to prove

Proposition (6.12). The image of ϵ_A in Det_X is

$$\prod_{n>0} (1-q^n)^{-2} . \xi_A .$$

When A is changed to A' by $g \in \Gamma$ we have

$$\xi_{A'} = (c\tau + d)^{-1} . \xi_A , \qquad\qquad (6.13)$$

and so, in the light of (6.12), the result (6.11) is equivalent to the
modularity property

$$\eta(\tau')^2 = u(g)^{-1}(c\tau + d)^{-1}\eta(t)^2 \qquad\qquad (6.14)$$

of the square of the Dedekind η-function, which is defined by

$$\eta(\tau) = e^{\pi i \tau/12} \prod_{n>0} (1-q^n) .$$

Indeed, (6.14) can be taken as a definition of the homomorphism u. The
existence and modularity of the η-function amount to the following
geometrical statement:

if L is a complex line equipped with a lattice Λ then there is a
canonical isomorphism $L^{\otimes 12} \cong \mathbb{C}$, i.e. a canonical map $f : L \to \mathbb{C}$ which
is homogeneous of degree 12; in particular L contains a distinguished
set $\mu_{12}^L = f^{-1}(1)$ of 12 points.

Applying this to the case $(L, \Lambda) = (\mathbb{C}, \mathbb{Z} + \tau\mathbb{Z})$ for $\mathrm{Im}(\tau) > 0$ gives us a 12-sheeted covering of the upper half-plane with an action of the group $SL_2(\mathbb{Z})$ on it; and considering how the sheets are permuted gives the homomorphism $u : SL_2(\mathbb{Z}) \rightarrow \mu_{12}$ which in fact describes the abelianization of $SL_2(\mathbb{Z})$.

At this point we could assume the properties of η and deduce (6.11) from (6.12). But it is obviously more satisfying to deduce the properties of η from general facts about the determinant line. I shall give an argument based on Mumford's theorem (5.9), but I should mention that Deligne has given a much more illuminating argument, which, however, it would require too long a digression to explain.

For my argument we consider alongside the line Det_X the determinant line $E_X^{(2)}$ of the $\bar{\partial}$-operator acting on forms of type $(2,0)$. From (5.9) we know that $E_X^{(2)} \cong \mathrm{Det}_X^{\otimes 13}$, and, more precisely, that $\mathrm{Det}_X^{\otimes 13} \otimes (E_X^{(2)})^*$ has a canonical element μ_X which is multiplicative in the sense that $\mu_{X \circ Y} = \mu_X \cdot \mu_Y$. But the proof of (5.9) shows that for an annulus A one has $\mu_A = \epsilon_A^{13} \tilde{\epsilon}_A^{-1}$, where $\tilde{\epsilon}_A$ is the standard element of $E_A^{(2)}$. (The reason for this is that all of these standard elements are characterized by multiplicativity with respect to the relation $D_\infty \circ A \circ D = D_\infty \circ D$ in the category $\mathcal{C}$. Now the map $A \rightarrow X$ takes μ_A to μ_X, and so the image of $\epsilon_A^{13} \tilde{\epsilon}_A^{-1}$ depends only on X. But Proposition (6.12) can be generalized to a statement about the determinant line $E_X^{(m)}$ of the $\bar{\partial}$-operator acting on $\Omega^{\otimes m}$ for any value of m. An annulus $E_A^{(m)}$ has the usual canonical element $\epsilon_A^{(m)}$, and for a torus X there is an element $\xi_\gamma^{(m)}$ for each cycle γ.

<u>Proposition (6.15)</u>. The image of $\epsilon_A^{(m)}$ in $E_X^{(m)}$ is

$$(-1)^m \, q^{\frac{1}{2}m(m-1)} \, \prod_{n>0} (1-q^n)^{-2} \, \xi^{(m)} \ .$$

On the other hand, when X is a torus $E_X^{(2)}$ can be identified with Det_X, for

$$E_X^{(2)} \cong H^0(X;\Omega^{\otimes 2})^* \otimes H^1(X;\Omega^{\otimes 2})$$

$$\cong H^0(X;\Omega^{\otimes 2})^* \otimes H^0(X;\Omega^{\otimes(-1)})^*$$

$$\cong H^0(X;\Omega^1)^*$$

$$= Det_X .$$

(Here the second isomorphism comes from Serre duality, and the third from the product $\Omega^{\otimes 2} \otimes \Omega^{\otimes(-1)} \to \Omega^1$.) Under this identification $\xi_A^{(2)}$ corresponds to ξ_A.

Putting together the isomorphisms $E_X^{(2)} \cong Det_X^{\otimes 13}$ and $E_X^{(2)} \cong Det_X$ gives us $Det_X^{\otimes 12} \cong \mathbb{C}$, by a map which takes

$$q^{-1} \prod (1-q^n)^{-24} \xi_A^{12} \to \eta(\tau)^{-24} \xi_A^{12}$$

to 1. This is precisely the statement that $\eta(\tau)^{24}$ is a modular form of weight 12.

It remains to prove (6.15), which, of course, includes (6.12). Using (6.4) we can identify $E_X^{(m)}$ with the determinant line of isomorphism

$$\pi_A : \Omega_{hol}^{\otimes m}(A) \to \Omega^{\otimes m}(S^1)$$

given by $\pi_A(f) = (\varphi_1^* f)_+ - (\varphi_0^* f)_-$, where $\varphi_0, \varphi_1 : S^1 \to A$ are the parametrizations of the ends of A. The operator π_A itself defines the element $\epsilon_A^{(m)}$, while $\xi_A^{(m)}$ is defined by the Fredholm operator $\tilde{\pi}$ such that $\tilde{\pi}(f) = \varphi_1^* f - \varphi_0^* f$, together with the obvious choice of isomorphism between its kernel and its cokernel. The ratio of $\xi_A^{(m)}$ to $\epsilon_A^{(m)}$ is therefore the determinant of $\tilde{\pi} \circ \pi_A^{-1}$ restricted to the subspace of $\Omega(S^1)$ spanned by all $z^k dz^m$ with $k \neq -m$. This operator is diagonal. It multiplies $z^k dz^m$ by $(1 - q^{k+m})$ if $k > 0$, and by $(1 - q^{-k-m})$ if $k < 0$. The determinant is therefore $(-1)^m q^{-\frac{1}{2}m(m-1)} \Pi (1-q^n)^2$, which proves (6.15).

§7. Spin structures: discrete coverings of $\mathcal{C}$

The categories $\mathcal{C}^{\text{spin}}$ and $\mathcal{C}^G$

The kernel of the universal central extension of $\text{Diff}^+(S^1)$ is $\mathbb{R} \oplus \mathbb{Z}$ [S2]. The coordinate functions on $\mathbb{R} \oplus \mathbb{Z}$, in the notation of (5.8) are (c,h). The determinant line accounts for the factor $\mathbb{R}$, while $\mathbb{Z}$ is the centre of the simply connected covering group of $\text{Diff}^+(S^1)$, i.e. of the contractible group of diffeomorphisms $\varphi : \mathbb{R} \to \mathbb{R}$ satisfying $\varphi(\theta + 2\pi) = \varphi(\theta) + 2\pi$. Having discussed the determinant extension it is natural to ask about extensions connected with the fundamental group. The most important one is the spin covering.

A spin structure on a circle S is a real line bundle L on S together with an isomorphism $L \otimes L \cong TS$. There are two possible choices for L: trivial or Möbius. I shall write S_P and S_A for the respective pairs (S,L): the letters stand for "periodic" and "antiperiodic".

A spin structure on a Riemann surface X is a holomorphic line bundle L with an isomorphism $L \otimes L \cong TX$, where now TX denotes the complex tangent line bundle. If X has a boundary then a spin structure L on X induces one on the boundary, for the positively oriented tangent vectors to the boundary of X have square-roots which form a real line in $L|\partial X$. Every surface possesses a non-empty finite set of spin structures: they are acted on simply transitively by the group $H^1(X;\mathbb{Z}/2)$, for two spin structures on X differ by tensoring with a real line bundle.

The category $\mathcal{C}^{\text{spin}}$ has objects $C_{n,n'}$ for $n,n' \geqslant 0$, where $C_{n,n'}$ is the union of n copies of S_A^1 and n' of S_P^1. The morphisms $C_{m,m'} \to C_{n,n'}$ are Riemann surfaces (X,L) with a spin structure and a given isomorphism $(\partial X, L | \partial X) \cong C_{n,n'} \cdot C_{m,m'}$. It is easy to see that there are no morphisms $C_{m,m'} \to C_{n,n'}$ unless $m' \equiv n'$ (mod 2), in which case the morphisms form a principal bundle over $\text{Mor}_{\mathscr{C}}(C_{m+m'}; C_{n+n'})$ with group $H^1(X, \partial X; \mathbb{Z}/2)$. In particular the identity component of the semigroup of endomorphisms of $C_{1,0}$ is a double covering of $\mathscr{A}$, and its Shilov boundary is a double covering of $\text{Diff}^+(S^1)$.

There is no obvious generalization of $\mathscr{C}^{\text{spin}}$. For any n one can construct an n-fold covering of the semigroup $\mathscr{A}$ by considering pairs (A, L) such that $L^{\otimes n} \cong TA$; but that does not work for general surfaces X because usually TX has no n^{th} root.

There is, nevertheless, an extension $\mathscr{C}^G$ of $\mathscr{C}$ associated to each finite group G. The objects of $\mathscr{C}^G$ are principal G-bundles over one-dimensional manifolds, and the morphisms are principal G-bundles over Riemann surfaces. Unlike $\mathscr{C}^{\text{spin}}$, the extension $\mathscr{C}^G$ is <u>split</u>: $\mathscr{C}$ itself is a subcategory of $\mathscr{C}^G$.

The principal G-bundles over S^1 correspond to the conjugacy classes of elements of G: I shall write them S^1_g for $g \in G$. The endomorphisms of S^1_g in $\mathscr{C}^G$ which cover an annulus A correspond non-canonically to the elements of the centralizer Z_g of g in G. (Choose a path γ joining the base-points of the ends of A, and assign to a G-bundle on A the monodromy along γ.) More precisely, the endomorphisms of S^1_g over $\mathscr{A}$ form the extension of $\mathscr{A}$ by Z_g defined by the homomorphism $\mathbb{Z} = \pi_1(\mathscr{A}) \to Z_g$ which takes 1 to g.

<u>The associated modular functors</u>

The categories $\mathscr{C}^{\text{spin}}$ and $\mathscr{C}^G$ give rise to modular functors in the sense of §5. Let us first consider $\mathscr{C}^G$. We shall denote the set of isomorphism classes of principal G-bundles on a 1-manifold S by $\mathscr{P}(S)$, and if P is a G-bundle on ∂X we shall denote by $\mathcal{P}(X;P)$ the set of isomorphism classes of pairs (Q,α), where Q is a G-bundle on X and α is an isomorphism $Q|\partial X \to P$. Thus if S is a union of components of ∂X the group Aut(P|S) of automorphisms of P|S acts on $\mathcal{P}(X;P)$. The patching-together property of G-bundles is expressed as follows. If $X = X_1 \cup X_2$ is a union of surfaces attached along $S_0 = \partial X_1 \cap \partial X_2$, and $S_i = (\partial X_i) - S_0$, then

$$\mathcal{P}(X;P) = \coprod_{P_0 \epsilon \mathcal{P}(S_0)} \mathcal{P}(X_1 ; P_1 \amalg P_0) \underset{\text{Aut}(P_0)}{\times} \mathcal{P}(X_2 ; P_2 \amalg P_0) ,$$

$$(7.1)$$

where $P_i = P|S_i$. There is a similar formula far attaching two edges of the same surface. We can now introduce a modular functor E whose set Φ of labels is the set of isomorphism classes of pairs (P,V), where $P \epsilon \mathcal{P}(S^1)$ and V is a complex irreducible representation of Aut(P). For a surface X with boundary circles $S_1,\ldots,S_k$ labelled with (P_i,V_i) we define

$$E(X;P,V) = \text{Map}_{\text{Aut}(P)} (\mathcal{P}(X;P);V) , \qquad (7.2)$$

where $P = P_1 \amalg \ldots \amalg P_k$ and $V = V_1 \otimes \ldots \otimes V_k$. The property (7.1) then translates into

$$E(X;P,V) = \bigoplus_{(P_0,V_0)} E(X_1;P_1 \amalg P_0, V_1 \otimes V_0^*) \otimes E(X_2;P_2 \amalg P_0, V_2 \otimes V_0) .$$

$$(7.3)$$

The Verlinde algebra A_G of this modular functor has arisen in another context in work of Lusztig [L]. Additively we have

$$A_G = \bigoplus_{[g]} R(C_g) \; ,$$

where $R(C_g)$ is the representation ring of the centralizer C_g of $g \in G$, and the sum is over the conjugacy classes of elements of G. Thus A_G can be identified with the equivariant K-group $K_G(G)$, where G acts on itself by conjugation. Analysis of (7.2) when X is a disc with two holes shows that the multiplication in A_G is given by the K_G direct image map induced by the multiplication $G \times G \to G$. (Thus if an element of $K_G(G)$ is regarded as a family of vector spaces $\{V_g\}_{g \in G}$ the product of $\{U_g\}$ and $\{V_g\}$ is $\{W_g\}$, where

$$W_g = \bigoplus_{g_1 g_2 = g} U_{g_1} \otimes V_{g_2} \; .)$$

In particular, if G is abelian the ring A_G is the group ring of $G \times \hat{G}$, where $\hat{G} = \mathrm{Hom}(G, \mathbb{T})$.

To make everything as explicit as possible let us observe that when X is a torus E(X) is the vector space of functions on the set of conjugacy classes of pairs of commuting elements (g_1, g_2) of G. Constructing X from an annulus gives the standard isomorphism $A_G \otimes \mathbb{C} \to E(X)$ which maps $\chi \in R(C_g)$ to the function

$$(g_1, g_2) \longmapsto \begin{cases} \chi(g_2) & \text{if } g_1 = g \\ 0 & \text{if } g_1 \text{ is not conjugate to } g \; . \end{cases}$$

If A_G is regarded as a space of functions on a subset Γ of $G \times G$ by this map the multiplication is given by $(f_1, f_2) \longmapsto f_1 * f_2$, where

$$(f_1 * f_2)(x,y) = \sum_{x_1 x_2 = x} f_1(x_1,y) f_2(x_2,y) .$$

The action of $SL_2(\mathbb{Z})$ in this description comes from its natural action on $\Gamma = \text{Hom}(\mathbb{Z}^2;G)$, and one can easily check Verlinde's theorem (5.12).

Another aspect of the general theory which is easy to see in this example is the vector $\psi_Y \in E(X)$ associated to a 3-manifold Y such that $\partial Y = X$. For ψ_Y is simply the function on the set $\mathcal{P}(X)$ of classes of G-bundles on X whose value at a bundle P is given by

$$\psi_Y(P) = \sum_{(Q,\alpha) \in \mathcal{P}(Y,P)} \frac{1}{|\text{Aut}(Q,\alpha)|} .$$

Little needs now to be said about the modular functor associated to $\mathcal{C}^{\text{spin}}$. Instead of $\mathcal{P}(S)$ and $\mathcal{P}(X;P)$ we have the corresponding sets of spin structures $\mathcal{S}(S)$ and $\mathcal{S}(X;\sigma)$, where σ is a spin structure on ∂X. The group of automorphisms of σ is $H^0(\partial X;\mathbb{Z}/2)$. It acts on $\mathcal{S}(X;\sigma)$, and there is a sewing property just like (7.1). We obtain a modular functor with four labels $A^{\pm}$ and $P^{\pm}$, corresponding to the two spin structures on S^1 and the two representations of the group (± 1) of automorphisms.

The definition of a field theory based on $\mathcal{C}^G$ or $\mathcal{C}^{\text{spin}}$ is clear. For $\mathcal{C}^{\text{spin}}$, for example, we should have a projective functor $(S,\sigma) \mapsto H(S,\sigma)$ from oriented 1-manifolds with spin structure, and a ray in $H(S,\sigma)$ for each Riemann surface X with $\partial X = S$ equipped with an element of $\mathcal{S}(X,\sigma)$. I shall not repeat the conditions to be satisfied, but let us notice that Aut(σ) will act on H(S,σ). If it were not for the projective nature of the functor we could say at once that such a theory defines a weakly conformal theory with respect to the modular functor described above. For each label (σ,V) for S we could define a space

$$H(S,(\sigma,V)) = \{H(S,\sigma) \otimes V\}^{Aut(\sigma)} \ ,$$

and when $\partial X = S$ we should have an Aut(σ)-equivariant map $\mathcal{G}(X,\sigma) \to H(S,\sigma)$ which would induce

$$\begin{aligned} E(X,(\sigma,V)) &= Map_{Aut(\sigma)}(\mathcal{G}(X,\sigma);V) \\ &= \{\mathbb{C}[\mathcal{G}(X,\sigma)] \otimes V\}^{Aut(\sigma)} \\ &\to H(S,(\sigma,V)) \ . \end{aligned}$$

But not much needs to be altered to take the projectiveness into account. A central extension of $\mathscr{C}^{spin}$ defines a line bundle L on $\mathcal{G}(X,\sigma)$ which is equivariant under Aut(σ), and we simply define a modified modular functor E for which E(X,(σ,V)) is the space of equivariant sections of L $\otimes$ V.

The homological description of spin structures

Besides the direct geometrical description already given we shall need to make use of two other ways of characterizing spin structures. The first is cohomological. If γ is a smooth closed curve in a closed surface X then a spin structure L on X is either Möbius or trivial on γ. Define $\sigma_L(\gamma) = +1$ or -1 accordingly. It turns out that σ_L defines a quadratic form on $H_X = H_1(X;F_2) \cong H^1(X,F_2)$ which is associated with the cup-product (or intersection), i.e.

$$\sigma_L(\gamma_1 + \gamma_2) = \sigma_L(\gamma_1) + \sigma_L(\gamma_2) + \gamma_1 \cdot \gamma_2 \ . \qquad (7.5)$$

Obviously the function σ_L describes the spin structure completely. This can alternatively be expressed by introducing the abelian group $\tilde{H}_X$ which is the extension of H_X by F_2 got by using the cup-product as a cocycle. The formula (7.5) means that σ_L is a <u>splitting</u> of the extension $\tilde{H}_X \rightarrow H_X$. A theorem of Atiyah [A1] asserts

<u>Proposition (7.6)</u>. Splittings of the extension $\tilde{H}_X$ correspond canonically to spin structures on X.

<u>Remarks</u>. The group $\tilde{H}_X$ is simply the group of units of the commutative ring $H^*(X;F_2)$. Another description is that $\tilde{H}_X = \tilde{KO}(X)$, and - because a spin structure is an orientation for KO-theory - the splitting associated to a spin structure is the corresponding Gysin map $\tilde{KO}(X) \rightarrow KO^{-2}(\text{point}) = Z/2$. Yet again, the set $\mathcal{S}_X$ of spin structures on X is an affine space of H_X, and there is a function $\sigma : \mathcal{S}_X \rightarrow F_2$ which takes L to the parity of the dimension of the space of sections of L*. A choice of spin structure identifies $\mathcal{S}_X$ with H_X and makes σ into the quadratic form σ_L.

Proposition (7.6) can be generalized to surfaces with parametrized boundaries. (In fact all we shall use of the parametrization is a choice of base-point on each boundary circle; and all that is really needed is a choice of a double covering of the boundary.) If S is a 1-manifold with a base-point in each component, let S_0 be the set of base-points, and let $H_S = H_1(S,S_0;F_2)$. Define $\tilde{H}_S = F_2 \oplus H_S$. Then spin structures on S can be identified with splittings of $\tilde{H}_S \rightarrow H_S$. On the other hand a spin structure L on a surface X can be described by a

function $\sigma_L : H_X \to F_2$, where $H_X = H_1(X, \partial_0 X; F_2)$ and $\partial_0 X$ is the set of

base-points in ∂X. (Here a spin-structure on X means a choice of L together with an identification of the fibre L_x with a fixed choice for each $x \in \partial_e X$.)

Proposition (7.7). For any surface X there is a canonical extension $\tilde{H}_X$

of H_X by F_2, and a canonical homomorphism $\tilde{H}_{\partial X} \to \tilde{H}_X$, such that spin

structures on X with a given restriction to ∂X correspond canonically

to splittings of $\tilde{H}_X$ extending a given splitting of $\tilde{H}_{\partial X}$.

Proof: The main point is to define $\tilde{H}_X$. Suppose that ∂X has k

components. Let Y be a standard sphere with k standard holes and a

chosen tree y linking the base-points of its boundary components. Let

$X* = X \cup Y$. The inclusion $(X, \partial_0 X) \to (X*, y)$ induces an isomorphism

$H_X \to H_1(X*, y) \cong H_{X*}$. But we know how to define $\tilde{H}_{X*}$ for the closed

surface X*, and it is now easy to deduce (7.7) from (7.6).

Spin structures and extensions of loop groups

The last way I shall mention of describing the spin structures on

a surface X is in terms of the group $\mathbb{C}_X^{\times}$ of holomorphic maps $X \to \mathbb{C}^{\times}$.

Proposition (7.8). For every spin structure on ∂X (with an even number

of periodic components) there is a central extension $\tilde{\mathbb{C}}_X^{\times}$ of $\mathbb{C}_X^{\times}$ by F_2

such that splittings of $\tilde{\mathbb{C}}_X^{\times}$ correspond precisely to spin structures on

X.

I shall give the proof in §12. (See also the end of §8.) At

this point I shall make just two remarks.

(i) The group of components of $C_X^\times$ is $H^1(X,Z)$, and if $\partial X = S^1$ then $\tilde{C}_X^\times$ is induced from $\bar{H}_X$ by the natural map $C_X^\times \to H^1(X;F_2) \cong H_X$. So the result is clear in this case.

(ii) Given a spin structure L on ∂X the group $C_{\partial X}^\times$ of smooth maps $\partial X \to C^\times$ acts on the polarized space $\Omega^{\frac{1}{2}}(\partial X;L)$ of $\frac{1}{2}$-forms on ∂X. This gives us a central extension $\tilde{C}_{\partial X}^\times$ of $C_{\partial X}^\times$ by $C^\times$. The extension $\tilde{C}_X^\times$ of (7.8) is a canonical subgroup of the restriction of $\tilde{C}_{\partial X}^\times$ to $C_X^\times$.

§8. The Grassmannian category: chiral fermions

Linear algebra

For clarity let us begin with finite dimensional vector spaces. A
linear map $T : V_0 \to V_1$ is completely determined by its graph W_T, a
subspace of $V_1 \oplus V_0$. It will be convenient in this section to define
the graph of T by

$$W_T = \{(Tv, -v) : v \in V_0\} .$$

One may ask whether the category of finite dimensional vector spaces
and linear maps is contained in a larger category in which the set of
morphisms $V_0 \to V_1$ is the Grassmannian manifold $Gr(V_1 \oplus V_0)$ of all
subspaces of $V_1 \oplus V_0$. If $W_{10} \in Tr(V_1 \oplus V_0)$, and $W_{21} \in Gr(V_2 \oplus V_1)$ one
would try to define the composite $W_{21} * W_{10}$ by

$$W_{21} * W_{10} = \{(v_2, -v_0) \in V_2 \oplus V_0 : \exists\, v_1 \in V_1 \quad \text{such that}$$

$$(v_2, -v_1) \in W_{21} \quad \text{and} \quad (v_1, -v_0) \in W_{10}\}$$

$$= pr_{20}((W_{21} \oplus V_0) \cap (V_2 \oplus W_{10})) , \tag{8.1}$$

where $pr_{20} : V_2 \oplus V_1 \oplus V_0 \to V_2 \oplus V_0$ is the projection.

If $\dim(W_{10}) = n_0 + a$ and $\dim(W_{21}) = n_1 + b$, where $n_i = \dim(V_i)$,
then generically $W_{21} * W_{10}$ has dimension $n_0 + a + b$. In fact,
$\dim(W_{21} * W_{10}) = n_0 + a + b$ if the following conditions are satisfied:

(i) $W_{21} \oplus W_{10} \rightarrow V_1$ is surjective, and

(8.2)

(ii) $W_{21} \oplus W_{10} \rightarrow V_2 \oplus V_1 \oplus V_0$ is injective.

Where these conditions do not hold the composition law is obviously badly discontinuous.

But we can do better. A subspace W of $V_1 \oplus V_0$ defines a ray in the exterior algebra $\Lambda(V_1 \oplus V_0)$, and hence, up to a scalar multiple, a linear map $T_W : \Lambda(V_0) \rightarrow \Lambda(V_1)$. For

$$\Lambda(V_1 \oplus V_0) \cong \Lambda(V_1) \otimes \Lambda(V_0)$$

$$\cong \Lambda(V_1) \otimes \Lambda(V_0)* \qquad (8.3)$$

$$\cong \text{Hom}(\Lambda(V_0) ; \Lambda(V_1)) .$$

The isomorphism $\Lambda^k(V_0) \rightarrow \Lambda^{n-k}((V_0)*$ used in (8.3) depends on the choice of an element of $\det(V_0)*$. Thus to get a specific map T_W we need not only W but also an element λ of $\det(W) \otimes \det(V_0)* = \det(V_0;W)$; we shall therefore write it $T_{W,\lambda}$. One readily verifies

<u>Proposition (8.4)</u>. If $W_{10} \subset V_1 \oplus V_0$ and $W_{21} \subset V_2 \oplus V_1$ then

$$T_{W_{21},\mu} \circ T_{W_{10},\lambda} = T_{W_{21}*W_{10},\mu \otimes \lambda}$$

if the conditions (8.2) are satisfied, where $\mu \otimes \lambda$ refers to the isomorphism $\det(V_1;W_{21}) \otimes \det(V_0;W_{10}) \cong \det(V_0;W_{21}*W_{10})$ induced by the exact sequence

$$0 \to W_{21} * W_{10} \to W_{21} \oplus W_{10} \to V_1 \to 0 .$$

If the conditions (8.2) are not satisfied, then $T_{W_{21},\mu} \circ T_{W_{10},\lambda} = 0$.

Corollary (8.5). (i) The category of finite dimensional vector spaces and linear maps is a subcategory of a category $\mathcal{V}$ which has the same objects, but in which a morphism from V_0 to V_1 is a pair (W,λ) with $W \in Gr(V_1 \oplus V_0)$ and $\lambda \in Det(V_0;W)$. (Here (W,λ) is regarded as independent of W if $\lambda = 0$, i.e. (W,λ) is really an element of $\Lambda(V_1 \oplus V_0) \otimes Det(V_0)*$.)

(ii) The exterior algebra functor $V \mapsto \Lambda(V)$ extends to $\mathcal{V}$, though as a functor into vector spaces rather than algebras. The morphism associated to (W,λ) raises degrees by $\dim(W) - \dim(V_0)$.

An endomorphism T of $\Lambda(V)$ has a trace and, more importantly, a supertrace $tr_S(T) = \Sigma(-1)^P tr(T|\Lambda^P)$. It is elementary to check

Proposition (8.6). If $(W,\lambda) : V \to V$ in $\mathcal{V}$ induces $T : \Lambda(V) \to \Lambda(V)$ then $tr_S(T)$ is the image of λ under

$$\Lambda(V \oplus V) \otimes Det(V)* \to \Lambda(V) \otimes Det(V)* \to \mathbb{C} ,$$

where the first map is induced by subtraction $V \oplus V \to V$.

Remark. If we used addition $V \oplus V \to V$ rather than subtraction we would obtain the trace instead of the supertrace.

More generally, a map $T : \Lambda(V_0 \oplus V) \to \Lambda(V_1 \oplus V)$ can be collapsed to a map $\check{T} : \Lambda(V_0) \to \Lambda(V_1)$ by taking the supertrace over $\Lambda(V)$.

Proposition (8.7). In this situation, if T is induced by (W, λ) in then $\check{T}$ is induced by $(\check{W}, \check{\lambda})$, where

$$\check{W} = \{(v_1, -v_0) \in V_1 \oplus V_0 : (v_1, v, -v_0, -v) \in W \text{ for some } v \in V\} ,$$

and $\check{\lambda}$ is the image of λ under

$$\Lambda(V_1 \oplus V \oplus V_0 \oplus V) \otimes \text{Det}(V_0 \oplus V)^* \cong \Lambda(V_1 \oplus V_0) \otimes \text{Det}(V_0)^* \otimes \Lambda(V \oplus V) \otimes \text{Det}(V)^*$$

$$\rightarrow \Lambda(V_1 \oplus V_0) \otimes \text{Det}(V_0)^* .$$

Polarized vector spaces and Fock spaces

We are not really interested in finite dimensional vector spaces. Instead we want to consider the category of _polarized_ topological vector spaces.

Definition (8.8). A polarization of a topological vector space E is a class of operators $J : E \rightarrow E$ such that $J^2 = 1$, any two differing by an operator of trace class.

Thus a polarized space E has a preferred class of decompositions $E = E^+ \oplus E^-$ into the ± 1 eigenspaces of J. The typical example is the space $\Omega^0(S^1)$ of smooth functions on the circle, with E^+ and E^- spanned by $\{e^{in\theta}\}$ for $n < 0$ and $n \geqslant 0$ respectively.

A polarized space has a (restricted) Grassmannian Gr(E) which consists of the -1 eigenspaces E^- of all allowable J's. If W_0 and W_1 are two subspaces belonging to Gr(E) we can define a canonical relative determinant line $\text{Det}(W_0; W_1)$. Holding W_0 fixed and letting W_1 vary we obtain a holomorphic line bundle Det_{W_0} on $\text{Gr}(E)$. These bundles are all isomorphic, but not canonically: an isomorphism $\text{Det}_{W_0} \to \text{Det}_{W_1}$ is the same thing as an element of $\text{Det}(W_0;W_1)$. For a discussion of all these facts I refer to Appendix B. Let us notice also that the line $\text{Det}(W_0;W_1)$ is naturally a vector subspace of $\mathfrak{F}_{W_0}(E)$, i.e. the bundle Det_{W_0} on $\text{Gr}(E)$ is a sub-bundle of the trivial bundle $\text{Gr}(E) \times \mathfrak{F}_{W_0}(E)$.

We can now repeat the discussion in this section using the category of polarized topological vector spaces and systematically replacing the exterior algebra by its analogue for polarized spaces, which is the <u>Fock space</u>.

<u>Definition (8.9)</u>. For $W \in \text{Gr}(E)$ the Fock space $\mathfrak{F}_W(E)$ is the dual of the space of holomorphic sections of Det_W^*.

Thus the projective space of $\mathfrak{F}_W(E)$ is independent of W, and an isomorphism $\mathfrak{F}_{W_0}(E) \to \mathfrak{F}_{W_1}(E)$ is given by an element of $\text{Det}(W_0;W_1)$. I recall from [PS] Chap. 10 that if $E = E^+ \oplus E^-$ is an allowable decomposition then we have a map

$$\Lambda((E^-)^* \oplus E^+) \to \mathfrak{F}_{E^-}(E)$$

which identifies the left-hand-side (interpreted algebraically) with a dense subspace of the Fock space.

The analogue of the finite dimensional isomorphism $\Lambda(V) \otimes \text{Det}(V)^* \cong \Lambda(V)^*$ is a bilinear map

$$\mathfrak{F}_{E^+}(\bar{E}) \times \mathfrak{F}_{E^-}(E) \to \mathbb{C} , \tag{8.10}$$

where $\tilde{E}$ denotes E with the reversed polarization-class, i.e. with J
replaced by -J. (For the definition see Appendix B.)

We now have

<u>Proposition (8.11)</u>. (i) There is a category $\mathcal{U}_{pol}$ whose objects are
pairs (E,J) consisting of a polarized topological vector space and a
choice of J, and whose morphisms $(E_0,J_0) \rightarrow (E_1,J_1)$ are pairs (W,λ)
with $W \in Gr(\tilde{E}_0 \oplus E_1)$ and $\lambda \in Det(E_0^+ \oplus E_1^-;W)$.

(ii) The Fock space is a functor from $\mathcal{U}_{pol}$ to Z-graded topological
vector spaces and trace-class maps. It has exactly the same properties
with respect to "sewing" and the supertrace as held in the finite
dimensional case. In particular, a morphism (W,λ) raises dimension by
the relative dimension dim$(W : E_0^+ \oplus E_1^-)$.

The category $\mathcal{U}_{pol}$ is thus formally analogous to the category $\mathcal{C}$
made from circles and Riemann surfaces, and the Fock space functor is
analogous to a field theory. To make the analogy complete we need the
hermitian structure. If E has a hermitian inner product, and the
polarization J is self-adjoint, then the Fock space $\mathcal{F}(E)$ inherits an
inner product. If E is positive-definite then so is $\mathcal{F}(E)$. But there
is another way to give $\mathcal{F}(E)$ an inner product, using the canonical
pairing (8.10). If E has a real structure, i.e. an operation of
complex conjugation, which interchanges E^+ and E^-, then $\mathcal{F}(\tilde{E}) \cong \overline{\mathcal{F}(E)}$,
and (8.10) becomes a hermitian form. If the conjugation exchanges E^+
and E^- only up to a finite dimensional discrepancy - more precisely, if
it anticommutes with J modulo trace class operators - then the hermitian

form is defined only up to a scalar multiple[1]. The inner product on $\mathcal{J}(E)$ coming from a real structure on E is hyperbolic, i.e. as far as possible from being positive definite.

Chiral fermion theories

We can now give our first examples of conformal field theories. For any integer α one has the space $\Omega^\alpha(S^1)$ of differential forms of degree α on the circle - i.e. expressions of the form $f(\theta)(d\theta)^\alpha$. This is polarized by $\Omega^\alpha = \Omega^\alpha_+ \oplus \Omega^\alpha_-$, where Ω^α_+ (resp. Ω^α_-) is spanned by $e^{in\theta}(d\theta)^\alpha$ for n < 0 (resp. n ⩾ 0). The class of the polarization is independent of the parametrization ([PS] p.91), so $S \mapsto \Omega^\alpha(S)$ is a functor from oriented 1-manifolds to $\mathcal{U}_{pol}$. Reversing the orientation reverses the polarization class, and so does complex conjugation in Ω^α. On the other hand for each Riemann surface X with boundary we have the space $\Omega^\alpha(X)$ of holomorphic α-forms $f(z)(dz)^\alpha$ on X, and (see [PS] (8.11.10)).

<u>Proposition (8.12)</u>. The space $\Omega^\alpha(X)$ belongs to the restricted Grassmannian $Gr(\Omega^\alpha(\partial X))$. Its dimension relative to $\Omega^\alpha_-(\partial X)$ is

$$d_\alpha(X) = (2\alpha-1)(g+m-1) ,$$

where m is the number of outgoing boundary circles.

Here $\Omega^\alpha_+(\partial X)$ means the sum of a copy of $\Omega^\alpha_+(S^1)$ (resp. $\Omega^\alpha_-(S^1)$) for each incoming (resp. outgoing) circle.

----
[1] (*) A better way to say this is: there is a hermitian form on $\mathcal{J}_{E^-}(E)$ with values in $Det(\overline{E}^-;E^+)$.

Corollary (8.13). Ω^α defines a functor from $\mathscr{C}$ to $\mathscr{U}_{pol}$, and so $S \mapsto \mathscr{F}(\Omega^\alpha(S))$ is a holomorphic conformal field theory. A surface X defines an operator which raises degrees by $d_\alpha(X)$.

The spaces Ω^α and $\Omega^{1-\alpha}$ are in duality, but the duality reverses the polarization, so the Fock spaces $\mathscr{F}(\Omega^\alpha)$ and $\mathscr{F}(\Omega^{1-\alpha})$ - and with them the field theories - are identical. By calculating the Lie algebra cocycle one finds that the theory $\mathscr{F}(\Omega^\alpha)$ has central charge $c = 12\alpha(1-\alpha)-2$. A physicist does this calculation in the following way.

Let L_p denote $e^{-ip\theta}d/d\theta$ in $\mathrm{Vect}_\mathbb{C}(S^1)$, and let $\psi_q = e^{-iq\theta}(id\theta)^\alpha$ in Ω^α. We have

$$L_p \cdot \psi_q = -i(q + \alpha p)\psi_{p+q} . \qquad (8.14)$$

Let $\omega = \psi_m \wedge \psi_{m-1} \wedge \psi_{m-2} \wedge \ldots$ in $\mathscr{F}(\Omega^\alpha)$. Then

$$L_p\omega = -i \sum_{k=0}^{p-1} (m - k + \alpha p)\omega_k$$

if $p > 0$, where ω_k is obtained from ω by replacing ψ_{m-k} by ψ_{m-k+p}. If $p < 0$ then $L_p\omega = 0$. Hence

$$[L_{-p}, L_p]\omega = \{(\alpha(1-\alpha) - 1/6)p^3 + (m(m+1) + 1/6)p\}\omega .$$

Comparing this with (5.6) we find

$$c = 12\alpha(1-\alpha) - 2 \quad \text{and} \quad h = \tfrac{1}{2}\alpha(1-\alpha) + \tfrac{1}{2}m(1+m) . \qquad (8.15)$$

The most interesting of the theories $\mathcal{J}(\Omega^{\alpha})$ is the one with $\alpha = -1$ or $\alpha = 2$, i.e. when $\Omega^{\alpha}(S^1)$ is the Lie algebra $\mathrm{Vect}(S^1)$ or its dual. This has $c = -26$. It is the theory of <u>BRS ghosts</u> which was mentioned at the end of §4. The grading of $\mathcal{J}(\Omega^2(S))$ is the <u>ghost number</u>, and $d_2(X) = 3(g+m-1)$ is the <u>ghost number anomaly</u>. We shall return to this theory in §9.

If α is an integer the space $\Omega^{\alpha}(S^1)$ has no inner product. But if $\alpha = \frac{1}{2}$ it has a natural positive definite inner product, and we can use this instead of the real structure to define an inner product on $\mathcal{J}(\Omega^{\frac{1}{2}})$. A new point arises, however, because to define $\Omega^{\frac{1}{2}}(S^1)$, and still more to define $\Omega^{\frac{1}{2}}(X)$ for a surface X, one must choose a square-root L of the tangent bundle TS^1 or TX, i.e. a <u>spin structure</u>. (See §7.) Thus we have

<u>Proposition (8.16)</u>. $\Omega^{\frac{1}{2}}$ defines a functor from $\mathcal{C}^{\mathrm{spin}}$ to $\mathcal{V}_{\mathrm{pol}}$, and so $(S,L) \mapsto \mathcal{J}(\Omega^{\frac{1}{2}}(S,L))$ is a positive-definite projective representation of $\mathcal{C}^{\mathrm{spin}}$, i.e. a weakly conformal field theory (cf. (5.2)), with central charge $c = 1$.

This theory is called the <u>charged chiral fermion</u>. It is the basic example of the structure we are studying. For surfaces X with spin structures which are Mobius (i.e. antiperiodic) on each boundary circle the dimension of $\Omega^{\frac{1}{2}}(X)$ relative to $\Omega^{\frac{1}{2}}_{+}(\partial X)$ is zero, and the associated operator preserves the grading. For the spin structure S^1_P, as there is no vacuum vector in $\mathcal{J}(\Omega^{\frac{1}{2}}(S^1_P))$, it is not obvious how to grade the Fock space. The correct procedure is to grade it by $\mathbb{Z} + \frac{1}{2}$, so that $\psi_0 \wedge \psi_{-1} \wedge \psi_{-2} \wedge \ldots$ has degree $+\frac{1}{2}$. The operators of the theory then preserve the grading in all cases. From the formulae (8.15) we find that $\mathrm{Diff}^+(S^1)$ acts on $\mathcal{J}(\Omega^{\frac{1}{2}}(S^1_A))$ with $(c,h) = (1,0)$, and on $\mathcal{J}(\Omega^{\frac{1}{2}}(S^1_P))$ with $(c,h) = (1,1/8)$.

Field operators

We have seen that when X is a surface with $\partial X = \overline{S}_0 \amalg S_1$ the space
$\Omega^\alpha(X)$, together with a point in its determinant line, defines an
operator

$$U_X : \mathcal{F}(\Omega^\alpha(S_0)) \to \mathcal{F}(\Omega^\alpha(S_1)) .$$

If $\mathfrak{f} = \Sigma\, n_i[z_i]$ is a divisor on X, i.e. a set $\{z_1, \ldots, z_k\}$ of points in
the interior of X equipped with integral multiplicities $\{n_1, \ldots, n_k\}$,
then it is natural to consider the space of holomorphic α-forms on X
which vanish to order $-n_i$ at z_i. (If n_i is positive this means that
the form is allowed to have a pole of order n_i at z_i). This space will
be denoted $\Omega^\alpha(X; \mathfrak{f})$. Because it belongs to $Gr(\Omega^\alpha(\partial X))$ it too defines an
operator $\mathcal{F}(\Omega^\alpha(S_0)) \to \mathcal{F}(\Omega^\alpha(S_1))$, at least up to a scalar multiple,
which raises degree by $d_\alpha(X) + \Sigma\, n_i$. To get a precise operator we must
choose an element of

$$\mathrm{Det}^\alpha(\mathfrak{f}) = \mathrm{Det}(\Omega^\alpha(X); \Omega^\alpha(X; \mathfrak{f})) .$$

(I shall assume that an element of $\mathrm{Det}(\Omega_+^\alpha(\partial X); \Omega^\alpha(X))$ has already been
chosen, and is kept fixed in what follows.) Now

$$\mathrm{Det}^\alpha(\mathfrak{f}) = \bigotimes_i \mathrm{Det}(P_{n_i}^\alpha(z_i)) , \qquad (8.17)$$

where $P_n^\alpha(z)$ denotes the space of principal parts at z of meromorphic
α-forms with n^{th} order poles if $n > 0$, while if $n < 0$ then it denotes the
dual of the space of $(-n-1)$-jets of holomorphic α-forms at z. Thus

$$\mathrm{Det}(P_n^\alpha(z)) \;=\; (T_z X)^{\otimes(-\alpha n + \frac{1}{2} n(n+1))}$$

for all n. This means that as the divisor $\mathfrak{f}$ varies we get not an operator-valued function on the complex manifold

$$\{(z_1,\ldots,z_k) \;\epsilon\; X^k \;:\; z_i \neq z_j\}$$

but a holomorphic operator-valued form of multidegree $(d_1,\ldots,d_k)$, where $d_i = -\alpha n_i + \frac{1}{2} n_i(n_i + 1)$. We write this operator

$$\psi(\mathfrak{f}) d\mathfrak{f}^\alpha = \psi^{(n_1)}(z_1)\psi^{(n_2)}(z_2) \;\ldots\; \psi^{(n_k)}(z_k) \;.\; \Pi \, dz_i^{\otimes d_i} \;. \tag{8.18}$$

Despite the notation it should not be regarded as a composition of operators associated to the different points z_i. It depends in an alternating way on the order of the points, providing z_i is assigned degree n_i: that follows from the graded nature of the isomorphism (8.17).

We have been assuming that the points z_i are distinct. As a point of the Grassmannian $\mathrm{Gr}(\Omega^\alpha(\partial X))$ the space $\Omega^\alpha(X;\mathfrak{f})$ behaves smoothly when the points come together (and their multiplicities are added appropriately), and so does the line $\mathrm{Det}^\alpha(\mathfrak{f})$. The isomorphism (8.17), however, breaks down and must be reconsidered. As an illustration let us consider the case of a divisor $\mathfrak{f} = [z_2]-[z_1]$ contained in the annulus

$$X = X_{ab} = \{z \;\epsilon\; \mathbb{C} \;:\; a \leqslant |z| \leqslant b\} \;.$$

For simplicity we shall assume that $\alpha = 0$. We can define a reference element of $\mathrm{Det}(\Omega^0_+(\partial X); \Omega^0(X))$ by means of the basis $\{f_k\}_{k\epsilon}$ of $\Omega^0(X)$, where

$$f_k(z) = (z/b)^k \quad \text{for} \quad k \geqslant 0$$
$$= (z/a)^k \quad \text{for} \quad k < 0 .$$

A good basis for $\mathrm{Det}^0(\mathfrak{H})$ is then represented by the basis $\{\varphi_\mathfrak{H} f_k\}_{k\epsilon\mathbb{Z}}$ of $\Omega^0(X; \mathfrak{H})$, where $\varphi_\mathfrak{H}(z) = (z-z_1)/(z-z_2)$. This behaves smoothly when $z_1 = z_2$. On the other hand, in terms of the isomorphism (8.17) the natural basis element $(dz_1)^0(dz_2)^{-1}$ of the right-hand side corresponds to the element of $\mathrm{Det}^0(\mathfrak{H})$ represented by the basis $\{g_k\}$ of $\Omega^0(X; \mathfrak{H})$, where

$$g_k(\mathfrak{H}) = (z-z_2)^{-1} - (z_1-z_2)^{-1} \quad \text{if} \quad k = 0$$
$$= (z^k-z_1^k)/b^k \qquad\qquad \text{if} \quad k > 0$$
$$= (z^k-z_1^k)/a^k \qquad\qquad \text{if} \quad k < 0 .$$

It is easy to check that the determinant of $\{\varphi_\mathfrak{H} f_k\}$ with respect to $\{g_k\}$ is z_1-z_2, and so we have

Proposition (8.19). The operator-valued form

$$\psi^*(z_1)\psi(z_2)(z_1-z_2)dz_2 \; : \; \mathfrak{F}(\Omega^0(S^1)) \to \mathfrak{F}(\Omega^0(S^1))$$

on the annulus X is holomorphic everywhere, and its value when $z_1 = z_2$ is simply the operator U_X associated to the annulus.

Here we have written, as is usual, $\psi(z)$ for $\psi^{(1)}(z)$, and $\psi*(z)$ for $\psi^{(-1)}(z)$. The proposition is usually stated as an "operator product expansion"[1]

$$\psi*(z_1)\psi(z_2) = \frac{1}{z_1 - z_2} + \text{(holomorphic)} .$$

It remains true when Ω^0 is replaced by Ω^α for any α, except that dz_2 becomes $(dz_1)^\alpha(dz_2)^{1-\alpha}$.

A similar, but easier, calculation can be performed for the positive divisor $\mathfrak{z} = [z_1] + [z_2] + \ldots + [z_n]$ on the same annulus X. The operator $\psi(z_1)\psi(z_2) \ldots \psi(z_n)$ corresponds to the basis for $\Omega^\alpha(X; \mathfrak{z})$ which consists of $(z-z_i)^{-1}$ for $i = 1, \ldots, n$ together with $\{f_k\}_{k\epsilon \mathbb{Z}}$ as above. On the other hand a basis for $\Omega^\alpha(X; \mathfrak{z})$ which is everywhere defined is given by $z^k h$ for $k = 0, 1, \ldots, n-1$ together with the $\{f_k\}$, where $h = \Pi(z-z_i)^{-1}$. The determinant of the first basis in terms of the second is a Vandermonde determinant equal to $\underset{i<j}{\Pi} (z_i - z_j)$. Thus we have

<u>Proposition (8.20)</u>. The operator $\psi(z_1)\psi(z_2) \ldots \psi(z_n)/\Pi(z_i - z_j)$ is holomorphic for all $z_1, \ldots, z_n$. Its value when $z_1 = z_2 = \ldots = z_n = z$ is usually denoted

$$C \; : \; \psi(z)\psi'(z)\psi''(z) \ldots \psi^{(n-1)}(z) \; : \; ,$$

where $C^{-1} = 1!2!3! \ldots (n-1)!$.

----

[1]For the omission of U_X from the following notation, see the remarks after (9.3).

It is an instructive exercise to translate the field operators just defined into their more usual description. The Fock space $\mathcal{F}(\Omega^\alpha(S^1))$ is a module over the exterior algebra of $\Omega^\alpha(S^1)$. As above we write ψ_k for the basis element $z^{-k-\alpha}(dz)^\alpha$ of $\Omega^\alpha(S^1)$, and also for the corresponding multiplication operator on the Fock space. The formal series

$$\hat{\psi}(w) = \sum_{k \in Z} w^{k+\alpha-1} \psi_k$$

does not define an operator in $\mathcal{F}(\Omega^0(S^1))$, because it is unbounded and may not converge. Nevertheless, let U_{ab} denote the endomorphism of $\mathcal{F}(\Omega^\alpha(S^1))$ defined by the annulus X_{ab}. This is a contraction operator which depends only on a/b, and satisfies $U_{ab}\psi_k = (a/b)^k\psi_k U_{ab}$. It is easy to see that if $a < |w| < b$ and $a \leqslant 1 \leqslant b$ the composite $U_{1b}\hat{\psi}(w)U_{a1}$ is a well-defined operator in $\mathcal{F}(\Omega^\alpha(S^1))$. We have

<u>Proposition (8.21)</u>. The operator-valued $(1-\alpha)$-form $\psi(w)dw^{1-\alpha}$ on the annulus X_{ab} is given by

$$\psi(w) = U_{1b}\hat{\psi}(w)U_{a1} \; .$$

<u>Proof</u>: This is simply a matter of unravelling the definitions. The Fock space $\mathcal{F} = \mathcal{F}(\Omega^\alpha(S^1))$ has a natural basis $\{\omega_S\}$, where S runs through an appropriate class of subsequences of Z. (Cf. [PS] Chap. 10.) The dual space $\mathcal{F}^*$ has a dual basis $\{\omega_{\overline{S}}\}$, where $\overline{S} = Z-S$. The operator U_{ab} multiplies ω_S by $(a/b)^{\ell(S)}$, and as an element of $\mathcal{F}^* \otimes \mathcal{F}$ we can write U_{ab} in the form

$$U_{ab} = \sum_S (a/b)^{\ell(S)} \omega_{\overline{S}} \otimes \omega_S \; . \tag{8.22}$$

Now $\Omega^{\alpha}(X_{ab};[w])$ is spanned by $(z-w)^{-1}dz^{\alpha}$ modulo $\Omega^{\alpha}(X_{ab})$. When this form is restricted to the ends of the annulus it becomes

on one end, and

$$\varphi = w^{-1} \sum_{k<0} b^{k+\alpha} w^{-k} \psi_{-k-\alpha}$$

$$\theta = -w^{-1} \sum_{k\geq 0} a^{k+\alpha} w^{-k} \psi_{-k-\alpha}$$

on the other. The operator $\psi(w)$, as an element of $\mathcal{F}^* \otimes \mathcal{F}$, is therefore given by multiplying the expression (8.22) by $1 \otimes \varphi - \theta \otimes 1$, and this amounts to the assertion of (8.21).

The bosonic description of $\mathcal{F}(\Omega^{\alpha})$

Each of the theories $\mathcal{F}(\Omega^{\alpha})$ has an alternative "bosonic" description. We shall explain this very briefly now, and shall return to it in §9 and §12. For an oriented 1-manifold S the group $\mathbb{C}_S^{\times}$ of smooth maps $S \to \mathbb{C}^{\times}$ acts on $\Omega^{\alpha}(S)$ by multiplication, and preserves the polarization class ([PS](6.3.1)). It therefore acts projectively on $\mathcal{F}(\Omega^{\alpha}(S))$. To give a bosonic description of the theory $\mathcal{F}(\Omega^{\alpha})$ means, in one interpretation, to construct it purely in terms of the representation theory of the groups $\mathbb{C}_S^{\times}$ and $\mathbb{C}_X^{\times}$, the group of holomorphic maps from a surface X to $\mathbb{C}^{\times}$.

Of course all the spaces $\Omega^{\alpha}(S)$ and $\Omega^{\frac{1}{2}}(S_{\sigma})$, with $\alpha \in \mathbb{Z}$ and σ a spin structure, are isomorphic as representations of $\mathbb{C}_S^{\times}$, and $\mathcal{F}(\Omega^{\alpha}(S^1))$ is the basic irreducible representation described in [PS] Chap. 10. Nevertheless one must beware of identifying the $\mathcal{F}(\Omega^{\alpha}(S))$ for different α, even as projective spaces, as the isomorphisms involve a choice of parameter on S. (At first sight this seems to contradict Schur's lemma, but that lemma does not apply to projective representations.) Concretely, one has a fixed representation H of $\mathbb{C}_S^{\times}$, but a different action of $\mathrm{Diff}^+(S^1)$ on H for each α. These actions are described on page 208 of [PS].

The bosonic description begins by prescribing a definite
projective multiplier on $\mathbb{C}_S^\times$, or equivalently an extension $\widetilde{\mathbb{C}}_S^\times$ of $\mathbb{C}_S^\times$ by
$\mathbb{C}^\times$ with a definite action of $\mathrm{Diff}^+(S^1)$ on it. Then $F_S = \mathcal{J}(\Omega^\alpha(S))$ is
constructed as the unique irreducible representation (of positive
energy) of $\widetilde{\mathbb{C}}_S^\times$. To complete the description one has only to give the
ray F_X in $F_{\partial X}$ corresponding to each surface X. From the Fock space
description we know that the ray is invariant under the subgroup $\mathbb{C}_X^\times$ of
$\mathbb{C}_{\partial X}^\times$, so it defines a homomorphism $\widetilde{\mathbb{C}}_X^\times \to \mathbb{C}^\times$, i.e. a splitting of the
induced central extension of $\mathbb{C}_X^\times$. From the bosonic point of view one
must <u>give</u> the splitting of the extension of $\mathbb{C}_X^\times$ directly. Then $\mathbb{C}_X^\times$ acts
on $F_{\partial X}$. One proves that there is a unique ray in $F_{\partial X}$ which is
pointwise fixed under $\mathbb{C}_X^\times$, and calls it F_X. This programme will be
carried out in §12. In §9 I shall try to explain the relation of the
representation theory to "bosonic fields".

Even spin structures and the real chiral fermion

There is an important variant of the linear algebra of this
section. In finite dimensions the exterior algebra functor $V \mapsto \Lambda(V)$
has an analogue which takes a vector space V with an inner product to
the spin representation of the orthogonal group O(V). To be precise,
let us consider real vector spaces V with non-degenerate quadratic
forms, not necessarily positive-definite. The spin representation of
O(V) is a mod 2 graded complex projective irreducible representation Δ
on which the orientation-reversing elements of O(V) act with degree 1.
Alternatively, Δ is an irreducible graded module for the complexified
Clifford algebra C(V). (See [ABS].) If dim(V) is odd then Δ is
uniquely determined up to isomorphism, but if dim(V) is even there are
two possibilities (which differ by reversing the grading), and a choice
of Δ corresponds to choosing an orientation of V.

The spin representation is best understood in terms of the
isotropic Grassmannian $\mathcal{J}(V)$ of all maximal isotropic subspaces of $V_{\mathbb{C}}$.
(Cf. [PS] Chap. 12.) If dim(V) is odd then $\mathcal{J}(V)$ is connected, but if
dim(V) is even $\mathcal{J}(V)$ has two connected components (for W ϵ $\mathcal{J}(V)$
defines a complex structure, and hence an orientation, on V). There is
a holomorphic line bundle Pf on $\mathcal{J}(V)$ - the Pfaffian bundle - and in
the even dimensional case the spin representation Δ is the space of
holomorphic sections of Pf*, and can be graded in two ways. In
particular, each W ϵ $\mathcal{J}(V)$ defines a ray Pf_W in W, which can also be
characterized by the fact that it is annihilated by the subspace W of
the Clifford algebra C(V). Indeed one can identify Δ with $Pf_W \otimes \Lambda(W*)$,
because $C(V)/\Lambda(W) \cong \Lambda(W*)$. When dim(V) is odd, however, Δ is the sum
of two copies of $\Gamma(Pf*)$.

Let us also recall from [ABS] that if Δ_i is an irreducible
$C(V_i)$-module for i = 1,2 then $\Delta_1 \otimes \Delta_2$ is an irreducible $C(V_1 \oplus V_2)$-
module unless both V_1 and V_2 are odd dimensional, in which case
$\Delta_1 \otimes \Delta_2$ is the sum of the two distinct irreducible $C(V_1 \oplus V_2)$-modules.

We can now define a category $\mathcal{U}^{orth}$ whose objects are pairs (V,Δ),
where Δ is an irreducible C(V)-module. A morphism $(V_0,\Delta_0) \to (V_1,\Delta_1)$ is
a pair (W,λ), where W is a maximal isotropic subspace of $V_0 \oplus V_1$ and
$\lambda \epsilon (\Delta_0^* \otimes \Delta_1)^{even}$ is annihilated by W. (Here V_0 denotes V_0 with its
quadratic form multiplied by -1.)

A morphism $(V_0,\Delta_0) \to (V_1,\Delta_1)$ in $\mathcal{U}^{orth}$ defines a linear map
$\Delta_0 \to \Delta_1$ of degree 0, and the group of automorphisms of (V,Δ) is the
complexification of $Spin^c(V)$. A general morphism $(V_0,\Delta_0) \to (V_1,\Delta_1)$
corresponds to a choice of isotropic subspaces P_0 and P_1 in $V_{0,\mathbb{C}}$ and
$V_{1,\mathbb{C}}$ together with an isometry $P_0^\perp/P_0 \to P_1^\perp/P_1$.

Now let us turn to polarized infinite dimensional real vector
spaces with quadratic forms. In this situation a polarization of V

means a class of skew transformations $J : V \to V$ such that $J^2 = -1$
modulo trace-class operators, two members of the class differing as
usual by trace-class operators. (Cf. (10.3).) The theory of
irreducible modules for the Clifford algebra $C(V)$ proceeds just as in
finite dimensions. There are two cases, according as dim(ker J) is
even or odd. We shall refer to V as even or odd correspondingly. In
either case there is an isotropic Grassmannian $\mathcal{J}(V)$ consisting of
maximal isotropic subspaces W of $V_{\mathbf{c}}$ which belong to the polarization
class (i.e. which are the (+i)-eigenspaces of allowable polarization
operators J). It is connected if V is odd, and has two components if V
is even. There are respectively one or two irreducible graded modules
for $C(V)$, and we can define a category $\mathcal{U}_{\text{pol}}^{\text{orth}}$ analogous to
$\mathcal{U}^{\text{orth}}$.

Finally we come to Riemann surfaces and their boundary circles.
We shall define a weakly conformal unitary field theory based on a
modular functor with three labels which I shall call the <u>even spin</u>
functor. The theory itself is called the <u>real chiral fermion</u>. It has
central charge $c = \frac{1}{2}$, and is the simplest example of a theory with
non-integral c.

For an oriented circle S with a spin structure L the space $\Omega^{\frac{1}{2}}(S;L)$
of sections of L* belongs to the category $\mathcal{U}_{\text{pol}}^{\text{orth}}$. It is even or odd
according as L is Möbius or trivial. A label for the circle S consists
of L together with an irreducible graded module for the Clifford
algebra of $\Omega^{\frac{1}{2}}(S;L)$. To an oriented 1-manifold $S = S_1 \amalg \ldots \amalg S_k$, where
the circle S_i is labelled (L_i, Δ_i), the theory assigns the Hilbert space
$\Delta = \Delta_1 \otimes \ldots \otimes \Delta_k$. For a surface X with $\partial X = S$ we consider all spin
structures L on X which reduce to $L = \amalg L_i$ on S. We define the modular
functor by

$$E(X;L,\Delta) = \bigoplus_L E(X;L) \subset \Delta \ ,$$

where $E(X;L)$ is the subspace of the even part of Δ which is annihilated by $\Gamma(L)$, i.e. by the boundary values of holomorphic sections of L. If all of the L_i are Möbius then Δ is an irreducible representation of the Clifford algebra of $\Omega^{\frac{1}{2}}(S;L)$, and the maximal isotropic subspace $\Gamma(L)$ annihilates a unique ray in Δ. The spin structure L is called <u>even</u> or <u>odd</u> relative to Δ according as this ray belongs to the even or odd part of Δ. Thus when L is purely Möbius we have

$$\dim E(X;L,\Delta) = \bigl|\{\text{even spin structures on X relative to } \Delta\}\bigr|.$$

If on the other hand L has 2q non-Möbius components then

$$\dim E(X;L,\Delta) = 2^{q-1}\bigl|\{\text{spin structures on X}\}\bigr|$$
$$= 2^{2g+q-1} \ ,$$

where g is the genus of X. (In these statements the set of spin structures on X means not the set $\mathcal{S}(X;L)$ of §7, but rather its quotient by $\mathrm{Aut}(L)$.)

In the Möbius case we observe that $E(X;L)$ is the Pfaffian line of L, which is the square-root of the determinant line of the $\bar{\partial}$-operator of L. This explains why the theory has $c = \frac{1}{2}$.

The Verlinde algebra generated by the three labels $A^{\pm}, P$ is described by the multiplication rules

$$A^+ = 1$$
$$(A^-)^2 = 1$$
$$P^2 = A^+ + A^-$$
$$A^{\pm}P = P \ .$$

§9. Field operators

Primary fields

We shall now describe how to reconstruct some of the usual formalism of quantum field theory from a functor of the type we are studying. Thus we begin from a vector space H, and have an operator $U_{X,\xi} : H^{\otimes m} \to H^{\otimes n}$ when X is a Riemann surface with $\partial X = C_n - C_m$, and $\xi \in Det_X$. We suppose that

$$U_{X,\lambda\xi} = \bar{\lambda}^{-\frac{1}{2}c_L} \lambda^{-\frac{1}{2}c_R} U_{X,\xi} ,$$

whhere (c_L, c_R) is the central charge.[1]

First, the morphism $C_0 \to C_1$ defined by the standard unit disc D and the canonical element $\epsilon_D \in Det_D$ gives us a map $\mathbb{C} \to H$, or equivalently a vector $\Omega \in H$. This is the <u>vacuum vector</u>.

The complex semigroup $\mathscr{A}$ acts projectively on H. The semigroup $\mathscr{E}_0$ of holomorphic embeddings $f : D \to \overset{\circ}{D}$ such that $f(0) = 0$ is a sub-semigroup of $\mathscr{A}$ over which the central extension is canonically split, for the standard element $\epsilon_f \in Det_f$ is characterized by $\epsilon_D \epsilon_f = \epsilon_D$. It therefore makes sense to look for eigenvectors of $\mathscr{E}_0$ in H. The possible homomorphisms $\mathscr{E}_0 \to \mathbb{C}^\times$ are given by $f \mapsto f'(0)^p \overline{f'(0)}^q$ with $p, q \in \mathbb{R}$ such that $p-q \in \mathbb{Z}$.

---

[1]The discussion applies with little change to a weakly conformal theory based on a modular functor with index set Φ rather than on the determinant line. The vacuum vector and the energy-momentum tensor lie in the space H_1 corresponding to $1 \in \Phi$.

Definition (9.1). A <u>primary field</u> of type (p,q) is a vector $\psi \in H$ such that

$$U(f)\psi = f'(0)^p \overline{f'(0)}^q \psi$$

for all $f \in \mathcal{C}_0$.

The reason for the terminology is the following. Suppose that X is a Riemann surface with m incoming and n outgoing boundary circles. We suppose an element $\xi \in \mathrm{Det}_X$ has been chosen, but we shall suppress it from the notation. Then for each primary field ψ of type (p,q) there is an operator-valued (p,q) form $\psi_X(z)dz^p d\bar{z}^q$ on X with values in the trace-class operators $H^{\otimes m} \to H^{\otimes n}$. To define ψ_X, choose a holomorphic embedding $f : D \to X$ with centre $z \in X$. Because $X_z = X - f(\overset{\circ}{D})$ is a morphism $C_{m+1} \to C_n$ it induces $U_{X_z} : H^{\otimes m} \otimes H \to H^{\otimes n}$, and $\psi_X(z)$ is defined by $\psi_X(z).\xi = U_{X_z}(\xi \otimes \psi)$. The condition of (9.1) implies that $\psi_X(z)dz^p d\bar{z}^q$ is a well-defined differential form on the interior of X. (It probably always has a <u>distributional</u> extension to the closed surface.)

One can also define multipoint fields. If $\psi_1,\ldots,\psi_k$ are primary fields of types (p_i,q_i) then there is a differential form $(\psi_1 \ldots \psi_k)_X$, usually written

$$\psi_1(z_1) \ldots \psi_k(z_k)dz_1^{p_1} \ldots dz_k^{p_k} d\bar{z}_1^{q_1} \ldots d\bar{z}_k^{q_k} \, ,$$

defined on the manifold

$$\{ (z_1,\ldots,z_k) \in \overset{\circ}{X}{}^k \; : \; z_i \neq z_j \quad \text{if } i \neq j \}$$

with values in the operators $H^{\otimes m} \to H^{\otimes n}$. It is defined by

$$(\psi_1 \ \ldots \ \psi_k)_X \cdot \xi = U_{X'}(\xi \otimes \psi_1 \otimes \ldots \otimes \psi_k) \ ,$$

where now X' is obtained from X by removing k discs around the points $z_1, \ldots, z_k$. If $\psi_1 = \ldots = \psi_k$ then the operator is symmetric with respect to permuting $z_1, \ldots, z_k$ (or skew-symmetric if the theory is mod 2 graded).

By their construction the operator-valued differential forms ψ_X have the following naturality property, which is usually referred to as the <u>Ward identity</u>. If $Y = Z_1 \circ X \circ Z_2$ is a union of surfaces then

$$\psi_Y | X = U_{Z_1} \circ \psi_X \circ U_{Z_0} \ . \tag{9.2}$$

We also have, for example,

$$(\psi_1 \ \ldots \ \psi_k)_X \circ (\psi_{k+1} \ \ldots \ \psi_m)_Y = (\psi_1 \ \ldots \ \psi_m)_{X \bullet Y}$$

when the surfaces X and Y are composable.

One can go on and define secondary, tertiary,... fields: the bigraded space H is filtered $H_0 \subset H_1 \subset H_2 \subset \ldots \subset H$, where H_0 is the primary fields, and $\mathcal{E}_0$ acts scalarly on H_k/H_{k-1} for each k. In fact <u>any</u> vector $\psi \in H$ gives rise to an operator-valued function on any surface; but when we use the notation $\psi_X(z)$ we must beware that if ψ is an r-ary field then $\psi_X(z)$ depends not only on z but also on the rth order jet of a local coordinate at z. Thus for any surface X and for any $\psi \in H$ the formulae

$$i \frac{d}{dz} \psi_X(z) = (L_1\psi)_X(z) \ , \qquad i \frac{d}{d\bar{z}} \psi_X(z) = (\bar{L}_1\psi)(z) \ ,$$

make sense and are valid, where L_1 and $\bar{L}_1$ are the usual Virasoro operators on H representing the vector field $id/dz = e^{-i\theta} d/d\theta$ in the left- and right-hand actions of $\text{Vect}(S^1)$ on H. These formulae can be regarded as an infinitesimal version of (9.2). They hold because

$$idU_{X_z} = U_{X_z} \circ (L_1 dz + \bar{L}_1 d\bar{z}) \ .$$

In particular, in a holomorphic theory each primary field ψ gives rise to a <u>holomorphic</u> operator-valued differential form ψ_X, and even if ψ is not primary ψ_X is a holomorphic function in the domain of a given local parameter. If $X = X_{ab}$ is the annulus $\{z : a \leqslant |z| \leqslant b\}$ and ψ is of type $(p,0)$ we can always write ψ_X as a Laurent series

$$\psi_X = U_{1b} \{ \sum_{k \in Z} \psi_k z^{k-p} dz^{\otimes p} \} \ U_{a1} \ , \qquad (9.3)$$

where ψ_k is an unbounded operator in H and U_{a1} and U_{1b} are the operators associated to the annuli X_{a1} and X_{1b}. The advantage of this notation is that ψ_k depends only on ψ and not on a,b. The reader should be warned that in usual terminology it is the unbounded operator $z) = \Sigma \ \psi_k z^{k-n}$ which is called the field operator, rather than my ψ_X. In terms of the ψ_k the Ward identity (9.2) becomes

$$[L_n, \psi_k] = -i(k + n - pn)\psi_{k+n} \ . \qquad (9.4)$$

In a holomorphic theory the operators ψ_k are completely characterized in terms of $\psi \ \epsilon$ H by the property (9.4) together with

$$\psi_p \Omega = \psi \; . \tag{9.5}$$

The energy-momentum tensor

The most important fields in any theory are the energy-momentum tensors T and $\overline{T}$. These are secondary fields of type $(2,0)$ and $(0,2)$ respectively: T transforms under $\mathcal{E}_0$ by

$$U_f . T = f'(0)^2 T + c_R S_f(0) \Omega \; , \tag{9.6}$$

where $S_f(0) = f'''(0)/f'(0) - 3/2(f''(0)/f'(0))^2$ is the Schwarzian derivative of f, and c_R is the central charge. It should be noticed that

$$f \; \longmapsto \; \begin{pmatrix} f'(0)^2 & c \, S_f(0) \\ 0 & 1 \end{pmatrix}$$

is the only two dimensional representation of $\mathcal{E}_0$ which combines the one dimensional representations $f \mapsto f'(0)^2$ and $f \mapsto 1$.

To define T and $\overline{T}$ we consider the variation of the vacuum map $U_D : \mathbb{C} \to H$ when the complex structure on the disc D is changed. An infinitesimal change of structure is an element of $V = \text{Vect}_{\mathbb{C}}(S^1)/\text{Vect}(D)$, so the variation is a map $V \to H$ which is $\mathbb{R}$-linear (but not $\mathbb{C}$-linear unless the theory is holomorphic). We can regard it as a $\mathbb{C}$-linear map $V \oplus \overline{V} \to H$. In V there is an eigenvector of $\mathcal{E}_0$ of type $(2,0)$ represented by $z^{-1}d/dz$. The image of this in H is denoted by T, and the image of the corresponding element of $\overline{V}$ by $\overline{T}$. (I use the traditional notation T, $\overline{T}$ with misgivings, as $\overline{T}$ is not

necessarily the complex conjugate of T.) In terms of the Virasoro
generators we have $T = iL_2\Omega$ and $\bar{T} = -i\bar{L}_2\Omega$. To obtain the transformation
properties of T and $\bar{T}$ under $\mathcal{E}_0$ we must notice that the map $V \to H$ is
not $\mathcal{E}_0$-equivariant. Because $Vect(S^1)$ acts projectively on H we
actually have an $\mathcal{E}_0$-equivariant map $\mathbb{C} \oplus V \to H$, where $\mathbb{C}$ is that of
the central extension of $Vect(S^1)$. Taking this into account we find
that T transforms by (9.6). (See [S2] (   ).)

On a Riemann surface X the vector T gives rise not to an operator-
valued quadratic differential $T(z)dz^2$, but rather to a projective
connection, i.e. when the local parameter is changed from z to ζ the
operator $T(z)dz^2$ becomes

$$T(\zeta)d\zeta^2 + c_R S_\zeta U_X ,$$

where S_ζ is the Schwarzian derivative of the change of parameters.

The significance of the energy-momentum tensor is that it
describes the variation of the operator $U_X : H^{\otimes m} \to H^{\otimes n}$ associated to
a surface X when the complex structure of X is changed. An
infinitesimal change of structure corresponds (see (4.1)) to an
element of $Vect_{\mathbb{C}}(\partial X)/Vect(X)$.

Proposition (9.7). (i) The energy-momentum tensor $T(z)dz^2$ is a
holomorphic projective connection on the interior of any surface. It
possesses a distributional boundary value. Similarly, $\bar{T}(z)d\bar{z}^2$ is
antiholomorphic.

(ii) For the infinitesimal deformation of X defined by a complex
vector field ξ along ∂X we have

$$\delta U_X = \int_{\partial X} \xi(z).T(z)dz + \int_{\partial X} \overline{\xi(z)}.\overline{T}(z)d\overline{z} .$$

Remark. If $T(z)dz^2$ were a quadratic differential it would pair with the vector field ξ to give a 1-form which could be integrated around ∂X. But ξ is really an element of the central extension of $Vect_{\mathbb{C}}(\partial X)$, for it represents a deformation of a surface equipped with a chosen point in its determinant line. Thus ξ pairs with a projective connection, for the projective connections are precisely the dual of the central extension of $Vect(\partial X)$. ([S2](p.335).) If ξ extends
also [S3]
holomorphically over X the integrals above vanish by Cauchy's theorem because T (resp. $\overline{T}$) is holomorphic (resp. antiholomorphic). Conversely, the fact that $\delta U_X = 0$ when ξ extends holomorphically, i.e. the fact that the theory is conformally invariant, implies that T is holomorphic.

Proof of (9.7). Consider the variation of U_X in the space E of trace-class operators $H^{\otimes m} \to H^{\otimes n}$ when the structure of X is changed. This is expressed by a real-linear map

$$Vect_{\mathbb{C}}(\partial X)/Vect(X) \to E \qquad (9.8)$$

when a section of the central extension of $Vect_{\mathbb{C}}(\partial X)$ has been chosen. The dual of $Vect_{\mathbb{C}}(\partial X)/Vect(X)$ is the space Ω of holomorphic quadratic differentials on X with distributional boundary values. So (9.8) corresponds to an element of $E \otimes_{\mathbb{R}} \Omega = E \otimes_{\mathbb{C}} (\Omega \oplus \overline{\Omega})$. This means that we have a formula

$$\delta U_X = \int_{\partial X} (\xi t_X + \overline{\xi} \overline{t}_X) \qquad (9.9)$$

for <u>some</u> naturally defined operator-valued objects $t_X, \bar{t}_X$. Applying
(9.9) when $X = D$ and $\xi = z^{-1}d/dz$ we find that $t_D(0) = T_D(0)$ and
$\bar{t}_D(0) = \bar{T}_D(0)$. We then deduce that $t_X(z) = T_X(z)$ and $\bar{t}_X(z) = \bar{T}_X(z)$ in
all cases by using the naturality property (9.2).

<u>Corollary (9.10)</u>. If X is an annulus $\{z : a \leqslant |z| \leqslant 1\}$ we can write the
energy-momentum tensor in terms of the Virasoro generators:

$$T(z) = i\{ \sum_{k \in Z} z^{k-2} L_k \} \circ U_X .$$

Infinitesimal automorphisms

An important role of field operators is to describe the
infinitesimal automorphisms and deformations of a theory. An
automorphism of a theory based on a vector space H evidently means an
invertible operator $A : H \to H$ which preserves the hermitian form and
satisfies

$$A^{\otimes n} \circ U_X = U_X \circ A^{\otimes m}$$

for each morphism X from C_m to C_n in $\mathcal{C}$. An <u>infinitesimal automorphism</u>
is accordingly a (densely defined) skew-hermitian operator δ in H such
that

$$\{ \sum_{i=1}^{n} 1 \otimes \ldots \otimes \delta \otimes \ldots \otimes 1 \} \circ U_X = U_X \circ \{ \sum_{j=1}^{m} 1 \otimes \ldots \otimes \delta \otimes \ldots \otimes 1 \}$$

$$(9.11)$$

where the factors δ occur in the ith and j^{th} places on the left and
right.

Proposition (9.12). In a holomorphic theory a real primary field ψ of type (1,0) defines an infinitesimal automorphism δ_ψ whose domain of definition includes $U_X H$ for any annulus X, and which is characterized by

$$\delta_\psi \circ U_X = \int_\gamma \psi_X(z)\,dz \; , \qquad (9.13)$$

where γ is any simply closed curve going once around the annulus.

Proof: The right-hand side is independent of γ because $\psi_X(z)\,dz$ is a holomorphic 1-form, and the formula (9.2) shows that δ_ψ is independent of X and satisfies (9.11).

In a theory which is not chiral the 1-form $\psi_X(z)\,dz$ defined by a primary field of type (1,0) need not be closed, i.e. holomorphic. In that case an infinitesimal automorphism is defined by a pair of real primary fields (ψ_L, ψ_R) of types (0,1) and (1,0) such that $L_1\psi_L = \bar{L}_1\psi_R$. Then the 1-form $\psi_L(z)\,d\bar{z} + \psi_R(z)\,dz$ is closed, and its integral replaces the right-hand side of (9.13).

In the literature it is assumed that <u>any</u> infinitesimal automorphism of a field theory is given by a primary field in the way described. This may well follow from our axioms; if it does not then another axiom should probably be added to ensure it. The idea of such an axiom would be to express the fact that the space H associated to a circle is, in some sense which I do not know how to make precise, a continuous tensor product of spaces associated to the infinitesimal elements of the circle.

For the rest of the discussion of infinitesimal automorphisms I shall confine myself to holomorphic theories. The infinitesimal automorphisms evidently form a Lie algebra, and there is an induced Lie algebra structure on the finite dimensional space of primary fields of type $(1,0)$, for if ψ_1 and ψ_2 are such fields then

$$[\psi_1, \psi_2] = \delta_{\psi_1} \psi_2 \ .$$

__Proposition (9.14)__. Let $\mathfrak{g}$ be the Lie algebra of real primary fields of type $(1,0)$ in a unitary holomorphic theory. Then the loop algebra $L\mathfrak{g}$ acts projectively on H, intertwining with the action of $\mathrm{Diff}^+(S^1)$. Conversely, if $L\mathfrak{g}$ acts in this way then $\mathfrak{g}$ is contained in the Lie algebra of primary fields of type $(1,0)$.

__Remark__. Because we are assuming the hermitian form on H is positive-definite $\mathfrak{g}$ will be the Lie algebra of a __compact__ group.

One consequence of Proposition (9.14) is that the field theories we shall construct in §11 from the loop groups of compact groups are genuinely different from one another. The proposition also gives a convenient criterion for deciding when the group of automorphisms of a theory is finite: there must be no primary fields of type $(1,0)$. But the most important positive application of the proposition is the "vertex-operator" construction of the basic representation of the loop group LG when G is simply laced: we shall return to this in §12.

Proposition (9.14) is deduced by one of the very characteristic arguments of conformal field theory from the following "operator-product expansion".

Proposition (9.15). If ψ_1 and ψ_2 are primary fields of type $(1,0)$ in a unitary holomorphic field theory then on any Riemann surface X we have

$$(\psi_1\psi_2)_X (z,\zeta) = \langle\psi_1,\psi_2\rangle U_X \frac{dzd\zeta}{(z-\zeta)^2} + \frac{1}{2\pi i}[\psi_1,\psi_2]_X\frac{d\zeta}{z-\zeta} + \varphi(z,\zeta) ,$$

where φ is holomorphic everywhere on $X \times X$, for some invariant inner product $\langle\ ,\ \rangle$ on the primary fields.

Proof of (9.14) using (9.15). Let $\psi_1,\ldots,\psi_n$ be a basis for $\mathfrak{g}$. We shall restrict ourselves to real analytic elements $\xi = \Sigma \xi_i\psi_i$ of the loop algebra $L\mathfrak{g}$. Then ξ is the boundary value of a holomorphic function ξ defined in an annulus X, and we can define the action δ_ξ of ξ on H by the formula

$$\delta_\xi \circ U_X = \int_\gamma \Sigma \xi_i\psi_{i,X}(z)dz ,$$

analogous to (9.13). Using the basic functoriality property (9.2) we can write the commutator $[\delta_\xi,\delta_\eta] \circ U_X$ in the form

$$\left\{\int_{\gamma_1} dz \int_{\gamma_2} d\zeta - \int_{\gamma_3} dz \int_{\gamma_2} d\zeta\right\}(\Sigma \xi_i(z)\eta_j(\zeta)(\psi_i\psi_j)_X(z,\zeta))$$

where $\gamma_1,\gamma_2,\gamma_3$ are three non-intersecting simple closed curves going once round the annulus, with γ_1 outside γ_2 and γ_2 outside γ_3. Let us first perform the integral over z, holding ζ fixed. Because the integrand is holomorphic for $z \neq \zeta$ and $\gamma_1 - \gamma_3$ is homologous to a small circle around ζ the result is the residue of the integrand at $z = \zeta$, which by (9.15) is

$$\sum \xi_i(\zeta)\eta_j(\zeta)[\psi_i,\psi_j]_X(\zeta) \;+\; \sum \langle\psi_i,\psi_j\rangle\xi_i'(\zeta)\eta_j(\zeta)U_X \;.$$

Integrating this around γ_2 gives

$$[\delta_\xi, \delta_\eta] \;=\; \delta_{[\xi,\eta]} \;+\; c(\xi,\eta)1 \;,$$

where

$$c(\xi,\eta) = \int \langle\xi'(\zeta),\eta(\zeta)\rangle d\zeta$$

is a cocycle defining a central extension of $L\mathfrak{g}$.

Proof of (9.15). First observe that the terms on the right-hand side of (9.16) make invariant sense, e.g. that the 2-form $dzd\zeta/(z-\zeta)^2$ on $X \times X$ is independent of the local parameter modulo holomorphic forms. Because of this we can assume that X is the standard disc D, and that $\zeta = 0$. The element $(\psi_1,\psi_2)_D(z,0)$ of H can be expanded

$$(\psi_1\psi_2)_D(z,0) = \sum_{k \in \mathbf{Z}} A_k z^k \tag{9.17}$$

with $A_k \in H$. Let R_α denote the automorphism of D which rotates it through α. Its action on $(\psi_1,\psi_2)_D(z,0)$ is given by

$$R_\alpha\{(\psi_1\psi_2)_D(z,0)\} = e^{2i\alpha}(\psi_1\psi_2)_D(e^{-i\alpha}z,0) \ .$$

Applying this to (9.17) we find that $R_\alpha A_k = e^{(2-k)i\alpha}A_k$, i.e. that A_k is an element of H of degree 2-k. In a unitary theory there are no fields of negative degree, and only the vacuum vector has degree 0. So

$$(\psi_1\psi_2)_D(z,0) = \lambda\Omega z^{-2} + A_1 z^{-1} + \text{(holomorphic)} \ ,$$

for some $\lambda \in \mathbb{C}$. But by definition we have

$$\int_{S^1} (\psi_1\psi_2)_D(z,0) = \delta_{\psi_1\psi_2}$$

$$= [\psi_1,\psi_2] \ ,$$

so $2\pi i A_1 = [\psi_1,\psi_2]$. The proof is completed by observing that $(\psi_1,\psi_2) \mapsto \lambda$ is necessarily an invariant inner product on $\mathfrak{g}$.

Infinitesimal deformations

When one has a continuously varying family of conformal field theories one may as well assume that the hermitian vector space H is fixed and that all the variation takes place in the operators U_X associated to surfaces X. It is simplest to think in terms of Definition (4.4). Then an infinitesimal deformation will be a rule which associates to each surface X a vector $\theta_X \in H_{\partial X}$ such that

(i) $\theta_{X \sqcup Y} = \theta_X \otimes \Omega_Y + \Omega_X \otimes \theta_Y$,

and

(ii) $\theta_X \mapsto \theta_{\check{X}}$ for each sewing map $X \to \check{X}$.

In analogy with (9.12) we have

<u>Proposition (9.18)</u>. In any theory a primary field θ of type (1,1)
defines an infinitesimal deformation by the formula

$$\Theta_X = \int_X \theta_X(z)\,dz \wedge d\bar{z} \; .$$

As with automorphisms it is usually assumed that any infinitesimal
deformation arises from a primary field in this way, but I do not know
a proof. Thus a chiral theory should be automatically rigid. The
space of deformations of the σ-model of a torus will be considered in
§10.

<u>Examples</u>

We shall consider some fields in the holomorphic fermionic
theories $\mathfrak{J}(\Omega^\alpha)$.

(i) The most obvious primary fields are the vectors

$$\psi^{(m)} = \psi_{m-\alpha} \wedge \psi_{m-\alpha-1} \wedge \psi_{m-\alpha-2} \wedge \cdots \in H \; .$$

Evidently $\psi^{(0)}$ is the vacuum vector Ω, with degree 0. In general $\psi^{(m)}$
has degree $\tfrac{1}{2}m(m-2\alpha+1)$, which is negative if m is between 0 and $2\alpha-1$.
These are exactly the fields $\psi^{(m)}$ which were described in a different
way in (8.20). That is obvious in the case of $\psi^{(1)} = \psi$ from the
characterization of ψ_X by means of (9.4) and (9.5); we shall not pursue
the general case here.

(ii) We saw in §8 that the theory $\mathfrak{F}(\Omega^{\frac{1}{2}})$ is Z-graded[1]. This means that the circle group $\mathbb{T}$ acts on it as a group of automorphisms. The infinitesimal generator is the primary field

$$J = \psi_{\frac{1}{2}} \wedge \psi_{-3/2} \wedge \psi_{-5/2} \wedge \cdots$$

$$= \psi_{\frac{1}{2}} \, \psi_{-\frac{1}{2}}^{*} \, \Omega$$

in $\mathfrak{F}(\Omega^{\frac{1}{2}}(S_A^1))$, which is called the <u>current</u>. (In the second expression for J we write $\psi_{\frac{1}{2}}$ for the operator of multiplication by $\psi_{\frac{1}{2}}$ on $\mathfrak{F}(\Omega^{\frac{1}{2}})$, and $\psi_{-\frac{1}{2}}^{*}$ for the antiderivation corresponding to the dual basis element $_{\text{to } \psi_{-\frac{1}{2}}}$.) The field J provides us with an action of the whole loop group $L\mathbb{T}$ on $\mathfrak{F}(\Omega^{\frac{1}{2}})$ extending the action of $\mathbb{T}$. (In physical language the grading or $\mathbb{T}$-action is the <u>charge</u>, and the action of the "current group" $L\mathbb{T}$ expresses the fact that charge is <u>local</u>.) To prove that the vector J really does generate the $\mathbb{T}$-action on $\mathfrak{F}(\Omega^{\frac{1}{2}})$ one can write the field $J(z)dz$ on an annulus in the form (9.3).

<u>Proposition (9.19)</u>. For J as above we have on the annulus $X = \{z : a \leqslant |z| \leqslant 1\}$ the relation

$$J(z)dz = \{\Sigma \, J_k z^{k-1} dz\} \circ U_X \,,$$

where

$$J_k = \sum_{r \in Z + \frac{1}{2}} \psi_{k+r} \, \psi_r^{*} \qquad \text{when } k \neq 0 \,,$$

---

[1] For simplicity I shall ignore the spin structure S_P^1 for the moment, though actually the discussion applies to it just as well.

and

$$J_0 = \sum_{r>0} \psi_r \psi_r^* - \sum_{r<0} \psi_r^* \psi_r \ .$$

One must check that the expressions for the J_k given here are well-defined operators on vectors of finite energy in H. Granting that, the proposition is easily proved by checking the conditions (9.4) and (9.5), using $[L_n, \psi_m] = -i(m + \frac{1}{2}n)\psi_{m+n}$. The operator J_0 is by definition the infinitesimal automorphism δ_J of H corresponding to J, and it obviously multiplies each basis vector of H by its Fock space degree.

(iii) In $\mathfrak{F}(\Omega^\alpha)$ for any value of α there is a vector $J = \psi_{1-\alpha} \psi_{-\alpha}^*$ of degree 1 analogous to the current in $\mathfrak{F}(\Omega^{\frac{1}{2}})$. When $\alpha \neq \frac{1}{2}$ it is not a primary field, for

$$L_{-1} J = i(1-2\alpha)\Omega \ .$$

This means that under the action of a holomorphic map $f : D \to \mathring{D}$ such that $f(0) = 0$ the vector J transforms by

$$U_f . J = f'(0)J - \tfrac{1}{2}(1-2\alpha)f''(0)f'(0)^{-1}\Omega$$

(cf. (9.6)), or equivalently that we have an operator-valued 1-form $J(z)dz$ in each coordinate patch on any surface X, but when one changes coordinates by $z = z(\zeta)$ the form changes to

$$J(z(\zeta))z'(\zeta)d\zeta \quad - \quad \tfrac{1}{2}(1-2\alpha) \; U_X(z''(\zeta)/z'(\zeta))d\zeta \; .$$

We can still write $J(z)dz$ in the form (9.19), and the operator J_0 in H defines the Fock space degree, which is still called "charge". But the fact that $J(z)dz$ is not a 1-form corresponds to the failure of the operators U_X to preserve charge which was pointed out in (8.13). From the present viewpoint the charge anomaly can be calculated as follows. Let us suppose that all the boundary circles of X are outgoing. We can choose a holomorphic connection in the holomorphic tangent bundle to X. It is given by a 1-form $\gamma(z)dz$ in each coordinate patch, but a change of coordinates $z = z(\zeta)$ replaces $\gamma(z)dz$ by

$$\gamma(z(\zeta))z'(\zeta)d\zeta \quad + \quad (z''(\zeta)/z'(\zeta))d\zeta \; .$$

Then

$$(J(z) \quad + \quad \tfrac{1}{2}(1-2\alpha)\gamma(z)U_X)dz$$

is a global holomorphic 1-form on X with values in $H(\partial X)$. Applying the operator J_0 to $U_X \in H(\partial X)$ gives, by definition,

$$\frac{1}{2\pi i} \int_{\partial X} J(z)z^{-1}dz \; .$$

By Cauchy's theorem this equals

$$\{\frac{1}{4\pi} \; (2\alpha-1) \int_{\partial X} \gamma(z)dz\}U_X \; .$$

We can suppose the connection γ arose from a trivialization of TX. Then $\int_{\partial X} \gamma(z)dz$ is the angle by which the tangent vector d/dz to ∂X rotates relative to the trivialization when one travels around ∂X. This angle is $4\pi(g + m - 1)$, where g is the genus of X and m the number of boundary circles. So the charge anomaly is $(2\alpha-1)(g + m - 1)$, in agreement with (8.12).

BRST cohomology

For a holomorphic theory H with central charge 26 we define the BRST cohomology in the following way. First tensor the theory H with the "ghost" theory $\mathfrak{F}(\Omega^2)$. The Fock space grading of $\mathfrak{F}(\Omega^2)$, which was called the "charge" in the preceding example, induces a grading of $H = H \otimes \mathfrak{F}(\Omega^2)$ which is now called the "ghost number". We shall show that H contains a primary field Q of degree 1 whose associated infinitesimal automorphism δ_Q raises the ghost number by 1 and satisfies $\delta_Q^2 = 0$. The cohomology of H with respect to the differential δ_Q is the BRST cohomology H_{BRST}. Now the Fock space $\mathfrak{F}(\Omega^2(S^1))$ is a renormalized version of the exterior algebra on $\Omega^2(S^1)$, which is the dual of the Lie algebra of vector fields $\text{Vect}(S^1) = \Omega^{-1}(S^1)$. The differential δ_Q is similarly, as we shall see, a renormalization of the standard differential of Lie algebra cohomology, and so $H_{BRST}(S^1)$ is a renormalized version of the cohomology of $\text{Vect}(S^1)$ with coefficients in the module $H(S^1)$. In fact one can say more: for $\mathfrak{F}(\Omega^2(S^1))$ is a module over the exterior algebra on $\Omega^2(S^1)$, and this makes $H_{BRST}(S^1)$ a module over the ordinary cohomology of $\text{Vect}(S^1)$ with coefficients in $\mathbb{C}$.

At the moment, however, I want just to point out the field-theoretic aspects of H_{BRST}. Because δ_Q is an infinitesimal automorphism the operator U_X associated to a surface X in the theory H commutes with δ_Q, i.e. it is a homomorphism of cochain complexes. It induces a map of the cohomology H_{BRST} which changes the degree by the ghost number anomaly. This is still not quite what we need. The surfaces X of a particular topological type α with m incoming and n outgoing circles form the moduli space $\mathscr{C}_\alpha$ of §4 whose tangent space at X is $\text{Vect}_{\mathbb{C}}(\partial X)/\text{Vect}(X)$. The operator U_X is really an element of $\hat{H}(\partial X) = H(\partial X) \otimes \mathfrak{I}(\Omega^2(\partial X))$, which is a module over the exterior algebra on $\text{Vect}_{\mathbb{C}}(\partial X)$, the dual space of $\Omega^2(\partial X)$. The vector U_X is annihilated by the subspace $\text{Vect}(X)$ of $\text{Vect}_{\mathbb{C}}(\partial X)$. We can therefore define for each p a holomorphic differential form ω_p of degree p on $\mathscr{C}_\alpha$, with values in $\hat{H}(\partial X)$, by

$$\omega_p(X; \xi_1, \ldots, \xi_p) = \xi_1 \xi_2 \cdots \xi_p U_X ,$$

where $\xi_i \in \text{Vect}_{\mathbb{C}}(\partial X)$. The fact that $\delta_Q U_X = 0$ - because δ_Q is an infinitesimal automorphism - has the following generalization.

<u>Proposition (9.20)</u>. We have

$$d\omega_{p-1} = -\delta_Q \omega_p ,$$

where d denotes the exterior derivative of forms on $\mathscr{C}_\alpha$. Alternatively expressed, if the boundary circles of X are regarded as incoming rather than outgoing, the forms $\{\omega_p\}$ define a map of cochain complexes

$$\hat{H}(\partial X) \rightarrow \Omega_{\text{hol}}(\mathscr{C}_\alpha) \tag{9.21}$$

Proof: An element $\xi \in \text{Vect}_{\mathbb{C}}(\partial X)$ acts on H in two ways: by the exterior multiplication used above, and also by its action as an element of the Lie algebra of Diff(∂X). I shall write L_ξ for the latter action, and (for this proof) i_ξ for the former. The two are related by the usual Cartan and naturality formulae

$$L_\xi = [\delta_Q, i_\xi] \ , \quad i_{[\xi,\eta]} = [L_\xi, i_\eta]$$

where [,] is the graded commutator.

I shall give the proof of (9.20) when p = 2. We have

$$d\omega_1(X;\xi,\eta) = L_\eta \omega_1(X;\xi) - L_\xi \omega_1(X;\eta) + \omega_1(X;[\xi,\eta])$$

$$= (L_\eta i_\xi - L_\xi i_\eta + i_{[\xi,\eta]})U_X$$

$$= (L_\eta i_\xi - i_\eta L_\xi)U_X$$

$$= \delta i_\eta i_\xi U_X$$

$$= -\delta\omega_2(X;\xi,\eta) \ .$$

The first line of this calculation is the definition of the exterior derivative, regarding ξ and η as vector fields on $\mathcal{C}_\alpha$. One can identify the Lie derivative L_ξ for forms on $\mathcal{C}_\alpha$ with the operator L_ξ on $\hat{H}$ because $X \mapsto U_X$ is equivariant with respect to the action of $\text{Vect}_{\mathbb{C}}(\partial X)$.

Let us now suppose that X is a surface with m boundary circles, all incoming. We readily check that the cochain map (9.21) raises degree by 3g-3, where g is the genus of X. It is also compatible with the action of $\text{Vect}_{\mathbb{C}}(\partial X)$ - by both kinds of operators L_ξ and i_ξ. The action of $\text{Vect}_{\mathbb{C}}(\partial X)$ is the infinitesimal version of the natural action

of the semigroup $\mathscr{A}^m$. Inside $\mathscr{A}$ there is the subsemigroup $\mathscr{E}$ of

holomorphic embeddings $f : D \to \overset{\circ}{D}$. The quotient space of $\mathscr{C}_\alpha$ by $\mathscr{E}^m$ is

the moduli space $\mathcal{M}_g$ of closed surfaces of genus g, and has complex

dimension 3g-3. Now the vacuum vector in $H(\partial X)$ is basic for $\mathscr{E}^m$ (i.e.

annihilated by L_ξ and i_ξ for $\xi \in \text{Vect}(D)$). Its image under (9.21) is

therefore a form on $\mathscr{C}_\alpha / \mathscr{E}^m = \mathcal{M}_g$ which is holomorphic, and of the top

dimension 3g-3. It is natural to call this the <u>partition form</u> of the

chiral theory H. A more physical theory in which both chiralities were

present would have a partition form of bidegree (3g-3,3g-3), and this

could be integrated over $\mathcal{M}_g$. That is the situation in string theory.

Besides the vacuum vector we can consider other classes in

$H_{BRST}(\partial X) = H_{BRST}(S^1)^{\otimes m}$. For the theories which arise in practice all

elements of $H_{BRST}(S^1)$ are represented by elements of $\hat{H}(S^1)$ which are

basic for the action of the semigroup

$$\mathscr{E}_0 = \{f \in \mathscr{E} : f(0) = 0\} .$$

(This is true whenever $H(S^1)$ is a free module over the enveloping

algebra $U(\mathfrak{a})$ of the Lie algebra $\mathfrak{a}$ of $\mathscr{E}_0$, which is spanned by the

Virasoro generators $\{L_k\}_{k \geqslant 1}$. Cf. [FGZ].) The quotient space $\mathscr{C}_\alpha / \mathscr{E}_0^m$

is the moduli space $\mathcal{M}_{g,m}$ of surfaces with m marked points. It has

complex dimension 3g-3+m. If we choose an element ψ_i of ghost number 1

in $H_{BRST}(S^1)$ for each boundary circle then the image

$$\langle \psi_1 \psi_2 \cdots \psi_m \rangle$$

of $\psi_1 \otimes \cdots \otimes \psi_m$ under the map (9.21) is a top dimensional holomorphic form on $\mathcal{M}_{g,m}$, well defined up to the addition of an exact form. For this reason the elements of ghost number 1 in H_{BRST} are regarded as (chiral) "physical states".

We now return to the definition of the primary field Q and the verification of its properties. We can write it explicitly

$$Q = J \otimes \psi_{-1}\Omega + \Omega \otimes (\psi_0\psi_{-1}\psi^*_{-2} - 3\psi_1)\Omega \qquad (9.22)$$

in $\hat{H} = H \otimes \mathcal{F}(\Omega^2)$. As the energies in $\mathcal{F}(\Omega^2)$ are bounded below by -1 we need only check that $L_{-1}Q = L_{-2}Q = 0$ to see that Q is primary, and that is easily done by using the relation

$$[L_k, \psi_m] = -i(m+2k)\psi_{m+k}$$

of (8.14), together with the formula $L_{-2}T = 13\Omega$ in H which expresses that H has central charge $c = 26$. To understand where (9.22) comes from, however, it is best to recall the formula for the differential δ in the cochain complex $M \otimes \wedge \mathfrak{g}^*$ of a Lie algebra $\mathfrak{g}$ with coefficients in a $\mathfrak{g}$-module M. Let $\{\xi_k\}$ be a basis for $\mathfrak{g}$, and $\{\alpha_k\}$ the dual basis of $\mathfrak{g}^*$. Then

$$\delta = \Sigma \xi_k \otimes e_k + \tfrac{1}{2}\Sigma 1 \otimes e_k\xi_k ,$$

where e_k is the operation of multiplication by α_k on $\wedge \mathfrak{g}^*$, and ξ_k denotes the action of ξ_k on either M or $\wedge \mathfrak{g}^*$. The analogous operator on $\hat{H} = H \otimes \mathcal{F}(\Omega^2)$, in the usual notation, is

$$\delta_Q = \sum_{k \in \mathbb{Z}} L_k \otimes \psi_{-k} + \tfrac{1}{2} \sum_{k \geqslant -1} 1 \otimes \psi_k L_{-k} + \tfrac{1}{2} \sum_{k < -1} 1 \otimes L_{-k}\psi_k ,$$

(9.23)

where we have renormalized by "adding the infinite term"

$$\tfrac{1}{2} \sum_{k < -1} 1 \otimes [L_{-k}, \psi_k] = (-\tfrac{1}{2} \sum_{k < -1} k) \cdot \psi_0$$

to ensure that $\delta_Q(\Omega \otimes \Omega) = 0$. In the light of (9.23) one finds, after a little experiment, that the sequence of operators

$$Q_m = \sum_{k \in \mathbb{Z}} L_{k+m} \otimes \psi_{-k} + \tfrac{1}{2} \sum_{k \geqslant -1} 1 \otimes \psi_k L_{-k+m}$$

$$+ \tfrac{1}{2} \sum_{k < -1} 1 \otimes L_{-k+m}\psi_k + 3/2\ m(m+1)\psi_m$$

satisfies the relations $[L_p, Q_m] = -im\ Q_{m+p}$ of (9.4) as well as $Q_0 = \delta_Q$. Finally, we obtain the expression (9.22) by setting

$$Q = Q_1(\Omega \otimes \Omega) .$$

To conclude we need to know that δ_Q raises the ghost number by 1, and also that $\delta_Q^2 = 0$. The first is obvious from (9.23). The second is equivalent to $\delta_Q Q = 0$, for δ_Q^2 is the graded commutator $[\delta_Q, \delta_Q]$. One knows a priori that δ_Q^2 is an infinitesimal automorphism of the theory, but I do not know a non-computational proof that it vanishes.

§10. The σ-model for a torus

The Hilbert space

We shall now construct the field theory corresponding to strings moving in a torus T. As was explained in the introduction, the essential point is to choose a vector space to play the role of the space of square-summable functions on LT. If A is a locally compact abelian group there is a simple group-theoretic characterization of the Hilbert space $H = L^2(A)$, and our strategy is to adopt this characterization as a definition in the case of LT.

In the finite dimensional case $L^2(A)$ possesses a unitary action of A by translations and a unitary action of the Pontrjagin dual group $\hat{A}$ by multiplication operators. These actions fit together to define an irreducible unitary representation of the Heisenberg group $(A \times \hat{A})^{\sim}$, which is the central extension of $A \times \hat{A}$ by T associated with the pairing $A \times \hat{A} \to T$, i.e. the extension whose cocycle c is given by

$$c((a_1,\alpha_1),(a_2,\alpha_2)) = <\alpha_1,a_2> . \qquad (10.1)$$

The space $L^2(A)$ is characterized - up to a scalar multiplication - as the unique faithful irreducible representation of the Heisenberg group.

To generalize this, let us begin with a Riemannian torus $T = \mathfrak{t}/\Lambda$, where $\mathfrak{t}$ is a finite dimensional real vector space with a positive inner product, and Λ is a lattice in $\mathfrak{t}$. The dual lattice

$$\Lambda^* = \{\xi \in \mathfrak{t} : <\xi,\eta> \in Z \text{ for all } \eta \in \Lambda\}$$

gives rise to the dual torus $T^* = \mathfrak{t}/\Lambda^*$.

The loop groups LT and LT* are in duality under the bilinear
pairing $<< , >>$: LT $\times$ LT* $\rightarrow$ T defined by

$$<<f,g>> = \tfrac{1}{2} \int_0^1 (<f,dg> - <df,g>) + \tfrac{1}{2}(<f(0),\Delta_g> - <\Delta_f,g(0)>), \quad (10.2)$$

where T is regarded as R/Z, and an element f of LT is regarded as a map
$f : R \rightarrow \mathfrak{t}$ satisfying

$$f(t + 1) = f(t) + \Delta_f$$

for some $\Delta_f \in \Lambda$. (Thus f is defined only modulo the addition of a
constant element of Λ.) The pairing $<< , >>$ is nondegenerate, and is
invariant under $Diff^+(S^1)$. The groups LT and LT* are thus essentially
Pontrjagin duals of each other.

We can now define a Hilbert space H which is a faithful
irreducible representation of the Heisenberg group $\tilde{\Pi}$ formed from
$\Pi =$ LT $\times$ LT* by using the pairing $<< , >>$. Because $\tilde{\Pi}$ is infinite
dimensional it has many faithful irreducible representations. To
single one out one must choose a positive polarization of Π.

Definition (10.3). (i) A polarization of a real vector space E with a
skew form S is an equivalence class of operators $J : E \rightarrow E$ such that

(a) $S(J\xi,J\eta) = S(\xi,\eta)$, and

(b) $J^2 = -1$ modulo trace class operators.

Two such operators J are equivalent if their difference is of trace
class.

(ii) A polarization J is positive if the quadratic form
$\xi \mapsto S(\xi,J\xi)$ is positive-definite on a subspace of E of finite
codimension.

(iii) If Π is an abelian Lie group with a skew pairing $c : \Pi \times \Pi \to T$, and $\pi_0(\Pi)$ and $\pi_1(\Pi)$ are finitely generated, then a polarization of Π is a polarization of its Lie algebra $\text{Lie}(\Pi)$.

In our case $\Pi = LT \times LT*$, and

$$LT = T \times \Lambda \times V$$
$$LT* = T* \times \Lambda* \times V \; ,$$

where V is the real vector space $(L\mathfrak{t})/\mathfrak{t}$. A polarization of Π is the same as a polarization of $V \oplus V$, and this space has a canonical polarization given by the decomposition $V_{\mathbb{C}} = V_{\mathbb{C}}^+ \oplus V_{\mathbb{C}}^-$ into positive and negative frequency. This gives us a definite irreducible unitary representation of $\widetilde{\Pi}$ on the Hilbert space

$$H = L^2(T \times \Lambda) \otimes S(V_{\mathbb{C}}^+ \oplus V_{\mathbb{C}}^+) \; .$$

<u>Remark</u>. A positive polarization of $V \oplus V$ is the same thing as a quadratic form $q : V \to \mathbb{C}$ such that $\text{Im } q$ is positive definite. In the present case $q(f) = i \sum n \langle a_{-n}, a_n \rangle$ if $f = \sum a_n e^{in\theta}$. One can identify $S(V_{\mathbb{C}}^+ \oplus V_{\mathbb{C}}^+)$ with a completion of the space of all functions on V of the form $f \mapsto p(f)e^{iq(f)}$, where $p : V \to \mathbb{C}$ is a polynomial, so it is a very natural candidate for $L^2(V)$.

Because $\text{Diff}^+(S^1)$ acts on $\widetilde{\Pi}$ by automorphisms, and preserves the polarization class, it acts on the irreducible representation H. (In principle the action could be projective, but in fact is not.) Orientation-reversing diffeomorphisms of S^1 reverse the polarization of

Π, so they give antiunitary maps $H \to \bar{H}$. We have therefore the data

to assign a Hilbert space $H_{\partial X}$ to the boundary of any Riemann surface X.

To complete the construction of a conformal field theory we must

associate to X a ray Ω_X in $H_{\partial X}$, and check the conditions (4.4) - but in

fact Ω_X will be defined in a way which makes the conditions manifest.

Before turning to these rays it is worth describing the action of

the conformal group $\mathrm{Conf}(S^1 \times R)$ on H. We recall from §3 that this

group is a covering group of $\mathrm{Diff}^+(S^1) \times \mathrm{Diff}^+(S^1)$, and that the

natural geometric action of $\mathrm{Diff}^+(S^1)$ is diagonal with respect to this

description. Let $\tilde{D}$ denote the simply connected covering group of

$\mathrm{Diff}^+(S^1)$, i.e. the diffeomorphisms $\varphi : R \to R$ such that

$\varphi(\theta + 2\pi) = \varphi(\theta) + 2\pi$. The group $\mathrm{Conf}(S^1 \times R)$ is a quotient of

$\tilde{D} \times \tilde{D}$, and the latter acts on Π by

$$(\varphi_1, \varphi_2)^{-1} \cdot (f,g) = \tfrac{1}{2}(\varphi_1^*(f+g) + \varphi_2^*(f-g), \ \varphi_1^*(f+g) - \varphi_2^*(f-g)) \ .$$

The representation H of Π is induced from the representation of its

identity component Π_0 on $H_{0,0} = S(V_{\mathbb{C}}^+ \oplus V_{\mathbb{C}}^+)$. Under Π_0 it breaks up

$$H = \bigoplus H_{\lambda,\mu} \ ,$$

where (λ,μ) runs through the group of components $\Lambda \times \Lambda^*$, and

$H_{\lambda,\mu} = H_{0,0}$ as a representation of $V \times V \subset \Pi_0$ but is acted on by $T \times T^*$

via the character (λ,μ). The summands $H_{\lambda,\mu}$ are acted on separately by

$\tilde{D} \times \tilde{D}$: the action on $H_{\lambda,\mu}$ is obtained by twisting the action on $H_{0,0}$ by

the crossed homomorphism $\epsilon_{\lambda,\mu} : \tilde{D} \times \tilde{D} \to \Pi_0$ associated to $(\lambda,\mu) \in \Pi$. To

see the action of $\tilde{D} \times \tilde{D}$ on $H_{0,0}$ we rewrite $S(V_{\mathbb{C}}^+ \oplus V_{\mathbb{C}}^+)$ as

$S(W_L) \otimes S(W_R)$, where W_L and W_R are the diagonal and antidiagonal subspaces of $V_{\mathbb{C}}^+ \oplus V_{\mathbb{C}}^+$. Then the left- and right-hand copies of $\eth$ act purely on the left- and right-hand factors $S(W_L)$ and $S(W_R)$. On each the representation is the standard metaplectic representation of $\eth$ described, for example, in [S2]. We thus obtain

Proposition (10.4). (i) The representation of $\mathrm{Conf}(S^1 \times \mathbb{R})$ on H has central charge (d,d), where d is the dimension of T.

(ii) The partition function of the theory, i.e. the trace of the action on H of the standard annulus A_q, is

$$
|\varphi(q)|^{2d} \prod_{\lambda,\mu} q^{\frac{1}{2}||\lambda+\mu||^2} \; \bar{q}^{\frac{1}{2}||\lambda-\mu||^2} ,
$$

where $\varphi(q) = \prod(1-q^n)^{-1}$.

The ray associated to a surface

The general method of prescribing a ray in the Heisenberg representation of $\tilde{\Pi}$ is to give a maximal isotropic subgroup P of $\Pi_{\mathbb{C}}$ and a suitable character $\theta : \tilde{P} \to \mathbb{C}^\times$ which splits the central extension of P induced by $\tilde{\Pi}_{\mathbb{C}}$. The ray is then the eigenvector of $\tilde{P}$ corresponding to the character θ: it exists and is unique providing P is positive and compatible with the polarization of Π in the following sense.

Definition (10.5). (i) P is an _isotropic_ subgroup of $\Pi_{\mathbb{C}}$ if the induced extension $\tilde{P}$ is abelian.

(ii) P is _positive_ if $\mathrm{Im}\, c(\bar{p},p) \geqslant 0$ for all $p \in P$, and $\mathrm{Im}\, c(\bar{p},p) > 0$ except on a compact subgroup of P.

(iii) P is <u>compatible</u> with the polarization of Π if the
endomorphism J_P of Lie $(Π_{\mathbb{C}})$ which is +i (resp. -i) on Lie(P) (resp.
Lie($\overline{P}$)) belongs to the polarization class of Π.

<u>Remarks</u>. (a) We write Im $c(\overline{p},p) \geqslant 0$ rather than $|c(\overline{p},p)| \leqslant 1$ because we
are writing $\mathbb{C}^X$ additively, i.e. as $\mathbb{C}/\mathbb{Z}$.

  (b) A character $\theta : \overline{P} \rightarrow \mathbb{C}^X = \mathbb{C}/\mathbb{Z}$ is the same thing as a map
$\theta : P \rightarrow \mathbb{C}/\mathbb{Z}$ satisfying

$$\theta(p_1 + p_2) = \theta(p_1) + \theta(p_2) + c(p_1, p_2) . \qquad (10.6)$$

We require it to satisfy $2\theta(p) = c(p,p)$.

We apply Definition (10.5) to the group $Π_{\partial X} = \text{Map}(\partial X; T) \times \text{Map}(\partial X; T*)$.
We associate to the surface X the group

$$P_X = \{(f,g) \in \text{Map}(X;T_{\mathbb{C}}) \times \text{Map}(X;T_{\mathbb{C}}^*) : dg = *idf\} .$$

Elements of P_X are determined by their restrictions to ∂X, so P_X is a
subgroup of $Π_{\partial X, \mathbb{C}}$. In studying the cocycle on $Π_{\partial X}$ it is convenient to
construct X from a plane polygon Y with 4g + 3k sides, where g is the
genus of X and k is the number of boundary circles. We shall label the
sides of Y cyclically

$$\alpha_1, \beta_1, \gamma_1, \delta_1, \alpha_2, \beta_2, \gamma_2, \delta_2, \ldots, \alpha_g, \beta_g, \gamma_g, \delta_g, \lambda_1, \sigma_1, \mu_1, \ldots, \lambda_k, \sigma_k, \mu_k, \qquad (10.7)$$

and identify γ_i with α_i^{-1}, δ_i with β_i^{-1}, and μ_i with λ_i^{-1}. Thus the sides
σ_i become the boundary of X.

If (f_1, g_1) and (f_2, g_2) belong to P_X then we find

$$c((f_1, g_1), (f_2, g_2)) = -i \int_X <df_1, *df_2> .$$

This is symmetric in (f_1, f_2), so P_X is isotropic. We also see that there is a canonical choice for the character θ satisfying (10.6), namely

$$\theta_X(f, g) = -\frac{1}{2} i \int_X <df, *df> .$$

Similarly, if $(f, g) \in P_X$ then

$$c((\overline{f}, \overline{g}), (f, g)) = i \int_X <df, *d\overline{f}> ,$$

which shows that P_X is positive. In fact we have

Proposition (10.8). P_X is a positive maximal isotropic subgroup of $\Pi_{\partial X, \mathbb{C}}$, and is compatible with the polarization.

Proof: (a) To show that P_X is maximal isotropic we consider an element $(f, g) \in \Pi_{\partial X, \mathbb{C}}$ which is in the commutant of P_X, i.e. such that $<<f_1, g>> = <<g_1, f>>$ for all $(f_1, g_1) \in P_X$. In particular we can take $f_1 = \varphi$, $g_1 = -i\varphi$, where φ is an arbitrary holomorphic function $X \to \mathbf{t}_{\mathbb{C}}$. We find

$$\int_{\partial X} <\varphi, dg - idf> = 0 .$$

This shows that dg - idf is the boundary value of a holomorphic differential on X. Similarly, taking $f_1 = \bar{\varphi}$, $g_1 = i\bar{\varphi}$ we find that dg + idf is the boundary value of an antiholomorphic differential. Putting the two facts together, there is a harmonic 1-form ω on X such that $df = \omega | \partial X$, $dg = *i\omega | \partial X$. Let F and G be indefinite integrals of ω and $*i\omega$ defined on Y, and such that F agrees with f and G with g at one vertex of ∂Y. To complete the proof we must show that $\alpha_i(F), \beta_i(F)$, and $\sigma_i^*(F)$ belong to Λ, and $\alpha_i(G), \beta_i(G)$, and $\sigma_i^*(G)$ belong to $\Lambda*$, where $\alpha_i(F)$ denotes $\int_{\alpha_i} dF$, etc., and $\sigma_i^*(F)$ denotes the constant difference between F and f along σ_i. But if $(f_1, g_1) \in P_X$ we calculate

$$<<f_1, g>> - <<g_1, f>> = \Sigma \{<\alpha_i(f_1), \beta_i(G)> - <\beta_i(f_1), \alpha_i(G)> + <\sigma_i(f_1), \sigma_i^*(G)>\}$$

$$- \Sigma \{<\alpha_i(g_1), \beta_i(F)> - <\beta_i(g_1), \alpha_i(F)> + <\sigma_i(g_1), \sigma_i^*(F)>\}$$

Now $\{\alpha_i(f_1), \beta_i(f_1), \sigma_i(f_1)\}$ describe the class of $f_1 : X \to T$ in $H^1(X; \Lambda)$. The proof is therefore completed by the observation that the group of components of P_X is $H^1(X; \Lambda) \oplus H^1(X; \Lambda*)$, a fact which follows easily from the theorem that any element of $H^1(X; \mathfrak{k})$ can be represented by $*d\varphi$ for some harmonic map $\varphi : X \to \mathfrak{k}$.

(b) The operator J_{P_X} in $\text{Map}(\partial X; \mathfrak{k}) \oplus \text{Map}(\partial X; \mathfrak{k})$ which corresponds to P_X is $(f, g) \mapsto (j_X g, j_X f)$, where $j_X f$ is defined by $d(j_X f) = *dF$, where $F : X \to \mathfrak{k}$ is the unique harmonic extension of f. (Thus j_X is well-defined only up to the addition of a finite rank operator.) For the standard polarization j_X is replaced by the Hilbert transform. But it is easy to check that j_X differs from the Hilbert transform by a smoothing operator on ∂X, which is certainly of trace class.

Generalized toral theories and their parameter space

The theory we have just constructed is manifestly symmetric with respect to the tori T and T*, i.e. dual tori give rise to the same string theory. But we can say considerably more. The Hilbert space was constructed as a projective representation of the loop group Π of the torus U = T × T*. To define the cocycle (10.2) we did not need to identify the Lie algebra $\mathfrak{t}$ with its dual: we used only the vector space $\mathfrak{u} = \mathfrak{t} \oplus \mathfrak{t}*$ with its natural indefinite inner product, and also the self-dual integral lattice Σ = Λ ⊕ Λ* in it[1]. To define the polarization and the rays Ω_X, however, we did use the identification $\mathfrak{t} - \mathfrak{t}^*$: essentially we need an orthogonal transformation J of $\mathfrak{u}$ such that (i) $J^2 = 1$ and (ii) $\langle \xi, J\xi \rangle \geq 0$, for the definition of the subgroup P_X can be written

$$P_X = \{ \varphi \in \text{Map}(X; U_{\mathbb{C}}) : d\varphi = *iJd\varphi \} \ .$$

We do not even need the inner product on $\mathfrak{u}$ to have signature 0: it can be positive definite, in which case we must have J = 1, so that P_X is simply the group of holomorphic maps X → $U_{\mathbb{C}}$. The fact that P_X is maximal isotropic is equivalent to the unimodularity of the lattice Σ, but to have a canonical splitting of its central extension we need Σ to be <u>even</u> (i.e. $\langle \lambda, \lambda \rangle \in 2\mathbb{Z}$) in addition. A lattice of the form Λ ⊕ Λ* is automatically even. Let us recall, however, that an even unimodular lattice can exist only if the signature p-q is divisible by 8. ([S7] Chapter 5, §2.2.)

---

[1]Strictly it is the commutator pairing (and hence the isomorphism class of the extension) and not the cocycle which is defined by $(\mathfrak{u}, \Sigma)$. The cocycle involves choosing an integral bilinear form B on $\mathfrak{u}$ such that $\langle \xi, \eta \rangle = B(\xi, \eta) + B(\eta, \xi)$. This exists only if the lattice is even. The general case is discussed in §12.

We now have a class of generalized toral theories parametrized by triples $(\mathcal{U},\Sigma,J)$, where Σ is an even unimodular lattice in the real inner-product space $\mathcal{U}$. If the inner product is indefinite the pair $(\mathcal{U},\Sigma)$ is determined ([S7] Chapter 5, Th. 6) up to isomorphism by the dimension and signature of $\mathcal{U}$, say $p + q$ and $p - q$. The automorphism group of $(\mathcal{U},\Sigma)$ is the discrete orthogonal group $\Gamma_{p,q} = O(\Sigma)$. The possible choices of J form the homogeneous space $O_{p,q}/O_p \times O_q$, so the parameter space of the theories is

$$M_{p,q} = \Gamma_{p,q} \backslash O_{p,q} / O_p \times O_q .$$

If $\mathcal{U}$ is positive definite, however, the parameter space is the discrete set of classes of even unimodular lattices.

The Hilbert space of the general toral theory breaks up $H = \bigoplus H_\sigma$, where σ runs through the lattice Σ. As before we have $H_0 = S(W_L) \otimes S(W_R)$ under $\mathcal{D} \times \mathcal{D}$, where now the left- and right-moving parts are associated to the splitting $\mathcal{U} = \mathcal{U}^+ \oplus \mathcal{U}^-$ into $+1$ and -1 eigenspaces of J. (Thus W_L (resp. W_R) is the positive- (resp. negative-) frequency part of $L\mathcal{U}_{\mathbb{C}} / \mathcal{U}_{\mathbb{C}}$.) In particular the theory has central charge (p,q), and is holomorphic if $\mathcal{U}$ is positive-definite. We shall discuss the positive-definite case in more detail in §12, without assuming that the lattice is unimodular. In general the splitting $\mathcal{U} = \mathcal{U}^+ \oplus \mathcal{U}^-$ is irrational with respect to the lattice Σ, and this prevents us factorizing the theory as a whole into left- and right-moving parts. When the splitting is rational the theory can be factorized as the product of a pair of weakly holomorphic theories based on a modular functor: this will be explained in §13.

It is interesting to consider the family of toral theories in the light of their infinitesimal automorphisms and deformations. From the formula (cf. (10.4))

$$\varphi(q)^p \varphi(\overline{q})^q \; \sum_{\sigma \in \Sigma} \; q^{<\sigma_+, J\sigma_+>} \; \overline{q}^{<\sigma_-, J\sigma_->}$$

for the partition function we see that for a generic lattice (i.e. when $<\sigma, J\sigma>$ is never integral) the fields of types $(1,0),(0,1)$, and $(1,1)$ are contained in H_0 and isomorphic to $\mathcal{u}_{\mathbb{C}}^+, \mathcal{u}_{\mathbb{C}}^-$ and $\mathcal{u}_{\mathbb{C}}^+ \otimes \mathcal{u}_{\mathbb{C}}^-$ respectively, and are all primary. Now $\mathcal{u}_{\mathbb{C}}^+ \oplus \mathcal{u}_{\mathbb{C}}^- = \mathcal{u}_{\mathbb{C}}$ is the complexified Lie algebra of the torus U, which is the obvious group of automorphisms of the theory. Similarly $\mathcal{u}_{\mathbb{C}}^+ \otimes \mathcal{u}_{\mathbb{C}}^-$ is the complexified tangent space of the parameter space $M_{p,q}$, which suggests that the toral theories form a complete component of the moduli space of all theories.

Finally we should mention that when $p = q = n$ the parameter space $M_{n,n}$ has as a covering space the moduli space of n-dimensional Riemannian tori T equipped with a translation-invariant 2-form ω. For if we write $U = T \times T^*$, as is always possible, and write $J : \mathfrak{t} \oplus \mathfrak{t}^* \rightarrow \mathfrak{t} \oplus \mathfrak{t}^*$ as a 2 × 2 matrix

$$\begin{pmatrix} a & b \\ c & d \end{pmatrix}$$

then $c : \mathfrak{t} \rightarrow \mathfrak{t}^*$ is a Riemannian metric on T and $\omega = ca : \mathfrak{t} \rightarrow \mathfrak{t}^*$ is a skew 2-form. One can easily check that c and ω can be prescribed arbitrarily, and determine b and d. In fact the moduli space of pairs (T,ω) is $GL_n(\mathbb{Z}) \backslash O_{n,n} / O_n \times O_n$, where $GL_n(\mathbb{Z})$ is the subgroup of $\Gamma_{n,n}$ which preserves the chosen decomposition $\Sigma = \Lambda \oplus \Lambda^*$.

From the point of view of string theory the theory corresponding
to (T,ω) is that of strings moving in a constant background field ω: it
is obtained by replacing the usual action functional for a map
$f : X \to T$ from a surface to T by

$$S(f) = \tfrac{1}{2} \int_X <df,*df> + \int_X f*\omega .$$

The term involving ω depends only on the homotopy class of f, and so
does not affect the classical equations of motion. From this
standpoint, however, there are surprising equivalences between the
theories for different (T,ω) coming from the fact that the true
parameter space is a quotient by $\Gamma_{n,n}$ rather than $GL_n(\mathbb{Z})$.

§12. The loop group of a torus

The circle

For the loop group of a torus one can describe the representations and the associated modular functors very explicitly. Let us begin with the loops in $\mathbb{C}^\times$ - in fact, to keep the functoriality as clear as possible, let us begin with the group $\mathbb{C}_S^\times$ of smooth maps from an arbitrary compact oriented 1-manifold to $\mathbb{C}^\times$.

We have already pointed out in §8 that $\mathbb{C}_S^\times$ acts by pointwise multiplication on the polarized space $\Omega^{\frac{1}{2}}(S_\sigma)$ of $\frac{1}{2}$-forms on S. Here σ is a spin structure on S, i.e. a choice of a square-root of T*S, and $\Omega^{\frac{1}{2}}(S_\sigma)$ denotes the sections of the complexification of σ. This action gives us a central extension $\tilde{\mathbb{C}}_{S,\sigma}^\times$ of $\mathbb{C}_S^\times$ by $\mathbb{C}^\times$: an element of the extended group is a pair (γ, ϵ) with $\gamma \in \mathbb{C}_S^\times$ and ϵ an element of $\mathrm{Det}(W; \gamma W)$ for some subspace W of $\Omega^{\frac{1}{2}}(S_\sigma)$ belonging to the restricted Grassmannian. The extensions corresponding to different spin structures are isomorphic, but not canonically.

The Heisenberg group $\tilde{\mathbb{C}}_{S,\sigma}^\times$ has a canonical irreducible representation $H_{S,\sigma}$. This can be realized (cf. [PS]Ch.9) on the space of holomorphic $\mathbb{C}$-valued functions on $\mathbb{C}_S^\times/P$, where P is any suitable maximal isotropic subgroup of $\mathbb{C}_S^\times$. (The realization depends on the choice of a splitting of the induced extension $\tilde{P}$ of P: for $H_{S,\sigma}$ is the representation of $\tilde{\mathbb{C}}_{S,\sigma}^\times$ holomorphically induced from the character $\tilde{P} \to \mathbb{C}^\times$ given by the splitting.)

If X is a Riemann surface such that $\partial X = S$ then - as we shall prove - the group $\mathbb{C}_X^\times$ of holomorphic maps $X \to \mathbb{C}^\times$ is a suitable maximal isotropic subgroup of $\mathbb{C}_S^\times$. The natural way to define a splitting of the extension over $\mathbb{C}_X^\times$ is to choose an spin structure σ_X on X compatible with σ. Then the space W of holomorphic sections of σ_X belongs to the restricted Grassmannian of $\Omega^{\frac{1}{2}}(S_\sigma)$ and satisfies $\gamma W = W$ for all $\gamma \in \mathbb{C}_X^\times$; so $\gamma \mapsto (\gamma, \mathrm{id}_W)$ is a splitting.

At this point we have essentially defined a conformal field theory based on the category $\mathscr{C}^{\mathrm{spin}}$. For the realization of $H_{S,\sigma}$ as $\mathrm{Hol}(\mathbb{C}_S^\times/\mathbb{C}_X^\times)$ gives us a canonical ray in $H_{S,\sigma}$ - consisting of the constant functions - for each surface X with $\partial X = S$. And $H_{S,\sigma}$ has a hermitian form, characterized by the property that the action of $\mathbb{C}_S^\times$ is unitary (in the sense that $U(\gamma)^* = U(\bar\gamma)$). The "sewing" property of the theory is obvious from this point of view.

This bosonic construction of the theory described by means of Fock spaces in §8 seems at first to have few advantages, for it entails proving that $\mathbb{C}_X^\times$ is a suitable maximal isotropic subgroup of $\mathbb{C}_{\partial X}^\times$; and in any case the theory of Heisenberg representations is not so elementary as that of Fock spaces. But the bosonic theory can be used in situations where there is no fermionic version, and it gives more information, as we shall see.

A central extension $\tilde A$ of an abelian group A by $\mathbb{C}^\times$ is determined up to non-canonical isomorphism by its commutator pairing

$$\ll , \gg : A \times A \to \mathbb{C}^\times,$$

defined by $\ll a_1, a_2 \gg = \tilde a_1 \tilde a_2 \tilde a_1^{-1} \tilde a_2^{-1}$, where $\tilde a_i \in \tilde A$ is a lift of a_i. The commutator pairing of the extension of $L\mathbb{C}^\times$ that we are interested in is

$$\ll f, g \gg = \frac{1}{2} \int_0^1 (f\,dg - g\,df) + \frac{1}{2}(f(0)\Delta_g - \Delta_f g(0)) . \qquad (12.1)$$

Here $\mathbb{C}^\times$ is regarded as $\mathbb{C}/\mathbb{Z}$, and elements of $L\mathbb{C}^\times$ as maps $f : \mathbb{R} \to \mathbb{C}$ such that $f(x + 1) = f(x) + \Delta_f$, where $\Delta_f \in \mathbb{Z}$ is the winding number. The pairing $\ll , \gg$ is invariant under $\text{Diff}^+(S^1)$.

We can obtain a definite central extension of $L\mathbb{C}^\times$ by giving a cocycle c, which must satisfy

$$c(f,g) - c(g,f) = \ll f,g \gg .$$

Unfortunately there is no choice of c which is invariant under $\text{Diff}^+(S^1)$, but the formula

$$c(f,g) = \tfrac{1}{2} \int_0^1 f dg - \tfrac{1}{2} \Delta_f g(0) \qquad (12.2)$$

defines an extension $\tilde{A}$ of the group A of maps $f : \mathbb{R} \to \mathbb{C}$ such that $f(x + 1) = f(x) + \Delta_f$ which is invariant under the universal covering group of $\text{Diff}^+(S^1)$, i.e. the diffeomorphisms φ of $\mathbb{R}$ satisfying $\varphi(x + 1) = \varphi(x) + 1$. Of course $L\mathbb{C}^\times = A/\mathbb{Z}$, so we get an extension of $L\mathbb{C}^\times$ by lifting the inclusion $\mathbb{Z} \to A$ to $\tilde{A}$. Because c vanishes on $\mathbb{Z} \times \mathbb{Z}$ the possible lifts simply correspond to homomorphisms $\mathbb{Z} \to \mathbb{C}^\times$, i.e. to elements $\sigma \in \mathbb{C}^\times$. It is easy to check that the double covering of $\text{Diff}^+(S^1)$ acts on the extension corresponding to $\sigma = 1$, while $\text{Diff}^+(S^1)$ itself acts when $\sigma = -1$. These are the two extensions which correspond to the two spin structures on the circle. Using them we can associate an extension $\tilde{L}\mathbb{C}^\times_{S,\sigma}$ of $L\mathbb{C}^\times_S$ to every oriented 1-manifold with a spin structure. In doing so it is important to notice the following point. From (12.1) we find that

$$<<f,g>> = \tfrac{1}{2} \Delta_f \Delta_g$$

if f and g have disjoint supports in S^1, in other words loops of
winding number 1 <u>anticommute</u> if they have disjoint supports. For a
disconnected 1-manifold it is therefore appropriate to look for
extensions of $\mathbb{C}_S^\times$ whose commutator is given by

$$<<f,g>> = \sum_i <<f_i,g_i>> + \tfrac{1}{2} \sum_{i \neq j} \Delta_{f_i} \Delta_{g_i} \; ,$$

where f_i, g_i are the restrictions of f,g to the ith component S_i of S.
We therefore define $\tilde{\mathbb{C}}_{S,\sigma}^\times$ so that it contains each $\tilde{\mathbb{C}}_{S_i,\sigma_i}^\times$ as a subgroup,
and satisfies

$$f.g = (-1)^{\Delta_f \Delta_g} g.f$$

when $f \in \tilde{\mathbb{C}}_{S_i,\sigma_i}$, $g \in \tilde{\mathbb{C}}_{S_j,\sigma_j}^\times$, and $i \neq j$. The natural way to achieve
this is to use the cocycle

$$c(f,g) = \sum c_i(f,g) + \tfrac{1}{2} \sum_{i>j} \Delta_{f_i} \Delta_{g_j} \; , \qquad (12.3)$$

where c_i is the cocycle defining $\tilde{\mathbb{C}}_{S_i}^\times$. Notice that this cocycle is
defined only on a covering group A_S of $\mathbb{C}_S^\times$, and does not depend on the
spin structure σ. The group $\tilde{\mathbb{C}}_S^\times$ is obtained as the quotient group of $\tilde{A}_S$
by the image of a lift of the inclusion $H^0(S;\mathbb{Z}) \to A_S$, and the lift <u>does</u>
depend on the spin structure.

Proposition (12.4).

(i)   For any Riemann surface X the subgroup $C_X^\times$ is a positive maximal isotropic subgroup of $C_{\partial X}^\times$, compatible with its polarization.

(ii) When restricted to $C_X^\times \subset C_{\partial X}^\times$ the cocycle (12.3) takes its values in $\{0, \frac{1}{2}\}$, so defines a canonical extension $C_{X,\sigma}^\times$ of $C_X^\times$ by $\mathbb{Z}/2$.

(iii) The extension $C_{X,\sigma}^\times$ splits, and its splittings correspond canonically to the spin structures on X which restrict to σ on ∂X.

Proof: This is similar to Proposition (10.7), and we shall use the same notation. Suppose that two elements f,g of $C_{\partial X}^\times$ are represented by smooth maps f,g : Y → $\mathbb{C}$, where Y is the plane polygon with sides $\alpha_1, \ldots, \mu_k$. After some manipulation we find

$$c(f,g) = -\tfrac{1}{2} \, f \cup g \; + \; \tfrac{1}{2} \int_Y df \wedge dg \, , \qquad (12.5)$$

where

$$f \cup g \;\; = \;\; \sum \{\alpha_i(f)\beta_i(g) - \beta_i(f)\alpha_i(g)\} \, .$$

If f and g are holomorphic then the integral over Y vanishes, and if f and g are well-defined on X then $c(f,g) \in \tfrac{1}{2}\mathbb{Z}$ and $c(f,g) - c(g,f) \in \mathbb{Z}$. So $C_X^\times$ is isotropic. It is positive because if f is holomorphic then

$$\text{Im } c(\bar{f}, f) \;\; = \;\; \frac{1}{2i} \int_Y df \wedge d\bar{f} \, > \, 0 \, .$$

The proof that $C_X^\times$ is maximal isotropic and compatible with the polarization is sufficiently like that of (10.7) to need no further comment.

[136]

Appendix B   Determinant lines

This appendix preserves the conventions of Appendix A, and all topological vector spaces are assumed to be allowable. An operator of determinant class means one of the form $1 + T$ where T is of trace class, and $\det(1 + T)$ is defined by (A.12).

Definition (B.1). An operator $T : E \to F$ between complete locally convex vector spaces is Fredholm if it is invertible modulo compact operators, or, equivalently, modulo operators of finite rank.

If T is Fredholm then it has closed range and finite dimensional kernel and cokernel. If E and F are Fréchet the converse is also true. (Cf. [S1]) The index of a Fredholm operator T is the integer $\dim(\ker T) - \dim(\mathrm{coker}\ T)$.

Definition (B.2).
(i)   If $T : E \to F$ is Fredholm of index 0 then Det_T is the line whose points are pairs $[\theta,\lambda]$, where $\lambda \in \mathbb{C}$ and $\theta : E \to F$ differs from T by a trace-class operator, subject to the equivalence relation generated by

$$[\theta\varphi,\lambda] \sim [\theta,(\det\varphi).\lambda]$$

when $\varphi : E \to E$ is of determinant class.
(ii)   If $T : E \to F$ is Fredholm of index n then $\mathrm{Det}_T = \mathrm{Det}_{\tilde{T}}$, where $\tilde{T} = T \oplus 0 : E \to F \oplus \mathbb{C}^n$ if $n > 0$, and $\tilde{T} = T \oplus 0 : E \oplus \mathbb{C}^{-n} \to F$ if $n < 0$.

<u>Remarks</u>. If T has index 0 one can always choose an <u>invertible</u> θ such that θ-T is of trace class. Then $\lambda \mapsto [\theta, \lambda]$ defines an isomorphism $\mathbb{C} \to \mathrm{Det}_T$. If θ is not invertible then $[\theta, \lambda] = 0$ for all λ.

<u>Definition (B.4)</u>. If $T : E \to F$ is Fredholm of index 0 then det(T) is the element [T,1] of the line Det_T. If index(T) $\neq 0$ then det(T) = 0 ϵ Det_T.

<u>Corollary (B.5)</u>. T is invertible if and only if det(T) $\neq 0$.

<u>Proposition (B.6)</u>. If dim(ker T) = p and dim(coker T) = q there is a canonical isomorphism

$$\mathrm{Det}_T \cong \Lambda^p(\ker T)^* \otimes \Lambda^q(\mathrm{coker}\ T) .$$

<u>Proof</u>: It is enough to prove this when p = q. Let $\alpha_1, \ldots, \alpha_p$ be a basis for (ker T)* and $\eta_1, \ldots, \eta_q$ be a basis for coker T. Let $\tilde{\alpha}_i : E \to \mathbb{C}$ be an extension of α_i. Then

$$[T + \Sigma \eta_i \otimes \tilde{\alpha}_i , 1] \longleftrightarrow \alpha_1 \wedge \ldots \wedge \alpha_p \otimes \eta_1 \wedge \ldots \wedge \eta_p$$

defines the isomorphism.

We shall now show that the lines Det_T depend holomorphically on T, i.e.

<u>Proposition (B.7)</u>. If $\{T_x : E_x \to F_x\}_{x \in X}$ is a holomorphic family of Fredholm operators then the lines Det_{T_x} form a holomorphic line bundle on X.

Here the meaning of 'holomorphic family' must be understood in the sense of [S1], i.e. $\{E_x\}$ and $\{F_x\}$ are holomorphic vector bundles on X in the weakest sense, but we assume that there exists a continuous parametrix $\{P_x : F_x \to E_x\}$ such that the <u>families</u> $\{T_xP_x - 1\}$ and $\{P_xT_x - 1\}$ are compact, i.e. are compact operators which depend continuously on x in the uniform topology.

<u>Proof of (B.7)</u>. We can assume the bundles $\{E_x\}$ and $\{F_x\}$ are trivial, and that the T_x have index 0. Then for each finite rank operator $t : E \to F$ the set

$$U_t = \{ x \in X : T_x + t \text{ is invertible} \}$$

is open in X. We trivialize the lines Det_{T_x} for $x \in U_t$ by $x \longmapsto [T_x + t, 1]$, and in the intersection $U_{t_0} \cap U_{t_1}$ the transition function is

$$x \mapsto \det((T_x + t_1)(T_x + t_0)^{-1})$$
$$= \det(1 + (t_1 - t_0)(T_x + t_0)^{-1}) ,$$

which is holomorphic.

The main general fact about determinant lines is

<u>Proposition (B.8)</u>. Let

$$
\begin{array}{ccccccccc}
0 & \to & E' & \to & E & \to & E'' & \to & 0 \\
 & & \downarrow T' & & \downarrow T & & \downarrow T'' & & \\
0 & \to & F' & \to & F & \to & F'' & \to & 0
\end{array}
$$

be a commutative diagram of topological vector spaces with exact rows and Fredholm columns. Then there is a canonical isomorphism

$$\text{Det}_T \cong \text{Det}_{T'} \otimes \text{Det}_{T''}$$

which depends holomorphically on T',T,T".

"Exact rows" means of course that E' and F' are topological subspaces of E and F, and that E" and F" have the quotient topology.

Proof: If t',t" are finite rank operators such that T' + t' and T" + t" are invertible then one can find t of finite rank such that T + t forms a commutative diagram with T' + t' and T" + t". The desired isomorphism is

$$[T + t, 1] \longleftrightarrow [T' + t', 1] \otimes [T" + t", 1] \ .$$

The determinant line, the restricted Grassmannian, and the central extension of GL_{res}

A polarized topological vector space E (in the sense of Definition (8.8)) has a restricted Grassmannian Gr(E) which consists of the (+1)-eigenspaces of the preferred involutions J which define the polarization. If W_0 and W_1 are two points of Gr(E) there is a preferred class of Fredholm operators T : $W_0 \to W_1$, namely those which differ from the inclusion $W_0 \to H$ by trace class operators.

<u>Definition (B.9)</u>. For W_0, W_1 in Gr(E) we write $Det(W_0; W_1)$ for Det_T,

where $T : W_0 \to W_1$ is such a Fredholm operator.

From (B.3) we know that $Det(W_0; W_1)$ does not depend on the choice

of T. If W_0 is held fixed then $\bigcup\limits_W Det(W_0; W)$ is a holomorphic line

bundle on Gr(E). This is the determinant line bundle of [PS](Chap.

10): there the chosen W_0 was called H_+.

For three spaces W_0, W_1, W_2 we clearly have

$$Det(W_0; W_2) \cong Det(W_0; W_1) \otimes Det(W_1; W_2) \ .$$

Now suppose that $g : E \to E$ belongs to the restricted general

linear group of E, i.e. that $gJg^{-1} - J$ is of trace class for a

preferred involution J.

<u>Definition (B.10)</u>. For $g \in GL_{res}(E)$ we define

$$Det_g = Det(W; gW) \ ,$$

where W is an element of Gr(E).

The line Det_g is independent of W, for if W_0 and W_1 are two

choices then g defines an isomorphism between $Det(W_0; W_1)$ and

$Det(gW_0; gW_1)$, and hence between $Det(W_0; gW_0)$ and

$$Det(W_0; gW_0) \otimes Det(W_0; W_1)^* \otimes Det(gW_0; gW_1) = Det(W_1; gW_1) \ .$$

Evidently $\mathrm{Det}_{g_1 g_2} \cong \mathrm{Det}_{g_1} \otimes \mathrm{Det}_{g_2}$, and so we can define a central extension of $GL_{res}(E)$ which consists of all pairs $(g.\lambda)$ with $\lambda \in \mathrm{Det}_g$. By its construction this group acts holomorphically on the line bundle $\bigcup_W \mathrm{Det}(W_0, W)$ for any choice of W_0.

Riemann surfaces

We conclude this appendix with a result on the determinant line of a Riemann surface.

Proposition (B.11). If a closed Riemann surface Z is the union of two surfaces X and Y which intersect in a 1-manifold S then Det_Z is canonically isomorphic to the determinant line of the map

$$\mathrm{Hol}(X) \oplus \mathrm{Hol}(Y) \;\to\; \Omega^0(S)$$
$$(\; f \;,\; g \;) \longmapsto (f|S)-(g|S) \; .$$

Proof: First consider the diagram

$$\begin{array}{ccccc}
\Omega^0(Z) & \to & \Omega^0(X) \oplus \Omega^0(Y) & \to & J^0(S) \\
\downarrow \bar{\partial} & & \downarrow & & \downarrow \\
\Omega^{01}(Z) & \to & \Omega^{01}(X) \oplus \Omega^{01}(Y) \oplus \Omega^0(S) & \to & J^{01}(S) \oplus \Omega^0(S) \; ,
\end{array}$$

where $J^0(S)$ (resp. $J^{01}(S)$) is the space of infinite jets of functions (resp. $(0,1)$-forms) on Z along S, the middle vertical map is $(f,g) \longmapsto (\bar{\partial}f, \bar{\partial}g, (f|S)-(g|S))$,

and the right-hand vertical map is $f \longmapsto (\bar{\partial}f, f|S)$.

The horizontal maps are defined in the obvious way to give short exact

sequences. Notice that an element f of $J^0(S)$ can be identified with

the sequence $\{f_k\}$ of smooth functions on S such that $f = \Sigma\ f_k y^k$, where

y is a coordinate on Z transverse to S. The right-hand vertical map is

then

$$\{f_k\} \quad\longmapsto\quad \{\ f_0\ ,\ f_0' + f_1\ ,\ f_1' + f_2\ ,\ldots\}\ ,$$

and is therefore an isomorphism. Thus by (B.8) we can identify Det_Z

with the determinant of the middle map. Proposition (8.11) is then

obtained from the diagram

$$
\begin{array}{ccccc}
\text{Hol}(X) \oplus \text{Hol}(Y) & \to & \Omega^0(X) \oplus \Omega^0(Y) & \to & \Omega^{01}(X) \oplus \Omega^{01}(Y) \\
\downarrow & & \downarrow & & \downarrow\ \text{id} \\
\Omega^0(S) & \to & \Omega^{01}(X) \oplus \Omega^{01}(Y) \oplus \Omega^0(S) & \to & \Omega^{01}(X) \oplus \Omega^{01}(Y)
\end{array}
$$

just as in the proof of (6.3).

Corollary (B.12). In the situation of (B.11) we have

$$\text{Det}_Z \cong \text{Det}_X \otimes \text{Det}_Y\ .$$

Proof: The right-hand side is the determinant of the map

$$\text{Hol}(X) \oplus \text{Hol}(Y) \;\to\; \Omega^0(S)$$
$$(\; f \;,\; g \;) \;\mapsto\; (f|S)_- - (g|S)_+ \;.$$

This differs from the map of (B.9) by a trace-class operator (cf. the proof of (6.4)).

Notice that (B.12) provides the proof of Proposition (6.5), which was omitted earlier.

References

[A1]   Atiyah, M.F., Riemann surfaces and spin structures. *Ann. Sci. Éc. Norm. Sup.* **4** (1971), 47–62.

[A2]   Atiyah, M.F., On framings of 3-manifolds. *Topology* **29** (1990), 1–7.

[ABS]   Atiyah, M.F., R. Bott, and A. Shapiro, Clifford modules. *Topology* **3** (Suppl. 1) (1964), 3–38.

[APS]   Atiyah, M.F., V.K. Patodi, and I.M. Singer, Spectral asymmetry and Riemannian geometry I. *Math. Proc. Camb. Phil. Soc.* **77** (1975), 43–69.

[BK]   Bakalov, B. and A. Kirillov, Jr., *Lectures on tensor categories and modular functors.* Amer. Math. Soc. University Lecture Series, Vol. 21, 2001.

[BPZ]   Belavin, A.A., A.M. Polyakov, and A.B. Zamolodchikov, Infinite conformal symmetry in two dimensional quantum field theory. *Nucl. Phys.* **B241** (1984), 333–380.

[BR]   Bowick, M.J. and S.G. Rajeev, The holomorphic geometry of closed bosonic string theory and Diff S^1/S^1. *Nucl. Phys.* **B293** (1987), 348–384.

[FGZ]   Frenkel, I.B., H. Garland, and G.J. Zuckerman, Semi-infinite cohomology and string theory. *Proc. Nat. Acad. Sci.* **83** (1986), 8442–8446.

[GV]   Gelfand, I.M. and N. Ya. Vilenkin, *Generalized Functions.* Vol. 4, Academic Press, 1964.

[H1]   Hitchin, N.J., Flat connections and geometric quantization. *Commun. Math. Phys.* **131** (1990), 347–380.

[H2]   Howe, R., The oscillator semigroup. *Proc. Symp. Pure Math.*, 48, 61–132.

[K]   Kontsevich, R., Rational conformal field theory and invariants of 3-dimensional manifolds. Marseille preprint CPT-88/p. 2189.

[LZ]   Lian, B.H. and G.J. Zuckerman, Semi-infinite cohomology and 2D gravity I. *Commun. Math. Phys.* **145** (1992), 561–593.

[L]   Lusztig, G., Leading coefficients of character values of Hecke algebras. *Proc. Symp. Pure Math.*, vol. 47, Amer. Math. Soc. 1987.

[MS]   Moore, G. and N. Seiberg, Polynomial equations for rational conformal field theories, *Phys. Lett.* **212B** (1988), 451–

[N1]   Neretin, A. Yu., A complex semigroup containing the group of diffeomorphisms of a circle. *Funkts. Anal. Prilozh.* **21** (1987), 82–83.

[N2]   Neretin, A. Yu., *Categories of symmetries and infinite-dimensional groups.* Lond. Math. Soc. Monographs **16**, Oxford Univ. Press 1996.

[PS]   Pressley, A. and G. Segal, *Loop Groups.* Oxford Univ. Press, 1986.

[Q]   Quillen, D., Determinants of Cauchy–Riemann operators on Riemann surfaces. *Funkts. Anal. Prilozh.* **19** (1985), 37–41.

576

[S1]   Segal, G.B., Fredholm complexes. *Quarterly J. Math.* Oxford **21** (1970),
         385–402.

[S2]   Segal, G.B., Unitary representations of some infinite dimensional groups.
         Commun. *Math. Phys.* **80** (1981), 301–342.

[S3]   Segal, G.B., The geometry of the KdV equation. *Int. J. Mod. Phys.* **A6**
         (1991), 2859–2869.

[S4]   Segal, G.B., Notes of lectures at Stanford 1996, available at
         http://www.cgtp.duke.edu/ITP99/segal

[S5]   Segal, G.B., Topological structures in string theory. *Phil. Trans. Roy. Soc.*
         London **A359** (2001), 1389–1398.

[S6]   Seiberg, N., Notes on quantum Liouville theory and quantum gravity. *Progr.*
         Theoret. Phys. Suppl. **102** (1991), 319–349.

[S7]   Serre, J.-P., *Cours d'arithmétique*. Presses Universitaires de France, 1970.

[T]   Tillmann, U., The classifying space of the $1 + 1$-dimensional cobordism
         category. *J. Reine Angew. Math.* **479** (1996), 67–75.

[V]   Verlinde, E., Fusion rules and modular trnsformations in 2 dimensional
         conformal field theory. *Nucl. Phys.* **B300** (1988), 360–376.

[W]   Witten, E., Quantum field theory and the Jones polynomial. *Commun. Math.*
         Phys. **121** (1989), 351–399.

[Z]   Zamolodchikov, A.B., Renormalization group and perturbation theory near
         fixed points in two-dimensional field theory. *Yad. Fiz.* **46** (1987), 1819–1831.